Fokker–Planck–Kolmogorov Equations

Mathematical
Surveys
and
Monographs
Volume 207

Fokker–Planck–Kolmogorov Equations

Vladimir I. Bogachev
Nicolai V. Krylov
Michael Röckner
Stanislav V. Shaposhnikov

American Mathematical Society
Providence, Rhode Island

2010 *Mathematics Subject Classification.* Primary 35-02, 35J15, 35K10, 60J35, 60J60.

For additional information and updates on this book, visit
www.ams.org/bookpages/surv-207

Library of Congress Cataloging-in-Publication Data

Fokker–Planck–Kolmogorov equations /Vladimir I. Bogachev, Nicolai V. Krylov, Michael Röckner, Stanislav V. Shaposhnikov.
pages cm. — (Mathematical surveys and monographs ; volume 207)
Includes bibliographical references and index.
ISBN 978-1-4704-2558-6 (alk. paper) | Softcover ISBN 978-1-4704-7009-8
1. Fokker–Planck equation. 2. Stochastic differential equations. I. Bogachev, V. I. (Vladimir Igorevich), 1961- II. Krylov, N. V. (Nicolai Vladimirovich). III. Röckner, Michael, 1956- IV. Shaposhnikov, Stanislav V.

QA274.23.F65 2015
515′.353—dc23

2015024922

Printed in the United States of America.
Reprinted by the American Mathematical Society, 2022.

∞ The paper used in this book is acid-free and falls within the guidelines
established to ensure permanence and durability.
Visit the AMS home page at `http://www.ams.org/`
11 10 9 8 7 6 5 4 3 2   27 26 25 24 23 22

Contents

Preface

This book gives a systematic presentation of the theory of Fokker–Planck–Kolmogorov equations, which are second order elliptic and parabolic equations for measures. This direction goes back to Kolmogorov's works [**527**], [**528**], [**529**] and a number of earlier works in the physics literature by Fokker [**377**], Smoluchowski [**863**], Planck [**781**], and Chapman [**235**]. One of our principal objects is the elliptic operator of the form

$$L_{A,b}f = \operatorname{trace}(AD^2 f) + \langle b, \nabla f\rangle, \quad f \in C_0^\infty(\Omega),$$

where $A = (a^{ij})$ is a mapping on a domain $\Omega \subset \mathbb{R}^d$ with values in the space of nonnegative symmetric linear operators on $\mathbb{R}^d$ and $b = (b^i)$ is a vector field on Ω. In coordinate form, $L_{A,b}$ is given by the expression

$$L_{A,b}f = a^{ij}\partial_{x_i}\partial_{x_j} f + b^i \partial_{x_i} f,$$

where we always assume that the summation is taken over all repeated indices.

With this operator $L_{A,b}$, we associate the weak elliptic equation

$$L_{A,b}^*\mu = 0 \tag{1}$$

for Borel measures on Ω, which is understood in the following sense:

$$\int_\Omega L_{A,b} f\, d\mu = 0 \quad \forall\, f \in C_0^\infty(\Omega), \tag{2}$$

where we assume that b^i, $a^{ij} \in L^1_{\mathrm{loc}}(\mu)$. If μ has a density ϱ with respect to Lebesgue measure, then ϱ is sometimes called "an adjoint solution" and the equation is called "an equation in double divergence form". We use the above term "weak elliptic equation for measures". The corresponding equation for the density ϱ is

$$\partial_{x_i}\partial_{x_j}(a^{ij}\varrho) - \partial_{x_i}(b^i\varrho) = 0.$$

If $A = \mathrm{I}$, we obtain the equation $\Delta\varrho - \operatorname{div}(\varrho b) = 0$.

Similarly, one can consider parabolic operators and parabolic Fokker–Planck–Kolmogorov equations for measures on $\Omega\times(0,T)$ of the type

$$\partial_t \mu = L_{A,b}^*\mu.$$

The corresponding equations for densities are

$$\partial_t \varrho(x,t) = \partial_{x_i}\partial_{x_j}\big(a^{ij}(x,t)\varrho(x,t)\big) - \partial_{x_i}\big(b^i(x,t)\varrho(x,t)\big), \tag{3}$$

and if we also have an initial distribution μ_0 in a suitable sense, then we arrive at the Cauchy problem for the Fokker–Planck–Kolmogorov equation. However, it is crucial that a priori Fokker–Planck–Kolmogorov equations are equations for measures, not for functions; this becomes relevant when the coefficients are singular or degenerate and, in particular, in the infinite-dimensional case, where no Lebesgue measure exists. It is also important that equation (1) is meaningful under very broad assumptions about A and b: only their local integrability with respect to the regarded solution μ is needed. These coefficients may be quite singular with respect to Lebesgue measure even if the solution admits a smooth density. For example, for an arbitrary infinitely differentiable probability density ϱ on $\mathbb{R}^d$, the measure $\mu = \varrho\, dx$ satisfies the above equation with $A = \mathrm{I}$ and $b = \nabla\varrho/\varrho$, where we set $\nabla\varrho(x)/\varrho(x) = 0$ whenever $\varrho(x) = 0$. This is obvious from the integration by parts formula

$$\int_{\mathbb{R}^d} [\Delta f + \langle\nabla\varrho/\varrho, \nabla f\rangle]\varrho\, dx = \int_{\mathbb{R}^d} \varrho\Delta f + \int_{\mathbb{R}^d} \langle\nabla\varrho, \nabla f\rangle\, dx = 0.$$

Since ϱ may vanish on an arbitrary proper closed subset of $\mathbb{R}^d$, the vector field b can fail to be locally integrable with respect to Lebesgue measure, but it is locally integrable with respect to μ. Also note that in general our solutions need not be more regular than the coefficients (unlike in the case of usual elliptic equations). For example, if $d = 1$ and $b = 0$, then for an arbitrary positive probability density ϱ, the measure $\mu = \varrho\, dx$ satisfies the equation $L^*_{A,0}\mu = 0$ with $A = \varrho^{-1}$.

In this general setting, a study of weak elliptic equations for measures on finite- and infinite-dimensional spaces was initiated in the 1990s in the papers of the first three authors. Actually, the infinite-dimensional case was even a starting point, which was motivated by investigations of infinite-dimensional diffusion processes and other applications in infinite-dimensional stochastic analysis (developed in particular in the works of Albeverio, Høegh-Krohn [**21**] as well as A.I. Kirillov [**511**]–[**516**]). It was realized in the course of these investigations that even infinite-dimensional equations with very nice coefficients often require results on finite-dimensional equations with quite general coefficients. For example, we shall see in Chapter 10 that the finite-dimensional projections μ_n of a measure μ satisfying an elliptic equation on an infinite-dimensional space satisfy elliptic equations whose coefficients are the conditional expectations of the original coefficients with respect to the σ-algebras generated by the corresponding projection operators. As a result, even for smooth infinite-dimensional coefficients, the only information about their conditional expectations is related to their integrability with respect to μ_n, not with respect to Lebesgue measure; in particular, no local boundedness is given.

The theory of elliptic and parabolic equations for measures is now a rapidly growing area with deep and interesting connections to many directions in real analysis, partial differential equations, and stochastic analysis. Let us briefly describe the probabilistic picture behind our analytic framework. Suppose that $\xi = (\xi_t^x)_{t\geqslant 0}$ is a diffusion process in $\mathbb{R}^d$ governed by the stochastic differential equation

$$d\xi_t^x = \sigma(\xi_t^x)dW_t + b(\xi_t^x)dt, \quad \xi_0 = x.$$

The basic concepts related to this equation are recalled in § 1.3. The generator of the transition semigroup $\{T_t\}_{t\geqslant 0}$ has the form $L_{A,b}$, where $A = \sigma\sigma^*/2$. The matrix $A = (a^{ij})$ in the operator $L_{A,b}$ will be called the *diffusion matrix* or *diffusion coefficient*; this differs from the standard form of the diffusion generator by the absence of the factor 1/2 in front of the second order derivatives, but is more convenient when one deals with equations. The vector field b is called the *drift coefficient* or just the *drift*. The transition probabilities of ξ satisfy the corresponding parabolic equation. Any invariant probability measure μ of ξ (if such exists) satisfies (1), where μ is called invariant for $\{T_t\}_{t\geqslant 0}$ if the following identity holds:

$$\int_{\mathbb{R}^d} T_t f\,d\mu = \int_{\mathbb{R}^d} f\,d\mu \quad \forall f \in C_b(\mathbb{R}^d). \tag{4}$$

Measures satisfying (1) are called *infinitesimally invariant*, because this equation has deep connections with invariance with respect to the corresponding operator semigroups. More precisely, if there is an invariant probability measure μ, then $\{T_t\}_{t\geqslant 0}$ extends to $L^1(\mu)$ and is strongly continuous. Let L be the corresponding generator with domain $D(L)$. Then (4) is equivalent to the equality

$$\int_{\mathbb{R}^d} Lf\,d\mu = 0 \quad \forall f \in D(L).$$

Under reasonable assumptions about A and b, the generator of the semigroup associated with the diffusion governed by the indicated stochastic equation coincides with $L_{A,b}$ on $C_0^\infty(\mathbb{R}^d)$. As we shall see, invariance of the measure in the sense of (4) is not the same as (2). The point is that the class $C_0^\infty(\mathbb{R}^d)$ may be much smaller than $D(L)$. What is important is that the equation is meaningful and can have solutions under assumptions that are much weaker than those needed for the existence of a diffusion, so that this equation can be investigated without any assumptions about the existence of semigroups. On the other hand, there exist very interesting and fruitful relations between equations (2) and (4). For example, if A and b are both Lipschitz and if A is nondegenerate, they are equivalent.

Letting $P(x,t,\,\cdot\,)$ be the corresponding transition probabilities (the distributions of ξ_t^x), the semigroup property reads

$$P(x,t+s,B) = \int_{\mathbb{R}^d} P(u,s,B)\,P(x,t,du), \tag{5}$$

or in the case where there exist densities $p(x,t,y)$,

$$p(x,t+s,y) = \int_{\mathbb{R}^d} p(u,s,y)\,p(x,t,u)\,du.$$

Identity (5) is called the Smoluchowski equation or the Chapman–Kolmogorov equation. In his seminal paper [**527**] Kolmogorov posed the following problems: find conditions for the existence and uniqueness of solutions to the Cauchy problem for (3) and investigate when (5) holds for these solutions. Now, 80 years later, these problems are still not completely solved. However, considerable progress has been achieved; results obtained and some related open problems are discussed in this book.

We shall consider the following problems.

1) Regularity of solutions of equation (2), for example, the existence of densities with respect to Lebesgue measure, the continuity and smoothness of these densities, and certain related estimates (such as L^2-estimates for logarithmic gradients of solutions). In particular, we shall see in Chapter 1 that the measure μ is always absolutely continuous with respect to Lebesgue measure on the set $\{\det A > 0\}$ and has a continuous density from the Sobolev class $W^{p,1}_{loc}$ with $p > d$ provided that the diffusion coefficients a^{ij} are in this class, $|b| \in L^p_{loc}(dx)$ or $|b| \in L^p_{loc}(\mu)$, and the matrix A is positive definite. Global properties of solutions of equations with unbounded coefficients are studied in Chapter 3, where certain global upper and lower estimates for the densities are obtained. We shall also obtain analogous results for parabolic equations in Chapters 6–8.

2) Existence of solutions to elliptic equation (2) and existence of invariant measures in the sense of (4) as well as relations between these two concepts are the subjects of Chapter 2 and Chapter 5. In particular, we shall see in Chapter 5 that under rather general assumptions, for a given probability measure μ satisfying our elliptic equation (2), one can construct a strongly continuous Markov semigroup $\{T_t^\mu\}_{t\geqslant 0}$ on $L^1(\mu)$ such that μ is $\{T_t^\mu\}_{t\geqslant 0}$-invariant and the generator of $\{T_t^\mu\}_{t\geqslant 0}$ coincides with $L_{A,b}$ on $C_0^\infty(\mathbb{R}^d)$. For this, an easy to verify condition is the existence of a Lyapunov function for $L_{A,b}$. In the general case (without any additional assumptions), a bit less is true, namely, μ is only subinvariant for $\{T_t\}_{t\geqslant 0}$. We shall see examples where this really occurs, i.e., where μ is not invariant. Existence of solutions to parabolic equations is addressed in Chapter 6.

3) Various uniqueness problems are considered in Chapters 4 and 5; in particular, uniqueness of invariant measures in the sense of (4) and uniqueness of solutions to (2) in the class of all probability measures. Related interesting problems concern uniqueness of associated semigroups $\{T_t^\mu\}_{t\geqslant 0}$ and the essential self-adjointness of the operator $L_{A,b}$ on $C_0^\infty(\mathbb{R}^d)$ in the case when it is symmetric. Parabolic analogues are considered in Chapter 9.

First, we concentrate on the elliptic case, to which Chapters 1–5 are devoted. In Chapters 6–9 similar problems are studied for parabolic equations; however, parabolic equations appear already in Chapter 5 in relation to semigroups generated by elliptic operators. Chapter 10 is devoted to a brief discussion of infinite-dimensional analogues of the problems listed in 1)–3). The results obtained so far in the infinite-dimensional setting apply to various particular situations, although they cover many concrete examples arising in applications such as stochastic partial differential equations, infinite particle systems, Gibbs measures, and so on. The main purpose of Chapter 10 is to give applications of finite-dimensional results and to demonstrate the universality of certain ideas, methods, and techniques. Finally, in Chapters 2, 6, and 9 we discuss degenerate equations and nonlinear equations for measures; important examples of such equations are Vlasov-type equations. We made some effort to minimize dependencies between the chapters; the proofs of a number of fundamental results are rather difficult and can be omitted without any loss of understanding of the rest.

Every chapter opens with some synopsis mentioning the chief problems and results discussed. The last section of each chapter includes some complementary subsections (the numbers in brackets within these internal contents refer to the corresponding page numbers) and also brief historical and bibliographic comments and exercises. In the Bibliography each item is provided with indication of all pages where it is cited. The Subject Index also includes special notations used.

Acknowledgments. The material in this book has been discussed with many colleagues; we are especially grateful to S. Albeverio, L. Ambrosio, A.-B. Cruzeiro, G. Da Prato, A. Eberle, D. Elworthy, F. Flandoli, M. Fukushima, B. Goldys, I. A. Ibragimov, A. I. Kirillov, A. V. Kolesnikov, V. A. Kondratiev, Yu. G. Kondratiev, V. V. Kozlov, G. Leha, V. A. Liskevich, A. Lunardi, P. Malliavin, I. Marshall, G. Metafune, J. van Neerven, D. Pallara, E. Pardoux, S. I. Pokhozhaev, Yu. V. Prohorov, E. V. Radkevich, A. Rhandi, G. Ritter, T. A. Shaposhnikova, I. Shigekawa, A. Shirikyan, A. N. Shiryaev, Z. Sobol, W. Stannat, N. S. Trudinger, A. Yu. Veretennikov, F. Y. Wang, J. Zabczyk, M. Zakai, T. S. Zhang, and V. V. Zhikov. The Russian Science Foundation grant 14-11-00196 is gratefully acknowledged.

CHAPTER 1

Stationary Fokker–Planck–Kolmogorov Equations

In this chapter we introduce principal objects related to elliptic equations for measures, an important example of which is the stationary Fokker–Planck–Kolmogorov equation for invariant probabilities of diffusion processes. Although our approach is purely analytic, some concepts related to diffusion processes are explained. Our principal problems are explained and in the rest of this chapter we present the results on existence of densities of solutions to elliptic equations for measures and their local properties such as Sobolev regularity. Thus, it turns out that under broad assumptions our equations for measures are reduced to equations for their densities. However, these equations have a rather special form, which leads to certain properties of solutions that are different from the case of general second order equations.

1.1. Background material

Throughout we shall use the following standard notation. The inner product and norm in $\mathbb{R}^d$ are denoted by $\langle\cdot,\cdot\rangle$ and $|\cdot|$, respectively. The diameter of a set Ω is $\operatorname{diam}\Omega = \sup_{x,y\in\Omega}|x-y|$. The open ball of radius r centered at a is denoted by $U(a,r)$ or $U_r(a)$. The unit matrix is denoted by I. The trace of an operator A is denoted by $\operatorname{tr} A$. The inequality $A\leqslant B$ for operators on $\mathbb{R}^d$ means the estimate $\langle Ah,h\rangle \leqslant \langle Bh,h\rangle$, where $h\in\mathbb{R}^d$, for their quadratic forms. In expressions like $a^{ij}x_iy_i$ and b^ix_i the standard summation rule with respect to repeated indices will be meant. Set $u^+ = \max(u,0)$, $u^- = -\min(u,0)$, i.e., $u = u^+ - u^-$.

Throughout "*positive*" means "larger than zero".

The class of all smooth functions with compact support lying in an open set $\Omega\subset\mathbb{R}^d$ is denoted by $C_0^\infty(\Omega)$; the classes of the type $C_b^k(\Omega)$, $C_0^k(\Omega)$ of functions with k continuous derivatives etc. are defined similarly; $C(\Omega)$ and $C_b(\Omega)$ are the classes of continuous and bounded continuous functions. The class of functions whose derivatives up to order k have continuous extensions to the closure of Ω is denoted by the symbol $C^k(\overline{\Omega})$. The support of a function f, i.e., the closure of the set $\{f\neq 0\}$, is denoted by $\operatorname{supp} f$.

A measure μ on a σ-algebra $\mathcal{A}$ in a space Ω is a function $\mu\colon\mathcal{A}\to\mathbb{R}^1$ that is countably additive: $\mu(A)=\sum_{n=1}^\infty \mu(A_n)$ whenever $A_n\in\mathcal{A}$ are pairwise disjoint and their union is A. Such a measure is automatically bounded and can be written as $\mu=\mu^+-\mu^-$, where the measures μ^+ and μ^-, called the positive and negative parts of μ, respectively, are nonnegative and concentrated on disjoint sets $\Omega^+\in\mathcal{A}$ and $\Omega^+\in\mathcal{A}$, respectively, such that $\Omega=\Omega^+\cup\Omega^-$. The measure

$$|\mu| := \mu^+ + \mu^-$$

is called the total variation of the measure μ. The variational norm or the variation of the measure μ is defined by the equality $\|\mu\| := |\mu|(\Omega)$. Let $\mathcal{M}(\Omega)$ be the class of all bounded measures on $(\Omega, \mathcal{A})$ and $\mathcal{P}(\Omega)$ the class of all *probability measures* on $(\Omega, \mathcal{A})$ (i.e., measures $\mu \geqslant 0$ with $\mu(\Omega) = 1$). The simplest probability measure is Dirac's measure δ_a at a point $a \in \Omega$, it equals 1 at the point a and 0 at the complement of a. If $\mu \geqslant 0$ and $\mu(\Omega) \leqslant 1$, then μ is a *subprobability* measure.

It is useful to admit also unbounded measures with values in $[0, +\infty]$ defined similarly. Such a measure is called σ-finite if the space is the union of countably many parts of finite measure. The classical Lebesgue measure on $\mathbb{R}^d$ provides an example. Lebesgue measure of a set Ω will be occasionally denoted by $|\Omega|$. For most of the results discussed below we need only the classical Lebesgue measure and other measures absolutely continuous with respect to it (see below).

We recall that the Borel σ-algebra $\mathcal{B}(E)$ is the smallest σ-algebra containing all open sets of a given space E. The term "a Borel measure μ" will normally mean a finite (possibly signed) countably additive measure on the σ-algebra of Borel sets; cases where infinite measures (say, locally finite measures) are considered will always be specified, except for Lebesgue measure. A Borel measure μ on a subset in $\mathbb{R}^d$ is called locally finite if every point has a neighborhood of finite $|\mu|$-measure.

A finite Borel measure μ on a topological space X is called Radon if, for every Borel set $B \subset X$ and every $\varepsilon > 0$, there is a compact set $K_\varepsilon \subset B$ such that $|\mu|(B \backslash K_\varepsilon) < \varepsilon$. By Ulam's theorem, on all complete separable metric spaces all finite Borel measures are Radon. Throughout we consider only Borel measures.

The integral of a function f with respect to a measure μ over a set A is denoted by the symbols

$$\int_A f(x)\, \mu(dx), \quad \int_A f\, d\mu.$$

For a nonnegative measure μ and $p \in [1, \infty)$, the symbols $L^p(\mu)$ or $L^p(\Omega, \mu)$ denote the space of equivalence classes of μ-measurable functions f such that the function $|f|^p$ is integrable. This space is equipped with the standard norm

$$\|f\|_p := \|f\|_{L^p(\mu)} := \left(\int_\Omega |f|^p\, d\mu \right)^{1/p}.$$

The notation $L^p(\Omega)$ always refers to the classical Lebesgue measure; sometimes we write $L^p(\Omega, dx)$ in order to stress this.

Let $L^\infty(\mu)$ denote the space of equivalence classes of bounded μ-measurable functions equipped with the norm $\|f\|_\infty := \inf_{g \sim f} \sup_x |g(x)|$.

A measure μ is called *separable* if $L^1(\mu)$ is separable (and then so are also all spaces $L^p(\mu)$ for $p < \infty$).

As usual, for $p \in [1, +\infty)$ we set

$$p' := \frac{p}{p-1}.$$

The classical *Hölder inequality* says that

$$\int_\Omega |fg|\, d\mu \leqslant \|f\|_p \|g\|_{p'}, \quad f \in L^p(\mu),\ g \in L^{p'}(\mu).$$

It yields the *generalized Hölder inequality*

$$\int_\Omega |f_1 \cdots f_n|\, d\mu \leqslant \|f_1\|_{p_1} \cdots \|f_n\|_{p_n}, \quad f_i \in L^{p_i}(\mu),\ p_1^{-1} + \cdots + p_n^{-1} = 1.$$

In addition, if $pq \geqslant p+q$, $f \in L^p(\mu)$ and $g \in L^q(\mu)$, then by Hölder's inequality $fg \in L^r(\mu)$ and

$$\|fg\|_r \leqslant \|f\|_p \|g\|_q \quad \text{if } r = pq/(p+q). \tag{1.1.1}$$

The integrability of a function with respect to a signed measure μ is understood as its integrability with respect to the total variation $|\mu|$ of the measure μ; the corresponding classes will be denoted by $L^p(\mu)$ or $L^p(|\mu|)$ and by $L^p(U,\mu)$ or $L^p(U,|\mu|)$ in the case where μ is restricted to a fixed set $U \subset \Omega$.

For a Radon measure μ, the class $L^1_{\mathrm{loc}}(\mu)$ consists of all functions that are integrable with respect to μ on all compact sets.

Let I_A denote the indicator function of the set A, i.e., $I_A(x) = 1$ if $x \in A$, $I_A(x) = 0$ if $x \notin A$.

A measure ν on a σ-algebra $\mathcal{A}$ is called absolutely continuous with respect to a measure μ on the same σ-algebra if the equality $|\mu|(A) = 0$ implies the equality $\nu(A) = 0$; notation: $\nu \ll \mu$. By the Radon–Nikodym theorem this is equivalent to the existence of a function ϱ integrable with respect to $|\mu|$ such that

$$\nu(A) = \int_A \varrho(x)\,\mu(dx), \quad A \in \mathcal{A}.$$

The function ϱ is called the *density* (the *Radon–Nikodym density*) of the measure ν with respect to the measure μ and is denoted by the symbol $d\nu/d\mu$. It is customary to write also

$$\nu = \varrho \cdot \mu \quad \text{or} \quad \nu = \varrho\mu.$$

If $\nu \ll \mu$ and $\mu \ll \nu$, then the measures ν and μ are *equivalent*; notation: $\nu \sim \mu$. This is equivalent to the following property: $\nu \ll \mu$ and $d\nu/d\mu \neq 0$ $|\mu|$-almost everywhere. The term "almost everywhere" is shortened as μ-a.e. (for a signed measure μ, the term "μ-a.e." is understood as "$|\mu|$-a.e.").

A sequence of Borel measures μ_n converges weakly to a Borel measure μ if for every bounded continuous function f one has

$$\lim_{n\to\infty} \int f\,d\mu_n = \int f\,d\mu.$$

A family $\mathcal{M}$ of Radon measures on a topological space X is called uniformly tight if for each $\varepsilon > 0$ there is a compact set $K_\varepsilon \subset X$ such that $|\mu|(X\backslash K_\varepsilon) < \varepsilon$ for all measures $\mu \in \mathcal{M}$. According to the Prohorov theorem, a bounded family of Borel measures on a complete separable metric space is uniformly tight precisely when every infinite sequence in it contains a weakly convergent subsequence (see Bogachev [**125**, Chapter 8]).

Given an open set $\Omega \subset \mathbb{R}^d$ and $p \in [1, +\infty)$, we denote by $W^{p,1}(\Omega)$ or $H^{p,1}(\Omega)$ the Sobolev class of all functions $f \in L^p(\Omega)$ whose generalized partial derivatives $\partial_{x_i} f$ are in $L^p(\Omega)$. A generalized (or Sobolev) derivative is defined by the equality (the integration by parts formula)

$$\int_U \varphi \partial_{x_i} f\,dx = -\int_U f \partial_{x_i} \varphi\,dx, \quad \varphi \in C_0^\infty(\Omega).$$

This space is equipped with the Sobolev norm

$$\|f\|_{p,1} := \|f\|_p + \sum_{i=1}^d \|\partial_{x_i} f\|_p.$$

We also use higher-order Sobolev classes $W^{p,k}(\Omega) = H^{p,k}(\Omega)$ with $k \in \mathbb{N}$, consisting of functions whose Sobolev partial derivatives up to order k are in $L^p(\Omega)$ and equipped with naturally defined norms $\|f\|_{p,k}$, and fractional Sobolev spaces $H^{p,r}(\Omega)$ with noninteger r (the definition is given in §1.8); the notation with the letter H will normally be used in the case of fractional or parabolic Sobolev classes.

The class $W^{\infty,k}(\Omega)$ consists of functions with bounded Sobolev derivatives up to order k; for example, $W^{\infty,1}(\Omega)$ is the class of bounded Lipschitzian functions. Let $\nabla f = (\partial_{x_1} f, \ldots, \partial_{x_d} f)$.

The class $W_0^{p,k}(\Omega)$ is defined as the closure of $C_0^\infty(\Omega)$ in $W^{p,k}(\Omega)$.

The space $C^{0,\delta}(\Omega)$ consists of Hölder continuous of order $\delta \in (0,1)$ functions f on Ω with finite norm

$$\|f\|_{C^{0,\delta}} := \sup_{x\in\Omega} |f(x)| + \sup_{x,y\in\Omega, x\neq y} |f(x) - f(y)|/|x-y|^\delta.$$

Symbols like $W_{\text{loc}}^{p,1}(\mathbb{R}^d)$, $W_{\text{loc}}^{p,1}(\Omega)$, $L_{\text{loc}}^{p}(\Omega,\mu)$ denote the classes of functions f such that ζf belongs to the corresponding class without the lower index "loc" for every $\zeta \in C_0^\infty(\mathbb{R}^d)$ or $\zeta \in C_0^\infty(\Omega)$, respectively.

Let $W^{p,-1}(\mathbb{R}^d)$ denote the dual space to $W^{p',1}(\mathbb{R}^d)$ with $p' = p/(p-1)$, $p > 1$.

Let us define weighted Sobolev spaces or classes. Let a nonnegative measure μ on $\mathbb{R}^d$ be given by a locally integrable density ϱ with respect to Lebesgue measure. The class $W^{p,k}(\mu)$ is defined as the completion of $C_0^\infty(\mathbb{R}^d)$ with respect to the Sobolev norm $\|\cdot\|_{p,k,\mu}$ defined similarly to $\|\cdot\|_{p,k}$, but with the measure μ in place of Lebesgue measure. If the density ϱ is continuous and positive, then $W^{p,k}(\mu)$ coincides with the class of functions $f \in W_{\text{loc}}^{p,k}(\mathbb{R}^d)$ with $\|f\|_{p,k,\mu} < \infty$. Weighted classes are used below only in a very few places, mostly the classes $W^{p,1}(\mu)$, moreover, in such cases the measure μ has some additional properties, for example, possessing a continuous positive density or a weakly differentiable density, so that the weighted Sobolev classes are well-defined (see, e.g., Bogachev [**126**, §2.6]).

We shall need the class $W_{\text{loc}}^{d+,1}(\Omega)$ consisting of all functions f on an open set Ω such that the restriction of f to each ball U with closure in Ω belongs to $W^{p_U,1}(U)$ for some $p_U > d$, and also the class $L_{\text{loc}}^{d+}(\Omega)$ defined similarly.

In the theory of Sobolev spaces and its applications a very important role is played by the following Sobolev embedding theorem (the case $p=1$ is called the Gagliardo–Nirenberg embedding theorem).

1.1.1. Theorem. (i) *If $p > d$ or $p = d = 1$, then one has the embedding*

$$W^{p,1}(\mathbb{R}^d) \subset C_b(\mathbb{R}^d) = C(\mathbb{R}^d) \cap L^\infty(\mathbb{R}^d).$$

Moreover, there exists a number $C(p,d) > 0$ such that

$$\|f\|_\infty \leqslant C(p,d)\|f\|_{p,1}, \quad f \in W^{p,1}(\mathbb{R}^d). \tag{1.1.2}$$

(ii) *If $p \in [1,d)$, then $W^{p,1}(\mathbb{R}^d) \subset L^{dp/(d-p)}(\mathbb{R}^d)$, hence $L^q(\mathbb{R}^d) \subset W^{p',-1}(\mathbb{R}^d)$ if $q = dp/(dp+p-d)$, $p > 1$. Moreover, there is a number $C(p,d) > 0$ such that*

$$\|f\|_{dp/(d-p)} \leqslant C(p,d)\|f\|_{p,1}, \quad f \in W^{p,1}(\mathbb{R}^d). \tag{1.1.3}$$

For any bounded domain Ω with Lipschitzian boundary analogous embeddings hold with some number $C(p,d,\Omega)$.

Note that $p' = qd/(d-q)$ in (ii). Actually in place of (1.1.3) the inequality

$$\|f\|_{dp/(d-p)} \leqslant C(p,d)\big\||\nabla f|\big\|_p \quad \forall f \in W^{p,1}(\mathbb{R}^d) \tag{1.1.4}$$

holds, which for $p = 1$ is called the Galiardo–Nirenberg inequality; it shows that an integrable function on $\mathbb{R}^d$ with an integrable gradient belongs in fact to the class $L^{d/(d-1)}(\mathbb{R}^d)$, hence also to all $L^p(\mathbb{R}^d)$ with $1 \leqslant p \leqslant d/(d-1)$. For functions with support in the unit ball U we obtain the inequality

$$\|f\|_p \leqslant C(p)\big\||\nabla f|\big\|_p, \quad f \in W_0^{p,1}(U). \tag{1.1.5}$$

Note also the *Poincaré inequality*

$$\|f - f_U\|_p \leqslant C(p)\big\||\nabla f|\big\|_p, \quad f \in W^{p,1}(U), \ f_U = \int_U f\,dx. \tag{1.1.6}$$

A function from the class $W^{d,1}(\mathbb{R}^d)$ need not be even locally bounded, but on every ball U it belongs to all $L^r(U)$.

For higher derivatives the following assertions are valid.

1.1.2. Corollary. *One has the following embeddings.*
(i) *If $kp < d$, then $W^{p,k}(\mathbb{R}^d) \subset L^{dp/(d-kp)}(\mathbb{R}^d)$.*
(ii) *If $kp > d$, then $W^{p,k}(\mathbb{R}^d) \subset C(\mathbb{R}^d) \cap L^\infty(\mathbb{R}^d)$.*
(iii) *$W^{1,d}(\mathbb{R}^d) \subset C(\mathbb{R}^d) \cap L^\infty(\mathbb{R}^d)$.*

Hölder norms of Sobolev functions admit the following estimates.

1.1.3. Theorem. *Let $rp > d$, let U be a ball of radius 1 in $\mathbb{R}^d$, and let $f \in W^{p,r}(U)$. Then f has a modification f_0 which satisfies Hölder's condition with exponent $\alpha = \min(1, r - d/p)$, and there exists $C(d,p,r) > 0$ such that for all $x, y \in U$ one has the inequality*

$$|f_0(x) - f_0(y)| \leqslant C(d,p,r)\|f\|_{p,r}|x-y|^\alpha. \tag{1.1.7}$$

If $f \in W_0^{p,r}(U)$, then

$$|f_0(x) - f_0(y)| \leqslant C(d,p,r)\|D^r f\|_{L^p(U)}|x-y|^\alpha, \tag{1.1.8}$$

where $\|D^r f\|_{L^p(U)}$ denotes the $L^p(U)$-norm of the real function

$$x \mapsto \sup_{|v_i| \leqslant 1} |D^r f(x)(v_1, \ldots, v_r)|.$$

A similar assertion is true for domains with sufficiently regular boundaries, but the constants will depend also on the domains.

Unlike the whole space, for a bounded domain $\Omega \subset \mathbb{R}^d$, one has the inclusion $L^p(\Omega) \subset L^r(\Omega)$ whenever $p > r$. This yields a wider spectrum of embedding theorems. We formulate the main results for a ball $U \subset \mathbb{R}^d$. Let us set $W^{q,0} := L^q$.

1.1.4. Theorem. (i) *Let $kp < d$. Then*

$$W^{p,j+k}(U) \subset W^{q,j}(U), \quad q \leqslant \frac{dp}{d-kp}, \ j \in \{0, 1, \ldots\}.$$

(ii) *Let $kp = d$. Then*

$$W^{p,j+k}(U) \subset W^{q,j}(U), \quad q < \infty, \ j \in \{0, 1, \ldots\}.$$

If $p = 1$, then $W^{j+d,1}(U) \subset C_b^j(U)$.
(iii) *Let $kp > d$. Then*

$$W^{p,j+k}(U) \subset C_b^j(U), \quad j \in \{0, 1, \ldots\}.$$

In addition, these embeddings are compact operators, with the exception of case (i) *with $q = dp/(d-kp)$.*

Proofs of all these classic results can be found in the book Adams, Fournier [**3**].

For $p > d$ and any function $f \in W^{p,1}(\mathbb{R}^d)$ with support in a ball of radius R one has the estimate

$$\|f\|_{L^\infty} \leqslant C(p, d, R)\big\||\nabla f|\big\|_p.$$

Neither this estimate nor (1.1.4) hold for functions on bounded domains (for example, for constant functions). Also a constant $C(p, d, R)$ cannot be taken independently of R (excepting the case $d = p = 1$), as simple computations with the functions $f_j(x) = \max(1 - |x|/j, 0)$ show.

Under broad assumptions about a set Ω in $\mathbb{R}^d$, the class $W_0^{p,k}(\Omega)$ (defined above as the closure of $C_0^\infty(\Omega)$ in $W^{p,k}(\Omega)$) admits the following description (see Adams, Fournier [**3**, Theorem 5.29 and Theorem 5.37]).

1.1.5. Theorem. *Let Ω be a bounded open set with smooth boundary. Then the class $W_0^{p,k}(\Omega)$ coincides with the set of functions in $W^{p,k}(\Omega)$ whose extensions by zero outside Ω belong to $W^{p,k}(\mathbb{R}^d)$.*

1.1.6. Corollary. *Let Ω be a bounded open set with smooth boundary. Suppose that $f \in W^{p,k}(\Omega)$, where $p > d$. If the continuous version of f vanishes on $\partial\Omega$ along with its derivatives up to order $k - 1$, then $f \in W_0^{p,k}(\Omega)$.*

Let U_R be an open ball of radius R. First we want to recall some simple properties of the space $W^{p,-1}(U_R)$, which is the dual of $W_0^{p',1}(U_R)$ for $p \in (1, \infty)$. It is known (see, e.g., Adams, Fournier [**3**, Chapter III, Theorem 3.12]) that every $u \in W^{p,-1}(U_R)$ can be written as

$$u = \partial_{x_i} f^i, \quad f^i \in L^p(U_R),\ i = 1, \dots, d, \tag{1.1.9}$$

and, for all representations (1.1.9), one has

$$\|u\|_{W^{p,-1}(U_R)} \leqslant \|f\|_{L^p(U_R)}. \tag{1.1.10}$$

By using scaling to control the norms of the embeddings, we arrive at the following well-known lemma (see, e.g., Gilbarg, Trudinger [**409**, Theorem 7.10]).

1.1.7. Lemma. (i) *Let $d' < r < \infty$ and $R > 0$. Then we have the continuous embedding $L^{rd/(r+d)}(U_R) \subset W^{r,-1}(U_R)$. In addition, there exists a number N independent of R such that*

$$\|u\|_{W^{r,-1}(U_R)} \leqslant N\|u\|_{L^{rd/(r+d)}(U_R)} \tag{1.1.11}$$

for all $u \in L^{rd/(r+d)}(U_R)$ and all $R > 0$.

(ii) *Let $1 < r < d'$ and $R > 0$. Then $L^1(U_R) \subset W^{r,-1}(U_R)$ and the embedding operator is bounded. In addition, there exists a number N independent of R such that*

$$\|u\|_{W^{r,-1}(U_R)} \leqslant NR^{1-d/r'}\|u\|_{L^1(U_R)} \tag{1.1.12}$$

for all $u \in L^1(U_R)$ and all $R > 0$.

(iii) *Let $r = d'$, $s > 1$, and $R > 0$. Then $L^s(U_R) \subset W^{r,-1}(U_R)$. In addition, there exists N independent of R such that*

$$\|u\|_{W^{r,-1}(U_R)} \leqslant NR^{2+d/s}\|u\|_{L^s(U_R)} \tag{1.1.13}$$

for all $u \in L^s(U_R)$ and all $R > 0$.

1.2. Elliptic equations

For convenience of later references we collect here a number of known results about second order elliptic equations. Throughout $\Delta = \partial_{x_1}^2 + \cdots + \partial_{x_d}^2$ is the Laplace operator. An elliptic operator (or a "nondivergence form elliptic operator") is an expression

$$L_{A,b,c}u = a^{ij}\partial_{x_i}\partial_{x_j}u + b^i\partial_{x_i}u + cu,$$

where a^{ij}, b^i and c are functions on $\mathbb{R}^d$, $A = (a^{ij})_{i,j\leqslant d}$, $b = (b^i)_{i=1}^d$ and the summation over repeated upper and lower indices is meant, moreover, $A = A^* \geqslant 0$. Such operators should be distinguished from "divergence form" operators

$$\mathcal{L}_{A,b,c}u = \partial_{x_i}(a^{ij}\partial_{x_j}u + b^i u) + cu,$$

to which it is customary to ascribe also more general operators

$$\mathcal{L}_{A,b,\beta,c}u = \partial_{x_i}(a^{ij}\partial_{x_j}u + b^i u) + \beta^i\partial_{x_i}u + cu, \quad \beta = (\beta^i).$$

As we shall see below, different forms of operators lead not only to different properties of solutions to the equations of the form $L_{A,b,c}u = f$ (a "direct" or "nondivergence form" equation) or $\mathcal{L}_{A,b,c}v = f$ (a "divergence form" equation), but even to different settings of problems. We note at once that our principal object — a stationary Fokker–Planck–Kolmogorov equation — is in general something third.

Nondivergence and divergence form equations are most often solved in Hölder classes (functions with Hölder continuous derivatives up to the second order) and in Sobolev classes. Let us mention the basic facts about Dirichlet problems on domains and about equations on the whole space.

A function u in the class $W_0^{p,1}(\Omega)$ on an open set Ω in $\mathbb{R}^d$ is called a solution of the equation

$$\mathcal{L}_{A,b,\beta,c}u = \nu, \quad \text{where } \nu \in W_0^{p,-1}(\Omega),\ p > 1,$$

if a^{ij}, b^i, β^i, c are measurable, $a^{ij}|\nabla u|, b^i u, \beta^i|\nabla u|, cu \in L^p(\Omega)$, and

$$\int_\Omega \bigl[-\langle A\nabla u - ub, \nabla\varphi\rangle + \varphi\langle\beta, \nabla u\rangle + cu\varphi\bigr]\, dx = \nu(\varphi)$$

for all functions $\varphi \in W_0^{p/(p-1),1}(\Omega)$ or, equivalently, for all $\varphi \in C_0^\infty(\Omega)$, where $\nu(\varphi)$ is the value of the functional ν at φ. In the case of bounded coefficients the required integrability conditions are automatically fulfilled.

Let Ω be a bounded domain in $\mathbb{R}^d$ with smooth boundary, let functions a^{ij} be Hölder continuous on the closure of Ω, and let the matrix $A(x)$ be symmetric and positive definite on $\overline{\Omega}$. It is known (see, for example, Gilbarg, Trudinger [**409**, Theorem 6.14], Krylov [**552**, Theorem 6.5.3]) that for every function $f \in C_0^\infty(\Omega)$ there is a function $u \in C^2(\overline{\Omega})$ such that $u = 0$ on $\partial\Omega$ and

$$a^{ij}\partial_{x_i}\partial_{x_j}u = f \quad \text{on } \Omega.$$

It is known (see [**409**, Lemma 9.17] or Krylov [**556**, Theorem 2, p. 242]) that for every $r > 1$ there is a number C_r independent of f such that

$$\|u\|_{W^{r,2}(\Omega)} \leqslant C_r\|f\|_{L^r(\Omega)}. \tag{1.2.1}$$

If A is merely continuous on $\overline{\Omega}$, then for any $f \in L^r(\Omega)$ the equation $L_{A,b,c}u = f$ with lower order terms $b^i, c \in L^\infty(\Omega)$ has a solution in the space $W^{r,2}(\Omega) \cap W_0^{r,1}(\Omega)$ if $c \leqslant 0$ (say, if $c = 0$). In this case also the indicated estimate holds.

The most general known conditions on the second order coefficients ensuring the solvability in Sobolev classes are formulated in terms of the class VMO consisting of locally integrable functions a on $\mathbb{R}^d$ for each of which there is a positive continuous function ω on $[0, +\infty)$ with $\omega(0) = 0$ such that

$$\sup_{z\in\mathbb{R}^d, r<R} r^{-2d} \int_{U_r(z)\times U_r(z)} |a(x) - a(y)|\, dx\, dy \leqslant \omega(R) \quad \forall R > 0.$$

This class contains all uniformly continuous functions, but includes also some locally unbounded functions. Note that the inclusion $W^{d,1}(\mathbb{R}^d) \subset VMO$ holds. About these conditions, see Dong [**308**], [**309**], Krylov [**557**] and the references given there. The following result is proved in Krylov [**555**].

1.2.1. Theorem. *Suppose that*

$$a^{ij} \in VMO, \ A(x) \geqslant \varepsilon \cdot \mathrm{I}, \ |a^{ij}| + |b^i| + |\beta^i| + |c| \leqslant K.$$

Then, for every $p > 1$, there are numbers λ_0 and M depending only on p, d, K, ε and a common for all a^{ij} function ω from the condition of the membership in VMO such that for all $\lambda \geqslant \lambda_0$ and $f, g^1, \ldots, g^d \in L^p(\mathbb{R}^d)$ the equations

$$L_{A,b,c}u - \lambda u = f \quad \textit{and} \quad \mathcal{L}_{A,b,\beta,c}v - \lambda v = f + \operatorname{div} g, \ g = (g^1, \ldots, g^d),$$

have unique solutions $u\in W^{p,2}(\mathbb{R}^d)$ and $v\in W^{p,1}(\mathbb{R}^d)$ and

$$\lambda\|u\|_p + \|u\|_{p,2} \leqslant M\|(L_{A,b,c} - \lambda)u\|_p, \tag{1.2.2}$$

$$\|v\|_{p,1} \leqslant M(\|f\|_p + \|g\|_p). \tag{1.2.3}$$

Thus, the operator $L_{A,b,c} - \lambda$ is an isomorphism between the spaces $W^{p,2}(\mathbb{R}^d)$ and $L^p(\mathbb{R}^d)$, $\mathcal{L}_{A,b,\beta,c} - \lambda$ is an isomorphism between $W^{p,1}(\mathbb{R}^d)$ and $W^{p,-1}(\mathbb{R}^d)$.

Estimate (1.2.3) means that $\|v\|_{p,1} \leqslant M\|\mathcal{L}_{A,b,\beta,c}v - \lambda v\|_{p,-1}$.

1.2.2. Corollary. *Under the conditions indicated in the theorem, for every ball U, whenever $\lambda \geqslant \lambda_0$, we have the estimate*

$$\|u\|_{W_0^{p,1}(U)} \leqslant M\|\mathcal{L}_{A,b,\beta,c}u - \lambda u\|_{W^{p,-1}(U)}, \quad u \in W_0^{p,1}(U). \tag{1.2.4}$$

PROOF. Note that by defining u by zero outside U we obtain a function in $W^{p,1}(\mathbb{R}^d)$ with the same norm, but the norm of $\mathcal{L}_{A,b,\beta,c}u$ will change, so for justifying (1.2.4) we use a different reasoning. We take a sequence of functions $u_n \in C_0^\infty(U)$ converging to u in $W^{p,1}(U)$. Set $\mathcal{L} = \mathcal{L}_{A,b,\beta,c}$. Then

$$\|u_n\|_{W^{p,1}(U)} = \|u_n\|_{W^{p,1}(\mathbb{R}^d)} \leqslant M\|\mathcal{L}u_n - \lambda u_n\|_{W^{p,-1}(\mathbb{R}^d)} = M\|\mathcal{L}u_n - \lambda u_n\|_{W^{p,-1}(U)},$$

where the left-hand side converges to $\|u_n\|_{W^{p,1}(U)}$ and the right-hand side converges to $M\|\mathcal{L}u - \lambda u\|_{W^{p,-1}(U)}$, which follows by the estimate

$$\|\mathcal{L}v\|_{W^{p,-1}(U)} \leqslant \big\||A\nabla v| + |\beta|\,|\nabla v| + |bv| + |cv|\big\|_{L^p(U)}$$

and the boundedness of the coefficients. □

For equations without lower order terms (or under some other additional conditions) one can take $\lambda = 0$. For the proof of the following result under more general conditions (in particular, with a bounded domain with C^1-boundary in place of a ball), see Auscher, Qafsaoui [**78**], Byun [**214**]. Let us derive it from the previous corollary.

1.2.3. Proposition. *Under the conditions on A from the theorem, for the operator $\mathcal{L}_A = \mathcal{L}_{A,0,0,0}$ and every ball U one can find $M > 0$ such that*

$$\|u\|_{W_0^{p,1}(U)} \leqslant M\|\mathcal{L}_A u\|_{W^{p,-1}(U)}, \quad u \in W_0^{p,1}(U). \tag{1.2.5}$$

Moreover, for any $f, g^1, \ldots, g^d \in L^p(U)$ the equation

$$\mathcal{L}_A u = f + \operatorname{div} g, \quad g = (g^1, \ldots, g^d)$$

has a unique solution in $W_0^{p,1}(U)$.

PROOF. First we observe that our estimate yields the existence of a solution taking into account that the number M in (1.2.4) according to Theorem 1.2.1 depends on A only through p, d, K, ε and the function ω. Indeed, we can approximate A in $L^p(U)$ by a sequence of smooth mappings A_k with common parameters indicated above. The sequence of solutions $u_k \in W_0^{p,1}(U)$ of the equations $\mathcal{L}_{A_k} u_k = f + \operatorname{div} g$ turns out to be bounded in $W_0^{p,1}(U)$, hence a subsequence $\{u_{k_n}\}$ converges weakly in $W_0^{p,1}(U)$ to some function u, which obviously will be a solution to $\mathcal{L}_A u = f + \operatorname{div} g$.

We now establish estimate (1.2.5). Suppose that it fails. Then there exist functions $u_n \in W_0^{p,1}(U)$ such that

$$\|u_n\|_{W_0^{p,1}(U)} = 1, \quad \|\mathcal{L}_A u_n)\|_{W^{p,-1}(U)} \leqslant 1/n.$$

We observe that $\{u_n\}$ converges in $W_0^{p,1}(U)$, since otherwise there is a subsequence $\{v_n\}$ with $\|v_n - v_k\|_{W_0^{p,1}(U)} \geqslant c > 0$, whence we obtain

$$\|(\mathcal{L}_A - \lambda_0)(v_n - v_k)\|_{W^{p,-1}(U)} \geqslant c/M,$$

hence $\|v_n - v_k\|_{W^{p,-1}(U)} \geqslant c/(2M\lambda_0)$ for sufficiently large n. This contradicts the compactness of the embedding $W_0^{p,1}(U) \subset W^{p,-1}(U)$. Thus, there is $u = \lim_{n\to\infty} u_n$ in $W_0^{p,1}(U)$. Then $\|u\|_{W_0^{p,1}(U)} = 1$, but $\mathcal{L}_A u = 0$, i.e., the integral of $\langle A\nabla u, \nabla\varphi\rangle$ vanishes for all functions $\varphi \in C_0^\infty(U)$, then also for all $\varphi \in W_0^{p',1}(U)$, whence it follows that $u = 0$. Indeed, if $p \geqslant 2$, then the integral of $\langle A\nabla u, \nabla u\rangle$ over U vanishes, which is only possible if $u = 0$, since $u \in W_0^{p,1}(U)$. The estimate proven for $p \geqslant 2$ yields also the existence of a solution, as observed above.

We can now complete our proof of (1.2.5) in the case $1 < p < 2$. It remains to show that $u = 0$ if $u \in W_0^{p,1}(U)$ and $\mathcal{L}_A u = 0$. As shown above, we can solve the equation $\mathcal{L}_A w = \operatorname{sign} u$ in $W_0^{p',1}(U)$. Then the integral of $|u|$ equals the integral of $-\langle A\nabla w, \nabla u\rangle$, which equals the vanishing integral of $-\langle\nabla w, A\nabla u\rangle$. □

1.2.4. Corollary. *Let the conditions on A indicated in Theorem* 1.2.1 *hold and $u \in W_0^{q,1}(U)$ for some $q > 1$. If*

$$\mathcal{L}_A u = f + \operatorname{div} g, \quad g = (g^1, \ldots, g^d),$$

where $f, g^i \in L^p(U)$ and $p > q$, then $u \in W_0^{p,1}(U)$.

PROOF. Let $w \in W_0^{p,1}(U)$ be a solution of the equation $\mathcal{L}_A u = f + \operatorname{div} g$, which exists by Proposition 1.2.3. Then the difference $v = u - w \in W_0^{q,1}(U)$ satisfies the homogeneous equation $\mathcal{L}_A v = 0$, but this equation has only zero solution in the class $W_0^{q,1}(U)$. Therefore, $u = w$ almost everywhere. □

Below we need the following technical assertion which follows from the previous proposition and embedding theorems.

1.2.5. Lemma. *Let p and q be two numbers satisfying the estimates $p \geqslant d$, $q \geqslant p'$, but not such that $p = d = q'$. Let $R_1 > 0$. Assume that the functions $a^{ij} \in W^{p,1}(U_{R_1})$ are continuous and $A \geqslant \lambda \cdot \mathrm{I}$ for some $\lambda > 0$. Then, there exist $N > 0$ and $R_0 > 0$ depending only on p, q, d, λ, R_1, the modulus of continuity of A, $\|a^{ij}\|_{W^{p,1}(U_{R_1})}$, and the rate of decreasing to zero of $\|\nabla a^{ij}\|_{L^d(U_R)}$ as $R \to 0$, such that for all $R < R_0$ and $\varphi \in W_0^{q,1}(U_R)$, one has*

$$f := a^{ij}\partial_{x_i}\partial_{x_j}\varphi \in W^{q,-1}(U_R) \quad \text{and} \quad \|\nabla\varphi\|_{L^q(U_R)} \leqslant N\|f\|_{W^{q,-1}(U_R)}. \tag{1.2.6}$$

PROOF. We may assume that $R_1 = 1$. Note $f \in W^{q,-1}(U_R)$. This follows from the fact that, for every bounded function $\zeta \in W^{p,1}(U_R)$, the operator $\psi \mapsto \zeta\psi$ is continuous on $W_0^{q',1}(U_R)$ by the estimate

$$\|\nabla(\zeta\psi)\|_{L^{q'}(U_R)} \leqslant C\|\nabla\psi\|_{L^{q'}(U_R)}.$$

Indeed, if $q' < d$, we have $|\nabla\zeta| \in L^p(U_R)$ and $\psi \in L^{q'd/(d-q')}(U_R)$. Hence by Hölder's inequality $|\psi\nabla\zeta| \in L^s(U_R)$, where

$$s = \frac{pq'd/(d-q')}{p + q'd/(d-q')} = \frac{pq'd}{pd - pq' + q'd} = q'\frac{pd}{pd - pq' + q'd} \geqslant q'.$$

In addition,

$$\begin{aligned}\|\psi\nabla\zeta\|_{L^s(U_R)} &\leqslant \|\psi\|_{L^{q'd/(d-q')}(U_R)}\|\nabla\zeta\|_{L^p(U_R)}\\ &\leqslant C\|\nabla\psi\|_{L^{q'}(U_R)}\|\nabla\zeta\|_{L^p(U_R)}.\end{aligned}$$

The case $q' > d$, where ψ is bounded and $q' \leqslant p$ (since $q \geqslant p'$), and the case $q' = d$, where $q' < p$ and ψ is in all $L^r(U_R)$, $r < \infty$, are similar. We note that

$$\partial_{x_i}(a^{ij}\partial_{x_j}\varphi) = f + \partial_{x_i}a^{ij}\partial_{x_j}\varphi =: f + g.$$

By the previous proposition

$$\|\nabla\varphi\|_{L^q(U_R)} \leqslant N_1\Bigl(\|g\|_{W^{q,-1}(U_R)} + \|f\|_{W^{q,-1}(U_R)}\Bigr), \tag{1.2.7}$$

where N_1 is independent of $R \in (0,1]$ and φ. Let $a = (a^j)$, $a^j = \partial_{x_i}a^{ij}$.

Now we consider three cases.

Case $q > d'$. By Lemma 1.1.7(i) and (1.1.1) we have

$$\|g\|_{W^{q,-1}(U_R)} \leqslant N_2\|a\|_{L^d(U_R)}\|\nabla\varphi\|_{L^q(U_R)},$$

which along with (1.2.7) yields

$$\|\nabla\varphi\|_{L^q(U_R)} \leqslant N_1N_2\|a\|_{L^d(U_R)}\|\nabla\varphi\|_{L^q(U_R)} + N_1\|f\|_{W^{q,-1}(U_R)}. \tag{1.2.8}$$

We emphasize that N_1 and N_2 are independent of R and f and note that, since $|\nabla a^{ij}| \in L^p(U_R)$ and $p \geqslant d$, we can choose R so small that

$$N_1N_2\|a\|_{L^d(U_R)} \leqslant 1/2.$$

For such an R, inequality (1.2.8) implies (1.2.6).

Case $p' < q \leqslant d'$. In that case it follows from $q > p'$ that, for r defined by $rd/(r+d) = pq/(p+q)$, we have $r > d' \geqslant q$. Therefore, for $R \in (0,1)$, we obtain

$$\|g\|_{W^{q,-1}(U_R)} \leqslant N_3R^{d(r-q)/(rq)}\|g\|_{W^{r,-1}(U_R)} \leqslant N\|a\|_{L^p(U_R)}\|\nabla\varphi\|_{L^q(U_R)},$$

and we can finish the proof as above.

Case $q = p' < d'$. As is easy to see, this is the only remaining case. By Lemma 1.1.7 and (1.1.1), for $R \in (0,1)$, we have

$$\|g\|_{W^{q,-1}(U_R)} \leqslant N_4 R^{1-d/p} \|g\|_{L^1(U_R)} \leqslant N_4 \|a\|_{L^p(U_R)} \|\nabla\varphi\|_{L^q(U_R)},$$

and the argument from the first case applies again to complete the proof. □

In § 1.7 we return to divergence form equations.

An important role in the theory of partial differential equations is played by various maximum principles. These principles can be of the following types:

1) the weak maximum principle asserts that, under certain conditions, if $Lu \leqslant 0$ in a domain Ω, $u|_{\partial\Omega} \geqslant 0$, then $u \geqslant 0$ in Ω; this maximum principle is discussed in Chapter 2; note that if $Lu = 0$ and $L1 = 0$, then the maximum and minimum of u are attained at the boundary;

2) the strong maximum principle asserts that, under certain conditions (see Gilbarg, Trudinger [**409**, § 3.2]), if $Lu = 0$ and u attains its minimum or maximum in the interior of Ω, then u is a constant. Let us give a precise formulation.

1.2.6. Theorem. *Suppose that* $L_{A,b}u \geqslant 0$ *in a connected open set* Ω, *where* $u \in C^2(\Omega)$, $c_1 \cdot \mathrm{I} \leqslant A(x) \leqslant c_2 \cdot \mathrm{I}$ *with constant* $c_1, c_2 > 0$, *and* b *is bounded. If* u *attains its maximum in the interior of* Ω, *then* u *is constant in* Ω.

1.3. Diffusion processes

Fokker–Planck–Kolmogorov equations arise naturally in the study of diffusion processes. Here we recall basic concepts and consider some examples. These concepts are not used in the main part of the book, but some acquaintance with them is useful for understanding the origins of the central problems of the book and the character of the most important applications of the presented analytical results.

First we define the concept of a Markov transition function on a measurable space $(X, \mathcal{X})$ in which all singletons belongs to $\mathcal{X}$. Suppose we are given a nonempty set $T \subset \mathbb{R}$. A function $(s,x,t,B) \mapsto P(s,x,t,B)$ defined for all $s,t \in T$ with $s \leqslant t$, $x \in X$ and $B \in \mathcal{X}$ is called a Markov transition function if

1) for all fixed s,t,x, the function $B \mapsto P(s,x,t,B)$ is a probability measure on $\mathcal{X}$ and for $s = t$ it is Dirac's measure at the point x;

2) for all fixed s,t,B, the function $x \mapsto P(s,x,t,B)$ is measurable with respect to $\mathcal{X}$;

3) whenever $s,t,u \in T$ and $s \leqslant t \leqslant u$, for all $x \in X$ and $B \in \mathcal{X}$ we have the equality

$$P(s,x,u,B) = \int_X P(t,y,u,B)\, P(s,x,t,dy), \tag{1.3.1}$$

called the *Chapman–Kolmogorov equation.*

A random process $\{\xi_t\}_{t\in T}$ with values in X is called a Markov process with the given transition function $P(s,x,t,B)$ if, for all $t,u \in T$ with $t \leqslant u$ and $B \in \mathcal{X}$, the function $P(t,\xi_t,u,B)$ serves as a conditional probability $P(\xi_u \in B|\mathcal{F}_{\leqslant t})$ with respect to the σ-algebra $\mathcal{F}_{\leqslant t}$ generated by the random elements ξ_s with $s \leqslant t$.

The quantity $P(s,x,t,B)$ can be interpreted as the probability of hitting the set B by the process at the time t under the condition that it is at the point x at the time $s \leqslant t$. So the measures $P(s,x,t,\,\cdot\,)$ are also called the transition probabilities of the process. It is also possible to consider Markov families $\{\xi_{s,x,t}\}$ for which $s \leqslant t$ and $\xi_{s,x,s} = x$. Certainly, in the general case there is no requirement that the process must be at a fixed point at the initial time. If $P(s,x,t,\,\cdot\,)$ depends on s,t through $t-s$, then the process is called *homogeneous*; in this case

$$P(s,x,t,\,\cdot\,) = P(0,x,t-s,\,\cdot\,) =: P(x,t-s,\,\cdot\,).$$

The one-dimensional distributions P_t of the process $\{\xi_t\}_{t\in T}$ are defined by the equality $P_t(B) := P(\xi_t \in B)$. A necessary and sufficient condition that a process be Markov with the given transition function is the equality

$$(1.3.2)\quad P((\xi_{t_1},\dots,\xi_{t_n}) \in C) = \int_X \cdots \int_X I_C(x_1,\dots,x_n) \times P(t_{n-1},x_{n-1},t_n,dx_n)\cdots P(t_1,x_1,t_2,dx_2)P_{t_1}(dx_1)$$

for all $C \in \mathcal{X}^n$ and $t_i \in T$ with $t_1 < \cdots < t_n$.

A somewhat more general concept is obtained if in place of the family of the σ-algebras $\mathcal{F}_{\leqslant t}$ we take an increasing family of σ-algebras $\mathcal{F}_t$ with the property that ξ_t is $\mathcal{F}_t$-measurable.

Let $U(x,\varepsilon) = \{y\colon |x-y| < \varepsilon\}$, $V(x,\varepsilon) = \{y\colon |x-y| > \varepsilon\}$. A Markov process with values in $\mathbb{R}^d$ with transition probabilities $P(s,x,t,B)$ is called a *diffusion process* or a *diffusion* (see Wentzell [**937**] or Gikhman, Skorokhod [**408**]) if there is a mapping $b\colon \mathbb{R}^d \times [0,+\infty) \to \mathbb{R}^d$, called the *drift coefficient*, and a mapping $(x,t) \mapsto A(x,t)$ with values in the space of symmetric operators on $\mathbb{R}^d$, called the *diffusion coefficient* or *diffusion matrix*, such that

(i) for all $\varepsilon > 0$, $t \geqslant 0$ and $x \in \mathbb{R}^d$ we have

$$\lim_{h\to 0} h^{-1} P\bigl(t,x,t+h,V(x,\varepsilon)\bigr) = 0,$$

(ii) for some $\varepsilon > 0$ and all $t \geqslant 0$, $x \in \mathbb{R}^d$ we have

$$\lim_{h\to 0} h^{-1} \int_{U(x,\varepsilon)} (y-x)\, P(t,x,t+h,dy) = b(x,t),$$

(iii) for some $\varepsilon > 0$ and all $t \geqslant 0$, $x,z \in \mathbb{R}^d$ we have

$$\lim_{h\to 0} h^{-1} \int_{U(x,\varepsilon)} \langle y-x, z\rangle^2\, P(t,x,t+h,dy) = 2\langle A(x,t)z, z\rangle.$$

If A and b do not depend on t, then the diffusion is homogeneous.

1.3.1. Proposition. *Suppose that relations* (i)–(iii) *hold locally uniformly in x and the functions a^{ij}, b^i are locally bounded. Then the transition probabilities satisfy the parabolic Fokker–Planck–Kolmogorov equation*

$$\partial_t \mu = \partial_{x_i}\partial_{x_j}(a^{ij}\mu) - \partial_{x_i}(b^i \mu)$$

in the sense of generalized functions (*see Chapter* 6). *If ν is a finite Borel measure on $\mathbb{R}^d$ and*

$$\mu_t(dx) = \int_{\mathbb{R}^d} P(0,y,t,dx)\,\nu(dy),$$

then the measure $\mu = \mu_t(dx)\,dt$ gives a solution to the Cauchy problem with the initial condition $\mu\big|_{t=0} = \nu$.

PROOF. We give a brief justification, see details in Wentzell [**937**, § 11.2] or Gikhman, Skorokhod [**408**, Chapter 1, § 1]. Let $f \in C_0^\infty(\mathbb{R}^d)$. Then

$$\frac{d}{dt}\int_{\mathbb{R}^d} f(y)\,P(s,x,t,dy)$$
$$= \lim_{h\to 0} h^{-1}\Bigl(\int_{\mathbb{R}^d} f(y)\,P(s,x,t+h,dy) - \int_{\mathbb{R}^d} f(z)\,P(s,x,t,dz)\Bigr).$$

By the Chapman–Kolmogorov equation the right-hand side equals

$$\lim_{h\to 0}\int_{\mathbb{R}^d} h^{-1}\int_{\mathbb{R}^d} (f(y)-f(z))\,P(t,z,t+h,dy)\,P(s,x,t,dz).$$

By using conditions (i)–(iii) and Taylor's expansion for f, we obtain

$$\lim_{h\to 0} h^{-1}\int_{\mathbb{R}^d} (f(y)-f(z))\,P(t,z,t+h,dy)$$
$$= a^{ij}(z,t)\partial_{z_i}\partial_{z_j} f(z) + b^i(z,t)\partial_{z_i} f(z).$$

Since convergence as $h\to 0$ is uniform in z, we have

$$\frac{d}{dt}\int_{\mathbb{R}^d} f(y)\,P(s,x,t,dy)$$
$$= \lim_{h\to 0}\int_{\mathbb{R}^d} \bigl(a^{ij}(z,t)\partial_{z_i}\partial_{z_j} f(z) + b^i(z,t)\partial_{z_i} f(z)\bigr)\,P(s,x,t,dz).$$

Thus, we have proved that the transition probabilities satisfy the indicated equation. In addition, for each function $\zeta \in C_0^\infty(\mathbb{R}^d)$, condition (i) gives the equality

$$\lim_{h\to 0}\int_{\mathbb{R}^d} \zeta(y)\,P(s,x,s+h,dy) = \zeta(x),$$

i.e., $P(s,x,t,dy)$ satisfies the condition $P\big|_{t=s} = \delta_x$. This proves the last assertion in the case where $\nu = \delta_x$. The general case follows by integration with respect to ν. □

In the case where the transition probabilities $P(s,x,t,dy)$ are given by densities $\varrho(s,x,t,y)$ with respect to Lebesgue measure, in the variables (y,t) they satisfy the above Fokker–Planck–Kolmogorov equation (also called the forward Kolmogorov equation)

$$\partial_t\varrho(s,x,t,y) = \partial_{y_i}\partial_{y_j}\bigl(a^{ij}(y,t)\varrho(s,x,t,y)\bigr) - \partial_{y_i}\bigl(b^i(y,t)\varrho(s,x,t,y)\bigr),$$

and in the variables (x,s) they satisfy the backward Kolmogorov equation

$$-\partial_s\varrho(s,x,t,y) = a^{ij}(x,s)\partial_{x_i}\partial_{x_j}\varrho(s,x,t,y) + b^i(x,s)\partial_{x_i}\varrho(s,x,t,y).$$

If A and b do not depend on t, then

$$P(s,x,t,dy) = P(0,x,t-s,dy)$$

under broad assumptions, i.e., the transition probabilities are determined by the probabilities

$$P(x,t,dy) = P(0,x,t,dy).$$

In case the latter have densities $\varrho(x,t,y)$, the backward Kolmogorov equation takes the form

$$\partial_t\varrho(x,t,y) = a^{ij}(x)\partial_{x_i}\partial_{x_j}\varrho(x,t,y) + b^i(x)\partial_{x_i}\varrho(x,t,y).$$

The differential operator

$$L_{A,b}\varphi = a^{ij}\partial_{x_i}\partial_{x_j}\varphi + b^i\partial_{x_i}\varphi$$

is called the *generator* of the given process. This terminology is connected with the fact that under suitable conditions the operators

$$T_t f(x) = \int_{\mathbb{R}^d} f(y)\, P(x,t,dy)$$

form a semigroup in a suitable functional space (semigroups will be discussed in Chapter 5). Then the forward Kolmogorov equation (which is the Fokker–Planck–Kolmogorov equation) reads $\partial_t T_t f = T_t L f$ and the backward equation becomes $\partial_t T_t f = L T_t f$. Say, if $A = \mathrm{I}$, $b^i \in C_b^\infty(\mathbb{R}^d)$, then the given formal relations have the usual meaning for $f \in C_b^\infty(\mathbb{R}^d)$.

In the case of a homogeneous process an important concept of a *stationary distribution* or *invariant measure* of the process (or of its transition semigroup) arises. This is a probability measure μ such that

$$\mu(B) = \int_{\mathbb{R}^d} P(y,t,B)\,\mu(dy) \quad \forall t \geqslant 0,\ B \in \mathcal{B}(\mathbb{R}^d).$$

It follows from what has been said that any stationary distribution satisfies the *stationary Fokker–Planck–Kolmogorov equation*

$$\partial_{x_i}\partial_{x_j}(a^{ij}\mu) - \partial_{x_i}(b^i\mu) = 0,$$

which will be the main object of study in Chapters 1–5.

Diffusion processes can be considered also in a broader sense, for example, one can consider almost surely continuous Markov processes in $\mathbb{R}^d$ such that their transition probabilities $P(s,x,t,dy)$ satisfy the Fokker–Planck–Kolmogorov equation with the initial condition $P\big|_{t=s} = \delta_x$. Such processes are called *quasi-diffusions*.

Since the distribution of a Markov process (or the family of its finite-dimensional distributions) is completely determined by its initial distribution and its transition probabilities, uniqueness of a probability solution to the Cauchy problem for the Fokker–Planck–Kolmogorov equation yields the weak uniqueness of the diffusion process whose transition probabilities satisfy this equation.

We recall that the conditional expectation of an integrable function ξ on a probability space $(\Omega, \mathcal{A}, \mu)$ with respect to a σ-algebra $\mathcal{B} \subset \mathcal{A}$ is a $\mathcal{B}$-measurable integrable function $\mathbb{E}_\mu[\xi|\mathcal{B}]$ such that

$$\int_\Omega \eta\xi\, d\mu = \int_\Omega \eta\mathbb{E}_\mu[\xi|\mathcal{B}]\, d\mu$$

for every bounded $\mathcal{B}$-measurable function η.

Jensen's inequality for the conditional expectation says that if V is a convex function and $V(\xi) \in L^1(\mu)$, then a.e.

$$V(\mathbb{E}_\mu[\xi|\mathcal{B}]) \leqslant \mathbb{E}_\mu[V(\xi)|\mathcal{B}].$$

A real or vector random process $\{\xi_t\}_{t\in T}$ with a directed index set T is called a *martingale* with respect to a family of σ-algebras $\mathcal{F}_t$ that is increasing in the sense that $\mathcal{F}_s \subset \mathcal{F}_t$ whenever $s \leqslant t$, provided that the element ξ_s is measurable with respect to $\mathcal{F}_s$, integrable and almost surely $\xi_s = \mathbb{E}[\xi_t|\mathcal{F}_s]$ for $t \geqslant s$, where $\mathbb{E}[\xi_t|\mathcal{F}_s]$ is the conditional expectation (the existence of the conditional expectation follows by the integrability).

One of the most important examples of a Markov process which is also a martingale (and one of the most important for applications processes) is the Wiener process.

1.3.2. Example. A Wiener process (or a Brownian motion) $\{w_t\}_{t\geqslant 0}$ is a real random process with the following properties:

(i) the trajectory $t \mapsto w_t(\omega)$ is continuous for every ω and $w_0 = 0$,

(ii) the random variables $w_{t_1}, w_{t_2} - w_{t_1}, \ldots, w_{t_n} - w_{t_{n-1}}$ are independent, whenever $0 \leqslant t_1 < t_2 < \cdots < t_n$,

(iii) for each t the random variable w_t is Gaussian with zero mean and variance t, i.e., $\mathbb{E}w_t^2 = t$.

A Wiener process in $\mathbb{R}^d$ is just a collection $(w_t^1, \ldots, w_t^d)$ of independent Wiener processes. The existence of Wiener processes is not straightforward and is proved in many textbooks (see, e.g., Wentzell [**937**]).

Let $\mathcal{F}_t$ be the σ-algebra generated by the variables w_s with $s \leqslant t$. Then with respect to this family the Wiener process is a martingale, since for $s \leqslant t$ the conditional expectation of $w_t - w_s$ with respect to $\mathcal{F}_s$ vanishes by condition (ii). In addition, the Wiener process is Markov with respect to the indicated family with the transition function $P(s, x, t, \,\cdot\,)$ defined as follows: if $s = t$, then this is Dirac's measure at the point x, if $s < t$, then this is the Gaussian measure with mean x and variance $t - s$, i.e., the measure with density $y \mapsto (2\pi)^{-1/2} \exp\bigl[-(y-x)^2/(2t-2s)\bigr]$. The Chapman–Kolmogorov equation is verified directly. The Markov property is verified by means of (1.3.2).

A diffusion process with a nonzero drift is not a martingale, which can be seen from the Itô equation (see below).

The most important way of constructing diffusion processes is solving stochastic differential equations. First we introduce the Itô integral.

Suppose we are given a Wiener process $\{w_t\}_{t\geqslant 0}$ and a process $\{\xi_t\}_{t\geqslant 0}$ measurable in (ω, t) such that the variable ξ_t is measurable with respect to the σ-algebra $\mathcal{F}_t$ generated by the variables w_s with $s \leqslant t$ (such a process is called adapted). Let also $T > 0$. Suppose that

$$\int_0^T \mathbb{E}|\xi_t|^2 \, dt < \infty.$$

The stochastic Itô integral

$$\int_0^T \xi_t \, dw_t$$

is defined as follows. If there are points $0 = t_1 < \cdots < t_n = T$ such that $\xi_t = \xi_{t_i}$ whenever $t_i \leqslant t < t_{i+1}$, then this integral is naturally defined as the sum $\sum_{i=1}^n \xi_{t_i}(\omega)\bigl(w_{t_{i+1}}(\omega) - w_{t_i}(\omega)\bigr)$. In the general case the function $\xi_t(\omega)$ is approximated in $L^2(P\otimes dt)$ by a sequence of functions of the indicated form and it is proved that the stochastic integrals of the approximations converge in $L^2(P)$; the limit is taken for the stochastic integral of the original process.

Similarly one defines the stochastic integral with respect to a Wiener process in $\mathbb{R}^d$ of a real-valued or operator-valued process.

Suppose now we are given Borel functions σ and b on $\mathbb{R}$. If

$$\xi_t = \xi_{t_0} + \int_{t_0}^t \sigma(\xi_s) \, dw_s + \int_{t_0}^t b(\xi_s) \, ds, \quad t_0 \leqslant t \leqslant T,$$

then the adapted process $\{\xi_t\}_{t\in[t_0,T]}$ is called a *strong solution* of the stochastic differential equation

$$d\xi_t = \sigma(\xi_t)dw_t + b(\xi_t)dt \tag{1.3.3}$$

on $[t_0,T]$ with the initial distribution ξ_{t_0}. This equation is a symbolic expression for the previous integral equation. Similarly stochastic equations are introduced for processes in $\mathbb{R}^d$, when b is a vector field, σ is an operator-valued mapping.

It is known (see Wentzell [**937**] or Krylov [**551**]) that in the case of Lipschitzian coefficients σ and b for any $\mathcal{F}_{t_0}$-measurable square-integrable random variable ξ_{t_0} this stochastic equation has a unique solution and this solution is a diffusion process with the drift b and the diffusion coefficient $\sigma\sigma^*/2$. There are also more subtle results (see Ikeda, Watanabe [**473**]). The following theorem on existence of a strong solution is proved in Gyöngy, Krylov [**432**] (we give its version for the whole space).

1.3.3. Theorem. *Suppose that $\sigma = (\sigma^{ij})_{i,j\leqslant d}$ and b are Borel mappings from $\mathbb{R}^d\times[0,+\infty)$ to the space of matrices on $\mathbb{R}^d$ and to $\mathbb{R}^d$, respectively, such that for every $k\in\mathbb{N}$ there are a positive function M_k integrable on $[0,k]$ and a number $\varepsilon_k>0$ such that*

$$|\sigma^{ij}(x,t)-\sigma^{ij}(y,t)|^2 \leqslant M_k(t)|x-y|^2,\ \ |b(x,t)|+|\sigma^{ij}(x,t)|\leqslant M_k(t),$$

$$A(x,t) := 2^{-1}\sigma(x,t)\sigma^*(x,t)\geqslant \varepsilon_k\cdot \mathrm{I}\quad \textit{if } |x|\leqslant k,\quad t\in[0,k],\ i,j\leqslant d.$$

Let $V\geqslant 0$ be a function on $\mathbb{R}^d\times[0,+\infty)$ with continuous first and second derivatives in x and a continuous derivative in t such that for some increasing sequence of bounded domains D_k covering $\mathbb{R}^d$ for each $T>0$ we have

$$L_{A,b}V(x,t)\leqslant M(t)V(t,x),\qquad \inf_{x\in\partial D_k,\, t\leqslant T} V(x,t)\to+\infty,$$

where M is a locally integrable function on $[0,+\infty)$. Then (1.3.3) *has a unique strong solution on $[0,+\infty)$.*

In Chapter 9 also the concept of weak solution will be mentioned and the related concept of martingale problem (which in turn is strongly related to Fokker–Planck–Kolmogorov equations).

Apart from the Wiener process, the *Ornstein–Uhlenbeck process* is very useful in applications. This process is given by the linear stochastic equation (scalar or vector)

$$d\xi_t = dw_t - 2^{-1}\xi_t\, dt.$$

It can be expressed via the Wiener process by the formula

$$\xi_t = e^{-t/2}\xi_0 + e^{-t/2}\int_0^t e^{s/2}\, dw_s.$$

For $\xi_0 = x_0$ the process $\widehat{\xi_t} = e^{-t/2}x_0 + e^{-t/2}w_{e^t-1}$ has the same finite-dimensional distributions as ξ_t (but does not satisfy the above stochastic equation). The generator of the Ornstein–Uhlenbeck process has the form $L/2$, where the operator

$$L\varphi(x) = \Delta\varphi(x) - \langle x, \nabla\varphi(x)\rangle$$

is called the *Ornstein–Uhlenbeck operator.*

We observe that the Wiener process has no stationary probability measures, but the standard Gaussian measure is invariant for the Ornstein–Uhlenbeck process and the Ornstein–Uhlenbeck semigroup (see also Examples 1.4.7, 5.1.1 and Exercise 5.6.57).

Let us also mention the *Itô formula*. If the process $\xi_t = (\xi_t^1, \dots, \xi_t^d)$ in $\mathbb{R}^d$ satisfies the equation $d\xi_t = \sigma(\xi_t)dw_t + b(\xi_t)dt$, then for any smooth function f the scalar process $f(\xi_t)$ satisfies the equation

$$\begin{aligned} df(\xi_t) &= \sum_{i=1}^d \partial_{x_i} f(\xi_t) d\xi_t^i + \frac{1}{2} \sum_{i,j=1}^d \partial_{x_i}\partial_{x_j} f(\xi_t) d\xi_t^i d\xi_t^j \\ &= \sum_{i,k=1}^d \partial_{x_i} f(\xi_t)\sigma^{ik}(\xi_t) dw_t^k + \sum_{i,k=1}^d \partial_{x_i} f(\xi_t)\sigma^{ik}(\xi_t) b^i(\xi_t)dt \\ &+ \frac{1}{2} \sum_{i,j,k=1}^d \partial_{x_i}\partial_{x_j} f(\xi_t)\sigma^{ik}(\xi_t)\sigma^{jk}(\xi_t)dt. \end{aligned}$$

The Itô formula will not be used in this book, but some acquaintance with it is useful for better understanding the methods of obtaining certain estimates. For example, the integral of the function f with respect to the solution of the parabolic Fokker–Planck–Kolmogorov equation at the moment t is usually the expectation of $f(\xi_t)$. For example, if $\sigma = \mathrm{I}$, then by the Itô formula this gives the expectation

$$\mathbb{E}f(\xi_0) + \mathbb{E}\int_0^t \Big[\langle \nabla f(\xi_t), b(\xi_t)\rangle + \frac{1}{2}\Delta f(\xi_t)\Big]\, dt.$$

Under the integral we have $Lf(\xi_t)$, $L = \Delta/2 + b\cdot\nabla$. If $Lf \leqslant C + Cf$, then the right-hand side is estimated by the integral of $C + C\mathbb{E}f(\xi_t)$ over $[0,t]$, which enables us to estimate $\mathbb{E}f(\xi_t)$ by means of the known Gronwall inequality (see Exercise 7.5.3).

1.4. Basic problems

Suppose we are given a locally finite Borel measure μ (possibly signed) on an open set $\Omega \subset \mathbb{R}^d$, a Borel function c on Ω, a Borel vector field $b = (b^i)$ on Ω, and a matrix-valued mapping $A = (a^{ij})_{i,j\leqslant d}$ on Ω such that the functions a^{ij} are Borel measurable. For $\varphi \in C^\infty(\Omega)$ let us set

$$L_{A,b}\varphi := \sum_{i,j\leqslant d} a^{ij}\partial_{x_i}\partial_{x_j}\varphi + \sum_{i\leqslant d} b^i \partial_{x_i}\varphi, \quad L_{A,b,c}\varphi = L_{A,b}\varphi + c\varphi.$$

We shall also consider the divergence form operators

$$\mathcal{L}_{A,b}\varphi := \sum_{i,j\leqslant d} \partial_{x_i}(a^{ij}\partial_{x_j}\varphi) + \sum_{i\leqslant d} b^i\partial_{x_i}\varphi$$

and the correspondingly defined operators $\mathcal{L}_{A,b,c}$.

1.4.1. Definition. We say that μ satisfies the equation

$$L^*_{A,b,c}\mu = 0 \tag{1.4.1}$$

in Ω if $a^{ij}, b^i, c \in L^1_{\mathrm{loc}}(|\mu|)$ and one has

$$\int_\Omega L_{A,b,c}\varphi(x)\,\mu(dx) = 0 \quad \forall \varphi \in C_0^\infty(\Omega). \tag{1.4.2}$$

For a given measure ν on Ω the equation

$$L^*_{A,b,c}\mu = \nu \tag{1.4.3}$$

is defined similarly as the identity $\displaystyle\int_\Omega L_{A,b,c}\varphi\, d\mu = \int_\Omega \varphi\, d\nu.$

For $c = 0$ we arrive at the equations $L^*_{A,b}\mu = 0$ and $L^*_{A,b}\mu = \nu$, the first of which is called the *stationary Fokker–Planck–Kolmogorov equation*; when $c \neq 0$ it is called the *Fokker–Planck–Kolmogorov equation with a potential.*

Equation (1.4.1) can be written as the equality

$$\partial_{x_i}\partial_{x_j}(a^{ij}\mu) - \partial_{x_i}(b^i\mu) + c\mu = 0$$

in the sense of generalized functions. If it is known in advance that the measure μ is given by a density ϱ of class $W^{1,1}_{\text{loc}}$ (which is true under broad assumptions, as we shall see below) and the functions a^{ij} are locally bounded and belong to $W^{1,1}_{\text{loc}}$ and the functions $\partial_{x_j}a^{ij}\varrho$ are integrable, then the integration by parts in (1.4.2) yields the equation

$$\partial_{x_i}(a^{ij}\partial_{x_j}\varrho) - \partial_{x_i}\big((b^i - \partial_{x_j}a^{ij})\varrho\big) + c\varrho = 0,$$

also understood in the sense of generalized functions.

If the coefficients are smooth and it is known in advance that the measure μ is given by a smooth density ϱ (which is true if the matrix $A(x)$ is not degenerate), then the double integration by parts in (1.4.2) yields the usual equation

$$a^{ij}\partial_{x_i}\partial_{x_j}\varrho + 2\partial_{x_i}a^{ij}\partial_{x_j}\varrho - b^i\partial_{x_i}\varrho - \partial_{x_i}b^i\varrho + c\varrho = 0.$$

Unlike the direct elliptic equations of the form $L_{A,b,c}u = 0$, the density of a solution of equation (1.4.1) even with Lipcshitzian coefficients and a nondegenerate matrix A may fail to belong to the second class $W^{1,2}_{\text{loc}}$ (simple examples are given below).

The equation

$$\mathcal{L}^*_{A,b,c}\mu = 0$$

is defined similarly, but it requires additional assumptions about either a^{ij} or μ (which will be made in appropriate places), because it is necessary to give meaning to the integral of $\mathcal{L}_{A,b,c}\varphi$ with respect to μ. For example, if $a^{ij} \in C^1(\Omega)$, then we write $\partial_{x_i}(a^{ij}\partial_{x_j}\varphi)$ as $\partial_{x_i}a^{ij}\partial_{x_j}\varphi + a^{ij}\partial_{x_i}\partial_{x_j}\varphi$ and use the previous definition.

Let us give a precise definition of a solution of the elliptic equation

$$\mathcal{L}^*_{A,b,c}\mu = 0 \tag{1.4.4}$$

for Borel measures μ on Ω, where $\mathcal{L}$ is an elliptic second order operator of divergence form

$$\mathcal{L}\varphi(x) := \partial_{x_i}\big(a^{ij}(x)\partial_{x_j}\varphi(x)\big) + b^i(x)\partial_{x_i}\varphi(x).$$

The interpretation of this equation is as usual: the functions a^{ij} and b^i must be integrable on every compact set in Ω with respect to the measure μ and, for every function $\varphi \in C_0^\infty(\Omega)$, we must have the equality

$$\int_\Omega \mathcal{L}_{A,b,c}\varphi\, d\mu = 0.$$

However, the latter can be understood in one of the following two ways.

(I) One has $a^{ij} \in W^{1,1}_{\text{loc}}(\Omega)$, the functions a^{ij}, $\partial_{x_i}a^{ij}$, and b^i are Borel measurable and locally integrable with respect to $|\mu|$, and

$$\int_\Omega [a^{ij}\partial_{x_i}\partial_{x_j}\varphi + \partial_{x_i}a^{ij}\partial_{x_j}\varphi + b^j\partial_{x_j}\varphi]\, d\mu = 0. \tag{1.4.5}$$

(II) The measure μ possesses a density ϱ in the class $W^{1,1}_{\text{loc}}(\Omega)$ such that the functions $a^{ij}\partial_{x_i}\varrho$ and $b^i\varrho$ are locally Lebesgue integrable and

$$\int_\Omega [-a^{ij}\partial_{x_i}\varrho\partial_{x_j}\varphi + b^i\partial_{x_i}\varphi\varrho]\, dx = 0. \tag{1.4.6}$$

Clearly, if the coefficients a^{ij} are locally Sobolev and the functions $\partial_{x_i} a^{ij} \varrho$ are locally integrable, then (1.4.6) can be written as (1.4.5).

Throughout we deal with the case where the matrix A is symmetric and nonnegative, but this is not needed for the definition (unlike for most of the results).

Probability solutions are those that are probability measures. *Integrable solutions* are those given by integrable densities (possibly signed).

In general, equation (1.4.1) can fail to have nonzero solutions in the class of bounded measures (take $\Omega = \mathbb{R}^1$, $A = 1$, $b = 0$, then the equation $\mu'' = 0$ means that the density of μ is linear), it can have many solutions even in the class of probability measures, and its solutions can be quite singular (e.g., if $A = 0$ and also $b = 0$, then any measure is a solution). However, even in the generality under consideration some positive information is available.

The one-dimensional case is much simpler than the multidimensional case.

1.4.2. Proposition. *Let $d = 1$ and let Ω be an interval (x_0, x_1). Suppose that $A > 0$ on Ω. Then, any measure μ satisfying the equation $L^*_{A,b,c}\mu = \nu$ is absolutely continuous with respect to Lebesgue measure and has a density ϱ of the form $\varrho = \varrho_0/A$, where ϱ_0 is absolutely continuous on every compact subinterval in Ω.*

If $c = 0$ and b/A is locally Lebesgue integrable, $x_2 \in (x_0, x_1)$ is fixed, then

$$\varrho(x) = A(x)^{-1} E(x)\Big(C_1 + \int_{x_0}^{x} \frac{C_2 + F(t)}{E(t)}\, dt\Big), \tag{1.4.7}$$

$$E(x) := \exp \int_{x_2}^{x} \frac{b(t)}{A(t)}\, dt, \quad F(x) := \nu\big((x_0, x)\big).$$

If $A = 1$, $c = 0$, $\nu = 0$, $\Omega = (-1, 1)$, and b is locally Lebesgue integrable on the interval $(-1, 1)$, then

$$\varrho(x) = \Big(k_1 + k_2 \int_0^x \exp\Big(-\int_0^s b(t)\, dt\Big)\, ds\Big) \exp \int_0^x b(t)\, dt, \tag{1.4.8}$$

where k_1 and k_2 are constants.

Proof. We have the identity

$$\int_\Omega (A\varphi'' + b\varphi' + c\varphi)\, d\mu = \int_\Omega \varphi\, d\nu \quad \forall \varphi \in C_0^\infty(\Omega),$$

which can be written as the equality

$$(A\mu)'' - (b\mu)' + c\mu = \nu$$

in the sense of distributions. Hence the distributional derivative of $(A\mu)' - b\mu$ is a locally bounded measure, i.e., $(A\mu)' - b\mu$ is a function of locally bounded variation. This shows that the distributional derivative of $A\mu$ is a locally bounded measure as well. Hence $A\mu$ is absolutely continuous and has a density ϱ_0. Therefore, μ is absolutely continuous. Now it is seen from our reasoning that the distributional derivative of $A\mu$ is a locally integrable function, so that ϱ_0 admits a locally absolutely continuous version. In the case $A = 1$, $c = 0$, $\nu = 0$, we arrive at the equation $\mu'' - (b\mu)' = 0$, whence $\mu' - b\mu = k_2$ for some constant k_2. If b is locally Lebesgue integrable, this equation can be explicitly solved. The general case reduces to this one by passing to the measure $A\mu$. □

Even in this simplest one-dimensional case we observe that a solution μ can fail to have a continuous density if A is positive but not continuous. We actually see that in the case of nondegenerate A (i.e., $\det A \neq 0$) the regularity of solutions is essentially the regularity of A. We shall see below that in higher dimensions the picture is similar, although the proofs involve much deeper techniques. Another simple observation is that without any assumptions of nondegeneracy on A we obtain that the measure $A\cdot\mu$ is absolutely continuous. A highly nontrivial analogue of this is true also in the multidimensional case.

Sometimes it is useful to construct an equation for which a given function is a solution.

1.4.3. Example. In the one-dimensional case for any two smooth functions f and g with everywhere nonzero Wronskian $W = f'g - fg'$ it is easy to write the equation $L^*_{1,b,c}\mu = 0$ with smooth coefficients for which they form a basis in the space of solutions. To this end we equate the determinant of the matrix with the rows (u, u', u''), (f, f', f''), (g, g', g'') to zero, which gives a second order equation $-Wu''+Au'+Bu = 0$ satisfied by f and g. Dividing by $-W$ we obtain the equation $u'' - (A/W)u' - (B/W)u = 0$, which can be written as $L^*_{1,b,c}u = 0$, $b = A/W$, $c = b' - B/W$. For nonzero c it is not always possible to find an explicit solution, but a new degree of freedom appears, which leads to some effects impossible in the case where $c = 0$.

Let us consider one more instructive example.

1.4.4. Example. Let $\varrho \in W^{1,1}_{\mathrm{loc}}(\mathbb{R}^d)$ and let $\mu = \varrho\, dx$. Then μ satisfies the equation $L^*_{1,b}\mu = 0$ with

$$b := \frac{\nabla\varrho}{\varrho}, \quad \text{where } b(x) := 0 \text{ whenever } \varrho(x) = 0.$$

Indeed, $|b|$ is locally $|\mu|$-integrable. For any $\varphi \in C_0^\infty(\mathbb{R}^d)$, by the integration by parts formula we have

$$\int [\Delta\varphi + \langle b, \nabla\varphi\rangle]\,\varrho\, dx = \int [-\langle\nabla\varphi, \nabla\varrho\rangle + \langle b, \nabla\varphi\rangle\varrho]\, dx = 0$$

since $b\varrho = \nabla\varrho$ almost everywhere due to the fact that $\nabla\varrho$ vanishes almost everywhere on the set $\{\varrho = 0\}$ (see Exercise 1.8.19).

1.4.5. Definition. Let $\varrho \in W^{1,1}_{\mathrm{loc}}(\mathbb{R}^d)$. The mapping $\nabla\varrho/\varrho$, where we set by definition $\nabla\varrho(x)/\varrho(x) := 0$ if $\varrho(x) = 0$, is called the *logarithmic gradient* of the measure μ or of the density ϱ.

In this example, we can even choose ϱ to be infinitely differentiable, but b can be quite singular with respect to Lebesgue measure. For instance, given a proper closed subset $Z \subset \mathbb{R}^d$, we can find a probability density $\varrho \in C^\infty(\mathbb{R}^d)$ with $Z = \{\varrho = 0\}$; in this way one can even obtain b that is not Lebesgue locally integrable on a closed set of positive Lebesgue measure. The simplest example of a singularity is this:

$$\varrho(x) = x^2\exp(-x^2/2)/\sqrt{2\pi}, \quad b(x) = x + 2x^{-1}. \tag{1.4.9}$$

In the case of smooth coefficients and nondegenerate A all solutions are smooth. This is a corollary of the following classical result (see Taylor [**894**, Chapter III],

Trèves [**898**, Chapter I]), which is often referred to as Weyl's regularity theorem for the equation

$$L_{A,b,c}\mu = \nu.$$

1.4.6. Theorem. *Suppose that* a^{ij}, b^i, $c \in C^\infty(\Omega)$ *and* $\det A > 0$. *If* μ *is a distribution on* Ω *such that* $L_{A,b,c}\mu \in C^\infty(\Omega)$, *then* $\mu \in C^\infty(\Omega)$.

Therefore, if a measure ν *on* Ω *has an infinitely differentiable density, then any measure* μ *on* Ω *satisfying the equation* $L^*_{A,b,c}\mu = \nu$ *possesses an infinitely differentiable density.*

The second assertion follows from the first one, since in the case of smooth coefficients the equation $L^*_{A,b,c}\mu = \nu$ can be written as the equality

$$\partial_{x_i}\partial_{x_j}(a^{ij}\mu) - \partial_{x_i}(b^i\mu) + c\mu = \nu$$

in the sense of distributions, which can be rewritten as

$$a^{ij}\partial_{x_i}\partial_{x_j}\mu + (\partial_{x_i}\partial_{x_j}a^{ij})\mu + 2\partial_{x_i}a^{ij}\partial_{x_j}\mu - (\partial_{x_i}b^i)\mu - b^i\partial_{x_i}\mu + c\mu = \nu,$$

i.e., as $L_{A,b_0,c_0}\mu = \nu$ with some smooth coefficients b_0 and c_0.

Explicitly solvable equations are rather rare, although there are important cases when they appear.

1.4.7. Example. Let μ be the standard Gaussian measure on $\mathbb{R}^d$, i.e., a probability measure with the standard Gaussian density

$$\varrho(x) = (2\pi)^{-d/2}\exp(-|x|^2/2).$$

Its logarithmic gradient has an especially simple form:

$$\frac{\nabla\varrho(x)}{\varrho(x)} = -x.$$

According to the previous example, μ satisfies the equation $L^*\mu = 0$, where L is the Ornstein–Uhlenbeck operator

$$L\varphi(x) = \Delta\varphi(x) - \langle x, \nabla\varphi(x)\rangle,$$

already encountered above. This operator plays an important role in analysis, probability theory, and the most diverse applications. We shall see below that any bounded measure σ on $\mathbb{R}^d$ satisfying the equation $L^*\sigma = 0$ has the form $\sigma = k\mu$, where k is a constant. It is worth noting that the operator L has an eigenbasis in $L^2(\gamma)$. For $d = 1$ an eigenbasis is formed by the Hermite–Chebyshev polynomials

$$H_0 = 1, \quad H_n(t) = \frac{(-1)^n}{\sqrt{n!}}e^{t^2/2}\frac{d^n}{dt^n}\left(e^{-t^2/2}\right), \quad n > 0.$$

Here $LH_n = -nH_n$. For $\mathbb{R}^d$ an eigenbasis is formed by the polynomials

$$H_{k_1,\dots,k_d}(x_1,\dots,x_d) = H_{k_1}(x_1)\cdots H_{k_d}(x_d), \quad k_i \geqslant 0.$$

Here

$$LH_{k_1,\dots,k_d} = -(k_1 + \cdots + k_d)H_{k_1,\dots,k_d}.$$

In this case the operator L is obviously symmetric in $L^2(\mu)$ on the domain of definition $C_0^\infty(\mathbb{R}^d)$, but this is not always true, as one can see from the following result.

1.4.8. Proposition. *Suppose that a nonnegative locally finite measure μ with a density $\varrho \in W^{p,1}_{\mathrm{loc}}(\mathbb{R}^d)$ satisfies the equation $L^*_{A,b}\mu = 0$, where $a^{ij} \in W^{p,1}_{\mathrm{loc}}(\mathbb{R}^d)$, $b^i \in L^p_{\mathrm{loc}}(\mathbb{R}^d)$, $p \geqslant 2$. Then the symmetry of the operator $L_{A,b}$ on domain $C_0^\infty(\mathbb{R}^d)$ in $L^2(\mu)$, i.e., the identity*

$$\int_{\mathbb{R}^d} \varphi L_{A,b}\psi \, d\mu = \int_{\mathbb{R}^d} \psi L_{A,b}\varphi \, d\mu, \quad \varphi, \psi \in C_0^\infty(\mathbb{R}^d),$$

is equivalent to the almost everywhere equality

$$A\nabla\varrho = \varrho b_0, \quad b_0^i := b^i - \sum_{j=1}^d \partial_{x_j} a^{ij}.$$

For $A = \mathrm{I}$ the symmetry is equivalent to the equality $\nabla\varrho = \varrho b$.

PROOF. Indeed, by the integration by parts formula the indicated identity is equivalent to the identity

$$\int_{\mathbb{R}^d} \langle \varphi\nabla\psi - \psi\nabla\varphi, A\nabla\varrho - \varrho b_0\rangle \varrho \, dx = 0,$$

which by the identity

$$\int_{\mathbb{R}^d} \langle \nabla(\varphi\psi), A\nabla\varrho - \varrho b_0\rangle \, dx = 0$$

that follows from the equation turns out to be equivalent to the relation

$$\int_{\mathbb{R}^d} \varphi\langle \nabla\psi, A\nabla\varrho - \varrho b_0\rangle \, dx = 0, \quad \varphi, \psi \in C_0^\infty(\mathbb{R}^d).$$

The latter is equivalent to the equality

$$\langle \nabla\psi, A\nabla\varrho - \varrho b_0\rangle = 0 \quad \text{a.e. for every function } \psi \in C_0^\infty(\mathbb{R}^d),$$

i.e., is the announced equality, since for ψ we can take a function that coincides with the coordinate function x_i on a given cube, which yields that $(A\nabla\varrho)^i = \varrho b_0^i$. □

Note that the symmetry of the operator $L_{A,b}$ on domain $C_0^\infty(\mathbb{R}^d)$ in $L^2(\mu)$ implies that $L^*_{A,b}\mu = 0$ under the much weaker assumption that $a^{ij}, b^i \in L^2_{\mathrm{loc}}(\mu)$, since for $\varphi \in C_0^\infty(\mathbb{R}^d)$ we can take $\psi \in C_0^\infty(\mathbb{R}^d)$ in such a way that $\psi = 1$ on the support of φ.

We shall see below that under broad assumptions any solution of the equation $L^*_{A,b,c}\mu = 0$ has the same smoothness as the diffusion coefficient A. However, even in the one-dimensional case it is easy to find an example where the smoothness of the solution does not exceed that of A.

1.4.9. Example. Let us take a probability measure μ with a smooth density that satisfies the equation $L^*_{\mathrm{I},b_0}\mu = 0$, e.g., let μ be the standard Gaussian measure and $b_0(x) = -x$. If now g is any Borel function with $1 \leqslant g \leqslant 2$, then the measure $g \cdot \mu$ satisfies the equation $L^*_{A,b}\mu = 0$ with $A = g^{-1}\mathrm{I}$ and $b = g^{-1}b_0$. In particular, in this way we can obtain an example, where A and b are Hölder continuous and A is uniformly nondegenerate, but the density of μ is not weakly differentiable and its Hölder order is not greater than that of A.

1.5. Existence of densities

We now turn to conditions for the existence of densities of solutions. Suppose that $A = (a^{ij})_{i,j=1}^d$ is a Borel measurable mapping on an open set $\Omega \subset \mathbb{R}^d$ with values in the space of nonnegative symmetric operators on $\mathbb{R}^d$. The main results of this section assert that under broad assumptions any solution of the equation $L_{A,b,c}^*\mu = 0$ has a density on the set where $\det A > 0$, and it is possible to estimate certain L^p-norms of the density.

For the proof we need the classical result following from the Riesz theorem and asserting that every linear functional Λ defined on a linear subspace E in $L^p(\lambda)$, where λ is a nonnegative σ-finite measure and $p \in [1, \infty)$, and satisfying the estimate

$$\Lambda(f) \leqslant C\|f\|_{L^p(\lambda)}, \quad f \in E,$$

is given by means of some function $g \in L^{p/(p-1)}(\lambda)$ in the form

$$\Lambda(f) = \int fg\, d\lambda, \quad f \in E.$$

In addition, we also need the following corollary of a very deep maximum principle due to A.D. Aleksandrov.

1.5.1. Theorem. *For every smooth positive function f on a uniformly convex smooth domain Ω (the principal curvatures of $\partial\Omega$ are separated from zero, e.g., Ω is a ball) there is a convex function $z \in C^2(\Omega) \cap C(\overline{\Omega})$ such that $z|_{\partial\Omega} = 0$ and*

$$\alpha^{ij}\partial_{x_i}\partial_{x_j} z \geqslant d|\det(\alpha^{ij})|^{1/d} f \quad \text{on } \Omega$$

for every nonnegative symmetric matrix (α^{ij}) and

$$\sup_{x\in\Omega} |z(x)| \leqslant C(d, \Omega)\|f\|_{L^d(\Omega)}.$$

PROOF. It is known (see Gilbarg, Trudinger [**409**, Theorem 17.23]) that for every smooth positive function f on a uniformly convex smooth domain Ω there exists a convex solution $z \in C^2(\Omega) \cap C(\overline{\Omega})$ of the Dirichlet problem

$$\det(D^2 z) = f^d, \quad z|_\Omega = 0. \tag{1.5.1}$$

Then by A.D. Aleksandrov's maximum principle (see [**409**, Theorem 9.1]) we have

$$\sup_{x\in\Omega} |z(x)| \leqslant C(d, \Omega)\|f\|_{L^d(\Omega)}.$$

If now $\alpha = (\alpha^{ij})$ is a nonnegative symmetric $d\times d$-matrix, we have

$$\alpha^{ij}\partial_{x_i}\partial_{x_j} z = \operatorname{tr}(\alpha D^2 z) \geqslant d|\det\alpha \det(D^2 z)|^{1/d} = d|\det\alpha|^{1/d} f,$$

since $\operatorname{tr}(AB) \geqslant d|\det(AB)|^{1/d}$ for any nonnegative symmetric $d\times d$-matrices, because $\operatorname{tr}(AB) = \operatorname{tr}(\sqrt{B}A\sqrt{B})$ and $\sqrt{B}A\sqrt{B}$ is a nonnegative matrix. $\square$

Here is one of the main results in this section.

1.5.2. Theorem. *Suppose that the matrix $A(x)$ is symmetric and nonnegative-definite for every x. Let μ be a locally finite Borel measure on Ω (possibly signed) such that $a^{ij} \in L^1_{\text{loc}}(\Omega, \mu)$, and for some $C > 0$ one has*

$$\int_\Omega a^{ij}\partial_{x_i}\partial_{x_j}\varphi\, d\mu \leqslant C(\sup_\Omega|\varphi| + \sup_\Omega|\nabla\varphi|) \tag{1.5.2}$$

for all nonnegative $\varphi \in C_0^\infty(\Omega)$. Then the following assertions are true.

(i) *If* μ *is nonnegative, then* $(\det A)^{1/d}\mu$ *has a density in* $L^{d'}_{\mathrm{loc}}(\Omega, dx)$.

(ii) *If* A *is locally Hölder continuous and* $\det A > 0$, *then* μ *has a density which belongs to* $L^r_{\mathrm{loc}}(\Omega, dx)$ *for every* $r \in [1, d')$.

PROOF. We shall start with case (ii) which is simpler. Let U_0 be a ball with compact closure in Ω and let $\zeta \in C_0^\infty(\Omega)$ be such that $0 \leqslant \zeta \leqslant 1$, $\zeta = 1$ on U_0 and the support of ζ belongs to a ball $U \subset \Omega$. Let us consider the measure $\nu = \zeta\,\mu$. By substituting $\zeta\psi$ in place of φ in (1.5.2), for every nonnegative smooth function ψ on Ω, we obtain

$$\int_U a^{ij}\partial_{x_i}\partial_{x_j}\psi\,d\nu \leqslant C_1(\sup_U|\psi| + \sup_U|\nabla\psi|), \tag{1.5.3}$$

where

$$C_1 = C + \bigl(C + 2d^2\sup_{i,j}\|a^{ij}\|_{L^1(U,\mu)}\bigr)\sup_U|\nabla\zeta| + \|a^{ij}\|_{L^1(U,\mu)}\sup_U|\partial_{x_i}\partial_{x_j}\zeta|$$

is independent of ψ. It is easily seen that (1.5.3) remains true for every nonnegative $\psi \in C^2(\overline{U})$. By considering the function $\psi + \sup|\psi|$, we arrive at the estimate

$$\Bigl|\int_U a^{ij}\partial_{x_i}\partial_{x_j}\psi\,d\nu\Bigr| \leqslant C_1(\sup_U|\psi| + \sup_U|\nabla\psi|) \quad \forall\,\psi \in C^2(\overline{U}). \tag{1.5.4}$$

Now let $r > d$. As we noted in § 1.2, for every $f \in C_0^\infty(U)$ there exists a function $u \in C^2(\overline{U})$ such that

$$a^{ij}\partial_{x_i}\partial_{x_j}u = f$$

on U and $u = 0$ on ∂U. Moreover, there exists a constant C_2 independent of f such that

$$\|u\|_{W^{r,2}(U)} \leqslant C_2\|f\|_{L^r(U)}.$$

By the Sobolev embedding theorem, we obtain

$$\sup_U|\nabla u| + \sup_U|u| \leqslant C_3\|f\|_{L^r(U)}.$$

Together with (1.5.4) this yields

$$\int_U f\,d\nu \leqslant C_1C_3\|f\|_{L^r(U)} \quad \forall\, f \in C_0^\infty(U). \tag{1.5.5}$$

Hence ν is absolutely continuous with $\nu = g\,dx$, $g \in L^{r'}(U)$.

Let us now consider case (i). The above reasoning does not work in this case even for bounded uniformly nondegenerate A, since the equation $a^{ij}\partial_{x_i}\partial_{x_j}u = f$ need not be solvable; for continuous A, the solution u of this equation is only in $W^{r,2}$ and not in C^2, hence one cannot pass from C_0^∞-functions to u in (1.5.4). In order to overcome this difficulty, we need the assumption that μ is nonnegative. As above, by considering a suitable function ζ, we arrive at estimate (1.5.4) for the measure $\nu = \zeta\mu$ on the open ball $U_{R_0}(x_0)$. Note that the support of the measure ν is contained in a ball $U_R(x_0)$ of radius $R = R_0 - 2r$, where $r > 0$. In that case, instead of solving the elliptic equation, we shall employ Theorem 1.5.1, according to which, for every nonnegative continuous function f on $\mathbb{R}^d$ vanishing outside the closed ball $\overline{U_{R_0}}(x_0)$, there exists a nonnegative continuous concave function z (the convex function from the theorem with the minus sign) on $\overline{U_{R_0}}(x_0)$ with the following property:

$$-\alpha^{ij}\partial_{x_i}\partial_{x_j}z \geqslant |\det(\alpha^{ij})|^{1/d}f$$

in $U_{R_0}(x_0)$ for every nonnegative matrix (α^{ij}) and

$$\sup_{U_{R_0}(x_0)} z \leqslant N\|f\|_{L^d(U_{R_0}(x_0))},$$

where N is independent of f and (α^{ij}). Let g be a fixed smooth probability density on $\mathbb{R}^d$ whose support is contained in the unit ball centered at the origin. For any locally integrable function v, we set

$$v_\varepsilon = v * g_\varepsilon, \quad g_\varepsilon(x) = \varepsilon^{-d} g(\varepsilon^{-1}x).$$

Then, for every nonnegative matrix (α^{ij}) and every $\varepsilon \in (0, r)$, one has the estimates

$$-\alpha^{ij}\partial_{x_i}\partial_{x_j} z_\varepsilon(x) \geqslant |\det(\alpha^{ij})|^{1/d} f_\varepsilon(x), \tag{1.5.6}$$

$$\sup_{U_R(x_0)} |z_\varepsilon| \leqslant N\|f_\varepsilon\|_{L^d(U_{R_0}(x_0))} \leqslant N\|f\|_{L^d(U_{R_0}(x_0))} \tag{1.5.7}$$

on $U_{R+r}(x_0)$, where N is independent of f, α^{ij}, and ε. Clearly, the functions z_ε are smooth, nonnegative and concave on $U_{R+r}(x_0)$ if $\varepsilon < r$. We observe that, for every nonnegative continuously differentiable concave function w on $U_{R+r}(x_0)$, one has

$$|\nabla w(x)| \leqslant r^{-1} \sup_{y \in U_{R+r}(x_0)} w(y) \quad \forall\, x \in U_R(x_0).$$

This estimate follows by considering the one-dimensional case. Together with (1.5.4), (1.5.6) applied to $\alpha^{ij} = a^{ij}(x)$ and (1.5.7), this yields the estimate

$$\int |\det(a^{ij})|^{1/d} f_\varepsilon \, d\nu \leqslant \left| \int a^{ij}\partial_{x_i}\partial_{x_j} z_\varepsilon \, d\nu \right| \leqslant C_1 \sup_{U_R(x_0)} \Bigl(|\nabla z_\varepsilon| + |z_\varepsilon| \Bigr)$$
$$\leqslant C_1 N (1 + r^{-1}) \|f\|_{L^d(U_{R_0}(x_0))}.$$

As in case (ii), we complete the proof. □

Notice that in assertion (ii) one cannot expect that the density of μ is continuous even for infinitely differentiable a^{ij}, which is seen if one takes $d = 1$, $\Omega = (-1, 1)$, $A = 1$ and $\mu(dx) = I_{(0,+\infty)}\, dx$.

We do not know whether assertion (i) remains true for signed measures.

1.5.3. Corollary. *Let μ be a locally finite (possibly signed) Borel measure on Ω and let a^{ij}, b^i, $c \in L^1_{\mathrm{loc}}(\Omega, \mu)$. Assume that*

$$\int_\Omega (L_{A,b}\varphi + c\varphi)\, d\mu \leqslant 0 \quad \textit{for all nonnegative } \varphi \in C_0^\infty(\Omega). \tag{1.5.8}$$

Then the following assertions are true.

(i) *If μ is nonnegative, then the locally finite measure* $(\det A)^{1/d}\mu$ *has a density in $L^{d'}_{\mathrm{loc}}(\Omega, dx)$.*

(ii) *If A is locally Hölder continuous and* $\det A > 0$, *then μ has a density which belongs to $L^r_{\mathrm{loc}}(\Omega, dx)$ for every $r \in [1, d')$.*

In particular, the above statements are true if (1.4.1) *holds.*

Proof. It suffices to note that, for every bounded open $\Omega_0 \subset \overline{\Omega}_0 \subset \Omega$, one has

$$\left| \int_{\Omega_0} (b^i \partial_{x_i}\varphi + c\varphi)\, d\mu \right| \leqslant \sup_{\Omega_0} |\nabla\varphi| \int_{\Omega_0} |b|\, d|\mu| + \sup_{\Omega_0} |\varphi| \int_{\Omega_0} |c|\, d|\mu|$$

for every smooth function φ with support in Ω_0. □

In assertion (ii) of this corollary one cannot expect the density of μ to be Hölder continuous, since for $d = 1$ and $A = 1$ one can take the measure μ with density

$$\exp \int_0^x b(t)\, dt$$

with a suitable function b (see Exercise 1.8.11).

The previous corollary has the following important generalization with the same proof concerned with the nonhomogeneous equation $L^*_{A,b,c}\mu = \nu$ with a measure on the right.

1.5.4. Corollary. *Let μ and ν be two locally finite (possibly signed) Borel measures on Ω and let a^{ij}, b^i, $c \in L^1_{\rm loc}(\Omega, \mu)$. Assume that*

$$\int_\Omega \Bigl[L_{A,b}\varphi + c\varphi\Bigr]\, d\mu = \int_\Omega \varphi\, d\nu \quad \textit{for all nonnegative } \varphi \in C_0^\infty(\Omega). \tag{1.5.9}$$

Then the following assertions are true.

(i) *If μ is nonnegative, then the locally finite measure* $(\det A)^{1/d}\mu$ *has a density in $L^{d'}_{\rm loc}(\Omega, dx)$.*

(ii) *If A is locally Hölder continuous and* $\det A > 0$, *then μ has a density which belongs to $L^r_{\rm loc}(\Omega, dx)$ for every $r \in [1, d')$.*

1.5.5. Remark. (i) Assertions (i) of Theorem 1.5.2, Corollary 1.5.3, and Corollary 1.5.4 for nonnegative measures extend to the case when μ is a σ-finite nonnegative Borel measure on Ω (not necessarily locally bounded). Indeed, (1.5.2), (1.5.8), and (1.5.9) make sense also for σ-finite μ provided that $a^{ij}, b^i, c \in L^1_{\rm loc}(\Omega, \mu)$. One can find a probability measure μ_0 such that $\mu = f\mu_0$, where f is a positive Borel function. Let

$$a_0^{ij} := fa^{ij},\ b_0^i := fb^i,\ c_0 := fc,\ A_0 = (a_0^{ij})_{i,j\leqslant d},\ b_0 = (b_0^i)_{i\leqslant d}.$$

Clearly, $a_0^{ij}, b_0^i, c_0 \in L^1_{\rm loc}(\mu_0)$ and μ_0 satisfies the hypotheses of the above mentioned assertions with A_0, b_0, and c_0 in place of A, b, and c. Hence the measure $(\det A_0)^{1/d}\mu_0$ has a density $\varrho \in L^{d'}_{\rm loc}(\Omega, dx)$. Since we have the equality $(\det A_0)^{1/d} = f(\det A)^{1/d}$, this means that $(\det A)^{1/d}\mu$ has the same density.

(ii) Assume that the hypotheses of Corollary 1.5.3(i) are fulfilled. Suppose that the ball $U_{R_1}(x_0)$ of radius $R_1 > 0$ centered at a point x_0 is contained in Ω. Then, for every $R < R_1$ and $r < d'$, there exists a number N depending only on R_1, R, r, d such that the density ϱ_A of $(\det A)^{1/d}\mu$ satisfies the estimate

$$\|\varrho_A\|_{L^{d'}(U_R)} \leqslant N\bigl\|1 + |b| + |c|\bigr\|_{L^1(U_{R_1},\mu)}.$$

In addition, for fixed d, the number N can be chosen as a locally bounded function of R_1, R, r. This follows from the proof of Theorem 1.5.2.

(iii) Assume that the hypotheses of Corollary 1.5.3(ii) are fulfilled. Let $U_{R_1}(x_0)$ belong to Ω. Then, for every $R < R_1$ and $r < d'$, there exists a number N depending only on R_1, R, r, d, $\inf_{U_{R_1}} \det A$, $\sup_{i,j} \sup_{U_{R_1}} |a^{ij}|$, and the Hölder norm of A on U_{R_1} such that the density ϱ of μ satisfies the estimate

$$\|\varrho\|_{L^r(U_R)} \leqslant N\bigl\|1 + |b| + |c|\bigr\|_{L^1(U_{R_1},\mu)}.$$

In addition, for fixed d, the number N can be chosen as a locally bounded function of the indicated quantities. This also follows from the proof of Theorem 1.5.2.

Let us consider an elliptic operator

$$Lu = a^{ij}\partial_{x_i}\partial_{x_j}u + b^i\partial_{x_i}u + cu,$$

where the coefficients a^{ij}, b^i and c are bounded Borel measurable functions on a domain $\Omega \subset \mathbb{R}^d$, the matrix $A(x) = \big(a^{ij}(x)\big)_{1\leqslant i,j\leqslant d}$ is symmetric and for some positive constants $\lambda > 0$ and $\gamma > 0$ one has

$$\lambda \mathrm{I} \leqslant A(x) \leqslant \lambda^{-1}\mathrm{I} \quad \forall x \in \Omega.$$

Moreover, we assume that for every ball $U(x_0, r) \subset \Omega$ we have

$$\sup_{x\in U(x_0,r)} \Big[r|b(x)| + r^2|c(x)|\Big] \leqslant \lambda^{-1}.$$

The following theorem was obtained in Bauman [**94**]. We say that a nonnegative Borel measure μ satisfies the inequality $L^*\mu \leqslant 0$ in Ω if

$$\int_\Omega L\varphi\, d\mu \leqslant 0 \quad \text{whenever } \varphi \in C_0^\infty(\Omega), \varphi \geqslant 0.$$

1.5.6. Theorem. *Suppose that $U(x_0, r) \subset \Omega$ and $0 < \sigma < \gamma < 1$. There exists a constant $C > 0$ depending only on γ, σ, λ, and d such that if a Borel measure μ is a nonnegative solution of the inequality $L^*\mu \leqslant 0$, then*

$$\mu\big(U(x_0, \gamma r)\big) \leqslant C\mu\big(U(x_0, \sigma r)\big).$$

PROOF. The theorem follows if we prove that there exists a number $\theta \in (0, 1)$ such that for all $\gamma \in (\theta, 1)$ there holds the estimate

$$\mu\big(U(x_0, \gamma r)\big) \leqslant C\mu\big(U(x_0, \theta r)\big),$$

where C depends only on γ, θ, λ, and d. Indeed, iterations of the above estimate will imply that

$$\mu\big(U(x_0, \gamma r)\big) \leqslant C^k\mu\big(U(x_0, \theta^k r/\gamma^{k-1})\big).$$

By choosing k so that $\theta^k < \sigma\gamma^{k-1}$ we obtain the assertion of the theorem. Changing variables we may assume that $r = 1$ and $x_0 = 0$.

Set $\varphi(u) = \exp(-u^{-1})$ if $u > 0$ and $\varphi(u) = 0$ if $u \leqslant 0$. We have

$$\begin{aligned} L\varphi(1 - |x|^2) &= -\varphi'(1 - |x|^2)\big(2\mathrm{tr}A(x) + 2\langle b(x), x\rangle\big) \\ &\quad - 4\varphi''(1 - |x|^2)\langle A(x)x, x\rangle + c(x)\varphi(1 - |x|^2) \\ &\geqslant \varphi(1 - |x|^2)(1 - |x|^2)^{-4}\Big(\langle A(x)x, x\rangle\big(4 - 2(1 - |x|^2)\big) \\ &\qquad - (1 - |x|^2)^2\big(2\mathrm{tr}A(x) + 2\langle b(x), x\rangle\big) + c(x)(1 - |x|^2)^4\Big). \end{aligned}$$

We can choose $\theta \in (0, 1)$ such that

$$\begin{aligned} \langle A(x)x, x\rangle\big(4 - 2(1 - |x|^2)\big) - (1 - |x|^2)^2\big(2\mathrm{tr}A(x) + 2\langle b(x), x\rangle\big) \\ + c(x)(1 - |x|^2)^4 \geqslant C_0 > 0 \end{aligned}$$

for every x with $\theta < |x| < 1$. Here C_0 depends only on λ, θ, and d. We have $L\varphi \geqslant 0$ on $U(0, 1) \setminus U(0, \theta)$ and for every $\gamma \in (\theta, 1)$ we have $L\varphi \geqslant C_1$ on $U(0, \gamma) \setminus U(0, \theta)$, where C_1 depends only on λ, θ, γ, and d. Note also that there exists a constant

$C_2 > 0$ depending only on λ and d such that $|L\varphi| \leqslant C_2$ on $U(0,1)$. Since $\mu \geqslant 0$ and $L^*\mu \leqslant 0$, we obtain

$$C_1\mu(U(0,\gamma)\setminus U(0,\theta)) \leqslant \int_{U(0,\gamma)\setminus U(0,\theta)} L\varphi\, d\mu \leqslant \int_{U(0,1)\setminus U(0,\theta)} L\varphi\, d\mu$$
$$\leqslant -\int_{U(0,\theta)} L\varphi\, d\mu \leqslant C_2\mu\big(U(0,\theta)\big).$$

Hence $\mu\big(U(0,\gamma)\big) \leqslant (C_1 + C_2)\mu\big(U(0,\theta)\big)$. □

The following theorem is a type of reverse Hölder's inequality. It follows immediately from Theorem 1.5.6 and Theorem 1.5.2.

1.5.7. Theorem. *Suppose that μ is a nonnegative solution of the inequality $L^*\mu \leqslant 0$ on Ω. Let $\gamma > 1$. Then μ has a density ϱ with respect to Lebesgue measure and there exists a constant $C > 0$ depending only on λ, γ, and d such that for every ball $U(x_0, r)$ with $U(x_0, \gamma r) \subset \Omega$, one has*

$$\biggl(\int_{U(x_0,r)} \varrho^{d/(d-1)}\, dx\biggr)^{(d-1)/d} \leqslant C|U(x_0,r)|^{-1/d} \int_{U(x_0,r)} \varrho\, dx.$$

The next interesting fact was discovered in Gehring [**403**].

1.5.8. Lemma. *Let Q be an arbitrary cube in $\mathbb{R}^d$ and let g be a nonnegative function in $L^q_{\rm loc}(\mathbb{R}^d)$ with $q > 1$. Suppose that for almost all $x \in Q$ the inequality*

$$|U|^{-1} \int_U g^q\, dx \leqslant C\biggl(|U|^{-1} \int_U g\, dx\biggr)^q$$

holds for every ball U centered at x. Then, there is a number $\varepsilon > 0$, depending only on q, C and d, such that $g \in L^p(Q)$ for every $p \in [q, q+\varepsilon)$ and

$$|Q|^{-1} \int_Q g^p\, dx \leqslant \frac{\varepsilon}{q+\varepsilon-p}\biggl(|Q|^{-1} \int_Q g^q\, dx\biggr)^{p/q}.$$

1.5.9. Corollary. *Let μ be a nonnegative solution of the inequality $L^*\mu \leqslant 0$ on Ω and let U be a ball with closure in Ω. Then μ has a density ϱ with respect to Lebesgue measure such that there is a number $\varepsilon > 0$, depending only on λ and U, with the property that $\varrho \in L^p(U)$ for every p in the interval $\big[1, d/(d-1)+\varepsilon\big)$.*

Recall that a Borel measure μ belongs to the Muckenhoupt class A_∞ if for every $\varepsilon > 0$ there exists a $\delta > 0$ such that $\mu(E) \leqslant \varepsilon\mu(U)$ whenever $U \subset \Omega$ is a ball and $E \subset U$ is a Borel set with $|E| \leqslant \delta|U|$.

1.5.10. Corollary. *Let μ be a nonnegative solution of the inequality $L^*\mu \leqslant 0$ on Ω. Let $\gamma > 1$. Then there exists a constant $C > 0$, depending only on λ, γ and d, such that for every ball $U(x_0, r)$ with $U(x_0, \gamma r) \subset \Omega$ and for every Borel set $E \subset U(x_0, r)$ we have*

$$\frac{\mu(E)}{\mu(U(x_0,r))} \leqslant C\biggl(\frac{|E|}{|U(x_0,r)|}\biggr)^{1/d},$$

in particular, μ belongs to A_∞ on every subdomain Ω' with $\overline{\Omega'} \subset \Omega$.

PROOF. Let ϱ be a density of the measure μ. Applying Hölder's inequality and Theorem 1.5.7 we obtain

$$\mu(E) = \int_{U(x_0,r)} I_E \varrho\, dx \leqslant |E|^{1/d} \Bigl(\int_{U(x_0,r)} \varrho^{d/(d-1)}\, dx \Bigr)^{(d-1)/d}$$
$$\leqslant C|E|^{1/d} |U(x_0,r)|^{-1/d} \mu(U(x_0,r)),$$

as required. □

1.5.11. Remark. (i) According to Coifman, Fefferman [**254**, Theorem V] and Muckenhoupt [**728**, Theorem 1], the last corollary implies that ϱ is in the Muckenhoupt class A_p for some $p > 1$ on every subset Ω' with $\overline{\Omega'} \subset \Omega$, i.e., there exists a constant $C_1 > 0$ such that

$$\Bigl(\frac{1}{|U|} \int_U \varrho\, dx \Bigr) \Bigl(\frac{1}{|U|} \int_U \varrho^{-1/(p-1)}\, dx \Bigr)^{p-1} \leqslant C_1,$$

for every ball $U \subset \Omega'$, where C_1 depends only on λ, d, and Ω'.

(ii) Moreover, for every $\varepsilon > 0$ there exists a constant β such that, for every ball $U \subset \Omega'$, Lebesgue measure of the set

$$\Bigl\{ y \in U\colon\ \beta |U|^{-1} \int_U \varrho\, dx \leqslant \varrho(y) \leqslant \beta^{-1} |U|^{-1} \int_U \varrho\, dx \Bigr\}$$

is not less than $(1-\varepsilon)|U|$. This assertion may be interpreted as a generalized Harnack principle.

(iii) There exist constants $C_2 > 0$ and $k > 1$ depending only on λ, d, and Ω' such that

$$\Bigl(\int_{\Omega'} |\varphi|^{kp} \varrho\, dx \Bigr)^{1/(kp)} \leqslant C_2 \Bigl(\int_{\Omega'} |\nabla \varphi|^p \varrho\, dx \Bigr)^{1/p}$$

for every $\varphi \in C_0^\infty(\Omega')$. See Fabes, Kenig, Serapioni [**347**] for a proof.

1.6. Local properties of densities

We now proceed to the regularity results. Throughout this section we assume that $A(x)$ is symmetric and positive and *$A(x)$ is continuous in x*. By the Sobolev embedding theorem, the continuity assumption is automatically satisfied for some version of A if $a^{ij} \in W^{p,1}_{loc}$, where $p > d$. In Theorem 1.4.6 we have already considered the case of smooth coefficients.

Let us consider the case where the coefficients are only Hölder continuous. The following result was proved in Sjögren [**861**].

1.6.1. Theorem. *Suppose that the coefficients a^{ij}, b^i, c are locally Hölder continuous in Ω and* $\det A > 0$. *Then any solution μ of the equation $L^*_{A,b,c}\mu = 0$ has a locally Hölder continuous density.*

Note that the solutions in [**861**] were a priori locally integrable functions, but by the above results the theorem remains true for measures. It would be interesting to study the case where only the coefficients a^{ij} are Hölder continuous. The continuity of all coefficients does not guarantee the Hölder continuity of a solution even if $d = 1$ and $A > 0$. However, it is not clear whether densities of solutions are continuous in the case where the coefficients are just continuous and A is uniformly elliptic. Without the requirement of uniform ellipticity, when A is just nondegenerate and continuous, one can construct a discontinuous probability solution on $\mathbb{R}^d$

with $d > 1$, using an example from Bauman [**95**] (which in turn employs a construction from Modica, Mortola [**720**]). In this example on a disc U in the plane a uniformly elliptic operator L_A with continuous A is such that there is a locally unbounded integrable function $\varrho \geqslant 0$ on $\mathbb{R}^2$ with $L_A^*(\varrho\, dx) = 0$. Taking a diffeomorphism $G\colon \mathbb{R}^2 \to U$, $G = (g^1, g^2)$ with a positive Jacobian, we obtain that the measure μ with the locally unbounded density $\varrho \circ G \det DG$ satisfies the equation $L_{Q,b}^*\mu = 0$ with continuous coefficients, where $Q = (q^{mk})$, $q^{mk} = a^{ij}\partial_{x_j} g^k \partial_{x_i} g^m$, and $b^k = a^{ij}\partial_{x_i}\partial_{x_j} g^k$.

We now proceed to the most difficult case where the diffusion coefficient is somewhat better than Hölder continuous, but is not smooth, and we want to have some Sobolev regularity of densities of solutions. One of the reasons why this is important is that, having established the Sobolev regularity of our solution, we can rewrite the equation $L_{A,b,c}^*\mu = 0$ for μ as a classical equation for its density ϱ in the sense of weak solutions: indeed, integrating by parts, we find that

$$\int_\Omega [a^{ij}\partial_{x_i}\varrho\partial_{x_j}\varphi + \partial_{x_i}a^{ij}\partial_{x_j}\varphi\varrho + b^i\partial_{x_i}\varphi\varrho + c\varrho]\, dx = 0 \quad \forall \varphi \in C_0^\infty(\Omega).$$

The difference between the main idea of the proofs in this section and that of the previous one is that now we verify that the solution determines a functional not on L^p, but on a negative Sobolev class, which gives the membership of the measure in a positive Sobolev class.

1.6.2. Theorem. *Let $d \geqslant 2$, $p \geqslant d$, $1 < q < \infty$, and $R_1 > 0$. Suppose that $a^{ij} \in W^{p,1}(U_{R_1})$ and $A \geqslant \lambda I$, where $\lambda > 0$. Then there exist numbers $R_0 > 0$ and $N_0 > 0$ with the following properties. Let $R < R_0$ and let μ be a measure of finite total variation on U_R such that for any $\varphi \in C_0^2(U_R) := C^2(\overline{U_R}) \cap \{u\colon u|_{\partial U_R} = 0\}$ we have the bound*

$$\left| \int_{U_R} a^{ij}\partial_{x_i}\partial_{x_j}\varphi\, d\mu \right| \leqslant N\|\nabla\varphi\|_{L^q(U_R)} \tag{1.6.1}$$

with a number N independent of φ. Furthermore, assume one of the following:

a) $p > d$ or

b) $p = d > q'$ and $\mu \in \bigcup_{r>1} L^r(U_R)$, where we identify μ with its density.

Then $\mu \in W_0^{q'\wedge p,1}(U_R)$ and

$$\|\mu\|_{W_0^{q'\wedge p,1}(U_R)} \leqslant N_0. \tag{1.6.2}$$

In addition, the radius R_0 can be taken such that it depends only on p, q, d, λ, R_1, $\|a^{ij}\|_{W^{p,1}(U_{R_1})}$, and the rate of decrease of $\|\nabla a^{ij}\|_{L^d(U_R)}$ as $R \to 0$, and N_0 depends on the same quantities and N.

PROOF. We break the proof into three cases.

Case $q \geqslant p'$ and $q \neq d'$. Take $f = (f^1, \ldots, f^d) \in C^2(\overline{U_R})$ and solve the equation

$$a^{ij}\partial_{x_i}\partial_{x_j}\varphi = \partial_{x_i} f^i$$

in U_R with zero boundary conditions. If a) holds, then $p > d$ and A is Hölder continuous in U_R and, by Hölder space theory (see § 1.2), there exists a unique solution $\varphi \in C_0^2(U_R)$ of our problem, which we can substitute into estimate (1.6.1). If b) holds, then, since A is continuous, $\partial_{x_i}\partial_{x_j}\varphi$ are summable to any power by

L^p-theory (see § 1.2), and, owing to $\mu \in \bigcup_{r>1} L^r(U_R)$, we again can substitute φ into inequality (1.6.1). By Lemma 1.2.5 and (1.6.1) we have

$$\left|\int_{U_R} \partial_{x_i} f^i \, \mu \, dx\right| \leqslant N \|\partial_{x_i} f^i\|_{W^{q,-1}(U_R)},$$

which implies our claim. We emphasize that we have established the inclusion of μ to $W_0^{q',1}(U_R)$, not just to $W^{q',1}(U_R)$, since the dual to $W^{q,-1}(U_R)$ is the former smaller space.

Case $q = d' \geqslant p'$. In this case by our assumptions we have $p > d$, so that Lemma 1.2.5 is still applicable.

Case $1 < q < p' < d'$. As is easy to see, this is the only remaining case. Observe that, of course, (1.6.1) is satisfied with $r = (p' + d')/2$ in place of q and, by the first case, we have $\mu \in W_0^{r',1}(U_R)$ if R is sufficiently small. Since $r' > d$, by the Sobolev embedding theorem, μ is bounded in U_R. Furthermore, we note that (1.6.1) means that

$$\varphi \mapsto \int_{U_R} a^{ij} \partial_{x_i} \partial_{x_j} \varphi \mu \, dx$$

is a linear functional defined on a dense subspace $C_0^2(U_R)$ of $W_0^{q,1}(U_R)$ and bounded in the $W_0^{q,1}(U_R)$-norm. By the duality between $W_0^{q,1}(U_R)$ and $W^{q',-1}(U_R)$, we have that

$$\int_{U_R} a^{ij} \partial_{x_i} \partial_{x_j} \varphi \mu \, dx = \int_{U_R} f \varphi \, dx,$$

where $f \in W^{q',-1}(U_R) \subset W^{p,-1}(U_R)$. Thus, μ is a generalized solution of the equation

$$\partial_{x_j}(a^{ij} \partial_{x_i} \mu) = f - \partial_{x_j}(\partial_{x_i} a^{ij} \mu) =: g.$$

Here $\partial_{x_i} a^{ij} \mu \in L^p(U_R)$, since μ is bounded, so that $g \in W^{p,-1}(U_R)$. Since μ belongs to $W_0^{r',1}(U_R)$ and $r' > d \geqslant 2$, we conclude that $\mu \in W_0^{p,1}(U_R)$ by Corollary 1.2.4, which is applicable, since $p > d$. □

1.6.3. Remark. The proof of this theorem actually shows that if μ has compact support in U_{R_1} and (1.6.1) holds for all $\varphi \in C_0^\infty(U_{R_1})$, then $\mu \in W_0^{q' \wedge p,1}(U_R)$ for some $R < R_1$. Moreover, even without the assumption of compactness of support, one can show that $\mu \in W_{\text{loc}}^{q' \wedge p,1}(U_R)$, but this requires some extra work (Exercise 1.8.16).

This theorem yields at once a certain low regularity of solutions to our elliptic equations.

1.6.4. Corollary. *Suppose that* $p > d \geqslant 2$, $a^{ij} \in W_{\text{loc}}^{p,1}(\Omega)$, $\det A > 0$, *and* μ *satisfies the equation* $L_{A,b}^* \mu = 0$, *where* $b \in L_{\text{loc}}^r(\mu)$ *for some* $r > 1$. *Then* μ *has a density in the class* $W_{\text{loc}}^{\alpha,1}(\Omega)$ *for each* $\alpha < dr/(dr - r + 1)$.

Proof. Let us take $\eta \in C_0^\infty(\Omega)$ with support in a ball $U \subset \Omega$ and $0 \leqslant \eta \leqslant 1$. Consider the measure $\mu_0 = \eta \cdot \mu$. We know that μ has a density in $L_{\text{loc}}^s(\Omega)$ with any $s < d/(d-1)$, which will be denoted also by μ. For every $\varphi \in C_0^\infty(U)$ we have

$$\eta a^{ij} \partial_{x_i} \partial_{x_j} \varphi = L_{A,b,c}(\eta\varphi) - \varphi a^{ij} \partial_{x_i} \partial_{x_j} \eta - 2a^{ij} \partial_{x_i} \varphi \partial_{x_j} \eta - \varphi b^i \partial_{x_i} \eta - \eta b^i \partial_{x_i} \varphi - c\eta\varphi.$$

Let $q = \alpha'$. By Hölder's inequality, the integral of $\eta b^i \partial_{x_i}\varphi$ with respect to μ is estimated by

$$\|\nabla\varphi\|_{L^q(U)}\|\eta b\mu\|_{L^\alpha(U)} \leqslant \|\nabla\varphi\|_{L^q(U)}\|\eta b\|_{L^r(\mu)}\|\mu\|_{L^s(U)}^{(r-\alpha)/(r\alpha)},$$

where $s = (r\alpha-\alpha)/(r-\alpha) < d(d-1)$, since $\alpha < dr/(dr-r+1)$, so $\|\mu\|_{L^s(U)}$ is finite. The integrals of the remaining terms are estimated similarly. In particular, the integral of $\eta c\varphi$ is estimated by $\|\varphi\|_\infty\|\eta c\|_{L^1(\mu)}$ and the norm $\|\varphi\|_\infty$ is estimated by $C\|\nabla\varphi\|_{L^q(U)}$, since $q > d$ due to the inequality $dr/(dr-r+1) < d/(d-1)$, which is readily verified. Therefore,

$$\int_U a^{ij}\partial_{x_i}\partial_{x_j}\varphi\, d\mu_0 \leqslant C(\eta, A, b, c)\|\nabla\varphi\|_q.$$

Hence $\eta\mu \in W_0^{\alpha,1}(U)$, which yields our assertion. □

More can be obtained if b is better integrable.

1.6.5. Theorem. *Let $\mu = \varrho\, dx$, $\varrho \in L^r_{\rm loc}(\Omega, dx)$, $a^{ij} \in W^{p,1}_{\rm loc}(\Omega)$, where $p > d$, $r \in (p', \infty)$. Suppose that the mapping A^{-1} is locally bounded and we are given functions*

$$\beta \in L^p_{\rm loc}(\Omega, dx) + L^p_{\rm loc}(\Omega, \mu) \quad \textit{and} \quad \gamma \in L^{pd/(p+d)}_{\rm loc}(\Omega, dx) + L^{pd/(p+d)}_{\rm loc}(\Omega, \mu)$$

such that for every $\varphi \in C_0^\infty(\Omega)$ we have

$$\biggl|\int_\Omega a^{ij}(x)\partial_{x_i}\partial_{x_j}\varphi(x)\,\mu(dx)\biggr| \leqslant \int_\Omega \bigl(|\varphi(x)|\,|\gamma(x)| + |\nabla\varphi(x)|\,|\beta(x)|\bigr)\,|\mu|(dx).$$

Then $\varrho \in W^{p,1}_{\rm loc}(\Omega)$.

PROOF. Let $\gamma = \gamma_1 + \gamma_2$, $\gamma_1 \in L^{pd/(p+d)}_{\rm loc}(\Omega, dx)$ and $\gamma_2 \in L^{pd/(p+d)}_{\rm loc}(\Omega, \mu)$. Let also $\beta = \beta_1 + \beta_2$, where $\beta_1 \in L^p_{\rm loc}(\Omega, dx)$ and $\beta_2 \in L^p_{\rm loc}(\Omega, \mu)$.

Note that we can assume that $r \neq p'd/(d-p')$, since otherwise we could just slightly decrease the number r. Since $r > p'$, we have $pr > p + r$. Then

$$q := \frac{pr}{pr-p-r} > 1, \quad q' = \frac{pr}{p+r} > 1.$$

According to (1.1.1), $\beta_1\varrho \in L^{q'}_{\rm loc}(\Omega)$ provided $\beta_1 \in L^p_{\rm loc}(\Omega)$. The same is true for $\beta_2 \in L^p_{\rm loc}(\Omega, \mu)$, since

$$|\beta_2|^{q'}|\varrho|^{q'} = |\beta_2|^{pr/(p+r)}|\varrho|^{r/(p+r)}|\varrho|^{(pr-r)/(p+r)},$$

where $|\beta_2|^{pr/(p+r)}|\varrho|^{r/(p+r)} \in L^s_{\rm loc}(\Omega)$ and $|\varrho|^{(pr-r)/(p+r)} \in L^{s'}_{\rm loc}(\Omega)$ with the number $s = (p+r)/r$.

Since $r \neq p'd/(d-p')$, we have $q \neq d$. If $q > d$, then $r > p'd/(d-p')$ and for every ball U_R with $\overline{U_R} \subset \Omega$ we have by Hölder's inequality

$$\|\gamma\varrho\|_{L^1(U_R)} \leqslant \|\gamma_1\|_{L^{pd/(p+d)}(U_R)}\|\varrho\|_{L^{p'd/(d-p')}(U_R)} + \|\gamma_2\|_{L^1(U_R,|\mu|)}.$$

Let $q < d$. Define k by $kdq'/(d+q') = r$, which gives

$$\frac{k'dq'}{d+q'} = \frac{pd}{p+d},$$

because $(d+q')/(dq') = 1/r + 1/p + 1/d$ due to $1/q' = 1/r + 1/p$. Hence by Hölder's inequality with the exponents k' and k we obtain

$$\|\gamma\varrho\|_{L^{dq'/(d+q')}(U_R)} \leqslant \|\gamma_1\|_{L^{pd/(p+d)}(U_R)}\|\varrho\|_{L^r(U_R)} + \|\gamma_2\|_{L^{pd/(p+d)}(U_R,|\mu|)}\|\varrho\|_{L^s(U_R)},$$

where $s = r(1-(d+q')/dq') < r$.

Observe that for every number $R > 0$ such that $U_R := U_R(x_0) \subset \Omega$, whenever $\eta \in C_0^\infty(U_R)$ and $\varphi \in C_0^2(U_R)$, one has

$$\begin{aligned}
&\left|\int_{U_R} a^{ij}\partial_{x_i}\partial_{x_j}\varphi(\eta\varrho)\,dx\right| \leqslant \left|\int_{U_R} a^{ij}\partial_{x_i}\partial_{x_j}(\varphi\eta)\varrho\,dx\right| \\
&\quad + \left|\int_{U_R} a^{ij}\partial_{x_i}\partial_{x_j}\eta(\varphi\varrho)\,dx\right| + 2\int_{U_R} \|A\|\,|\nabla\eta|\,|\nabla\varphi|\,|\varrho|\,dx \\
&\leqslant N_1\int_{U_R}(|\varphi|+|\nabla\varphi|)|\beta\varrho|\,dx \leqslant N_2\|\beta\varrho\|_{L^{q'}(U_R)}\|\nabla\varphi\|_{L^q(U_R)} = N_3\|\nabla\varphi\|_{L^q(U_R)},
\end{aligned} \tag{1.6.3}$$

where the constants N_1, N_2, and N_3 are independent of φ. The last inequality above is due to the estimate $\|\varphi\|_{L^{dq/(d-q)}(U_R)} \leqslant N\|\nabla\varphi\|_{L^q(U_R)}$ if $q < d$ and the estimate $\|\varphi\|_{L^\infty(U_R)} \leqslant N\|\nabla\varphi\|_{L^q(U_R)}$ if $q > d$ with some constant N. It follows by Theorem 1.6.2 that $\eta\varrho \in W_0^{q'\wedge p,1}(U_R)$ if R is small enough, and, since we can take any point as x_0 and $q' < p$, we have

$$\varrho \in W_{\rm loc}^{q',1}(\Omega). \tag{1.6.4}$$

Moreover, if $q' < d$, then by the Sobolev embedding $\varrho \in L_{\rm loc}^{r_1}(\Omega)$ with

$$r_1 = q'd/(d-q') = \frac{prd}{(p+r)d-pr}.$$

The inequality $q' < d$ is equivalent to $r < pd/(p-d)$. Thus, on the interval

$$\bigl(p/(p-1), pd/(p-d)\bigr)$$

we obtain a mapping $T\colon r \mapsto r_1$ with the property that if $\varrho \in L_{\rm loc}^r(\Omega)$, then we have $\varrho \in L_{\rm loc}^{r_1}(\Omega)$. It is easy to see that

$$\frac{r_1}{r} = \frac{pd}{pd-r(p-d)} \geq \frac{pd}{pd-p'(p-d)} = \frac{d'}{p'} > 1,$$

where the first inequality is due to $pr > p + r$ and the second one is due to $p > d$. Hence after finitely many applications of T to the given number r we will come to

$$s \in \bigl(p/(p-1), pd/(p-d)\bigr)$$

such that $t = T(s) \geq pd/(p-d)$ and $\mu \in L^t(U_R)$. Actually, without loss of generality we may assume that $t > pd/(p-d)$, since otherwise we could just slightly decrease the initial point r (and increase the number of iterations of T). This shows that we could assume from the very beginning that $r > pd/(p-d)$ that is $q' > d$. In that case (1.6.4) implies that the function μ is locally bounded, which shows that (1.6.3) is true with $q' = p$. Now it only remains to apply again Theorem 1.6.2. □

1.6.6. Remark. The condition on the density of μ in Theorem 1.6.5 can be replaced by the condition that $\beta, \gamma \in L_{\rm loc}^1(\Omega, \mu)$. This follows by Theorem 1.5.2.

1.6.7. Corollary. *Let μ be a locally finite Borel measure on Ω satisfying the equation $L_{A,b,c}^*\mu = 0$. Let A^{-1} be locally bounded in Ω with $a^{ij} \in W_{\rm loc}^{p,1}(\Omega)$, where $p > d$, and let either*

(i) $b^i \in L_{\rm loc}^p(\Omega, dx)$, $c \in L_{\rm loc}^{pd/(p+d)}(\Omega, dx)$

or

(ii) $b^i \in L_{\rm loc}^p(\Omega, \mu)$, $c \in L_{\rm loc}^{pd/(p+d)}(\Omega, \mu)$. *Then μ has a density in $W_{\rm loc}^{p,1}(\Omega)$ that is locally Hölder continuous.*

PROOF. It suffices to take $\beta = |b|$, $\gamma = |c|$ and apply Theorem 1.6.5. □

1.6.8. Corollary. *Let μ be a locally finite Borel measure on U_R. Suppose that the mapping A^{-1} is locally bounded on U_R with $a^{ij} \in W^{p,1}_{\text{loc}}(U_R)$, where $p > d$, $\partial_{x_i} a^{ij} \in L^p_{\text{loc}}(\mu)$, and b^i, $c \in L^p_{\text{loc}}(\mu)$. Suppose that*

$$\int_{U_R} \Bigl[a^{ij}\partial_{x_i}\partial_{x_j}\varphi + \partial_{x_i}a^{ij}\partial_{x_j}\varphi + b^i\partial_{x_i}\varphi + c\varphi\Bigr]\, d\mu = 0 \quad \forall \varphi \in C_0^\infty(U_R).$$

Then μ has a density in $W^{p,1}_{\text{loc}}(U_R)$ that is locally Hölder continuous.

Corollary 1.6.7 can be generalized as follows.

1.6.9. Corollary. *Let $p > d$, let $a^{ij} \in W^{p,1}_{\text{loc}}(\Omega)$, b^i, f^i, $c \in L^p_{\text{loc}}(\Omega)$, and let A^{-1} be locally bounded in Ω. Assume that μ is a locally finite Borel measure on Ω such that b^i, $c \in L^1_{\text{loc}}(\Omega, \mu)$ and, for every function $\varphi \in C_0^\infty(\Omega)$, one has*

$$\int_\Omega \Bigl[a^{ij}\partial_{x_i}\partial_{x_j}\varphi + b^i\partial_{x_i}\varphi + c\varphi\Bigr]\, d\mu = \int_\Omega f^i \partial_{x_i}\varphi\, dx.$$

Then μ has a density in $W^{p,1}_{\text{loc}}(\Omega)$.

It is easily seen that in Corollary 1.6.7 one cannot omit the hypotheses that A^{-1} is locally bounded and $a^{ij} \in W^{p,1}_{\text{loc}}$. Indeed, if A and b vanish at a point x_0, then Dirac's measure at x_0 satisfies our elliptic equation. In particular, if it is not given in advance that μ is absolutely continuous, then one cannot take an arbitrary Lebesgue version of A. We have already seen in Example 1.4.9 that a solution may fail to be more regular than A. Also, the condition $p > d$ is essential for the membership of μ in a Sobolev class even if $A = \mathrm{I}$ (see the example below). However, if μ is a probability measure on $\mathbb{R}^d$, then the condition $|b| \in L^2(\mu)$ implies that $\mu = \varrho\, dx$ with $\varrho \in W^{1,1}(\mathbb{R}^d)$ and $|\nabla\varrho|^2/\varrho \in L^1(\mathbb{R}^d)$ (see § 3.1).

1.6.10. Example. Let $d > 3$ and

$$L^*F = \Delta F - \operatorname{div}(Fb) - F = \Delta F + \alpha\partial_{x_i}(x_i|x|^{-2}F) - F,$$

where $\alpha = d - 3$ and

$$b(x) = -\alpha x|x|^{-2} = \nabla(|x|^{-\alpha})/|x|^{-\alpha}.$$

Then the function $F(x) = (e^r - e^{-r})r^{-(d-2)}$, $r = |x|$, is locally Lebesgue integrable and $L^*F = 0$ in the sense of distributions, but F does not belong to $W^{2,1}_{\text{loc}}(\mathbb{R}^d)$. Here $|b| \in L^{d-\varepsilon}_{\text{loc}}(\mathbb{R}^d)$ for all $\varepsilon > 0$. In a similar way, if the term $-F$ is omitted in the equation above, then the function $F(x) = r^{-(d-3)}$ has the same properties.

PROOF. Observe that $\partial_{x_i}F$, $\partial_{x_i}\partial_{x_j}F$ are locally Lebesgue integrable. Hence the equation $L^*F = 0$ follows easily from the equation

$$f'' + \frac{(d-1+\alpha)}{r} f' + \alpha\frac{d-2}{r^2} f - f = 0$$

on $(0, \infty)$, which is satisfied for the function $f(r) = (e^r - e^{-r})r^{-(d-2)}$. It remains to note that F, ∇F, and ΔF are locally Lebesgue integrable, since $f(r)r^{d-1}$, $f'(r)r^{d-1}$, and $f''(r)r^{d-1}$ are locally bounded, but ∇F is not Lebesgue square-integrable at the origin. If $d \geqslant 6$, then F is also not Lebesgue square-integrable at the origin. In the case without the term $-F$ in the equation similar calculations show that $F(x) = r^{-(d-3)}$ has the same properties. □

1.7. Regularity of solutions to divergence type equations

Here we present several important results related to divergence form equations. First we consider the equation

$$\partial_{x_i}(a^{ij}\partial_{x_j}u) = 0. \tag{1.7.1}$$

Suppose that $\lambda_1 \cdot \mathrm{I} \leqslant A(x) \leqslant \lambda_2 \cdot \mathrm{I}$. What we call its solution can be defined in different ways depending on the properties of the coefficients a^{ij} (see § 1.4). If the functions a^{ij} are merely measurable and locally bounded, then we require that the solution must satisfy the inclusion $u \in W^{1,1}_{\mathrm{loc}}$ and the identity

$$\int_{\mathbb{R}^d} a^{ij}\partial_{x_i}\varphi\partial_{x_j}u\,dx = 0 \quad \forall\,\varphi \in C_0^\infty(\mathbb{R}^d).$$

In this case E. De Giorgi (see De Giorgi [**286**]) showed that any solution in $W^{2,1}_{\mathrm{loc}}$ is locally Hölder continuous. J. Serrin (see Serrin [**842**]) constructed an example showing that the membership in $W^{p,1}_{\mathrm{loc}}$ with $p < 2$ does not ensure the local boundedness of a solution, and his conjecture that in the case of Hölder continuous coefficients any solution in $W^{1,1}_{\mathrm{loc}}$ belongs automatically to $W^{2,1}_{\mathrm{loc}}$ was proved by H. Brezis (see Brezis [**207**], Ancona [**48**]), even under somewhat weaker assumptions: the solution must belong to the class of functions of bounded variation and the coefficients must be Dini continuous. In addition, in the case of merely continuous a^{ij}, Brezis proved that any solution in $W^{p,1}_{\mathrm{loc}}$ with some $p > 1$ belongs to all $W^{q,1}_{\mathrm{loc}}$ with $q < \infty$. However, in Jin, Maz'ya, Van Schaftingen [**492**] an example was constructed showing that for $p = 1$ this is not true.

Let now all functions a^{ij} be locally Lipschitzian. Then equation (1.7.1) can be written as $L^*_{A,b}u = 0$ with $b^i = \sum_{j=1}^d \partial_{x_j}a^{ij}$, hence a priori solutions from L^1_{loc} are admissible. In this situation, in Zhang, Bao [**955**] the conjecture of Brezis was proved that all solutions belong to all classes $W^{q,2}_{\mathrm{loc}}$ with $q < \infty$; the problem was to prove the inclusion in $W^{2,1}_{\mathrm{loc}}$, then the classical results increase the regularity.

Let us consider a general divergence form equation. Set

$$\mathcal{L}u = \partial_{x_i}\big(a^{ij}\partial_{x_j}u - b^i u\big) + \beta^i\partial_{x_i}u + cu, \tag{1.7.2}$$

where functions a^{ij}, b^i, β^i and c are measurable on a bounded open set $\Omega \subset \mathbb{R}^d$.

Suppose that

$$\lambda_1 \cdot \mathrm{I} \leqslant A(x) \leqslant \lambda_2 \cdot \mathrm{I}, \quad \lambda_1, \lambda_2 > 0,$$

$g = |c| + \sum_{i=1}^d \big[|a^i|^2 + |b^i|^2\big] \in L^s(\Omega)$, $f \in L^s(\Omega)$, $g^i \in L^{2s}(\Omega)$, $s > d/2$. We shall say that a function $u \in W^{2,1}_{\mathrm{loc}}(\Omega)$ satisfies the equation

$$\mathcal{L}u = f + \mathrm{div} g$$

if we have the equality

$$\int_\Omega \Big[\langle A\nabla u, \nabla\varphi\rangle\,dx + u\langle b, \nabla\varphi\rangle + \langle \beta, \nabla u\rangle\varphi + cu\varphi\Big]\,dx = \int_\Omega \big[\langle g, \nabla\varphi\rangle - f\varphi\big]\,dx$$

for every function $\varphi \in C_0^\infty(\Omega)$. For such solutions the following important result of Trudinger [**904**] holds (its first assertion was proved already in the first edition of the book Ladyzhenskaya, Ural'tseva [**577**, Theorem 14.1]).

1.7.1. Theorem. *Any solution u has a locally Hölder continuous version and for every ball $U_R(z)$ in Ω and $r < R$ one has the inequality*

$$\sup_{x,y\in U_r(z)} |u(x)-u(y)| \leqslant C_1 r^{\alpha}\Bigl(\sup_{U_R(z)}|u| + \|f\|_{L^s(\Omega)} + \max_i \|g^i\|_{L^{2s}(\Omega)}\Bigr),$$

where the numbers C_1 and $\alpha > 0$ depend only on $d, s, R, \lambda_1, \lambda_2, \|g\|_{L^s(\Omega)}$. If $f = 0$ and $g = 0$ and $u \geqslant 0$, then u satisfies Harnack's inequality

$$\sup_{x\in U_r(z)} u(x) \leqslant C_2 \inf_{x\in U_r(z)} u(x),$$

where the number C_2 depends only on the same quantities as C_1.

For solutions of the equation $L^*_{A,b,c}\mu = 0$ we obtain the following.

1.7.2. Corollary. *Let μ be a nonnegative locally finite Borel measure on a domain Ω in $\mathbb{R}^d$ satisfying the equation $L^*_{A,b,c}\mu = 0$. Let A^{-1} be locally bounded in Ω with $a^{ij} \in W^{p,1}_{\rm loc}(\Omega)$, where $p > d$, and let $b^i \in L^p_{\rm loc}(\Omega, dx)$, $c \in L^{p/2}_{\rm loc}(\Omega, dx)$. Then the continuous density ϱ of μ has the following property: for every compact set K contained in a connected open set U with compact closure in Ω, one has*

$$\sup_K \varrho \leqslant C \inf_K \varrho,$$

where the number C depends only on the quantities $\|a^{ij}\|_{W^{p,1}(U)}$, $\|b\|_{L^p(U)}$, $\|c\|_{L^p(U)}$, $\inf_U \det A$, and K. In particular, ϱ does not vanish in U if it is not identically zero in U.

The dependence of C on the indicated quantities will be studied in Chapter 3. The assumption that $b^i \in L^p_{\rm loc}(\Omega, dx)$ in Theorem 1.7.2 cannot be replaced by the alternative assumption from Corollary 1.6.7 that $b^i \in L^p_{\rm loc}(\Omega, \mu)$. Indeed, it suffices to take $b = \nabla\varrho/\varrho$ such that ϱ is a probability density which has zeros, but $|b| \in L^p(\mu)$ (for example, we can take ϱ which behaves like $\exp(-x^{-2})$ in a neighborhood of the origin).

1.7.3. Proposition. *Suppose that the hypotheses of the previous corollary are fulfilled and Ω is connected. Let μ be some positive measure on Ω satisfying the equation $L^*_{A,b,c}\mu = 0$. Then, any other solution μ_0 can be written as $\mu_0 = f\cdot\mu$, where $f \in W^{p,2}_{\rm loc}(\Omega)$.*

PROOF. Suppose first that $d > 1$. Then $p > 2$. We know that μ and μ_0 have continuous densities ϱ and ϱ_0, respectively, in the class $W^{p,1}_{\rm loc}(\Omega)$ and that ϱ has no zeros in Ω. Set $f = \varrho_0/\varrho$. Then $\mu_0 = f\cdot\mu$ and $f \in W^{p,1}_{\rm loc}(\Omega)$. We have $a^i := \sum_{j=1}^d \partial_{x_j} a^{ij} \in L^p_{\rm loc}(\Omega)$. Set $a := (a^i)$. Let us verify that f satisfies the elliptic equation

$$a^{ij}\varrho\partial_{x_i}\partial_{x_j} f + \langle \nabla f, 2\varrho a + 2A\nabla\varrho - \varrho b\rangle = 0 \tag{1.7.3}$$

in the sense of weak solutions in the class $W^{p,1}_{\rm loc}(\Omega)$, i.e., in the sense of the identity

$$\int_\Omega \bigl[-\varphi\partial_{x_i}(a^{ij}\varrho)\partial_{x_j} f - \langle \varrho A\nabla f, \nabla\varphi\rangle + \langle \nabla f, 2\varrho a + 2A\nabla\varrho - \varrho b\rangle\varphi\bigr]\, dx = 0$$

for all $\varphi \in C_0^\infty(\Omega)$. This will yield the desired inclusion $f \in W^{p,2}_{\rm loc}(\Omega)$, since we have $\varrho a^{ij} \in W^{p,1}_{\rm loc}(\Omega)$, ϱA is nondegenerate, $a\varrho$, $b\varrho$, $c\varrho \in L^p_{\rm loc}(\Omega)$. In order to establish equality (1.7.3) we observe that the equality $L^*_{A,b,c}\mu_0 = L^*_{A,b,c}\mu = 0$ and the integration by parts formula give the identities

$$\int [-\partial_{x_i}(a^{ij}\varrho f)\partial_{x_j}\varphi + \langle f\varrho b, \nabla\varphi\rangle + c\varrho f\varphi]\, dx = 0, \tag{1.7.4}$$

$$\int [-\partial_{x_i}(a^{ij}\varrho)\partial_{x_j}\varphi + \langle \varrho b, \nabla\varphi\rangle + c\varrho\varphi]\, dx = 0 \tag{1.7.5}$$

for all $\varphi \in C_0^\infty(\Omega)$. Since $a^{ij}, \varrho, f \in W^{p,1}_{\text{loc}}(\Omega)$ and $p > 2$, it follows that equality (1.7.5) remains true for all functions φ of the form $\varphi = f\psi$ with $\psi \in C_0^\infty(\Omega)$. This yields the identity

$$\int [-\partial_{x_i}(a^{ij}\varrho)f\partial_{x_j}\varphi - \partial_{x_i}(a^{ij}\varrho)\varphi\partial_{x_j}f + \langle \varrho b, f\nabla\varphi\rangle + \langle \varrho b, \varphi\nabla f\rangle + c\varrho f\varphi]\, dx = 0$$

for all $\varphi \in C_0^\infty(\Omega)$. Subtracting this identity from (1.7.4) and differentiating the products by the Leibniz formula we arrive at (1.7.3). In the case $d = 1$ this reasoning does not apply if $p < 2$, but in this case a simple direct proof works: we have $(Af\varrho)' = f\varrho b + \psi$ and $(A\varrho)' = b\varrho + k$, where ψ is the indefinite integral of $cf\varrho$ and k is constant. Then $f' = (\psi - kf)(A\varrho)^{-1}$. □

In the rest of this section we discuss a priori estimates of solutions on a bounded domain $\Omega \subset \mathbb{R}^d$, which will play an important role in the proofs of the theorems on existence and regularity of solutions. The simplest a priori estimate is obtained by substituting in the identity defining the equation $\mathcal{L}u = f + \operatorname{div} g$ the function $u\varphi$, where the function $\varphi \in C_0^\infty(\Omega)$ equals 1 on Ω' and $\overline{\Omega'} \subset \Omega$. Suppose for simplicity that $b = \beta = c = 0$ and $f, g^i \in L^2(\Omega)$. Then we immediately obtain

$$\|u\|_{W^{2,1}(\Omega')} \leqslant C\bigl(\|u\|_{L^2(\Omega)} + \|f\|_{L^2(\Omega)} + \|g\|_{L^2(\Omega)}\bigr).$$

The next result generalizes this estimate to the case of the operator $\mathcal{L}$ of the form (1.7.2) with all coefficients and the space $W^{p,1}$.

For functions on Ω we shall write that $a^{ij} \in VMO$ if a^{ij} extends to all of $\mathbb{R}^d$ as a function in VMO. We recall that the membership of a function a^{ij} in the class VMO is expressed in terms of a certain function ω denoted below by the symbol ω_A (see § 1.2). This condition is weaker than the uniform continuity. In the case of uniformly continuous coefficients we can take for ω_A the modulus of continuity. For Hölder continuous coefficients, ω_A can be easily expressed in terms of their Hölder norms.

Suppose that the coefficients a^{ij}, b, β and c satisfy the following conditions with some numbers $\lambda_1, \lambda_2 > 0$:

$$\begin{aligned}
&a^{ij} \in VMO,\ a^{ij} = a^{ji},\ \ \lambda_1 \cdot \mathrm{I} \leqslant A(x) \leqslant \lambda_2 \cdot \mathrm{I},\\
&b^i, \beta^i \in L^d(\Omega),\ c \in L^{d/2}(\Omega) \text{ if } 2 \leqslant q < d,\\
&b^i, \beta^i \in L^s(\Omega),\ c \in L^{s/2}(\Omega) \text{ with some } s > d \text{ if } 2 \leqslant q = d,\\
&b^i, \beta^i \in L^q(\Omega),\ c \in L^{dq/(d+q)}(\Omega) \text{ if } q > d.
\end{aligned} \tag{1.7.6}$$

As above, let $U(x, r)$ denote the ball of radius r centered at x and let $|U(x, r)|$ be its volume.

For $\gamma \geqslant 1, r > 0$ and $\eta \in L^\gamma(\Omega)$ we set

$$\Theta_{\gamma,\eta}(r) = \sup_{x\in\Omega} \biggl(\int_{U(x,r)\cap\Omega} |\eta(y)|^\gamma\, dy\biggr)^{1/\gamma}.$$

We recall Young's inequality. If $x > 0$, $y > 0$, $\varepsilon > 0$, $\delta > 0$, $\gamma > 0$, and $\delta^{-1} + \gamma^{-1} = 1$, then

$$xy \leqslant \varepsilon \frac{x^\delta}{\delta} + \varepsilon^{1-\gamma} \frac{y^\gamma}{\gamma}. \tag{1.7.7}$$

1.7.4. Theorem. *Suppose that the coefficients of the operator $\mathcal{L}$ satisfy conditions (1.7.6) and $q \geqslant 2$. Let $g \in L^q(\Omega)$, $f \in L^p(\Omega)$, where $p = dq/(d+q)$ if $q \neq d$ and $p > d/2$ if $q = d$. If a function $u \in W^{q,1}(\Omega)$ is a solution to the equation $\mathcal{L}u = f + \operatorname{div} g$ on Ω, then for every open set Ω' with compact closure in Ω we have the estimate*

$$\|u\|_{W^{q,1}(\Omega')} \leqslant C\big(\|u\|_{L^1(\Omega)} + \|g\|_{L^q(\Omega)} + \|f\|_{L^p(\Omega)}\big),$$

where the constant C depends only on $\Omega, \Omega', d, \lambda_1, \lambda_2, q$, ω_A and on the rate of convergence of the quantities $\Theta_{d,b}(r)$, $\Theta_{d,\beta}(r)$, $\Theta_{d/2,c}(r)$ to zero as $r \to 0$ in case $2 \leqslant q < d$ and on the number s and the norms $\|b^i\|_{L^s}$, $\|\beta^i\|_{L^s}$, $\|c\|_{L^{s/2}}$ in case $q = d$, and, finally, on the norms $\|b^i\|_{L^q}$, $\|\beta^i\|_{L_q}$, $\|c\|_{L^{dq/(d+q)}}$ in case $q > d$.

Proof. Let $U = U(a,r)$ be a ball with closure in Ω and $0 < r < 1$ and let a function ζ in $C_0^\infty(U)$ be such that $0 \leqslant \zeta \leqslant 1$, $\zeta(x) > 0$ in U and $\zeta(x) = 1$ on the twice smaller ball $U(a, r/2)$, and also

$$J(s) := \sup_x |\nabla\zeta(x)|\zeta(x)^{-s} + \sup_{i,j,x} |\partial_{x_i}\partial_{x_j}\zeta(x)|\zeta(x)^{-s} < \infty, \quad 0 < s < 1.$$

These conditions can be easily ensured by taking $\zeta(x) = \psi(|x|/r)$, $\psi \in C_0^\infty(\mathbb{R})$, $0 \leqslant \psi \leqslant 1$, $\psi(y) = 0$ if $|y| \geqslant 1$, $\psi(y) > 0$ if $|y| < 1$, $\psi(y) = 1$ if $|y| \leqslant 1/2$ and $\psi(y) = \exp\big[(y^2-1)^{-1}\big]$ near the points -1 and 1. We shall use as a cut-off function only this ζ.

1. We first prove the theorem for $q = 2$. To this end, we shall estimate the norm $\|u\zeta\|_{W^{2,1}(U)}$. Substituting in the identity defining the equation $\mathcal{L}u = f + \operatorname{div} g$ the function $\varphi = u\zeta^2$ and integrating, we have

$$\int_\Omega \big[a^{ij}u_{x_j}u_{x_i}\zeta^2\,dx + 2a^{ij}u_{x_j}u\zeta\zeta_{x_i} + b^i u_{x_i}u\zeta^2 + 2b^i u^2\zeta_{x_i}\zeta \\ + g^i u_{x_i}\zeta^2 + 2g^i u\zeta\zeta_{x_i} + c^i u_{x_i}u\zeta^2 + hu^2\zeta^2 + fu\zeta^2\big]\,dx = 0.$$

Using inequality (1.7.7) with $\alpha = \beta = 2$ and a sufficiently small $\varepsilon > 0$ and the first condition in (1.7.6), we find that

$$\|\zeta\nabla u\|_{L^2}^2 \leqslant C_1\big(\|u\nabla\zeta\|_{L^2}^2 + \|bu\zeta\|_{L^2}^2 + \|\beta u\zeta\|_{L^2}^2 \\ + \|\sqrt{|c|}u\zeta\|_{L^2}^2 + \|g\zeta\|_{L^2}^2 + \|fu\zeta^2\|_{L^1}\big), \tag{1.7.8}$$

where $C_1 = C_1(d, m, M)$. Let us estimate the summands in the right-hand side of this inequality separately. Set $t = d$ if $d > 2$ and $t = (s+2)/2$ if $d = 2$. It is clear that $2 < t < s$ if $n = d$. In order to estimate $\|\eta u\zeta\|_{L^2}$, where η equals b, c or $\sqrt{|h|}$, we apply Hölder's inequality with the exponent $t/2$ and the Sobolev embedding $W^{2,1} \subset L^{2t/(t-2)}$. We have

$$\|\eta u\zeta\|_2 \leqslant \|\eta\|_t \|u\zeta\|_{2t/(t-2)} \leqslant C(d,t)\|\eta\|_t\|u\zeta\|_{2,1}, \tag{1.7.9}$$

where $\|\eta\|_t = \Theta_{d,\eta}(r)$ if $t = d$, $\|\eta\|_t \leqslant |U(a,r)|^{1-t/s}\|\eta\|_s$ if $2 < t < s$. Applying Hölder's inequality with the exponents p, $p' = p(p-1)^{-1}$, where $p = 2d(d+2)^{-1}$

if $d > 2$ and $p > 1$ if $d = 2$, inequality (1.7.7) with $\gamma = \delta = 2$ and the Sobolev embedding $W^{2,1} \subset L^{p'}$, we obtain

$$\|fu\zeta^2\|_1 \leqslant \|f\zeta\|_p\|u\zeta\|_{p'} \leqslant C(d,p)\|f\zeta\|_p\|u\zeta\|_{2,1}$$
$$\leqslant \varepsilon^2\|u\zeta\|_{2,1}^2 + \varepsilon^{-2}C(d,p)^2\|f\zeta\|_p^2.$$

Let us estimate $\|u\nabla\zeta\|_{L^2}$. Let $l = 4(d+2)^{-1}$ if $d > 2$ and $0 < l < 1$ if $d = 2$. By the Sobolev embedding theorem $W^{2,1} \subset L^{(2-l)/(1-l)}$. Applying Hölder's inequality, we arrive at the inequalities

$$\|u\nabla\zeta\|_2 = \|u\zeta^{(2-l)/2}\zeta^{-(2-l)/2}\nabla\zeta\|_2 \leqslant J(1-l/2)\biggl(\int_U |u\zeta|^{2-l}|u|^l\,dx\biggr)^{1/2}$$
$$\leqslant J(1-l/2)\|u\|_1^{l/2}\|u\zeta\|_{(2-l)/(1-l)}^{(2-l)/2} \leqslant J(1-l/2)C(n,l)\|u\|_1^{l/2}\|u\zeta\|_{2,1}^{(2-l)/2}.$$

Young's inequality (1.7.7) with $\alpha = 2/l$ and $\beta = 2/(2-l)$ gives

$$\|u\nabla\zeta\|_2 \leqslant \varepsilon\|u\zeta\|_{2,1} + \varepsilon^{-l/(2-l)}C(d,l)\|u\|_{L^1}.$$

According to estimate (1.7.9) with $\eta = 1$, we have

$$\|u\zeta\|_{L^2(U(a,r))} \leqslant C(d,t)|U(a,r)|^{1/t}\|u\zeta\|_{2,1}.$$

Substituting the obtained estimates in (1.7.8) and taking into account that by the Leibniz formula

$$\|u\zeta\|_{2,1} \leqslant \|u\nabla\zeta\|_2 + \|\zeta\nabla u\|_2 + \|u\zeta\|_2,$$

we find that

$$\|u\zeta\|_{W^{2,1}(U)} \leqslant C_2\varepsilon^{-1}\bigl(\|u\|_{L^1(U)} + \|g\|_{L^2(U)} + \|f\|_{L^p(U)}\bigr)$$
$$+ C_2\bigl(\|b\|_{L^t(U)} + \|\beta\|_{L^t(U)} + \|c\|_{L^{t/2}(U)}^{1/2} + |U(a,r)|^{1/t} + \varepsilon\bigr)\|u\zeta\|_{W^{2,1}(U)},$$

where $C_2 = C_2(d,p,t,m,M)$. Choosing $r > 0$ and $\varepsilon > 0$ such that

$$C_2\Bigl(\|b\|_{L^t(U(a,r))} + \|c\|_{L^t(U(a,r))} + \|h\|_{L^{t/2}(U(a,r))}^{1/2} + |U(a,r)|^{1/t} + \varepsilon\Bigr) < \frac{1}{2},$$

we arrive at the estimate $\|u\zeta\|_{2,1} \leqslant 2C_2\varepsilon^{-1}\bigl(\|u\|_1 + \|g\|_2 + \|f\|_p\bigr)$. By using a smooth partition of unity associated with a finite covering of the domain Ω' by balls of radius r with closure in Ω, we obtain the required estimate.

2. Let us consider the case $2 < q < d$. As above, we start with an estimate of the quantity $\|u\zeta\|_{W^{q,1}(U)}$. We may assume that A is extended to the whole space $\mathbb{R}^d$ with preservation of all conditions. By Theorem 1.2.1, for a sufficiently large number $\lambda = \lambda(d,q,m,M,\omega_A) > 0$ there is a number $N = N(d,q,m,M,\omega_A) > 1$ such that for every function $w \in W^{q,1}(\mathbb{R}^d)$ with compact support the generalized function $\mathcal{A}w = \partial_{x_i}(a^{ij}\partial_{x_j}w) - \lambda w$ satisfies the inequality

$$\|w\|_{W^{q,1}(\mathbb{R}^d)} \leqslant N\|\mathcal{A}w\|_{W^{q,-1}(\mathbb{R}^d)}. \tag{1.7.10}$$

The function $w = \zeta u$ satisfies the equation

$$\mathcal{A}w = -\lambda\zeta u + a^{ij}\partial_{x_i}\zeta\partial_{x_j}u + \partial_{x_i}(a^{ij}u\partial_{x_j}\zeta) - \zeta\partial_{x_i}(b^i u)$$
$$+ \beta^i\partial_{x_i}u\zeta + cu\zeta - \zeta\partial_{x_i}g^i + f\zeta. \tag{1.7.11}$$

According to (1.7.10) it suffices to estimate the norm of the right-hand side in $W^{q,-1}(U)$ through the parameters indicated in the theorem. Let us rewrite the right-hand side of the last equality in the following form convenient for our later use:

$$\text{(1.7.12)}\quad \mathcal{A}w = \partial_{x_i}(a^{ij}u\partial_{x_j}\zeta - b^i u\zeta - g^i\zeta) - \lambda\zeta u + a^{ij}\partial_{x_i}\zeta\partial_{x_j}u + b^i u\partial_{x_i}\zeta + \beta^i\partial_{x_i}(u\zeta) - \beta^i u\partial_{x_i}\zeta + cu\zeta + g^i\partial_{x_i}\zeta + f\zeta.$$

We recall that $\|\partial_{x_i}\eta^i\|_{q,-1} \leqslant \|\eta\|_q$ for every vector function $\eta \in L^q$ and by the embedding theorem we have $L^p \subset W^{q,-1}$ if $p = dq(d+q)^{-1}$. Therefore, the norm of the right-hand side is estimated by

$$\text{(1.7.13)}\quad \|a^{ij}u\partial_{x_j}\zeta\|_q + \|b^i u\zeta\|_q + \|g^i\zeta\|_q + C(d,p)\big[\lambda\|\zeta u\|_p + \|a^{ij}\partial_{x_i}\zeta\partial_{x_j}u\|_p + \|b^i u\partial_{x_i}\zeta\|_p + \|\beta^i\partial_{x_i}(u\zeta)\|_p + \|\beta^i u\partial_{x_i}\zeta\|_p + \|hu\zeta\|_p + \|g^i\partial_{x_i}\zeta\|_p + \|f\zeta\|_p\big],$$

where $C = C(d,p)$ is the constant from the embedding theorem. We observe that

$$\|a^{ij}u\partial_{x_j}\zeta\|_q \leqslant M\|u\nabla\zeta\|_q, \quad \|g^i\partial_{x_i}\zeta\|_p \leqslant \|\nabla\zeta\|_d\|g\|_q.$$

Applying Hölder's inequality with the exponents d/q, $d/(d-q)$ and the exponents $(d+q)/q$, $(d+q)/d$, we obtain

$$\|b^i u\zeta\|_q + C(d,p)\big[\|b^i u\partial_{x_i}\zeta\|_p + \|c^i\partial_{x_i}(u\zeta)\|_p + \|\beta^i u\partial_{x_i}\zeta\|_p\big] \leqslant C(d,q,p)\big[(\|b\|_d + \|c\|_d)\|u\zeta\|_{q,1} + (\|b\|_d + \|\beta\|_d)\|u\nabla\zeta\|_q\big].$$

Hölder's inequality with the exponents $(d+q)/2q$, $(d+q)/(d-q)$ and the Sobolev embedding theorem give

$$\lambda\|\zeta u\|_p + \|cu\zeta\|_p \leqslant C(d,q)\Big(\lambda|U(a,r)|^{2q/(d+q)} + \|c\|_{d/2}\Big)\|u\zeta\|_{q,1}.$$

It remains to estimate $\|u\nabla\zeta\|_{L^q}$. Let $l = q^2/(dq+q-d)$. We observe that $0<l<1$ and $(q-l)/(1-l) = dq/(d-q)$. By the Sobolev embedding theorem and Hölder's inequality we have

$$\|u\nabla\zeta\|_q = \|\zeta^{1-l/q}u\zeta^{-1+l/q}\nabla\zeta\|_q \leqslant J(1-l/q)\Big(\int_U |u\zeta|^{q-l}|u|^l\,dx\Big)^{1/q}$$
$$\leqslant J(1-l/q)\|u\|_1^{l/q}\|u\zeta\|_{(q-l)/(1-l)}^{(q-l)/q} \leqslant J(1-l/q)C(d,l,q)\|u\|_1^{l/q}\|\zeta u\|_{q,1}^{(q-l)/q}.$$

According to Young's inequality (1.7.7) with $\alpha = q/l, \beta = q/(q-l)$, we obtain

$$\|u\nabla\zeta\|_q \leqslant \varepsilon\|u\zeta\|_{q,1} + C(d,l,q,\varepsilon)\|u\|_1.$$

Similarly to the case $q=2$ we pick $\varepsilon > 0$ and $r > 0$ sufficiently small. We obtain

$$\|u\zeta\|_{q,1} \leqslant C_3(\|\nabla u\nabla\zeta\|_p + \|u\|_1 + \|g\|_q + \|f\|_p).$$

By using a partition of unity we arrive at the estimate

$$\|u\|_{W^{q,1}(\Omega')} \leqslant C_4\big(\|u\|_{W^{p,1}(\Omega'')} + \|u\|_{L^1(\Omega)} + \|g\|_{L^q(\Omega)} + \|f\|_{L^p(\Omega)}\big)$$

for any domains $\Omega' \subset \Omega'' \subset \Omega$ with $\overline{\Omega'} \subset \Omega''$, $\overline{\Omega''} \subset \Omega$. The constant C_4 depends only on $\Omega', \Omega'', \Omega$ and the quantities indicated in the hypotheses of the theorem. Let us set

$$p_k = dp_{k-1}(d+p_{k-1}),\ p_0 = q,\ p_1 = p.$$

Since $p_k/p_{k-1} \leqslant d/(d+1)$, we can find a natural number K such that $p_K \leqslant 2$. Let $\{\Omega_k\}_{0\leqslant k\leqslant K}$ be a family of open sets such that $\Omega' \subset \Omega_k \subset \Omega$ and $\overline{\Omega_k} \subset \Omega_{k+1}$. Then, as shown above, whenever $0 \leqslant k \leqslant K$, we have the estimate

$$(1.7.14)\quad \|u\|_{W^{p_{k-1},1}(\Omega_{k-1})} \leqslant C(k)\big(\|u\|_{W^{p_k,1}(\Omega_k)} + \|u\|_{L^1(\Omega)} + \|g\|_{L^q(\Omega)} + \|f\|_{L^p(\Omega)}\big).$$

Since $p_k \leqslant 2$ for $k = K$, we estimate $\|u\|_{W^{p_K,1}}$ via $\|u\|_{W^{2,1}}$ and use the already established estimate for the case $q = 2$. Repeatedly applying inequality (1.7.14) for each k, in finitely many steps we obtain the required estimate for $\|u\|_{W^{p_0,1}(\Omega')}$, which completes the proof in the case $2 < q < d$.

Let us consider the case $q > d$. As in the previous case, the norm of the right-hand side in equality (1.7.11) is estimated by (1.7.13). Applying condition (1.7.6) and Hölder's inequality with the exponents $(d+q)/q$ and $(d+q)/d$, we obtain the following inequalities:

$$\begin{aligned}\|a^{ij}u\partial_{x_j}\zeta\|_q &+ C(d,p)\big[\|b^i u\partial_{x_i}\zeta\|_p + \|\beta^i u\partial_{x_i}\zeta\|_p\big]\\ &\leqslant C(d,p)(M + \|b\|_d + \|\beta\|_d)\|u\nabla\zeta\|_q\\ &\leqslant C(d,p)\big(M + |U(a,r)|^{(q-d)/d}\|b\|_q + |U(a,r)|^{(q-d)/d}\|\beta\|_q\big)\|u\nabla\zeta\|_q.\end{aligned}$$

Let us estimate $\|u\nabla\zeta\|_q$. Since $q > d$, by the Sobolev embedding theorem

$$\|u\zeta\|_{L^\infty(U(a,r))} \leqslant C(n,p)r^{(q-d)/q}\|u\zeta\|_{W^{q,1}(U(a,r))}.$$

Then

$$\begin{aligned}\|u\nabla\zeta\|_q &= \Big(\int_U |u\zeta|^{q-1}|u|\,|\nabla\zeta|^q\zeta^{1-q}\,dx\Big)^{1/q}\\ &\leqslant C(d,q)J\big((q-1)/q\big)\|u\|_{q,1}^{(q-1)/q}\|u\|_1^{1/q} \leqslant \varepsilon\|u\|_{q,1} + \varepsilon^{-1}\widetilde{C}(d,q)\|u\|_1.\end{aligned}$$

Applying again the Sobolev embedding theorem, we obtain

$$\begin{aligned}\|b^i u\zeta\|_q &+ C(d,p)\big[\lambda\|u\zeta\|_p + \|cu\zeta\|_p\big]\\ &\leqslant C(d,q,p)r^{(q-d)/q}(1 + \|b\|_q + \|c\|_p)\|u\zeta\|_{q,1}.\end{aligned}$$

It remains to observe that

$$\|\beta^i\partial_{x_i}(u\zeta)\|_p \leqslant |U(a,r)|^{(q-d)/q}\|\beta\|_q\|u\|_{q,1},\quad \|g^i\partial_{x_i}\zeta\|_p \leqslant \|g\|_q\|\nabla\zeta\|_d.$$

Similarly to the previous cases we pick $\varepsilon > 0$ and $r > 0$ sufficiently small and obtain that

$$\|u\zeta\|_{q,1} \leqslant C_5\big(\|\nabla u\nabla\zeta\|_p + \|u\|_1 + \|g\|_q + \|f\|_p\big).$$

By using a partition of unity, we arrive at the estimate

$$\|u\|_{W^{q,1}(\Omega')} \leqslant C_6\big(\|u\|_{W^{p,1}(\Omega'')} + \|u\|_{L^1(\Omega)} + \|g\|_{L^q(\Omega)} + \|f\|_{L^p(\Omega)}\big)$$

for any domains $\Omega' \subset \Omega'' \subset \Omega$ with $\overline{\Omega'} \subset \Omega''$, $\overline{\Omega''} \subset \Omega$. Note that $p < d$ and we can use the previous step for estimating the norm $\|u\|_{p,1}$. The proof in the case $q = d$ is similar. □

It is clear that the norm $\|u\|_{L^1(\Omega)}$ can be replaced (by Hölder's inequality) with the norm $\|u\|_{L^q(\Omega)}$.

The next assertion describes the dependence of the constant from the theorem on the diameter of the domain in the simplest case where $b = \beta = 0$ and $c = f = 0$.

1.7.5. Corollary. *Let $\Omega = U(z,R)$, where $R<1$, and let $u\in W^{q,1}_{\rm loc}(U(z,R))$ satisfy the equation $\partial_{x_i}(a^{ij}\partial_{x_i}u-g^i)=0$, where $q>d$ and* (1.7.6) *is fulfilled. Then, whenever $\lambda<1$, we have*

$$\|u\|_{W^{q,1}(U(z,\lambda R))}\leqslant R^{-1}C\bigl(\|u\|_{L^q(U(z,R))}+R\|g\|_{L^q(U(z,R))}\bigr),$$

where $C=C(d,q,\alpha,\omega_A,\lambda)$.

PROOF. Let us change the coordinates $x=z+Ry$. Then the obtained function $v(y)=u(z+Ry)$ satisfies in $U(0,1)$ the equation

$$\partial_{y_j}\bigl(a^{ij}(z+Ry)\partial_{y_i}v(y)+Rg^i(z+Ry)\bigr)=0.$$

By assumption, $a^{ij}\in VMO$ with the function ω_A, so we may assume that the function ω_A does not change, since $R<1$. Then for the function v we have

$$\|v\|_{W^{q,1}(U(0,\lambda))}\leqslant C(d,q,\alpha,\|A\|_{C^{0,\delta}},\lambda)\bigl(\|v\|_{L^q(U(0,1))}+R\|g\|_{L^q(U(0,1))}\bigr).$$

Making the inverse change of coordinates and taking into account that $R<1$ we obtain

$$\begin{aligned}\|v\|_{W^{q,1}(U(0,\lambda))}&=R^{-d/q}\|u\|_{L^q(U(z,\lambda R))}+R^{1-d/q}\|\nabla u\|_{L^q(U(z,\lambda R))}\\&\geqslant R^{1-d/q}\|u\|_{W^{q,1}(U(z,\lambda R))}.\end{aligned}$$

Similarly, we have

$$\begin{aligned}\|v\|_{L^q(U(0,1))}+R\|h\|_{L^q(U(0,1))}&=R^{-d/q}\|u\|_{L^q(U(z,R))}+R^{1-d/q}\|g\|_{L^q(U(z,R))}\\&=R^{-d/q}\bigl(\|u\|_{L^q(U(z,R))}+R\|g\|_{L^q(U(z,R))}\bigr).\end{aligned}$$

On the basis of these estimates we obtain

$$\|u\|_{W^{q,1}(U(z,\lambda R))}\leqslant C(d,q,\alpha,\|A\|_{C^{0,\delta}},\lambda)R^{-1}\bigl(\|u\|_{L^q(U(z,R))}+R\|g\|_{L^q(U(z,R))}\bigr),$$

as required. □

From Theorem 1.7.4 we deduce an estimate on the whole domain Ω for the function $u\in W^{q,1}_0(\Omega)$ in the case where the boundary $\partial\Omega$ is sufficiently regular.

Recall that the boundary $\partial\Omega$ is of class C^1 if it can be locally made flat by a diffeomorphism of class C^1.

1.7.6. Corollary. *Suppose that the coefficients of the operator $\mathcal{L}$ satisfy conditions* (1.7.6) *and $q\geqslant 2$. Let also $g\in L^q(\Omega)$, $f\in L^p(\Omega)$, where $p=dq/(d+q)$ if $q\neq d$ and $p>d/2$ if $q=d$. If Ω is a bounded domain with boundary of class C^1 and a function $u\in W^{q,1}_0(\Omega)$ is a solution of the equation $\mathcal{L}u=f+{\rm div} g$ in Ω, then*

$$\|u\|_{W^{q,1}_0(\Omega)}\leqslant C\bigl(\|u\|_{L^1(\Omega)}+\|g\|_{L^q(\Omega)}+\|f\|_{L^p(\Omega)}\bigr),$$

where the constant C depends only on the same quantities as in Theorem 1.7.4.

PROOF. It suffices to consider the following situation: $u\in W^{q,1}(K)$, where K is a cube one of the faces of which belongs to the hyperplane $\{x_d=0\}$ and the cube itself belongs to the open half-space $\{x_d>0\}$. Suppose that $u\psi\in W^{q,1}_0(\Omega)$ for every function $\psi\in C^\infty(\overline{K})$ vanishing in a neighborhood of every face excepting $x_d=0$. This actually means that $u=0$ when $x_d=0$. Set $y=(x_1,\ldots,x_{d-1})$. For any $x_d<0$ we set $u(y,x_d)=-u(y,-x_d)$ and

$$A(y,x_d)=A(y,-x_d),\ b(y,x_d)=-b(y,-x_d),\ \beta(y,x_d)=-\beta(y,-x_d),$$

$$c(y,x_d)=c(y,-x_d),\ f(y,x_d)=f(y,-x_d),\ g^i(y,x_d)=-g^i(y,-x_d).$$

Let K' be the cube obtained by reflecting K with respect to $\{x_d = 0\}$. The function u belongs to $W_0^{1,q}(K \cup K')$ and satisfies the equation $\mathcal{L}u = f + \mathrm{div}g$ in $K \cup K'$. Let K_1 be a cube inside of K such that all faces of K_1 are strictly inside of K, excepting one face belonging to the hyperplane $\{x_d = 0\}$. By Theorem 1.7.4 we have

$$\|u\|_{W^{q,1}(K_1)} \leqslant C\bigl(\|u\|_{L^1(K\cup K')} + \|g\|_{L^q(K\cup K')} + \|f\|_{L^p(K\cup K')}\bigr).$$

Thus, we obtain the required estimate up to the boundary $\{x_d = 0\}$. □

Existence of solutions to divergence form equations is discussed in Chapter 2, the increasing of the Sobolev regularity is considered in § 1.8(ii).

1.8. Complements, comments, and exercises

1.8(i). Fractional Sobolev classes

In the case where the diffusion matrix A is infinitely differentiable a somewhat more special result holds in terms of the scale of fractional Sobolev classes. Given $p \in (1, +\infty)$ and $s \in \mathbb{R}$, we set

$$H^{p,s}(\mathbb{R}^d) := (I - \Delta)^{-s/2}\bigl(L^p(\mathbb{R}^d)\bigr), \quad \|f\|_{p,s} = \|(I-\Delta)^{s/2} f\|_p,$$

where the operator $(I - \Delta)^{-s/2}$ is applied in the sense of the space of tempered distributions $\mathcal{S}'(\mathbb{R}^d)$; it can be defined via the Fourier transform by using the operator of multiplication by the function $(1 + |x|^2)^{-s/2}$. If $s \geqslant 0$, then the space $H^{p,s}(\mathbb{R}^d)$ is continuously embedded into $L^p(\mathbb{R}^d)$. For $s \in \mathbb{N}$ the class $H^{p,s}(\mathbb{R}^d)$ coincides with $W^{p,s}(\mathbb{R}^d)$ and the respective norms are equivalent. The class $H^{p,s}_{\mathrm{loc}}(\mathbb{R}^d)$, where $s \in \mathbb{R}$, $p > 1$, consists of all functions f such that $\varphi f \in H^{p,s}(\mathbb{R}^d)$ for all $\varphi \in C_0^\infty(\mathbb{R}^d)$.

In the proofs below we use the following well-known lemma. Let Ω be a domain in $\mathbb{R}^d$ and A a mapping on Ω with values in the space of positive symmetric operators on $\mathbb{R}^d$.

1.8.1. Lemma. *Suppose that $a^{ij} \in C^\infty(\Omega)$ and* $\det A > 0$.

(i) *Let $r \in (-\infty, \infty)$ and $p > 1$. If u is a distribution such that $L_A u \in H^{p,r}_{\mathrm{loc}}(\Omega)$, then $u \in H^{p,r+2}_{\mathrm{loc}}(\Omega)$; also if $u \in H^{p,r}_{\mathrm{loc}}(\Omega)$, then $\partial_{x_i} u \in H^{p,r-1}_{\mathrm{loc}}(\Omega)$, $1 \leqslant i \leqslant d$.*

(ii) *We have $H^{p,1}_{\mathrm{loc}}(\Omega) \subset L^{dp/(d-p)}_{\mathrm{loc}}(\Omega)$ and $L^p_{\mathrm{loc}}(\Omega) \subset H^{dp/(d-p),-1}_{\mathrm{loc}}(\Omega)$ whenever $1 < p < d$, and $H^{p,1}_{\mathrm{loc}}(\Omega) \subset C^{1-d/p}_{\mathrm{loc}}(\Omega)$ if $p > d$, so that in the latter case all functions in $H^{p,1}_{\mathrm{loc}}(\Omega)$ are locally bounded. In addition, whenever $q > p > 1$, we have the inclusion $L^p_{\mathrm{loc}}(\Omega) \subset H^{q,d/q-d/p}_{\mathrm{loc}}(\Omega)$.*

(iii) *If μ is a locally bounded Radon measure on Ω, then $\mu \in H^{p,-m}_{\mathrm{loc}}(\Omega)$, whenever $p > 1$ and $m > d(1 - 1/p)$.*

PROOF. Assertion (i) is well-known: its first statement is a well-known elliptic regularity result (see Taylor [**894**, Chapter III]) and the second statement follows from the boundedness of Riesz's transforms. Assertion (ii) is just the Sobolev embedding theorem (mentioned in § 1.1 for $H^{p,1}$). Assertion (iii) follows from this embedding theorem, because for any regular sub-domain U of Ω one has the embedding $H^{q,m}(U) \subset C(\overline{U})$ if $qm > d$, whence by duality we obtain that the space $H^{q/(q-1),-m}(U) = [H_0^{q,m}(U)]^*$ contains all finite measures on U. □

We formulate the following result for $d > 1$ just because the case $d = 1$ is elementary and has already been discussed. In addition, we include in the formulation some assertions which follow also from the already mentioned results (but the proof we give is direct and does not use the results above).

1.8.2. Theorem. *Under the same assumptions about A as in the lemma, let $d \geqslant 2$ and let μ and ν be Radon measures on Ω (possibly signed). Let a mapping $b = (b^i)\colon \Omega \to \mathbb{R}^d$ and a function $c\colon \Omega \to \mathbb{R}$ be such that $|b|$, $c \in L^1_{\rm loc}(\Omega, \mu)$. Suppose that $L^*_{A,b,c}\mu = \nu$. Then the following assertions are true.*

(i) *One has $\mu \in H^{p,1-d(p-1)/p-\varepsilon}_{\rm loc}(\Omega)$ for any $p \geqslant 1$ and any $\varepsilon > 0$. Here $1 - d(p-1)/p > 0$ if $p \in \bigl[1, d/(d-1)\bigr)$ and, in particular, μ admits a density $F \in L^p_{\rm loc}(\Omega)$ for any $p \in \bigl[1, d/(d-1)\bigr)$.*

(ii) *If $|b| \in L^\gamma_{\rm loc}(\Omega,\mu)$, $c \in L^{\gamma/2}_{\rm loc}(\Omega,\mu)$ and $\nu \in L^{d/(d-\gamma+2)}_{\rm loc}(\Omega)$ where $d \geqslant \gamma > 1$, then $F := d\mu/dx \in H^{p,1}_{\rm loc}(\Omega)$ for any $p \in [1, d/(d-\gamma+1))$. In particular, $F \in L^p_{\rm loc}(\Omega)$ for any $p \in \bigl[1, d/(d-\gamma)\bigr)$, where we set $d/(d-\gamma) := \infty$ if $\gamma = d$.*

(iii) *If $\gamma > d$ and either*

(a) *$|b| \in L^\gamma_{\rm loc}(\Omega)$ and $c, \nu \in L^{\gamma d/(d+\gamma)}_{\rm loc}(\Omega)$,*

or

(b) *$|b| \in L^\gamma_{\rm loc}(\Omega,\mu)$, $c \in L^{\gamma d/(d+\gamma)}_{\rm loc}(\Omega,\mu)$, and $\nu \in L^{\gamma d/(d+\gamma)}_{\rm loc}(\Omega)$,*

then μ admits a density $F \in H^{\gamma,1}_{\rm loc}(\Omega)$. In particular, $F \in C^{1-d/\gamma}_{\rm loc}(\Omega)$.

Proof. (i) We have in the sense of distributions

$$L_A\mu = \partial_{x_i}\bigl((b^i - \partial_{x_j}a^{ij})\mu\bigr) - \partial_{x_i}a^{ij}\partial_{x_j}\mu - c\mu + \nu \tag{1.8.1}$$

on Ω. Here by Lemma 1.8.1(iii) the right-hand side belongs to $H^{p,-m-1}_{\rm loc}(\Omega)$ if $m > d(1-1/p)$. By Lemma 1.8.1(i) we conclude $\mu \in H^{p,-m+1}_{\rm loc}(\Omega)$, which leads to the result after substituting $m = d(1-1/p) + \varepsilon$.

Before we prove (ii) and (iii), we need some preparations. Fix $p_1 > 1$ and assume that $F = d\mu/dx \in L^{p_1}_{\rm loc}(\Omega)$. Such a number p_1 exists by (i). Set

$$r := r(p_1) := \frac{\gamma p_1}{\gamma - 1 + p_1} \tag{1.8.2}$$

and observe that owing to the inequalities $1 < \gamma$ and $p_1 > 1$, we have $1 < r < \gamma$. Next, starting with the formula

$$|bF|^r = (|b||F|^{1/\gamma})^r |F|^{r-r/\gamma}$$

and using Hölder's inequality (with $s = \gamma/r > 1$ and $t := s/(s-1) = \gamma/(\gamma - r)$ and the relations $|b|\,|F|^{1/\gamma} \in L^\gamma_{\rm loc}(\Omega)$ and $F \in L^{p_1}_{\rm loc}(\Omega)$, we obtain that $b^iF \in L^r_{\rm loc}(\Omega)$. By Lemma 1.8.1(i) one has

$$b^iF \in H^{r,0}_{\rm loc}(\Omega), \quad (b^iF)_{x_i} \in H^{r,-1}_{\rm loc}(\Omega). \tag{1.8.3}$$

(ii) Set

$$q := q(p_1) := \frac{\gamma p_1}{\gamma - 2 + 2p_1} \vee 1, \tag{1.8.4}$$

and note that $q > 1 \Leftrightarrow \gamma > 2 \Leftrightarrow q < \gamma/2$, in particular, $q < \gamma$ in any case. Hence repeating the above argument with the triple c, $\gamma/2$, q in place of $|b|$, γ, r, we obtain that

$$cF \in L^q_{\mathrm{loc}}(\Omega). \tag{1.8.5}$$

Fix $p_1 > 1$ such that $F := \frac{d\mu}{dx} \in L^{p_1}_{\mathrm{loc}}(\Omega)$ and let r, q be as in (1.8.2), (1.8.4), respectively. Since $\gamma \leqslant d$, we have $q < d$, which by (1.8.5) and assertions (ii) and (iii) of Lemma 1.8.1 implies that $cF \in H^{dq/(d-q),-1}_{\mathrm{loc}}(\Omega)$ if $q > 1$ and that $cF \in H^{s,-1}_{\mathrm{loc}}(\Omega)$ for any $s \in (1, d/(d-1))$ if $q = 1$.

It turns out that if $p_1 < d/(d-\gamma)$, then

$$cF \in H^{r,-1}_{\mathrm{loc}}(\Omega). \tag{1.8.6}$$

Indeed, if $q > 1$, then (1.8.6) follows from the fact that if $p_1 \in \big(1, d/(d-\gamma)\big)$, then the inequality $r \leqslant dq/(d-q)$ holds. If $q = 1$, then $\gamma \leqslant 2$ and (1.8.6) follows from the fact that $r < d/(d-\gamma+1) \leqslant d/(d-1)$ for $p_1 < d/(d-\gamma)$.

Finally, by Lemma 1.8.1 (ii) we obtain that $\nu \in H^{d/(d-\gamma+1),-1}_{\mathrm{loc}}(\Omega)$ if $\gamma > 2$ and $\nu \in H^{s,-1}_{\mathrm{loc}}(\Omega)$ for any $s \in \big(1, d/(d-1)\big)$ if $\gamma \leqslant 2$. In the same way as above, $\nu \in H^{r,-1}_{\mathrm{loc}}(\Omega)$ whenever $1 < p_1 < d/(d-\gamma)$. This along with (1.8.3) and (1.8.6) shows that the right-hand side of (1.8.1) is now in $H^{r,-1}_{\mathrm{loc}}(\Omega)$. By Lemma 1.8.1(i) we have

$$\mu \in H^{r,1}_{\mathrm{loc}}(\Omega) \tag{1.8.7}$$

and by Lemma 1.8.1(ii) we have $F \in L^{p_2}_{\mathrm{loc}}(\Omega)$, where

$$p_2 := \frac{dr}{d-r} = \frac{d\gamma p_1}{d\gamma - d + (d-\gamma)p_1} =: f(p_1).$$

Thus, we obtain

$$p_1 \in \left(1, \frac{d}{d-\gamma}\right) \text{ and } F \in L^{p_1}_{\mathrm{loc}}(\Omega) \Longrightarrow F \in L^{f(p_1)}_{\mathrm{loc}}(\Omega).$$

One can easily check that $p_2 = f(p_1) > p_1$ if $p_1 < d/(d-\gamma)$, and that the only positive solution of the equation $q = f(q)$ is $q = d/(d-\gamma)$. Therefore, by taking p_1 in $\big(1, d/(d-1)\big)$, which is possible by (i), and by defining $p_{k+1} = f(p_k)$ we obtain an increasing sequence of numbers $p_k \uparrow d/(d-\gamma)$, which implies that $F \in L^p_{\mathrm{loc}}(\Omega)$ for any $p < d/(d-\gamma)$.

But as $p_k \nearrow d/(d-\gamma)$, the sequence of numbers $r(p_k)$ defined according to equality (1.8.2) increases to the limit

$$\frac{\gamma d/(d-\gamma)}{\gamma - 1 + d/(d-\gamma)} = \frac{d}{d-\gamma+1}\,.$$

By (1.8.7) this proves (ii).

(iii) First we consider case (b). By the last assertion in (ii) we have $F \in L^{p_1}_{\mathrm{loc}}(\Omega)$ for any finite $p_1 > 1$. Let $r := r(p_1)$ be defined as in (1.8.2). Then $1 < r < \gamma$ and inclusions (1.8.3) hold. Set

$$q := q(p_1) := \frac{\frac{d\gamma}{d+\gamma}\, p_1}{\frac{d\gamma}{d+\gamma} - 1 + p_1}\,. \tag{1.8.8}$$

If $2 \leqslant d < \gamma$, then $d\gamma/(d+\gamma) > 1$. Therefore, since $p_1 > 1$, it follows that $1 < q < d\gamma/(d+\gamma)$. Hence repeating the arguments that led to (1.8.3) with the triple

c, $\frac{d\gamma}{d+\gamma}$, q in place of $|b|$, γ, r, we obtain $cF \in L^q_{\rm loc}(\Omega)$, thus, $cF \in H^{dq/(d-q),-1}_{\rm loc}(\Omega)$ by assertion (ii) of the lemma. Observe that, as $p_1 \to \infty$, we have

$$r \uparrow \gamma, \quad q \uparrow \frac{d\gamma}{d+\gamma}, \quad \frac{dq}{d-q} \uparrow \gamma.$$

Therefore, combining this with our assumption that ν is contained in the class $L^{d\gamma/(d+\gamma)}_{\rm loc}(\Omega)$, which by assertion (ii) of the lemma is contained in $H^{\gamma,-1}_{\rm loc}(\Omega)$, and by taking p_1 large enough, we see that the right-hand side of (1.8.1) is in $H^{\gamma-\varepsilon,-1}_{\rm loc}(\Omega)$ for any $\varepsilon \in (0, \gamma-1)$. By Lemma 1.8.1(ii) we conclude that $F \in H^{\gamma-\varepsilon,1}_{\rm loc}(\Omega)$. Since $\gamma > d$, the function F is locally bounded. Now we see that above we can take $p_1 = \infty$ and therefore the right-hand side of (1.8.1) is in $H^{\gamma,-1}_{\rm loc}(\Omega)$, which by assertion (i) of the lemma gives us the desired result.

In the remaining case (a) we take $p_1 > \gamma/(\gamma-1)$ and assume that $F \in L^{p_1}_{\rm loc}(\Omega)$. Then instead of (1.8.2) and (1.8.8) we define

$$r := r(p_1) := \frac{\gamma p_1}{\gamma + p_1}, \quad q := q(p_1) := \frac{\frac{d\gamma}{d+\gamma} p_1}{\frac{d\gamma}{d+\gamma} + p_1} \vee 1 \tag{1.8.9}$$

and observe that, since $p_1 > \gamma/(\gamma-1)$, we have $r > 1$, which (because of the relation $p_1^{-1} + \gamma^{-1} = r^{-1}$) allows us to apply Hölder's inequality starting with $|bF|^r = |b|^r|F|^r$ to conclude that (1.8.3) holds. Since $c \in L^1_{\rm loc}(\Omega, \mu)$,

$$\frac{d\gamma}{d+\gamma} > 1 \quad \text{and} \quad \left(\frac{d\gamma}{d+\gamma}\right)^{-1} + p_1^{-1} = q^{-1},$$

we also have that $cF \in L^q_{\rm loc}(\Omega)$. Obviously, $q < d$. As in part (ii) this yields that $cF \in H^{dq/(d-q),-1}_{\rm loc}(\Omega)$ if $q > 1$ and $cF \in H^{s,-1}_{\rm loc}(\Omega)$ for any $s \in (1, d/(d-1))$ if $q = 1$. We assert that (1.8.6) holds (with $r = r(p_1)$ as in (1.8.9)) for all $p_1 > \gamma/(\gamma-1)$, $p_1 \neq d\gamma/(d\gamma - d - \gamma)$.

Indeed, if $q > 1$, then $dq/(d-q) = r$. If $q = 1$, then $p_1 \leqslant d\gamma/(d\gamma - d - \gamma)$. But since $p_1 \neq d\gamma/(d\gamma - d - \gamma)$, we have $p_1 < d\gamma/(d\gamma - d - \gamma)$, which is equivalent to the inequality $r < d/(d-1)$.

Thus, since $\nu \in L^{d\gamma/(d+\gamma)}_{\rm loc}(\Omega) \subset H^{\gamma,-1}_{\rm loc}(\Omega) \subset H^{r,-1}_{\rm loc}(\Omega)$, because $r < \gamma$, assertion (i) in the lemma yields the following:

$$\Bigl(p_1 > \frac{\gamma}{\gamma-1},\ p_1 \neq \frac{d\gamma}{d\gamma - d - \gamma},\ F \in L^{p_1}_{\rm loc}(\Omega)\Bigr) \Longrightarrow F \in H^{r,1}_{\rm loc}(\Omega). \tag{1.8.10}$$

If $r < d$, then the latter in turn implies by assertion (ii) in Lemma 1.8.1 that $F \in L^{p_2}_{\rm loc}(\Omega)$. Let us summarize what has been shown:

$$\begin{aligned} &\Bigl(p_1 > \tfrac{\gamma}{\gamma-1},\ p_1 \neq \tfrac{d\gamma}{d\gamma-d-\gamma},\ r := \tfrac{\gamma p_1}{\gamma+p_1} < d,\ F \in L^{p_1}_{\rm loc}(\Omega)\Bigr) \\ &\qquad \Longrightarrow F \in L^{p_2}_{\rm loc}(\Omega), \end{aligned} \tag{1.8.11}$$

where

$$p_2 := \frac{dr}{d-r} = \frac{d\gamma p_1}{d\gamma - (\gamma-d)p_1} > \frac{d\gamma}{d\gamma - (\gamma - d)} p_1.$$

Also, note that $\gamma/(\gamma-1) < d/(d-1) < \frac{d\gamma}{\gamma d - d - \gamma}$, so that by (i) we can find a number p_1 to start with. Then starting with p_1 close enough to $d/(d-1)$, by iterating (1.8.11) we always increase p by some factor greater than $d\gamma/(d\gamma - (\gamma-d)) > 1$. While doing this, we can obviously choose the first p so that the iterated numbers

p will never be equal to $d\gamma/(d\gamma - d - \gamma)$ and the corresponding numbers r will not coincide with d. After several steps we shall come to the situation where $r > d$, and then we conclude from (1.8.10) that F is locally bounded (one cannot keep iterating (1.8.11) infinitely because of the restriction $r < d$). As in case (b), we can now easily complete the proof. □

Example 1.6.10 shows that assertion (iii) of this theorem may fail if $\gamma > d$ is replaced by $\gamma = d - \varepsilon$. Then F does not even need to be in $H^{2,1}_{\mathrm{loc}}(\Omega)$.

1.8(ii). Increasing the Sobolev regularity of solutions

The following result of Morrey is known (see Morrey [**723**, Theorem 5.5.3, p. 154], where A is continuous) about raising the integrability of a solution of the full equation with the operator $\mathcal{L}_{A,b,\beta,c}$ on a domain $\Omega \subset \mathbb{R}^d$, $d > 1$, where $a^{ij} \in VMO$ and $A(x) \geqslant \delta \cdot \mathrm{I}$ with some $\delta > 0$.

1.8.3. Theorem. *Let $d/(d-1) \leqslant q < r$ and let $u \in W^{q,1}_{\mathrm{loc}}(\Omega)$ satisfy the equation*

$$\mathcal{L}_{A,b,\beta,c} u = f + \operatorname{div} g, \quad g = (g^1, \dots, g^d),$$

where $f \in L^{dr/(d+r)}_{\mathrm{loc}}(\Omega)$, $g^i \in L^r_{\mathrm{loc}}(\Omega)$, $b^i, \beta^i \in L^s_{\mathrm{loc}}(\Omega)$, $c \in L^t_{\mathrm{loc}}(\Omega)$, $s = d$ and $t = d/2$ if $r < d$, $s > d$ and $t > d/2$ if $r = d$, $s = r$ and $t = dr/(d+r)$ if $r > d$. Then $u \in W^{r,1}_{\mathrm{loc}}(\Omega)$.

PROOF. 1. Under the assumptions of the theorem for every ball $U(x_0, \varepsilon)$ in Ω of a sufficiently small radius ε there is a function $v \in W^{r,1}_0\big(U(x_0,\varepsilon)\big)$ satisfying the equation

$$\partial_{x_i}(a^{ij}\partial_{x_j} v - b^i v) + \beta^i \partial_{x_i} v = f + \partial_{x_i} g^i. \tag{1.8.12}$$

It suffices to establish the existence in the case of smooth coefficients, since approximating the coefficients by smooth functions and using the estimate

$$\|u\|_{W^{r,1}_0(U(x_0,\varepsilon))} \leqslant N(\|g\|_{L^r(U(x_0,\varepsilon))} + \|f\|_{L^{dr/(d+r)}(U(x_0,\varepsilon))}),$$

one can construct a sequence of smooth functions converging to the solution. However, it is important that the radius ε of the ball $U(x_0, \varepsilon)$ could depend only on ω_A and the quantities δ, $\|\beta\|_{L^s}$, and $\|b\|_{L^s}$ and be independent of the smoothness of the coefficients.

For constructing a solution in the case of smooth coefficients it suffices to show that for a sufficiently small ball $U(x_0, \varepsilon)$ the solution of the homogeneous equation ($f = g^i = 0$) in the class $W^{2,1}_0\big(U(x_0,\varepsilon)\big)$ must be zero. Then, by the Fredholm alternative (see Proposition 2.1.4), there exists a solution of the nonhomogeneous equation on this ball. Thus, let $v \in W^{2,1}_0\big(U(x_0,\varepsilon)\big)$ and

$$\partial_{x_i}(a^{ij}\partial_{x_j} v - b^i v) + \beta^i \partial_{x_i} v = 0.$$

Multiplying by v and integrating by parts we obtain

$$\varepsilon \int_{U(x_0,\varepsilon)} |\nabla v|^2\, dx \leqslant \int_{U(x_0,\varepsilon)} |b + \beta|\, |v|\, |\nabla v|\, dx.$$

We recall that by the Sobolev inequality $\|v\|_{L^{2d/(d-2)}(U(x_0,\varepsilon))} \leqslant c(d)\|\nabla v\|_{L^2(U(x_0,\varepsilon))}$, where $c(d)$ does not depend on ε. Applying Hölder's inequality, we obtain

$$\int_{U(x_0,\varepsilon)} |b + \beta|\, |v|\, |\nabla v|\, dx \leqslant c(d)\|b + \beta\|_{L^d(U(x_0,\varepsilon))}\|\nabla v\|^2_{L^2(U(x_0,\varepsilon))}.$$

Choosing ε so small that $c(d)\|b+\beta\|_{L^d(U(x_0,\varepsilon))} \leqslant \delta/2$, we obtain the inequality

$$\int_{U(x_0,\varepsilon)} |\nabla v|^2\, dx \leqslant 0.$$

Therefore, $v = 0$ on $U(x_0, \varepsilon)$.

Thus, we have shown uniqueness of a solution in $W_0^{2,1}\big(U(x_0,\varepsilon)\big)$, hence also in $W_0^{r,1}\big(U(x_0,\varepsilon)\big)$ with $r \geqslant 2$. We observe once again that for constructing a solution in the smooth case this uniqueness is enough. If $r \leqslant 2$, then some additional reasoning is required.

We show that the solution v is unique. It suffices to show that the solution v of the homogeneous equation must be zero. To this end we solve the adjoint equation with the right-hand side $f = \operatorname{sign} v$ and $g = 0$, i.e.,

$$\partial_{x_i}(a^{ij}\partial_{x_i} w - \beta^i w) + b^i w = \operatorname{sign} v.$$

Since the right-hand side is bounded, there is a solution $w \in W_0^{m,1}\big(U(x_0,\varepsilon)\big)$ with m as close to d as we wish (so $w \in L^p\big(U(x_0,\varepsilon)\big)$ for any p). Multiplying this equation by v and integrating by parts we conclude that $v = 0$. Thus, on a sufficiently small ball $U(x_0,\varepsilon)$ there is a unique solution $v \in W_0^{r,1}\big(U(x_0,\varepsilon)\big)$ of equation (1.8.12).

2. Let $r < d$ and $\zeta \in C_0^\infty(\Omega)$, where the support of ζ belongs to a ball U of a sufficiently small radius indicated above. Then

(1.8.13)
$$\begin{aligned}\partial_{x_i}\big(a^{ij}\partial_{x_j}(u\zeta) - b^i(u\zeta)\big) + \beta^i\partial_{x_i}(u\zeta) &= \partial_{x_i}(a^{ij}u\partial_{x_j}\zeta - g^i\zeta) + a^{ij}\partial_{x_i}\zeta\partial_{x_j}u \\ &\quad + b^i u\partial_{x_i}\zeta - \beta^i u\partial_{x_i}\zeta + cu\zeta + g^i\partial_{x_i}\zeta + f\zeta.\end{aligned}$$

Set

$$\widetilde{f} = a^{ij}\partial_{x_i}\zeta\partial_{x_j}u + b^i u\partial_{x_i}\zeta - \beta^i u\partial_{x_i}\zeta + cu\zeta + g^i\partial_{x_i}\zeta + f\zeta,\ \widetilde{g}^i = a^{ij}u\partial_{x_j}\zeta - g^i\zeta.$$

Then

$$\partial_{x_i}\big(a^{ij}\partial_{x_j}(u\zeta) - b^i(u\zeta)\big) + \beta^i\partial_{x_i}(u\zeta) = \widetilde{f} + \partial_{x_i}\widetilde{g}^i.$$

We observe that $\widetilde{f} \in L^q(U)$ and $\widetilde{g}^i \in L^{dq/(d-q)}(U)$. Then by the previous step we have $u\zeta \in W_0^{dq/(d-q),1}(U)$. Note that $dq/(d-q) \geqslant q(d-1)/(d-2)$, since $q \geqslant d/(d-1)$. Repeating this reasoning on a smaller ball we again increase the smoothness of our solution until we obtain the inclusion to the class $W^{r,1}$. The cases $r = d$ and $r > d$ are similar. □

1.8(iii). Renormalized solutions

Let $\Omega \subset \mathbb{R}^d$. Suppose we are given a nonnegative nonzero function $W \in L^1_{\text{loc}}(\Omega)$ satisfying the equation

(1.8.14)
$$\partial_{x_i}\partial_{x_j}(a^{ij}W) = 0,$$

where $A = A^*$ is infinitely differentiable, $\lambda^{-1}\cdot \mathrm{I} \leqslant A(x) \leqslant \lambda\cdot \mathrm{I}$, $x \in \Omega$. A renormalized with respect to W solution of equation (1.8.14) is a function w such that we have $wW \in L^1_{\text{loc}}(\Omega)$ and wW in place of W satisfies (1.8.14). It turns out that the renormalized solutions possess many nice properties. For example, the maximum principle holds for them. Indeed, the function w satisfies the equation

$$a^{ij}\partial_{x_i}\partial_{x_j}w + 2W^{-1}\partial_{x_j}(a^{ij}W)\partial_{x_i}w = 0,$$

for the solutions of which the classical strong maximum principle holds. In the next theorem we have collected some typical results from the papers Bauman [**94**], Escauriaza [**341**], [**342**].

1.8.4. Theorem. *Let w be a renormalized with respect to W solution of equation* (1.8.14). *Then the following assertions are true.*

(i) *Harnack's inequality. Let $w \geqslant 0$. For every ball $U(x_0, r)$ with $U(x_0, 5r) \subset \Omega$ one has the inequality*

$$\sup_{U(x_0,r)} w(x) \leqslant C(\lambda, d) \inf_{U(x_0,r)} w(x).$$

(ii) *Hölder's continuity. There is a number $\alpha = \alpha(\lambda, d) \in (0, 1]$ such that for every ball $U(x_0, r) \subset \Omega$ one has the estimate*

$$|w(x) - w(y)| \leqslant C(\lambda, d)\Big(\frac{|x-y|}{r}\Big)^{\alpha} \sup_{U(x_0,r)} |w(z)|.$$

(iii) *Boundedness of solutions. For every ball $U(x_0, r) \subset \Omega$ one has the estimate*

$$\sup_{U(x_0,r/2)} |w(x)| \leqslant C(\lambda, d)\Big(\int_{U(x_0,r)} |w| W\, dx\Big)\Big(\int_{U(x_0,r)} W\, dx\Big)^{-1}.$$

The constants $C(\lambda, d)$ and $\alpha(\lambda, d)$ depend only on λ and d and are independent of the smoothness of A.

Assertion (i) is obtained in [**94**, Theorem 4.4]. Assertion (ii) is a corollary of (i) and the fact that any constant is a renormalized solution. Finally, (iii) is proved in [**341**, Theorem 2.3]. In [**342**], these results are used for obtaining some estimates on the Green's function.

1.8(iv). Generalizations of the maximum principle of A.D. Aleksandrov and k-Hessians

For the proof of the existence of densities in § 1.5 we have used the maximum principle of A.D. Aleksandrov (see Theorem 1.5.1). Here we consider some of its generalizations. Let Ω be a convex bounded domain in $\mathbb{R}^d$ with boundary of class C^1, e.g., a ball. It is known (see Gilbarg, Trudinger [**409**, Theorem 8.15]) that if $u \in C^2(\Omega) \bigcap C_0(\overline{\Omega})$ satisfies the equation $\Delta u = f$, then

$$\sup_{\Omega} |u| \leqslant C(d, \Omega, p)\|f\|_{L^p(\Omega)}$$

for any $p > d/2$. According to Aleksandrov's maximum principle, if a convex function $u \in C^2(\Omega) \bigcap C_0(\overline{\Omega})$ satisfies the equation $\det D^2 u = f$, then

$$\sup_{\Omega} |u| \leqslant C(d, \Omega)\|f\|^{1/d}_{L^1(\Omega)}.$$

It turns out that some intermediate estimates hold.

Let $S_k(u)$ be the sum of the principal k-minors in the matrix $D^2 u$. In this case $S_1(u) = \Delta u = \mathrm{tr} D^2 u$ and $S_d(u) = \det D^2 u$. The expression $S_k(u)$ is called the k-Hessian. A function $u \in C^2(\Omega) \bigcap C_0(\overline{\Omega})$ is called k-admissible if $S_j(u) \geqslant 0$ for all indices $j \leqslant k$. The set of all k-admissible functions will be denoted by $\Phi_0^k(\Omega)$. Let

$$\|u\|_{\Phi_0^k(\Omega)} = \Big(\int_{\Omega} -u S_k(u)\, dx\Big)^{1/(k+1)}.$$

The next result was proved in Wang [**934**, Theorem 5.1].

1.8.5. Theorem. *Let* $u \in \Phi_0^k(\Omega)$.

(i) *If* $1 \leqslant k < d/2$, *then* $\|u\|_{L^{p+1}(\Omega)} \leqslant C\|u\|_{\Phi_0^k(\Omega)}$ *for each* $p+1 \in [1, k^*]$, *where we set* $k^* = d(k+1)/(d-2k)$.

(ii) *If* $k = d/2$, *then* $\|u\|_{L^p(\Omega)} \leqslant C\|u\|_{\Phi_0^k(\Omega)}$ *for all* $p \geqslant 1$.

(iii) *If* $d/2 < k \leqslant d$, *then* $\|u\|_{L^\infty(\Omega)} \leqslant C\|u\|_{\Phi_0^k(\Omega)}$.

The number C *depends only on* d, k, *and* $\operatorname{diam}\Omega$.

Note that in the case $k = d/2$ in Tian, Wang [**895**] the following was proved: there are numbers $\alpha(d) > 0$ and $C(d, \Omega) > 0$ such that

$$\int_\Omega \exp\Bigl(\alpha\Bigl(\frac{|u|}{\|u\|_{\Phi_0^k(\Omega)}}\Bigr)^\beta\Bigr)\,dx \leqslant C(d,\Omega), \quad \text{where } 1 \leqslant \beta \leqslant (d+2)/d.$$

By using Moser's iterations, one can derive from this result analogs of Aleksandrov's maximum principle for k-Hessians. The next corollary is proved in [**934**, Theorem 5.5].

1.8.6. Corollary. *Let* $u \in \Phi_0^k(\Omega)$ *and* $S_k(u) = f$. *Then*

$$\sup_\Omega |u| \leqslant C\|f\|_{L^p(\Omega)}^{1/k},$$

where $p > d/(2k)$ *if* $k \leqslant d/2$ *and* $p = 1$ *if* $k > d/2$.

Let $\sigma_k(\lambda) = \sum_{1\leqslant i_1<i_2<\ldots<i_k\leqslant d} \lambda_{i_1}\cdots\lambda_{i_k}$ be the kth elementary symmetric polynomial. Set

$$\Gamma_k = \{\lambda \in \mathbb{R}^d\colon \sigma_j(\lambda) \geqslant 0, j = 1, 2, \ldots, k\}, \quad \Gamma_k^* = \{\lambda \in \mathbb{R}^d\colon \langle\lambda, \mu\rangle \geqslant 0\,\forall \mu \in \Gamma_k\}.$$

Let A be a symmetric matrix and let $\lambda(A)$ be the vector of eigenvalues of A. Suppose that $\lambda(A) \in \Gamma_k^*$. Set

$$\varrho_k^*(A) = \inf\bigl\{\langle\lambda(A), \mu\rangle\colon \mu \in \Gamma_k, \sigma_k(\mu) \geqslant 1\bigr\}$$

For example, $\lambda(A) \in \Gamma_1^*$ precisely when $A = g\cdot \mathrm{I}$, where g is a nonnegative number. In this case $\varrho_k^*(A) = g$.

The next related result is proved in Kuo, Trudinger [**565**].

1.8.7. Corollary. *Let* $u \in C^2(\Omega)\bigcap C_0(\overline{\Omega})$ *and* $\operatorname{tr}(AD^2u) \geqslant f$. *Suppose that* $\lambda(A) \in \Gamma_k^*$ *and* $\varrho_k^*(A) > 0$. *Then*

$$\sup_\Omega u \leqslant C\|f/\varrho_k^*(A)\|_{L^q(\Omega)},$$

where $q = k$ *if* $k > d/2$ *and* $q > d/2$ *if* $k \leqslant d/2$. *The number* C *depends only on* d, k, q, *and* Ω.

Comments

Bibliographic materials related to the outstanding Russian mathematician Andrey Nikolaevich Kolmogorov and the great German physicist, a Nobel prize winner Max Planck, whose names are in the title of this book, can be found in Kolmogorov [**530**] and Klyauc, Frankfurt [**517**], where some additional references are given. Adriaan Daniël Fokker (17.VIII.1887 – 24.IX.1972) is a Dutch physicist, a member of the Royal Dutch Academy. He was born in the island of Java, a Dutch colony at the time, in 1904 – 1905 was a student at the Polytechnic school in Delft, in 1906 – 1913 studied physics at the University of Leiden with H. Lorentz, on

October 24, 1913, defended his thesis "Over Brown'sche bewegingen in het stralingsveld, en waarschijnlijkheids-beschouwingen in de stralingstheorie", the main results of which were published in his paper [**377**] (where the stationary equation was considered). Later he continued his studies with Albert Einstein. After his military service in World War I he worked in 1917 – 1918 as an assistant of H. Lorentz and P. Ehrenfest, in 1923 – 1927 he was Professor at the High Technical School in Delft, in 1928 – 1955 he was curator of the physical cabinet of the Teylers Museum in Haarlem and was Professor at the University of Leiden. Fokker's main scientific works are devoted to radiation, X-rays, physics of electron, general relativity, gravitation, the theory of fluctuations, and the theory of gyroscope. The equation bearing his name was needed for establishing the distribution law of the average energy of a rotating electric dipole, for this purpose Fokker developed a method first used by Einstein [**329**] for describing the Brownian motion. In addition, Fokker was an expert in music theory, he was enthusiast of the 31 equal temperament, his 31-tone equal-tempered organ, which was installed in Teyler's Museum in Haarlem in 1951, is called the Fokker organ. It is worth mentioning that Kolmogorov and Planck also were amateurs and connoisseurs of music, moreover, Planck was playing at a professional level.

There is an extensive literature on the theory of Sobolev spaces, see Adams [**2**], Adams, Fournier [**3**], Besov, Il'in, Nikolskiĭ [**112**], Bogachev [**126**], Brezis [**208**], Burenkov [**212**], Evans, Gariepy [**344**], Gol'dshteĭn, Reshetnyak [**412**], Haroske, Triebel [**440**], Krylov [**556**], Kufner, Sändig [**558**], Leoni [**602**], Maz'ja [**682**], [**683**], Runst, Sickel [**828**], Stein [**877**], Triebel [**899**], [**900**], and Ziemer [**965**], where numerous additional references can be found. In [**126**], [**558**] and Zhikov [**961**] also weighted Sobolev classes are considered.

Many books are devoted to the general theory of elliptic second order linear partial differential equations, see Agmon [**5**], Agmon, Douglis, Nirenberg [**6**], Bers, John, Schechter [**109**], Borsuk [**198**], Borsuk, Kondratiev [**199**], Chen, Wu [**237**], Demengel, Demengel [**293**], Egorov, Kondratiev [**325**], Gilbarg, Trudinger [**409**], Han, Lin [**438**], Hörmander [**461**], Helffer [**447**], Kenig [**501**], Kondrat'ev, Landis [**533**], Koshelev [**537**], Kozlov, Maz'ya, Rossmann [**541**], Kresin, Maz'ya [**542**], Krylov [**552**], [**556**], Ladyzhenskaya, Ural'tseva [**577**], Landis [**581**], Lions, Magenes [**618**], Maugeri, Palagachev, Softova [**681**], Maz'ya, Rossmann [**685**], Miranda [**713**], Nazarov, Plamenevsky [**737**], Oleĭnik, Radkevič [**757**], Radkevich [**798**], Rempel, Schulze [**803**], Sauvigny [**834**], Shimakura [**857**], Shishmarev [**859**], Stampacchia [**870**], Stroock [**882**], Troianiello [**901**], Volpert [**924**], Wu, Yin, Wang [**944**], and also Garroni, Menaldi [**400**].

However, Fokker–Planck–Kolmogorov equations have significant specific features and so far have not become the subject of a separate exposition, although some of their aspects are discussed in depth in books with probabilistic motives, see Gihman, Skorokhod [**407**], Krylov [**549**], Kushner [**570**], Soize [**865**] (where also explicit solutions are considered), Stroock [**882**], Stroock, Varadhan [**884**]. These specific features are connected, on the one hand, with the fact that such equations by their nature are equations with respect to measures (sometimes they are called "double divergence form" equations and their solutions are called "adjoint solutions", in the case of irregular coefficients they cannot be written as divergence form equations or as direct equations), and, on the other hand, with unusual for the classical theory classes of solutions (say, integrable on the whole space, but without any restrictions on growth or smoothness class). A study of such equations goes back to Kolmorogov's works [**527**], [**528**], [**529**] and a series of earlier works in physics by Fokker [**377**], Smoluchowski [**863**], Planck [**781**], and Chapman [**235**], where equations for probability densities were considered (see also Hostinský [**466**]). The informative survey Fuller [**394**] lists also Lord Rayleigh and L. Bachelier among predecessors.

Connections between elliptic operators and diffusions are discussed in the books Bass [**91**] and Pinsky [**780**].

Traditionally, second order elliptic equations are solved in Hölder classes or in Sobolev classes depending on the properties of the coefficients. It is more convenient to consider the major problems of this book in Sobolev classes even in the case of smooth coefficients, which is due to the significant role of various a priori estimates in terms of quantities like Sobolev norms.

The existence of densities of solutions under minimal assumptions is based on the A.D. Aleksandrov estimates (see [**34**]–[**37**]). Unlike direct elliptic equations and divergence form equations, Fokker–Planck–Kolmogorov equations can have solutions whose regularity is not higher than the regularity of the diffusion coefficients. There are many works devoted to generalizations of Aleksandrov's estimates, see Cabré [**215**], Kuo, Trudinger [**565**]. Elliptic inequalities of the type $L^*u \geqslant 0$ were considered long ago, see, for example, Littman [**628**], [**629**].

Theorem 1.7.4 makes precise the statement of a result formulated by Ch. Morrey in his book [**723**, p. 156] not quite correctly (with $\Omega' = \Omega$). The assertion given there with $\Omega' = \Omega$ would be false, for example, for the Laplace equation on a ball. A proof of Morrey's estimate with an investigation of the dependence of the constant on the coefficients was given in Shaposhnikov [**843**] with the same inaccuracy as in [**723**]. Actually, the reasoning in Shaposhnikov [**843**] yields exactly the estimate we give, as explained in Shaposhnikov [**846**], and an estimate with $\Omega = \Omega'$ is possible only for solutions with zero boundary values on a domain with a sufficiently regular boundary (see Corollary 1.7.6). We observe that in the existing applications of Morrey's theorem actually only the presented correct statement was used, although in some papers it was formulated with the indicated inaccuracy (see, for example, Bogachev, Röckner [**160**] and Bogachev, Krylov, Röckner [**152**]). The proof of the corrected statement was given in Bogachev, Röckner, Shaposhnikov [**165**] and Shaposhnikov [**846**], where even a more general fact is proved.

Various results which can be regarded as results about properties of densities of solutions to elliptic Fokker–Planck–Kolmogorov equations with coefficients of a rather general form were obtained in the books cited above and also in the papers Krylov [**547**], Sjögren [**861**], Bauman [**94**], [**95**], Escauriaza [**341**], [**342**], Escauriaza, Kenig [**343**], Fabes, Stroock [**348**], Gushchin [**430**], Maz'ya, McOwen [**684**]; note also a more abstract approach of Herve [**452**]. A systematic study of the whole complex of these problems was initiated in Bogachev, Röckner [**157**], [**158**], [**160**], Bogachev, Krylov, Röckner [**149**], [**152**] and continued by many authors. Stationary Fokker–Planck–Kolmogorov equations with various special restrictions on the coefficients are considered in Arapostathis, Borkar [**55**], Bensoussan [**105**], Noarov [**745**]–[**749**] and in the works cited on concrete occasions in the subsequent chapters. Fokker–Planck–Kolmogorov equations can be also considered on more singular manifolds such as fractals and metric measure spaces.

Exercises

1.8.8. Suppose that probability measures μ and ν on the real line satisfy the equations $L^*_{1,b_1,c_1}\mu = 0$ and $L^*_{1,b_2,c_2}\nu = 0$. Show that the measure $\sigma = \mu\otimes\nu$ satisfies the equation $L^*_{\mathrm{I},b,c}\sigma = 0$ with $b(x,y) = \bigl(b_1(x), b_2(y)\bigr)$, $c(x,y) = c_1(x) + c_2(y)$.

1.8.9. Let a measure μ on $\mathbb{R}^d$ satisfy the equation $L^*_{A,b}\mu = 0$. Write down the equation to which the measure $\nu = \mu\circ F^{-1}$ satisfies, where $F = (F^i)$ is a diffeomorphism of $\mathbb{R}^d$ with inverse $G = (G^i)$.

HINT: Make the change of variable $x = G(y)$ in the integral identity, substituting $\varphi = \psi\circ F$, obtain the equation $L^*_{Q,h}\nu = 0$ with the matrix $Q = (q^{mk})$, where $q^{mk}(y) = a^{ij}\bigl(G(y)\bigr)\partial_{x_i}F^k\bigl(G(y)\bigr)\partial_{x_j}F^m\bigl(G(y)\bigr)$, and the drift $h = (h^k)$, where $h^k(y) = a^{ij}\bigl(G(y)\bigr)\partial_{x_i}\partial_{x_j}F^k\bigl(G(y)\bigr) + b^i\bigl(G(y)\bigr)\partial_{x_i}F^k\bigl(G(y)\bigr)$.

1.8.10. Suppose that a probability measure μ on the real line satisfies the equation $L^*_{1,b}\mu = 0$ on $(-\infty, 0)$ and on $(0, +\infty)$ with some continuous function b on the real line. Is it true that μ satisfies this equation on the whole real line?

1.8.11. Give an example of an absolutely continuous function on $[0,1]$ that does not satisfy the Hölder condition of any order.

1.8.12. (Bogachev, Röckner [**160**]) Let $A_k = (a_k^{ij})$ be a sequence of continuous mappings on $\mathbb{R}^d$ with values in the space of symmetric matrices and let $b_k = (b_k^i)$ be a sequence of Borel vector fields on $\mathbb{R}^d$. Suppose that for every ball $U_r \subset \mathbb{R}^d$ of radius r there exist numbers $c_r > 0$, $\alpha_r > 0$, and $p = p_r > d$ such that

$$A_k \geqslant c_r\mathrm{I}, \quad \|a_k^{ij}\|_{W^{p,1}(U_r)} + \|b_k^i\|_{L^p(U_r)} \leqslant \alpha_r \quad \text{for all } i,j,k.$$

Assume that there are probability measures μ_k on $\mathbb{R}^d$ such that $L^*_{A_k,b_k}\mu_k = 0$. Then the measures μ_k have continuous strictly positive densities that are uniformly Hölder continuous on every ball. If the sequence $\{\mu_k\}$ is uniformly tight, then it has compact closure in the variation norm, and every measure μ in its closure has a continuous strictly positive density of class $W^{p,1}(U_r)$ for every $r > 0$.

HINT: It follows from our hypotheses and Theorem 1.6.5 that the measures μ_k have continuous densities f_k. Since the functions f_k are probability densities, we obtain by (1.6.2) that, for every $r > 0$, the sequence $\{f_k\}$ is bounded in $W^{p,1}(U_r)$. By the Sobolev embedding theorem $\{f_k\}$ is uniformly Hölder continuous on U_r, in particular, has compact closure with respect to the sup-norm. If $\{\mu_k\}$ is uniformly tight, then some subsequence $\{\mu_{k_i}\}$ converges weakly to some probability measure μ. Passing to a subsequence once again we may assume that the functions f_{k_i} converge uniformly on compact sets and are uniformly bounded in $W^{p,1}(U_r)$ for each $r > 0$. Hence μ has a density $f \in W^{p,1}(U_r)$. Then we obtain a continuous and strictly positive version of f. Therefore, the probability measures μ_{k_i} converge to μ in the variation norm. This reasoning applies to any subsequence in $\{\mu_k\}$, whence we obtain the desired conclusion.

1.8.13. (Bogachev, Röckner [**160**]) The assertion of the previous exercise can be generalized as follows. Let Ω be an open set in $\mathbb{R}^d$ that is the union of increasing open sets Ω_k such that the closure of Ω_k is compact and contained in Ω_{k+1}. Let μ_k be probability measures on Ω_k satisfying the equations $L^*_{A_k,b_k}\mu_k = 0$ on Ω_k, where each A_k is a continuous mapping on Ω_k with values in the set of nonnegative symmetric matrices, the mappings A_k are uniformly bounded on compact sets in the

$W^{p,1}$-norm with some $p > 1$, the mappings A_k^{-1} are uniformly bounded on compact sets, and Borel vector fields b_k on the sets Ω_k are uniformly bounded in the $L^p(\mathbb{R}^d)$-norm on compact sets. Then the analogue of the assertion of the previous exercise is true. The same is true for Riemannian manifolds of dimension d.

1.8.14. Let $(X, \mathcal{B}, \mu)$ be a probability space, where the measure μ is separable, and let $(S, \mathcal{S})$ be a measurable space. Suppose that for each $s \in S$ we are given a μ-integrable function ξ_s such that for every set $B \in \mathcal{B}$ the integral of ξ_s over B is an $\mathcal{S}$-measurable function of s. Prove that for every $s \in S$ one can choose a version of ξ_s such that the function $(x, s) \mapsto \xi_s(x)$ is $\mathcal{B}\otimes\mathcal{S}$-measurable.

HINT: If μ is Lebesgue measure on $[0,1]$ and $f_s \in L^2(\mu)$, then we can take the standard trigonometric basis $\{e_n\}$ in $L^2(\mu)$ and set $\xi_s(x) = \sum_{n=1}^\infty (\xi_s, e_n)e_n(x)$ at the points of convergence, by using the Carleson theorem on convergence of this series almost everywhere and making ξ_s zero outside; the set of convergence belongs to $\mathcal{B}\otimes\mathcal{S}$, which ensures the $\mathcal{B}\otimes\mathcal{S}$-measurability of the obtained version. In the general case we can use the existence of a Schauder basis $\{\varphi_n\}$ in $L^1(\mu)$ consisting of $\mathcal{B}$-measurable functions with the property that for each $f \in L^1(\mu)$ the series $f = \sum_{n=1}^\infty c_n(f)\varphi_n$ converges μ-a.e.; the coefficients $c_n(f)$ are continuous linear functionals on $L^1(\mu)$, hence they are represented as the integrals of $f\psi_n$ with certain $\psi_n \in L^\infty(\mu)$, which ensures the measurability of $s \mapsto c_n(f_s)$. Finally, a Schauder basis with the indicated property can be constructed as follows: by the separability of the measure there is a countable collection of sets $B_n \in \mathcal{B}$ the linear span of the indicators of which is dense in $L^1(\mu)$; the σ-algebra $\mathcal{B}_n$ generated by $B_1, \ldots, B_n$ can be also generated by a partition of X into finitely many disjoint parts $B_{n,1}, \ldots, B_{n,k_n}$, the conditional expectations of f with respect to $\mathcal{B}_n$ converge to f in norm and almost everywhere by the martingale convergence theorem, and these conditional expectations can be represented as partial sums of a series in Haar-type functions constructed by means of the indicated decreasing partitions. Another construction can be found in the hint to Exercise 6.10.71 in Bogachev [**125**].

1.8.15. Let $\xi\colon [0,1] \to L^1(\mu)$ be a continuous mapping, where μ is a probability measure on a measurable space $(\Omega, \mathcal{B})$. Prove that there is a function $\eta\colon \Omega\times\mathbb{R} \to \mathbb{R}$, called a measurable modification of ξ, such that it is measurable with respect to $\mathcal{B}\otimes\mathcal{B}(\mathbb{R})$ and for each $t \in \mathbb{R}$ the equality $\eta(\omega, t) = \xi(t)(\omega)$ holds for μ-a.e. ω.

HINT: Use the previous exercise considering μ on the σ-algebra generated by $\xi(t), t \in \mathbb{Q}$; see also Neveu [**742**, § III.4].

1.8.16. Justify Remark 1.6.3.

1.8.17. Prove that if on the closed ball in $\mathbb{R}^d$ a uniformly Hölder continuous sequence of functions converges in measure, then it converges uniformly.

1.8.18. Let $u \in W^{p,2}(U) \cap W_0^{p,1}(U)$, where the set $U \subset \mathbb{R}^d$ is bounded and open and $p > d$. By using Corollary 1.1.6 show that for all $c > 0$ and $\alpha > 1$ the function $w := (u + c)^\alpha - c^\alpha$ belongs to $W_0^{p,2}(U)$.

1.8.19. Let $f \in W^{1,1}_{\rm loc}(\mathbb{R}^d)$. Show that $\nabla f(x) = 0$ a.e. on $f^{-1}(0)$.

1.8.20. Prove that the diffusion process on the real line given by the equation $d\xi_t = dw_t + f(\xi_t)dt$, where $f'(x) + f(x)^2 = ax^2 + bx + c$, has an invariant probability measure only under the condition that $f(x) = \alpha x + \beta$, $\alpha < 0$.

HINT: See Zeitouni [**953**].

CHAPTER 2

Existence of Solutions

This chapter is devoted to constructing solutions to elliptic equations for measures on the whole space, i.e., stationary Fokker–Planck–Kolmogorov equations. We first discuss the existence of positive locally bounded solutions to divergence form equations, which is closely connected with maximum principles. Then we study the conditions for global integrability of positive solutions and also establish a priori estimates of solutions by means of Lyapunov functions, which is one of the most important methods of constructing probability solutions (i.e., solutions that are probability measures).

One of the main results here is Theorem 2.4.1, according to which a positive local solution to the equation $L^*_{A,b}\mu = 0$ exists under rather general assumptions, namely, A is locally nondegenerate and either a^{ij} and b^i are locally bounded or the coefficients a^{ij} belong to the local Sobolev class $W^{d+,1}_{\mathrm{loc}}$ and $b^i \in L^{d+}_{\mathrm{loc}}$. There is a probability solution if there exists a Lyapunov function V such that $L_{A,b}V \leqslant -1$ outside of a ball.

2.1. The maximum principle and the Dirichlet problem

In this section we discuss connections between the existence of solutions on a bounded domain and maximum principles for the elliptic operator and its adjoint, moreover, we consider a general divergence form operator.

Let $\Omega \subset \mathbb{R}^d$ be a bounded domain with C^∞-smooth boundary. Set

$$\mathcal{L}u = \partial_{x_i}\big(a^{ij}\partial_{x_j}u - b^i u\big) + \beta^i\partial_{x_i}u + cu, \tag{2.1.1}$$

where the functions a^{ij}, b^i, β^i and c are measurable in Ω.

In this section we assume that the coefficients satisfy the following conditions: there are numbers $\lambda_1, \lambda_2 > 0$ and $p > d$ such that

$$\begin{aligned} &\lambda_1 \cdot \mathrm{I} \leqslant A(x) \leqslant \lambda_2 \cdot \mathrm{I},\ a^{ij} = a^{ji},\ x \in \Omega, \\ &|b|, |\beta| \in L^p(\Omega),\ c \in L^s(\Omega),\ s = p/2, \end{aligned} \tag{2.1.2}$$

where in case $d = 1$ the number s is an arbitrary number greater than 1.

For the sequel the following estimates will be useful, which follow by Hölder's inequality and the Sobolev embedding theorem. Let $h \in L^d(\Omega)$, $g \in L^s(\Omega)$, where $s = d/2$ if $d > 2$ and s is an arbitrary number greater than 1 if $d \leqslant 2$, $u, v \in W^{2,1}(\Omega)$. Then

$$\begin{aligned} &\big\|hu|\nabla v|\big\|_{L^1(\Omega)} \leqslant C(d,\Omega)\|h\|_{L^d(\Omega)}\|u\|_{W^{2,1}(\Omega)}\|v\|_{W^{2,1}(\Omega)}, \\ &\|guv\|_{L^1(\Omega)} \leqslant C(d,s,\Omega)\|g\|_{L^s(\Omega)}\|u\|_{W^{2,1}(\Omega)}\|v\|_{W^{2,1}(\Omega)}. \end{aligned} \tag{2.1.3}$$

Let us consider on $W^{2,1}(\Omega)$ the following bilinear form $\mathcal{L}(u,v)$:

$$\mathcal{L}(u,v) = \int_\Omega \big[\langle A\nabla u, \nabla v\rangle\, dx + u\langle b, \nabla v\rangle + \langle \beta, \nabla u\rangle v + cuv\big]\, dx. \tag{2.1.4}$$

We observe that according to inequalities (2.1.3) all expressions in this definition are meaningful. If all coefficients are smooth, then obviously $\mathcal{L}(u,v) = -\langle \mathcal{L}u, v\rangle_{L^2(\Omega)}$ for $u, v \in W_0^{2,1}(\Omega)$.

2.1.1. Definition. We shall say that a function $u \in W^{2,1}(\Omega)$ satisfies the inequality $\mathcal{L}u \geqslant 0$ if $\mathcal{L}(u,v) \leqslant 0$ for all nonnegative functions $v \in W_0^{2,1}(\Omega)$.

We shall say that the operator $\mathcal{L}$ satisfies the maximum principle if, for every function $u \in W_0^{2,1}(\Omega)$, the inequality $\mathcal{L}u \geqslant 0$ implies the inequality $u \leqslant 0$ in Ω.

In the latter case we can replace the condition $u \in W_0^{2,1}(\Omega)$ with the conditions $u \in W^{2,1}(\Omega)$ and $u^+ = \max\{u, 0\} \in W_0^{2,1}(\Omega)$. Indeed, the inequality $\mathcal{L}u \geqslant 0$ implies the inequality $\mathcal{L}u^+ \geqslant 0$ (see Exercise 2.5.16).

Note that here, certainly, it is not asserted that every solution with any boundary value achieves its maximum or minimum on the boundary of the domain (so we deal with the weak maximum principle).

2.1.2. Definition. Let $f, g^1, \ldots, g^d \in L^2(\Omega)$. We shall say that a function $u \in W^{2,1}(\Omega)$ is a solution of the equation $\mathcal{L}u = f + \operatorname{div} g$ if we have the identity

$$\mathcal{L}(u,v) = -\langle f, v\rangle_{L^2(\Omega)} + \langle g^i, \partial_{x_i} v\rangle_{L^2(\Omega)} \quad \forall\, v \in W_0^{2,1}(\Omega).$$

We observe that if the operator $\mathcal{L}$ satisfies the maximum principle, then the equation $\mathcal{L}u = 0$ has only zero solution in the class $W_0^{2,1}(\Omega)$. Indeed, such a solution must be simultaneously nonnegative and nonpositive.

2.1.3. Lemma. *Let condition* (2.1.2) *be fulfilled. Then*

(i) $\mathcal{L}(u,v) \leqslant N_1 \|u\|_{W^{2,1}(\Omega)} \|v\|_{W^{2,1}(\Omega)}$,

(ii) $\mathcal{L}(u,u) \geqslant \frac{\lambda_1}{2} \|u\|^2_{W^{2,1}(\Omega)} - N_2 \|u\|^2_{L^2(\Omega)}$, *where* $N_1 > 0$ *and* $N_2 > 0$ *are numbers that depend only on the coefficients of* $\mathcal{L}$ *and the domain* Ω.

PROOF. Estimate (i) follows by (2.1.3) with the functions $h = |b - \beta|$ and $g = c$. Let us prove (ii). Let $h = |b - \beta|$. We observe that

$$\big\| hu|\nabla u| \big\|_{L^1(\Omega)} \leqslant \frac{\lambda_1}{8} \|\nabla u\|^2_{L^2(\Omega)} + \frac{2}{\lambda_1} \|uh\|^2_{L^2(\Omega)}.$$

Let $M > 0$ be such that

$$\left(\int_{h>M} |h|^d\, dx \right)^{2/d} \leqslant \frac{\lambda_1^2}{16}.$$

Then

$$\|uh\|^2_{L^2(\Omega)} \leqslant M \|u\|^2_{L^2(\Omega)} + \frac{\lambda_1^2}{16} \|u\|^2_{W^{2,1}(\Omega)}.$$

Therefore,

$$\big\| hu|\nabla u| \big\|_{L^1(\Omega)} \leqslant \frac{\lambda_1}{4} \|\nabla u\|^2_{L^2(\Omega)} + \left(\frac{2M}{\lambda_1} + \frac{\lambda_1}{4} \right) \|u\|_{L^2(\Omega)}.$$

Similarly we estimate $\|cu^2\|_{L^1(\Omega)}$. □

2.1.4. Proposition. *Suppose that conditions* (2.1.2) *are fulfilled on* Ω *and that the operator* $\mathcal{L}$ *satisfies the maximum principle on* Ω. *Then, for any functions* $f, g^1, \ldots, g^d \in L^2(\Omega)$, *there exists a solution* $u \in W_0^{2,1}(\Omega)$ *of the Dirichlet problem* $\mathcal{L}u = f + \mathrm{div} g$, $u|_{\partial\Omega} = 0$, *where* $g = (g^i)$ *and the equality* $u|_{\partial\Omega} = 0$ *is understood as the inclusion* $u \in W_0^{2,1}(\Omega)$.

PROOF. According to Lemma 2.1.3, for a sufficiently large $\sigma > 0$, the bilinear form $\mathcal{L}_\sigma(u,v)$ corresponding to the operator $\mathcal{L}_\sigma u = \mathcal{L}u - \sigma u$ satisfies the inequality

$$\mathcal{L}_\sigma(u,v) \leqslant \widetilde{N}_1 \|u\|_{W_0^{2,1}(\Omega)} \|v\|_{W_0^{2,1}(\Omega)}, \quad \mathcal{L}_\sigma(u,u) \geqslant \frac{\lambda}{2} \|u\|^2_{W_0^{2,1}(\Omega)}.$$

By the Lax–Milgram lemma (see Gilbarg, Trudinger [**409**, Theorem 5.8]) that asserts the surjectivity of a bounded operator T in a Hilbert space for which $(Tx, x) \geqslant \varepsilon(x,x)$ with some $\varepsilon > 0$, there is a continuous operator $\mathcal{L}_\sigma^{-1}$ from $W^{2,-1}(\Omega) := \big(W_0^{2,1}(\Omega)\big)^*$ to $W_0^{2,1}(\Omega)$ such that the equality $\mathcal{L}_\sigma u = F$ is equivalent to the equality $u = \mathcal{L}_\sigma^{-1} F$ for every $F \in W^{2,-1}(\Omega)$.

The identity embedding operator $I\colon W_0^{2,1}(\Omega) \to W^{2,-1}(\Omega)$ is compact. Suppose that $F \in W^{2,-1}(\Omega)$. Then the equation $\mathcal{L}u = F$ can be written in the form $u + \sigma\mathcal{L}_\sigma^{-1} I u = \mathcal{L}_\sigma^{-1} F$. Now our assertion follows by the Fredholm alternative, since the operator $\sigma\mathcal{L}_\sigma^{-1} I$ is compact and by the maximum principle the zero solution of the equation $\mathcal{L}u = 0$ is the only solution in $W_0^{1,2}(\Omega)$. □

2.1.5. Remark. Let $f \in L^{p/2}(\Omega)$, $g^1, \ldots, g^d \in L^p(\Omega)$. According to Trudinger [**904**, Theorem 4.1], the solution u constructed in the lemma satisfies the estimate

$$\sup_{x\in\Omega} |u(x)| \leqslant N\big(\|f\|_{L^{p/2}(\Omega)} + \|g\|_{L^p(\Omega)}\big).$$

2.1.6. Corollary. *Suppose that in addition to conditions* (2.1.2) *we are given that* $c \in L^{pd/(p+d)}(\Omega)$. *If* $a^{ij} \in VMO$, $f \in L^{pd/(p+d)}(\Omega)$, *and* $g^i \in L^p(\Omega)$, *then the solution* u *constructed in Proposition* 2.1.4 *belongs to the class* $W_0^{p,1}(\Omega)$ *and we have the estimate*

$$\|u\|_{W_0^{p,1}(\Omega)} \leqslant N\big(\|g\|_{L^p(\Omega)} + \|f\|_{L^{pd/(p+d)}(\Omega)}\big).$$

PROOF. Consider first the case of the operator

$$\mathcal{L}u = \partial_{x_i}(a^{ij}\partial_{x_j} u) + \beta^i \partial_{x_i} u.$$

Let $\{a_n^{ij}\}, \{\beta_n^i\}, \{f_n\}, \{g_n^i\} \subset C^\infty(\overline{\Omega})$, $A_n \geqslant 2^{-1}\lambda \cdot \mathrm{I}$, let $\omega_{A_n} \leqslant \omega_A$ for the corresponding functions from the inclusion in VMO (see § 1.2), and let

$$\lim_{n\to\infty} \|a_n^{ij} - a^{ij}\|_{L^p(\Omega)} = 0, \quad \lim_{n\to\infty} \|\beta_n^i - \beta^i\|_{L^p(\Omega)} = 0,$$

$$\lim_{n\to\infty} \|g_n^i - g^i\|_{L^p(\Omega)} = 0, \quad \lim_{n\to\infty} \|f_n - f\|_{L^{dp/(d+p)}(\Omega)} = 0.$$

Let $\mathcal{L}_n$ denote the elliptic operator with the smooth coefficients a_n^{ij}, β_n^i. It is known (see Gilbarg, Trudinger [**409**, Theorem 8.14]) that the corresponding Dirichlet problem $\mathcal{L}_n u = f_n + \mathrm{div} g_n$, $u|_{\partial\Omega} = 0$, where $g_n = (g_n^1, \ldots, g_n^d)$, has a unique classical solution $u_n \in C^\infty(\overline{\Omega})$. By Theorem 1.7.4 this solution satisfies the estimate

$$\|u_n\|_{W_0^{p,1}(\Omega)} \leqslant N\big(\|u_n\|_{L^1(\Omega)} + \|g_n\|_{L^p(\Omega)} + \|f_n\|_{L^{pd/(p+d)}(\Omega)}\big),$$

where N does not depend on n. From this estimate and Remark 2.1.5 we obtain the uniform boundedness of $\|u_n\|_{W_0^{p,1}(\Omega)}$. Hence we can choose a subsequence $\{u_{n_k}\}$ converging weakly in $W^{p,1}(\Omega)$, which converges to a function u in $W_0^{p,1}(\Omega)$. The same subsequence converges weakly in $W_0^{2,1}(\Omega)$. By the embedding theorem (recall that here $p > d$) this sequence uniformly converges to u. Again applying the estimate from Theorem 1.7.4 to the difference $u_n - u_m$, we conclude that the sequence $\{u_n\}$ is Cauchy in $W_0^{p,1}(\Omega)$. Therefore, $\{u_n\}$ converges to u in $W_0^{p,1}(\Omega)$. Clearly, u is a solution of the equation $\mathcal{L}u = f + \mathrm{div} g$ in the class $W_0^{p,1}(\Omega)$. By Theorem 2.1.8 proven below, the solution u is the unique solution in $W_0^{2,1}(\Omega)$, hence it coincides with the one constructed in the proposition.

We now consider the general operator $\mathcal{L}$ with all coefficients a^{ij}, b^i, β^i and c. Let us rewrite the equation $\mathcal{L}u = f + \mathrm{div} g$ in the form

$$\partial_{x_i}(a^{ij}\partial_{x_j}u) + \beta^i\partial_{x_i}u = f - cu + \partial_{x_i}(g^i + b^i u).$$

Remark 2.1.5 implies the inclusions $cu \in L^{pd/(p+d)}(\Omega)$ and $b^i u \in L^p(\Omega)$. As shown above, there exists a solution $w \in W_0^{p,1}(\Omega)$ of the equation

$$\partial_{x_i}(a^{ij}\partial_{x_j}w) + \beta^i\partial_{x_i}w = f - cu + \partial_{x_i}(g^i + b^i u).$$

Since w is the unique solution in the class $W_0^{2,1}(\Omega)$, it coincides almost everywhere with u. Therefore, $u \in W_0^{p,1}(\Omega)$. The estimate for u follows by Theorem 1.7.4. □

Set

$$\mathcal{L}^*u = \partial_{x_i}\big(a^{ij}\partial_{x_j}u - \beta^i u\big) + b^i\partial_{x_i}u + cu, \tag{2.1.5}$$

i.e., the coefficients b and β are simply interchanged.

2.1.7. Lemma. *If the operator $\mathcal{L}$ satisfies the maximum principle on Ω, then the operator $\mathcal{L}^*$ also satisfies the maximum principle on Ω.*

PROOF. Let $u \in W_0^{2,1}(\Omega)$ and $\mathcal{L}^*u \geqslant 0$. We show that $u \leqslant 0$. It suffices to verify that the integral of fu is nonpositive if $f \in C_0^\infty(\Omega)$, $f \geqslant 0$. Since on Ω we have the maximum principle for $\mathcal{L}$, by Proposition 2.1.4 there exists a solution $v \in W_0^{2,1}(\Omega)$ of the Dirichlet problem $\mathcal{L}v = -f$, $v|_{\partial\Omega} = 0$. By the maximum principle $v \geqslant 0$, since $\mathcal{L}v \leqslant 0$. By the definition of a solution the integral of fu equals $\mathcal{L}^*(u, v)$, which is nonnegative, since $\mathcal{L}^*u \geqslant 0$. □

Very general sufficient conditions for the validity of the maximum principle are obtained in Trudinger [**905**], where the following result is proved.

2.1.8. Theorem. *Suppose that condition* (2.1.2) *is fulfilled and for some number $\gamma \in [0, 1]$ one has the inequality*

$$\int_\Omega \big[cv + \langle \gamma b + (1-\gamma)\beta, \nabla v\rangle\big]\, dx \leqslant 0 \quad \textit{if } v \in C_0^\infty(\Omega),\ v \geqslant 0. \tag{2.1.6}$$

Then the operator $\mathcal{L}$ of the form (2.1.1) *satisfies the maximum principle.*

PROOF. The case $d = 1$ is Exercise 2.5.17. Further we deal with $d \geqslant 2$. It follows from our assumptions that (2.1.6) remains true for all nonnegative functions $v \in W_0^{2,1}(\Omega)$. By Lemma 2.1.7 it suffices to prove our assertion only for $\gamma \leqslant 1/2$. Suppose that $u \in W_0^{2,1}(\Omega)$ and $\mathcal{L}u \geqslant 0$. Then $u^+ \in W_0^{2,1}(\Omega)$. Let $f \in C_b^1([0, +\infty))$ be a function such that $f(0) = 0$, $f \geqslant 0$ and also $\sup_t f(t)/t < \infty$. Then we have $v := f(u^+) \in W_0^{2,1}(\Omega)$ and almost everywhere

$$\nabla f(u^+) = f'(u^+)\nabla u^+, \quad \langle A\nabla u, \nabla u^+\rangle = \langle A\nabla u^+, \nabla u^+\rangle,$$

since $\nabla u^+ = 0$ almost everywhere on the set $\{u < 0\}$. Hence

$$\int_\Omega f'(u^+)\langle A\nabla u^+, \nabla u^+\rangle\, dx = \int_\Omega \bigl[f'(u^+)\langle b, \nabla u^+\rangle u^+ + \langle \beta, \nabla u^+\rangle f(u^+) + cu^+ f(u^+)\bigr]\, dx.$$

Applying inequality (2.1.6), we obtain

$$\int_\Omega f'(u^+)\langle A\nabla u^+, \nabla u^+\rangle\, dx \leqslant \int_\Omega \langle b-\beta, \nabla u^+\rangle \bigl[\gamma f(u^+) - (1-\gamma) f'(u^+)u^+\bigr]\, dx.$$

Picking a function f for different values of γ, we can show that $u^+ = 0$. Namely, let $\gamma \in (0, 1/2]$. Let us set $f(t) = (t+\varepsilon)^p - \varepsilon^p$, where $\varepsilon > 0$ and $p = \gamma/(1-\gamma)$. We observe that $0 < p \leqslant 1$. Then

$$\begin{aligned} p\int_\Omega (u^+ + \varepsilon)^{p-1}\langle A\nabla u^+, \nabla u^+\rangle\, dx &\leqslant \frac{\gamma\varepsilon}{p}\int_\Omega (u^+ + \varepsilon)^{p-1}|b-\beta|\,|\nabla u^+|\, dx + \gamma\varepsilon^p \int_\Omega |b-\beta|\,|\nabla u^+|\, dx \\ &\leqslant \Bigl(\frac{\gamma\varepsilon^{2-p}}{p} + \gamma\varepsilon\Bigr)\int_\Omega |b-\beta|\,|\nabla u^+|\, dx. \end{aligned}$$

Letting $\varepsilon \to 0$, we obtain that u^+ is a constant. The inclusion $u^+ \in W_0^{2,1}(\Omega)$ gives now the equality $u^+ = 0$ a.e. on Ω.

We now consider the case where $\gamma = 0$. Let $\varepsilon > 0$. Set $f(t) = t/(t+\varepsilon)$. We have

$$\begin{aligned} \int_\Omega (u^+ + \varepsilon)^{-2}\langle A\nabla u^+, \nabla u^+\rangle\, dx &\leqslant \int_\Omega \langle \beta - b, \nabla u^+\rangle u^+ (u^+ + \varepsilon)^{-2}\, dx \\ &\leqslant \int_\Omega |\beta - b|\,|\nabla u^+|(u^+ + \varepsilon)^{-1}\, dx. \end{aligned}$$

Letting $w_\varepsilon = \ln(1 + \varepsilon^{-1}u^+)$, we obtain

$$\begin{aligned} \int_\Omega |\sqrt{A}\nabla w_\varepsilon|^2\, dx &\leqslant \int_\Omega |\beta - b|\,|\nabla w_\varepsilon|\, dx \\ &\leqslant \Bigl(\int_\Omega |\sqrt{A}\nabla w_\varepsilon|^2\, dx\Bigr)^{1/2}\Bigl(\int_\Omega |\sqrt{A^{-1}}(\beta - b)|^2\, dx\Bigr)^{1/2}. \end{aligned}$$

Hence

$$\int_\Omega |\sqrt{A}\nabla w_\varepsilon|^2\, dx \leqslant \int_\Omega |\sqrt{A^{-1}}(\beta - b)|^2\, dx.$$

Since $w \in W_0^{2,1}(\Omega)$, by the Poincaré inequality we have

$$\int_\Omega |w_\varepsilon|^2\, dx \leqslant C\int_\Omega |\sqrt{A^{-1}}(\beta - b)|^2\, dx,$$

where C does not depend on ε. Letting $\varepsilon \to 0$, we conclude that $u^+ = 0$ a.e. by Fatou's theorem, since $w_\varepsilon(x) \to +\infty$ if $u^+(x) > 0$. □

Let us give one more simple sufficient condition for the validity of the maximum principle.

2.1.9. Corollary. *Let condition* (2.1.2) *be fulfilled. If there exists a function* $V \in W^{p,2}(\Omega)$, *where* $p > d$ *and* $p = 2$ *if* $d = 1$, *such that* $\inf_\Omega V > 0$ *and* $\mathcal{L}V \leqslant 0$, *then* $\mathcal{L}$ *satisfies the maximum principle.*

PROOF. We observe that by the embedding theorem the function V has a continuous positive bounded version on Ω. We shall deal with this version. Let $u \in W_0^{2,1}(\Omega)$ and $\mathcal{L}u \geqslant 0$. Set $w = u/V$. Clearly, $w \in W_0^{2,1}(\Omega)$. Let

$$\mathcal{L}_V w = \partial_{x_i}\big(Va^{ij}\partial_{x_j}w + (a^{ij}\partial_{x_j}V - Vb^i)w\big) + V\beta^i\partial_{x_i}w + (\beta^i\partial_{x_i}V + cV)w.$$

We observe that by assumption $\mathcal{L}V \leqslant 0$, i.e., we have the inequality

$$\partial_{x_i}(a^{ij}\partial_{x_j}V - Vb^i) + (\beta^i\partial_{x_i}V + cV) \leqslant 0.$$

By Theorem 2.1.8 with $\gamma = 1$ the operator $\mathcal{L}_V$ satisfies the maximum principle. Since $\mathcal{L}_V w = \mathcal{L}u \geqslant 0$, we have $w \leqslant 0$. Hence $u \leqslant 0$. □

2.1.10. Remark. Let us single out the following two cases in which the hypotheses of Theorem 2.1.8 are obviously satisfied: (i) $\gamma = 1$, $b = 0$ and $c \leqslant 0$, (ii) $\gamma = 0$, $\beta = 0$ and $c \leqslant 0$. When we deal with elliptic operators of the form $Lu = a^{ij}\partial_{x_i}\partial_{x_j}u + b^i\partial_{x_i}u$, where $a^{ij} \in W^{p,1}_{\mathrm{loc}}$, we can write the formal adjoint L^* as

$$L^*u = \partial_{x_i}\partial_{x_j}(a^{ij}u) - \partial_{x_i}(b^iu) = \partial_{x_i}\big(a^{ij}\partial_{x_j}u - (b^i - \partial_{x_j}a^{ij})u\big),$$

which is $\mathcal{L}_{A,b_0}u$ with a new drift b_0 with components $b^i - \partial_{x_j}a^{ij}$. Therefore, case (ii) above applies and yields that L^* satisfies the maximum principle. We shall see in the next section that this ensures the existence of positive solutions (but not necessarily integrable) to the equation $L^*u = 0$.

2.2. Positive solutions of divergence type equations

Here we prove that maximum principles for the elliptic operator $\mathcal{L}$ of divergence form (2.1.1) and the operator $\mathcal{L}^*$ of the form (2.1.5) are equivalent to the existence of positive solutions for them. We recall (see Theorem 1.7.2 and also Chapter 3) that for every nonnegative solution $u \in W^{2,1}(\Omega)$ of the equation $\mathcal{L}u = 0$ (or the equation $\mathcal{L}^*u = 0$) the following Harnack inequality is fulfilled: for every open ball U with $\overline{U} \subset \Omega$ there exists a number $C(U) > 0$ such that

$$\sup_U u \leqslant C(U)\inf_U u.$$

If the continuous version vanishes at least at one point, then it is identically zero. Finally, if a nonnegative solution u belongs to the class $C(\Omega)$ and at least at one point in U is greater than zero, then in U it is separated from zero by some constant.

Now we can prove a theorem on the existence of positive solutions. For the definition of the class VMO, see p. 8, 37.

2.2.1. Theorem. *Suppose in addition to conditions* (2.1.2), *where the bounded domain* Ω *has smooth boundary, that* $c \in L^{pd/(p+d)}(\Omega)$ *and* $a^{ij} \in VMO$. *Then, the following assertions are equivalent on the domain* Ω:

(i) *the operator* $\mathcal{L}$ *satisfies the maximum principle;*
(ii) *there exists* $u \in W^{p,1}(\Omega)$ *with* $u > 0$ *and* $\mathcal{L}u = 0$;
(iii) *there exists* $u \in W^{p,1}(\Omega)$ *with* $u > 0$ *and* $\mathcal{L}u \leqslant 0$;
(iv) *the operator* $\mathcal{L}^*$ *satisfies the maximum principle;*
(v) *there exists* $u \in W^{p,1}(\Omega)$ *with* $u > 0$ *and* $\mathcal{L}^*u = 0$;
(vi) *there exists* $u \in W^{p,1}(\Omega)$ *with* $u > 0$ *and* $\mathcal{L}^*u \leqslant 0$.

PROOF. The equivalence of assertions (i) and (iv) follows by Lemma 2.1.7. We show that the implications (i)⇒(ii) and (iv)⇒(v) follow by Proposition 2.1.4 and Corollary 2.1.6. Indeed, let $v \in W_0^{p,1}(\Omega)$ be a solution of the equation $\mathcal{L}v = \operatorname{div} b - c$. Then $u = v + 1$ is a solution of the equation $\mathcal{L}u = 0$. In addition, $u \geqslant 0$, since $\mathcal{L}u^- \geqslant 0$ by Exercise 2.5.16 and $u^- \in W_0^{2,1}(\Omega)$. Since $u - 1 \in W_0^{p,1}(\Omega)$, we can choose a version of u in $C(\overline{\Omega})$. Therefore, $u > 0$ in Ω by Harnack's inequality. The case of $\mathcal{L}^*$ is similar.

The implications (ii)⇒(iii) and (v)⇒(vi) are obvious. The implications (iii)⇒(i) and (vi)⇒(iv) follow by Corollary 2.1.9. □

2.2.2. Example. Assertions (iii) and (vi) enable us to obtain easily sufficient conditions for the validity of the maximum principle and the existence of a positive solution. Say, the condition $-\operatorname{div} b + c \leqslant 0$, sufficient for the maximum principle, simply means that the function $V = 1$ satisfies the inequality $\mathcal{L}V \leqslant 0$.

Let now $v(x) = \exp(\alpha|x|^2/2)$. For the validity of the maximum principle and the existence of a positive solution it suffices to have the estimate

$$c(x) \leqslant -\alpha^2\langle A(x)x, x\rangle - \alpha \operatorname{tr} A(x) - \alpha\langle b(x), x\rangle.$$

In particular, if $d = 1$, $A = 1$, $b(x) = -x$, then for sufficient conditions with $\alpha = -1$ we obtain $c(x) \leqslant 1 - 2x^2$, and for $\alpha = 1$ we need $c(x) \leqslant -1 + 2x^2$.

Consider now the operator $\mathcal{L}$ on all of $\mathbb{R}^d$. Suppose that on every ball U the coefficients of $\mathcal{L}$ satisfy conditions (2.1.2), where in addition $c \in L^{pd/(p+d)}(U)$ and $a^{ij} \in VMO$ (we recall that the latter holds if a^{ij} are continuous).

2.2.3. Corollary. *The following assertions are equivalent:*
(i) *on every ball $\mathcal{L}$ satisfies the maximum principle;*
(ii) *there exists $u \in W_{\mathrm{loc}}^{p,1}(\mathbb{R}^d)$ with $u > 0$ and $\mathcal{L}u = 0$;*
(iii) *there exists $u \in W_{\mathrm{loc}}^{p,1}(\mathbb{R}^d)$ with $u > 0$ and $\mathcal{L}u \leqslant 0$;*
(iv) *on every ball $\mathcal{L}^*$ satisfies the maximum principle;*
(v) *there exists $u \in W_{\mathrm{loc}}^{p,1}(\mathbb{R}^d)$ with $u > 0$ and $\mathcal{L}^*u = 0$;*
(vi) *there exists $u \in W_{\mathrm{loc}}^{p,1}(\mathbb{R}^d)$ with $u > 0$ and $\mathcal{L}^*u \leqslant 0$.*

PROOF. It suffices to explain how to construct a positive solution on all of $\mathbb{R}^d$. Let U_n be the ball of radius n centered at the origin. By the previous theorem there exists a positive solution of the equation $\mathcal{L}u_n = 0$. Let $\varrho_n = u_n/u_n(0)$. Then $\varrho_n(0) = 1$. By Harnack's inequality, for every ball U with compact closure in U_n there exists a constant $C(U) > 0$ such that

$$C(U)^{-1}\varrho_n(0) \leqslant \inf_U \varrho_n(x), \quad \sup_U \varrho_n(x) \leqslant C(U)\varrho_n(0).$$

In addition, the norm in $W^{p,1}(U)$ of the solution ϱ_n is estimated by a number independent of n. Therefore, $\{\varrho_n\}$ has a subsequence weakly converging in $W^{p,1}$ and uniformly (by the embedding theorem) on all balls to some positive function ϱ, which obviously satisfies the equation $\mathcal{L}\varrho = 0$. Note also that we could pick a function u_n with the boundary condition $u_n|_{\partial U_n} = 1$. □

Thus, we obtained conditions ensuring the existence of a positive solution, which mean essentially that the differential operator in question must satisfy the maximum principle. For constructing probability solutions, i.e., positive solutions with finite integral, which becomes one after normalization, we need a priori estimates with Lyapunov functions.

2.3. Lyapunov functions and a priori estimates

In this section we establish certain general a priori estimates which will be useful in proving the existence of solutions. These estimates provide bounds on the integrals of certain given functions with respect to unknown solutions. They involve the concept of Lyapunov function, which we now explain.

2.3.1. Definition. A real-valued function V on a topological space is called compact if all the sets $\{V \leqslant c\}$ are compact.

A function V is *quasi-compact* if the space can be represented as the union of increasing compact sets $\{V \leqslant c_k\}$ for some increasing sequence of numbers c_k.

Clearly, a compact function is quasi-compact. A continuous function V on $\mathbb{R}^d$ is compact precisely when $\lim\limits_{|x|\to\infty} V(x) = +\infty$. Any even continuous function on the real line which is increasing on $[0, +\infty)$ to some number c but does not assume the value c is quasi-compact; if $c < \infty$, then it is not compact. A *Lyapunov function* for an operator L is a quasi-compact function V with a suitable estimate for LV.

2.3.2. Theorem. *Let $V \in C^2(\mathbb{R}^d)$ and let U be a bounded set of the form $U := \{V < c\}$ for some c. Suppose that μ is a nonnegative Borel measure on U satisfying the equation $L^*_{A,b}\mu = 0$. Suppose that there exist a Borel function $\Psi \in L^1(\mu)$ and a Borel function $\Phi \geqslant 0$ such that*

$$LV \leqslant \Psi - \Phi \quad \mu\text{-a.e. on } U.$$

Then

$$\int_U \Phi\, d\mu \leqslant \int_U \Psi\, d\mu. \tag{2.3.1}$$

If μ is a nonnegative measure on $\mathbb{R}^d$ satisfying the equation $L^*_{A,b}\mu = 0$ on the whole space and $V \in C^2(\mathbb{R}^d)$ is a quasi-compact function with $LV \leqslant \Psi - \Phi$ on $\mathbb{R}^d$, then

$$\int_{\mathbb{R}^d} \Phi\, d\mu \leqslant \int_{\mathbb{R}^d} \Psi\, d\mu. \tag{2.3.2}$$

The assumption that $V \in C^2(\mathbb{R}^d)$ can be replaced by $V \in W^{d,2}_{\rm loc}(\mathbb{R}^d)$ if, for each ball U in $\mathbb{R}^d$, there are numbers $m = m(U) > 0$ and $M = M(U) > 0$ such that $m \cdot \mathrm{I} \leqslant A(x) \leqslant M \cdot \mathrm{I}$ for all $x \in U$ and $b^i \in L^d_{\rm loc}(\mathbb{R}^d)$.

Proof. Adding a constant we may assume that $V \geqslant 0$. It suffices to prove our assertion for bounded Φ. Indeed, once this is done, we consider the functions $\Phi_k = \min(\Phi, k)$, for which we have $LV \leqslant \Psi - \Phi_k$, and then apply Fatou's theorem. Assuming that Φ is bounded, given $\varepsilon > 0$, we can find a compact set $K \subset U$ such that $\mu(U) < \mu(K) + \varepsilon$ and

$$\int_U \Phi\, d\mu < \int_K \Phi\, d\mu + \varepsilon.$$

By the continuity of V there is a number $r < c$ such that

$$K \subset \{V < r\} \subset \{V \leqslant r\} \subset U.$$

This number can be taken with the property that $\mu(x\colon V(x) = r) = 0$. Hence there is a number $r_1 < r$ such that $K \subset \{V < r_1\}$ and

$$\int_{\{r_1 \leqslant V \leqslant r\}} |\langle \nabla V, b\rangle|\, d\mu < \varepsilon.$$

Let us take a function $\varphi \in C^2(\mathbb{R})$ such that $\varphi(t) = t$ if $t \leqslant r_1$, $\varphi(t) = (r_1 + r)/2$ if $t \geqslant r$, $0 \leqslant \varphi' \leqslant 1$, and $\varphi'' \leqslant 0$. Then

$$L_{A,b}(\varphi \circ V) = \varphi''(V)\langle A\nabla V, \nabla V\rangle + \varphi'(V)L_{A,b}V,$$

which vanishes outside the set $\{V \leqslant r\}$ (which is compact by the boundedness of U), because $\varphi \circ V$ is constant outside $\{V \leqslant r\}$, and on $\{V \leqslant r_1\}$ we have the identity $L_{A,b}(\varphi \circ V) = L_{A,b}V$. Since $\varphi \circ V \in C^2(U)$ and $\varphi \circ V$ is constant outside the set $\{V \leqslant r\}$, by approximating $\varphi \circ V$ by functions from $C_0(U)$ uniformly with the first and second order derivatives we have

$$\int_U L_{A,b}(\varphi \circ V)\, d\mu = 0.$$

Taking into account that $\varphi''(V) \leqslant 0$ and $0 \leqslant \varphi'(V) \leqslant 1$, we obtain that

$$\begin{aligned}\int_{\{V\leqslant r_1\}} [\Psi - \Phi]\, d\mu \geqslant \int_{\{V\leqslant r_1\}} L(\varphi \circ V)\, d\mu &= -\int_{\{r_1<V\leqslant r\}} L(\varphi \circ V)\, d\mu \\ &\geqslant -\int_{\{r_1<V\leqslant r\}} |\langle \nabla V, b\rangle|\, d\mu \geqslant -\varepsilon,\end{aligned}$$

whence

$$\int_{\{V\leqslant r_1\}} \Phi\, d\mu \leqslant \int_{\{V\leqslant r_1\}} \Psi\, d\mu + \varepsilon.$$

Since $\varepsilon > 0$ was arbitrary and the numbers r and r_1 can be chosen as close to c as we like, the desired estimate is proven. The assertion concerning the whole space follows at once since we can take increasing sets $\{V < c_k\}$ whose union is $\mathbb{R}^d$.

A similar reasoning applies in case $V \in W^{d,2}_{\text{loc}}(\mathbb{R}^d)$ if it is known additionally that $m \cdot \mathrm{I} \leqslant A \leqslant M \cdot \mathrm{I}$ on each ball and $b^i \in L^d_{\text{loc}}(\mathbb{R}^d)$. The fact that the integral of $L_{A,b}\varphi\, \varrho$ vanishes follows by the fact that by the Sobolev embedding theorem $\partial_{x_i} V \in L^r_{\text{loc}}(\mathbb{R}^d)$ for all numbers $r < \infty$, and nondegeneracy of A yields the inclusion $\varrho \in L^{d/(d-1)}_{\text{loc}}(\mathbb{R}^d)$ for the density ϱ of the measure μ, as shown in § 1.5. □

We emphasize that in the main part of this theorem nothing is required from A and b except for the requirements in the definition of solutions (i.e., the local μ-integrability). It should be also observed that for signed solutions the conclusion of this theorem is false (see Example 4.1.3).

2.3.3. Corollary. *Let μ be a locally bounded nonnegative measure on $\mathbb{R}^d$ satisfying the equation $L^*_{A,b}\mu = 0$ and let $V \in C^2(\mathbb{R}^d)$ be a quasi-compact function such that there are numbers C and R with*

(2.3.3) $$L_{A,b}V(x) \leqslant -C \quad \textit{if } |x| > R.$$

Then μ is bounded and

(2.3.4) $$\mu(x\colon\ |x| > R) \leqslant C^{-1} \int_{|x|\leqslant R} |L_{A,b}V|\, d\mu < \infty.$$

The assumption that $V \in C^2(\mathbb{R}^d)$ can be replaced by $V \in W^{d,2}_{\text{loc}}(\mathbb{R}^d)$ if $b^i \in L^d_{\text{loc}}(\mathbb{R}^d)$ and for each ball U in $\mathbb{R}^d$ there are $m = m(U) > 0$ and $M = M(U) > 0$ such that $m \cdot \mathrm{I} \leqslant A(x) \leqslant M \cdot \mathrm{I}$ whenever $x \in U$.

PROOF. Estimate (2.3.2) applies with $\Phi = |L_{A,b}V| I_U$ and $\Psi = C I_{\mathbb{R}^d\setminus U}$, where we set $U = \{x\colon\ |x| \leqslant R\}$. □

2.3.4. Corollary. *Suppose that a probability measure μ on $\mathbb{R}^d$ satisfies the equation $L^*_{A,b}\mu = 0$, where $M := \sup_x \|A(x)\| < \infty$. Suppose also that there are numbers $p \in \mathbb{N}$ and $R > 0$ such that*

$$\langle b(x), x\rangle \leqslant -1 - M(2p+d) \quad \text{if } |x| > R. \tag{2.3.5}$$

Then

$$\int_{\mathbb{R}^d} \langle x, x\rangle^p \,\mu(dx) \leqslant \int_{\{|x|\leqslant R\}} \langle x, x\rangle^p \Big[(2p+d)M + 1 + \big|\langle x, b(x)\rangle\big|\Big]\,\mu(dx). \tag{2.3.6}$$

PROOF. Letting $V(x) = 2^{-1}(p+1)^{-1}\langle x, x\rangle^{p+1}$, we have

$$\begin{aligned} L_{A,b}V(x) &= 2p\langle x, x\rangle^{p+1} \sum_{i,j=1}^d a^{ij}(x)x_i x_j + \langle x, x\rangle^p \operatorname{tr} A(x) + \langle x, x\rangle^p \langle x, b(x)\rangle \\ &\leqslant \langle x, x\rangle^p \Big[2p\|A(x)\| + d\|A(x)\| + \langle x, b(x)\rangle\Big], \end{aligned}$$

which is estimated by $-\langle x, x\rangle^p$ whenever $|x| > R$. In addition, $|L_{A,b}V(x)|$ does not exceed $\langle x, x\rangle^p\big[(2p+d)M + |\langle x, b(x)\rangle|\big]$, so that (2.3.2) applies with $\Phi(x) = \langle x, x\rangle^p$, $\Psi(x) = \langle x, x\rangle^p\big[(2p+d)M + 1 + |\langle x, b(x)\rangle|\big] I_{\{|x|\leqslant R\}}$. □

The following natural question arises: when does a compact function V exist such that $L_{A,b}V(x) \leqslant -C$ for sufficiently large $|x|$?

Let us consider several related examples.

Constructing radial Lyapunov functions. Set

$$L_{I,b}u = \Delta u + \langle b, \nabla u\rangle.$$

Suppose that there is a function $\beta \in L^\infty_{\mathrm{loc}}\big((0, +\infty)\big)$ such that

$$\frac{d - 1 + \langle x, b(x)\rangle}{|x|} \leqslant \beta(|x|) \quad \text{whenever } |x| > r_0 > 0. \tag{2.3.7}$$

Let us explain how to find a function $W \in W^{\infty,2}_{\mathrm{loc}}\big([r_0, +\infty)\big)$ and a number $\gamma > 0$ such that $W \geqslant 0$, $W' \geqslant 0$, $\lim\limits_{r\to+\infty} W(r) = +\infty$ and

$$W''(r) + \beta(r)W'(r) = -\gamma \quad \text{whenever} \quad r > r_0.$$

Set

$$h(r) = \int_{r_0}^r \exp\Big(\int_{r_0}^t \beta(s)\,ds\Big)\,dt.$$

Then

$$W'(r) = \frac{C_1 - \gamma h(r)}{h'(r)} \quad \text{and} \quad W(r) = C_2 + \int_{r_0}^r \frac{C_1 - \gamma h(t)}{h'(t)}\,dt.$$

Let $h(+\infty) = h_\infty < \infty$, $C_2 = 0$, $C_1 = 1$ and $\gamma = 1/(2h_\infty)$. Then

$$W(r) \geqslant \frac{1}{2}\int_{r_0}^r \frac{1}{h'(t)}\,dt.$$

It remains to observe that the condition $h(+\infty) < \infty$ implies the equality

$$\int_{r_0}^{+\infty} \frac{1}{h'(t)}\,dt = +\infty.$$

Since $L_{I,b}W(|x|) = W''(|x|) + |x|^{-1}W'(|x|)\big(\langle b(x), x\rangle + d - 1\big)$, we arrive at the following assertion.

2.3.5. Theorem. *Suppose that there is a function $\beta \in L^\infty_{\rm loc}((0,+\infty))$ such that (2.3.7) holds and the function h defined above satisfies the condition $h(+\infty) < +\infty$. Then the function V defined by the equality*

$$V(x) = h(+\infty)W(|x|) = \int_{r_0}^{|x|} \frac{2h(+\infty) - h(t)}{h'(t)}\, dt \qquad \text{if } |x| > r_0,$$

belongs to $W^{\infty,2}_{\rm loc}(\mathbb{R}^d)$ and is a weak Lyapunov function for the operator $L_{\rm I,b}$, i.e., $\lim\limits_{|x|\to+\infty} V(x) = +\infty$ and $L_{\rm I,b}V(x) \leqslant -1$ if $|x| > r_0$.

For example, if $\langle b(x), x\rangle = -(1 + 2\mathrm{sign}\, \sin|x|)|x|$, then

$$\beta(r) = \frac{d-1}{r} - (1 + 2\mathrm{sign}\, \sin r).$$

The condition $h(+\infty) < \infty$ in this case takes the form

$$\int_{r_0}^{+\infty} r^{d-1} \exp\Bigl(-\int_{r_0}^{r} (1 + 2\mathrm{sign}\, \sin s)\, ds\Bigr)\, dr < +\infty$$

and is obviously fulfilled. This theorem shows that under the assumptions of Theorem 3 in Noarov [**747**], where $A = \mathrm{I}$, $b \in C^\infty$, and the function

$$R^{d-1} \exp \int_0^R t^{-1} \max_{|x|=t} \langle b(x), x\rangle\, dt$$

is integrable on $(0, +\infty)$, there is a radial Lyapunov function. Here one can take, for example, the function $\beta(t) = \sup_{|x|=t}\langle b(x), x\rangle$.

Dissipativity outside a strip. It is well known that in the case of $L_{\rm I,0} = \Delta$ there is no Lyapunov function. Thus, even if $A = \mathrm{I}$ and b is globally bounded, there might be no Lyapunov functions. However, it turns out that if outside a strip (sufficiently narrow) b is dissipative and is bounded on the strip itself, then one can find a Lyapunov function. Let us consider for simplicity the case $d = 2$. Let $b(x,y) = \bigl(b^1(x,y), b^2(x,y)\bigr)$ be such that $|b| \leqslant 1$ and

$$b^1(x,y)x + b^2(x,y)y \leqslant -\sqrt{x^2+y^2} \qquad \text{whenever } |y| > \alpha.$$

We find $\alpha > 0$ such that the operator $L_{\rm I,b}$ possesses a Lyapunov function. Set $V(x,y) = x^2\psi(y) + \gamma(x^2+y^2)^{3/2}$, where the function $\psi \geqslant 0$ is such that

$$\psi''(0) + b^2(x,0)\psi'(0) = -2, \quad \psi''(y) + b^2(x,y)\psi'(y) \leqslant 2 - \delta, \quad \delta > 0.$$

Set $\psi(y) = \pi/2 - \mathrm{arctg}\,(y^2)$. We have

$$\psi''(y) + b^2(x,y)\psi'(y) = 2\Bigl(-\frac{1}{1+y^2} + \frac{2y^2}{(1+y^2)^2} - \frac{b^2(x,y)y}{1+y^2}\Bigr).$$

It is clear that the first condition on ψ is fulfilled. Since $|b| \leqslant 1$, we have

$$\psi''(y) + b^2(x,y)\psi'(y) \leqslant 2\Bigl(1 - \frac{1}{1+y^2}\Bigr),$$

so the second condition on ψ is fulfilled. Let $2 - \delta < 3\gamma < 2$. Then

$$\begin{aligned} L_{\rm I,b}V(x,y) = x^2\bigl(\psi''(y) + b^2(x,y)\psi'(y)\bigr) + 2\psi(y) + 2b^1(x,y)x\psi(y) \\ + 9\gamma\sqrt{x^2+y^2} + 3\gamma\sqrt{x^2+y^2}\bigl(b^1(x,y)x + b^2(x,y)y\bigr). \end{aligned}$$

Let $\alpha > 0$ be such that $\psi''(y) + b^2(x,y)\psi'(y) < -3\gamma$ whenever $|y| \leqslant \alpha$. Then $L_{I,b}V(x,y) \leqslant c_1 - c_2x^2$ if $|y| \leqslant \alpha$ for all sufficiently large $|x|$,

$$L_{I,b}V(x,y) \leqslant c_3 - c_4(x^2+y^2) \quad \text{if } |y| > \alpha$$

for all sufficiently large x^2+y^2. Thus, a Lyapunov function is constructed. Therefore, the existence theorem in the paper Noarov [**749**] under the stated condition of dissipativity outside a strip, where this situation was first considered, is also covered by the general theorem with a Lyapunov function.

The absence of Lyapunov functions under existence of a unique probability solution. We now show that the existence of a quasi-compact Lyapunov function is not necessary for the existence of a unique probability solution. Let $d = 1$, $A = \mathrm{I}$ and

$$b(x) = -x + e^{x^2/2}.$$

Then $\varrho(x) = e^{-x^2/2}$ is the unique probability solution of the obtained equation $(\varrho' - b\varrho)' = 0$. However, there is no function $V \in W^{1,2}(\mathbb{R})$ such that $V \geqslant 0$, $\lim_{|x|\to\infty} V(x) = +\infty$ and for all sufficiently large x the inequality $L_{1,b}V(x) \leqslant -\gamma$ holds with some number $\gamma > 0$. Indeed, suppose that for $|x| > R > 0$ we have $V''(x) + b(x)V'(x) \leqslant -\gamma$. Let $x > R$ and

$$h(x) = \int_R^x \exp\Bigl(\int_0^s b(t)\,dt\Bigr)\,ds.$$

Then for $x > R$ we have

$$V'(x) \leqslant \bigl(c - \gamma h(x)\bigr)/h'(x), \quad c = V'(R)h'(R),$$

which follows by the inequality $(V'h')' \leqslant \gamma h'$, which holds by the equality $bh' = h''$ and the supposed inequality for V. We observe that $h(+\infty) = +\infty$, because $b(x) \geqslant \delta e^{x^2/4}$ if $x > 0$. Hence there is a number $R_1 > R$ such that for all $x > R_1$ we obtain that $c - \gamma h(x) \leqslant 0$. Therefore, $V'(x) \leqslant 0$, which is impossible for a quasi-compact function.

A quasi-compact function V of class C^2 (or $W^{1,2}_{\mathrm{loc}}(\mathbb{R}^d)$) for which

$$L_{A,b}V \leqslant -C$$

outside a ball will be called a *weak Lyapunov function*; if we have

$$\lim_{|x|\to\infty} L_{A,b}V(x) = -\infty,$$

then V will be called a *strong Lyapunov function.*

It is easy to give an example of a weak Lyapunov function that is not strong. For example, let $A = 1$, $b(x) = -x^3$, and $V(x) = |x|^{-1}$ if $|x| \geqslant 1$. We are unaware of examples of operators for which there is a weak Lyapunov function, but there is no strong one. If the diffusion coefficient A is bounded and an unbounded weak Lyapunov function V is Lipschitzian, then V^2 is a strong Lyapunov function, since

$$L_{A,b}(V^2) = 2VL_{A,b}V + 2\langle A\nabla V, \nabla V\rangle \leqslant -V|L_{A,b}V|$$

outside a ball. See also Proposition 5.3.9 and Example 5.3.10(ii).

We observe that in the presence of a strong Lyapunov function the solutions possess certain additional useful properties (see Chapters 4 and 5). For example, if $\mu \in \mathcal{P}(\mathbb{R}^d)$ and $L^*_{A,b}\mu = 0$, then the existence of a strong Lyapunov function implies the existence and uniqueness of a Markov strongly continuous semigroup $\{T_t\}_{t\geqslant 0}$

on $L^1(\mu)$ with the generator extending $L_{A,b}$ such that the measure μ is invariant for it (Proposition 5.3.9 gives the converse under broad assumptions).

2.4. Construction of solutions to stationary Fokker–Planck–Kolmogorov equations

Here we obtain sufficient conditions for the existence of positive and probability solutions μ on the whole space $\mathbb{R}^d$ to the stationary Fokker–Planck–Kolmogorov equation $L^*_{A,b}\mu = 0$. The case where the coefficient c is not zero is discussed in § 2.5(i). We present sufficient conditions for the existence of solutions expressed in terms of Lyapunov functions, hence verifiable explicitly in terms of the coefficients of our operators. We consider two methods of proving existence: the first one gives nonnegative global solutions under rather general assumptions without Lyapunov functions and then provides a probability solution in the presence of a Lyapunov function, while the second method (which was historically the first) employs Lyapunov functions from the very beginning; this second method is presented in exercises.

Throughout this section $A(x) = \bigl(a^{ij}(x)\bigr)_{1\leqslant i,j\leqslant d}$ is a nonnegative symmetric matrix for each x.

Let $U_R = \{x \in \mathbb{R}^d\colon |x| \leqslant R\}$. We recall that $W^{d+,1}_{\rm loc}(\mathbb{R}^d)$ denotes the class of all functions on $\mathbb{R}^d$ whose restrictions to the balls U_R belong to $W^{p,1}(U_R)$ for some order $p = p_R > d$. Similarly, $L^{d+}_{\rm loc}(\mathbb{R}^d)$ denotes the class of measurable functions whose restrictions to U_R belong to $L^p(U_R)$ with some $p = p_R > d$.

2.4.1. Theorem. *Let $d \geqslant 2$. Assume that for every ball $U \in \mathbb{R}^d$ there exist numbers $m_U > 0$ and $M_U > 0$ such that*

$$m_U \cdot \mathrm{I} \leqslant A(x) \leqslant M_U \cdot \mathrm{I} \quad \forall x \in U. \tag{2.4.1}$$

(i) *Suppose that $a^{ij} \in W^{d+,1}_{\rm loc}(\mathbb{R}^d)$ and $b^i \in L^{d+}_{\rm loc}(\mathbb{R}^d)$. Then there exists a positive solution $\varrho \in W^{d+,1}_{\rm loc}(\mathbb{R}^d)$ to the equation $L^*_{A,b}(\varrho\, dx) = 0$. Moreover, if $a^{ij} \in W^{p,1}_{\rm loc}(\mathbb{R}^d)$, $b^i \in L^p_{\rm loc}(\mathbb{R}^d)$ for some $p > d$, then $\varrho \in W^{p,1}_{\rm loc}(\mathbb{R}^d)$.*

(ii) *Suppose that $a^{ij}, b^i \in L^\infty_{\rm loc}(\mathbb{R}^d)$. Then there exists a nonnegative solution $\varrho \in L^{d/(d-1)}_{\rm loc}(\mathbb{R}^d)$ to the equation $L^*_{A,b}(\varrho\, dx) = 0$ such that*

$$\int_{U_1} \varrho\, dx = 1.$$

PROOF. (i) Set $b^i_0 = b^i - \partial_{x_j} a^{ij}$, $L = L_{A,b}$. Then our equation can be written in the divergence form

$$\partial_{x_i}\bigl(a^{ij}\partial_{x_j}\varrho - b^i_0\varrho\bigr) = 0. \tag{2.4.2}$$

By Theorem 2.1.8 with $\gamma = 0$ the maximum principle holds for this equation on every ball, hence our assertion follows from Corollary 2.2.3.

(ii) Let $\omega \in C^\infty_0(\mathbb{R}^d)$, $\omega \geqslant 0$ and $\|\omega\|_{L^1(\mathbb{R}^d)} = 1$. Set $\omega_k(x) = k^d\omega(kx)$ and $a^{ij}_k = a^{ij} * \omega_k$, $b^i_k = b^i * \omega_k$. Then $a^{ij}_k, b^i_k \in C^\infty(\mathbb{R}^d)$. It is readily verified that these coefficients a^{ij}_k satisfy condition (2.4.1) for each k. According to what has been proved above, there exists a nontrivial solution ϱ_k to the equation $L^*_k(\varrho_k\, dx) = 0$, where

$$L_k = a^{ij}_k\partial_{x_i}\partial_{x_j} + b^i_k\partial_{x_i}.$$

Since the coefficients a_k^{ij}, b_k^i are smooth, we have $\varrho_k \in C^\infty(\mathbb{R}^d)$. Multiplying by a suitable constant we may assume that

$$\int_{U_1} \varrho_k \, dx = 1.$$

By Theorem 1.5.6 for every $R > 1$ there exists a number $C_1(R)$ that does not depend on k such that

$$\int_{U_R} \varrho_k \, dx \leqslant C_1(R) \int_{U_1} \varrho_k \, dx = C_1(R).$$

By Remark 1.5.5(ii) there exists a number $C_2(R)$ such that

$$\int_{U_R} \varrho_k^{\frac{d}{d-1}} \, dx \leqslant C_2(R) \int_{U_R} \varrho_k \, dx \leqslant C_1(R)C_2(R).$$

Therefore, passing to a subsequence we may assume that $\{\varrho_k\}$ converges weakly in $L^{d/(d-1)}(U)$ for every ball U to some function ϱ. It is clear that $\varrho \geqslant 0$ and

$$\int_{U_1} \varrho \, dx = 1.$$

We observe that the functions a_k^{ij} and b_k^i converge to a^{ij} and b^i, respectively, in $L^d(U)$ on every ball U. For every function $\varphi \in C_0^\infty(\mathbb{R}^d)$ we have

$$\int_{\mathbb{R}^d} \varrho L\varphi \, dx = \int_{\mathbb{R}^d} (\varrho - \varrho_k) L\varphi \, dx + \int_{\mathbb{R}^d} \varrho_k (L - L_k)\varphi \, dx.$$

By Hölder's inequality

$$\Bigl|\int_{\mathbb{R}^d} \varrho_k (L - L_k)\varphi \, dx\Bigr| \leqslant C_3 \|\varrho_k\|_{L^{d/(d-1)}(U)} \sum_{i,j} \bigl(\|a^{ij} - a_k^{ij}\|_{L^d(U)} + \|b^i - b_k^i\|_{L^d(U)}\bigr),$$

where U is a ball containing the support of φ. Note also that $L\varphi \in L^d(U)$. Therefore, letting $k \to \infty$, we obtain the equality

$$\int_{\mathbb{R}^d} \varrho L\varphi \, dx = 0,$$

which completes the proof. □

We observe that in the case $d = 1$ the existence of a positive solution to the equation $L_{A,b}^*(\varrho \, dx) = 0$ follows easily from condition (2.4.1) along with the inclusion $b \in L_{\mathrm{loc}}^1(\mathbb{R})$. Indeed, the function

$$\varrho(x) = a(x)^{-1} \exp\biggl(\int_0^x \frac{b(s)}{a(s)} \, ds\biggr)$$

is such a solution. This is why in the theorem we considered only the case $d \geqslant 2$.

2.4.2. Corollary. *If in the situation of Theorem* 2.4.1 *there are a quasi-compact function* $V \in W_{\mathrm{loc}}^{d,2}(\mathbb{R}^d)$ *and numbers* $C > 0$ *and* $R > 0$ *such that*

$$L_{A,b}V(x) \leqslant -C \quad \text{if } |x| > R, \tag{2.4.3}$$

then there is a probability density ϱ *satisfying the equation* $L_{A,b}^*(\varrho \, dx) = 0$.

PROOF. A positive solution exists by Theorem 2.4.1. By Corollary 2.3.3 every positive solution ϱ belongs to $L^1(\mathbb{R}^d)$; after normalization this solution will be a probability measure. □

2.4.3. Corollary. *The assertion of the previous corollary is true if A is a locally Lipschitzian uniformly bounded mapping to the space of positive operators on $\mathbb{R}^d$ and $b = (b^i)$ is a mapping such that $b^i \in L^{d+}_{\rm loc}(\mathbb{R}^d)$ and there exist positive numbers R and C such that*

$$\langle b(x), x\rangle \leqslant -C - \sup_y \operatorname{tr} A(y) \quad \text{if} \quad |x| \geqslant R.$$

PROOF. It suffices to take $V(x) = \langle x, x\rangle$. □

For example, this corollary obviously applies to the situation where $A = \mathrm{I}$ and $b(x) = -k(x)x$, where $k \in L^{d+}_{\rm loc}(\mathbb{R}^d)$ is such that for some $C > 0$

$$k(x)|x|^2 \geqslant d + C, \quad \text{whenever} \quad |x| \geqslant R.$$

More generally, if $A = \mathrm{I}$, then it suffices to have a weaker relation

$$\limsup_{|x|\to\infty} \langle x, x\rangle^{\gamma-1}\Bigl[2(\gamma-1) + d + \langle b(x), x\rangle\Bigr] = -\infty$$

for some $\gamma \geqslant 1$ (then the function $V(x) = \langle x, x\rangle^\gamma$ can be used).

2.4.4. Corollary. *Let $A = (a^{ij})$ be a continuous mapping on $\mathbb{R}^d$ with values in the space of nonnegative symmetric linear operators on $\mathbb{R}^d$ and let b be a Borel vector field on $\mathbb{R}^d$. Suppose that there exists a quasi-compact function $V \in C^2(\mathbb{R}^d)$ with (2.4.3). Then the following assertions are true.*

(i) *If b is continuous, then there exists a probability measure μ satisfying the equation $L^*_{A,b}\mu = 0$.*

(ii) *If $\det A > 0$ and b is locally bounded, then there exists a probability measure μ that has a density of class $L^{d/(d-1)}_{\rm loc}(\mathbb{R}^d)$ and satisfies the equation $L^*_{A,b}\mu = 0$.*

PROOF. (i) It is readily seen that there exists a sequence of smooth mappings $A_k = (a^{ij}_k)$ with values in the space of nondegenerate nonnegative symmetric operators on $\mathbb{R}^d$ converging to A uniformly on compact sets such that for all sufficiently large k we have

$$L_{A_k,b}V(x) \leqslant -C/2 \quad \text{if } |x| > R.$$

To this end, we find smooth nondegenerate mappings A_k such that the following estimate holds for each $m \in \mathbb{N}$:

$$\sup_{x\in U_m} |a^{ij}(x) - a^{ij}_k(x)| \leqslant k^{-1}\Bigl(\sup_{x\in U_m} |\partial_{x_i}\partial_{x_j} V(x)| + 1\Bigr)^{-1}.$$

Then $\lim\limits_{k\to\infty} \sup_x |a^{ij}(x)\partial_{x_i}\partial_{x_j}V(x) - a^{ij}_k(x)\partial_{x_i}\partial_{x_j}V(x)| = 0$. There are probability measures μ_k with positive continuous densities ϱ_k solving the equations

$$L^*_{A_k,b}\mu_k = 0.$$

We renormalize these solutions in order to have the equality $\mu_k(U) = 1$, where U is the ball of radius R centered at the origin. According to Theorem 2.3.2 the remormalized sequence $\{\mu_k\}$ is still bounded, since the functions $L_{A_k,b}V$ are uniformly bounded on the ball U. Hence it contains a subsequence that converges weakly on every ball in $\mathbb{R}^n$. The measure μ obtained in the limit is bounded and nonnegative, but is not identically zero, because $\mu(U) = 1$. Since the coefficients are continuous we obtain the equality $L^*_{A,b}\mu = 0$. It is remains to renormalize μ to get to a probability measure on $\mathbb{R}^d$. Case (ii) follows by assertion (ii) in the theorem. □

2.4.5. Remark. It is obvious that in the case of nondegenerate A the Borel measurability of b can be replaced by its Lebesgue measurability and that our solution does not depend on a Lebesgue equivalent version of b, because any solution has a density.

Let us observe that (2.4.3) holds if there are positive numbers c_1 and c_2 such that

$$LV(x) \leqslant c_1 - c_2 V(x) \quad \text{and} \quad \lim_{|x|\to\infty} V(x) = +\infty.$$

It should be noted that for $A = I$ the estimate $\langle x, b(x)\rangle \leqslant -c < -d$ outside some ball is sufficient, but cannot be replaced by the condition that $\lim\limits_{|x|\to\infty} \langle x, b(x)\rangle = -d$. Indeed, let $d = 1$, $A = 1$, and $b(x) = -\operatorname{sign} x/(1+|x|)$. Then $xb(x) \to -1$ as $|x| \to \infty$. Assume that μ is a probability measure satisfying the equation $L^*_{1,b}\mu = 0$. Then μ has a locally absolutely continuous density ϱ such that $\varrho'' + (b\varrho)' = 0$ in the sense of distributions, whence we have $\varrho' + b\varrho = 0$, since $b\varrho$ is integrable (it suffices to integrate over the intervals $[t_k, s_k]$, where $t_k \to -\infty$, $s_k \to +\infty$, $\varrho(t_k) \to 0$, and $\varrho(s_k) \to 0$). In addition, $\varrho > 0$, so that $(\ln \varrho)' = -\ln(1+|x|)'$, whence we obtain that $\varrho(x) = c/(1+|x|)$, which is a nonintegrable function.

It is worth noting that in the survey Bogachev, Krylov, Röckner [**155**] after Remark 1.4.4 it is mistakenly claimed that (2.4.3) is not sufficient for existence of a probability solution, but the example given there (which is the above example) is actually related not to (2.4.3), but to the inequality $\langle b(x), x\rangle \leqslant c < 0$.

It is clear that the presented existence results can be modified for divergence form operators. For example, if the functions a^{ij} are Lipschitzian and

$$\langle b(x), x\rangle \leqslant c_1 - c_2|x| \ln|x|$$

outside of some ball, then the extra term in the drift which appears when we write the equation in nondivergence form does not destroy condition (2.4.3).

Note also that in the case where b is a gradient, more special sufficient conditions are available that guarantee the existence of a probability measure whose logarithmic gradient is b (see Exercise 2.5.25 and Bogachev [**126**, § 7.5]), but even in this case the general sufficient conditions presented above turn out to be very efficient.

In Exercise 2.5.24 we present a slightly different method of constructing solutions under a stronger condition on the Lyapunov function. This method combines in a sense the reasoning from the previous two theorems.

2.5. Complements, comments, and exercises

2.5(i). Solvability of equations with potentials

The next theorem gives a sufficient condition for the existence and the global integrability of positive solutions in the case $c \neq 0$; thus we obtain a sufficient condition for the existence of a probability solution. In this case the global integrability is connected not with estimates for $L_{A,b,c}V$, but rather with the behaviour of the potential c.

2.5.1. Theorem. *Let $a^{ij} \in W^{p,1}_{loc}(\mathbb{R}^d)$, where $p > d$. Assume that the continuous version of A is nondegenerate, $b^i \in L^p_{loc}(\mathbb{R}^d)$, $c \in L^{pd/(p+d)}_{loc}(\mathbb{R}^d)$, $V \in C^2(\mathbb{R}^d)$, and also $\lim\limits_{|x|\to+\infty} V(x) = +\infty$. Suppose that*

$$L_{A,b,c}V(x) \leqslant 0 \quad \forall x \in \mathbb{R}^d$$

and $c(x) \geqslant c_0 > 0$ when $|x| \geqslant R$ for some $R > 0$. Then a positive solution of the equation $L^*_{A,b,c}(\varrho\,dx) = 0$ exists and every positive solution belongs to $L^1(\mathbb{R}^d)$. Hence there exists a probability solution.

PROOF. The existence of a positive solution follows from Corollary 2.2.3. Let ϱ be a positive solution of the equation $L^*_{A,b,c}(\varrho dx)=0$. Let $\zeta \in C^2(\mathbb{R})$, $\zeta(0) = 1$, $\zeta(t) = 0$ if $t \geqslant 1$, $\zeta' \leqslant 0$ and $\zeta'' \geqslant 0$. Set $\zeta_N(x) = \zeta\bigl(V(x)/N\bigr)$. Then the function ζ_N has compact support and

$$\begin{aligned} L_{A,b,c}\zeta_N(V) = N^{-1}\zeta'(V/N)L_{A,b,c}V + N^{-2}\zeta''(V/N)\langle A\nabla V, \nabla V\rangle \\ + c\bigl(\zeta_N - N^{-1}\zeta'(V/N)V\bigr). \end{aligned}$$

We can assume (increasing R) that $V(x) > 0$ if $|x| \geqslant R$. Since $\zeta' \leqslant 0$, $\zeta'' \geqslant 0$ and $L_{A,b,c}V(x) \leqslant 0$ and $c(x) > 0$ whenever $|x| \geqslant R$, for such x we obtain the inequality

$$L_{A,b,c}\zeta_N(x) \geqslant c(x)\zeta_N(x).$$

It follows from the equation that the integral of $L_{A,b,c}\zeta_N\varrho$ vanishes, so

$$\int_{|x|\geqslant R} c\zeta_N\varrho\,dx \leqslant -\int_{|x|<R} L_{A,b,c}\zeta_N\varrho\,dx.$$

Therefore,

$$c_0\int_{|x|\geqslant R} \zeta_N\varrho\,dx \leqslant -\int_{|x|<R} L_{A,b,c}\zeta_N\varrho\,dx.$$

Since the right-hand side tends to the integral of $c\varrho$ over the ball $\{|x| < R\}$ and $\zeta_N \to \zeta(0) = 1$ as $N \to \infty$, we have $\varrho \in L^1(\mathbb{R}^d)$. □

In this theorem the condition $c(x) \geqslant c_0 > 0$ cannot be replaced with the condition $c(x) \geqslant 0$ or the condition $c(x) \leqslant c_0 < 0$. For example $\varrho(x) = e^x$ satisfies the equation $\varrho'' - \varrho = 0$, where $c = -1$ and $LV(x) < 0$ with the function $V(x) = |x|$ if $|x| \geqslant 1$.

As an example of application of this theorem we consider the following situation. Let $A = \mathrm{I}$, $b(x) = -x$, $V(x) = |x|^2 + 1$, $c(x) \leqslant (2|x|^2 - 2d)/(|x|^2 + 1)$, and, in addition, $c(x) \geqslant 1$ outside some ball. Then there exists a probability solution of the equation $L^*_{\mathrm{I},b,c}\mu = 0$.

2.5(ii). Solvability of equations on manifolds

Most of the results in § 2.4 with the same justifications extend to the case of a Riemannian manifold M of dimension d and an elliptic operator L on M that in local coordinates satisfies the conditions in Theorem 2.4.1. For example, the following fact is true.

2.5.2. Theorem. *Suppose that there exists a function $V \in C^2(M)$ such that $LV \leqslant -C < 0$ outside some compact set and for every N there exists a compact set K_N such that $V \geqslant N$ outside K_N. Then the equation $L^*\mu = 0$ has a probability solution.*

In the manifold case the existence of Lyapunov functions is not always easy to verify. Let us mention some sufficient conditions for the existence of infinitesimally invariant measures which can be easier to verify in applications. For the proofs, based on similar ideas, see Bogachev, Röckner, Wang [**179**]. Let M be a smooth complete connected Riemannian manifold of dimension d. Let λ_M denote the Riemannian volume on M. The Ricci tensor will be denoted by Ric.

Let $o \in M$ be a fixed point and let $\varrho(x) := \varrho(x, o)$ be the Riemannian distance. The cut-locus set $\mathrm{cut}(o)$ of o consists of all points x with the property that there is a unit vector $V \in T_oM$ such that $t = \varrho(o, \exp[tV])$ precisely when $t \in [0, \varrho(o, x)]$, where exp is the exponential mapping (see Kobayashi, Nomizu [**524**], Gallot, Hulin, Lafontaine [**397**]).

Let us consider the equation $L_Z^*\mu = 0$ on M for $L_Z\varphi = \Delta + \langle Z, \nabla\varphi\rangle$, where Δ is the Laplace–Beltrami operator on M, i.e., $\Delta\varphi = \mathrm{div}\,\nabla\varphi$.

2.5.3. Theorem. *Suppose that a measurable vector field Z on M is such that $|Z| \in L^\alpha_{\mathrm{loc}}(\lambda_M)$, where $\alpha > d$, and there is a function $F \in C^2[0,\infty)$ such that*

$$\lim_{r\to\infty} F(r) = \infty \quad \text{and} \quad \lim_{r\to\infty} \sup_{\{\varrho\geqslant r\}\backslash \mathrm{cut}(o)} [F'(\varrho)L_Z\varrho + F''(\varrho)] = -\infty, \tag{2.5.1}$$

where $\sup\varnothing = -\infty$. *Then there exists a probability measure μ satisfying the equation $L_Z^*\mu = 0$ and having a density $\varrho \in W^{\alpha,1}_{\mathrm{loc}}(\lambda_M)$.*

2.5.4. Example. Let $\mathrm{Ric} \geqslant -k$, where $k \geqslant 0$ is a constant. If $|Z| \in L^\alpha_{\mathrm{loc}}(\lambda_M)$, $\alpha > d$, and $\lim\limits_{r\to\infty} \sup_{\{\varrho\geqslant r\}\backslash \mathrm{cut}(o)} \langle Z, \nabla\varrho\rangle < -\sqrt{k(d-1)}$, then the conclusion of Theorem 2.5.3 is true. It suffices to take $F(r) = r^2$ and observe that

$$\lim_{r\to\infty} \sup_{\{\varrho\geqslant r\}\backslash \mathrm{cut}(o)} \Delta\varrho \leqslant \sqrt{k(d-1)}.$$

Let us give a sufficient condition for the existence of solutions in terms of the curvature. For a vector field Z of class C^1 we set

$$k(r) = \inf_{\{\varrho\geqslant r\}\backslash \mathrm{cut}(o)} \Bigl\{\mathrm{Ric}(\nabla\varrho, \nabla\varrho) - \langle \nabla_{\nabla\varrho} Z, \nabla\varrho\rangle\Bigr\}.$$

Let us also introduce the function

$$\overline{k}(r) = \inf_{\{\varrho\geqslant r\}\backslash \mathrm{cut}(o)} \Bigl\{-\langle \nabla_{\nabla\varrho} Z, \nabla\varrho\rangle\Bigr\}.$$

2.5.5. Theorem. *Suppose that the vector field Z of class C^1 is such that*

$$\int_0^\infty k(r)\,dr = \infty. \tag{2.5.2}$$

Then there exists a probability solution μ of the equation $L_Z^*\mu = 0$. If the Ricci curvature is bounded from below, then this is true if (2.5.2) *is fulfilled for $\overline{k}$ in place of k.*

Certainly, for manifolds many distinctions appear. For example, the Laplace equation $\Delta\mu = 0$ on $\mathbb{R}^d$ has no nonconstant positive solutions and no nonzero integrable solutions. There are connected noncompact manifolds on which such solutions exist, see Kuz'menko, Molĉanov [**575**], Chung [**248**], Li, Schoen [**605**], and Grigoryan [**423**]. In the case of a disconnected manifold it can happen that there is a probability solution in one component without nonzero integrable solutions in some other component.

Let us mention a result of the paper Bogachev, Röckner, Zhang [**182**] on the existence of probability solutions of the equation $\mathcal{L}^*_{A,b}\nu = 0$ on M in terms of perturbations of symmetric operators. Let μ be a probability measure on M with a positive density $\varrho_\mu \in W^{p,1}_{\mathrm{loc}}(M)$, $p > d$. Let a Borel vector field v on M be such that $|v| \in L^2(\mu)$ and there exist numbers $\varepsilon \in (0,1)$, $C_1 > 0$, and $C_2 > 0$ such that

$$\left|\int_M \langle v, \nabla f\rangle g\, d\mu\right| \leqslant C_2 \|f\|_{W^{2,1}(\mu)} \|g\|_{W^{2,1}(\mu)}, \quad f, g \in C_0^\infty(M),$$

$$\int_M \langle v, \nabla f\rangle f\, d\mu \geqslant -\varepsilon \int_M |\nabla f|^2\, d\mu - C_1 \int_M f^2\, d\mu, \quad f \in C_0^\infty(M).$$

The indicated conditions are fulfilled if b is bounded and also if μ satisfies the logarithmic Sobolev inequality (see also (5.6.5))

$$\int_M f^2 \ln\bigl(|f|/\|f\|_{L^2(\mu)}\bigr)\, d\mu \leqslant \lambda \int_M |\nabla f|^2\, d\mu, \quad f \in C_0^\infty(M),$$

and, in addition, $\exp(\alpha|b|^2) \in L^1(\mu)$ for $\alpha > \lambda + 1$.

2.5.6. Theorem. *Let $c_1 \cdot \mathrm{I} \leqslant A \leqslant c_2 \cdot \mathrm{I}$ for some $c_1, c_2 > 0$, and suppose that in local coordinates the matrix elements of A belong to $W^{p,1}_{\mathrm{loc}}$. Suppose that the embedding $W^{2,1}(\mu) \to L^2(\mu)$ is compact. Then there exists a positive function $\psi \in L^2(\mu)$ for which the measure $\nu = \psi \cdot \mu$ satisfies the equation $\mathcal{L}^*_{A,b}\nu = 0$.*

The required hypotheses are fulfilled if μ satisfies the logarithmic Sobolev inequality and $\exp(\alpha|b|^2) \in L^1(\mu)$ for $\alpha > \lambda + 1$.

2.5(iii). The absence of integrable solutions

We now show that the inequality $LV > 0$, opposite to the required one for the existence of probability solutions, under broad assumptions yields the absence of nonzero solutions of bounded variation.

2.5.7. Theorem. *Let $b \in L^p_{\mathrm{loc}}(\mathbb{R}^d)$, where $p > d$, and suppose that there is a positive function $V \in C^2(\mathbb{R}^d)$ such that $V(x) \to +\infty$ as $|x| \to +\infty$ and for some numbers $C > 0$, $R > 0$ and almost all x with $V(x) > R$ we have the inequality $L_{\mathrm{I},b}V(x) > 0$ and $|\nabla V(x)|^2 \leqslant CV(x)$. Then the equation $L^*_{\mathrm{I},b}\mu = 0$ has no nonzero solutions of bounded variation.*

PROOF. Let u be an integrable solution: the measure $\mu = u\,dx$ satisfies the equation $L^*\mu = 0$, $L = L_{\mathrm{I},b}$. According to Exercise 2.5.16, for all nonnegative functions $\varphi \in C_0^\infty(\mathbb{R}^d)$ we have the inequality

$$\int_{\mathbb{R}^d} L\varphi\, d|\mu| \geqslant 0.$$

This inequality remains valid also for all functions $\varphi \geqslant 0$ of class $W^{\infty,2}(\mathbb{R}^d)$ with compact support. Let us define a function z_N for $N > R + 1$ as follows:

$$z_N(t) = 0 \text{ if } t \in (-\infty, R] \cup [2N, +\infty), \quad z_N(t) = t - R \text{ if } t \in (R, R+1),$$

$$z_N(t) = 1 \text{ if } t \in [R+1, N], \quad z_N(t) = N^{-1}(2N - t) \text{ if } t \in (N, 2N).$$

Set

$$\zeta_N(t) = \int_0^t z_N(s)\, ds.$$

Then ζ_N is a nonnegative nondecreasing function and $\zeta_N(t) \equiv \zeta_N(2N)$ for $t > 2N$. Set $\varphi(x) = \zeta_N(2N) - \zeta_N\big(V(x)\big)$. It is readily seen that this function is nonnegative and has compact support. Hence

$$\int_{\mathbb{R}^d} \zeta_N'\big(V(x)\big)LV(x)\,d|\mu| + \int_{\mathbb{R}^d} \zeta_N''\big(V(x)\big)|\nabla V(x)|^2\,d|\mu| \leqslant 0.$$

It is easy to deduce from this estimate and the properties of $\zeta_N' = z_N$ that

$$\int_{R+1<V<N} LV\,d|\mu| \leqslant \frac{1}{N}\int_{N<V\leqslant 2N} |\nabla V|^2\,d|\mu|.$$

By hypothesis $|\nabla V(x)|^2 \leqslant CV(x)$ on the set $\{V > R\}$. Therefore,

$$\int_{R+1<V<N} LV\,d|\mu| \leqslant 2C|\mu|\big(x\colon\ N < V(x) \leqslant 2N\big).$$

Since $LV(x) > 0$ on the set $\{V > R\}$, letting $N \to +\infty$ we obtain that

$$|\mu|\big(x\colon V(x) > R+1\big) = 0.$$

Thus, the solution μ vanishes outside some ball $U(0, R_0)$ and hence satisfies the Dirichlet problem with zero boundary condition on the ball $U(0, R_0)$. This Dirichlet problem has only zero solution. Therefore, the measure μ is identically zero. □

Let us consider the following example: $V(x) = |x|^2/2$. According to the proven theorem, if $\langle b(x), x\rangle > -d$, then our equation has no nonzero integrable solutions. Note that this condition is sharp in the sense that the estimate $\langle b(x), x\rangle \geqslant -d - \varepsilon$, $\varepsilon > 0$ is not sufficient. Indeed, let us consider the measure given by a density ϱ which coincides with the radial function $r^{-d-\varepsilon}$ outside the unit ball, and let $b = \nabla\varrho/\varrho$, $A = \mathrm{I}$. It is readily seen that we have the equality $\langle b(x), x\rangle = -d - \varepsilon$.

Note that the condition $|\nabla V|^2 \leqslant CV$ cannot be omitted, i.e., it is not enough to admit arbitrary functions V with $LV > 0$. Indeed, take $V(x) = |x|^3$. Then $LV > 0$ if b is such that $\langle b(x), x\rangle > -d - 5$, which, as one can see from the previous example with $\varrho = r^{-d-5}$ outside the unit ball, can be fulfilled also in the case where there is a nonzero integrable solution.

2.5(iv). Nonlinear equations

Here we briefly discuss nonlinear elliptic equations for measures. We consider the case where nonlinearities arise due to dependence of coefficients of the linear differentiable operator on the solutions itself. The classical stationary Vlasov equations in plasma theory give an important motivation. Suppose that we are given a set $\mathcal{P}_0$ of probability measures on $\mathbb{R}^d$ and for every measure μ in this set we are given a Borel mapping $A(\mu, \cdot)$ from $\mathbb{R}^d$ to the space of symmetric nonnegative operators on $\mathbb{R}^d$ and also a Borel vector field $b(\mu)$ on $\mathbb{R}^d$. Let $A(\mu, x) = \big(a^{ij}(\mu, x)\big)_{i,j\leqslant d}$, $b(\mu, x) = \big(b^i(\mu, x)\big)_{i\leqslant d}$. We consider the operator

$$L_\mu f(x) := a^{ij}(\mu, x)\partial_{x_j}\partial_{x_i} f(x) + b^i(\mu, x)\partial_{x_i} f(x). \tag{2.5.3}$$

For a fixed measure μ we can solve the linear equation $L_\mu^*\sigma = 0$ with respect to the measure σ. Here we are interested in the nonlinear equation

$$L_\mu^*\mu = 0$$

in the class $\mathcal{P}_0$. Naturally, nonlinearities appear in the case of a nontrivial dependence of the coefficients on the solution μ. For example, if the coefficient A is independent of μ (say, $A = \mathrm{I}$), then a nonlinearity may arise due to dependence of b on μ. An important example of such dependence is given by vector fields arising in the Vlasov equations and having the form of the convolution $b_0 * \mu$, where b_0 is a fixed vector field (see Kozlov [**539**]). Rather difficult problems arise already in the case $d = 1$, which is in sharp contrast with linear equations.

For constructing solutions to nonlinear equations we need the following result from Chapter 1: if a probability measure σ on $\mathbb{R}^d$ satisfies the equation $L^*_{A,b}\sigma = 0$ and for every $R > 0$ there exist positive numbers $\varepsilon(R)$ and $M(R)$ such that on the ball U_R of radius R centered at the origin we have the estimate

$$A(x) \geqslant \varepsilon(R) \cdot \mathrm{I}, \ \|a^{ij}\|_{W^{p,1}(U_R)} \leqslant M(R), \ \|b^i\|_{L^p(U_R)} \leqslant M(R), \tag{2.5.4}$$

where $p > d$ is a fixed number, then for every $R > 0$ there exists a number $K\bigl(R, \varepsilon(R), M(R)\bigr) > 0$ such that the density ϱ of the measure σ satisfies the condition

$$|\varrho(x) - \varrho(y)| \leqslant K\bigl(R, \varepsilon(R), M(R)\bigr)|x - y|^\alpha, \quad x, y \in U_R, \tag{2.5.5}$$

where $\alpha := 1 - d/p$.

Below we assume that we are given some positive functions ε and M on $(0, +\infty)$ for which a function K is chosen according to conditions (2.5.4) and (2.5.5).

Let $\mathcal{P}_0$ denote the class of all probability measures on $\mathbb{R}^d$ possessing densities satisfying (2.5.5).

Suppose that the following conditions are fulfilled:

(H1) for every measure $\mu \in \mathcal{P}_0$ we are given Borel mappings $A(\mu, \cdot)$ and $b(\mu, \cdot)$ on $\mathbb{R}^d$ with values in the spaces of operators on $\mathbb{R}^d$ and $\mathbb{R}^d$ itself, respectively, such that estimate (2.5.4) is fulfilled uniformly in μ;

(H2) if a sequence of measures $\mu_n \in \mathcal{P}_0$ is such that their densities locally uniformly converge to the density of a measure $\mu \in \mathcal{P}_0$, then the functions $a^{ij}(\mu_n, \cdot)$ and $b^i(\mu_n, \cdot)$ converge in L^1 on every ball (with the usual Lebesgue measure) to the functions $a^{ij}(\mu, \cdot)$ and $b^i(\mu, \cdot)$, respectively;

(H3) we are given a Borel function $\Psi \geqslant 0$ on $\mathbb{R}^d$ with $\lim\limits_{|x|\to\infty} \Psi(x) = +\infty$ and a number $C \geqslant 0$ such that, for every measure $\mu \in \mathcal{P}_0$, there exists a quasi-compact function $V_\mu \in C^2(\mathbb{R}^d)$ for which $L_\mu V_\mu(x) \leqslant C - \Psi(x)$.

2.5.8. Theorem. *Let conditions* (H1)–(H3) *be fulfilled. Then there exists a measure $\mu \in \mathcal{P}_0$ such that $L^*_\mu \mu = 0$.*

The proof is based on the following two lemmas.

2.5.9. Lemma. *The set*

$$\mathcal{K} := \biggl\{\mu \in \mathcal{P}_0 \colon \int_{\mathbb{R}^d} \Psi \, d\mu \leqslant C\biggr\}$$

is convex and compact in the Banach space of measures on $\mathbb{R}^d$ with the variation norm.

PROOF. Suppose that we are given a sequence of measures $\mu_n \in \mathcal{K}$. By estimate (2.5.5) it contains a subsequence of measures μ_{n_k} whose densities converge uniformly on compact sets. Since the integrals of Ψ against the measures μ_n are uniformly bounded and $\lim\limits_{|x|\to\infty} \Psi(x) = +\infty$, the whole sequence $\{\mu_n\}$ is uniformly tight. Therefore, passing to a further subsequence, we can assume that the sequence $\{\mu_{n_k}\}$ converges weakly to some measure μ. It follows from the locally uniform convergence of the densities of these measures that the measure μ also possesses a density ϱ and the latter satisfies condition (2.5.5). In addition, it is seen from the results in § 2.3 that the integral of Ψ against the measure μ does not exceed C, i.e., $\mu \in \mathcal{K}$. Since μ_n and μ are probability measures, convergence of densities implies convergence in variation, i.e., convergence of densities in $L^1(\mathbb{R}^d)$, see Bogachev [**125**, Theorem 2.8.9]. □

We now define a mapping $T\colon \mathcal{K} \to \mathcal{K}$ as follows:

$$\nu = T(\mu) \quad \Longleftrightarrow \quad L_\mu^* \nu = 0.$$

The mapping T is well-defined, since under our conditions for every measure μ in the class of probability measures there is a unique solution ν of the indicated equation (the uniqueness is proved in Chapter 5). This solution belongs to $\mathcal{K}$, since it satisfies condition (2.5.5) by (2.5.4) and condition (H1). In addition, by condition (H3) for the operator L_μ, the following estimate holds (see § 2.3):

$$\int_{\mathbb{R}^d} \Psi \, d\nu \leqslant C.$$

2.5.10. Lemma. *The mapping T is continuous when $\mathcal{K}$ is equipped with the variation norm.*

Proof. We show that if measures $\mu_n = \varrho_n dx$, $\mu = \varrho dx$ belong to $\mathcal{K}$ and $\|\mu_n - \mu\| \to 0$, then $\|T(\mu_n) - T(\mu)\| \to 0$. By the compactness of the set $\mathcal{K}$ we can find a subsequence $\{T(\mu_{n_k})\}$ converging in variation to some measure $\nu \in \mathcal{K}$. It is clear from what has been said above that the densities f_k of the measures $T(\mu_{n_k})$ locally uniformly converge to the density f of the measure ν. We show that ν satisfies the equation $L_\mu^*\nu = 0$, which by uniqueness of a solution will give the equality $\nu = T(\mu)$, from which the relation $\|T(\mu_n) - T(\mu)\| \to 0$ follows, since our reasoning is true for every subsequence in $\{T(\mu_n)\}$. For every $\varphi \in C_0^\infty(\mathbb{R}^d)$ we have the identity

$$\int_{\mathbb{R}^d} \bigl[a^{ij}(\mu_{n_k}, x)\partial_{x_j}\partial_{x_i}\varphi f_k(x) + \langle b(\mu_{n_k}, x), \nabla\varphi(x)\rangle f_k(x)\bigr]\, dx = 0.$$

Let S be the support of φ. Since the functions $a^{ij}(\mu_{n_k}, x)$ and $b(\mu_{n_k}, x)$ converge on S in $L^1(S)$ to the functions $a^{ij}(\mu, x)$ and $b(\mu, x)$, respectively, and the functions f_k converge uniformly on S to f, we obtain the following equality:

$$\int_{\mathbb{R}^d} \bigl[a^{ij}(\mu, x)\partial_{x_j}\partial_{x_i}\varphi f(x) + \langle b(\mu, x), \nabla\varphi(x)\rangle f(x)\bigr]\, dx = 0.$$

This means that $L_\mu^*\nu = 0$. Thus, T is continuous. □

Now the application of the Schauder fixed point theorem completes the proof of Theorem 2.5.8.

We proceed to degenerate equations, including the fully degenerate equation $L_\mu^*\mu = 0$ with $A = 0$, i.e., the equation

$$\operatorname{div}\bigl(b(\mu, \cdot)\mu\bigr) = 0,$$

where the divergence is understood in the sense of generalized functions. This equation is called the *continuity equation* (or the *transport equation*, but the latter term is more often used for nondivergence form equations).

Suppose that for every measure $\mu \in \mathcal{P}(\mathbb{R}^d)$ we are given a vector field $b(\mu, \cdot)$ on $\mathbb{R}^d$ and there are constants $C, R > 0$ such that

$$\langle b(\mu, x), x\rangle \leqslant -C \quad \text{if } |x| \geqslant R \tag{2.5.6}$$

for all μ, and for every ball U there holds the inequality

$$\sup_{\mu} \sup_{x\in U} |b(\mu, x)| < \infty. \tag{2.5.7}$$

Suppose also that for every measure $\mu \in \mathcal{P}(\mathbb{R}^d)$ we are given a continuous mapping $A(\mu, \cdot)$ with values in the space of nonnegative symmetric operators on $\mathbb{R}^d$ such that

$$\sup_{x,\mu} \operatorname{tr} A(\mu, x) \leqslant M_0 < C/3. \tag{2.5.8}$$

In addition, suppose that if measures μ_n converge weakly to a measure μ, then the mappings $b(\mu_n, \cdot)$ and $A(\mu_n, \cdot)$ converge to $b(\mu, \cdot)$ and $A(\mu, \cdot)$, respectively, uniformly on compact sets.

Let us now consider the nonlinear elliptic equation with the additional term $\varepsilon\delta_{ij}$ at a^{ij}, where $\varepsilon > 0$, i.e., in place of $A(\mu, \cdot)$ we take the operator $A(\mu, \cdot) + \varepsilon \cdot \mathrm{I}$. Denote the corresponding operator by $L_{\mu,\varepsilon}$; the corresponding equation has the form $L^*_{\mu,\varepsilon}\mu = 0$.

We observe that by (2.5.6)–(2.5.8), taking the function $V(x) = \langle x, x\rangle^2$ for all measures, for some positive constants C_1, C_2, ε_0 simultaneously for all measures μ we obtain the estimate

$$\begin{aligned} L_{\mu,\varepsilon}V(x) = 4\langle x, x\rangle[\operatorname{tr} A(\mu, x) + \varepsilon d] + 8\langle A(\mu, x)x, x\rangle + 8\varepsilon\langle x, x\rangle \\ + 4\langle x, x\rangle\langle b(\mu, x), x\rangle \leqslant C_1 - C_2\langle x, x\rangle, \quad \varepsilon \in (0, \varepsilon_0). \end{aligned}$$

Therefore, for every fixed $\varepsilon \in (0, \varepsilon_0)$ we obtain a probability measure μ_ε satisfying the equation $L^*_{\mu_\varepsilon,\varepsilon}\mu_\varepsilon = 0$. Letting $\mu_n := \mu_{1/n}$, we obtain a sequence of probability measures for which

$$\sup_n \int_{\mathbb{R}^d} \langle x, x\rangle \, \mu_n(dx) < \infty.$$

Hence this sequence is uniformly tight and contains a subsequence converging weakly to some probability measure μ. Without loss of generality we can assume that the whole sequence is weakly convergent. Clearly, under our assumptions, for every function $\varphi \in C_0^\infty(\mathbb{R}^d)$, the equality

$$\int_{\mathbb{R}^d} L_{\mu_n,1/n}\varphi \, d\mu_n = 0$$

admits passage to the limit as $n \to \infty$, which shows that the integral of $L_\mu\varphi$ against the measure μ vanishes. Therefore, we have proved the following result.

2.5.11. Theorem. *Under conditions* (2.5.6)–(2.5.8) *there exists a solution of the equation* $L^*_\mu\mu = 0$.

In particular, under the assumptions made, the first order equation

$$\operatorname{div}\big(b(\mu, \cdot)\mu\big) = 0$$

is solvable.

In place of the whole space of probability measures we could consider only its convex compact part $\mathcal{P}_0$ consisting of all measures μ for which

$$\int_{\mathbb{R}^d} \langle x, x\rangle\, \mu(dx) \leqslant C_1/C_2.$$

Then the same reasoning yields the following more general result.

2.5.12. Theorem. *Let the mappings $A(\mu,\,\cdot\,)$ and $b(\mu,\,\cdot\,)$ be defined only for measures $\mu \in \mathcal{P}_0$ and satisfy the conditions of the previous theorem. Then in the class $\mathcal{P}_0$ there exists a solution of the equation $L_\mu^*\mu = 0$.*

For justification it suffices to observe that for $\varepsilon \leqslant \varepsilon_0$ the solution μ_ε of the equation $L^*_{\mu,\varepsilon}\mu_\varepsilon = 0$ satisfies the estimate

$$\int_{\mathbb{R}^d} \Psi\, d\mu_\varepsilon \leqslant C_1/C_2,$$

which means that $\mu_\varepsilon \in \mathcal{P}_0$.

Let us give some examples of existence of solutions of equations with $A = 0$ and drift coefficients of different types.

2.5.13. Example. Let b have the form $b(x,\mu) = k(\mu)b(x)$, where $k(\mu)$ is a real coefficient and b does not depend on μ and is continuous. Let the coefficient $k(\mu)$ be uniformly separated from zero and bounded, i.e., $0 < c_1 \leqslant k(\mu) \leqslant c_2 < \infty$ for all μ. Suppose also that for some $C, R > 0$ we have the estimate

$$\langle b(x), x\rangle \leqslant -C \quad \text{whenever } |x| \geqslant R.$$

Then the existence of a solution to the continuity equation is ensured if the function k is continuous on the space of probability measures with weak topology.

2.5.14. Example. Let $b(x,\mu) = -x + b_0(x,\mu)$, where the field b_0 is uniformly bounded and depends continuously on the pair (x,μ). Then

$$\langle -x + b_0(x,\mu), x\rangle \leqslant M|x| - \langle x, x\rangle$$

for some $M > 0$, which ensures the required condition and gives the solvability of the continuity equation.

2.5.15. Example. We now consider the field

$$b(x,\mu) = \int_{\mathbb{R}^d} b_0(x,y)\mu(dy),$$

where b_0 is a bounded continuous vector field such that for some $C, R > 0$ we have the estimate $\langle b_0(x,y), x\rangle \leqslant -C$ for all x with $|x| \geqslant R$ and all y. Then there exists a solution of the continuity equation. Unfortunately, this example does not apply to vector fields of the form $b_0(x-y)$ leading to convolutions $b_0 * \mu$ arising in the stationary Vlasov equation (see Kozlov [**539**]).

Some results on the uniqueness of solutions of the considered equations will be presented in § 4.5(iv). In Chapter 6 we discuss nonlinear parabolic equations for measures.

Comments

Maximum principles are discussed in the books Fraenkel [**383**], Kresin, Maz'ya [**542**], Protter, Weinberger [**792**], Pucci, Serrin [**794**], Sperb [**868**]; see also the paper Bony [**195**]. Connections of the maximum principle on a bounded domain with the existence of a solution to the boundary value problem and also with the maximum principle for the adjoint operator were studied in the papers Chicco [**239**] (divergence form operators), Peetre, Rus [**769**] (the case of smooth coefficients), Bogachev, Kirillov, Shaposhnikov [**145**].

The method of Lyapunov functions is a classical tool for the proof of existence of solutions in diverse problems in the theory of ordinary differential equations and partial differential equations, and also in the theory of dynamic systems. Its application to constructing diffusions and their stationary distributions and in this way to the proof of solvability of the corresponding Fokker–Planck–Kolmogorov equations goes back to R.Z. Hasminskii [**444**], [**502**] and was later used by many authors (see, for example, Miyahara [**716**], Zakai [**949**], [**950**]). The Hasminskii theorem on the existence of stationary distributions was generalized to the case of non-Lipschitz coefficients in Bogachev, Röckner [**160**]. The results presented in this chapter are borrowed (with some reinforcements) from the papers Bogachev, Kirillov, Shaposhnikov [**144**], [**145**], Bogachev, Krylov, Röckner [**155**], Bogachev, Röckner, Shaposhnikov [**171**], Bogachev, Röckner, Wang [**179**], Bogachev, Röckner, Zhang [**182**]. In Chapters 4 and 5 we shall see that the existence of a Lyapunov function not only gives the existence of a probability solution, but ensures its uniqueness and also invariance with respect to the associated Markov semigroup. In Chapter 5 we shall also present some results on existence of invariant measures of general Markov semigroups.

In the paper Bensoussan [**105**], the existence of a probability solution for the equation $L^*_{\mathrm{I},b}\mu = 0$ is established under condition (2.4.3) and the additional assumption that $|\nabla V|^2/V \leqslant k$ for some number $k > 0$. However, in this case we can take a new Lyapunov function $W = V^\alpha$ with a sufficiently large α for which $L_{\mathrm{I},b}W(x) \to -\infty$ as $|x| \to \infty$. On existence of solutions to stationary equations, see also Birrell, Herzog, Wehr [**117**], Liu, Shen [**630**].

In § 2.5(iv) our exposition follows the paper Tonoyan [**897**].

Exercises

2.5.16. Prove that for every function $u \in W^{2,1}(\Omega)$ the inequality $\mathcal{L}u \geqslant 0$ implies the inequalities $\mathcal{L}|u| \geqslant 0$ and $\mathcal{L}u^+ \geqslant 0$.

HINT: In the case of smooth coefficients for every smooth function f it is easily verified that

$$\mathcal{L}f(u) = f'(u)\mathcal{L}u + f''(u)|\sqrt{A}\nabla u|^2 + (c - \mathrm{div} b)\big(f(u) - uf'(u)\big).$$

For convex f, dropping the nonnegative term with f'' and substituting formally the function $f(u) = |u|$, we arrive at once at the inequality $\mathcal{L}|u| \geqslant 0$. By means of smooth approximations this can be easily justified.

2.5.17. Prove the maximum principle in the one-dimensional case.

2.5.18. Show that the equation $L^*_{A,b,c}\mu = 0$ has no probability solutions on the real line if $A = 1$, $b(x) = -x$, $c = 1$.

2.5.19. Show that the equation $L^*_{A,b,c}\mu = 0$ has a probability solution on the real line in the case where $A = 1$, $b(x) = -x$ if $|x| > 1$, $b(x) = x^2 - 1$ if $|x| \leqslant 1$, $c(x) = 0$ if $|x| > 1$, $c(x) = -2x$ if $|x| \leqslant 1$. This solution has a Gaussian density outside $[-1, 1]$ and a constant density in $(-1, 1)$.

2.5.20. Let μ be a probability measure on $\mathbb{R}^d$ with $L^*_{A,b}\mu = 0$, where A is uniformly bounded and $\langle b(x), x\rangle \leqslant c_1 - c_2|x|^r$. Prove that $\exp(\varepsilon|x|^r) \in L^1(\mu)$ for sufficiently small ε (indicate explicitly an admissible value ε via $c_2, r, \sup_x \|A(x)\|$).

2.5.21. Let Ω be a bounded domain in $\mathbb{R}^d$ with smooth boundary,

$$Lu = a^{ij}\partial_{x_i}\partial_{x_j} + b^i\partial_{x_i}u + cu, \quad L^*u = \partial_{x_i}(a^{ij}\partial_{x_j}) - \partial_{x_i}(b^i u) + cu,$$

where $a^{ij} \in W^{p,1}(\Omega)$, $b^i \in L^p(\Omega)$, $c \in L^{p/2}(\Omega)$, $p > d$. By using the Sobolev embedding theorem and Hölder's inequality show that the operators L and L^* map continuously $W_0^{2,1}(\Omega)$ to $W^{2,-1}(\Omega)$.

2.5.22. (Bogachev, Röckner [**160**]) Suppose we are given mappings $A_k = (a_k^{ij})$ on $\mathbb{R}^d$ with values in the space of positive symmetric linear operators on $\mathbb{R}^d$ and Borel vector fields $b_k = (b_k^i)$ on $\mathbb{R}^d$ such that for every $R > 0$ there exist numbers $\alpha_R > d$, $\beta_R > 0$ and $\gamma_R > 0$ for which

$$\kappa_R := \sup_k \int_{|x|\leqslant R} \Bigl[|b_k(x)|^{\alpha_R} + \sum_{i,j=1}^d \|\nabla a_k^{ij}(x)\|^{\alpha_R}\Bigr]\, dx < \infty, \tag{2.5.9}$$

$$\beta_R \mathrm{I}\cdot \leqslant A_k(x) \leqslant \gamma_R \cdot \mathrm{I} \quad \forall k,\ \forall x \in \{y\colon\ |y| \leqslant R\}. \tag{2.5.10}$$

Suppose also that there exists a quasi-compact (for example, just compact) continuous function $V \in W^{2,1}_{\mathrm{loc}}(\mathbb{R}^d)$ such that

$$\lim_{|x|\to\infty} \sup_k L_{A_k,b_k} V(x) = -\infty. \tag{2.5.11}$$

Let $\{\mu_k\}$ be a sequence of probability measures on $\mathbb{R}^d$ satisfying the equations $L^*_{A_k,b_k}\mu_k = 0$. Set $U_R := \{V < R\}$. Then the following assertions are true.

(i) the measures μ_k have densities ϱ_k such that for every fixed $R > 1$ the functions $\varrho_k|_{U_R}$ are uniformly bounded, uniformly Hölder continuous and uniformly norm bounded in $W^{\alpha_R,1}(U_{R-1})$.

(ii) For every $R > 0$ we have $\inf\limits_k \inf\limits_{x\in U_R} \varrho_k(x) > 0$.

(iii) The sequence $\{\mu_k\}$ is relatively weakly compact on $\mathbb{R}^d$.

In addition, suppose that for every k the probability measure μ_k is defined on $U_k = \{V < c_k\}$, where $c_k \uparrow \infty$, and satisfies our equation $L^*_{A_k,b_k}\mu_k = 0$ only in U_k. Then all assertions (i)–(iii) remain true with the following changes: in assertions (i) and (ii) we consider only those k for which $c_k > R$.

2.5.23. Prove that in the previous exercise the sequence $\{\mu_k\}$ has compact closure in variation.

2.5.24. By using the two previous exercises and assuming the existence of a strong Lyapunov function, construct a probability solution as a limit of positive solutions on balls.

2.5.25. Let $b = \nabla G$, where $G \in W^{1,1}_{\mathrm{loc}}(\mathbb{R}^d)$ and $|\nabla G| \exp G \in L^1_{\mathrm{loc}}(\mathbb{R}^d)$. Suppose that we are given a function $V \in C^2(\mathbb{R}^d)$ such that the sets $\{V \leqslant c\}$ are compact, there exist numbers $c_k \to +\infty$ for which the sets $V^{-1}(c_k)$ are Lipschitzian surfaces, and $LV := \Delta V + \langle \beta, \nabla V\rangle \leqslant -1$ outside some compact set. Prove that $\exp G \in L^1(\mathbb{R}^d)$, i.e., b is the logarithmic gradient of the probability measure with density $\mathrm{const}\cdot \exp G$.

HINT: See Bogachev [**126**, Proposition 7.5.2].

CHAPTER 3

Global Properties of Densities

This chapter is concerned with the properties of densities of solutions to elliptic equations for measures such as their membership in Sobolev classes on the whole space, boundedness on the whole space, and two-sided estimates.

In case $A = \mathrm{I}$ one of the main results in this chapter (Theorem 3.1.2) says that the density ϱ of a probability solution μ to the equation $L^*_{A,b}\mu = 0$ with $|b| \in L^2(\mu)$ belongs to the Sobolev class $W^{1,1}$ and $\sqrt{\varrho} \in W^{2,1}$, i.e., $|\nabla\varrho|^2/\varrho$ is integrable on $\mathbb{R}^d$. If $|b| \in L^p(\mu)$ with $p > d$, then $\varrho \in W^{p,1}$ and ϱ is uniformly bounded (Corollary 3.2.5). Analogous results hold for nonconstant a^{ij} from certain Sobolev classes (Theorem 3.2.4). Moreover, similar results extend to nonhomogeneous equations, which enables us to obtain upper bounds on the solutions densities (see § 3.3). These bounds have the form $\varrho(x) \leqslant C\Phi(x)^{-1}$, where Φ is a given function satisfying certain simple conditions expressed in terms of A and b (a typical result is Theorem 3.3.1). On the other hand, in § 3.4 and § 3.5 we consider Harnack's inequality, which yields lower bounds and positivity of the densities of solutions.

3.1. Square integrability of logarithmic gradients

An important role will be played below by square integrability of the logarithmic gradients of densities ϱ of probability solutions to elliptic equations with respect to these solutions, i.e., the inclusion $\sqrt{\varrho} \in W^{2,1}(\mathbb{R}^d)$, hence also $\varrho \in W^{1,1}(\mathbb{R}^d)$, which can be obtained under rather weak global conditions on the drift (moreover, under dimension invariant conditions). It is interesting that, unlike Example 1.6.10, the assumptions about the order of integrability of the drift may be considerably weakened in the case of the global integrability.

The following lemma will be employed below to control the norms of convolutions.

3.1.1. Lemma. *Let μ and ν be two Borel probability measures on $\mathbb{R}^d$ and let $f \in L^2(\mu)$. Let g denote the Radon–Nikodym density of $(f\mu) * \nu$ with respect to the convolution $\mu * \nu$, where*

$$\mu * \nu\,(A) = \int_{\mathbb{R}^d} \mu(A - x)\,\nu(dx).$$

Then

$$\int_{\mathbb{R}^d} g^2\,d(\mu * \nu) \leqslant \int_{\mathbb{R}^d} f^2\,d\mu.$$

More generally, $\|g\|_{L^p(\mu*\nu)} \leqslant \|f\|_{L^p(\mu)}$ if $f \in L^p(\mu)$, $p \geqslant 1$.

PROOF. Let $\Lambda\colon \mathbb{R}^d\times\mathbb{R}^d \to \mathbb{R}^d$ be defined by

$$\Lambda(x,y) = x+y.$$

Then, letting $\overline{f}\colon \mathbb{R}^d\times\mathbb{R}^d \to \mathbb{R}^1$, $\overline{f}(x,y) = f(x)$, $(x,y)\in\mathbb{R}^d\times\mathbb{R}^d$, we obtain that

$$g\circ\Lambda = \mathbb{E}_{\mu\otimes\nu}[\bar{f} \mid \sigma(\Lambda)],$$

where on the right we have the conditional expectation of f with respect to the product of μ and ν, given the σ-algebra generated by Λ. Indeed, we have to verify that for any bounded $\mathcal{B}$-measurable function φ the integrals of $\varphi(x+y)g(x+y)$ and $f(x)\varphi(x+y)$ with respect to $\nu\otimes\nu$ coincide. This is true, since the second integral is the integral of φ against the measure $(f\mu)*\nu$, but the latter coincides with the integral of φg against $\mu*\nu$. Hence Jensen's inequality for conditional expectations implies the assertion. □

For $a^{ij}\in L^1_{\text{loc}}(\mathbb{R}^d)$ set

$$a := (a^1,\dots,a^d), \quad a^j := \sum_{i=1}^d \partial_{x_i} a^{ij}.$$

3.1.2. Theorem. *Suppose that a mapping A with values in the space of symmetric matrices is uniformly bounded, uniformly Lipschitzian and that there is a number $\alpha>0$ such that $A(x)\geqslant \alpha\cdot\mathrm{I}$. Let μ be a Borel probability measure on $\mathbb{R}^d$ satisfying the equation $L^*_{A,b}\mu=0$, where $|b|\in L^2(\mu)$. Then*

(i) *we have $\mu=\varrho\,dx$, where $\varrho=\varphi^2$ and $\varphi\in W^{2,1}(\mathbb{R}^d)$, hence $\varrho\in W^{1,1}(\mathbb{R}^d)$ and also $\varrho\in L^{d/(d-2)}(\mathbb{R}^d)$ if $d>2$;*

(ii) *we have*

$$\int_{\mathbb{R}^d}\Bigl|\frac{\nabla\varrho}{\varrho}\Bigr|^2\varrho\,dx = 4\int_{\mathbb{R}^d}|\nabla\varphi|^2\,dx \leqslant \frac{1}{\alpha^2}\int_{\mathbb{R}^d}|b-a|^2\,d\mu;$$

(iii) *the mapping $\nabla\varrho/\varrho$ coincides μ-a.e. with the orthogonal projection of the vector field $A^{-1}(b-a)$ onto the closure of the set $\{\nabla u|\ u\in C_0^\infty(\mathbb{R}^d)\}$ in the space $L^2(\mu,\mathbb{R}^d)$ equipped with the inner product $\langle\cdot,\cdot\rangle_2$ defined by*

$$\langle F,G\rangle_2 := \int_{\mathbb{R}^d}\langle AF,G\rangle\,d\mu.$$

Therefore,

$$\int_{\mathbb{R}^d}\Bigl|\frac{\sqrt{A}\nabla\varrho}{\varrho}\Bigr|^2\varrho\,dx \leqslant \int_{\mathbb{R}^d}|A^{-1/2}(b-a)|^2\,d\mu. \tag{3.1.1}$$

In particular, if $A=\mathrm{I}$, then these assertions hold with $\alpha=1$ and $a=0$.

PROOF. Let g be the standard Gaussian density on $\mathbb{R}^d$. Set

$$f_\varepsilon(x) := g_\varepsilon*\mu(x) = \int g_\varepsilon(x-y)\,\mu(dy), \quad g_\varepsilon(x)=\varepsilon^{-d}g(x/\varepsilon),$$

where the integral is taken over $\mathbb{R}^d$ as well as the integrals below. Then each f_ε is strictly positive, infinitely differentiable, and tends to zero at infinity together with all its derivatives by Lebesgue's dominated convergence theorem. Furthermore,

$$\int f_\varepsilon\,dx = 1 \quad\text{and}\quad f_\varepsilon\,dx\to\mu \quad \text{weakly as } \varepsilon\to 0.$$

Assertion (i) is proved by the following claim: there exist a sequence of numbers $\varepsilon_n>0$ with $\varepsilon_n\to 0$ and a function $\varrho\in L^1(\mathbb{R}^d)$ such that $f_{\varepsilon_n}\to\varrho$ in $L^1(U)$ for every ball U in $\mathbb{R}^d$) and $\varphi := \sqrt{\varrho} \in W^{2,1}(\mathbb{R}^d)$. This claim, however, follows from the estimate

$$\int |\nabla \sqrt{f_\varepsilon}|^2 \, dx \leqslant const. \tag{3.1.2}$$

Indeed, if (3.1.2) holds, then for $\varphi_n := \sqrt{f_{1/n}}$ we have

$$\int (\varphi_n^2 + |\nabla \varphi_n|^2) \, dx \leqslant const.$$

Since the embedding $W^{2,1}(U) \subset L^2(U)$ is compact for every open ball U, there exist a function $\varphi \in W^{2,1}(\mathbb{R}^d)$ and a subsequence $\{n_k\}$ such that $\varphi_{n_k} \to \varphi$ weakly in $W^{2,1}(\mathbb{R}^d)$ and strongly in $L^2(U)$. We take $\varrho := \varphi^2$. So it remains to prove estimate (3.1.2). For any $u \in C_0^\infty(\mathbb{R}^d)$, integrating by parts and using Fubini's theorem we obtain that

$$\begin{aligned}
&-\int \partial_{x_i} u \big(\partial_{x_j}[(a^{ij}\mu) * g_\varepsilon] - \partial_{x_i}[(b^i\mu) * g_\varepsilon]\big) \, dx \\
&= \int \partial_{x_i}\partial_j u [(a^{ij}\mu) * g_\varepsilon] \, dx + \int \partial_{x_i} u [(b^i\mu) * g_\varepsilon] \, dx \\
&= \iint \partial_{x_i}\partial_j u(x) \, g_\varepsilon(x-y) \, a^{ij}(y) \, \mu(dy) \, dx \\
&\quad + \iint \partial_{x_i} u(x) \, g_\varepsilon(x-y) b^i(y) \, \mu(dy) \, dx \\
&= \int a^{ij} \, \partial_{x_i}\partial_{x_j}(u * g_\varepsilon) \, d\mu + \int b^i \, \partial_{x_i}(u * g_\varepsilon) \, d\mu = \int L(u * g_\varepsilon) \, d\mu = 0,
\end{aligned} \tag{3.1.3}$$

since $u * g_\varepsilon \in C_b^\infty(\mathbb{R}^d)$. It is readily seen that for any function $h \in W^{1,1}_{\text{loc}}(\mathbb{R}^d)$ with $|\nabla h| \in L^2(\mathbb{R}^d)$ there exist functions $u_n \in C_0^\infty(\mathbb{R}^d)$, $n \in \mathbb{N}$, such that we have $|\nabla h - \nabla u_n| \to 0$ in $L^2(\mathbb{R}^d)$. Since for any $c > 0$ the function $\ln(f_\varepsilon + c)$ is bounded and $\nabla \ln(f_\varepsilon + c) = \nabla f_\varepsilon/(f_\varepsilon + c) \in L^2(\mathbb{R}^d)$, (3.1.3) implies that

$$\int \frac{\partial_{x_i} f_\varepsilon}{f_\varepsilon + c} \partial_{x_j}[(a^{ij}\mu) * g_\varepsilon] \, dx = \int \frac{\partial_{x_i} f_\varepsilon}{f_\varepsilon + c} [(b^i\mu) * g_\varepsilon] \, dx. \tag{3.1.4}$$

So far we only used that each a^{ij} is bounded. The Lipschitzness was not required.

Note that for any bounded function a that is Lipschitzian with constant λ, whenever $1 \leqslant j \leqslant d$, we have

$$\partial_{x_j}[(a\mu) * g_\varepsilon](x) = a(x) \, \partial_{x_j} f_\varepsilon(x) + \int \partial_{x_j} g_\varepsilon(x-y)[a(y) - a(x)] \, \mu(dy) \tag{3.1.5}$$

and also

$$\begin{aligned}
\Big| \int \partial_{x_j} g_\varepsilon(x-y)[a(y) - a(x)] \, \mu(dy) \Big| &\leqslant \lambda \int |\partial_{x_j} g_\varepsilon(x-y)| \, |y-x| \, \mu(dy) \\
&\leqslant \lambda \int \varepsilon^{-d} \frac{|x-y|^2}{\varepsilon^2} \, g\Big(\frac{x-y}{\varepsilon}\Big) \, \mu(dy) = \lambda(\mu * q_\varepsilon)(x),
\end{aligned} \tag{3.1.6}$$

where we set $q_\varepsilon(x) := \varepsilon^{-d} g(x/\varepsilon)\,|x/\varepsilon|^2$, $x \in \mathbb{R}^d$. Using (3.1.5) we can rewrite equality (3.1.4) as

$$\int \frac{a^{ij}\,\partial_{x_i} f_\varepsilon\,\partial_{x_j} f_\varepsilon}{f_\varepsilon + c}\,dx = -\int \frac{\partial_{x_i} f_\varepsilon}{f_\varepsilon + c}(x) \int \partial_{x_j} g_\varepsilon(x-y)[a^{ij}(y) - a^{ij}(x)]\,\mu(dy)\,dx + \int \frac{\partial_{x_i} f_\varepsilon}{f_\varepsilon + c}(x)[(b^i\mu) * g_\varepsilon]\,dx,$$

whose right-hand side by (3.1.6) and the Cauchy–Bunyakovsky inequality is dominated by

$$d^{3/2}\lambda\Bigl(\int \frac{|\nabla f_\varepsilon|^2}{f_\varepsilon + c}\,dx\Bigr)^{1/2}\Bigl(\int \frac{(\mu * q_\varepsilon)^2}{f_\varepsilon + c}\,dx\Bigr)^{1/2} + \Bigl(\int \frac{|\nabla f_\varepsilon|^2}{f_\varepsilon + c}\,dx\Bigr)^{1/2}\Bigl(\int \frac{\sum_{i=1}^d [(b^i\mu) * g_\varepsilon]^2}{f_\varepsilon + c}\,dx\Bigr)^{1/2}.$$

By virtue of Lemma 3.1.1 (or a direct calculation using the Cauchy–Bunyakovsky inequality) we have that

$$\int \frac{[(b^i\mu) * g_\varepsilon]^2}{f_\varepsilon}\,dx \leqslant \int |b^i|^2\,d\mu, \quad 1 \leqslant i \leqslant d, \qquad \int \frac{(\mu * q_\varepsilon)^2}{f_\varepsilon}\,dx \leqslant \int |x|^4\,g(x)\,dx.$$

Using that $a^{ij}x_i x_j \geqslant \alpha|x|^2$ for all $x = (x_1, \ldots, x_d) \in \mathbb{R}^d$, we see from the last three inequalities that

$$\int \frac{|\nabla f_\varepsilon|^2}{f_\varepsilon + c}\,dx \leqslant \alpha^{-2}\Bigl[\Bigl(\int |b|^2\,dx\Bigr)^{1/2} + d^{3/2}\lambda\Bigl(\int |x|^4 g(x)\,dx\Bigr)^{1/2}\Bigr]^2.$$

Since $|\nabla\sqrt{f_\varepsilon}|^2 = \frac{1}{4}|\nabla f_\varepsilon|^2\,f_\varepsilon^{-1}$, by Fatou's lemma this implies (3.1.2). The second part of assertion (i) follows by the Gagliardo–Nirenberg embedding theorem (see Theorem 1.1.1(ii)).

(ii) We now use the following fact (see Bogachev [**126**, Theorem 2.6.11]): there exist functions $f_n \in C_0^\infty(\mathbb{R}^d)$, $n \in \mathbb{N}$, such that

$$\int \Bigl|\frac{\nabla\varrho}{\varrho} - \nabla f_n\Bigr|^2\,d\mu \to 0. \tag{3.1.7}$$

Since

$$\int L f_n\,\varrho\,dx = 0$$

and $\varrho \in W^{1,1}(\mathbb{R}^d)$, we can integrate by parts to obtain that

$$\int a^{ij}\,\partial_{x_i} f_n\,\frac{\partial_{x_j}\varrho}{\varrho}\,\varrho\,dx = -\int \partial_{x_j} a^{ij}\partial_{x_i} f_n\,\varrho\,dx + \int b^i\partial_{x_i} f_n\,\varrho\,dx,$$

where we used that $\partial_{x_j}\varrho = 0$ a.e. on $\{\varrho = 0\}$. By (3.1.7) we can let $n \to \infty$, which yields

$$\int a^{ij}\,\frac{\partial_{x_i}\varrho}{\varrho}\,\frac{\partial_{x_j}\varrho}{\varrho}\,\varrho\,dx = -\int \partial_{x_j} a^{ij}\,\frac{\partial_{x_i}\varrho}{\varrho}\,\varrho\,dx + \int b^i\,\frac{\partial_{x_i}\varrho}{\varrho}\,\varrho\,dx.$$

Using the uniform ellipticity of A and the Cauchy–Bunyakovsky inequality we get

$$\int \Bigl|\frac{\nabla\varrho}{\varrho}\Bigr|^2\,d\mu \leqslant \alpha^{-1}\Bigl(\int \Bigl|\frac{\nabla\varrho}{\varrho}\Bigr|^2\,d\mu\Bigr)^{1/2}\Bigl(\int |b - a|^2\,d\mu\Bigr)^{1/2},$$

which implies assertion (ii).

(iii) By (3.1.7) we already know that $\nabla\varrho/\varrho$ is in the closure of the gradients of smooth compactly supported functions in $L^2(\mu,\mathbb{R}^d)$. Hence it suffices to check that for each $u\in C_b^\infty(\mathbb{R}^d)$ we have the equality

$$\int\Bigl\langle A\nabla u, A^{-1}(b-a)-\frac{\nabla\varrho}{\varrho}\Bigr\rangle\,d\mu=0.$$

But integrating the left-hand side by parts we see that the left-hand side is equal to $\displaystyle\int L_{A,b}u\,d\mu$, which vanishes by assumption. □

3.1.3. Corollary. *Let a probability measure μ on $\mathbb{R}^d$ satisfy the equation $L_{\mathrm{I},b}^*\mu=0$, where $|b|\in L^2(\mu)$. The operator $L_{\mathrm{I},b}$ on domain $C_0^\infty(\mathbb{R}^d)$ in $L^2(\mu)$ is symmetric precisely when b coincides μ-a.e. with the logarithmic gradient of μ.*

PROOF. If $\mu=\varrho\,dx$ and $b=\nabla\varrho/\varrho$, then for any $\varphi,\psi\in C_0^\infty(\mathbb{R}^d)$ we have

$$\begin{aligned}\int[\Delta\varphi+\langle b,\nabla\varphi\rangle]\psi\,d\mu&=\int[-\langle\nabla\varphi,\nabla\psi\rangle\varrho-\langle\nabla\varphi,\nabla\varrho\rangle\psi+\langle\nabla\varrho,\nabla\varphi\rangle\psi]\,dx\\&=-\int\langle\nabla\varphi,\nabla\psi\rangle\,d\mu,\end{aligned}$$

which shows that $L_{\mathrm{I},b}$ is symmetric. Conversely, suppose that $L_{\mathrm{I},b}$ is symmetric. Then

$$\int L_{\mathrm{I},b}\varphi\,\psi\varrho\,dx=\int[-\langle\nabla\varphi,\nabla\psi\rangle\varrho-\langle\nabla\varphi,\nabla\varrho\rangle\psi+\langle b,\nabla\varphi\rangle\psi\varrho]\,dx$$

for all $\varphi,\psi\in C_0^\infty(\mathbb{R}^d)$. Since this expression coincides with the integral of $L_{\mathrm{I},b}\psi\,\varphi\varrho$, we arrive at the identity

$$\int\psi\langle\nabla\varphi,b-\nabla\varrho/\varrho\rangle\,\varrho\,dx=\int\varphi\langle\nabla\psi,b-\nabla\varrho/\varrho\rangle\,\varrho\,dx,$$

which yields the identity

$$\int\varphi\langle\nabla\psi,b-\nabla\varrho/\varrho\rangle\,\varrho\,dx=0,$$

because $b-\nabla\varrho/\varrho$ is orthogonal to $\nabla(\psi\varphi)$ in $L^2(\mu,\mathbb{R}^d)$. Using that $b-\nabla\varrho/\varrho$ is μ-square-integrable, we conclude that the last identity is true for any coordinate function $\varphi(x)=x_i$, which shows that $b-\nabla\varrho/\varrho=0$ almost everywhere with respect to μ. We recall that if $\varrho\in W_{\mathrm{loc}}^{p,1}(\mathbb{R}^d)$ and $p\geqslant 2$, then our assertion has already been proved in Proposition 1.4.8. □

Our next global elliptic regularity result employs a certain uniform local condition on A. For $a^{ij}\in W_{\mathrm{loc}}^{1,1}(\mathbb{R}^d)$ we set

$$\Theta_A(x):=\sum_{j=1}^d\Bigl|\sum_{i=1}^d\partial_{x_i}a^{ij}(x)\Bigr|.$$

For given $p>1$ and $\gamma>0$ let

$$q=q(d,p,\gamma)=\begin{cases}d & \text{if } \ p>d/(d-1),\\ d+\gamma & \text{if } \ p=d/(d-1),\\ p'=p/(p-1)\ \ (>d) & \text{if } \ p<d/(d-1).\end{cases}$$

We introduce the following condition on A and $p>1$: $a^{ij}\in W^{1,1}_{\rm loc}(\mathbb{R}^d)$ and

$$\lim_{r\to 0}\sup_{z\in\mathbb{R}^d}\int_{U(z,r)}\Theta_A^q(x)\,dx=0, \tag{3.1.8}$$

where q is defined above (in the case $p=d/(d-1)$ this equality must be fulfilled with $q=d+\gamma$ for some $\gamma>0$).

We observe that this condition is weaker than

$$\lim_{r\to 0}\sup_{z\in\mathbb{R}^d}\int_{U(z,r)}\widetilde{\Theta}_A^q(x)\,dx=0, \tag{3.1.9}$$

where

$$\widetilde{\Theta}_A(x):=\sum_{j=1}^d\sum_{i=1}^d|\partial_{x_i}a^{ij}(x)|.$$

It is clear that if there is $p_0>d>1$ such that

$$\sup_{z\in\mathbb{R}^d}\int_{U(z,1)}\sum_{i,j=1}^d|\nabla a^{ij}(x)|^{p_0}\,dx<\infty, \tag{3.1.10}$$

then A satisfies condition (3.1.8) (as well as (3.1.9)) for any $p\in(1,p_0)$ and is uniformly continuous (even uniformly Hölder continuous) on all of $\mathbb{R}^d$. In particular, both properties hold if A is uniformly Lipschitzian.

In the rest of this section it will be more convenient to formulate our results for divergence form equations

$$\mathcal{L}^*_{A,b}\mu=0,$$

the interpretation of which is explained in § 1.4 (see (1.4.4)). It is worth noting that although in most of our results we assume that $a^{ij}\in W^{p,1}_{\rm loc}(\mathbb{R}^d)$, hence one can write $\mathcal{L}_{A,b}$ as L_{A,b_0} with $b_0^i:=b^i+\partial_{x_j}a^{ij}$, the case of $\mathcal{L}_{A,b}$ does not always reduce to that of $L_{A,b}$, because the global integrability conditions on $|b|$ and $|\nabla a^{ij}|$ are different. In some situations, it is easier to deal with divergence form operators, in others the standard form is more convenient. In the manifold case, usually divergence form operators lead to more natural geometric objects. Apparently, the most natural setting for most of such problems should appeal to the geometry related to A and to weighted Sobolev spaces. However, the corresponding techniques, in particular, embedding theorems, are less developed than the classical Sobolev theory.

The next result is a partial generalization of Theorem 3.1.2; we weaken the assumptions about A, but add an extra local condition on b. That extra condition is not needed if we know in advance that μ has a locally bounded density in $W^{2,1}_{\rm loc}(\mathbb{R}^d)$. Unlike most of other results in this section, this theorem is false for signed measures.

3.1.4. Theorem. *Suppose that $\mu\in\mathcal{P}(\mathbb{R}^d)$ and $\mathcal{L}^*_{A,b}\mu=0$, where the mapping A is continuous, $\det A>0$, $a^{ij}\in W^{d+,1}_{\rm loc}(\mathbb{R}^d)$, and $|b|\in L^{d+}_{\rm loc}(\mu)$. Suppose in addition that $|A^{-1/2}b|\in L^2(\mu)$ and that*

$$\liminf_{r\to\infty}\int_{r\leqslant|x|\leqslant 2r}\Bigl[r^{-2}\|A(x)\|+r^{-1}\Theta_A(x)\Bigr]\,\mu(dx)=0. \tag{3.1.11}$$

Then μ has a density $\varrho\in W^{d+,1}_{\rm loc}(\mathbb{R}^d)$ such that

$$\int_{\mathbb{R}^d}\Bigl|\frac{\sqrt{A}\nabla\varrho}{\varrho}\Bigr|^2\,d\mu\leqslant\int_{\mathbb{R}^d}|A^{-1/2}b|^2\,d\mu. \tag{3.1.12}$$

Under the additional assumption that $A \geqslant \varepsilon \cdot \mathrm{I}$, $\varepsilon > 0$, *one has* $\sqrt{\varrho} \in W^{2,1}(\mathbb{R}^d)$, $\varrho \in L^{d/(d-2)}(\mathbb{R}^d)$ *if* $d > 2$ *and* $\varrho \in L^s(\mathbb{R}^d)$ *for all* $s \in [1, \infty)$ *if* $d = 2$.

PROOF. The desired estimate is obtained by formally substituting the function $\varphi = \ln \varrho$ in the identity

$$\int_{\mathbb{R}^d} \mathcal{L}_{A,b}\varphi \, d\mu = 0 \tag{3.1.13}$$

and integrating by parts in the term with $\mathcal{L}_A$. In order to justify this straightforward procedure, we need a certain local regularity of μ, hence some better local integrability of b. In order to avoid this extra assumption, a smoothing procedure by convolutions has been used above (in Bogachev, Röckner, Wang [**179**] in the manifold case, the heat semigroup was employed); however, smoothing requires certain global conditions on A. By the local theory, we know that μ has a continuous density $\varrho \in W^{d+,1}_{\mathrm{loc}}(\mathbb{R}^d)$. Hence (3.1.13) with $L = \mathcal{L}_{A,b}$ extends to $\varphi \in W^{2,1}_0(\mathbb{R}^d)$. Integrating by parts we obtain

$$\int_{\mathbb{R}^d} \langle A\nabla\varphi, \nabla\varrho\rangle \, dx = \int_{\mathbb{R}^d} \langle b, \nabla\varphi\rangle \varrho \, dx.$$

We fix a function $\zeta \in C_0^\infty(\mathbb{R}^d)$ such that $\zeta(x) = 1$ if $|x| \leqslant 1$ and for any $j \in \mathbb{N}$ set $\zeta_j(x) := \zeta(x/j)$. Letting $\varrho_{k,\delta} := \min\bigl(k, \max(\varrho, \delta)\bigr)$ for small $\delta > 0$ and large $k > 0$, $\Omega_{k,\delta} := \{\delta < \varrho < k\}$, and taking $\varphi = \zeta_j^2 \ln \varrho_{k,\delta}$ we arrive at the equality

$$\begin{aligned}\int_{\mathbb{R}^d} \Bigl\langle A\nabla\varrho, \frac{\nabla\varrho}{\varrho}\Bigr\rangle \zeta_j^2 I_{\Omega_{k,\delta}} \, \varrho \, dx + 2\int_{\mathbb{R}^d} \langle A\nabla\varrho, \nabla\zeta_j\rangle \ln \varrho_{k,\delta}\zeta_j \, dx \\ = \int_{\mathbb{R}^d} I_{\Omega_{k,\delta}} \Bigl\langle b, \frac{\nabla\varrho}{\varrho}\Bigr\rangle \zeta_j^2 \varrho \, dx + 2\int_{\mathbb{R}^d} \langle b, \nabla\zeta_j\rangle\zeta_j \ln \varrho_{k,\delta} \, \varrho \, dx.\end{aligned}$$

Integrating by parts in the second term on the left, we find that

$$\begin{aligned} S_{j,k,\delta} &:= \int_{\Omega_{k,\delta}} \Bigl\langle A\nabla\varrho, \frac{\nabla\varrho}{\varrho}\Bigr\rangle \zeta_j^2 \, \varrho \, dx \\ &= 2\int_{\Omega_{k,\delta}} \Bigl\langle \frac{\nabla\varrho}{\varrho}, A\nabla\zeta_j\Bigr\rangle \zeta_j \varrho \, dx + 2\int_{\mathbb{R}^d} \operatorname{div}(\zeta_j A\nabla\zeta_j) \ln \varrho_{k,\delta} \, \varrho \, dx \\ &\quad + \int_{\Omega_{k,\delta}} \Bigl\langle b, \frac{\nabla\varrho}{\varrho}\Bigr\rangle \zeta_j^2 \varrho \, dx + 2\int_{\mathbb{R}^d} \langle b, \nabla\zeta_j\rangle\zeta_j \ln \varrho_{k,\delta} \, \varrho \, dx \\ &\leqslant \sqrt{S_{j,k,\delta}}\Bigl(2\|I_{\Omega_{k,\delta}}\sqrt{A}\nabla\zeta_j\|_{L^2(\mu,\mathbb{R}^d)} + \|A^{-1/2}b\|_{L^2(\mu,\mathbb{R}^d)}\Bigr) + R_{j,k,\delta}, \end{aligned}$$

where

$$R_{j,k,\delta} := 2\int_{\mathbb{R}^d} \operatorname{div}(\zeta_j A\nabla\zeta_j) \ln \varrho_{k,\delta} \, \varrho \, dx + 2\int_{\mathbb{R}^d} \langle b, \nabla\zeta_j\rangle\zeta_j \ln \varrho_{k,\delta} \, \varrho \, dx.$$

Keeping k and δ fixed, we observe that, given $\varepsilon > 0$, for all sufficiently large numbers j of the form $j = r_l$ with $r_l \to \infty$ chosen according to (3.1.11), the quantity $R_{j,k,\delta}$ can be made smaller than ε in absolute value. Indeed, it follows by the hypotheses and the estimates

$$\sup_x |\nabla\zeta_j(x)| \leqslant j^{-1} \sup_x |\nabla\zeta(x)|, \quad \sup_x |\partial_{x_i}\partial_{x_m}\zeta_j(x)| \leqslant j^{-2} \sup_x |\partial_{x_i}\partial_{x_m}\zeta(x)|$$

that for all $j = r_l$ the first term in the expression for $R_{j,k,\delta}$ can be estimated by

$$M_{k,\delta} r_l^{-2} \int_{\{r_l \leqslant |x| \leqslant 2r_l\}} \|A(x)\| \, \mu(dx) + M_{k,\delta} r_l^{-1} \int_{\{r_l \leqslant |x| \leqslant 2r_l\}} \Theta_A(x) \, \mu(dx),$$

where $M_{k,\delta} = 2M(\ln k - \ln \delta)$ with a constant M that depends on the maxima of the first and second derivatives of ζ. Similarly, by the Cauchy inequality and the estimate $|b(x)| \leqslant \|A^{1/2}(x)\| \, |A^{-1/2}(x)b(x)|$ the second term is majorized by

$$r_l^{-1} M_{k,\delta} \|A^{-1/2} b\|_{L^2(\mu,\mathbb{R}^d)} \biggl(\int_{r_l \leqslant |x| \leqslant 2r_l} \|A(x)\| \, \mu(dx) \biggr)^{1/2}.$$

Therefore, for all $l > l(k,\delta)$, one has $S_{r_l,k,\delta} \leqslant \|A^{-1/2} b\|_{L^2(\mu,\mathbb{R}^d)} + \varepsilon$. This yields that the integrals of $|\sqrt{A} \nabla \varrho / \varrho|^2$ over the sets $\Omega_{k,\delta}$ against μ are majorized by the same quantity. Letting $k \to \infty$ and $\delta \to 0$, we see that $|\sqrt{A} \nabla \varrho / \varrho| \in L^2(\mu)$ and obtain the desired bound. □

3.1.5. Remark. (i) Condition (3.1.11) is fulfilled if

$$|\nabla a^{ij}(x)| \leqslant C_0 + C_1 |x|.$$

If μ is known to have finite first moment, i.e., $|x| \in L^1(\mu)$, then quadratic growth of $|\nabla a^{ij}|$ is allowed.

(ii) Condition (3.1.11) can be replaced by the assumption that for some $r > 0$ one has

$$\liminf_{R \to \infty} \int_{R \leqslant |x| \leqslant R+r} \Bigl[\|A(x)\| + \Theta_A(x) \Bigr] \, \mu(dx) = 0.$$

This condition is weaker on the part of Θ_A, but is stronger on the part of $\|A\|$; for uniformly bounded A, it is weaker.

(iii) Note also that if A is uniformly bounded and satisfies (3.1.8), then equality (3.1.11) is ensured by the assumption that

$$\liminf_{r \to \infty} r^{d-1} \mu(\{|x| \geqslant r\}) = 0,$$

which is fulfilled, e.g., if $|x|^{d-1} \in L^1(\mu)$. The latter can be effectively verified in terms of A and b by the Lyapunov function method.

Estimate (3.1.12) can be regarded as the estimate

$$\int \Bigl\langle \frac{\nabla \varrho}{\varrho}, \frac{\nabla \varrho}{\varrho} \Bigr\rangle \, d\mu \leqslant \int \langle b, b \rangle \, d\mu$$

with respect to the Riemannian geometry generated by A. Such an estimate holds indeed for a broad class of Riemannian manifolds (see § 3.7(i)).

In the situation of Theorem 3.1.4, we do not know whether the natural estimate (3.1.12) holds without any extra local assumptions on b and without (3.1.11). However, there is an important special case when (3.1.11) is not needed.

3.1.6. Theorem. *Let A be continuous,* $\det A > 0$, $a^{ij} \in W^{d+,1}_{loc}(\mathbb{R}^d)$, *and* $|b| \in L^{d+}_{loc}(\mathbb{R}^d)$. *Suppose that there exists a quasi-compact function $V \in W^{2,2}_{loc}(\mathbb{R}^d)$ such that*

$$\mathcal{L}_{A,b} V(x) \to -\infty \quad \textit{as } |x| \to +\infty.$$

Assume also that there are numbers $c_1, c_2 > 0$ such that

$$\mathcal{L}_{A,b} V \leqslant c_1 - c_2 |A^{-1/2} b|^2$$

outside some ball. Then there exists a measure $\mu \in \mathcal{P}(\mathbb{R}^d)$ *with a positive density* $\varrho \in W^{d+,1}_{\rm loc}(\mathbb{R}^d)$ *such that* $\mathcal{L}^*_{A,b}\mu = 0$ *and* $|\sqrt{A}\nabla\varrho|^2/\varrho \in L^1(\mathbb{R}^d)$.

If, in addition, there exists a positive Borel function θ *on* $[0,+\infty)$ *such that* $\lim\limits_{t\to\infty}\theta(t) = +\infty$ *and*

$$\mathcal{L}_{A,b}V \leqslant c_1 - c_2\theta(|A^{-1/2}b|)|A^{-1/2}b|^2$$

outside some ball, then $|A^{-1/2}b| \in L^2(\mu)$ *and* (3.1.12) *holds.*

PROOF. We recall (see Remark at the end of the proof of Corollary 2.2.3 and Exercise 2.5.22) that the existence of μ can be shown by taking positive functions $f_j \in W^{p,1}(U_j)$, $U_j = U(0,j)$, $p = p(U_j) > d$, such that

$$\operatorname{div}(A\nabla f_j - f_j b) = 0, \quad f_j|_{\partial U_j} = 1.$$

Then μ is obtained as a weak limit of $\|f_j\|^{-1}_{L^1(U_j)} f_j \cdot dx$. By multiplying the above equation by $\ln f_j$ (which vanishes on ∂U_j) and integrating by parts we obtain

$$\int_{U_j}\Bigl\langle A\nabla f_j, \frac{\nabla f_j}{f_j}\Bigr\rangle\,dx = \int_{U_j}\langle b, \nabla f_j\rangle\,dx = \int_{U_j}\langle A^{-1/2}b, A^{1/2}\nabla f_j\rangle\,dx.$$

The same is true for the normalized probability densities $\varrho_j := f_j\|f_j\|^{-1}_{L^1(U_j)}$, whence by the Cauchy inequality we find that

$$\int_{U_j}\Bigl\langle A\nabla \varrho_j, \frac{\nabla \varrho_j}{\varrho_j}\Bigr\rangle\,dx \leqslant \int_{U_j}|A^{-1/2}b|^2\,\varrho_j\,dx.$$

The right-hand side is uniformly bounded in j due to the estimate

$$\mathcal{L}_{A,b}V \leqslant c_1 - c_2|A^{-1/2}b|^2$$

outside some ball, because it follows from § 2.1 that the integrals of the functions $|\mathcal{L}_{A,b}V|\varrho_j$ over U_j are uniformly bounded. This yields the integrability of $|\sqrt{A}\nabla\varrho|^2/\varrho$, since $\varrho_j \to \varrho$ locally uniformly. The existence of a function θ with the properties mentioned in the formulation yields the uniform boundedness of the integrals of the functions $I_{U_j}\theta(|A^{-1/2}b|)|A^{-1/2}b|^2\varrho_j$, which ensures convergence of the integrals of $I_{U_j}|A^{-1/2}b|^2\varrho_j$ to the integral of $|A^{-1/2}b|^2\varrho$ (see Exercise 3.7.22). □

3.2. Global Sobolev regularity

Theorem 3.1.2 increases the global integrability of ϱ, but obviously cannot ensure the uniform boundedness of ϱ. The latter will be obtained below from the inclusion $\varrho \in W^{p,1}(\mathbb{R}^d)$ with $p > d$ under additional assumptions, but already now we can use local estimates from § 1.6 to establish a uniform bound on ϱ. Along with equations with the operator $L_{A,b}$ we shall consider equations with $\mathcal{L}_{A,b}$, their definition is given in § 1.4. Below $U(x,1)$ is the ball of radius 1 centered at x.

3.2.1. Theorem. *Suppose that* μ *is a probability measure on* $\mathcal{B}(\mathbb{R}^d)$ *such that* $L^*_{A,b}\mu = 0$, *where* A *and* A^{-1} *are uniformly bounded,* A *is Lipschitzian on every ball of radius* 1 *with a constant independent of the center of the ball, and* b *satisfies the following condition with some* $p > d$:

$$\text{either } \sup\nolimits_{x\in\mathbb{R}^d}\|b\|_{L^p(U(x,1))} < \infty \quad \text{or } \sup\nolimits_{x\in\mathbb{R}^d}\|b\|_{L^p(U(x,1),\mu)} < \infty.$$

Then the continuous version ϱ *of the density of* μ *is uniformly bounded.*

If $|b| \in L^p(\mu)$, *then* $\varrho \in W^{p,1}(\mathbb{R}^d)$.

PROOF. It follows from § 1.6 that the maximum of ϱ on $U(z,1)$ is estimated by a number that depends only on $\sup_x \Big[\|A(x)\|+\|A(x)^{-1}\|\Big]$, the Lipschitzian constant for A, and $\|b\|_{L^p(U(z,1))}$ (or $\|b\|_{L^p(U(z,1),\mu)}$ in the second case), hence is majorized by a number independent of z. If $|b| \in L^p(\mu)$, then the $W^{p,1}$-norm of ϱ on a cube K with unit edge length is estimated by $C(\|\varrho\|_{L^1(Q)} + \|b\|_{L^p(Q,\mu)})$, where Q is the cube with the same center and doubled edge, and C does not depend on the center. Hence the $W^{p,1}$-norm of ϱ is finite on the whole space. □

The last assertion will be strengthened below. The simplest example where the hypothesis of the theorem is fulfilled is bounded b.

We now study nonhomogeneous equations

$$L^*_{A,b}\mu = \nu \quad \text{and} \quad \mathcal{L}^*_{A,b}\mu = \nu.$$

We recall (see § 1.1) that, according to the Sobolev embedding theorem, for every number $p > d/(d-1)$, there is a number $C(d,p)$ such that

$$\|\varphi\|_{p,-1} \leqslant C(d,p)\|\varphi\|_{L^s(\mathbb{R}^d)}, \quad s = \frac{pd}{p+d} > 1, \tag{3.2.1}$$

for all $\varphi \in L^s(\mathbb{R}^d)$. If $1 < p < d/(d-1)$, then

$$\|\varphi\|_{p,-1} \leqslant C(d,p)\|\varphi\|_{L^1(\mathbb{R}^d)}. \tag{3.2.2}$$

In the case $p = d/(d-1)$, that is $p' = d$, for any number $r \in (d,+\infty)$ one has $\kappa := d/r \in (0,1)$, i.e.,

$$1 - \frac{n}{p'} = \kappa - \frac{d}{r},$$

and by the Sobolev embedding theorem for fractional classes $H^{r,\kappa}(\mathbb{R}^d)$ one obtains

$$W^{p',1}(\mathbb{R}^d) \subset H^{r,\kappa}(\mathbb{R}^d) \subset L^r(\mathbb{R}^d).$$

Therefore, $L^{r'}(\mathbb{R}^d) \subset W^{p,-1}(\mathbb{R}^d)$ with

$$\|\varphi\|_{p,-1} \leqslant C(d,d',r')\|\varphi\|_{L^{r'}(\mathbb{R}^d)}. \tag{3.2.3}$$

The next theorem about solutions of direct and divergence form equations is very close to the results in § 1.2, but is placed here, because it is often used in this section.

3.2.2. Theorem. *Let $\varepsilon \cdot \mathrm{I} \leqslant A(x)$ for some $\varepsilon > 0$. Let $p > 1$ and let $u \in W^{p,1}_{\mathrm{loc}}(\mathbb{R}^d)$ be such that*

$$u \cdot \operatorname{tr} A \in L^p(\mathbb{R}^d), \quad L_A u \in W^{p,-1}(\mathbb{R}^d).$$

Suppose that A satisfies condition (3.1.8) *for p with the corresponding exponent $q = q(d,p)$ and that the functions $a^{ij} \in W^{q,1}_{\mathrm{loc}}(\mathbb{R}^d)$ are uniformly continuous. Then we have $u \in W^{p,1}(\mathbb{R}^d)$. The same is true in the case of $\mathcal{L}_A$.*

If $u \in W^{p,2}_{\mathrm{loc}}(\mathbb{R}^d)$, $L_A u \in L^p(\mathbb{R}^d)$, $|\nabla u| \operatorname{tr} A \in L^p(\mathbb{R}^d)$, then $u \in W^{p,2}(\mathbb{R}^d)$.

PROOF. First of all we show that under our assumptions $L_A u$ is well-defined as an element of the negative space $W^{p,-1}_{\mathrm{loc}}(\mathbb{R}^d)$, hence our hypothesis that $L_A u$ belongs to $W^{p,-1}(\mathbb{R}^d)$ is meaningful. Indeed, one has $\partial_{x_i}\partial_{x_j} u \in W^{p,-1}_{\mathrm{loc}}(\mathbb{R}^d)$. In addition, a^{ij} is a multiplier of the class $W^{p,-1}_{\mathrm{loc}}(\mathbb{R}^d)$, since for any function $\varphi \in W^{p',1}_0(\mathbb{R}^d)$ one has $a^{ij}|\nabla\varphi| \in L^{p'}(\mathbb{R}^d)$ and $\varphi|\nabla a^{ij}| \in L^{p'}(\mathbb{R}^d)$. The latter inclusion follows easily by the Sobolev embedding theorem. Indeed, if $p > d/(d-1)$, then $p' < d$ and we obtain

$\varphi \in L^{p'd/(d-p')}(\mathbb{R}^d)$, hence by Hölder's inequality with the exponent $r = d/(d-p')$ (so that $r' = d/p'$) and our choice of q, one obtains the integrability of $|\varphi|^{p'}|\nabla a^{ij}|^{p'}$. The cases $p = d/(d-1)$ and $p < d/(d-1)$ are similar.

Without loss of generality we may assume that $2\varepsilon < C_0$, where the number C_0 is chosen as follows: if A_0 is a constant symmetric matrix with $2^{-1}\varepsilon\cdot \mathrm{I} \leqslant A_0 \leqslant 2\varepsilon\cdot \mathrm{I}$, then

$$\|\psi\|_{p,1} \leqslant C_0\|L_{A_0}\psi\|_{p,-1}$$

for every $\psi \in W_0^{p,1}(\mathbb{R}^d)$ having support in a ball of radius 1. Let $\delta \in (0, \varepsilon/2)$ be so small that $1 - \big(C(d,p)+1\big)C_0\delta > 1/2$. Let $r \in (0, 1/2)$ be a number such that $\|A(x) - A(y)\| < \delta$, whenever $|x-y| < 2r$ and

$$\int_{U(z,2r)} \Theta_A(x)^q\, dx \leqslant \delta^q \quad \text{for all } z \in \mathbb{R}^d.$$

Let us show that there is $C > 0$ such that, whenever $v \in W_0^{p,1}(\mathbb{R}^d)$ has support in a cube B of diameter $2r$, we have

$$\|v\|_{p,1} \leqslant C\|L_A v\|_{p,-1}. \tag{3.2.4}$$

Let $A_0 = A(x_0)$, where x_0 is the center of B. Then $L_{A_0}v = L_A v + L_{A_0-A}v$, hence

$$\|v\|_{p,1} \leqslant C_0\|L_A v\|_{p,-1} + C_0\|L_{A_0-A}v\|_{p,-1}.$$

Let us estimate the last term on the right. Given $\varphi \in W^{p',1}(\mathbb{R}^d)$, we have

$$\int \varphi L_{A_0-A}v\, dx = -\int \langle (A_0 - A)\nabla v, \nabla\varphi\rangle\, dx - \int \partial_{x_i}(a_0^{ij} - a^{ij})\partial_{x_j} v\, \varphi\, dx.$$

The first integral on the right is majorized by $\delta\|\nabla v\|_{L^p(B)}\|\nabla\varphi\|_{L^{p'}(B)}$. The second one is estimated by

$$\|\varphi\|_{p',1}\big\|\Theta_A|\nabla v|\big\|_{p,-1}.$$

By (3.2.1) and Hölder's inequality for $p > d/(d-1)$ one has

$$\begin{aligned}\big\||\nabla v|\Theta_A\big\|_{p,-1} &\leqslant C(d,p)\big\||\nabla v|\Theta_A\big\|_{L^s(\mathbb{R}^d)}\\ &\leqslant C(d,p)\|\Theta_A\|_{L^d(B)}\|v\|_{p,1} \leqslant \delta C(d,p)\|v\|_{p,1}.\end{aligned}$$

Hence

$$\|L_{A_0-A}v\|_{p,-1} \leqslant \delta\big(1 + C(d,p)\big)\|v\|_{p,1}.$$

This proves (3.2.4) with $C = 2C_0$ for $p > d/(d-1)$.

If $p < d/(d-1)$, then $p' > d$ and by (3.2.2) and Hölder's inequality

$$\begin{aligned}\big\|\Theta_A|\nabla v|\big\|_{p,-1} &\leqslant C(d,p)\big\|\Theta_A|\nabla v|\big\|_{L^1(\mathbb{R}^d)}\\ &\leqslant C(d,p)\|v\|_{p,1}\|\Theta_A\|_{L^{p'}(B)} \leqslant \delta C(d,p)\|v\|_{p,1},\end{aligned}$$

which again leads to (3.2.4). Finally, if $p = d/(d-1)$, then we apply (3.2.3) with any $r > d$. For r' sufficiently close to 1 we use Hölder's inequality and obtain (3.2.4).

It is easily seen that $\mathbb{R}^d$ can be covered by finitely many collections $\mathcal{F}_1, \ldots, \mathcal{F}_N$ of cubes with the following property: each $\mathcal{F}_i$ consists of a sequence of cubes K_j of diameter r such that the cubes Q_j with the same centers and twice bigger diameters are disjoint. Let us fix one of these collections and take functions $\zeta_j \in C_0^\infty(\mathbb{R}^d)$ such that

$$0 \leqslant \zeta_j \leqslant 1,\ \sup_{j,x}\Big[|\nabla\zeta_j(x)| + \|D^2\zeta_j(x)\|\Big] := M < \infty,\ \operatorname{supp}\zeta_j \subset Q_j,\ \zeta_j|_{K_j} = 1.$$

We have

$$L_A(\zeta_j u) = \zeta_j L_A u + 2\partial_{x_k}(a^{ik}\partial_{x_i}\zeta_j u) - 2u\partial_{x_k}(a^{ik}\partial_{x_i}\zeta_j) + uL_A\zeta_j$$
$$= \zeta_j L_A u + 2\partial_{x_k}(a^{ik}\partial_{x_i}\zeta_j u) - 2u\partial_{x_k}a^{ik}\partial_{x_i}\zeta_j - uL_A\zeta_j.$$

Therefore,

$$\|\zeta_j u\|_{p,1} \leqslant C\Big[\|\zeta_j L_A u\|_{p,-1} + 2M\sum_{i,k=1}^{d}\|a^{ik}u\|_{L^p(Q_j)} + 2\|u\partial_{x_k}a^{ik}\partial_{x_i}\zeta_j\|_{p,-1} + \|uL_A\zeta_j\|_{p,-1}\Big]. \tag{3.2.5}$$

Let us estimate separately each term on the right. The second term is estimated as follows:

$$2M\sum_{i,k=1}^{d}\|a^{ik}u\|_{L^p(Q_j)} \leqslant 2Md^2\|u\mathrm{tr}\, A\|_{L^p(Q_j)}.$$

For the last term one has

$$\|uL_A\zeta_j\|_{p,-1} \leqslant \|uL_A\zeta_j\|_{L^p(Q_j)} \leqslant Md^2\|u\mathrm{tr}\, A\|_{L^p(Q_j)}.$$

Let us consider the third term. If $d/(d-1) < p$, one has

$$2\|u\partial_{x_k}a^{ik}\partial_{x_i}\zeta_j\|_{p,-1} \leqslant 2C(p,d)\|u\partial_{x_k}a^{ik}\partial_{x_i}\zeta_j\|_{L^s(Q_j)}$$

with $s = pd/(p+d)$. By Hölder's inequality with the exponents $(p+d)/d$ and $(p+d)/p$ and condition (3.1.8) we find that

$$\|u\partial_{x_k}a^{ik}\partial_{x_i}\zeta_j\|_{L^s(Q_j)} \leqslant \|u\|_{L^p(Q_j)}\|\partial_{x_k}a^{ik}\partial_{x_i}\zeta_j\|_{L^d(Q_j)} \leqslant const\|u\|_{L^p(Q_j)}.$$

Note also that

$$\|u\|_{L^p(Q_j)} \leqslant \varepsilon^{-1}\|u\mathrm{tr}\, A\|_{L^p(Q_j)}.$$

If $1 < p < d/(d-1)$, we have $p' > d$, hence $W^{p',1}(\mathbb{R}^d) \subset L^\infty(\mathbb{R}^d)$. In addition, $q = p'$, so that $\|\Theta_A\|_{L^{p'}(Q_j)} \leqslant M_1 < \infty$, where M_1 is a constant independent of j. Therefore,

$$\|u\partial_{x_k}a^{ik}\partial_{x_i}\zeta_j\|_{p,-1} \leqslant C(d,p)\|u\partial_{x_k}a^{ik}\partial_{x_i}\zeta_j\|_{L^1(Q_j)}$$
$$\leqslant MC(d,p)\|u\|_{L^p(Q_j)}\|\Theta_A\|_{L^{p'}(Q_j)} \leqslant MM_1C(d,p)\varepsilon^{-1}\|u\mathrm{tr}\, A\|_{L^p(Q_j)}.$$

In the remaining case $p = d/(d-1)$ we apply (3.2.3) and find that

$$\|u\partial_{x_k}a^{ik}\partial_{x_i}\zeta_j\|_{p,-1} \leqslant C(d,d',r')\|u\partial_{x_k}a^{ik}\partial_{x_i}\zeta_j\|_{L^{r'}(\mathbb{R}^d)}.$$

We choose the number r' sufficiently close to 1 and estimate the right-hand side by Hölder's inequality through $const\|u\|_{L^p(Q_j)}$, which is possible, since $q = d+\gamma > d$ and the quantities $\|\Theta_A\|_{L^q(Q_j)}$ are uniformly bounded.

Taking into account that $\zeta_j|_{K_j} = 1$, we obtain from (3.2.5) and the foregoing estimates that

$$\||\nabla u|\|^p_{L^p(K_j)} \leqslant C_2\big[\|\zeta_j L_A u\|^p_{p,-1} + \|u\mathrm{tr}\, A\|^p_{L^p(Q_j)}\big], \tag{3.2.6}$$

where C_2 is a constant independent of j. Let us observe that there is a constant N_0 such that for every $\nu \in W^{p,-1}(\mathbb{R}^d)$ one has

$$\sum_{j=1}^{\infty}\|\zeta_j\nu\|^p_{p,-1} \leqslant N_0\|\nu\|^p_{p,-1}. \tag{3.2.7}$$

Indeed, one has $\nu = g_0 - \sum_{i=1}^d \partial_{x_i} g_i$, where $g_i \in L^p(\mathbb{R}^d)$. To see this, it suffices to write $\nu = g_0 - \Delta g_0$, $g_0 \in W^{p,1}(\mathbb{R}^d)$, and set $g_i := \partial_{x_i} g_0$. Then

$$N_1 \sum_{i=0}^d \|g_i\|_{L^p(\mathbb{R}^d)} \leqslant \|\nu\|_{p,-1} \leqslant N_2 \sum_{i=0}^d \|g_i\|_{L^p(\mathbb{R}^d)} \tag{3.2.8}$$

with some constants N_1 and N_2 independent of ν, because $I - \Delta$ is an isomorphism between $W^{p,1}(\mathbb{R}^d)$ and $W^{p,-1}(\mathbb{R}^d)$. Since $\|\zeta_j g_0\|_{p,-1} \leqslant \|\zeta_j g_0\|_{L^p(\mathbb{R}^d)}$ and

$$\|\zeta_j \partial_{x_i} g_i\|_{p,-1} = \|\partial_{x_i}(\zeta_j g_i) - g_i \partial_{x_i} \zeta_j\|_{p,-1} \leqslant (1+M)\|g_i\|_{L^p(Q_j)},$$

by (3.2.8) we find that

$$\begin{aligned}\sum_{j=1}^\infty \|\zeta_j \nu\|_{p,-1}^p &\leqslant (n+1)^p \sum_{j=1}^\infty \Bigl[\|\zeta_j g_0\|_{p,-1}^p + \sum_{i=1}^d \|\zeta_j \partial_{x_i} g_i\|_{p,-1}^p\Bigr] \\ &\leqslant (1+M)^p (d+1)^p \sum_{j=1}^\infty \sum_{i=0}^d \|g_i\|_{L^p(Q_j)}^p \leqslant (1+M)^p (d+1)^p \sum_{i=0}^d \|g_i\|_{L^p(\mathbb{R}^d)}^p \\ &\leqslant (1+M)^p (d+1)^p N_1^{-p} \|\nu\|_{p,-1}^p.\end{aligned}$$

Therefore, estimate (3.2.6) yields

$$\||\nabla u|\|_{L^p(\mathbb{R}^d)}^p \leqslant NC_2 \|u \operatorname{tr} A\|_{L^p(\mathbb{R}^d)}^p + N_0 C_2 \|L_A u\|_{p,-1}^p.$$

Clearly, $u \in L^p(\mathbb{R}^d)$ by our assumptions that $u \operatorname{tr} A \in L^p(\mathbb{R}^d)$ and $A \geqslant \varepsilon \cdot \mathrm{I}$. So, the proof in the case of L_A is complete.

The case of $\mathcal{L}_A$ is similar. We only note that $\mathcal{L}_A u \in W^{p,-1}_{\mathrm{loc}}(\mathbb{R}^d)$, because $\partial_{x_i} a^{ij} \partial_{x_j} u \in L^{pq/(p+q)}_{\mathrm{loc}}(\mathbb{R}^d)$, and that

$$\mathcal{L}_A(\zeta_j u) = \zeta_j \mathcal{L}_A u + 2\partial_{x_i}(a^{ik} u \partial_{x_k} \zeta_j) - u \partial_{x_i}(a^{ik} \partial_{x_k} \zeta_j).$$

The $W^{p,-1}$-norm of the right-hand side is estimated along the same lines as in the first case. The last assertion of the theorem is proved similarly by using the obvious equality $L_A(\zeta_j u) = \zeta_j L_A u + u L_A \zeta_j + 2\langle A\nabla\zeta_j, \nabla u\rangle$. □

We emphasize that in Theorem 3.2.2 no boundedness of A is assumed. Part of the hypotheses of Theorem 3.2.2 is the integrability of the solution to a power bigger than 1. Since we intend to apply this theorem to measures, it is desirable to ensure such an integrability for the densities of measures satisfying equations of the type $L^*\mu = \nu$. The next three theorems serve this purpose. Their proofs use Theorem 1.2.1 on isomorphisms defined by the operators $L_A - \lambda$ and $\mathcal{L}_A - \lambda$. We note at once that this theorem gives the uniqueness of a solution to the equation $\mathcal{L}_A F - F = G \in W^{p,-1}$ in the class $W^{p,1}$, but not in broader classes of distributions to which a priori can belong our measure μ.

3.2.3. Theorem. *Let $\mu \in \mathcal{M}(\mathbb{R}^d)$ be such that $L_{A,b}^*\mu = \nu \in W^{p,-1}(\mathbb{R}^d)$ for some $p \in \bigl(1, \frac{d}{d-1}\bigr)$ and $|b| \in L^1(|\mu|)$. Suppose that the functions a^{ij} belong to VMO, for example, are uniformly continuous, and $c_1 \cdot \mathrm{I} \leqslant A(x) \leqslant c_2 \cdot \mathrm{I}$ for some constants $c_1, c_2 > 0$. Then μ has a density in $L^r(\mathbb{R}^d)$ for every $r \in [1, p]$.*

In the case of the operator $\mathcal{L}_{A,b}$ the same is true under the additional assumption that $a^{ij} \in W^{1,1}_{\mathrm{loc}}(\mathbb{R}^d)$ for all i, j and $\partial_{x_i} a^{ij} \in L^1(|\mu|)$ for every j.

PROOF. Let us consider the case of $L_{A,b}$. We know that μ has a density ϱ, hence

$$|b|\varrho \in L^1(\mathbb{R}^d) \subset W^{p,-1}(\mathbb{R}^d),$$

because $p < d/(d-1)$. For any test function φ we have

$$\begin{aligned}\Bigl|\int (L_A\varphi - \varphi)\, d\mu\Bigr| &= \Bigl|-\int \varphi\, d\mu - \int \langle b, \nabla\varphi\rangle\, d\mu + \nu(\varphi)\Bigr| \\ &\leqslant C \sup_x \bigl[|\varphi(x)| + |\nabla\varphi(x)|\bigr] + C\|\varphi\|_{p',1}, \end{aligned} \tag{3.2.9}$$

where C is a number independent of φ. If we fix a ball $U \subset \mathbb{R}^d$, then for all $\varphi \in C_0^\infty(U)$ we have

$$\|\varphi\|_{p',1} \leqslant C(U) \sup_{x\in U} \bigl[|\varphi(x)| + |\nabla\varphi(x)|\bigr],$$

hence (3.2.9) yields

$$\Bigl|\int (L_A\varphi - \varphi)\, d\mu\Bigr| \leqslant C\bigl(1 + C(U)\bigr) \sup_{x\in U} \bigl[|\varphi(x)| + |\nabla\varphi(x)|\bigr]$$

for all $\varphi \in C_0^\infty(U)$. From § 1.5 we know that $\varrho \in L^r_{\mathrm{loc}}(\mathbb{R}^d)$ for any $r < d/(d-1)$. Hence estimate (3.2.9) extends to all $\varphi \in W_0^{p',2}(\mathbb{R}^d)$. Note that $p' > d$, hence we have a continuous embedding of $W^{p',2}(\mathbb{R}^d)$ to $C_b^1(\mathbb{R}^d)$. Let $\psi \in C_0^\infty(\mathbb{R}^d)$. There is a function $\varphi \in W^{p',2}(\mathbb{R}^d)$ such that $L_A\varphi - \varphi = \psi$ and

$$\sup_x \bigl[|\varphi(x)| + |\nabla\varphi(x)|\bigr] + \|\varphi\|_{p',1} \leqslant M\|\varphi\|_{p',2} \leqslant M_0\|\psi\|_{L^{p'}(\mathbb{R}^d)},$$

where M and M_0 are independent of ψ (see Theorem 1.2.1). Let us take a sequence of functions $\zeta_k \in C_0^\infty(\mathbb{R}^d)$ such that $0 \leqslant \zeta_k \leqslant 1$, $\zeta_k(x) = 0$ if $|x| \geqslant k+1$, $\zeta_k(x) = 1$ if $|x| \leqslant k$, and

$$\sup_k \sup_x \bigl[|\nabla\zeta_k(x)| + \|D^2\zeta_k(x)\|\bigr] = M_1 < \infty.$$

We have

$$\begin{aligned} L_A(\zeta_k\varphi) - \zeta_k\varphi &= \zeta_k(L_A\varphi - \varphi) + 2\langle A\nabla\zeta_k, \nabla\varphi\rangle + \varphi L_A\zeta_k \\ &= \zeta_k\psi + 2(\langle\nabla\zeta_k, \nabla\varphi\rangle + \varphi L_A\zeta_k. \end{aligned}$$

Therefore,

$$\begin{aligned} \int \zeta_k\psi\, d\mu &= \int (L_A(\zeta_k\varphi) - \zeta_k\varphi)\, d\mu - 2\int \langle A\nabla\zeta_k, \nabla\varphi\rangle\, d\mu - \int \varphi L_A\zeta_k\, d\mu \\ &\leqslant C \sup_x \bigl[|\zeta_k(x)\varphi(x)| + |\nabla(\zeta_k(x)\varphi(x))|\bigr] + C\|\zeta_k\varphi\|_{p',1} \\ &\quad + (2c_2M_1M_0 + M_0M_1n^2c_2)\|\psi\|_{L^{p'}(\mathbb{R}^d)}|\mu|\bigl(\{k \leqslant |x| \leqslant k+1\}\bigr). \end{aligned}$$

The last term on the right converges to zero as $k \to \infty$. Therefore, we obtain

$$\begin{aligned} \int \psi(x)\varrho(x)\, dx &\leqslant C \sup_x \bigl[|\varphi(x)| + M_1|\varphi(x)| + |\nabla\varphi(x)|\bigr] + C\|\varphi\|_{p',1} \\ &\leqslant M_0C(1 + M_1)\|\psi\|_{L^{p'}(\mathbb{R}^d)}, \end{aligned}$$

which yields the inclusion $\varrho \in L^p(\mathbb{R}^d)$. Since $\varrho \in L^1(\mathbb{R}^d)$, one has the inclusion $\varrho \in L^r(\mathbb{R}^d)$ for all $r \in [1, p]$. In the case of the operator $\mathcal{L}_{A,b}$ it suffices to note that $\mathcal{L}^*_{A,b}\mu = L^*_{A,b_0}\mu$ with $b_0^i = b^i - \partial_{x_j}a^{ij} \in L^1(|\mu|)$. □

3.2.4. Theorem. *Let $\mu \in \mathcal{M}(\mathbb{R}^d)$ be such that $\mathcal{L}_{A,b}^*\mu = \nu$. Suppose that*

(a) *$A \geqslant \varepsilon \cdot \mathrm{I}$ with some $\varepsilon > 0$, $a^{ij} \in W^{\alpha,1}_{\mathrm{loc}}(\mathbb{R}^d)$, and either $|b| \in L^{\alpha}_{\mathrm{loc}}(\mathbb{R}^d)$ or $|b| \in L^{\alpha}_{\mathrm{loc}}(|\mu|)$, where $\alpha > d$,*

(b) *$|b|, \operatorname{tr} A \in L^{\beta}(|\mu|)$, where $\beta > 1$,*

(c) *A satisfies (3.1.8) for the number β in (b) and is uniformly continuous.*

Assume also that the density ϱ of μ belongs to $L^{\beta_0}(\mathbb{R}^d)$ for some $\beta_0 > 1$, which is automatically the case in (i)–(iii) below if A is uniformly bounded and $\partial_{x_i} a^{ij} \in L^{\beta}(|\mu|)$ for each j.

(i) *Let $1 < \beta < d$ and let $\nu \in W^{\theta,-1}$ for all numbers $\theta \in \bigl(1, \frac{d}{d-\beta+1}\bigr)$. Then $\varrho \in W^{r,1}(\mathbb{R}^d)$ for all $r \in \bigl(1, \frac{d}{d-\beta+1}\bigr)$. Moreover, if the measure μ is nonnegative, then this inclusion is true for $r = \frac{d}{d-\beta+1}$.*

(ii) *Let $\beta = d$ and $\nu \in W^{\theta,-1}(\mathbb{R}^d)$ for all $\theta \in (1, d)$. Then $\varrho \in W^{r,1}(\mathbb{R}^d)$ for all $r \in (1, d)$.*

(iii) *Let $d < \beta \leqslant \alpha$ and $\nu \in W^{\theta,-1}(\mathbb{R}^d)$ for all $\theta \in (1, \beta]$. Then we have $\varrho \in W^{r,1}(\mathbb{R}^d)$ for any $r \in (1, \beta]$. In particular, $\varrho \in L^{\infty}(\mathbb{R}^d)$.*

The same is true in the case of the operator $L_{A,b}$ provided that one has, in addition, $\partial_{x_i} a^{ij} \in L^{\beta}(|\mu|)$ for each j.

PROOF. (i) Suppose we know that $\varrho \in L^{\beta_k}(\mathbb{R}^d)$ for some $\beta_k \geqslant \beta_0$. Let

$$p_k = \frac{\beta\beta_k}{\beta + \beta_k - 1}. \tag{3.2.10}$$

By Hölder's inequality with the exponent $t = \beta/p_k$ we obtain

$$|\varrho b| = |\varrho|^{1-1/\beta} |\varrho|^{1/\beta} |b| \in L^{p_k}(\mathbb{R}^d).$$

Hence $\partial_{x_i}(b^i \varrho) \in W^{p_k,-1}(\mathbb{R}^d)$. By the same reasoning $\partial_{x_i}(b^i \varrho) \in W^{s,-1}(\mathbb{R}^d)$ for every $s \in (1, p_k]$. Hence $\mathcal{L}_A \varrho \in W^{s,-1}(\mathbb{R}^d)$ for each $s \in (1, p_k]$. Note that we have $\varrho \operatorname{tr} A \in L^{s}(\mathbb{R}^d)$ for each $s \in (1, p_k]$, which follows as above by Hölder's inequality from our assumption that $\operatorname{tr} A \in L^{\beta}(|\mu|)$. Theorem 3.2.2 yields $\varrho \in W^{s,1}(\mathbb{R}^d)$ whenever $s \in (1, p_k]$. By the Sobolev embedding theorem $\varrho \in L^{\beta_{k+1}}(\mathbb{R}^d)$ with

$$\beta_{k+1} = \frac{d\beta\beta_k}{d(\beta + \beta_k - 1) - \beta\beta_k}.$$

Starting from β_0 and iterating we obtain a sequence $\{\beta_k\}$ that is increasing and, as one can easily verify, converges to $d/(d - \beta)$. By using (3.2.10) once again we conclude that $\varrho \in W^{r,1}(\mathbb{R}^d)$ for all $r \in \bigl(1, d/(d - \beta + 1)\bigr)$. Other cases are similar. We only note that in the case $\beta > d$ we first obtain that $\varrho \in W^{p,1}(\mathbb{R}^d)$ with any number $p \in (d, \beta)$, which yields that $\varrho \in L^{\infty}(\mathbb{R}^d)$, so $|\varrho b| \in L^{\beta}(\mathbb{R}^d)$. To this end we apply (ii), which gives $\varrho \in L^{p}(\mathbb{R}^d)$ for all $p < \infty$. Then $\varrho|b|, \varrho \operatorname{tr} A \in L^{s}(\mathbb{R}^d)$ for all $s < \beta$, hence $\varrho \in W^{s,1}(\mathbb{R}^d)$.

Let us note that if A is bounded and $\partial_{x_i} a^{ij} \in L^{\beta}(|\mu|)$ for each j, then, in each of the cases (i)–(iii), Theorem 3.2.3 yields that $\varrho \in L^{\beta_0}(\mathbb{R}^d)$ with some $\beta_0 > 1$. The case of $L_{A,b}$ reduces to that of $\mathcal{L}_{A,b_0}$ as above. □

3.2.5. Corollary. *Suppose that $\mu \in \mathcal{P}(\mathbb{R}^d)$ satisfies the equation $\mathcal{L}_{A,b}^*\mu = 0$ and there is a number $\alpha > d$ such that $|b|, \operatorname{tr} A \in L^{\alpha}(\mu)$, $a^{ij} \in W^{\alpha,1}_{\mathrm{loc}}(\mathbb{R}^d)$ are uniformly continuous, conditions (3.1.8) and (3.1.11) hold, and $A \geqslant \varepsilon \cdot \mathrm{I}$, $\varepsilon > 0$. Then $\mu = \varrho\, dx$, where $\varrho \in W^{\alpha,1}(\mathbb{R}^d)$. In particular, $\varrho \in L^{\infty}(\mathbb{R}^d)$.*

In particular, the conclusion holds true if we have (3.1.10) *and, in addition,*

$$A \geqslant \varepsilon \cdot \mathrm{I},\ |b| \in L^\alpha(\mu),\ \operatorname{tr} A \in L^\alpha(\mu),\ (1+|x|)^{-1}\Theta_A \in L^1(\mu).$$

The case $\beta = 1$ in the above theorem has not been studied so far. In particular, it is unknown whether $\varrho \in W^{1,1}(\mathbb{R}^d)$, i.e., $|\nabla\varrho/\varrho| \in L^1(\mu)$, if $|b| \in L^1(\mu)$ and μ is a probability measure satisfying the equation $L^*_{\mathrm{I},b}\mu = 0$. If $|b| \in L^2(\mu)$, then this is true by Theorem 3.1.2. In addition, it is unknown whether in the same case we have $|\nabla\varrho/\varrho| \in L^p(\mu)$ if $|b| \in L^p(\mu)$ with some exponent $p \neq 2$. Additional sufficient conditions for this are given in Theorem 3.4.5. So, without any additional assumptions such as the existence of Lyapunov functions or bounds on the coefficients, the inclusion $\varrho \in W^{1,1}(\mathbb{R}^d)$ is established so far (even for $A = \mathrm{I}$) only under the assumption that $|b| \in L^2(\mu)$. It is clear that ϱ may not belong to the class $W^{p,2}$ unless we require certain regularity of b.

3.2.6. Theorem. *Suppose that* $\alpha \geqslant 2d$, $A \geqslant \varepsilon \cdot \mathrm{I}$ *with some number* $\varepsilon>0$, $a^{ij}, b^i \in W^{\alpha,1}_{\mathrm{loc}}(\mathbb{R}^d)$, *$A$ is uniformly continuous and satisfies condition* (3.1.8) *for* α. *Let* $\mu \in \mathcal{P}(\mathbb{R}^d)$ *satisfy* $\mathcal{L}^*_{A,b}\mu = \nu$, *where* $\nu \in L^r(\mathbb{R}^d)$ *for all* $r \in (1,\alpha]$. *Assume also that* $|b|, \operatorname{div} b \in L^\alpha(\mu)$ *and* $\operatorname{tr} A, |\nabla a^{ij}| \in L^\theta(\mu)$ *for some* $\theta \geqslant \max(2d^2,\alpha)$. *Then* $\mu = \varrho\,dx$, *where* $\varrho \in W^{r,2}(\mathbb{R}^d)$ *for all* $r \in (1,\alpha/2)$.

If $\alpha > 2d$, *then we have* $\varrho \in W^{\alpha/2,2}(\mathbb{R}^d)$ *and* $|\nabla\varrho| \in L^\infty(\mathbb{R}^d)$.

If, in addition, $a^{ij}, |\nabla a^{ij}| \in L^\infty(\mathbb{R}^d)$, *then these assertions are true for any solution μ from the class of all measures of bounded variation.*

PROOF. We consider the case $d > 1$, since the case $d = 1$ is simple and can be verified directly. From Theorem 1.1.1 we obtain $\nu \in W^{\theta,-1}(\mathbb{R}^d)$ for $\theta \in (1,\alpha]$. Suppose first that μ is a probability measure. By case (iii) of the previous theorem μ has a bounded positive continuous density ϱ in $W^{\beta,1}(\mathbb{R}^d)$ with $1 < \beta \leqslant \alpha$, i.e., $|\nabla\varrho| \in L^\beta(\mathbb{R}^d)$ if $1 < \beta \leqslant \alpha$. In addition, it is known from the local theory (and follows from Theorem 1.2.1) that $\varrho \in W^{\alpha,2}_{\mathrm{loc}}$. Therefore,

$$L_A\varrho = -\partial_{x_i} a^{ij}\partial_{x_j}\varrho + \varrho \operatorname{div} b + \langle b, \nabla\varrho\rangle + \nu.$$

Since ϱ is bounded, we obtain that $\varrho\operatorname{div} b \in L^s(\mathbb{R}^d)$ if $s \leqslant \alpha$. Note that $\Theta_A \in L^\theta(\mu)$ and $\|A(x)\| \leqslant c_1 + c_2|x|$ by the uniform continuity of A. Hence by Theorem 3.1.4 we have $|\nabla\varrho|/\varrho \in L^2(\mu)$. Letting

$$\gamma := \frac{1}{2}\theta\frac{\alpha+2}{\alpha+\theta}$$

and noting that

$$\frac{1}{2} + \frac{\gamma}{\theta} + \frac{\gamma-1}{\alpha} = 1,\ |\nabla a^{ij}|\varrho^{1/2} \in L^\theta(\mathbb{R}^d),\ |\nabla\varrho| \in L^\alpha(\mathbb{R}^d),$$

we obtain from the generalized Hölder inequality (see p. 2) that

$$|\nabla a^{ij}|^\gamma|\nabla\varrho|^\gamma = |\nabla a^{ij}|^\gamma\varrho^{1/2}|\nabla\varrho|^{\gamma-1}|\nabla\varrho|\varrho^{-1/2} \in L^1(\mathbb{R}^d).$$

Hence $|\partial_{x_i} a^{ij}\partial_{x_j}\varrho|^\gamma \in L^1(\mathbb{R}^d)$.

Note that $\gamma \geqslant d$ and the equality is only possible if $\alpha = 2d$ and $\theta = 2d^2$. Indeed, the inequality $\gamma \geqslant d$ is equivalent to

$$\theta \geqslant 2d\frac{\alpha}{\alpha+2-2d} = 2d\Big(1+\frac{2d-2}{\alpha+2-2d}\Big).$$

As $\alpha \geqslant 2d$, the right-hand side attains its maximum $2d^2$ at $\alpha = 2d$.

In addition, $\langle b, \nabla\varrho\rangle \in L^p(\mathbb{R}^d)$, where $1 \leqslant 3\alpha/(2\alpha+2) < p \leqslant \alpha/4+1/2$ (note that $3\alpha/(2\alpha+2) < \alpha/4+1/2$ as $\alpha > 2$). Indeed, we have by the Cauchy inequality

$$|\langle b, \nabla\varrho\rangle|^p \leqslant |b|^p|\nabla\varrho|^{p-1}\varrho^{1/2}|\nabla\varrho|\varrho^{-1/2} \in L^1(\mathbb{R}^d),$$

since $|\nabla\varrho|\varrho^{-1/2} \in L^2(\mathbb{R}^d)$, the function ϱ is bounded and $|b|^{2p}|\nabla\varrho|^{2p-2}\varrho \in L^1(\mathbb{R}^d)$. The latter inclusion is easily verified by Hölder's inequality with conjugated exponents $t = \alpha/(\alpha+2-2p)$ and $s = \alpha/(2p-2) > 1$ by using the fact that

$$|b|^{2pt}\varrho^t \in L^1(\mathbb{R}^d),\ |\nabla\varrho|^\alpha \in L^1(\mathbb{R}^d), \quad \text{where } 2pt = 2p\alpha/(\alpha+2-2p) \leqslant \alpha.$$

The latter is true by the estimate $4p \leqslant \alpha+2$. The same estimates hold for $\operatorname{tr} A$ in place of $|b|$.

By the last assertion of Theorem 3.2.2 we obtain $\varrho \in W^{p,2}(\mathbb{R}^d)$. If A is uniformly bounded, then also Theorem 1.2.1 applies. Moreover, one has $\varrho \in W^{s,2}(\mathbb{R}^d)$ for any $s \in (1,p]$, which is proved by similar estimates. Under the condition $\alpha/4+1/2 \geqslant d$ we obtain at once by the Sobolev embedding theorem that the inclusion $|\nabla\varrho| \in L^s(\mathbb{R}^d)$ holds for all $s \in (1,+\infty)$.

If $p = \alpha/4+1/2 < d$, then we show that one can increase the order p to $p \geqslant d$ by several iterations. Namely, from the inclusion $|\nabla\varrho| \in L^q(\mathbb{R}^d)$ we deduce the inclusion

$$\varrho \in W^{r,2}(\mathbb{R}^d) \quad \text{with } 1 < r \leqslant t,\ \frac{1}{t} = \Bigl(1-\frac{2}{\alpha}\Bigr)\frac{1}{q}+\frac{2}{\alpha}.$$

To this end, we observe that by the generalized Hölder inequality with three exponents w, r, r, where $1/w+1/r+1/r = 1$, i.e., $w = r/(r-2)$, we have

$$|b|^t|\nabla\varrho|^t = |b|^t\varrho^{1/r}|\nabla\varrho|^{2/r}\varrho^{-1/r}|\nabla\varrho|^{t-2/r} \in L^1(\mu),$$

since $|b|^{rt}\varrho, |\nabla\varrho|^2/\varrho, |\nabla\varrho|^{(rt-2)/(r-2)} \in L^1(\mathbb{R}^d)$, because for the indicated value of t we have $rt = \alpha$ and $(rt-2)/(r-2) = q$. The same is true for $\operatorname{tr} A$ in place of $|b|$, so that the last assertion of Theorem 3.2.2 is again applicable. Now we can take $q_0 = \alpha/4+1/2$ and set

$$p_n = \Bigl(1-\frac{2}{\alpha}\Bigr)\frac{1}{q_n}+\frac{2}{\alpha}, \qquad \frac{1}{q_{n+1}} = \frac{1}{p_{n+1}}-\frac{1}{d}.$$

As shown above, the inclusion $|\nabla\varrho| \in L^{q_n}(\mathbb{R}^d)$ yields the inclusion $\varrho \in W^{p_n,2}(\mathbb{R}^d)$. If we admit that $p_n < d$ for all indices n, then by the Sobolev embedding theorem $|\nabla\varrho| \in L^{q_{n+1}}(\mathbb{R}^d)$ for all n. Since $q_0 \geqslant d/2$, we have $1/q_{n+1} \leqslant 1/q_n$, i.e., the numbers q_n increase to some l, but this is only possible if $l = \infty$, since otherwise we have $1/l = (1-2/\alpha)/l + 2/\alpha - 1/d < (1-2/\alpha)/l$, which is impossible.

Thus, $\varrho \in W^{r,2}(\mathbb{R}^d)$ for all $r < \alpha/2$. If $\alpha > 2d$, then this implies the inclusion $|\nabla\varrho| \in L^\infty(\mathbb{R}^d)$, so that we finally obtain that $\varrho \in W^{\alpha/2,2}(\mathbb{R}^d)$.

In the case where $a^{ij}, |\nabla a^{ij}| \in L^\infty(\mathbb{R}^d)$, some estimates above simplify and we do not need the integrability of $|\nabla\varrho|^2/\varrho$, so that our reasoning applies to signed measures. □

3.3. Upper estimates for densities

We now turn to pointwise bounds of solutions. The idea is simple: in order to show that $|\varrho(x)| \leqslant C\Psi(x)$ for some positive function Ψ, one has to consider the measure μ_0 with density ϱ/Ψ and verify that this measure satisfies an equation of the type considered in Theorem 3.2.4. Here we continue to study equations with divergence form operators $\mathcal{L}_{A,b}$.

3.3.1. Theorem. *Suppose that μ is a probability measure satisfying the equation $\mathcal{L}^*_{A,b}\mu = 0$, where A and b satisfy hypotheses* (a)–(b) *of Theorem* 3.2.4 *for some number $\alpha = \beta > d$. Let $\Phi \in W^{1,1}_{\rm loc}(\mathbb{R}^d)$ be a positive function such that for some $\theta > d$ and all $j = 1, \ldots, d$ one has*

$$\Phi \in L^1(\mu), \ |\nabla\Phi| \in L^\theta(\mu), \ \partial_{x_i}a^{ij} \in L^d(\mu).$$

Then the density ϱ of μ satisfies $\varrho(x) \leqslant C\Phi(x)^{-1}$ with some constant C.

PROOF. We may assume that $d < \theta \leqslant \beta$. Let us consider the bounded positive measure $\mu_0 = \Phi \cdot \mu$. It is easily verified that

$$\mathcal{L}^*_{A,b}\mu_0 = -\langle b, \nabla\Phi\rangle\varrho - \partial_{x_i}a^{ij}\partial_{x_j}\Phi\varrho + 2\partial_{x_i}(a^{ij}\partial_{x_j}\Phi\varrho).$$

Now it suffices to show that the right-hand side belongs to the negative space $W^{s,-1}(\mathbb{R}^d)$ for all $s \in (1, \theta]$. Clearly, $a^{ij}\partial_{x_j}\Phi\varrho \in L^s(\mathbb{R}^d)$ whenever $s \leqslant \theta$ by the boundedness of a^{ij} and ϱ and the inclusion $|\nabla\Phi| \in L^\theta(\mu)$. In addition, $L^{d\theta/(d+\theta)}(\mathbb{R}^d) \subset W^{\theta,-1}(\mathbb{R}^d)$, so it remains to verify that $\langle b, \nabla\Phi\rangle\varrho$ and $\partial_{x_i}a^{ij}\partial_{x_j}\Phi\varrho$ belong to $L^{d\theta/(d+\theta)}(\mathbb{R}^d)$, which by the boundedness of ϱ reduces to proving that $\langle b, \nabla\Phi\rangle$ and $\partial_{x_i}a^{ij}\partial_{x_j}\Phi$ belong to $L^{d\theta/(d+\theta)}(\mu)$. The latter is true by Hölder's inequality, since $\partial_{x_i}a^{ij} \in L^d(\mu)$ and $|\nabla\Phi| \in L^\theta(\mu)$. The same reasoning shows that $\langle b, \nabla\Phi\rangle\varrho$ and $\partial_{x_i}a^{ij}\partial_{x_j}\Phi\varrho$ belong to $W^{s,-1}(\mathbb{R}^d)$ for any s with $d/(d-1) < s \leqslant \theta$. Finally, these functions belong also to $W^{s,-1}(\mathbb{R}^d)$ if $1 < s \leqslant d/(d-1)$, because $\langle b, \nabla\Phi\rangle$ and $\partial_{x_i}a^{ij}\partial_{x_j}\Phi$ belong to all $L^r(\mu)$ with $1 \leqslant r \leqslant d\theta/(d+\theta)$ and ϱ is bounded. By Theorem 3.2.4 one has $\Phi\varrho \in W^{\theta,1}(\mathbb{R}^d)$. Hence $\Phi\varrho \in L^\infty(\mathbb{R}^d)$. □

We shall now see that the existence of polynomial or exponential moments of solutions yields the corresponding decay of densities at infinity.

3.3.2. Example. Let $\mu \in \mathcal{P}(\mathbb{R}^d)$ satisfy the equation $\mathcal{L}^*_{A,b}\mu = 0$, where A is uniformly Lipschitzian and A and A^{-1} are uniformly bounded. Furthermore, assume that $|b| \in L^p(\mu)$ for some $p > d$.

(i) If $\Phi \in W^{1,1}_{\rm loc}(\mathbb{R}^d)$ is a positive function such that

$$\Phi \in L^1(\mu), \ |\nabla\Phi| \in L^\theta(\mu), \qquad \theta > d,$$

then $\varrho(x) \leqslant C\Phi(x)^{-1}$.

(ii) Let $k > 1$ and suppose that $|x|^r \in L^1(\mu)$ for some $r > (k-1)d$. Then, letting $\Phi(x) = |x|^k$, we obtain $\varrho(x) \leqslant C|x|^{-k}$.

(iii) Suppose that $\exp(\alpha|x|^\beta) \in L^1(\mu)$, where $\alpha, \beta > 0$. Then, for any positive number $\kappa < \beta/d$, there is a number $C > 0$ such that $\varrho(x) \leqslant C\exp(-\kappa|x|^\beta)$. To ensure that $|b| \in L^p(\mu)$, it is enough to have $|b(x)| \leqslant C + C\exp(\alpha_0|x|^\beta)$ with some number $\alpha_0 < \alpha/d$.

It is surprising that the above estimates, very rough at first glance, are in fact sufficiently sharp. Below we shall see that there are lower bounds of the same order.

In a similar manner one obtains upper bounds on $|\nabla\varrho|$.

3.3.3. Proposition. *Suppose that in the situation of Theorem* 3.3.1 *we have additionally $\Phi \in W^{1,2}_{\rm loc}(\mathbb{R}^d)$ and*

$$\varrho|\nabla\Phi| \in L^\infty(\mathbb{R}^d), \quad |b|, |\nabla\Phi|, \partial_{x_i}a^{ij}\partial_{x_j}\Phi, L_A\Phi, |A\nabla\Phi| \in L^r(\mu), \ r > 2d.$$

Then $|\nabla\varrho(x)| \leqslant C\Phi(x)^{-1}$.

PROOF. The measure μ_0 satisfies the equation with the right-hand side

$$-\langle b, \nabla\Phi\rangle\varrho + \partial_{x_i} a^{ij}\partial_{x_j}\Phi\varrho + 2L_A\Phi\varrho + 2\langle A\nabla\Phi, \nabla\varrho\rangle.$$

We know that ϱ and $\nabla\varrho$ are bounded. Hence the right-hand side is in $L^s(\mathbb{R}^d)$ for any $s \in (1, r/2]$, which yields $\Phi\varrho \in W^{\theta,2}(\mathbb{R}^d)$, $1 < \theta < r/2$. Therefore, $\nabla(\Phi\varrho)$ is bounded, whence the claim follows. □

3.3.4. Example. Let $\mu \in \mathcal{P}(\mathbb{R}^d)$ satisfy the equation $\mathcal{L}^*_{A,b}\mu = 0$, where A is uniformly Lipschitzian, A and A^{-1} are uniformly bounded, and $|b|, \operatorname{div} b \in L^p(\mu)$ for some $p > 2d$.

(i) Let $\Phi(x) = |x|^k$, $k \geqslant 1$, and let $|x|^m \in L^1(\mu)$, where $m > 2d(k-1)$. Then

$$|\nabla\varrho(x)| \leqslant C(1 + |x|)^{-k}.$$

(ii) Let $\Phi(x) = \exp\bigl(K|x|^\beta\bigr)$ and suppose that $\exp\bigl(M|x|^\beta\bigr) \in L^1(\mu)$ for some number $M > 2dK$. Then

$$|\nabla\varrho(x)| \leqslant C\exp\bigl(-K|x|^\beta\bigr).$$

By using the method of Lyapunov functions, one can give effective conditions for the existence of polynomial or exponential moments for μ. For example, if we have $A(x) \leqslant \Lambda\cdot \mathrm{I}$ and $\langle b(x), x\rangle \leqslant -K < -\Lambda d$ outside some ball, then letting $V(x) = \langle x, x\rangle^\gamma$ with $1 < \gamma < 1 + (K - \Lambda d)/2$, we obtain outside some ball

$$L_{A,b}V(x) \leqslant 2\gamma\langle x, x\rangle^{\gamma-1}\bigl[\Lambda d + 2(\gamma-1) + \langle b(x), x\rangle\bigr] \leqslant -\kappa|x|^{2\gamma-2},$$

where $\kappa > 0$. Hence $|x|^{2\gamma-1} \in L^1(\mu)$. Stronger decay of $\langle b(x), x\rangle$, e.g., the estimate $\langle b(x), x\rangle \leqslant c_1 - c_2|x|^r$, implies the exponential integrability (see Exercise 2.5.20). Certainly, the required integrability of the coefficients can be also deduced from such estimates provided we know certain bounds on the coefficients.

3.4. Harnack's inequality and lower estimates for densities

In this section we discuss lower bounds for densities of solutions to elliptic equations of the form

$$\mathcal{L}^*_{A,b}\mu = 0 \tag{3.4.1}$$

for Borel measures μ on $\mathbb{R}^d$, where $\mathcal{L}$ is an elliptic second order operator of divergence form

$$\mathcal{L}\varphi(x) := \partial_{x_i}\bigl(a^{ij}(x)\partial_{x_j}\varphi(x)\bigr) + b^i(x)\partial_{x_i}\varphi(x).$$

The interpretation of this equation is explained in § 1.4. The divergence form of operators is used only for convenience of formulations; under our standard assumptions about A all the main results of this section can be easily rewritten for the operator $L_{A,b}$.

Lower estimates are deduced from Harnack's inequality, before which we make precise the dependence of constants on the coefficients of the equation. Unlike Harnack's inequality formulated in § 1.7 for general divergence form equations, here we discuss the simplest case of bounded coefficients. Note also that among many possible ways of proving this inequality we have chosen Moser's method of iterations. This is explained by its simplicity and applicability with minor changes to the case where the boundedness of coefficients is replaced with their integrability with respect to the solution. This case is discussed in the next section. In addition, in the last section of this chapter (see § 3.7(ii)) we discuss another approach to the proof of Harnack's inequality.

Let Ω be a bounded domain in $\mathbb{R}^d$. We shall assume that there are constants $\gamma \geqslant 0$ and $\alpha > 0$ such that

$$\sum_{i,j} |a^{ij}(x)|^2 \leqslant \gamma^2 \quad \text{and} \quad A(x) \geqslant \alpha \cdot \mathrm{I} \quad \text{for all } x \in \Omega. \tag{3.4.2}$$

Let a mapping $b\colon \Omega \to \mathbb{R}^d$ be measurable and

$$\sup_{x\in\Omega} |b(x)| \leqslant B < \infty.$$

Suppose that a nonnegative function $u \in W^{2,1}(\Omega)$ satisfies the equation

$$\partial_{x_i}(a^{ij}\partial_{x_j}u - b^i u) = 0, \tag{3.4.3}$$

i.e., for every function $\varphi \in C_0^1(\Omega)$ we have the equality

$$\int_\Omega \partial_{x_i}\varphi(a^{ij}\partial_{x_j}u - b^i u)\, dx = 0. \tag{3.4.4}$$

According to the results in § 1.7, the function u has a locally Hölder continuous version, in particular, this version is locally bounded on Ω.

The proof of the following important result, which is often called the John–Nirenberg lemma, can be found in the book Gilbarg, Trudinger [**409**, Lemma 7.21]; see also our discussion in § 3.7(iii), p. 120. For any integrable function u we set

$$u_\Omega = |\Omega|^{-1}\int_\Omega u\, dx,$$

where $|\Omega|$ is the volume of Ω. Let $U(a,r)$ be the ball of radius r centered at a.

3.4.1. Lemma. *Let Ω be a convex domain and let $v \in W^{1,1}(\Omega)$ be a function such that there exists $K > 0$ such that, for every ball $U(x_0,R)$, one has the inequality*

$$\int_{\Omega\cap U(x_0,R)} |\nabla v|\, dx \leqslant K R^{d-1}.$$

Then there exist positive numbers σ_0 and C depending only on d such that

$$\int_\Omega \exp\Bigl(\frac{\sigma}{K}|v - v_\Omega|\Bigr)\, dx \leqslant C(\operatorname{diam}\Omega)^d,$$

where $\sigma = \sigma_0|\Omega|(\operatorname{diam}\Omega)^{-d}$, $\operatorname{diam}\Omega = \sup_{x,y\in\Omega}|x-y|$.

3.4.2. Theorem. *Let $\theta = 1 + 4\delta$, $0 < \delta \leqslant 3$. If $U(y,\theta R) \subset \Omega$, then for the continuous version of u the following Harnack inequality is true*:

$$\sup_{x\in U(y,R)} u(x) \leqslant C \inf_{x\in U(y,R)} u(x), \tag{3.4.5}$$

where

$$C = \exp\bigl\{c(d)\delta^{-1}\bigl(\gamma\alpha^{-1} + B\alpha^{-1}R\bigr)\bigr\},$$

and the number $c(d)$ depends only on d.

PROOF. We may assume that $d \geqslant 3$. The case $d \leqslant 2$ reduces to the case $d = 3$ if we regard u as a function on $\mathbb{R}^3$ and set $b^3 = 0$, $a^{33} = 1$, $a^{32} = a^{23} = a^{31} = a^{13} = 0$.

1. Let $R = 1, y = 0$, $\eta \in C_0^1(\Omega)$ and $u_\varepsilon = u + \varepsilon$, where $\varepsilon > 0$. Let also $\beta \in (-\infty, +\infty)$. It is readily seen that equality (3.4.4) remains valid also for the function $\varphi = \eta^2 u_\varepsilon^\beta$, whence we obtain

$$\beta \int_\Omega \langle A\nabla u, \nabla u_\varepsilon\rangle u_\varepsilon^{\beta-1}\eta^2\,dx = -2\int_\Omega \langle A\nabla u, \nabla\eta\rangle u_\varepsilon^\beta \eta\,dx$$
$$+\beta\int_\Omega \langle ub, \nabla u_\varepsilon\rangle u_\varepsilon^{\beta-1}\eta^2\,dx + 2\int_\Omega \langle ub, \nabla\eta\rangle u_\varepsilon^\beta\eta\,dx.$$

Applying the Cauchy inequality and taking into account our assumptions about the coefficients, we obtain

$$\int_\Omega |\nabla u_\varepsilon| u_\varepsilon^{\beta-1}\eta^2\,dx \leqslant C\int_\Omega (\eta + |\nabla\eta|)^2 u_\varepsilon^{\beta+1}\,dx, \tag{3.4.6}$$

where

$$C = 4\Bigl(\frac{\gamma}{|\beta|\alpha} + \frac{B}{\alpha}\Bigr)^2 + \frac{4B}{|\beta|\alpha}.$$

Since $\alpha \leqslant \gamma$, we have

$$C \leqslant 8\Bigl(\frac{\gamma}{|\beta|\alpha} + \frac{B}{\alpha}\Bigr)^2.$$

Hence we shall use estimate (3.4.6) with the larger constant

$$C = 8\bigl(\gamma|\beta|^{-1}\alpha^{-1} + B\alpha^{-1}\bigr)^2.$$

Let $w = u_\varepsilon^{(\beta+1)/2}$ if $\beta \neq -1$ and let $w = \ln u_\varepsilon$ if $\beta = -1$. If $\beta = -1$, then

$$\int_\Omega |\nabla w|^2\eta^2\,dx \leqslant C_1\int_\Omega (\eta + |\nabla\eta|)^2\,dx, \tag{3.4.7}$$

where

$$C_1 = 8\Bigl(\frac{\gamma}{\alpha} + \frac{B}{\alpha}\Bigr)^2.$$

If $\beta \neq -1$, then

$$\int_\Omega |\nabla w|^2\eta^2\,dx \leqslant C_2\int_\Omega (\eta + |\nabla\eta|)^2 w^2\,dx, \tag{3.4.8}$$

where

$$C_2 = 2(\beta+1)^2\Bigl(\frac{\gamma}{|\beta|\alpha} + \frac{B}{\alpha}\Bigr)^2.$$

2. Set

$$F(p, r) = \Bigl(\int_{U(0,r)} u_\varepsilon^p\,dx\Bigr)^{1/p}, \quad p \neq 0.$$

We now find numbers

$$p_0 = p_0(d, \delta, \alpha, \gamma, B) > 0, \quad C_3 = C_3(d, \delta, \alpha, \gamma, B) > 0$$

such that

$$F(p_0, 1 + 3\delta) \leqslant C_3 F(-p_0, 1 + 3\delta).$$

Let $r \leqslant 1$, $U_{r+\delta r}$ an arbitrary ball of radius $r + \delta r$ contained in $U(0, \theta)$, and let U_r be the ball of radius r with the same center as $U_{r+\delta r}$. Let us take a smooth function η such that $\eta(x) = 1$ if $x \in U_r$, $\eta(x) = 0$ if $x \notin U_{r+\delta r}$, moreover, $0 \leqslant \eta \leqslant 1$ and $|\nabla\eta| \leqslant 2(\delta r)^{-1}$. According to (3.4.7) with $\beta = -1$ we obtain

$$\int_{U_r} |\nabla w|\,dx \leqslant (2r)^{d/2}\Bigl(\int_{U_r} |\nabla w|^2\,dx\Bigr)^{1/2} \leqslant K_0(d)\delta^{-1}C_1^{1/2}r^{d-1}.$$

By Lemma 3.4.1 there exist numbers $C(d)$ and $K(d)$ depending only on d such that the following inequality is fulfilled:

$$\int_{U(0,1+3\delta)} \exp\bigl(p_0|w - w_{U(0,1+3\delta)}|\bigr)\,dx \leqslant K(d), \quad p_0 := C(d)\delta C_1^{-1/2}.$$

Therefore,

$$\int_{U(0,1+3\delta)} e^{p_0 w}\,dx \int_{U(0,1+3\delta)} e^{-p_0 w}\,dx$$
$$\leqslant \int_{U(0,1+3\delta)} e^{p_0 w - w_{U(0,1+3\delta)}}\,dx \int_{U(0,1+3\delta)} e^{w_{U(0,1+3\delta)} - p_0 w}\,dx \leqslant K(d)^2.$$

Recalling that $w = \ln u_\varepsilon$, we obtain the estimate

$$F(p_0, 1+3\delta) \leqslant C_3 F(-p_0, 1+3\delta),$$

where

$$C_3 := K(d)^{2/p_0} = \exp\bigl(c_1(d)\delta^{-1}\alpha^{-1}(\gamma + B)\bigr),$$

and the number $c_1(d)$ depends only on d. The numbers p_0 and C_3 are found.

3. It is known from the theory of integral (see Exercise 3.7.25 or Bogachev [**125**, Exercise 4.7.45]), that

$$F(+\infty, 1) := \sup_{U(0,1)} u_\varepsilon = \lim_{p\to+\infty} F(p,1), \quad F(-\infty, 1) := \inf_{U(0,1)} u_\varepsilon = \lim_{p\to-\infty} F(p,1).$$

We find numbers $C_4 = C_4(d,\delta,\alpha,\gamma,B) > 0$ and $C_5 = C_5(d,\delta,\alpha,\gamma,B) > 0$ such that

$$F(+\infty, 1) \leqslant C_4 F(p_0, 1+3\delta) \leqslant C_3 F(-p_0, 1+3\delta) \leqslant C_5 F(-\infty, 1).$$

By the triangle inequality we have

$$\Bigl(\int_\Omega |\nabla(\eta w)|^2\,dx\Bigr)^{1/2} \leqslant \Bigl(\int_\Omega |\eta\nabla w|^2\,dx\Bigr)^{1/2} + \Bigl(\int_\Omega |w\nabla\eta|^2\,dx\Bigr)^{1/2}.$$

According to estimate (3.4.8) with $\beta \neq -1$ we obtain

$$\Bigl(\int_\Omega |\nabla(\eta w)|^2\,dx\Bigr)^{1/2} \leqslant 2(C_2+1)^{1/2}\Bigl(\int_\Omega (\eta + |\nabla\eta|)^2 w^2\,dx\Bigr)^{1/2}. \tag{3.4.9}$$

Let η be a smooth function such that $0 \leqslant \eta \leqslant 1$, $\eta(x) = 1$ if $|x| \leqslant r_1$, $\eta(x) = 0$ if $|x| \geqslant r_2$, $|\nabla\eta| \leqslant 2/(r_2 - r_1)$, where $1 < r_1 < r_2 \leqslant 1+3\delta$. By the Sobolev embedding theorem inequality (3.4.9) gives the estimate

$$\Bigl(\int_{U(0,r_1)} w^{2d/(d-2)}\,dx\Bigr)^{(d-2)/2d} \leqslant \frac{C_0(d)(C_2+1)^{1/2}}{r_2 - r_1}\Bigl(\int_{U(0,r_2)} w^2\,dx\Bigr)^{1/2},$$

where the number $C_0(d)$ depends only on the dimension d. Let $\beta + 1 > 0$. Since $w = u_\varepsilon^{(\beta+1)/2}$, we have

$$F\Bigl(\frac{d}{d-2}(\beta+1), r_1\Bigr) \leqslant \Bigl(\frac{C_0(d)(C_2+1)^{1/2}}{r_2 - r_1}\Bigr)^{2/(\beta+1)} F(\beta+1, r_2).$$

Set $q = d/(d-2) > 1$. We shall apply the obtained estimate to the numbers r_1 and r_2 of the form

$$r_1 = r_m = 1 + \delta 2^{-m}, \quad r_2 = r_{m-1} = 1 + \delta 2^{-m+1},$$

but we shall take for $\beta+1$ the numbers $q^m p$, where $m=0,1,2,\dots$ and the number p is defined by the equality

$$p = d(d-1)^{-1}\big((d-2)/d\big)^k,$$

in which k is an integer number such that

$$\frac{\ln p_0 - \ln d + \ln(d-1)}{\ln(d-2)-\ln d} < k < \frac{\ln p_0 - \ln d + \ln(d-1)}{\ln(d-2)-\ln d} + 2.$$

We observe that the following inequality is valid:

$$(d-2)^2 d^{-2} p_0 < p < p_0.$$

In addition, for all nonnegative integer numbers m we have the estimate

$$|q^m p - 1| = |d(d-1)^{-1}(1-2/d)^{k-m} - 1| \geqslant (d-1)^{-1},$$

which is obvious from the consideration of the cases $k \geqslant m$ and $k < m$. It follows from this estimate that $C_2 \leqslant 2(d-1)^2(\beta+1)^2\alpha^{-2}(\gamma+B)^2$, whence by the relation $p < p_0$ we obtain the inequality

$$C_2 \leqslant 2(d-1)^2 q^{2m} p_0^2 \Big(\frac{\gamma}{\alpha} + \frac{B}{\alpha}\Big)^2 = 2\delta^2 C(d)^2 (d-1)^2 q^{2m}.$$

Thus, we arrive at the relation

$$F(q^{m+1}p, r_m) \leqslant \big(Q(d)\delta^2\big)^{2m/(q^m p)} F(q^m p, r_{m-1}),$$

where the number $Q(d)$ depends only on d. This relation gives the estimate

$$F(q^m p, r_m) \leqslant \big(Q(d)\delta^2\big)^{S/p} F(p, 1+\delta), \quad S := S(d) := 2\sum_{m=1}^{\infty} m q^{-m}.$$

By Hölder's inequality $F(p,1+\delta) \leqslant \exp(c_2(d)/p_0) F(p_0, 1+3\delta)$. Finally, we obtain the inequality $F(+\infty,1) \leqslant C_4 F(p_0, 1+3\delta)$, where

$$C_4 = \exp\big\{c_3(d)\delta^{-1}\alpha^{-1}(\gamma+B)\big\}.$$

If $\beta+1<0$, then in the same way we have

$$F(\beta+1, r_2) \leqslant \left(\frac{C_0(d)(C_2+1)^{1/2}}{r_2-r_1}\right)^{2/(|\beta+1|)} F\Big(\frac{d}{d-2}(\beta+1), r_1\Big).$$

Repeating the previous reasoning, we obtain the estimate

$$F(-p_0, 1+3\delta) \leqslant C_5 F(-\infty, 1), \quad C_5 = \exp\big\{c_4(d)\delta^{-1}\alpha^{-1}(\gamma+B)\big\}.$$

Along with our previous estimates this gives the estimate (3.4.5) for u_ε in place of u with a constant of the required form, independent of ε in the case $R=1$, $y=0$. Letting $\varepsilon \to 0$, we obtain the desired estimate also for u. The case $R \neq 1$ reduces to the considered one by the change of variables $x = y + Rz$. The function $v(z) = u(y+zR)$ satisfies the equation

$$\partial_{z_i}\big(a^{ij}\partial_{z_j} v - Rb^i v\big) = 0.$$

Hence the number B must be replaced with BR and the constants γ and α can be left without change. □

We observe that in proof we never used the symmetry of the matrix A (but a priori properties of u were important).

In the next theorem we assume that a nonnegative locally bounded measure μ on $\mathbb{R}^d$ has a density $\varrho \in W^{2,1}_{\rm loc}(\mathbb{R}^d)$. Let the measure μ satisfy equation (3.4.1) on $\mathbb{R}^d$ in the sense of equality (1.4.6), i.e., in case (II), where

$$\mathcal{L} = \partial_{x_i}(a^{ij}\partial_{x_j}) + b^i\partial_{x_i},$$

the matrix-valued mapping $A = (a^{ij})_{1\leqslant i,j\leqslant d}$ is Lebesgue measurable, condition (3.4.2) is fulfilled locally, i.e., the functions $\|A(x)\|$ and $\|A(x)^{-1}\|$ are locally bounded, the coefficient $b = (b^i)_{i\leqslant d}$ is a measurable locally bounded vector field. We recall that in the case where $a^{ij} \in W^{d+,1}_{\rm loc}(\mathbb{R}^d)$ and $b^i \in L^{d+}_{\rm loc}(\mathbb{R}^d)$ and $A(x)^{-1}$ is locally bounded, every solution of (3.4.1) in the sense of (1.4.5) automatically has a density in $W^{d+,1}_{\rm loc}(\mathbb{R}^d)$, hence satisfies also (1.4.6).

Let V be a continuous increasing function on $[0,\infty)$ with $V(0) > 0$.

3.4.3. Theorem. *Let $|b(x)| \leqslant V(|x|/\theta)$, where $\theta > 1$. Set*

$$\alpha(r) := \sup_{|x|\leqslant r} \|A(x)^{-1}\|, \quad \gamma(r) := \sup_{|x|\leqslant r} \|A(x)\|.$$

Then there exists a positive number $K(d)$ depending only on d such that the continuous version of the function ϱ satisfies the inequality

$$\varrho(x) \geqslant \varrho(0)\exp\bigl\{-K(d)(\theta-1)^{-1}\alpha(\theta|x|)^{-1}\bigl(\gamma(\theta|x|) + V(|x|)|x|\bigr)\bigr\}.$$

In particular, if $\|A(x)\| \leqslant \gamma$ and $\|A(x)^{-1}\| \leqslant \alpha$, then there exists a positive number $K = K(d,\alpha,\gamma,\theta)$ such that the continuous version of ϱ satisfies the inequality

$$\varrho(x) \geqslant \varrho(0)\exp\bigl\{-K\bigl(1 + V(|x|)|x|\bigr)\bigr\}.$$

PROOF. Let us fix x. Let us consider the ball $U(0,\theta|x|)$. Theorem 3.4.2 gives the desired estimate, since $\sup_{z\in U(0,\theta|x|)} |b(z)| \leqslant V(|x|)$. □

3.4.4. Example. Suppose that

$$\sup_x \Bigl[\|A(x)\| + \|A(x)^{-1}\|\Bigr] < \infty.$$

If, for some numbers $c_1, c_2 > 0$, for almost all x we have the estimate

$$|b(x)| \leqslant c_1|x|^\beta + c_2,$$

then there exists a constant K such that the following inequality holds:

$$\varrho(x) \geqslant \varrho(0)\exp\bigl\{-K\bigl(1 + |x|^{\beta+1}\bigr)\bigr\}.$$

If we have

$$\sup_{x,i,j}\Bigl[\|A(x)\| + \|A(x)^{-1}\| + |\nabla a^{ij}(x)|\Bigr] < \infty,$$

$$|b(x)| \leqslant c_1|x|^\beta + c_2, \quad \limsup_{|x|\to\infty} |x|^{-\beta-1}(b(x),x) < 0,$$

then we obtain the following two-sided estimate:

$$\exp\bigl\{-K_1\bigl(1+|x|^{\beta+1}\bigr)\bigr\} \leqslant \varrho(x) \leqslant \exp\bigl\{-K_2\bigl(1+|x|^{\beta+1}\bigr)\bigr\}.$$

For example, if $A = \mathrm{I}$ and $b(x) = -x$, then the Gaussian measure with density $\varrho(x) = \exp(-|x|^2/2)$ is a solution. The previous results ensure the bounds

$$\exp\bigl(-K_1(1+|x|^2)\bigr) \leqslant \varrho(x) \leqslant \exp\bigl(-K_2(1+|x|^2)\bigr)$$

with some numbers $K_1, K_2 > 0$, which gives a sufficiently adequate description of the decay at infinity, although it does not yield the sharp asymptotics.

It should be noted that the hypothesis that $\limsup\limits_{|x|\to\infty} |x|^{-\beta-1}\langle b(x), x\rangle < 0$ is only needed to ensure the integrability of $\exp\bigl(M|x|^\beta\bigr)$ and can be replaced by the latter (however, its advantage is that it is expressed explicitly in terms of A and b).

We recall (see Example 3.3.2) that for upper bounds it is enough to have that $|b| \in L^p(\mu)$ with some $p > d$ and $\exp\bigl(M|x|^\beta\bigr) \in L^1(\mu)$. It turns out that also lower bounds can be obtained under similar conditions of μ-integrability of certain functions (see the next section).

By using the obtained estimates we can give effectively verified conditions for the membership of the logarithmic gradient $\nabla\varrho/\varrho$ of the measure μ in $L^p(\mu)$. In the case $p = 2$ and $A = \mathrm{I}$ a simple sufficient condition is $|b| \in L^2(\mu)$.

In the next theorem and its corollaries we suppose that μ is a probability measure on $\mathbb{R}^d$ with a continuous positive density ϱ satisfying the elliptic equation (1.4.4) in the sense of (1.4.5), i.e., we deal with Case (I). In particular, the weighed Sobolev class $W^{p,1}(\mu)$ is well-defined.

3.4.5. Theorem. *Let $a^{ij} \in C^{0,\delta}(\mathbb{R}^d) \cap W^{p_0,1}_{loc}(\mathbb{R}^d)$, $\alpha \cdot \mathrm{I} \leqslant A \leqslant \widetilde{\alpha} \cdot \mathrm{I}$, where $\alpha, \widetilde{\alpha}, \delta > 0$ and $p_0 > d$, and let $\lim\limits_{r\to 0} \sup\limits_x \|\partial_{x_i} a^{ij}\|_{L^d(U(x,r))} = 0$ (the latter holds if A is Lipschitzian). Let $\Phi \in W^{1,1}_{loc}(\mathbb{R}^1)$ be a positive continuous function increasing on $[0, +\infty)$ such that $\Phi(N + 1) \leqslant C\Phi(N)^{1+\varepsilon}$ for some $C, \varepsilon > 0$, and let the functions $\Phi(|x|)$ and $\Phi'(|x|)^{p_1}$ with some $p_1 > d$ be integrable with respect to μ on $\mathbb{R}^d$. Suppose also that there exist numbers $C_0 > 0$, $\theta > 1$, $p > 1$, and $\gamma \in [0, 1/d)$ such that*

$$|b(x)| \leqslant C_0\Phi(|x| - \theta)^\gamma, \quad |\nabla a^{ij}(x)|^d \leqslant C_0\Phi(|x|),$$

$$\sum_{N=1}^{\infty} N^{d-1}\Phi(N)^{-q} < \infty, \quad \textit{where } q := 1 - \gamma(p + \varepsilon d) > 0.$$

Then $\ln\varrho \in W^{p,1}(\mu)$ and $|\nabla\varrho|/\varrho \in L^p(\mu)$.

PROOF. We may assume that $\theta \leqslant 3$. As we know from Chapter 1, under the stated assumptions μ has a positive continuous density $\varrho \in W^{p_0,1}_{loc}(\mathbb{R}^d)$. By Theorem 3.3.1, under our assumptions, which ensure the hypotheses of that theorem, the density ϱ is estimated as follows:

$$\varrho(x) \leqslant C_1\Phi(|x|)^{-1}.$$

Since the series of $N^{d-1}\Phi(N)^{-q}$ converges, one has $\Phi(|x|)^{-q} \in L^1(\mathbb{R}^d)$, hence we have $\varrho^q \in L^1(\mathbb{R}^d)$ and $\ln\varrho \in L^p(\mu)$, because $0 < q < 1$ and ϱ is bounded. For any fixed $x \in \mathbb{R}^d$ we have

$$B := \sup_{z\in U(x,\theta)} |b(z)| \leqslant C_0 \sup_{z\in U(x,\theta)} \Phi(|x| - \theta)^\gamma \leqslant C_0\Phi(|x|)^\gamma.$$

Let $0 < r < \min\{1/B, 1\}$. Then $rB < 1$. Hence by Theorem 3.4.2 there exists a constant $K = K(\alpha, \gamma, d, \theta, \delta)$ such that for every $y \in U(x, \theta^{1/2}r)$ one has the inequality

$$K^{-1}\varrho(x) \leqslant \varrho(y) \leqslant K\varrho(x). \tag{3.4.10}$$

Let us estimate the following integral:

$$I := \int_{U(x,r)} \frac{|\nabla \varrho(y)|^p}{\varrho(y)^{p-1}}\, dy.$$

According to inequality (3.4.10) we have

$$I \leqslant \varrho(x)^{1-p} K^{p-1} \int_{U(x,r)} |\nabla \varrho(y)|^p\, dy. \tag{3.4.11}$$

In addition, we have the estimate

$$\int_{U(x,\theta^{1/2}r)} |b|^p \varrho^p\, dy \leqslant B^p \int_{U(x,\theta^{1/2}r)} \varrho^p\, dy.$$

In the case $p > d$, by Corollary 1.7.5 with $h^i = b^i\varrho$ and inequality (3.4.10), we obtain the estimate

$$\begin{aligned}\int_{U(x,r)} |\nabla \varrho(y)|^p\, dy &\leqslant C_2 r^{-p} \int_{U(x,\theta^{1/2}r)} [1+|b(y)|^p]\varrho(y)^p\, dy\\ &\leqslant C_2 r^{-p}(1+r^p B^p)\|\varrho\|^p_{L^p(U(x,\theta^{1/2}r))}\\ &\leqslant C_2' r^{d-p}(1+r^pB^p)K^p\varrho(x)^p,\end{aligned}$$

where the numbers C_2 and C_2' are independent of x and r. The same estimate holds in the case $p \leqslant d$. Indeed, since it is true for $p_0 > d$, Hölder's inequality and the elementary verified inequality $(1+t^{p_0})^p \leqslant (1+t^p)^{p_0}$ yield that

$$\begin{aligned}\|\nabla\varrho\|^p_{L^p(U(x,r))} &\leqslant \|\nabla\varrho\|^{p/p_0}_{L^{p_0}(U(x,r))} C(d) r^{d(p_0-p)/p_0}\\ &\leqslant C_3 r^{(d-p_0)p/p_0} r^{d(p_0-p)/p_0}(1+r^pB^p)\varrho(x)^p\\ &= C_3 r^{d-p}(1+r^pB^p)\varrho(x)^p,\end{aligned}$$

where C_3 does not depend on r and x. Therefore, from (3.4.11) we obtain

$$I \leqslant C_4 r^{d-p}(1+r^pB^p)\varrho(x) \leqslant 2C_4 r^{d-p}\varrho(x),$$

where C_4 is independent of x and r. Let us take $r = (2C_0)^{-1}\Phi(|x|)^{-\gamma}$, which is possible due to the above estimate on B. By the bound on ϱ given at the beginning of the proof, the following inequality holds:

$$\int_{U(x,r)} \frac{|\nabla\varrho(y)|^p}{\varrho(y)^{p-1}}\, dy \leqslant C_5 \Phi(|x|)^{\gamma p - \gamma d - 1},$$

where C_5 does not depend on x and r. Set

$$Q_N = U(0, N+1)\backslash U(0,N), \quad N > 1.$$

It is easily verified by induction on d that there exists a constant C_6, which depends only on the dimension d, such that Q_N can be covered by balls $U(x,r)$ of radius $r = (2C_0)^{-1}\Phi(N+1)^{-\gamma}$, whose total number does not exceed $C_6 C_0^d N^{d-1}\Phi(N+1)^{d\gamma}$. According to what has been proved above we obtain

$$I_N := \int_{Q_N} \frac{|\nabla\varrho(y)|^p}{\varrho(y)^{p-1}}\, dy \leqslant C_5C_6C_0^d N^{d-1}\Phi(N+1)^{d\gamma}\Phi(N)^{\gamma p - \gamma d - 1}.$$

Since $\Phi(N+1) \leqslant C\Phi(N)^{1+\varepsilon}$, one has the inequality

$$N^{d-1}\Phi(N+1)^{d\gamma}\Phi(N)^{\gamma p-\gamma d-1} \leqslant CN^{d-1}\Phi^{-q}.$$

Hence $\sum_{N=1}^\infty I_N < \infty$. Therefore, $|\nabla \ln\varrho|^p\varrho \in L^1(\mathbb{R}^d)$. □

3.4.6. Corollary. *Let* $a^{ij} \in C^{0,\delta}(\mathbb{R}^d) \cap W^{p_0,1}_{loc}(\mathbb{R}^d)$, $\alpha \cdot \mathrm{I} \leqslant A \leqslant \widetilde{\alpha} \cdot \mathrm{I}$, *where* $\alpha, \widetilde{\alpha}, \delta > 0$ *and* $p_0 > d > 1$, *and let* $\lim\limits_{r\to 0} \sup\limits_x \|\partial_{x_i} a^{ij}\|_{L^d(U(x,r))} = 0$. *Let* $p > 1$ *be given. Suppose that for some* $M > 0$ *and* $\beta > 0$ *the function* $\exp(M|x|^\beta)$ *is integrable with respect to the measure* μ. *Let*

$$|b(x)| \leqslant C_0 \exp\bigl(\kappa|x|^\beta\bigr),\ |\nabla a^{ij}(x)| \leqslant C_0 \exp\bigl(\kappa|x|^\beta\bigr),\ \textit{where } 0 < \kappa d \max(p,d) < M.$$

Then $\ln \varrho \in W^{p,1}(\mu)$. *In particular, if for every* $\kappa > 0$ *there is a number* $C(\kappa)$ *such that*

$$|b(x)| + |\nabla a^{ij}(x)| \leqslant C(\kappa) \exp\bigl(\kappa|x|^\beta\bigr),$$

then $\ln \varrho \in W^{p,1}(\mu)$ *for all* $p \in [1, +\infty)$.

PROOF. Clearly, there is a sufficiently small number $\varepsilon_0 > 0$ such that letting $\gamma := \max(p,d)^{-1} - \varepsilon_0$ one has $\gamma > 0$ and $\kappa\gamma^{-1} < Md^{-1}$. Let us take for Φ the function $\Phi(r) = \exp(M_0|r|^\beta)$, where M_0 is chosen in the interval $(\kappa\gamma^{-1}, Md^{-1})$. Since $p\gamma < 1$, there is $\varepsilon > 0$ such that $q = 1 - \gamma(p + \varepsilon d) > 0$. Set $\theta = 2$. We observe that

$$\exp\bigl(\kappa r^\beta\bigr) \leqslant C_1 \exp\bigl(\gamma M_0 |r-2|^\beta\bigr),$$

since $\kappa < \gamma M_0$. Hence $|b(x)| \leqslant C_0 C_1 \Phi(|x| - 2)^\gamma$. It is also clear that for some $p_1 > d$ the function $\Phi'(|x|)^{p_1}$ is integrable with respect to μ, since $M_0 d < M$. It is easily seen that all other assumptions of the above theorem are fulfilled as well. □

Taking $\Phi(r) = r^\beta + 1$, one proves the following result.

3.4.7. Corollary. *Let* $a^{ij} \in C^{0,\delta}(\mathbb{R}^d) \cap W^{p_0,1}_{loc}(\mathbb{R}^d)$, $\alpha \cdot \mathrm{I} \leqslant A \leqslant \widetilde{\alpha} \cdot \mathrm{I}$, *where* $\alpha, \widetilde{\alpha}, \delta > 0$ *and* $p_0 > d > 1$, *and let* $\lim\limits_{r\to 0} \sup\limits_x \|\partial_{x_i} a^{ij}\|_{L^d(U(x,r))} = 0$. *Suppose that for some* $\beta > d$ *the function* $|x|^{\beta d}$ *is integrable with respect to the measure* μ *and that*

$$|b(x)| \leqslant C_0 + C_0|x|^{\beta\gamma},\ |\nabla a^{ij}(x)| \leqslant C_0 + C_0|x|^{\beta/d}, \textit{where } \gamma < d^{-1}, \gamma < 1 - d\beta^{-1}.$$

Then $\ln \varrho \in W^{p,1}(\mu)$ *for every* $p \in \Bigl[1, (\beta - d)\beta^{-1}\gamma^{-1}\Bigr)$.

PROOF. Let $\Phi(r) = |r|^\beta + 1$. Then $d - \beta + \beta\gamma p + \beta\gamma\varepsilon d < 0$ for any sufficiently small $\varepsilon > 0$, since $p < (\beta - d)\beta^{-1}\gamma^{-1}$; this ensures convergence of the series of $N^{d-1}\Phi(N)^{-q}$ with $q = 1 - \gamma(p + \varepsilon d)$. Note that the bound $\gamma < 1 - d\beta^{-1}$ yields the inequality $(\beta - d)\beta^{-1}\gamma^{-1} > 1$. The remaining assumptions of the above theorem are easily verified. □

3.5. Positivity of densities

We now consider lower bounds without assumptions on the growth of the drift coefficient, but using instead a certain integrability of the drift with respect to the regarded solution. Until the end of this section we shall assume that the matrix $A(x) = \bigl(a^{ij}(x)\bigr)_{1\leqslant i,j\leqslant d}$ is symmetric and satisfies the following condition:

(G1) for some $p > d$ the functions a^{ij} belong to the class $W^{p,1}_{loc}(\mathbb{R}^d)$ and there exist numbers $m, M > 0$ such that for all $x, y \in \mathbb{R}^d$ we have

$$m \cdot \mathrm{I} \leqslant A(x) \leqslant M \cdot \mathrm{I}.$$

If in addition to Condition (G1) we have $b^i \in L^p_{\mathrm{loc}}(\mu)$ (or $b^i \in L^p_{\mathrm{loc}}(\mathbb{R}^d)$), then μ is given by a continuous density $\varrho \in W^{p,1}_{\mathrm{loc}}(\mathbb{R}^d)$, which we shall deal with. Therefore, equation (1.4.4) can be written as the equality

$$\partial_{x_i}(a^{ij}\partial_{x_j}\varrho) - \partial_{x_i}(b^i\varrho) = 0,$$

understood in the weak sense.

The method of obtaining lower bounds discussed above is not applicable in the case of locally unbounded b. We shall now show that without any restrictions on the growth of b one can obtain estimates of the form

$$\varrho(x) \geqslant e^{-f(c_1|x|+c_2)}, \tag{3.5.1}$$

where c_1, c_2 are some positive numbers and the function $f \in C^2\big([0,\infty)\big)$ satisfies the following conditions:

(H1) $f(z) > 0$, $f'(z) > 0$, $f''(z) > 0$ if $z > 0$;

(H2) the function $e^{-f(z)}$ is convex (that is, $(e^{-f})'' \geqslant 0$) on the set $z > z_0$ for some $z_0 \geqslant 0$ and decreases to 0 as $z \to +\infty$.

For obtaining estimate (3.5.1) it suffices to require the following conditions in addition to (G1):

(G2) $|b|\exp\big(\psi(|b|)\big) \in L^p(\mu)$ for some $p > \max\{2, d\}$, where ψ is a nonnegative strictly increasing continuous function mapping $[0,\infty)$ onto $[0,\infty)$ such that for some number $N > 0$ one has

(H3) $\psi^{-1}(z) \leqslant Nf'\big(f^{-1}(z)\big)$ for all $z > 0$.

Let us give several typical examples of such functions f and ψ; in these examples for a given function f we construct ψ. If $f(z) = e^z$, then one can take $\psi(z) = \delta \cdot z$, where $\delta > 0$. Then (H1), (H2) and (H3) are fulfilled. In this case, as we shall see in Example 3.5.3, requiring also that $\exp(\delta|b|) \in L^1(\mu)$, we obtain the estimate

$$\varrho(x) \geqslant \exp\big(-\widetilde{c_2}\exp(\widetilde{c_1}|x|)\big).$$

If $f(z) = z^{r/(r-1)}$ with $r > 1$, then $\psi(z) = \delta \cdot z^r$ is suitable. Then

$$\varrho(x) \geqslant \widetilde{c_2}\exp\big(-\widetilde{c_1}|x|^{r/(r-1)}\big).$$

It follows from (3.5.1) that the solution density has no zeros under a condition weaker than the exponential integrability of $|b|$ (sufficiency of the latter condition was proved in Bogachev, Mayer-Wolf [**156**]). For example, if we set $f(z) = e^{e^z}$ and $\psi(z) = \delta \cdot \dfrac{z}{|\ln z|^\kappa}$ for $z > 2$ and $0 < \kappa < 1$, then we obtain a condition that is sufficient for positivity, but is weaker than the exponential integrability of $|b|$. If $d = 1$, $A = 1$, and $b = \varrho'/\varrho$, then this new sufficient condition for positivity is close to the one obtained in Scheutzow, von Weizsäcker [**836**], and in a sense the latter cannot be improved (it is given below in Theorem 3.5.5).

According to condition (G2), we have $|b| \in L^p(\mu)$ for some $p > d$. Hence, as we already know, the measure μ is given by a continuous density $\varrho \in W^{p,1}_{\mathrm{loc}}(\mathbb{R}^d)$ such that $\|\varrho\|_{L^\infty(\mathbb{R}^d)} < \infty$.

We fix a cube Q with edge of unit length.

3.5.1. Theorem. *Let $\mathcal{L}^*_{A,b}\mu = 0$, where $\mu = \varrho\,dx$ is a probability measure, the coefficients a^{ij}, b^i satisfy conditions* (G1), (G2), *and let conditions* (H1), (H2), *and* (H3) *be fulfilled. Then there exist numbers* $C > 0, \alpha > 0$ *and* $\Lambda \in (0,1)$, *independent of the center of* Q, *such that* $\Lambda\|\varrho\|_\infty < 1/2$ *and one has*

$$\sup_{x\in Q} \exp\Bigl(f^{-1}(|\ln(\Lambda\varrho)|)\Bigr) \leqslant C\biggl(\int_Q \exp\Bigl(-\alpha f^{-1}(|\ln(\Lambda\varrho)|)\Bigr)\,dx\biggr)^{-1/\alpha}. \tag{3.5.2}$$

Moreover, the numbers C *and* α *depend only on* f *and the following quantities:*

$$p,\ N,\ m,\ M,\ d,\ \|\varrho\|_{L^\infty(\mathbb{R}^d)},\ \int_{\mathbb{R}^d} |b|^p e^{p\psi(|b|)}\varrho\,dx.$$

The proof is given in § 3.6. It follows from (3.5.2) that for every measurable set $\Omega \subset Q$ of positive measure we have the estimate

$$\sup_{x\in Q} u(x) \leqslant C|\Omega|^{-1}\sup_{x\in\Omega} u(x), \quad u := \exp\Bigl(\alpha f^{-1}(|\ln(\Lambda\varrho)|)\Bigr).$$

It is this estimate that we actually use below. To see that it holds, we observe that

$$\biggl(\int_Q u^{-1}\,dx\biggr)^{-1} \leqslant \biggl(\int_\Omega u^{-1}\,dx\biggr)^{-1} \leqslant |\Omega|^{-1}\sup_{x\in\Omega} u(x),$$

because $u^{-1} \geqslant \bigl(\sup_{x\in\Omega} u(x)\bigr)^{-1}$ on Ω, hence the integral of u^{-1} over Ω is not less than $|\Omega|\bigl(\sup_{x\in\Omega} u(x)\bigr)^{-1}$, which leads to the announced estimate.

3.5.2. Theorem. *Let* $\mu = \varrho\,dx$ *be a solution of equation* (1.4.4), *where the coefficients* a^{ij}, b^i *satisfy conditions* (G1), (G2), *and let conditions* (H1), (H2), *and* (H3) *be fulfilled. Then there exist numbers* $c_1 > 0$ *and* $c_2 > 0$ *such that* (3.5.1) *holds.*

PROOF. Let $u = \exp(\alpha f^{-1}(|\ln(\Lambda\varrho)|))$, where α and Λ are numbers from condition (3.5.2). Let us fix $x \in \mathbb{R}^d$. Let $N = [|x|] + 1$ and $x_i = ix/N$. Then $x_0 = 0$, $x_N = x$ and $|x_i - x_{i-1}| \leqslant 1$. Let Q_i denote the cube with the center at the point x_i and unit edge parallel to the vector x. For every i we have $x_{i-1} \in Q_i$, $|Q_i \cap Q_{i-1}| = 1/2$. Therefore, by the estimate above we have

$$\sup_{Q_i} u(x) \leqslant C|Q_i\cap Q_{i-1}|^{-1}\sup_{Q_i\cap Q_{i-1}} u(x) \leqslant 2C\sup_{Q_{i-1}} u(x).$$

We obtain the inequality

$$\sup_{Q_i} u(x) \leqslant 2C\sup_{Q_{i-1}} u(x).$$

Applying this inequality for all i starting with $i = N$, we find that

$$u(x) = u(x_N) \leqslant (2C)^N \sup_{Q_0} u(x) \leqslant (2C)^N \sup_{|x|\leqslant 2} u(x).$$

Since $N = [|x|] + 1 \leqslant |x| + 1$, for some $\lambda_1 > 0$ and $\lambda_2 > 0$ we have

$$u(x) \leqslant \exp(\lambda_1|x| + \lambda_2),\ x \in \mathbb{R}^d.$$

Taking into account that $\varrho = \Lambda^{-1}e^{-f(\alpha^{-1}\ln u)}$ due to the estimate $\Lambda\varrho < 1/2$ and recalling that $\Lambda^{-1} \geqslant 1$ and the function f is increasing, we obtain the desired estimate. □

This result gives lower bounds for the density of the stationary measure of the diffusion process with the diffusion coefficient $\sqrt{2A}$ and drift b if this process exists.

3.5.3. Example. Let condition (G1) hold and let a number $r > 1$ be given.

(i) In order to obtain the estimate

$$\varrho(x) \geqslant \widetilde{c_2} \exp\big(-\widetilde{c_1}|x|^{r/(r-1)}\big), \tag{3.5.3}$$

it suffices to have $\exp(\delta|b|^r) \in L^1(\mu)$ for some $\delta > 0$.

Indeed, the function $\psi(z) = \delta z^r/(2p)$ satisfies condition (H3) with

$$f(z) = z^{r/(r-1)}.$$

There exists a number $C > 0$ such that $|z| \leqslant C\exp\big(\delta|z|^r/(2p)\big)$. Then

$$|b|^p \exp\big(\delta|b|^r/2\big) \leqslant C^p \exp\big(\delta|b|^r\big),$$

hence $|b|\exp\big(\delta|b|^r/(2p)\big) \in L^p(\mu)$, that is, condition (G2) is fulfilled.

(ii) In order to obtain the estimate

$$\varrho(x) \geqslant \exp\big(-\widetilde{c_2}\exp(\widetilde{c_1}|x|)\big), \tag{3.5.4}$$

it suffices that $\exp(\delta|b|) \in L^1(\mu)$ for some $\delta > 0$.

Indeed, whenever $0 < \delta_1 < \delta$, the functions $\psi(z) = \delta_1 \cdot z$ and $f(z) = e^z$ satisfy condition (H3) with $N = 1/\delta_1$ and (G2) is fulfilled as well.

3.5.4. Example. Suppose that $\mu = \varrho\, dx$ is a probability measure on $\mathbb{R}^d$ such that $\varrho \in W^{1,1}_{\mathrm{loc}}(\mathbb{R}^d)$. Then, as we know, μ obviously satisfies equation (1.4.4) with $A = \mathrm{I}$ and $b = \nabla\varrho/\varrho$, where $b(x) := 0$ if $\varrho(x) = 0$. Therefore, to get estimate (3.5.3) it suffices that $\exp(\delta|\nabla\varrho/\varrho|^r) \in L^1(\mu)$ for some $\delta > 0$, and for estimate (3.5.4) it suffices to have $\exp(\delta|\nabla\varrho/\varrho|) \in L^1(\mu)$ for some $\delta > 0$.

For $d = 1$ the assertion in the last example was obtained in Nualart [**751**] (where in the case $r = 1$ the formulation contains a minor inaccuracy: $\widetilde{c_1}$ is replaced by 1; but the function $\varrho(x) = \exp\big(-\exp(2|x|)\big)$ shows that one cannot get rid of $\widetilde{c_1}$). For $d > 1$ and $r = 1$ the assertion of the last example is given in [**126**, Exercise 6.8.4]; in [**656**] the case $r > 1$ is considered. However, the methods of [**751**] and [**656**] employ in a very essential way the fact that b is the logarithmic gradient $\nabla\varrho/\varrho$.

We now compare the sufficient condition for positivity established above with that of Scheutzow, Weizsäcker [**836**], where in the case $d = 1$ the positivity of ϱ in terms of the integrability of ϱ'/ϱ was investigated and the following final result was obtained. We recall that in our setting this corresponds to the elliptic equation with $d = 1$, $A = 1$ and $b = \varrho'/\varrho$.

3.5.5. Theorem. *Let ψ_0 be an increasing positive convex function on the half-line $[0,\infty)$. Then the following conditions are equivalent:*

(1) *if a locally absolutely continuous probability density ϱ satisfies the condition*

$$\int_{-\infty}^{\infty} \psi_0\big(|\varrho'(t)/\varrho(t)|\big)\varrho(t)\, dt < \infty, \tag{3.5.5}$$

then $\varrho(t) > 0$ for almost all points t with respect to Lebesgue measure;

(2) *for every $r > 0$, we have*

$$\int_r^{\infty} \frac{\ln\psi_0(t)}{t^2}\, dt = +\infty. \tag{3.5.6}$$

As observed in Scheutzow, Weizsäcker [**836**], for every $r > 0$ and every homeomorphism $h\colon [r,\infty) \to [R,\infty)$, the following equality is fulfilled:

$$\int_R^\infty \frac{1}{h^{-1}(s)}\,ds = \int_r^\infty \frac{h(t)}{t^2}\,dt - \frac{R}{r}, \tag{3.5.7}$$

that is, either both integrals are infinite, or both are finite and the indicated equality holds. Indeed, if h is continuously differentiable on $[r,T]$ and $h' > 0$, then

$$\int_R^T \frac{1}{h^{-1}(s)}\,ds = \int_r^{h^{-1}(T)} \frac{h'(t)}{t}\,dt = \int_r^{h^{-1}(T)} \frac{h(t)}{t^2}\,dt + \frac{T}{h^{-1}(T)} - \frac{h(r)}{r}.$$

By approximation the left and right sides of the equality remain equal in the general case. Now, if the left side of (3.5.7) is finite, then one can find an increasing sequence $T_n \to +\infty$ such that $T_n/h^{-1}(T_n) \to 0$, which yields that the right side is finite as well and (3.5.7) holds. Conversely, if the right side is finite, then there is a sequence $T_n \to +\infty$ such that $h(t_n)/t_n \to 0$, hence the sequence $T_n = h(t_n)$ works.

Thus, condition (3.5.6) can be written in the form

$$\int_R^\infty \frac{1}{(\ln\psi_0)^{-1}(t)}\,dt = +\infty, \quad R > 0. \tag{3.5.8}$$

3.5.6. Theorem. *Suppose that a probability measure $\mu = \varrho\,dx$ on $\mathbb{R}^d$ satisfies the equation $\mathcal{L}^*_{A,b}\mu = 0$, where the coefficients a^{ij} satisfy condition* (G1). *Suppose also that we are given a strictly increasing continuously differentiable function $\psi\colon [0,\infty) \to [0,\infty)$ such that condition* (G2) *holds and, for some $\alpha > \psi(0)$, the following conditions are fulfilled:*

$$\psi'(t) \geqslant \frac{1}{t} \quad \textit{if } t \in [\alpha, +\infty), \tag{3.5.9}$$

$$\int_\alpha^{+\infty} \frac{dt}{\psi^{-1}(t)} = +\infty. \tag{3.5.10}$$

Let the function $f\colon [0,+\infty) \to [\alpha,+\infty)$ be defined by the relation

$$f^{-1}(z) = \int_\alpha^z \frac{dt}{\psi^{-1}(t)}, \quad z \geqslant \alpha. \tag{3.5.11}$$

Then there exist numbers $c_1, c_2 > 0$ such that for the continuous version of ϱ estimate (3.5.1) *holds. In particular, the continuous version of ϱ is strictly positive.*

Proof. Set $g(y) = \psi^{-1}(y)$ on $[\psi(0), +\infty)$. Then (3.5.11) leads to the equality

$$t = \int_\alpha^{f(t)} \frac{dy}{g(y)}. \tag{3.5.12}$$

It is easily seen from this equality that $f(t)$ is unboundedly increasing as $t \to +\infty$. Note also that (3.5.12) is true for all $t \geqslant 0$ due to condition (3.5.10), since the function f must be defined on $[0, +\infty)$. In addition, the function f defined in this way satisfies the differential equation

$$g(z) = f'\bigl(f^{-1}(z)\bigr), \tag{3.5.13}$$

which is verified by differentiating equality (3.5.11). Therefore, we obtain that condition (H3) is fulfilled with $N = 1$.

Equality (3.5.13) yields the relation

$$f'(t) = g\bigl(f(t)\bigr),$$

which after differentiation gives the equality

$$f''(t) = g'\big(f(t)\big)\, f'(t).$$

We observe that since the function $\psi\colon [0,\infty) \to [0,\infty)$ is strictly increasing, the inverse function g is also nonnegative and increasing. We deduce from this that $f'(t) > 0$, $f''(t) > 0$. Thus, condition (H1) is fulfilled.

We now consider the expression

$$\begin{aligned}\big(e^{-f(t)}\big)'' &= \big(-f'(t)e^{-f(t)}\big)' = -f''(t)e^{-f(t)} + \big(f'(t)\big)^2 e^{-f(t)}\\ &= e^{-f(t)}\Big(\big(f'(t)\big)^2 - f''(t)\Big).\end{aligned}$$

We show that under our restrictions $\big(f'(t)\big)^2 - f''(t) \geqslant 0$. Substituting the expressions for f' and f'' obtained above and letting $z = f(t)$, we find that

$$\big(f'(t)\big)^2 - f''(t) = g(z)\cdot\big(g(z) - g'(z)\big),$$

where for any point $s = \psi^{-1}(z) \geqslant \alpha$, that is, for $z \geqslant \psi(0)$, we have

$$g(z) - g'(z) = \psi^{-1}(z) - \frac{1}{\psi'\big(\psi^{-1}(z)\big)} = s - \frac{1}{\psi'(s)} \geqslant 0$$

by condition (3.5.9). Hence condition (H2) holds as well. Thus, we have verified all conditions required for the validity of the desired estimate. □

3.5.7. Example. A typical example of a function ψ satisfying the conditions of the theorem is the function $\psi_\varepsilon(t) = \varepsilon t/\ln t$ on $[3,+\infty)$ linearly extended to $[0,3)$, where $\varepsilon > 0$. Hence, if condition (G1) is fulfilled and $\exp\big(\psi_\varepsilon(|b|)\big) \in L^1(\mu)$ with some $\varepsilon > 0$, then the continuous version of the density ϱ has no zeros.

We recall that condition (3.5.10) is equivalent to the condition

$$\int_R^{+\infty} \frac{\psi(t)}{t^2}\, dt = +\infty. \tag{3.5.14}$$

Note that the only condition in this theorem without exact analogs among the conditions in Theorem 3.5.2 and Theorem 3.5.5 is (3.5.9), that is, the estimate $\psi'(t) \geqslant 1/t$ for sufficiently large t (though, the continuous differentiability of ψ is not required in those theorems either). Nevertheless, this condition is close to the convexity of the function ψ_0 in Theorem 3.5.5. An analog of ψ_0 in our conditions is the function $e^{p\psi}$ with from (G2). However, the convexity of $e^{p\psi}$ is not enough for obtaining estimate (3.5.9). It suffices to require the convexity of the function $e^{\delta\psi}$ for some number $\delta \in (0,1)$. Indeed, if, for some $\delta \in (0,1)$, the function $e^{\delta\psi(t)}$ is convex down-wards on $(t_0,+\infty)$ for some $t_0 \geqslant 0$, that is, $(e^{\delta\psi(t)})'' \geqslant 0$ as $t > t_0$, then

$$\psi''(t) + \delta\big(\psi'(t)\big)^2 \geqslant 0.$$

Let $u := \psi'$. Then

$$\frac{u'}{u^2} + \delta \geqslant 0.$$

Therefore,

$$-\frac{1}{u(t)} + \delta t \geqslant -\frac{1}{u(t_0)} + \delta t_0,$$

whence

$$u(t) \geqslant \frac{1}{\delta t + C}, \quad C = \frac{1}{u(t_0)} - \delta t_0.$$

Since $0 < \delta < 1$, there exists $t_1 > t_0$ such that for all $t > t_1$ we have

$$\psi'(t) = u(t) \geqslant \frac{1}{t}.$$

Note that for practical applications, condition (3.5.9) actually does not impose any restriction, since the main condition (3.5.10) (or the equivalent condition (3.5.14)), which can be written in the form

$$\int_R^{+\infty} \frac{\psi'(y)}{y}\, dy = +\infty,$$

is fulfilled for any function $\psi'(y)$ which behaves like $(\ln y)^{-1}$, but is broken for $\psi'(y) = (\ln y)^{-1-\varepsilon}$ with $\varepsilon > 0$. Certainly, formally speaking, condition (3.5.9) does not follow from (3.5.10). We do not know whether (3.5.9) can be omitted. This is concerned with the fact that this condition of technical character is used in the verification of the validity of (H2). Considering the equality $\psi^{-1}(z) = Nf'\bigl(f^{-1}(z)\bigr)$ for all values of N instead of only $N = 1$ gives no advantages.

Condition (G2) employed in Theorem 3.5.2 and Theorem 3.5.6 is quite analogous to condition (3.5.5) in Theorem 3.5.5: if it were not for the multiplier $|b|$ in the expression $|b| \exp \psi(|b|) \in L^p(\mu)$ from (G2) (we recall that here p is any fixed number greater than $\max(d, 2)$), we could assume that the functions ψ_0 from (3.5.5) and ψ from (G2) are related by the equality $\ln \psi_0 = p\psi$, so in that case condition (3.5.6) would coincide with (3.5.14), equivalent to (3.5.10). Actually, if we set $\ln \psi_0 = p\psi$, then (G2) is a bit stronger than (3.5.5). However, the difference is almost vanishing for close to optimal functions of the type $\psi(t) = t/\ln t$. Moreover, suppose that a strictly increasing continuously differentiable function ψ satisfies conditions (3.5.9) and (3.5.14), but in place of (G2) we are only given the μ-integrability of $\exp\bigl(p\varphi(|b|)\bigr)$. If we were not obliged to ensure condition (3.5.9), we could take a new strictly increasing continuously differentiable function $\psi_1 = \lambda\psi$, where $0 < \lambda < 1$, for which both (3.5.14) and (G2) are fulfilled. Thus, the most notable difference between condition (3.5.6) and conditions (3.5.9), (3.5.14) is that (3.5.6) is stable under multiplication of $\ln \psi_0$ by any number $0 < \lambda < 1$ (certainly, condition (3.5.5) is also stable, although the convexity of ψ_0 may be destroyed), but when multiplying ψ by a number $0 < \lambda < 1$ in the general case we preserve only condition (3.5.14). Yet, this drawback can be eliminated in this way: in place of (3.5.9) we shall require the following stronger (but for practical purposes also nonrestrictive) condition:

$$\forall C > 0\ \exists \alpha_C > 0\colon \quad \psi'(t) \geqslant \frac{C}{t} \quad \text{whenever } t \geqslant \alpha_C. \tag{3.5.15}$$

It is clear that this condition is stable under multiplication of ψ by any number $\varepsilon > 0$. In addition, it yields that $\psi(t) \geqslant C \ln t + \psi(\alpha_C) - \ln \alpha_C$ for all $t \geqslant \alpha_C$.

Note also that if for every $\delta \in (0, 1)$ there exists $t_\delta > 0$ such that the function $e^{\delta\psi(t)}$ is convex for $t \geqslant t_\delta$, then (3.5.15) is satisfied. Now, on the basis of what has been said above, we obtain the following sufficient condition for the positivity of densities (its L^1-condition differs from the L^p-condition in (G2) due to (3.5.15)).

3.5.8. Corollary. *Suppose that a probability measure $\mu = \varrho\, dx$ on $\mathbb{R}^d$ satisfies the equation $\mathcal{L}_{A,b}^*\mu = 0$, where the coefficients a^{ij} satisfy condition* (G1). *Assume that there is an increasing continuously differentiable function $\psi\colon [0, \infty) \to [0, \infty)$*

such that conditions (3.5.14) *and* (3.5.15) *are fulfilled and that*

$$\exp\bigl(\psi(|b|)\bigr) \in L^1(\mu).$$

Then the continuous version of ϱ is strictly positive and if the function f is defined by relation (3.5.11), *then, for some $c_1, c_2 > 0$, estimate* (3.5.1) *holds.*

For example, the function $\psi(t) = \varepsilon t/\ln t$ on the half-line satisfies conditions (3.5.14) and (3.5.15) for all $\varepsilon > 0$.

Let us also give a local version of our main result, a useful distinction of which is that condition (G2) can hold only locally. Its proof is also given in § 3.6.

3.5.9. Theorem. *Let a nonnegative measure $\mu = \varrho\,dx$ on an open ball $U \subset \mathbb{R}^d$ satisfy the equation $\mathcal{L}^*_{A,b}\mu = 0$, where the coefficients a^{ij} satisfy condition* (G1). *In addition, suppose that there exists an increasing continuously differentiable function $\psi\colon [0,\infty) \to [0,\infty)$ such that*

$$|b| \exp\bigl(\psi(|b|)\bigr) \in L^p(U,\mu) \quad \textit{for some } p > \max(2,d)$$

and that conditions (3.5.9) *and* (3.5.10) *are satisfied. Then the continuous version of the density ϱ is either identically zero on U or it has no zeros in U.*

It follows from this theorem that if μ is a nonzero nonnegative Borel measure on a ball U and $L^*_{\mathrm{I},b}\mu = 0$, where $\exp(\varepsilon|b|) \in L^1_{\mathrm{loc}}(\mu)$ for some number $\varepsilon > 0$, then μ has a continuous strictly positive density.

3.6. Justification of the results on positivity

Here we justify the results of the previous section left there without proof.

Let $V = e^f/f'$. Since $(e^{-f})'' = [(f')^2 - f'']e^{-f} \geqslant 0$ on $[z_0, +\infty)$, we have

$$V' = [(f')^2 - f'']e^{-f}(f')^{-2} \geqslant 0 \quad \text{on } [z_0, +\infty).$$

In addition, V increases to $+\infty$, since the function $1/V = f'e^{-f}$ cannot be separated from zero on $[0, +\infty)$. It follows from conditions (H1) and (H3) that $f'(y) \to +\infty$ as $y \to +\infty$. Therefore, there exists $y_0 > \max\{z_0, 1\}$ such that $f'(y) \geqslant 1$ and $V(y) \geqslant e^{\psi(0)}$ whenever $y \geqslant y_0$. Let $\tau_0 := \exp\{-f(\ln y_0)\}$. Then $0 < \tau_0 < 1$.

For $\tau \in (0, \tau_0)$ and $q \geqslant 0$ we put

$$h_q(\tau) := -\int_\tau^{\tau_0} \bigl|V\bigl(f^{-1}(|\ln s|)\bigr)\bigr|^2 \exp\{2qf^{-1}(|\ln s|)\}\,ds.$$

3.6.1. Lemma. *Suppose that conditions* (H1), (H2), (H3) *are fulfilled and that $\tau \in (0, \tau_0)$. Then*

(i) *the inequality $V(y)\psi^{-1}\bigl(\ln V(y)\bigr) \leqslant Ne^{f(y)}$ is fulfilled for $y > y_0$;*

(ii) *there exists a number $N_1 > 0$ such that*

$$\frac{1}{V(y)} \int_{y_0}^{y} V(s)\,ds \leqslant N_1, \quad y > y_0,$$

and, in addition,

$$h_q^2(\tau)/h_q'(\tau) \leqslant N_1^2 \exp\bigl(2qf^{-1}(|\ln \tau|)\bigr).$$

PROOF. Applying condition (H3) and taking into account that $V(y) \leqslant e^{f(y)}$ for all $y > y_0$, we obtain

$$V(y)\psi^{-1}\bigl(\ln V(y)\bigr) = \frac{1}{f'(y)} e^{f(y)} \psi^{-1}\bigl(f(y)\bigr) \leqslant Ne^{f(y)}.$$

Assertion (i) is proven. Let us prove (ii). Integrating by parts we find that

$$\int_{y_0}^{y} \frac{e^{f(s)}}{f'(s)}\,ds = \frac{e^{f(y)}}{\big(f'(y)\big)^2} - \frac{e^{f(y_0)}}{\big(f'(y_0)\big)^2} + 2\int_{y_0}^{y} \frac{f''(s)e^{f(s)}}{\big(f'(s)\big)^3}\,ds.$$

Since the function e^f/f' increases on $(y_0, +\infty)$, we have

$$\int_{y_0}^{y} \frac{f''(s)e^{f(s)}}{\big(f'(s)\big)^3}\,ds \leqslant \frac{e^{f(y)}}{f'(y)}\int_{y_0}^{y} \frac{f''(s)}{\big(f'(s)\big)^2}\,ds = \frac{e^{f(y)}}{f'(y)}\Big(\frac{1}{f'(y_0)} - \frac{1}{f'(y)}\Big).$$

This gives the first estimate in (ii). In order to obtain the second one, we change variables $y = f^{-1}(|\ln\tau|)$, $y_0 = f^{-1}(|\ln\tau_0|)$ and use the monotonicity of the function e^{2qy} on the real line. This yields the inequality

$$\frac{h_q^2(\tau)}{h_q'(\tau)} = \frac{e^{-2qy}}{V^2(y)}\Big(\int_{y_0}^{y} e^{2qs}V(s)\,ds\Big)^2 \leqslant e^{2qy}\Big(\frac{1}{V(y)}\int_{y_0}^{y} V(s)\,ds\Big)^2,$$

which along with the previous estimate gives the desired inequality. □

By (G2) we have $|b| \in L^p(\mu)$ for some $p > d$. Hence, as we know, the measure μ is given by a continuous density $\varrho \in W^{p,1}_{\rm loc}(\mathbb{R}^d)$ with $\|\varrho\|_\infty = \|\varrho\|_{L^\infty(\mathbb{R}^d)} < \infty$. For any $k \in \mathbb{N}$ we put

$$\varrho_k = \Lambda\varrho + 1/k, \quad \xi_k := f^{-1}(|\ln\varrho_k|),$$

$$\Lambda := \min\{\tau_0(2\|\varrho\|_\infty)^{-1}, 1\}. \tag{3.6.1}$$

Then $\Lambda\varrho < 1/2$ and $\varrho_k < \tau_0$ for all natural numbers $k > 1/(2\tau_0)$. Hence substituting ϱ_k in place of τ in h_q and letting $y = \xi_k$ we have all assertions of Lemma 3.6.1.

3.6.2. Lemma. *Let $\mu = \varrho\,dx$ be a solution of the equation $\mathcal{L}^*_{A,b}\mu = 0$, where the coefficients a^{ij}, b^i satisfy conditions* (G1), (G2), *and let conditions* (H1), (H2), *and* (H3) *be fulfilled. Let $s > 1$, $s' = s/(s-1)$, where $2s' \leqslant p$. Suppose we are given a function $\eta \in C_0^2(Q)$, where Q is a cube of edge length 2, and let $|\eta| \leqslant 1$. Then the following estimate holds:*

$$\int_Q |\nabla\varrho|^2 h_q'(\varrho_k)\eta^2\,dx \leqslant N_2\Big[\int_Q |\nabla\eta|^2\exp\{2q\xi_k\}\,dx + \Big(\int_Q \exp\{2sq\xi_k\}\eta^2\,dx\Big)^{1/s}\Big],$$

where N_2 is a number depending only on the following quantities:

$$s,\ N,\ N_1,\ \tau_0,\ m,\ M,\ d,\ \|\varrho\|_{L^\infty(\mathbb{R}^d)},\ \int_{\mathbb{R}^d} |b|^{2s'}\exp\{2s'\psi(|b|)\}\varrho\,dx.$$

PROOF. For every function $\varphi \in W_0^{1,2}(Q)$ we have

$$\int_Q \langle A\nabla\varrho, \nabla\varphi\rangle\,dx = \int_Q \langle b, \nabla\varphi\rangle\varrho\,dx.$$

Substituting $\varphi = h_q(\varrho_k)\eta^2$, we obtain

$$\int_Q \langle A\nabla\varrho, \nabla\varrho\rangle h_q'(\varrho_k)\eta^2\,dx = I + J + L,$$

$$I = -\int_Q 2\langle A\nabla\varrho, \nabla\eta\rangle h_q(\varrho_k)\eta\,dx, \quad J = \int_Q 2\langle b, \nabla\eta\rangle\varrho h_q(\varrho_k)\eta\,dx,$$

$$L = \int_Q \langle b, \nabla\varrho\rangle h_q'(\varrho_k)\eta^2\varrho\,dx.$$

Let us estimate every term separately. Let $\varepsilon > 0$. Then

$$I \leqslant \varepsilon \int_Q |\nabla \varrho|^2 h_q'(\varrho_k)\eta^2\,dx + \varepsilon^{-1}M^2 \int_Q |\nabla \eta|^2 \frac{h_q^2(\varrho_k)}{h_q'(\varrho_k)}\,dx.$$

By estimate (ii) in Lemma 3.6.1 we have

$$I \leqslant \varepsilon \int_Q |\nabla \varrho|^2 h_q'(\varrho_k)\eta^2\,dx + \varepsilon^{-1}(MN_1)^2 \int_Q |\nabla \eta|^2 e^{2qf^{-1}(\varrho_k)}\,dx.$$

We estimate J as follows:

$$J \leqslant \int_Q |\nabla \eta|^2 \frac{h_q^2(\varrho_k)}{h_q'(\varrho_k)}\,dx + \int_Q |b|^2 \varrho^2 h_q'(\varrho_k)\eta^2\,dx.$$

The first term is estimated in the same way as above. Let us consider the second term. By Hölder's inequality with exponents s' and s we have

$$\int_Q |b|^2\varrho^2 h_q'(\varrho_k)\eta^2\,dx \leqslant \biggl(\int_Q |b|^{2s'}\varrho^{2s'}V(\xi_k)^{2s'}\eta^2\,dx\biggr)^{1/s'} \biggl(\int_Q \exp\{2qs\xi_k\}\eta^2\,dx\biggr)^{1/s}.$$

Let us estimate the first factor. Since ψ is increasing, for all $\alpha \geqslant 1$ and $\beta \geqslant 0$ we have

$$\alpha\beta \leqslant \alpha\psi^{-1}(\ln\alpha) + \beta e^{\psi(\beta)}.$$

This inequality and estimate (i) in Lemma 3.6.1 yield

$$|b|\varrho V(\xi_k) \leqslant \varrho\bigl[|b|e^{\psi(|b|)} + V(\xi_k)\psi^{-1}(\ln V(\xi_k))\bigr] \leqslant |b|e^{\psi(|b|)}\varrho + N/\Lambda,$$

where Λ is defined by (3.6.1). By using the inequalities $(x+y)^{2s'} \leqslant 2^{2s'}(x^{2s'}+y^{2s'})$ and $\eta^2 \leqslant 1$, we obtain that

$$\int_Q |b|^{2s'}\varrho^{2s'}V(\xi_k)^{2s'}\eta^2\,dx \leqslant 4^{s'}\|\varrho\|_{L^\infty(\mathbb{R}^d)}^{2s'-1}\int_{\mathbb{R}^d}|b|^{2s'}e^{2s'\psi(|b|)}\varrho\,dx + (2N/\Lambda)^{2s'}|Q|.$$

Therefore, there exists a number $C_1 > 0$ depending only on the quantities indicated in the lemma such that

$$J \leqslant C_1\biggl[\int_Q |\nabla\eta|^2 e^{2q\xi_k}\,dx + \biggl(\int_Q e^{2sq\xi_k}\eta^2\,dx\biggr)^{1/s}\biggr].$$

It remains to estimate the term L. We have

$$L \leqslant \varepsilon \int_Q |\nabla\varrho|^2 h_q'(\varrho_k)\eta^2\,dx + 4\varepsilon^{-1}\int_Q |b|^2\varrho^2 h_q'(\varrho_k)\eta^2\,dx.$$

Estimating here the second term in the same way as above, we obtain

$$L \leqslant \varepsilon \int_Q |\nabla\varrho|^2 h_q'(\varrho_k)\eta^2\,dx + 4\varepsilon^{-1}C_1\biggl(\int_Q e^{2sq\xi_k}\eta^2\,dx\biggr)^{1/s}.$$

We observe that

$$m\int_Q |\nabla\varrho|^2 h_q'(\varrho_k)\eta^2\,dx \leqslant \int_Q \langle A\nabla\varrho, \nabla\varrho\rangle h_q'(\varrho_k)\eta^2\,dx.$$

Collecting the obtained estimates and letting $\varepsilon = m/3$ we find that

$$\int_Q |\nabla\varrho|^2 h_q'(\varrho_k)\eta^2\,dx \leqslant N_2\biggl[\int_Q |\nabla\eta|^2 e^{2q\xi_k}\,dx + \biggl(\int_Q e^{2sq\xi_k}\eta^2\,dx\biggr)^{1/s}\biggr],$$

which completes the proof. □

PROOF OF THEOREM 3.5.1. Let $d > 2$. Without loss of generality we may assume that

$$Q = \prod_{i=1}^{d}\Bigl[x_i^0 - \frac{1}{2}, x_i^0 + \frac{1}{2}\Bigr], \quad Q_n = \prod_{i=1}^{d}\Bigl[x_i^0 - \frac{1}{2} - \frac{1}{2^{n+1}}, x_i^0 + \frac{1}{2} + \frac{1}{2^{n+1}}\Bigr].$$

1. We observe that

$$\int_{Q_0} |\nabla\xi_k|^2\eta^2\,dx = \int_{Q_0} |\nabla\varrho|^2 V^2(\xi_k)\eta^2\,dx = \int_{Q_0} |\nabla\varrho|^2 h_0'(\varrho_k)\eta^2\,dx.$$

Let $s = p/(p-2)$. Then $s < d/(d-2)$ and $2s' = p$. Lemma 3.6.2 with $q = 0$ gives

$$\int_{Q_0} |\nabla\xi_k|^2\eta^2\,dx \leqslant N_2\Bigl[\int_{Q_0} |\nabla\eta|^2\,dx + \Bigl(\int_{Q_0}\eta^2\,dx\Bigr)^{(p-2)/p}\Bigr].$$

Let us take two balls $U(y,r) \subset U(y,2r) \subset Q_0$. Let $\eta(x) = 1$ if $x \in U(y,r)$ and $\eta(x) = 0$ if $x \notin U(y,2r)$. Suppose also that $|\eta| \leqslant 1$ and $|\nabla\eta| \leqslant c_1 r^{-1}$ with some constant c_1. Substituting η in the above estimate we find that

$$\int_{B(y,r)} |\nabla\xi_k|^2\,dx \leqslant C_0 r^{d-2}.$$

Here the number C_0 depends only on the parameters indicated in the theorem, but does not depend on y, r, k. Therefore, for every ball $U(y,r)$, by the Cauchy–Bunyakovsky inequality we obtain the estimate

$$\int_{U(y,r)} |\nabla\xi_k|\,dx \leqslant (C_0\omega_d)^{1/2} r^{d-1},$$

where ω_d is the volume of the d-dimensional unit ball. Applying Lemma 3.4.1 we obtain that there exist constants $\alpha > 0$ and $L > 0$ such that

$$\int_{Q_1} \exp\Bigl(\alpha|\xi_k - (\xi_k)_{Q_1}|\Bigr)\,dx \leqslant L.$$

Then

$$\int_{Q_1} e^{\alpha\xi_k}\,dx \int_{Q_1} e^{-\alpha\xi_k}\,dx \leqslant \Bigl(\int_{Q_1} \exp\{\alpha|\xi_k - (\xi_k)_{Q_1}|\}\,dx\Bigr)^2 \leqslant L^2. \tag{3.6.2}$$

2. We observe that for all $\eta \in C_0^2(Q_1)$ one has the equality

$$\int_{Q_1} |\nabla e^{q\xi_k}|^2\eta^2\,dx = q^2\int_{Q_1} |\nabla\varrho|^2 V^2(\xi_k) e^{2q\xi_k}\eta^2\,dx = q^2\int_{Q_1} |\nabla\varrho|^2 h_q'(\varrho_k)\eta^2\,dx.$$

Applying Lemma 3.6.2 with $q > 0$ we obtain

$$\int_{Q_1} |\nabla e^{q\xi_k}|^2\eta^2\,dx \leqslant q^2 N_2\Bigl[\int_{Q_1} |\nabla\eta|^2 e^{2q\xi_k}\,dx + \Bigl(\int_{Q_1} e^{2sq\xi_k}\eta^2\,dx\Bigr)^{1/s}\Bigr].$$

According to the Leibniz formula $\nabla(e^{q\xi_k}\eta) = \eta\nabla e^{q\xi_k} + e^{q\xi_k}\nabla\eta$. Then

$$\int_{Q_1} |\nabla(e^{q\xi_k}\eta)|^2\,dx \leqslant q^2 N_2\Bigl[\int_{Q_1} |\nabla\eta|^2 e^{2q\xi_k}\,dx + \Bigl(\int_{Q_1} e^{2sq\xi_k}\eta^2\,dx\Bigr)^{1/s}\Bigr].$$

Suppose that a smooth function $\eta = \eta_n$ vanishes outside Q_n and equals 1 on the cube Q_{n+1}. Let $|\eta_n| \leqslant 1$ and $|\nabla\eta_n| \leqslant c_2 2^{n+1}$ for some constant c_2 independent of n. Applying Hölder's inequality with exponents s and s' we find that

$$\int_{Q_1} |\nabla(e^{q\xi_k}\eta)|^2\,dx \leqslant (q^2+1)C_1^n\Bigl(\int_{Q_n} e^{2sq\xi_k}\,dx\Bigr)^{1/s}.$$

By the Sobolev embedding theorem we obtain

$$\Bigl(\int_{Q_{n+1}} |e^{q\xi_k}|^{2d/(d-2)}\,dx\Bigr)^{(d-2)/d} \leqslant (q^2+1)C_2^n\Bigl(\int_{Q_n} e^{2sq\xi_k}\,dx\Bigr)^{1/s}.$$

For any measurable set E and $t \neq 0$ we put

$$F(t,E) := \Bigl(\int_E e^{t\xi_k}\,dx\Bigr)^{1/t}, \quad F(+\infty,E) = \sup_{x\in E} e^{\xi_k}.$$

Therefore,

$$F\Bigl(\frac{2qd}{d-2}, Q_{n+1}\Bigr) \leqslant \bigl((q^2+1)C_2\bigr)^{n/q} F(2qs, Q_n).$$

Set $p_n = 2qs$ and $p_{n+1} = ds^{-1}(d-2)^{-1}p_n, p_1 = \alpha$. For $s = p/(p-2)$ we obtain $s < d/(d-2)$, $\lambda = ds^{-1}(d-2)^{-1} > 1$, $p_n = \alpha\lambda^n$, $p_n \to +\infty$,

$$F(p_{n+1}, Q_{n+1}) \leqslant C_3^{n\lambda^{-n}} F(p_n, Q_n).$$

Since $0 < \lambda < 1$, one has $\sum_{n=1}^\infty n\lambda^{-n} < \infty$. Hence there exists $C_4 > 0$ such that

$$F(p_{n+1}, Q_{n+1}) \leqslant C_3^\theta F(\alpha, Q_1) \leqslant C_4 F(\alpha, Q_1), \quad \theta = \sum_{n=1}^\infty n\lambda^{-n}.$$

It is known that $F(+\infty, Q) = \lim\limits_{t\to\infty} F(t,Q)$. Therefore, as $n \to +\infty$, we obtain

$$F(+\infty, Q) \leqslant C_4 F(\alpha, Q_1).$$

Hence by (3.6.2) the inequality $F(\alpha, Q_1) \leqslant L^2 F(-\alpha, Q_1)$ is true. Letting $k \to \infty$, we obtain

$$\sup_{x\in Q} \exp\Bigl(f^{-1}(|\ln(\Lambda\varrho)|)\Bigr) \leqslant C_4 L^2 \Bigl(\int_{Q_1} \exp\bigl(-\alpha f(|\ln \Lambda\varrho|)\bigr)\Bigr)^{-1/\alpha}.$$

It remains to observe that replacing Q_1 by E increases the right-hand side. Thus, (3.5.2) is proven if $d > 2$. The cases $d = 1$ and $d = 2$ are even simpler, because in the Sobolev inequality in place of the exponent $2d(d-2)^{-1}$ one can take any number $r > 1$. □

3.6.3. Remark. The numbers C and α in the formulation of the proven theorem depend only on the following quantities:

$$p,\ N,\ N_1,\ \tau_0,\ m,\ M,\ d,\ \|\varrho\|_{L^\infty(\mathbb{R}^d)},\ \int_{\mathbb{R}^d} |b|^p e^{p\psi(|b|)}\varrho\,dx.$$

Proof of Theorem 3.5.9. We can assume that $\varrho \leqslant 1/2$. Let the function f be defined by (3.5.11). It follows from the proof of Theorem 3.5.1 that, for every ball V with closure in U, there exist numbers $C > 0$, $\Lambda > 0$, and $\gamma > 0$ such that for every $k > 1$ and $\varrho_k = \Lambda\varrho + 1/k$ we have

$$\sup_{x\in V} \exp\Bigl(f^{-1}(|\ln(\varrho_k)|)\Bigr) \leqslant C\Bigl(\int_V \exp\bigl(-\gamma f^{-1}(|\ln \varrho_k|)\,dx\bigr)\Bigr)^{-1/\gamma}.$$

If the continuous version of the density ϱ does not vanish at least at one point of such ball V, then the integral

$$\int_V \exp\Bigl(-\gamma f^{-1}(|\ln \Lambda\varrho|)\Bigr)\,dx$$

does not vanish (where $f^{-1}(+\infty) := +\infty$) and for every $k > 1$ one has the estimate

$$\int_V \exp(-\gamma f^{-1}(|\ln \Lambda\varrho|)\,dx \leqslant \int_V \exp\Bigl(-\gamma f^{-1}(|\ln \varrho_k|)\Bigr)\,dx.$$

Therefore, for every $x \in V$ we have the inequality

$$\exp\Bigl[f^{-1}\Bigl(\bigl|\ln(\varrho_k(x))\bigr|\Bigr)\Bigr] \leqslant D, \quad D = C\Bigl(\int_V \exp(-\gamma f^{-1}(|\ln \Lambda\varrho|))\,dx\Bigr)^{-1/\gamma}.$$

Hence for ϱ we obtain

$$\Lambda^{-1}\exp(-f(\ln D)) - \Lambda^{-1}k^{-1} \leqslant \varrho(x) \leqslant \Lambda^{-1}\exp\bigl(f(\ln D)\bigr) - \Lambda^{-1}k^{-1}.$$

Letting k increase to infinity, we obtain the inequality

$$\Lambda^{-1}\exp\bigl(-f(\ln D)\bigr) \leqslant \varrho(x) \leqslant \Lambda^{-1}\exp\bigl(f(\ln D)\bigr),$$

which yields immediately the positivity of ϱ in V, hence also in U. □

3.7. Complements, comments, and exercises

(i) Estimates of solutions on manifolds (119). (ii) More on Harnack's inequality (119). (iii) The John–Nirenberg lemma (120). (iv) Differentiability of solutions with respect to a parameter (122). Comments (127). Exercises (128).

3.7(i). Estimates of solutions on manifolds

The next theorem was proved in Bogachev, Röckner, Wang [**179**], [**181**] for a broad class of Riemannian manifolds.

3.7.1. Theorem. *Let M be a Riemannian manifold with Riemannian volume measure λ such that the Ricci curvature is bounded from below and the Riemannian volumes of balls of any fixed positive radius are bounded away from zero. Let μ be a Borel probability measure on M such that $L^*\mu = 0$, where $Lf = \Delta f + \langle b, \nabla f\rangle$ and $|b| \in L^2(\mu)$. Then $\mu = \varrho\cdot\lambda$, where $\sqrt{\varrho} \in W^{2,1}(M)$ and*

$$\int_M \frac{|\nabla\varrho|^2}{\varrho^2}\,d\mu \leqslant \int_M |b|^2\,d\mu.$$

If in place of $|b| \in L^2(\mu)$ we have $|b| \in L^2(\lambda)$, then

$$\int_M \frac{|\nabla\varrho|^2}{\varrho^2}\,d\lambda \leqslant \int_M |b|^2\,d\lambda.$$

Actually, the technical conditions imposed on the manifold M in [**179**] are even broader and are expressed in terms of the heat semigroup. However, as noted in [**181**, Remark 2.5(ii)], this estimate may fail for general Riemannian manifolds even if $b = 0$. Namely, according to Chung [**248**], Grigoryan [**423**], Li, Schoen [**605**], there exist complete connected Riemannian manifolds on which there are nonconstant positive integrable harmonic functions; such a function defines a measure satisfying our equation with $b = 0$, and the above estimates fail for it.

3.7(ii). More on Harnack's inequality

The proof of Harnack's inequality given in § 3.4 can be slightly simplified. It turns out that it suffices to prove this inequality for $b = 0$, and then by adding a fictitious variable to reduce the general case to this one. The idea is borrowed from Kruzhkov [**543**] (see also Oleĭnik, Kružkov [**756**]).

Let $u \in W^{2,1}(\Omega)$ satisfy the equation

$$\partial_{x_i}(a^{ij}\partial_{x_j} u) = 0, \tag{3.7.1}$$

where

$$\sum_{i,j} |a^{ij}(x)|^2 \leqslant \gamma^2, \quad \langle A(x)h, h\rangle \geqslant \alpha |h|^2.$$

Here we *do not assume* that the matrix A is symmetric! Then

$$\sup_{U(z,r)} u(x) \leqslant C(d,\alpha,\gamma) \inf_{U(z,r)} u(x) \tag{3.7.2}$$

if $U(z,3r) \subset \Omega$, where $U(z,r)$ is the ball of radius r centered at z, and

$$C(d,\alpha,\gamma) = \exp\Bigl(C(d)\frac{\gamma}{\alpha}\Bigr).$$

Let now $u \in W^{2,1}(\Omega)$ satisfy the equation

$$\partial_{x_i}(a^{ij}\partial_{x_i} u - b^i u) = 0.$$

Set $B = \sup_\Omega |b(x)|$. Then the function $v(x,y) = yu(x)$ satisfies in $\Omega \times [1,2]$ the equation

$$\partial_{x_i}(a^{ij}y^{-1}\partial_{x_i} v - b^i \partial_y v) + \partial_y\bigl((\gamma + B)\partial_y v\bigr) = 0.$$

We observe that this is an equation of type (3.7.1) with the matrix

$$\widetilde{A}(x,y) = \widetilde{A}(x) = \begin{pmatrix} A(x) & b(x) \\ 0 & (\gamma + B) \end{pmatrix},$$

where

$$\sum_{i,j} |\widetilde{a}^{ij}|^2 \leqslant 4(\gamma + B)^2, \quad \langle \widetilde{A}h, h\rangle \geqslant \frac{\alpha}{2}|h|^2.$$

Therefore, u satisfies Harnack's inequality (3.7.2) with the constant

$$C(d,r,\gamma,\alpha,B) = \exp\Bigl(C(d)\alpha^{-1}(\gamma + Br)\Bigr).$$

Its dependence on r is the result of a simple scaling. In the same manner we can add the dependence on other coefficients in the full divergence form equation.

In addition, if we are not interested in an explicit form of dependence of the constant on γ and α, then its dependence on B can be easily obtained by scaling.

3.7(iii). The John–Nirenberg lemma

We recall that in our proof of Harnack's inequality the crucial role was played by Lemma 3.4.1. However, this lemma is a special case of the following more general assertion due to F. John and L. Nirenberg [**493**]. As above, let

$$v_\Omega = |\Omega|^{-1} \int_\Omega v\, dx.$$

3.7.2. Theorem. *Let v be a measurable function on the ball $U_0 \in \mathbb{R}^d$. Suppose that there is a number $K > 0$ such that for every ball $U \subset U_0$ we have the inequality*

$$|U|^{-1} \int_U |v - v_U|\, dx \leqslant K.$$

Then

$$\bigl|\{x \in U_0\colon |v(x) - v_{U_0}| > \alpha\}\bigr| \leqslant K_1(d) e^{-K_2(d)\alpha/K} |U_0|$$

for all $\alpha > 0$, where $K_1(d)$ and $K_2(d)$ depend only on d. This estimate yields that

$$\int_{U_0} \exp\bigl(c|v - v_{U_0}|\bigr)\, dx \leqslant C(d, K, U_0, c) < \infty \tag{3.7.3}$$

for every number $c < K_2(d)/K$.

The assertion of the lemma follows from this theorem by the aid of the Poincaré inequality (see (1.1.6)):

$$\int_{U(x_0,R)} |v - v_{U(x_0,R)}|\, dx \leqslant C(d) R \int_{U(x_0,R)} |\nabla v|\, dx \leqslant KC(d)R^d.$$

We observe that for the proof of Harnack's inequality we actually need the following weaker assertion, whose proof is relatively easy.

3.7.3. Lemma. *Let a function $v \in W^{2,1}(3U)$, where $U = U(0,1)$ is the unit ball, be such that for every nonnegative function $\psi \in C_0^\infty(3U)$ we have the inequality*

$$\int_{3U} |\nabla v|^2 \psi^2\, dx \leqslant K \int_{3U} \bigl[|\psi|^2 + |\nabla\psi|^2\bigr]\, dx. \tag{3.7.4}$$

Then there exist numbers $c > 0$ and M, depending only on d and K, but not on v, such that

$$\int_U \exp\bigl(c|v - v_{2U}|\bigr)\, dx \leqslant M.$$

PROOF. We observe that (3.7.4) is invariant with respect to adding a constant to v. Hence we shall assume that $v_{2U} = 0$. Set $R_n = 1 + 2^{-n}$, $U_n = U(0, R_n)$. Let ψ_n be a function such that $\psi_n \in C_0^\infty(U_{n-1})$, $\psi_n(x) = 1$ if $x \in U_n$ and $0 \leqslant \psi_n \leqslant 1$, $|\nabla\psi_n| \leqslant C_1 2^n$. By the Poincaré inequality

$$\int_{U_0} |v|^2\, dx \leqslant K|3U|(1 + 2C_1). \tag{3.7.5}$$

Let now $|v|_k = \min\{|v|, k\}$, where $k \in \mathbb{N}$. Let us substitute in (3.7.4) the function $\psi = \bigl(|v|_k^m + (8Km)^m\bigr)\psi_m$. We obtain the inequality

$$\int_{U_0} |\nabla v|^2 \bigl(|v|_k^{2m} + (8Km)^{2m}\bigr)\psi_m^2\, dx$$
$$\leqslant 4K \int_{U_0} \Bigl[m^2 |\nabla |v|_k|^2 |v|_k^{2m-2}\psi_m^2 + \bigl(|v|_k^{2m} + (8Km)^{2m}\bigr)\bigl(\psi_m^2 + |\nabla\psi_m|^2\bigr)\Bigr]\, dx.$$

By Young's inequality we have $m^2|v|_k^{2m-2} \leqslant \varepsilon(m-1)m^{-1}|v|_k^{2m} + \varepsilon^{-m+1}m^{2m-1}$. Let us set $\varepsilon = (8K)^{-1}$. Then

$$\int_{U_0} |\nabla v|^2 \bigl(|v|_k^{2m} + (8Km)^{2m}\bigr)\psi_m^2\, dx$$
$$\leqslant 8K \int_{U_0} \bigl(|v|_k^{2m} + (8Km)^{2m}\bigr)\bigl(\psi_m^2 + |\nabla\psi_m|^2\bigr)\, dx.$$

Applying the Sobolev inequality, we obtain the following estimate:

$$\left(\int_{U_m} |v|_k^{(2m+2)d/(d-2)}\,dx\right)^{(d-2)/d} \leqslant 4^m C_2 \int_{U_{m-1}} |v|_k^{2m+2}\,dx + C_2(8Km)^{2m+2}.$$

Let $p_m = 2m+2$, $p_{m+1} = \beta p_m$, $0 < p_1 \leqslant 2$ and $\beta = d/(d-2)$. Set

$$I(p,U) = \left(\int_U |v|_k^p\,dx\right)^{1/p}.$$

Then the last estimate can be rewritten as follows:

$$I(p_{m+1}, U_m) \leqslant C_3^{m\beta^{-m}} I(p_m, U_{m-1}) + C_4\beta^m.$$

Repeating this estimate for smaller m and applying (3.7.5), we find that

$$I(p_{m+1}, U_m) \leqslant C_5(1 + p_{m+1}).$$

Thus, applying if necessary Hölder's inequality, we obtain

$$\int_{U(x_0,R_0/2)} |v|_k^m\,dx \leqslant C_6^m(1+m)^m,$$

where C_6 does not depend on m. Therefore, for sufficiently small $\delta > 0$ we have

$$\int_{U(x_0,R_0/2)} \exp(\delta|v|_k)\,dx = \sum_{m=0}^{\infty} \int_{U(x_0,R_0/2)} \frac{(\delta|v|_k)^m}{m!}\,dx < \infty.$$

Letting $k \to \infty$, we complete the proof. □

An assertion similar to this lemma is proved in Trudinger [**903**]. In the parabolic case this approach is used in Lieberman [**611**, Lemma 6.20].

3.7(iv). Differentiability of solutions with respect to a parameter

Here we give broad sufficient conditions for the differentiability of solutions to stationary Fokker–Planck–Kolmogorov equations

$$\partial_{x_i}\partial_{x_j}(a_\alpha^{ij}\mu_\alpha) - \partial_{x_i}(b_\alpha^i\mu_\alpha) = 0$$

with respect to a parameter. These conditions are expressed in terms of Lyapunov functions and apply to unbounded coefficients. They become especially simple in the case where a_α^{ij} and $\partial_\alpha a_\alpha^{ij}$ are uniformly bounded and $\partial_{x_i}a_\alpha^{ij}$, $\partial_\alpha\partial_{x_i}a_\alpha^{ij}$, b_α^i, $\partial_\alpha b_\alpha^i$ have at most polynomial growth: just the relation $\lim\limits_{|x|\to\infty} \sup_\alpha \langle b_\alpha(x), x\rangle = -\infty$ for the drift coefficient b_α is needed. For the proofs, see Bogachev, Shaposhnikov, Veretennikov [**183**] (these results were announced in [**184**]).

Suppose that for every $\alpha \in [0,1]$ we are given a second order elliptic operator

$$L_\alpha\varphi = a_\alpha^{ij}\partial_{x_i}\partial_{x_j}\varphi + b_\alpha^i\partial_{x_i}\varphi$$

with coefficients satisfying certain conditions specified below. Suppose also that for each α there is a unique probability measure μ_α satisfying the stationary Fokker–Planck–Kolmogorov equation

$$L_\alpha^*\mu_\alpha = 0. \tag{3.7.6}$$

We assume that we deal with real coefficients a_α^{ij} and b_α^i on $\mathbb{R}^d$, Borel measurable in (x,α) and satisfying the following conditions:

the matrices $A_\alpha(x) = \big(a_\alpha^{ij}(x)\big)_{i,j\leqslant d}$ are symmetric and for every ball $U \subset \mathbb{R}^d$ we have

$$\sup_\alpha \|a_\alpha^{ij}\|_{W^{p,1}(U)} \leqslant M_1(U) < \infty, \quad \sup_\alpha \|b_\alpha^i\|_{L^p(U)} \leqslant M_2(U) < \infty, \tag{3.7.7}$$

where $p = p(U) > d$, and for all x we have

$$A_\alpha(x) \geqslant c_0 \mathrm{I}, \quad c_0 > 0, \tag{3.7.8}$$

where I is the unit operator and c_0 is a constant (independent of U).

Recall that a family $\mathcal{M}$ of probability measures is uniformly tight if, for each number $r > 0$, there is a compact set K such that $\mu(\mathbb{R}^d\backslash K) \leqslant r$ for all $\mu \in \mathcal{M}$.

A criterion for the uniform tightness (see Exercise 3.7.28) is the existence of a locally bounded Borel function $W \geqslant 0$ such that $\lim\limits_{|x|\to\infty} W(x) = +\infty$ and

$$\sup_{\mu\in\mathcal{M}} \int_{\mathbb{R}^d} W\, d\mu < \infty.$$

The case of continuity is much easier and here we have the following result (in which (3.7.8) is replaced by a local bound).

3.7.4. Proposition. *Suppose that* (3.7.7) *holds,* $\inf_{\alpha,x\in U} \det A_\alpha(x) > 0$ *for every ball* U *and that the family of measures* μ_α *(that are unique probability solutions to the corresponding equations* (3.7.6)*) is uniformly tight. Assume also that, for every ball* U*, the restrictions of* a_α^{ij} *and* b_α^i *to* U *are continuous in* α *in* $L^1(U)$*. Then, one can choose densities* ϱ_α *of* μ_α *such that the function* $\varrho_\alpha(x)$ *will be jointly continuous. In addition, the mapping* $\alpha \mapsto \varrho_\alpha$ *with values in* $L^1(\mathbb{R}^d)$ *is continuous, i.e., the mapping* $\alpha \mapsto \mu_\alpha$ *is continuous in the variation norm.*

A sufficient condition for the uniform tightness of the measures μ_α is the existence of a Lyapunov function V such that $V(x) \to +\infty$ and $\sup_\alpha L_\alpha V(x) = -\infty$ as $|x| \to \infty$. Certainly, this condition ensures also the existence and uniqueness of solutions. In order to have the local continuity in L^1 it is enough to have the usual continuity of the coefficients in α along with their uniform integrability on balls.

Recall that a mapping $\alpha \mapsto f_\alpha$ from $(0,1)$ to $L^p(U)$ is differentiable if there is a mapping $\alpha \mapsto g_\alpha$ from $(0,1)$ to $L^p(U)$ such that $(f_{\alpha+s} - f_\alpha)/s \to g_\alpha$ in $L^p(U)$ as $s \to 0$ for each fixed $\alpha \in (0,1)$. If $\alpha \mapsto g_\alpha$ is continuous, then f_α is said to be continuously differentiable in L^p.

Suppose that for every ball U there is a number $p_0 = p_0(U) > d$ such that the mappings

(3.7.9) $\quad \alpha \mapsto a_\alpha^{ij}|_U$ and $\alpha \mapsto b_\alpha^i|_U$ are continuously differentiable in $L^{p_0}(U)$.

Note that this condition is fulfilled if, in addition to (3.7.7), the functions a_α^{ij}, $\partial_{x_k} a_\alpha^{ij}$ and b_α^i are differentiable in α and their derivatives in α are continuous in α and locally bounded in both variables.

Set

$$A_\alpha^i := (a_\alpha^{i1}, \dots, a_\alpha^{id}), \quad \operatorname{div} A_\alpha := (\operatorname{div} A_\alpha^1, \dots, \operatorname{div} A_\alpha^d),$$
$$B_\alpha := \partial_\alpha b_\alpha = (\partial_\alpha b_\alpha^1, \dots, \partial_\alpha b_\alpha^d),$$
$$S_\alpha := \partial_\alpha A_\alpha = (\partial_\alpha a_\alpha^{ij})_{i,j\leqslant d}, \quad R_\alpha^i = \partial_\alpha \partial_{x_j} a_\alpha^{ij}, \quad R_\alpha := (R_\alpha^1, \dots, R_\alpha^d) = \operatorname{div} S_\alpha.$$

We assume that

$$\sup_{\alpha,x} \|S_\alpha(x)\| \leqslant \lambda_0 < \infty. \tag{3.7.10}$$

This condition is obviously fulfilled if $A_\alpha(x)$ does not depend on α or is uniformly Lipschitzian in α.

Condition (3.7.10) implies that in (3.7.9) actually a stronger condition on the diffusion coefficient is fulfilled: the functions $\alpha \mapsto a_\alpha^{ij}|_U$ are continuously differentiable in every $L^{p_1}(U)$ with $p_1 < \infty$. In particular, we can take $p_1 = p_1(U) > \dfrac{dp}{p-d}$, where $p = p(U)$ is the number from (3.7.7).

3.7.5. Theorem. *Let conditions* (3.7.7)–(3.7.10) *hold. Let* $V \in C^2(\mathbb{R}^d)$ *and let* $W \in L^1_{\rm loc}(\mathbb{R}^d)$ *be such that* $\lim\limits_{|x|\to\infty} V(x) = +\infty$, $\lim\limits_{|x|\to\infty} W(x) = +\infty$ *and*

$$\sup_\alpha L_\alpha V(x) \leqslant -W(x) \quad \textit{if} \ |x| \geqslant R \tag{3.7.11}$$

for some $R > 0$. *Assume that for some numbers* $C_V > 0, m \geqslant 1$ *we have*

$$\sup_\alpha \Bigl(|A_\alpha^{-1/2}(b_\alpha - \operatorname{div} A_\alpha)|^2 + |A_\alpha^{-1/2}(\partial_\alpha b_\alpha - \partial_\alpha \operatorname{div} A_\alpha)|^2 + |L_\alpha V|\Bigr) \leqslant C_V + C_V V^m W, \tag{3.7.12}$$

Finally, assume that for some $\varepsilon < 1/(4m+1)$ *there is a ball outside of which*

$$\sup_\alpha \langle A_\alpha \nabla V, \nabla V\rangle \leqslant \varepsilon V W. \tag{3.7.13}$$

Then $\partial_\alpha \varrho_\alpha$ *exists and for each* $\alpha \in (0,1)$ *satisfies the equation*

$$L_\alpha^* \partial_\alpha \varrho_\alpha = \operatorname{div}(B_\alpha \varrho_\alpha - R_\alpha \varrho_\alpha - S_\alpha \nabla \varrho_\alpha). \tag{3.7.14}$$

In addition, the mapping $\alpha \mapsto \varrho_\alpha$ *with values in* $L^1(\mathbb{R}^d)$ *is differentiable.*

Finally, if the diffusion matrix A *does not depend on* α, *then* (3.7.12) *can be replaced by the bound* $\sup_\alpha(|A^{-1/2}\partial_\alpha b_\alpha|^2 + |L_\alpha V|) \leqslant C_V + C_V V^m W$.

The theorem employs seven conditions (3.7.7)–(3.7.13) (or four global conditions (3.7.10)–(3.7.13) once we fix our local assumptions), but if $A_\alpha = \mathrm{I}$, $|b_\alpha|$ and $|\partial_\alpha b_\alpha|$ have polynomial bounds, then, by taking $V(x) = |x|^2$, it suffices to have only one condition that $\lim\limits_{|x|\to\infty} \sup_\alpha \langle b_\alpha(x), x\rangle = -\infty$.

Let us briefly comment on the hypotheses of this theorem.

3.7.6. Remark. (i) Condition (3.7.11) ensures the existence and uniqueness of probability solutions to (3.7.6) for each α. It also ensures the uniform boundedness of the integrals of W with respect to the measures μ_α; moreover, for each $k < 4m+1$ the integrals of $V^k W$ against μ_α are uniformly bounded.

(ii) Note also that if A is constant (independent of α) and nondegenerate, then (3.7.8) and (3.7.10) are fulfilled (along with the first condition in (3.7.7)) and that we have $R_\alpha = S_\alpha = 0$.

(iii) It follows from (3.7.8) that (3.7.12) is ensured by the estimate

$$\sup_\alpha \Bigl(|b_\alpha - \operatorname{div} A_\alpha|^2 + |\partial_\alpha b_\alpha - \partial_\alpha \operatorname{div} A_\alpha|^2 + |L_\alpha V|\Bigr) \leqslant C_V + C_V V^m W.$$

However, for growing diffusion coefficients the operators $A_\alpha^{-1/2}$ in (3.7.12) can help. Certainly, for uniformly bounded A_α both estimates are equivalent.

3.7.7. Corollary. *Suppose that* A_α, A_α^{-1} *and* $\partial_\alpha A_\alpha$ *are uniformly bounded, condition* (3.7.9) *holds and that*

$$|\partial_{x_k} a_\alpha^{ij}(x)| + |\partial_\alpha \partial_{x_k} a_\alpha^{ij}(x)| + |b_\alpha^i(x)| + |\partial_\alpha b_\alpha^i(x)| \leqslant C + C|x|^k \quad \forall\, x, \alpha$$

for some constants C and k. Assume also that we have the equality

$$\lim_{|x|\to\infty} \sup_{\alpha} \langle b_\alpha(x), x\rangle = -\infty.$$

Then $\varrho_\alpha(x)$ is differentiable in α and $\partial_\alpha \varrho_\alpha(x)$ satisfies the equation indicated in the theorem.

3.7.8. Corollary. *Suppose that the operator norms of A_α, A_α^{-1} and $\partial_\alpha A_\alpha$ are uniformly bounded, condition (3.7.9) holds and that*

$$|\partial_{x_k} a_\alpha^{ij}(x)| + |\partial_\alpha \partial_{x_k} a_\alpha^{ij}(x)| + |b_\alpha^i(x)| + |\partial_\alpha b_\alpha^i(x)| \leqslant C \exp(q|x|^\beta) \quad \forall\, x, \alpha$$

for some positive numbers C, q, and β. Assume also that there is a number

$$\gamma > (9 \sup_{\alpha,x} \|A_\alpha(x)\| + 1/4) q\beta$$

such that outside of some ball we have

$$\sup_{\alpha} \langle b_\alpha(x), x\rangle \leqslant -\gamma |x|^\beta.$$

Then $\varrho_\alpha(x)$ is differentiable in α and $\partial_\alpha \varrho_\alpha(x)$ satisfies the equation indicated in the theorem.

3.7.9. Example. Let $A = \mathrm{I}$, $b(x) = -x + h_\alpha(x)$, $h_\alpha(x)$ is continuously differentiable in α, where $\sup_{\alpha,x} |h_\alpha(x)| < \infty$, $|\nabla h_\alpha(x)| \leqslant C \exp(q|x|^2)$, $q < 1/20$. Then probability solutions μ_α to the corresponding equations (3.7.6) exist, are unique and have densities ϱ_α differentiable in α.

3.7.10. Example. Let $a_\alpha^{ij}(x)$ and $b_\alpha^i(x)$ be of class C_b^1 in both variables, let $A_\alpha^{-1}(x)$ be uniformly bounded, and let $\sup_\alpha \langle b_\alpha(x), x\rangle \to -\infty$ as $|x| \to \infty$. Then $\varrho_\alpha(x)$ is continuously differentiable in both variables. This case was considered in Pardoux, Veretennikov [**764**] and Veretennikov [**916**].

Note that in applications of these results to stationary distributions of diffusions governed by stochastic equations $d\xi_{\alpha,t} = \sigma_\alpha(\xi_{\alpha,t}) dw_t + b_\alpha(\xi_{\alpha,t}) dt$ the hypotheses must be checked for the matrices $A_\alpha = \sigma_\alpha \sigma_\alpha^*/2$.

Let us consider examples showing that certain additional assumptions, besides smoothness of our coefficients, are needed to guarantee even the continuity of densities with respect to the parameter.

3.7.11. Example. One can find a bounded function $b_\alpha(x)$ on the plane of class C^∞ in both variables such that the integral

$$J_\alpha = \int_{-\infty}^{+\infty} \exp \int_0^x b_\alpha(y)\, dy\, dx$$

exists, but is not continuous at $\alpha = 0$. It is not difficult to give explicit examples of such functions; it suffices to take a positive integrable smooth function g such that g'/g is bounded (say, $(1+x^2)^{-1}$) and set $g(\alpha, x) = g(x) + \alpha g(\alpha x)$; in this case the integral in x is not continuous in α at the origin. Then the probability density

$$\varrho_\alpha(x) = J_\alpha^{-1} \exp \int_0^x b_\alpha(y)\, dy, \quad b_\alpha(x) = \partial_x g(\alpha, x)/g(\alpha, x),$$

satisfies the equation $\varrho_\alpha'' - (b_\alpha \varrho_\alpha)' = 0$, but $\varrho_\alpha(x)$ is discontinuous in α at $\alpha = 0$ for all x. A bit more involved example (see the next example) provides bounded $b_\alpha(x)$ that is Lipschitzian in α. It is also worth noting that if we consider our equation with a parameter as an equation with an extra variable (or pass to a system of equations), then we obtain a degenerate equation.

3.7.12. Example. Let us give an explicit example (suggested by I. Yaroslavtsev) of a uniformly bounded function $b_\alpha(x)$ with bounded $\partial_\alpha b_\alpha(x)$ such that the probability solution $\varrho_\alpha(x)$ to the equation $L^*_\alpha(\varrho_\alpha dx) = 0$ is not continuous in α.

Set $b_\alpha(x) = -x - 1$ if $x < 0$, $\alpha \in (-1,1)$ and define $b_\alpha(x)$ for $x \geqslant 0$, $\alpha \in (-1,1)$ as follows. Let $b_0(x) = -(x+1)^{-1/2}$. We now construct three functions φ_1, φ_2 and φ_3 on $[0,+\infty)$. Set $\varphi_1(x) = 2(x+1)^{-1/2}$. Next, on the closed set where $\alpha \leqslant \varphi_1(x)$ we let $b_\alpha(x) = -(x+1)^{-1/2} + \alpha$. Clearly, on the curve $\bigl(x, \varphi_1(x)\bigr)$ our function equals $(x+1)^{-1/2}$. Next, we take a decreasing function φ_2 on $(0,+\infty)$ such that for all $\alpha > 0$ one has

$$\int_0^{\varphi_2^{-1}(\alpha)} \exp \int_0^x b_\alpha(y)\,dy\,dx = 2 + \int_0^\infty \exp \int_0^x b_0(y)\,dy\,dx.$$

Note that $\varphi_2^{-1}(\alpha)$ is determined by the previously defined values of b. The values $\varphi_2^{-1}(\alpha)$ are well-defined. Indeed, once $\alpha > 0$ is fixed, we have

$$b_\alpha(x) = -(x+1)^{-1/2} + \alpha \text{ if } x < 4\alpha^{-2} - 1,\ b_\alpha(x) = (x+1)^{-1/2} \text{ if } 4\alpha^{-2} - 1 \leqslant x,$$

hence the integral in the left-hand side of the expression above taken over $[0,+\infty)$ is infinite and this integral equals $2-\alpha < 2$ when taken over $[0, 4\alpha^{-2}-1]$, which enables one to find suitable $\varphi_2^{-1}(\alpha) > 4\alpha^{-2} - 1$. Note that φ_2 is continuous decreasing and $\varphi_2(x) \to 0$ as $x \to \infty$.

On the set $\varphi_1(x) < \alpha \leqslant \varphi_2(x)$ we put $b_\alpha(x) = (x+1)^{-1/2}$. Finally, we take

$$\varphi_3(x) = \varphi_2(x) + 2(x+1)^{-1/2}$$

and on the set $\varphi_2(x) < \alpha < \varphi_3(x)$ we put

$$b_\alpha(x) = (x+1)^{-1/2} - \bigl(\alpha - \varphi_2(x)\bigr),$$

and on the set where $\alpha \geqslant \varphi_3(x)$ we put $b_\alpha(x) = -(x+1)^{-1/2}$. It is clear that $|b_\alpha(x)| \leqslant 1$, b is continuous and $|\partial_\alpha b_\alpha(x)| \leqslant 1$, more precisely, in the interiors of the sets bounded by the three curves defined above $\partial_\alpha b_\alpha(x)$ is 1, 0 and -1 and again 0, respectively. The corresponding solution $\varrho_\alpha(x)$ is discontinuous at $\alpha = 0$. This property can be retained by smoothing b and making it differentiable in α everywhere with a uniformly bounded derivative in α.

It is instructive to see which conditions of the theorem cannot be ensured in this example. Here Corollary 3.7.7 almost applies with $V(x) = x^2$ and for any fixed parameter α we have $b_\alpha(x)x < x^{-1/2}/2$ outside of some interval, but there is no uniformity in α.

3.7.13. Remark. It follows from L^1-differentiability that

$$\int_{\mathbb{R}^d} \partial_\alpha \varrho_\alpha(x)\,dx = 0.$$

3.7.14. Remark. (i) The main theorem can be combined with the results of § 1.7 on Sobolev regularity of nonhomogeneous equations in order to ensure the differentiability of $\alpha \mapsto \varrho_\alpha$ with values in $W^{r,1}(\mathbb{R}^d)$. For example, suppose that in the main theorem A_α, A_α^{-1}, ∇a^{ij}, and $\partial_\alpha A_\alpha$ are uniformly bounded and

$$|b_\alpha|^p + |\partial_\alpha b_\alpha|^p + |\partial_\alpha \mathrm{div}\, A_\alpha|^p \leqslant C_V + C_V V^m W$$

with some $p > d$. Then we have differentiability in the Sobolev space $W^{r,1}(\mathbb{R}^d)$ for any order $r < p$.

(ii) Condition (3.7.10) is essential in the proof in [**183**] in estimating the integral of the expression $|A_\alpha^{-1/2} S_\alpha \nabla \varrho_\alpha|^2/\varrho_\alpha$ and similar integrals for certain approximations, since we have an a priori bound just for the integral of $|\nabla \varrho_\alpha|^2/\varrho_\alpha$, so that a growing $\|S_\alpha\|$ could destroy this estimate. We do not know whether this condition can be removed. However, under assumptions similar to those used in Corollary 3.7.8 it is proved in § 3.2 that there is a bound for the integral of $|\nabla \varrho_\alpha|^p/\varrho_\alpha$ with a sufficiently large $p > d$. This enables us to replace (3.7.10) by an exponential bound and make our condition on $\partial_\alpha A_\alpha$ closer to that of the condition on $\partial_\alpha b_\alpha$.

Comments

Theorem 3.1.4, partially generalizing Theorem 3.1.2 obtained in Bogachev, Krylov, Röckner [**149**], strengthens Theorem 3.1 from the paper Metafune, Pallara, Rhandi [**694**], where A^{-1} is bounded and $|b| \in L^2(\mu)$; in addition, in [**694**] the same local assumptions are made as in Theorem 3.1.4 along with some condition which is slightly stronger than (3.1.11).

The idea of obtaining upper estimates of densities of solutions to stationary Fokker–Planck–Kolmogorov equations by means of uniform estimates of densities of solutions of nonhomogeneous equations (see § 3.3) was used in Metafune, Pallara, Rhandi [**694**] in the case of exponential functions. This method was further developed in Bogachev, Krylov, Röckner [**154**], where the presented results were obtained. Case (iii) in Example 3.3.2 gives the estimate from [**694**] under somewhat weaker assumptions. See also Fornaro, Fusco, Metafune, Pallara [**380**] for related results.

The results presented in § 3.4 were obtained in Bogachev, Röckner, Shaposhnikov [**165**]; they generalize the results from Metafune, Pallara, Rhandi [**694**], where the theory of nonlinear equations was applied (in particular, the known results of S.N. Bernstein), which required certain additional assumptions about A. The first general result for $p>2$ was established in [**694**] (its special case had been earlier considered in Cruzeiro, Malliavin [**260**]). The condition from [**165**] strengthens this results, since does not require the differentiability of the drift coefficient and assumes a lower regularity of the diffusion coefficient (in [**694**] it was assumed that $a^{ij} \in C^3(\mathbb{R}^d)$ and $b \in C^2(\mathbb{R}^d)$). This weakening of conditions on the coefficients became possible, because, unlike [**694**], the proof in [**165**] does not use the methods of the theory of nonlinear equations.

The modern form of a Harnack type elliptic inequality was obtained in the classical work Moser [**724**]. Harnack's inequality for the direct elliptic equation under very broad assumptions was established by Safonov [**830**]. Harnack's inequality is the main subject of the book DiBenedetto, Gianazza, Vespri [**301**]. The positivity of densities of solutions of stationary equations was studied in Agafontsev, Bogachev, Shaposhnikov [**4**], Shaposhnikov [**844**].

Our presentation in § 3.5 follows Bogachev, Röckner, Shaposhnikov [**167**].

The global regularity of solutions of elliptic equations for measures with coefficients of a general form was also studied in Bogachev, Krylov, Röckner [**153**]. A condition for the differentiability of the density of an invariant measure with respect to a parameter was obtained in Pardoux, Veretennikov [**764**].

On diffusion operators on manifolds, see also Bratteli, Robinson [**205**], Li [**606**], Robinson [**808**]. Obviously, the local positivity result proves above extends to the manifold case without any changes. Some other results involving global conditions can also natural analogs for manifolds, while in certain cases the geometry of manifolds can play a role, as is the case, for example, for square integrability of logarithmic gradients of densities of solutions or Liouville-type theorems.

Exercises

3.7.15. Give an example of a smooth probability density ϱ on the real line such that it has a zero, but the function ϱ'/ϱ belongs to all classes $L^p(\varrho\,dx)$ whenever $1 \leqslant p < \infty$.

3.7.16. Can it happen that the equation $L^*_{1,b}\mu = 0$ with a continuous periodic drift coefficient b on the real line has a probability solution?

3.7.17. Let $h \in W^{1,1}_{\text{loc}}(\mathbb{R}^d)$ and $|\nabla h| \in L^2(\mathbb{R}^d)$. Prove that there is a sequence of functions $u_n \in C_0^\infty(\mathbb{R}^d)$ such that $|\nabla h - \nabla u_n| \to 0$ in the norm of $L^2(\mathbb{R}^d)$.

HINT: Use the Poincaré inequality.

3.7.18. Investigate which estimates for the density can be obtained from the results in this chapter in the case of bounded b; consider the density $\varrho(x) = \exp(-|x|)$ satisfying the equation with $A = 1$, $b(x) = -\operatorname{sign} x$.

3.7.19. A probability measure on the real line with a density ϱ satisfies the equation $L^*_{1,b}\mu = 0$ with a bounded smooth drift b. Must the derivatives of ϱ be bounded?

3.7.20. Suppose that a probability measure μ on $\mathbb{R}^d$ has a density $\varrho \in W^{p,1}_{\text{loc}}(\mathbb{R}^d)$ and satisfies the equation $L^*_{A,b}\mu = 0$, where the mapping $A = (a^{ij})$ is bounded, $a^{ij} \in W^{p,1}_{\text{loc}}(\mathbb{R}^d)$, $b^i \in L^p_{\text{loc}}(\mathbb{R}^d)$, $p > d$. Prove that

$$\int_{\mathbb{R}^d} \psi(f) L_{A,b} f \, d\mu = -\int_{\mathbb{R}^d} \psi'(f) \langle A\nabla f, \nabla f\rangle \, d\mu \tag{3.7.15}$$

for all locally Lipschitzian functions ψ on the real line and all functions $f \in C_b^2(\mathbb{R}^d)$ such that $f|b|, \langle b, \nabla f\rangle \in L^1(\mu)$. In particular,

$$\int_{\mathbb{R}^d} f L_{A,b} f \, d\mu = -\int_{\mathbb{R}^d} \langle A\nabla f, \nabla f\rangle \, d\mu. \tag{3.7.16}$$

The condition $f \in C_b^2(\mathbb{R}^d)$ can be replaced with the conditions $f \in W^{p,2}_{\text{loc}}(\mathbb{R}^d)$, $\|f\|_\infty < \infty$ combined with the boundedness of $D^2 f$ outside a ball.

HINT: If $f \in C_0^2(\mathbb{R}^d)$, $\psi \in C^2(\mathbb{R})$, then

$$\int_{\mathbb{R}^d} \psi(f) L_{A,b} f \, d\mu = -\int_{\mathbb{R}^d} \psi'(f) \langle A\nabla f, \nabla f\rangle \, d\mu + \int_{\mathbb{R}^d} \psi(f) \partial_{x_i} f [\varrho b^i - \partial_{x_j}(a^{ij}\varrho)] \, dx$$

by the integration by parts formula, where the second integral vanishes by the equation, since $\psi(f)\nabla f$ has compact support and is the gradient of the function $\Psi(f) \in C_0^2(\mathbb{R}^d)$, where $\psi = \Psi'$ and $\Psi(0) = 0$. By approximation the equality extends to locally Lipschitzian functions ψ (that are Lipschitzian on the range of f). In the general case the equality is obtained by passing to the limit in the equalities for $f\varphi_n$, where $\varphi_n(x) = \varphi(x/n)$, $\varphi \in C_0^\infty(\mathbb{R}^d)$, $\varphi = 1$ on the unit ball, since $|\langle b, \nabla(f\varphi_n)\rangle| \leqslant C|\langle b, \nabla f\rangle| + C|fb|$.

3.7.21. Suppose that a continuous function $\psi > 0$ is strictly increasing on the set $[0, +\infty)$, $\psi([0, +\infty)) = [0, +\infty)$. Prove that for all $\alpha \geqslant 1$ and $\beta \geqslant 0$ we have

$$\alpha\beta \leqslant \alpha\psi^{-1}(\ln \alpha) + \beta e^{\psi(\beta)}.$$

HINT: Justify and apply the inequality $ab \leqslant a\psi(a) + b\psi^{-1}(b)$.

3.7.22. Suppose that probability densities ϱ_j on a probability space (Ω, μ) converge almost everywhere to a probability density ϱ, measurable functions g_j are uniformly bounded and almost everywhere converge to a function g, $f \geqslant 0$ is a measurable function such that for some positive Borel function θ on $[0, +\infty)$ with $\lim\limits_{t\to\infty} \theta(t) = +\infty$ we have

$$\sup_j \int_\Omega f\theta(f)\varrho_j \, d\mu < \infty.$$

Prove that

$$\int_\Omega f g_j \varrho_j \, d\mu \to \int_\Omega f g \varrho \, d\mu.$$

HINT: If f is bounded, then this is true, since $\|\varrho - \varrho_j\|_1 \to 0$ (see Bogachev [**125**, Theorem 2.8.9]); in the general case for a sufficiently large N the integrals of $|f - \min(f, N)|\varrho$ and $|f - \min(f, N)|\varrho_j$ are small due to the condition on the function θ.

3.7.23. Modifying the reasoning from the proof of Lemma 3.7.3 prove that if, for every nonnegative function $\psi \in C_0^\infty(3U)$ and some function $v \in W^{2,1}(3U)$, we have the inequality

$$\int_{3U} |\nabla v|^2 \psi^2 \, dx \leqslant K \int_{3U} |\nabla \psi|^2 \, dx + K\biggl(\int_{3U} \psi^2 \, dx\biggr)^{(p-2)/p},$$

where $p > d$, then there exist numbers $c > 0$ and M for which the conclusion of the indicated lemma is true.

3.7.24. Prove that the function

$$e^{-x^2} \int_0^x e^{s^2} \, ds$$

is not integrable on $\mathbb{R}$.

3.7.25. Let μ be a probability measure and let $f \in L^\infty(\mu)$. Prove the equality $\lim\limits_{p\to\infty} \|f\|_p = \|f\|_\infty$.

HINT: Observe that this is obvious for the function f with $\|f\|_\infty = 1$ and finitely many values.

3.7.26. Let $\varrho \in W^{1,1}_{\rm loc}(\mathbb{R}^d)$, $\varrho \geqslant 0$ and $\exp(\varepsilon|\nabla\varrho/\varrho|) \in L^1_{\rm loc}(\varrho\, dx)$ with some $\varepsilon > 0$, where $\nabla\varrho/\varrho(x) = 0$ if $\varrho(x) = 0$. Without using the results of this chapter prove that ϱ has a continuous density which either vanishes identically or has no zeros.

HINT: Let U be a ball; from the inequality $x^k e^{-x} \leqslant k^k e^{-k}$ derive the estimate

$$\varepsilon^k \int_U |\partial_{x_i}\varrho/\varrho|^k \varrho \, dx \leqslant k^k e^{-k} \int_U \exp(\varepsilon|\partial_{x_i}\varrho/\varrho|)\varrho \, dx \leqslant k^k e^{-k} M$$

with a number M independent of k; next, for the functions $f_k = \varrho^{1/k}$ obtain the estimate $\|\partial_{x_i} f_k\|_{L^k(U)} \leqslant \varepsilon^{-1}e^{-1}M^{1/k}$; conclude by the embedding theorem that $\{f_k\}$ contains a uniformly convergent subsequence, but its limit can be only 0 or 1; see Bogachev, Mayer-Wolf [**156**].

3.7.27. Let a nonnegative measure μ have a locally bounded continuous density $\varrho \in W^{2,1}_{loc}(\mathbb{R}^d)$ and satisfy the inequality

$$\int_{\mathbb{R}^d} \mathcal{L}_{A,b,c} u \, d\mu \leqslant 0 \quad \text{whenever } u \in C_0^\infty(\mathbb{R}^d), \ u \geqslant 0.$$

Suppose that condition (G1) from § 3.5 is fulfilled, $c \leqslant 0$ and

$$\exp(\delta|b|^2), \ \exp(\delta|c|) \in L^1_{loc}(\mu)$$

for some $\delta > 0$. Prove that either $\varrho = 0$ or ϱ has no zeros.

HINT: See Bogachev, Röckner, Shaposhnikov [**166**].

3.7.28. Show that a family $\mathcal{M}$ of Borel probability measures on $\mathbb{R}^d$ is uniformly tight precisely when there exists a locally bounded Borel function $W \geqslant 0$ such that $\lim\limits_{|x|\to\infty} W(x) = +\infty$ and

$$\sup_{\mu\in\mathcal{M}} \int_{\mathbb{R}^d} W \, d\mu < \infty.$$

HINT: This condition is sufficient by the Chebyshev inequality, since the sets $\{W \leqslant R\}$ are bounded; in the opposite direction, if $\mathcal{M}$ is uniformly tight, then there are increasing compact sets K_n such that $\mu(\mathbb{R}^d \backslash K_n) \leqslant 2^{-n}$ for all $\mu \in \mathcal{M}$ and the union of K_n is the whole space; one can take the function W defined by $W = n$ on $K_{n+1} \backslash K_n$.

3.7.29. Suppose that a sequence of probability measures μ_n on $\mathbb{R}^d$ with densities of class $W^{1,1}(\mathbb{R}^d)$ is uniformly tight and that the sequence of densities is bounded in the Sobolev space $W^{1,1}(\mathbb{R}^d)$. Show that it has a subsequence converging in the variation norm.

HINT: Use the compactness of the embedding $W^{1,1}(U) \subset L^1(U)$ for every ball.

3.7.30. Suppose that a sequence of probability measures μ_n on $\mathbb{R}^d$ with densities ϱ_n of class $W^{1,1}(\mathbb{R}^d)$ is uniformly tight and $L^*_{I,b_n}\mu_n = 0$, where b_n are Borel vector fields such that

$$\sup_n \int_{\mathbb{R}^d} |b_n|^2 \, d\mu_n < \infty.$$

Show that it has a subsequence converging in the variation norm.

HINT: Use that

$$\sup_n \int_{\mathbb{R}^d} \frac{|\nabla \varrho_n(x)|^2}{\varrho_n(x)} \, dx < \infty$$

and the previous exercise.

CHAPTER 4

Uniqueness Problems

Given an elliptic operator $L = L_{A,b}$, we address the following problems. When is a probability solution to the equation $L^*\mu = 0$ unique? When does a given probability solution have the property that any solution of bounded variation is a multiple of it? Which dimension can have the simplex of probability solutions $\mathcal{P}$?

It turns out that even for $A = \mathrm{I}$ and infinitely differentiable b on $\mathbb{R}^d$ with $d > 1$ a probability solution is not always unique and the simplex of all probability solutions can be infinite-dimensional (but it is not known whether this dimension can be finite greater than one); this is discussed in § 4.2. It may also happen that a probability solution is unique, but there is no uniqueness for signed solutions in the sense that the dimension of the space of solutions has dimension greater than one. Nevertheless, it is still unknown whether nontrivial signed solutions can exist when there is no probability solution. However, there are broad conditions for uniqueness of probability solutions and for uniqueness of signed solutions. One of the main uniqueness results in this chapter is Theorem 4.1.6 with several alternative assumptions: in addition to our usual local assumptions about the coefficients A and b, we need that either $a^{ij}(x)/(1+|x|^2)$ and $b^i(x)/(1+|x|)$ are in $L^1(\mu)$ or there exists a Lyapunov function V such that $LV(x) \leqslant CV(x)$. These two uniqueness properties are not equivalent, and signed solutions are discussed in § 4.3. In the next chapter we continue the study of uniqueness looking at semigroups associated with our elliptic operators.

Let $\mathcal{M}_{\mathrm{ell}}^{A,b}$ be the set of all probability solutions of the equation $L_{A,b}^*\mu = 0$ and let $\#\mathcal{M}_{\mathrm{ell}}^{A,b}$ be its cardinality.

4.1. Uniqueness conditions

The explicit expression for the solutions enables us to investigate completely the uniqueness problem in the one-dimensional case (see § 1.4). We recall that even if $d = 1$ and we have $A = 1$, there might be many probability solutions for singular b (see the example described in (1.4.9)). Let us consider locally integrable b.

4.1.1. Example. Suppose that $\Omega = \mathbb{R}^1$, A is locally absolutely continuous and positive, and b is locally Lebesgue integrable. If

$$\int_{-\infty}^{0} \frac{1}{\sqrt{A(s)}}\, ds = \infty, \quad \int_{0}^{+\infty} \frac{1}{\sqrt{A(s)}}\, ds = \infty, \tag{4.1.1}$$

then $\#\mathcal{M}_{\mathrm{ell}}^{A,b} \leqslant 1$.

PROOF. Let (4.1.1) hold. Suppose that there exist two different probability measures satisfying our elliptic equation. It is clear from Proposition 1.4.2 that the space of all solutions is two-dimensional. If we set

$$\Psi(x) = \exp\Bigl(\int_0^x \frac{b(s) - A'(s)}{A(s)}\, ds\Bigr) = \frac{A(0)}{A(x)} \exp \int_0^x \frac{b(s)}{A(s)}\, ds,$$

then this space is spanned by the functions Ψ and

$$\Phi(x) := \Psi(x) \int_0^x A(s)^{-1}\Psi(s)^{-1}\, ds.$$

The existence of two linearly independent solutions implies that Ψ and Φ are integrable on $\mathbb{R}^1$. We observe that by (4.1.1) the function $(A\Psi)^{-1}$ is not integrable at $+\infty$ and $-\infty$, since

$$\Bigl(\int \frac{1}{\sqrt{A(s)}}\, ds\Bigr)^2 \leqslant \int \Psi(s)\, ds \int \frac{1}{A(s)\Psi(s)}\, ds.$$

Since $\Phi(x) < 0$ if $x < 0$ and $\Phi(x) > 0$ if $x > 0$, at least one of the two different measures from $\mathcal{M}_{\rm ell}^{A,b}$ has the form $C_1\Psi + C_2\Phi$ with $C_1 > 0$ and $C_2 \neq 0$. This leads to a contradiction, because Φ is the product of Ψ and a function with infinite limits of different signs at $-\infty$ and $+\infty$. $\square$

It is easily seen that if (4.1.1) does not hold, then $\mathcal{M}_{\rm ell}^{A,b}$ may contain infinitely many elements. For example, let

$$A(x) = \exp(2x^2), \quad b(x) = 2x\exp(2x^2).$$

Then $\bigl[b(s) - A'(s)\bigr]A(s)^{-1} = -2s$ and $\Psi(x) = \exp(-x^2)$. Hence one element from $\mathcal{M}_{\rm ell}^{A,b}$ has the density $\sqrt{\pi}^{-1}\exp(-x^2)$ and the function

$$C\exp(-x^2)\int_0^x \exp(-s^2)\, ds + 2C\exp(-x^2),$$

where C is a normalization constant, is the density of another one. Moreover, the following statement is true.

4.1.2. Proposition. *Let $A > 0$ be a locally absolutely continuous function on the real line such that* (4.1.1) *does not hold. Then there exists a locally integrable function b such that $\#\mathcal{M}_{\rm ell}^{A,b} = \infty$.*

PROOF. If both integrals in (4.1.1) are finite, then, setting $b = A'/2$, we can take $\Psi = A^{-1/2}$, which after multiplication by a constant gives one element from the set $\mathcal{M}_{\rm ell}^{A,b}$. This is seen from (1.4.7), since

$$\exp\int_0^x \frac{b(s) - A'(s)}{A(s)}\, ds = \frac{\sqrt{A(0)}}{\sqrt{A(x)}} = \sqrt{A(0)}\Psi(x).$$

Taking a constant C bigger than the integral of $A^{-1/2}$, we obtain another element of $\mathcal{M}_{\rm ell}^{A,b}$ in the form

$$C_0\Psi(x)\Bigl[\int_0^x \frac{1}{A(s)\Psi(s)}\, ds + C\Bigr],$$

where $C_0 > 0$ is a normalization constant. The second factor above is positive and bounded, since $A(s)\Psi(s) = \sqrt{A(s)}$. The construction is similar if only one of the integrals in (4.1.1) is finite, say, the integral over $(-\infty, 0]$ is less than 1. Namely, let us assume for simplicity that $A(0) = 1$ and set

$$b(x) = A'(x)/2, \quad \Psi(x) = A(x)^{-1/2} \quad \text{if } x \leqslant 0.$$

Now we have two possibilities: A^{-1} may be integrable on $[0,+\infty)$ or not. In the first case, we set

$$\Psi(x) = e^{-x}A(x)^{-1}, \quad x > 0.$$

This gives us one positive integrable solution of our equation with the coefficient b on $(0,+\infty)$ determined from the equation

$$\exp\int_0^x \frac{b(s)-A'(s)}{A(s)}\,ds = \Psi(x). \tag{4.1.2}$$

Since A is positive and A' is locally integrable, the function b is also locally integrable. A linearly independent positive integrable solution is obtained by setting

$$\begin{aligned}\Psi_1(x) &= \frac{1}{\sqrt{A(x)}}\Bigl[\int_0^x \frac{1}{\sqrt{A(s)}}\,ds + 1\Bigr], \quad x \leqslant 0,\\ \Psi_1(x) &= \frac{1+e^{-x}}{2A(x)}, \quad x > 0.\end{aligned} \tag{4.1.3}$$

If A^{-1} is not integrable on $[0,+\infty)$, then we set

$$\Psi(x) = \exp(-x^2)\exp\Bigl(-\int_0^x \exp(s^2)\frac{1}{A(s)}\,ds\Bigr), \quad x \geqslant 0, \tag{4.1.4}$$

and determine b from (4.1.2). Again Ψ is a positive integrable solution. A linearly independent positive integrable solution is obtained by (4.1.3) on $(-\infty,0]$ and

$$\Psi_1(x) = \Psi(x)\Bigl(\int_0^x \frac{1}{A(s)\Psi(s)}\,ds + 1\Bigr), \quad x > 0.$$

The integrability of Ψ_1 on $[0,+\infty)$ follows from the following relation:

$$\Psi(x)\int_0^x \frac{1}{A(s)\Psi(s)}\,ds = \exp(-x^2) - \exp(-x^2)\exp\Bigl(-\int_0^x \exp(s^2)\frac{1}{A(s)}\,ds\Bigr).$$

In order to verify this relation, we observe that by (4.1.4) it is equivalent to the equality

$$\int_0^x \frac{1}{A(s)\Psi(s)}\,ds = \exp\int_0^x \exp(s^2)\frac{1}{A(s)}\,ds - 1.$$

It remains to note that both sides of the last equality coincide at $x=0$ and the differentiation of both sides gives the identity

$$\frac{1}{A(x)\Psi(x)} = \exp(x^2)\frac{1}{A(x)}\exp\int_0^x \exp(s^2)\frac{1}{A(s)}\,ds,$$

which is valid by (4.1.4). □

The following example shows that even in the one-dimensional case there are smooth gradient-type drifts for which there is a unique solution in the class of probability measures, but there are also nonzero signed solutions in the class of bounded measures.

4.1.3. Example. Let $\varrho(x) = (1+4x^4)^{-1}\exp(-x^4)$, $A(x)=1$,

$$b(x) = \frac{\varrho'(x)}{\varrho(x)} = -4x^3 - 16x^3(1+4x^4)^{-1}.$$

Then the density $u(x) = x(1+4x^4)^{-1} = x\exp(x^4)\varrho(x)$ with respect to Lebesgue measure defines a bounded signed measure which is also a solution. In addition, $b \in L^p(\varrho\,dx)$ for any $p < \infty$. Indeed,

$$u'(x) = x\exp(x^4)\varrho'(x) + \bigl(\exp(x^4) + 4x^4\exp(x^4)\bigr)\varrho(x),$$

hence $u'(x) - b(x)u(x) = 1$.

Therefore, the problems of uniqueness in the class of probability measures and in the class of all bounded measures (in the sense of its dimension) are not equivalent.

We now turn to the multidimensional case $d > 1$ and assume that $L = L_{A,b}$ satisfies our standard local assumptions

(4.1.5) $$a^{ij} \in W^{d+,1}_{\rm loc}(\mathbb{R}^d),\ |\det A|^{-1} \in L^\infty_{\rm loc}(\mathbb{R}^d),\ b^i \in L^{d+}_{\rm loc}(\mathbb{R}^d).$$

Suppose that the equation

(4.1.6) $$L^*\mu = 0$$

has a probability solution μ, which, as we know, possesses a positive continuous density ϱ in the class $W^{d+,1}_{\rm loc}(\mathbb{R}^d)$. Suppose that a locally finite measure ν (not necessarily a probability measure) satisfies equation (4.1.6) too. Then ν is also given by a continuous density ζ with respect to Lebesgue measure and $\zeta/\varrho \in W^{d+,2}_{\rm loc}(\mathbb{R}^d)$ according to Proposition 1.7.3. Set

$$v = \zeta/\varrho = d\nu/d\mu.$$

The function v satisfies the equation

(4.1.7) $$\operatorname{div}\bigl(\varrho A\nabla v - hv\bigr) = 0,$$

where

$$h^i = b^i - \beta^i_\mu,\quad \beta^i_\mu = \sum_{j=1}^d \Bigl(\partial_{x_j}a^{ij} + a^{ij}\frac{\partial_{x_j}\varrho}{\varrho}\Bigr).$$

We observe that $\operatorname{div} h = 0$ in the sense that

$$\int_{\mathbb{R}^d} \langle h(x), \nabla\psi(x)\rangle\,dx = 0 \quad \forall\,\psi \in C_0^\infty(\mathbb{R}^d).$$

Let $L^1_+(\mu)$ denote the convex subset of $L^1(\mu)$ consisting of all nonnegative functions. Let us formulate our questions in terms of the function v.

When is any solution v from the class $L^1_+(\mu)$ constant?

When is any solution v from the class $L^1(\mu)$ constant?

Which dimension can have the simplex of solutions v to equation (4.1.7) belonging to the set $L^1_+(\mu)$?

Therefore, our uniqueness problems for equation (4.1.6) are reduced to proving Liouville-type theorems for equation (4.1.7) in the classes $L^1(\mu)$ and $L^1_+(\mu)$. We emphasize that in the setting of this section the existence of a probability solution is assumed a priori. In § 4.3 we consider signed solutions without this assumption. We do not know whether nonzero signed integrable solutions can exist without probability solutions (this is impossible on the real line, see Exercise 4.5.22).

To answer the first two questions we need an auxiliary lemma.

Let f be a smooth bounded function on $[0,+\infty)$. As above, set $v^+ = \max\{v, 0\}$, $v^- = -\min\{v, 0\}$.

4.1.4. Lemma. *Suppose that v is a solution to equation* (4.1.7). *Then, for every function $\psi \in C_0^\infty(\mathbb{R}^d)$, one has the following equality:*

$$\int_{\mathbb{R}^d} |\sqrt{A}\nabla v^+|^2 f''(v^+)\psi\, d\mu = \int_{\mathbb{R}^d} f(v^+)L\psi\, d\mu - f'(0)\int_{\mathbb{R}^d} v^- L\psi\, d\mu.$$

PROOF. Since v satisfies equation (4.1.7), we have

$$\int_{\mathbb{R}^d} \langle \varrho A\nabla v - hv, \nabla\varphi\rangle\, dx = 0 \quad \forall \varphi \in C_0^\infty(\mathbb{R}^d).$$

Clearly, this identity extends to functions $\varphi \in W^{2,1}(\mathbb{R}^d)$ with compact support. Take $\varphi = f'(v^+)\psi$. We obtain the equality

$$\int_{\mathbb{R}^d} \varrho|\sqrt{A}\nabla v^+|^2 f''(v^+)\psi\, dx$$
$$= -\int_{\mathbb{R}^d} \varrho\langle A\nabla v, \nabla\psi\rangle f'(v^+)\, dx - \int_{\mathbb{R}^d} \langle h, \nabla v\rangle f'(v^+)\psi\, dx.$$

Let us consider the first summand on the right-hand side of this equality. Writing the function v as $v = v^+ - v^-$ and integrating by parts, we find that

$$-\int_{\mathbb{R}^d} \varrho\langle A\nabla v, \nabla\psi\rangle f'(v^+)\, dx$$
$$= \int_{\mathbb{R}^d} \varrho\, \mathrm{div}(A\nabla\psi) f(v^+)\, dx + \int_{\mathbb{R}^d} \langle A\nabla\varrho, \nabla\psi\rangle f(v^+)\, dx$$
$$- \int_{\mathbb{R}^d} \varrho v^- f'(v^+)\mathrm{div}(A\nabla\psi)\, dx - \int_{\mathbb{R}^d} v^- \langle A\nabla\varrho, \nabla\psi\rangle f'(v^+)\, dx.$$

Acting similarly with the second term, we obtain

$$-\int_{\mathbb{R}^d} \langle h, \nabla v\rangle f'(v^+)\psi\, dx = \int_{\mathbb{R}^d} \langle h, \nabla\psi\rangle f(v^+)\, dx - \int_{\mathbb{R}^d} \langle h, \nabla\psi\rangle v^- f'(v^+)\, dx.$$

Let us sum these equalities and use the identity

$$h^i = (b^i - \beta_\mu^i)\varrho = b^i\varrho - \sum_{j=1}^d \partial_{x_j}(a^{ij}\varrho).$$

This yields that

$$\int_{\mathbb{R}^d} \varrho|\sqrt{A}\nabla v^+|^2 f''(v^+)\psi\, dx = \int_{\mathbb{R}^d} \varrho f(v^+)L\psi\, dx - \int_{\mathbb{R}^d} \varrho v^- f'(v^+)L\psi\, dx.$$

It remains to observe that $v^- f'(v^+) = v^- f'(0)$ almost everywhere. □

4.1.5. Remark. In the case $v \geqslant 0$ we obtain that $v^+ = v$, $v^- = 0$ and

$$\int \varrho|\sqrt{A}\nabla v|^2 f''(v)\psi\, dx = \int \varrho f(v)L\psi\, dx.$$

4.1.6. Theorem. *Let conditions* (4.1.5) *be fulfilled and suppose that a function $v \in L_+^1(\mu)$ is a solution to equation* (4.1.7). *Suppose also that at least one of the following conditions is satisfied:*

(i) $\dfrac{|a^{ij}|}{1+|x|^2}, \dfrac{|b^i|}{1+|x|} \in L^1(\mu)$, *for example*, $a^{ij}, b^i \in L^1(\mu)$,

(ii) $a^{ij}, b^i - \beta_\mu^i \in L^1(\mu)$,

(iii) *there exists a positive function* $V \in C^2(\mathbb{R}^d)$ *such that* $V(x) \to +\infty$ *as* $|x| \to +\infty$ *and for some number* $C > 0$ *and all* x *we have*

$$LV(x) \leqslant CV(x).$$

Then the function v *is constant. In particular, the solution* μ *is the only probability solution to the equation* $L^*\mu = 0$.

PROOF. (i) Let ζ be a nonnegative function on the real line such that $\zeta(s) = 1$ if $|s| \leqslant 1$, $\zeta(s) = 0$ if $|s| > 2$, $\zeta'(s) \leqslant 0$ if $s > 0$, and there is a number $M > 0$ such that for all s

$$|\zeta(s)| \leqslant M, \quad |\zeta'(s)| \leqslant M, \quad |\zeta''(s)| \leqslant M.$$

Set $\psi_N(x) = \zeta(|x|/N)$. Then

$$|\nabla\psi_N(x)| \leqslant N^{-1}M, \quad |\partial_{x_i}\partial_{x_j}\psi_N(x)| \leqslant N^{-2}M.$$

Let us apply Lemma 4.1.4 to the function $f(t) = -\mathrm{arctg}\, t$, for which we have $f''(t) = 2t/(1+t^2)^2$. Taking into account the remark after that lemma we obtain

$$2\int_{\mathbb{R}^d} |\sqrt{A}\nabla v|^2 v(1+v^2)^{-2}\psi_N\, d\mu = \int_{\mathbb{R}^d} f(v)L\psi_N\, d\mu$$
$$\leqslant \pi M N^{-2}\int_{S_N}\sum_{i,j=1}^d |a^{ij}|\, d\mu + \pi M N^{-1}\int_{S_N}\sum_{i=1}^d |b^i|\, d\mu,$$

where $S_N = \{x\colon N \leqslant |x| \leqslant 2N\}$. As $N \to \infty$, the right-hand side tends to zero by the μ-integrability of $a^{ij}(x)/(1+|x|^2)$ and $b^i(x)/(1+|x|)$. Therefore, $|\nabla v(x)| = 0$ a.e. In case (ii) the reasoning is completely analogous.

(iii) Let $\zeta \in C^\infty([0,+\infty))$ be a nonnegative function such that $\zeta(0) = 1$, $\zeta(s) = 0$ if $s > 1$, $\zeta' \leqslant 0$ and $\zeta'' \geqslant 0$, i.e., the function ζ is convex and decreasing. Let $M = \max|\zeta'|$. Set

$$f(t) = -\mathrm{arctg}\, t, \quad \psi_N(x) = \zeta\bigl(N^{-1}V(x)\bigr).$$

Then

$$L\psi_N = N^{-1}\zeta'(V/N)LV + N^{-2}\zeta''(V/N)\langle A\nabla V, \nabla V\rangle.$$

We observe that $f(v)L\psi_N \leqslant 2^{-1}\pi M N^{-1}V$. Lemma 4.1.4 yields the inequality

$$2\int_{\mathbb{R}^d} |\sqrt{A}\nabla v|^2 v(1+v^2)^{-2}\psi_N\, d\mu \leqslant 2^{-1}\pi M N^{-1}\int_{\{V<N\}} V\, d\mu.$$

Let γ be any number in $(0,1)$. Then

$$N^{-1}\int_{\{V<N\}} V\, d\mu \leqslant \gamma + \mu(\{x\colon \gamma N < V < N\}).$$

Hence, letting $N \to \infty$, we obtain

$$2\int_{\mathbb{R}^d} |\sqrt{A}\nabla v|^2 v(1+v^2)^{-2}\, d\mu \leqslant 2^{-1}\pi M\gamma.$$

Letting $\gamma \to 0$, we conclude that $\nabla v = 0$ a.e., so $v = \mathrm{const}$. □

4.1.7. Remark. We observe that we never used in the proof that $v \in L^1(\mu)$, only the estimate $v \geqslant 0$ was employed. Therefore, if either of conditions (i), (ii), or (iii) holds, we obtain that any nonnegative solution to equation (4.1.7) is constant, in particular, any nonnegative solution to the equation $L^*\mu = 0$ is proportional to the probability solution μ (of course, we assume that a probability solution exists).

(ii) It is clear from the proof that in assertion (i) it suffices to have a weaker condition

$$\lim_{k\to\infty} \int_{k\leqslant|x|\leqslant k+1} \Big[k^{-2}\|A(x)\| + k^{-1}|b(x)|\Big]\,\mu(dx) = 0$$

or a similar condition with $b - \beta_\mu$ in place of b in assertion (ii).

(iii) Another proof of this theorem will be given in Chapter 5 (see Theorem 5.3.3, Proposition 5.3.9 and Examples 5.5.1, 5.5.3).

It follows from the theorem that, for example, if $A = \mathrm{I}$, $b^i \in L^p_{\mathrm{loc}}(\mathbb{R}^d)$, $p > d$, and $\mu = \varrho\,dx$ is a probability solution of the equation $L^*_{\mathrm{I},b}\mu = 0$ such that we have $|b - \nabla\varrho/\varrho| \in L^1(\mu)$, then μ is the only probability solution. However, the converse is not true, which follows from Exercise 4.5.27, i.e., the inclusion $|b-\nabla\varrho/\varrho| \in L^1(\mu)$ does not follow from the fact that the measure $\mu = \varrho\,dx$ is a unique probability solution to our equation (where $|b| \in L^p_{\mathrm{loc}}(\mathbb{R}^d)$ with $p > d$). We recall that it is unknown whether it is true that $|\nabla\varrho/\varrho| \in L^1(\mu)$ if $\mu = \varrho\,dx$ is a solution to the equation $L^*_{\mathrm{I},b}\mu = 0$, where $|b| \in L^1(\mu)$ and $|b| \in L^p_{\mathrm{loc}}(\mathbb{R}^d)$ with $p > d$.

4.1.8. Example. Let $V(x) = \ln\big(\ln(1 + |x|)\big)$ if $|x| > 1$. Then, whenever $|x| > 1$, we have

$$LV(x) = -\frac{\langle A(x)x, x\rangle}{|x|^2(1+|x|)^2\ln(1+|x|)}\Big(1 + \frac{1}{\ln(1+|x|)} + \frac{1+|x|}{|x|}\Big) + \frac{\operatorname{tr} A(x)}{|x|(1+|x|)\ln(1+|x|)} + \frac{\langle b(x), x\rangle}{|x|(|x|+1)\ln(|x|+1)}.$$

Therefore, in case of a locally bounded A and b, condition (iii) in Theorem 4.1.6 is fulfilled if, for some number $C > 0$, whenever $|x| \geqslant 1$ we have the estimates

$$\operatorname{tr} A(x) \leqslant C|x|^2\ln(1+|x|)\ln\big(\ln(1+|x|)\big), \quad \langle b(x), x\rangle \leqslant C|x|^2\ln(1+|x|)\ln\big(\ln(1+|x|)\big)$$

The second bound holds if we have $|b(x)| \leqslant C|x|\ln(1+|x|)\ln\big(\ln(1+|x|)\big)$. However, even for $A = \mathrm{I}$ the estimate $|b(x)| \leqslant c + c|x|\big(\ln(1+|x|)\big)^r$ with $r > 1$ is not sufficient: see Example 4.2.1.

Note that if we use the Lyapunov function $V_1(x) = \ln(1+|x|^2)$, then we obtain

$$LV_1(x) = 2(1+|x|^2)^{-1}\operatorname{tr} A(x) - 4(1+|x|^2)^{-2}\langle A(x)x, x\rangle + 2(1+|x|^2)^{-1}\langle b(x), x\rangle,$$

which will lead to a slightly worse sufficient condition.

4.1.9. Theorem. *Let conditions* (4.1.5) *be fulfilled and suppose that a function* $v \in L^1(\mu)$ *is a solution to equation* (4.1.7). *Suppose also that there exists a function* $V \in C^2(\mathbb{R}^d)$ *such that* $V(x) \to +\infty$ *as* $|x| \to +\infty$ *and, for some number* $C > 0$ *and all* x, *one has the inequalities*

$$LV(x) \geqslant -C, \quad |\sqrt{A(x)}\nabla V(x)| \leqslant C. \tag{4.1.8}$$

Then, the function v *is constant, in particular, any solution of the equation* $L^*\sigma = 0$ *in the class of bounded measures is proportional to the given probability solution* μ. *In place of* (4.1.8) *it suffices to have just*

$$LV(x) \geqslant -CV, \quad |\sqrt{A(x)}\nabla V(x)| \leqslant CV. \tag{4.1.9}$$

PROOF. Let $v \in L^1(\mu)$ be a solution to equation (4.1.7). Set

$$f(t) = (1+t)^{-1} \quad \text{and} \quad \psi(x) = \zeta\bigl(N^{-1}V(x)\bigr),$$

where a nonnegative function ζ is such that $\zeta(x) = 1$ if $|x| \leqslant 1$, $\zeta(x) = 0$ if $|x| > 2$, $\zeta' \leqslant 0$ if $x > 0$, and there exists a number $M > 0$ such that for all x one has the estimates

$$|\zeta(x)| \leqslant M, \quad |\zeta'(x)| \leqslant M, \quad |\zeta''(x)| \leqslant M.$$

Applying Lemma 4.1.4 and taking into account that $(1+v^+)^{-1} \leqslant 1$ and $\zeta' \leqslant 0$, we obtain the inequality

$$2\int \varrho|\sqrt{A}\nabla v^+|^2(1+v^+)^{-3}\psi\,dx \\ \leqslant (MCN^{-1} + MC^2N^{-2})\Bigl(\int_{V>N} \varrho\,dx + \int_{V>N} |\sigma|\,dx\Bigr).$$

Letting $N \to +\infty$ we obtain that $v^+ = const$. Considering $-\sigma$ we conclude that $v^- = const$, hence $v = const$ and $\sigma = const \cdot \varrho$. Finally, from (4.1.9) we can pass to (4.1.8) by replacing V with a new function $\ln(V + M)$, where $M = 1 - \min V$. □

Let us note an important difference between the hypotheses of this theorem and the previous one: here $LV \geqslant -CV$, while there we have had $LV \leqslant CV$. The latter does not guarantee uniqueness of an integrable solution, which is seen from Example 4.1.3, where one can take $V(x) = x^2$. Thus, the behavior of large positive values of LV is responsible for the uniqueness of probability solutions, and the behavior of large in absolute value negative values is responsible for the uniqueness of integrable solutions.

4.1.10. Example. Let $V(x) = \ln\bigl(\ln(1+|x|)\bigr)$ if $|x| > 1$. Then, whenever $|x| > 1$, we have

$$|\sqrt{A(x)}\nabla V(x)|^2 = \langle A(x)\nabla V(x), \nabla V(x)\rangle = \frac{\langle A(x)x, x\rangle}{|x|^2(|x|+1)^2\ln^2(|x|+1)}.$$

Using the expression for $LV(x)$ from Example 4.1.8, we conclude that in order to ensure the hypotheses in Theorem 4.1.9 it suffices to have the estimates

$$\langle A(x)x, x\rangle \leqslant C|x|^4\ln(1+|x|)\ln\bigl(\ln(1+|x|)\bigr),$$
$$\operatorname{tr} A(x) + \langle b(x), x\rangle \geqslant -C|x|^2\ln(1+|x|)\ln\bigl(\ln(1+|x|)\bigr)$$

for all $x \in \mathbb{R}^d$ with $|x| \geqslant 1$ and some number $C > 0$.

Now we consider the case of unit diffusion matrix and a gradient drift coefficient. Our equation in this case takes the form

$$\Delta\mu - \operatorname{div}(b\mu) = 0, \tag{4.1.10}$$

where $b = \nabla\Phi$ for some function Φ. Our question is this: when does the probability solution μ have the form $Ce^{\Phi}\,dx$?

It should be noted that even in the one-dimensional case a probability solution may fail to have such a form. Indeed, let

$$\Phi(x) = -\frac{x^2}{2} + \int_0^x e^{s^2/2}\,ds.$$

Then our equation has the unique probability solution $(2\pi)^{-1/2}e^{-x^2/2}\,dx$, but the function e^Φ is not integrable.

Let us give a sufficient condition which guarantees that $\mu = Ce^\Phi\,dx$.

4.1.11. Theorem. *Let $\Phi \in W^{d+,1}_{\rm loc}(\mathbb{R}^d)$. Suppose that a probability measure $\mu = \varrho\,dx$ satisfies equation (4.1.10). If $e^\Phi \in L^1(\mathbb{R}^d)$, then $\varrho = Ce^\Phi$ for some constant C. So the measure $\mu = Ce^\Phi\,dx$ is the only probability solution.*

PROOF. Set $w = \varrho e^{-\Phi}$. Then

$$\mathrm{div}(e^\Phi \nabla w) = 0,$$

which can be written as an integral identity

$$\int_{\mathbb{R}^d} e^\Phi \langle \nabla w, \nabla\varphi\rangle\,dx = 0, \quad \varphi \in C_0^\infty(\mathbb{R}^d). \tag{4.1.11}$$

We shall consider the case $d > 1$, since in the case $d = 1$ our equation has a unique probability solution and this solution is Ce^Φ. Hence (4.1.11) remains valid for all functions $\varphi \in W_0^{2,1}(\mathbb{R}^d)$. We apply this integral identity to $\varphi = w^{-1}\zeta_k^2$, where

$$\zeta_k(x) = \zeta(x/k),\ \zeta \in C_0^\infty(\mathbb{R}^d),\ \zeta \geqslant 0,\ \zeta(x) = 1 \text{ if } |x| < 1.$$

We obtain

$$\int_{\mathbb{R}^d} e^\Phi |\nabla w|^2 w^{-2}\zeta_k^2\,dx = 2\int_{\mathbb{R}^d} e^v w^{-1}\langle \nabla w, \nabla\zeta_k\rangle\zeta_k\,dx.$$

Applying the Cauchy inequality and taking into account that $|\nabla\zeta_k| \leqslant Ck^{-1}$, we find that

$$\int_{\mathbb{R}^d} e^\Phi |\nabla w|^2 w^{-2}\zeta_k^2\,dx \leqslant 4k^{-2}\int_{\mathbb{R}^d} e^v\,dx.$$

Letting $k \to +\infty$, we arrive at the equality

$$\int_{\mathbb{R}^d} e^\Phi |\nabla w|^2 w^{-2}\,dx = 0,$$

which yields that $|\nabla w| = 0$. Thus, w is constant and $\varrho = Ce^\Phi$. □

Applying Theorem 4.1.6 and Remark 4.1.7, we arrive at the following result.

4.1.12. Theorem. *Let $\Phi \in W^{p,1}_{\rm loc}(\mathbb{R}^d)$, where $p > d$. Suppose that a probability measure $\mu = \varrho\,dx$ satisfies equation (4.1.10) and either of the following conditions is fulfilled:*

(i) $|\nabla\Phi| \in L^1(\mu)$,

(ii) *there exists a positive function $V \in C^2(\mathbb{R}^d)$ such that $V(x) \to +\infty$ as $|x| \to +\infty$ and, for some number $C > 0$ and all x, one has the estimates*

$$LV(x) \leqslant C, \quad |\nabla V(x)| \leqslant C$$

or even weaker estimates

$$LV(x) \leqslant CV, \quad |\nabla V(x)| \leqslant CV.$$

Then, there exists a constant C such that $\varrho = Ce^\Phi$.

PROOF. The case $d = 1$ is considered as above. In the case $d \geqslant 2$ it suffices to observe that $e^\Phi\,dx$ is a nonnegative solution and apply Remark 4.1.7. □

Finally, we observe that the set of measures $\mathcal{M}^{A,b}_{\rm ell}$ is convex. Some observations about its extreme points are contained in Exercises 4.5.20 and 4.5.21.

4.2. Cases of nonuniqueness

In the next section and also in the next chapter we return to uniqueness problems for solutions of elliptic equations in the classes of bounded measures and probability measures. Here we present some negative results in the case of $A = \mathrm{I}$ and infinitely differentiable b. It is easy to construct such examples for singular drifts b (see (1.4.9)). In dimension 1, the singularity of b is the only reason for nonuniqueness (see Proposition 4.1.1). However, in any dimension $d > 1$ there are examples of nonuniqueness with smooth b. It is not easy to find such examples, since the existence results based on Lyapunov functions always produce unique solutions. For several years the problem remained open until the following explicit example was constructed in Bogachev, Röckner, Stannat [**177**], [**178**].

4.2.1. Example. Let

$$b^i(x) = -x_i - 2x_{\sigma(i)} e^{(x_i^2 - x_{\sigma(i)}^2)/2},$$

where $\sigma\colon \{1,\dots,d\} \to \{1,\dots,d\}$ is one-to-one and $\sigma(i) \neq i$. Then our equation has at least two solutions: one is the standard Gaussian measure μ on $\mathbb{R}^d$ and another is the measure $\nu = v \cdot \mu$ with

$$v(x) = c_d \sum_{i=1}^d \int_{-\infty}^{x_i} e^{-s^2/2}\, ds,$$

where c_d is a normalizing constant. More generally, let $f \in C^2(\mathbb{R}^1)$ be bounded, with $f, f' > 0$ and $f', f'' \in L^1(\mathbb{R}^1)$. Define $b = (b^i)\colon \mathbb{R}^d \to \mathbb{R}^d$ by

$$b^i(x) := \frac{f''(x_i)}{f'(x_i)} + 2\frac{f''(x_{\sigma(i)})}{f'(x_{\sigma(i)})}, \quad x = (x_1, \dots, x_d) \in \mathbb{R}^d,$$

and set

$$\mu := c_1 \prod_{i=1}^d f'(x_i)\, dx, \quad \nu := c_2 \sum_{i=1}^d f(x_i) \cdot \mu,$$

where $c_1, c_2 > 0$ are normalizing constants. Then μ and ν are two different elements in $\mathcal{M}_{\mathrm{ell}}^{\mathrm{I},b}$. In particular, let us take here a smooth function f with $f'(s) = c|s|^{-1}(\ln s)^{-r}$ outside of $[-2, 2]$, where $r > 1$. This gives an example of nonuniqueness in which $|b(x)| \leqslant c + c|x|\bigl(\ln(1 + |x|)\bigr)^r$ with $r > 1$. More generally, one can take any positive integrable function f' with bounded f'', which gives an example of nonuniqueness with $|b(x)| \leqslant C/f'(|x|)$.

Proof. All these assertions are verified directly, we only show that the measures μ and ν are solutions. We obviously have $\beta_\mu^i(x) = f''(x_i)/f'(x_i)$. To show that $\mu \in \mathcal{M}_{\mathrm{ell}}^{\mathrm{I},b}$, we have to prove that the vector field $b - \beta^\mu$ has zero divergence with respect to μ. Since $\sigma(i) \neq i$, the integration by parts formula yields that for each $\varphi \in C_0^\infty(\mathbb{R}^d)$ we have

$$\sum_{i=1}^d \int \frac{f''(x_{\sigma(i)})}{f'(x_i)} \partial_{x_i}\varphi(x)\, \mu(dx) = c_1 \sum_{i=1}^d \int \cdots \int \Bigl(\int \varphi(x)\, dx_i \Bigr)$$
$$\times f''(x_{\sigma(i)}) f'(x_1) \cdots \widehat{f'(x_i)} \cdots f'(x_d)\, dx_1 \cdots \widehat{dx_i} \cdots dx_d = 0,$$

where the symbol "$\widehat{\quad}$" indicates that the corresponding term is omitted. To show that also $\nu \in \mathcal{M}_{\rm ell}^{\rm I,b}$, we first observe that $\beta_\nu(x)$ has the components

$$\frac{f'(x_i)}{\sum_{j=1}^d f(x_j)} + \frac{f''(x_i)}{f'(x_i)}.$$

Hence we have to show that the vector field with components

$$\frac{2f''(x_{\sigma(i)})}{f'(x_i)} - \frac{f'(x_i)}{\sum_{j=1}^d f(x_j)}$$

has zero divergence with respect to ν. For any function $\varphi \in C_0^\infty(\mathbb{R}^d)$, integrating by parts and using the equality

$$\partial_{x_i} \sum_{j=1}^d f(x_j) f''(x_{\sigma(i)}) = f'(x_i) f''(x_{\sigma(i)}),$$

we obtain

$$\begin{aligned}
&\sum_{i=1}^d \int \Bigl(2\frac{f''(x_{\sigma(i)})}{f'(x_i)} - \frac{f'(x_i)}{\sum_{j=1}^d f(x_j)}\Bigr)\partial_{x_i}\varphi(x)\,\nu(dx)\\
&= c_2 \sum_{i=1}^d \int \Bigl(2\sum_{j=1}^d f(x_j)\,\frac{f''(x_{\sigma(i)})}{f'(x_i)} - f'(x_i)\Bigr)\,\partial_{x_i}\varphi(x)\,\mu(dx)\\
&= -2c_1c_2 \sum_{i=1}^d \int\cdots\int\Bigl(\int f'(x_i) f''(x_{\sigma(i)})\varphi(x) - f'(x_i) f''(x_i)\varphi(x)\,dx\Bigr)\\
&\qquad\qquad \times f'(x_1)\cdots \widehat{f'(x_i)}\cdots f'(x_d)\,dx_1\cdots \widehat{dx_i}\cdots dx_d\\
&= -2c_1c_2 \sum_{i=1}^d \Bigl(\int f''(x_{\sigma(i)})\varphi(x)\,\mu(dx) - \int f''(x_i)\varphi(x)\,\mu(dx)\Bigr) = 0,
\end{aligned}$$

which yields the desired conclusion. □

However, even in this explicit example it remained unknown whether there exist other linearly independent probabilistic solutions.

Until the end of this section we assume that $A = \mathrm{I}$ and $b^i \in C^\infty(\mathbb{R}^d)$ for each $i = 1, \ldots, d$. Then any solution of the equation $L_{I,b}^*\mu = 0$ has a density $\varrho \in C^\infty(\mathbb{R}^d)$ and the equation can be written as the following equation for ϱ:

$$\mathrm{div}\bigl(\nabla\varrho - b\varrho\bigr) = 0.$$

Let us set

$$L := L_{\mathrm{I},b}, \quad a := b\varrho - \nabla\varrho.$$

Then $a \in C^\infty(\mathbb{R}^d, \mathbb{R}^d)$ and

$$\mathrm{div}\, a = 0.$$

If ϱ is a probability density, then we know that $\varrho > 0$, hence the coefficient b is expressed in the following way:

$$b = \frac{\nabla\varrho}{\varrho} + \frac{a}{\varrho}. \tag{4.2.1}$$

Let us seek another solution of the equation $L^*\mu = 0$ in the form $\nu = v\cdot\mu$. The measure $\nu = v\cdot\mu$ satisfies the same equation if and only if the function v satisfies the equation

$$L_\mu v := \operatorname{div}\big(\varrho\nabla v - av\big) = 0. \tag{4.2.2}$$

Certainly, every constant will be a solution to equation (4.2.2). We would like to find a sufficient condition for the existence of a bounded positive solution which is not constant. Let us introduce the following bilinear skew-symmetric form on the domain $C_0^\infty(\mathbb{R}^d)$:

$$[f,g] := \int_{\mathbb{R}^d} \langle a, \nabla f\rangle g\,dx.$$

Note that $[f,g]$ is defined if g is bounded and $\langle a,\nabla f\rangle$ is integrable, but it can fail to be skew-symmetric. The next theorem gives sufficient conditions for the existence of a bounded positive solution to (4.2.2) which is not constant.

4.2.2. Theorem. *Assume that there exists a function $\varphi \in C_b^2(\mathbb{R}^d)$ such that we have $\langle a,\nabla\varphi\rangle \in L^1(\mathbb{R}^d)$,*

$$[\varphi, 1] = 0, \quad \textit{and} \quad [\varphi,\varphi] < 0. \tag{4.2.3}$$

Then equation (4.2.2) *has a bounded positive solution which is not constant.*

Proof. Set $M := \|\varphi\|_{C_b^2}$. There exists a solution $v_n \in W^{2,1}(U_n)$ of the Dirichlet problem

$$L_\mu v_n = 0, \quad v_n|_{\partial U_n} = \varphi$$

in the ball U_n of radius n centered at zero. Moreover, $v_n - \varphi \in W_0^{2,1}(U_n)$. For every ball $U \subset U_n$, one has $v \in C_b^2(U)$. According to the maximum principle, we have

$$\sup_{U_n} |v_n| \leqslant \sup_{\partial U_n} |\varphi| \leqslant M.$$

Set $u_n := v_n - \varphi$. One has $\sup_{U_n} |u_n| \leqslant 2M$ and

$$L_\mu u_n = -L_\mu\varphi, \quad u_n|_{\partial U_n} = 0.$$

Since div $a = 0$, we have

$$\mathcal{L}_\mu(u_n+\varphi) = \operatorname{div}\big(\varrho\nabla(u_n+\varphi)\big) - \langle a, \nabla(u_n+\varphi)\rangle.$$

Multiplying the equation $\mathcal{L}_\mu(u_n+\varphi) = 0$ by u_n and integrating by parts, we obtain

$$\int_{U_n} |\nabla u_n|^2 \varrho\,dx = I + J + K,$$

where

$$I = -\int_{U_n} \langle a, \nabla u_n\rangle u_n\,dx, \quad J = -\int_{U_n} \langle \nabla u_n, \nabla\varphi\rangle\varrho\,dx,$$

$$K = -\int_{U_n} \langle a, \nabla\varphi\rangle u_n\,dx.$$

Let us consider the terms I and J. Since div $a = 0$, we obtain $I = 0$. Applying the inequality $xy \leqslant (x^2+y^2)/2$ to the product $|\nabla\varphi|\cdot|\nabla u_n|$, we find that

$$J \leqslant \frac{1}{2}\int_{U_n} |\nabla u_n|^2\varrho\,dx + \frac{1}{2}\int_{U_n} |\nabla\varphi|^2\varrho\,dx.$$

Thus we arrive at the following estimate:

$$\int_{U_n} |\nabla u_n|^2 \varrho\, dx \leqslant \int_{U_n} |\nabla \varphi|^2 \varrho\, dx - 2\int_{U_n} \langle a, \nabla\varphi\rangle u_n\, dx.$$

Let us extend the function u_n by zero outside U_n. Since $u_n \in W_0^{2,1}(U_n)$, we obtain a function from $W_0^{2,1}(\mathbb{R}^d)$. Consequently, for each natural number n we have the inequality

$$\int_{\mathbb{R}^d} |\nabla u_n|^2 \varrho\, dx \leqslant \int_{\mathbb{R}^d} |\nabla \varphi|^2 \varrho\, dx - 2\int_{\mathbb{R}^d} \langle a, \nabla\varphi\rangle u_n\, dx. \tag{4.2.4}$$

Taking into account the estimates

$$\sup_{x\in\mathbb{R}^d} |\nabla\varphi(x)| \leqslant M, \quad \sup_{x\in\mathbb{R}^d} |u_n(x)| \leqslant 2M,$$

we deduce from (4.2.4) the estimate

$$\int_{\mathbb{R}^d} |\nabla u_n|^2 \varrho\, dx \leqslant M^2 + 4M\|\langle a, \nabla\varphi\rangle\|_{L^1(\mathbb{R}^d)}. \tag{4.2.5}$$

For every ball U_k there is a number $C_1(k) > 0$ such that $\varrho(x) \geqslant C_1(k)$ for all $x \in U_k$. Hence (4.2.5) implies the estimate $\|\nabla u_n\|_{L^2(U_k)} \leqslant C_2(k)$, where the number $C_2(k)$ does not depend on n. Since $\sup_{U_k} |u_n| \leqslant 2M$, one has $\|u_n\|_{L^2(U_k)} \leqslant C_3(k)$, where $C_3(k)$ does not depend on n. Consequently, for every ball U_k, we have the estimate $\|u_n\|_{W^{2,1}(U_k)} \leqslant C_4(k)$ uniformly in n. Set $k = 1$. Let us choose a subsequence $\{u_{n_{1,j}}\}$ weakly converging in $W^{2,1}(U_1)$. Next we choose a further subsequence that is weakly converging in $W^{2,1}(U_2)$. Continuing in this way and using the diagonal method we obtain a subsequence weakly converging in $W^{2,1}(U_k)$ for each k. Hence we can assume that $\{u_n\}$ itself has this property. We obtain a function u on the whole space such that, for each k, its restriction to U_k is the weak limit in $W^{2,1}(U_k)$ of the restrictions of u_n. Therefore, $\{u_n\}$ converges to u in the norm of $L^2(U_k)$ for every ball U_k. Let $k > 1$ be an arbitrary natural number. We fix the ball U_k, a natural number $l > 2 + d/2$ and numbers $k - 1 = s_l < s_{l-1} < \ldots < s_0 = k$. Let U_{s_i} denote the ball of radius s_i centered at the point $x = 0$. Clearly, one has $U_{k-1} \subset U_{s_i} \subset U_k$. There exists a number $N = N(k)$ such that for every $n > N$ the function u_n satisfies the equation $\mathcal{L}_\mu u_n = -\mathcal{L}_\mu\varphi$ on the ball U_k. Let $n, m > N$. Then we have $\mathcal{L}_\mu(u_n - u_m) = 0$ on U_k. According to Gilbarg, Trudinger [**409**, Theorem 8.10], we have the estimate

$$\|u_n - u_m\|_{W^{2,i+1}(U_{s_{i+1}})} \leqslant C(s_i, s_{i+1}, \varrho, a)\|u_n - u_m\|_{W^{2,i}(U_{s_i})},$$

where $W^{2,0}(U_{s_0}) := L^2(U_k)$. If the sequence $\{u_n\}$ converges in $W^{2,i}(U_{s_i})$, then, according to the previous estimate, it converges to u in $W^{2,i+1}(U_{s_{i+1}})$. Applying this reasoning for $0 \leqslant i \leqslant l-1$ and using the Sobolev embedding theorem $W^{2,l}(U_{k-1}) \subset C_b^2(U_{k-1})$, $2 < l - d/2$, we obtain that the sequence $\{u_n\}_{n\geqslant N}$ converges uniformly to u in $C_b^2(U_{k-1})$. Hence

$$u \in C_b^2(U_{k-1}), \quad \sup_{U_{k-1}} |u| \leqslant 2M, \quad \mathcal{L}_\mu u = -\mathcal{L}_\mu\varphi$$

on the ball U_{k-1}. Since k is arbitrary, for every ball U there exists a number $N = N(U)$ such that for every $n > N$ one has $u_n \in C_b^2(U)$ and the sequence

$\{u_n\}_{n>N}$ converges to u in $C_b^2(U)$ and, furthermore, $\mathcal{L}_\mu u = -\mathcal{L}_\mu \varphi$. According to Fatou's theorem and Lebesgue's theorem inequality (4.2.4) implies the estimate

$$\int_{\mathbb{R}^d} |\nabla u|^2 \varrho\, dx \leqslant \int_{\mathbb{R}^d} |\nabla \varphi|^2 \varrho\, dx - 2\int_{\mathbb{R}^d} \langle a, \nabla\varphi\rangle u\, dx.$$

Hence the function u does not coincide with the function $-\varphi + c$ for any constant c, because otherwise the last estimate and the condition

$$\int_{\mathbb{R}^d} \langle a, \nabla\varphi\rangle\, dx = 0$$

would immediately give the inequality

$$\int_{\mathbb{R}^d} \langle a, \nabla\varphi\rangle \varphi\, dx \geqslant 0,$$

contradicting the hypothesis of the theorem. Set $v := u + \varphi + 3M$. Finally, note that $v \geqslant 0$, $\mathcal{L}_\mu v = 0$ and that v is not constant. □

Obviously, by multiplying v by a positive constant we obtain a probability measure $v \cdot \mu$ satisfying the equation $L^*(v \cdot \mu) = 0$.

4.2.3. Remark. For the verification of the conditions of Theorem 4.2.2 it is useful to keep in mind the following expressions for $[\varphi, \varphi]$ and $[\varphi, 1]$. Let Ω_n be increasing bounded domains in $\mathbb{R}^d$ with piecewise smooth boundaries such that $\mathbb{R}^d = \bigcup_{n=1}^\infty \Omega_n$. Since div $a = 0$, we have

$$[\varphi, \varphi] = \int_{\mathbb{R}^d} \langle a, \nabla\varphi\rangle \varphi\, dx = \lim_{n\to\infty} \int_{\Omega_n} \langle a, \nabla\varphi\rangle \varphi\, dx = \lim_{n\to\infty} \frac{1}{2} \int_{\partial\Omega_n} \langle a, \nu_n\rangle \varphi^2\, ds,$$

$$[\varphi, 1] = \int_{\mathbb{R}^d} \langle a, \nabla\varphi\rangle\, dx = \lim_{n\to\infty} \int_{\Omega_n} \langle a, \nabla\varphi\rangle\, dx = \lim_{n\to\infty} \int_{\partial\Omega_n} \langle a, \nu_n\rangle \varphi\, ds,$$

where ν_n is the outward normal on $\partial\Omega_n$. Consequently, in order to ensure condition (4.2.3) it is enough to have

$$\lim_{n\to\infty} \int_{\partial\Omega_n} \langle a, \nu_n\rangle \varphi^2\, ds < 0, \qquad \lim_{n\to\infty} \int_{\partial\Omega_n} \langle a, \nu_n\rangle \varphi\, ds = 0.$$

To get an example of equation $L^*\mu = 0$ with at least two different probability solutions, it is sufficient to do the following. First of all, we find a smooth vector field a with div $a = 0$ and a function φ satisfying the conditions of Theorem 4.2.2. Next, we fix an arbitrary infinitely differentiable positive function ϱ with $\|\varrho\|_{L^1(\mathbb{R}^d)} = 1$. Finally, we take the vector field b given by (4.2.1). Then the equation $L^*\mu = 0$ with this coefficient b has at least two different probability solutions: one is the measure $\mu = \varrho\, dx$ and another is the measure $\nu = c_1 v \cdot \mu$, where c_1 is a normalizing constant, and the function v is a (non constant) solution of equation (4.2.2), which exists by Theorem 4.2.2. Let us present some examples of a and φ such that div $a = 0$ and conditions of Theorem 4.2.2 are fulfilled.

4.2.4. Example. Let $d = 2$. Let us take odd functions q, ψ, $\sigma \in C_b^2(\mathbb{R}^1)$ such that

$$q,\ q\psi,\ \sigma' \in L^1(\mathbb{R}^1), \qquad \lim_{n\to\infty} \sigma(n) = 1,$$

and $q\psi \geqslant 0$ does not vanish identically. Obviously, this is possible. Let

$$a(x, y) := \bigl(0, -q(x)\bigr), \quad \varphi(x, y) := \psi(x) + \sigma(y).$$

Clearly, we have $\operatorname{div} a = 0$. Let us verify the conditions of Theorem 4.2.2 using Remark 4.2.3. Let Ω_n be the square with vertices at the points (n,n), $(-n,n)$, $(n,-n)$, and $(-n,-n)$. Then

$$\int_{\partial\Omega_n} \langle a,\nu_n\rangle \varphi^2\,ds = -\bigl(\sigma(n)-\sigma(-n)\bigr)\int_{-n}^{n} q(x)\bigl(2\psi(x)+\sigma(n)+\sigma(-n)\bigr)\,dx,$$

$$\int_{\partial\Omega_n} \langle a,\nu_n\rangle \varphi\,ds = -\bigl(\sigma(n)-\sigma(-n)\bigr)\int_{-n}^{n} q(x)\,dx.$$

Consequently,

$$[\varphi,\varphi] = \lim_{n\to\infty}\frac{1}{2}\int_{\partial\Omega_n} \langle a,\nu_n\rangle \varphi^2\,ds = -2\int_{-\infty}^{+\infty} q(x)\psi(x)\,dx < 0,$$

$$[\varphi,1] = \lim_{n\to\infty}\int_{\partial\Omega_n} \langle a,\nu_n\rangle \varphi\,ds = -2\int_{-\infty}^{+\infty} q(x)\,dx = 0.$$

Hence the conditions of Theorem 4.2.2 are fulfilled. Therefore, choosing an arbitrary strictly positive smooth probability density ϱ, we can construct a drift b (as explained in the remark above) such that the corresponding equation $L^*\mu = 0$ is satisfied by at least two probability measures, one of which is the given probability measure $\mu = \varrho\,dx$.

4.2.5. Example. Let $d = 2$ and let functions q, ψ, $\sigma \in C_b^2(\mathbb{R}^1)$ not vanish identically and satisfy the following conditions:

$$q,\ \sigma' \in L^1(\mathbb{R}^1),\ q > 0,\ \lim_{n\to\infty}\sigma(n) = 1,\ \lim_{n\to\infty}\sigma(-n) = 0.$$

Set

$$a(x,y) := \bigl(0, -q(x)\bigr), \quad \varphi(x,y) := \psi(x)\sigma(y).$$

Again, $\operatorname{div} a = 0$. We calculate $[\varphi,1]$ and $[\varphi,\varphi]$ by using Remark 4.2.3. Let Ω_n be the square with vertices at the points (n,n), $(-n,n)$, $(n,-n)$, and $(-n,-n)$. Then

$$\int_{\partial\Omega_n} \langle a,\nu_n\rangle \varphi^2\,ds = -\bigl(\sigma^2(n)-\sigma^2(-n)\bigr)\int_{-n}^{n} q(x)\psi^2(x)\,dx,$$

$$\int_{\partial\Omega_n} \langle a,\nu_n\rangle \varphi\,ds = -\bigl(\sigma(n)-\sigma(-n)\bigr)\int_{-n}^{n} q(x)\psi(x)\,dx.$$

Consequently,

$$[\varphi,\varphi] = -\frac{1}{2}\int_{-\infty}^{+\infty} q(x)\psi^2(x)\,dx, \quad [\varphi,1] = -\int_{-\infty}^{+\infty} q(x)\psi(x)\,dx.$$

To satisfy the conditions of Theorem 4.2.2, it is sufficient to require the orthogonality of the functions ψ and 1 in $L^2(\mathbb{R}^1, q\,dx)$. Again, Remark 4.2.3 enables us to construct an equation $L^*\mu = 0$ with different probability solutions, one of which is any given measure $\mu = \varrho\,dx$ with a smooth positive density.

It is easy to extend the last example to the case $d \geqslant 2$.

4.2.6. Example. For $x \in \mathbb{R}^d$ set $x' := (x_1, x_2, \ldots, x_{d-1})$. Suppose that q, $\psi \in C_b^2(\mathbb{R}^{d-1})$ and $\sigma \in C_b^2(\mathbb{R}^1)$ do not vanish identically and satisfy the following conditions:

$$q \in L^1(\mathbb{R}^{d-1}),\ \sigma' \in L^1(\mathbb{R}^1),\ q > 0,\ \lim_{n\to\infty}\sigma(n) = 1,\ \lim_{n\to\infty}\sigma(-n) = 0.$$

Set

$$a^i(x) := 0 \ \text{ if } 1 \leqslant i \leqslant d-1, \ \ a^d(x) := -q(x'), \ \ \varphi(x) := \psi(x')\sigma(x_d).$$

Then $\operatorname{div} a = 0$ and

$$[\varphi, \varphi] = -\frac{1}{2}\int_{\mathbb{R}^{d-1}} q(x')\psi^2(x')\, dx', \ [\varphi, 1] = -\int_{\mathbb{R}^{d-1}} q(x')\psi(x')\, dx'.$$

To satisfy the conditions of Theorem 4.2.2 it is sufficient to require the orthogonality of the functions ψ and 1 in $L^2(\mathbb{R}^{d-1}, q\, dx')$.

We fix a with $\operatorname{div} a = 0$ and two different functions φ_1 and φ_2 satisfying the conditions of Theorem 4.2.2, and we construct two solutions v_1 and v_2 according to this theorem. This theorem guarantees that $1, v_1$ and $1, v_2$ are pairs of linearly independent functions. Under what conditions on φ_1 and φ_2 will the three functions 1, v_1 and v_2 be linearly independent? The following theorem answers this question and gives some additional information.

4.2.7. Theorem. *Let $n \in \mathbb{N}$ be given. Assume that there exist functions $\varphi_1, \varphi_2, \ldots, \varphi_{n+1} \in C_b^2(\mathbb{R}^d)$ satisfying the conditions of Theorem 4.2.2. Let functions $v_1, v_2, \ldots, v_{n+1}$ be solutions of equation (4.2.2) generated by these functions according to Theorem 4.2.2. Assume that the functions $1, v_1, \ldots, v_n$ are linearly independent and that for every vector $\alpha = (\alpha_1, \ldots, \alpha_n) \in \mathbb{R}^n$ the following inequality holds:*

$$\Bigl[\varphi_{n+1} - \sum_{k=1}^{n} \alpha_k \varphi_k, \varphi_{n+1} - \sum_{k=1}^{n} \alpha_k \varphi_k\Bigr] < 0. \tag{4.2.6}$$

Then the functions $1, v_1, \ldots, v_n, v_{n+1}$ are linearly independent.

Proof. Suppose the contrary. Then $v_{n+1} = \sum_{k=1}^{n} \alpha_k v_k + \gamma$, where γ, α_k are some numbers. Since $v_i = u_i + \varphi_i + C_i$ whenever $1 \leqslant i \leqslant n+1$, one has

$$u_{n+1} - \sum_{k=1}^{n} \alpha_k u_k = -\Bigl(\varphi_{n+1} - \sum_{k=1}^{n} \alpha_k \varphi_k\Bigr) + C.$$

Let $u = u_{n+1} - \sum_{k=1}^{n} \alpha_k u_k$ and $\varphi = \varphi_{n+1} - \sum_{k=1}^{n} \alpha_k \varphi_k$. As in Theorem 4.1.6, for each $1 \leqslant i \leqslant n+1$, we construct a sequence of functions $v_{i,l}$ and, consequently, the sequence $u_{i,l} = v_{i,l} - \varphi_i$. Repeating the reasoning from Theorem 4.1.6, we choose a subsequence of functions u_{1,l_j} with the following property: for every ball $U \subset \mathbb{R}^d$, there exists a number $N = N(U)$ such that, for every $j > N$, one has $u_{1,l_j} \in C_b^2(U)$ and the sequence $\{u_{1,l_j}\}_{j>N}$ converges to u_1 in $C_b^2(U)$. Suppose we have already chosen a sequence u_{i,l_j} with the following property: for each ball $U \subset \mathbb{R}^d$, there exists a number $N = N(U, i)$ such that for every $j > N$ one has $u_{i,l_j} \in C_b^2(U)$ and the sequence $\{u_{i,l_j}\}_{j>N}$ converges to u_i in $C_b^2(U)$. Then we apply the reasoning from Theorem 4.1.6 and choose a subsequence $\{u_{i+1,l_{j_k}}\}$ in $\{u_{i+1,l_j}\}_{j\geqslant 1}$ which converges to u_{i+1}. If $i = n$, then we obtain a sequence $\{u_{n+1,z_j}\}_{j\geqslant 1}$ converging to the function u_{n+1}. Moreover, whenever $1 \leqslant i \leqslant n$, the sequence $\{u_{i,z_j}\}_{j\geqslant 1}$ converges to u_i as well. Set

$$u^j := u_{n+1,z_j} - \sum_{k=1}^{n} \alpha_k u_{k,z_j}.$$

Note that $\{u^j\}$ converges to u. In addition, $\mathcal{L}_\mu u^j = -\mathcal{L}_\mu \varphi$ and $u^j(x) = 0$ if $x \in \partial U_{z_j}$. Multiplying the equality $\mathcal{L}_\mu(u^j + \varphi) = 0$ by u^j and acting as above in the justification of estimate (4.2.4) we find that

$$\int_{\mathbb{R}^d} |\nabla u^j|^2 \varrho\, dx \leqslant \int_{\mathbb{R}^d} |\nabla \varphi|^2 \varrho\, dx - 2\int_{\mathbb{R}^d} \langle a, \nabla\varphi\rangle u^j\, dx.$$

Letting $j \to +\infty$, we deduce the estimate

$$\int_{\mathbb{R}^d} |\nabla u|^2 \varrho\, dx \leqslant \int_{\mathbb{R}^d} |\nabla \varphi|^2 \varrho\, dx - 2\int_{\mathbb{R}^d} \langle a, \nabla\varphi\rangle u\, dx.$$

Since $u = -\varphi + C$, we obtain $[\varphi, \varphi] \geqslant 0$, which contradicts (4.2.3). □

4.2.8. Remark. Let $\Phi = (\Phi_{ij})_{i,j\leqslant n}$ and $h = (h_i)_{i\leqslant n}$, where

$$\Phi_{ij} = ([\varphi_i, \varphi_j] + [\varphi_j,\ \varphi_i])/2,\ h_i = [\varphi_i, \varphi_{n+1}] + [\varphi_{n+1}, \varphi_i],\ 1 \leqslant i, j \leqslant n.$$

Let $h_0 := [\varphi_{n+1}, \varphi_{n+1}]$. Then inequality (4.2.6) can be written as

$$\langle \Phi\alpha, \alpha\rangle - \langle h, \alpha\rangle + h_0 < 0.$$

Consequently, to ensure condition (4.2.6) it is enough to have the following: the matrix Φ is strictly negative and

$$4h_0 < \langle \Phi^{-1}h, h\rangle. \tag{4.2.7}$$

In particular, if $n = 1$, then inequality (4.2.7) has the following very simple form:

$$\bigl([\varphi_1, \varphi_2] + [\varphi_2, \varphi_1]\bigr)^2 < 4[\varphi_1, \varphi_1][\varphi_2, \varphi_2].$$

4.2.9. Remark. To verify condition (4.2.6) it is useful to keep in mind the following expression for $[\varphi_i, \varphi_j] + [\varphi_j, \varphi_i]$. Let Ω_n be increasing domains with piecewise smooth boundaries and $\mathbb{R}^d = \bigcup_{n=1}^\infty \Omega_n$. Since $\operatorname{div} a = 0$, we have

$$\begin{aligned}[\varphi_i, \varphi_j] + [\varphi_j, \varphi_i] &= \int_{\mathbb{R}^d} \langle a, \nabla\varphi_i\rangle \varphi_j\, dx + [\varphi_j, \varphi_i] \\ &= \lim_{n\to\infty} \int_{\Omega_n} \langle a, \nabla\varphi_i\rangle \varphi_j\, dx + [\varphi_j, \varphi_i] = \lim_{n\to\infty} \int_{\partial\Omega_n} \langle a, \nu_n\rangle \varphi_i\varphi_j\, ds,\end{aligned}$$

where ν_n is the outward normal on $\partial\Omega_n$.

4.2.10. Remark. Suppose that for a given smooth vector field a with $\operatorname{div} a = 0$ there exist functions $\varphi_1, \ldots, \varphi_{n+1}$ such that the conditions of Theorem 4.2.7 hold. Assume that we are given a strictly positive infinitely differentiable function ϱ with $\|\varrho\|_{L^1(\mathbb{R}^d)} = 1$. Then the equation $L^*\mu = 0$ with the coefficient b that is expressed via a and ϱ by formula (4.2.1) has at least $n+1$ linearly independent probability solutions, one of which is the measure $\mu = \varrho\, dx$, and n other solutions are the measures $\nu_i = c_i v_i \cdot \mu$, where c_i are normalizing constants and the functions v_i are nonconstant solutions of equation (4.2.2) generated by the functions φ_i.

Let us present an explicit example of a, φ_1, and φ_2 such that the conditions of Theorem 4.2.7 are fulfilled.

4.2.11. Example. Let us set $x' := (x_1, x_2, \ldots, x_{d-1})$. Let $\sigma \in C_b^2(\mathbb{R}^1)$ and $q, \psi_1, \psi_2 \in C_b^2(\mathbb{R}^{d-1})$ not vanish identically. Assume that

$$q \in L^1(\mathbb{R}^{d-1}),\ q > 0,\ \sigma' \in L^1(\mathbb{R}^1),\ \lim_{n\to\infty} \sigma(n) = 1,\ \lim_{n\to\infty} \sigma(-n) = 0.$$

Let

$$a^k(x) := 0 \ \text{ if } 1 \leqslant k \leqslant d-1, \ \ a^d(x) := -q(x'),$$
$$\varphi_1(x) := \psi_1(x')\sigma(x_d), \quad \varphi_2(x) := \psi_2(x')\sigma(x_d).$$

Then $\operatorname{div} a = 0$ and, whenever $1 \leqslant i, j \leqslant 2$, we have

$$[\varphi_i, \varphi_j] + [\varphi_j, \varphi_i] = -\int_{\mathbb{R}^{d-1}} \psi_i(x')\psi_j(x')q(x')\,dx',$$

$$[\varphi_i, 1] = -\int_{\mathbb{R}^{d-1}} \psi_i(x')q(x')\,dx'.$$

To satisfy the conditions of Theorem 4.2.7 it is enough to require the orthogonality of the functions $1, \psi_1, \psi_2$ in $L^2(\mathbb{R}^{d-1}, q\,dx')$.

This example can be easily extended to the case of an arbitrary number of functions φ_i. Moreover, we can give an example of an equation $L^*\mu = 0$ which has a countable sequence of linearly independent probability solutions. In particular, the space of solutions to such an equation in the class of bounded measures is infinite-dimensional. It is enough to find a sequence of positive bounded solutions $\{v_i\}_{i\geqslant 1}$ to (4.2.2) such that the functions $1, \{v_i\}_{i\geqslant 1}$ are linearly independent. According to Theorem 4.2.7 and Remark 4.2.8, it suffices to find a vector field a with $\operatorname{div} a = 0$ and a sequence of functions $\{\varphi_i\}$ satisfying the conditions of Theorem 4.2.2 such that, for each n, the functions $\varphi_1, \varphi_2, \ldots, \varphi_{n+1}$ satisfy condition (4.2.7).

4.2.12. Example. For $x \in \mathbb{R}^d$ set $x' := (x_1, x_2, \ldots, x_{d-1})$. Let functions $q \in C_b^2(\mathbb{R}^{d-1})$ and $\sigma \in C_b^2(\mathbb{R}^1)$ be such that

$$q \in L^1(\mathbb{R}^{d-1}),\ q > 0,\ \sigma' \in L^1(\mathbb{R}^1),\ \lim_{n\to\infty} \sigma(n) = 1,\ \lim_{n\to\infty} \sigma(-n) = 0.$$

Let $\psi_i \in C_b^2(\mathbb{R}^{d-1})$, where $i \in \mathbb{N}$. Set

$$a^k(x) := 0 \ \text{ if } 1 \leqslant k \leqslant d-1, \quad a^d(x) := -q(x'), \quad \varphi_i(x) := \psi_i(x')\sigma(x_d).$$

Then, for any $i, j \geqslant 1$, we have

$$[\varphi_i, \varphi_j] + [\varphi_j, \varphi_i] = -\int_{\mathbb{R}^{d-1}} \psi_i(x')\psi_j(x')q(x')\,dx',$$

$$[\varphi_i, 1] = -\int_{\mathbb{R}^{d-1}} \psi_i(x')q(x')\,dx'.$$

Let $1, \{\psi_i\}_{i\geqslant 1}$ be an orthonormal system in $L^2(\mathbb{R}^{d-1}, q\,dx')$. Then, for each n, condition (4.2.7) holds for the functions $1, \varphi_1, \ldots, \varphi_{n+1}$ because the matrix Φ is diagonal with $\Phi_{ii} = -1/2$, hence is negative, $h = 0$, and $h_0 = [\psi_{n+1}, \psi_{n+1}] < 0$.

4.2.13. Example. We have seen in Example 4.2.1 that there is an equation $L^*\mu = 0$ that has at least two different probability solutions. We can now show that it actually has a countable sequence of linearly independent solutions that are probability measures. We have

$$b^k(x) = -x_k - 2x_{\sigma(k)} \exp\Bigl((x_k^2 - x_{\sigma(k)}^2)/2\Bigr), \quad \varrho(x) = (2\pi)^{-d/2} \exp(-|x|^2/2),$$

where $\sigma\colon \{1, \ldots, d\} \to \{1, \ldots, d\}$ is one-to-one such that $\sigma(k) \neq k$. Then

$$a^k(x) = -2(2\pi)^{d/2} x_{\sigma(k)} \exp\Bigl(-x_{\sigma(k)}^2 - 2^{-1}\sum_{i\neq k} x_i^2\Bigr).$$

Let $\mathbb{R}^{d-1}_+ := \mathbb{R}^{d-1}\bigcap\{x_{\sigma(d)} > 0\}$ and $x' := (x_1, x_2, \dots, x_{d-1})$. Let us take functions $\omega, \psi_i^{**} \in C_0^\infty(\mathbb{R}^{d-1}_+)$, where $i \in \mathbb{N}$, with disjoint supports such that

$$-\int_{\mathbb{R}^{d-1}_+} \omega(x')a^d(x')\,dx' = 1.$$

Let

$$\psi_i^*(x') := \psi_i^{**}(x') + \omega(x')\int_{\mathbb{R}^{d-1}_+} \psi_i^{**}(y')a^d(y')\,dy', \quad x' \in \mathbb{R}^{d-1}_+.$$

Note that $\operatorname{div} a = 0$ and

$$\int_{\mathbb{R}^{d-1}_+} \psi_i^*(x')a^d(x')\,dx' = 0 \text{ for each } i \geqslant 1.$$

It is easy to see that the functions ψ_i^* are linearly independent. We apply the orthogonalization process in the space $L^2(\mathbb{R}^{d-1}_+, -a^d(x')\,dx')$ to the system $\{\psi_i^*\}_{i\geqslant 1}$ and obtain functions $\{\psi_i\}_{i\geqslant 1}$. We observe that the obtained functions have the following properties: $\lim\limits_{|x'|\to\infty} \psi_i(x') = 0$ and

$$\int_{\mathbb{R}^{d-1}_+} \psi_i(x')a^d(x')\,dx' = 0, \int_{\mathbb{R}^{d-1}_+} \psi_i(x')\psi_j(x')a^d(x')\,dx' = 0 \text{ for all } i, j \geqslant 1.$$

Let us extend ψ_i to the whole space $\mathbb{R}^{d-1}$ by zero outside $\mathbb{R}^{d-1}_+$. Clearly, we obtain functions from $C_0^\infty(\mathbb{R}^{d-1})$, since ω and ψ^{**} are of compact support in $\mathbb{R}^{d-1}_+$. Let σ be a smooth function such that $\sigma' \in L^1(\mathbb{R}^1)$, $\lim\limits_{n\to\infty} \sigma(n) = 1$, and $\lim\limits_{n\to\infty} \sigma(-n) = 0$. Set $\varphi_i(x) := \psi_i(x')\sigma(x_d)$. According to Theorem 4.2.2, we can construct nonconstant solutions corresponding to the functions φ_i. Similarly to the previous example, we obtain a sequence of solutions $1, v_1, v_2, \dots$ that are linearly independent.

Finally, we present one more sufficient condition for the existence of a nonconstant positive bounded solution of equation (4.2.2).

4.2.14. Proposition. *Assume that there exists a function $\varphi \in C_b^2(\mathbb{R}^d)$ such that*

$$2\sup_{x\in\mathbb{R}^d} |\varphi(x)| \int_{\mathbb{R}^d} |L_\mu\varphi|\,dx < \int_{\mathbb{R}^d} |\nabla\varphi|^2\varrho\,dx.$$

Then there exists a bounded positive solution of equation (4.2.2) *which is not constant.*

PROOF. We argue as in the proof of Theorem 4.1.6. Let $M := \sup_{x\in\mathbb{R}^d} |\varphi(x)|$. As in the proof of Theorem 4.1.6, we construct a sequence of functions v_n and the corresponding functions $u_n = v_n - \varphi$. Multiplying the equality $\mathcal{L}_\mu u_n = -\mathcal{L}_\mu\varphi$ by u_n and integrating by parts we obtain the following estimate:

$$\int_{\mathbb{R}^d} |\nabla u_n|^2\varrho\,dx \leqslant 2M\int_{\mathbb{R}^d} |\mathcal{L}_\mu\varphi|\,dx = \int_{\mathbb{R}^d} |\nabla\varphi|^2\varrho\,dx - c_0, \tag{4.2.8}$$

where

$$c_0 = \int_{\mathbb{R}^d} |\nabla\varphi|^2\varrho\,dx - 2M\int_{\mathbb{R}^d} |\mathcal{L}_\mu\varphi|\,dx > 0.$$

Note that c_0 does not depend on n. As in the proof of Theorem 4.1.6, passing to a subsequence, we can assume that, for every ball U, there exists a number

$N = N(U)$ such that, for every $n > N$, one has $u_n \in C_b^2(U)$ and the sequence $\{u_n\}_{n>N}$ converges to u in $C_b^2(U)$ and $\mathcal{L}_\mu u = -\mathcal{L}_\mu \varphi$. According to Fatou's theorem,

$$\int_{\mathbb{R}^d} |\nabla u|^2 \varrho\, dx \leqslant \int_{\mathbb{R}^d} |\nabla \varphi|^2 \varrho\, dx - c_0.$$

Hence the function u does not coincide with the function $-\varphi + c$ for any constant c. Indeed, otherwise the above estimate immediately implies that $c_0 \leqslant 0$, which contradicts the inequality $c_0 > 0$. Set $v := u + \varphi + 3M$. Note that $v \geqslant 0$, $\mathcal{L}_\mu v = 0$ and that v is not constant. □

The following example demonstrates an application of this proposition.

4.2.15. Example. Let $d = 2$ and let $\varrho_1, \varrho_2 \in C^\infty(\mathbb{R}^1) \bigcap L^1(\mathbb{R}^1)$ be positive functions such that $\|\varrho_1\|_{L^1} = \|\varrho_2\|_{L^1} = 1$. Set $\varrho(x,y) := \varrho_1(x)\varrho_2(y)$,

$$\varphi(x,y) := \int_{-\infty}^{x} \varrho_1(s)\, ds + \int_{-\infty}^{y} \varrho_2(s)\, ds,$$

$$a^1(x,y) := -2\varrho_2'(y)\varrho_2(y) + c(y)\varrho_2(y), \ a^2(x,y) := -2\varrho_1'(x)\varrho_2(x) + d(x)\varrho_1(x).$$

Then div $a = 0$ and

$$L_\mu \varphi(x,y) = \bigl(d(x) + c(y)\bigr)\varrho(x,y).$$

To satisfy the conditions of Proposition 4.2.14 it is enough to have the following estimate:

$$4\int_{\mathbb{R}^2} |d(x) + c(y)| \varrho(x,y)\, dx\, dy < \int_{\mathbb{R}^2} \Bigl[\varrho_1(x)^2 + \varrho_2(y)^2\Bigr] \varrho(x,y)\, dx\, dy.$$

4.2.16. Remark. Under the assumptions of Proposition 4.2.14 we have

$$\int_{\mathbb{R}^d} \operatorname{div}(\varrho \nabla \varphi)\, dx = 0.$$

Hence the estimate in the condition of that proposition can be replaced by the following one:

$$L_\mu \varphi \geqslant 0 \quad \text{and} \quad \int_{\mathbb{R}^d} |\nabla \varphi|^2 \varrho\, dx + 2 \sup_{x \in \mathbb{R}^d} |\varphi(x)| \int_{\mathbb{R}^d} \langle a, \nabla \varphi \rangle\, dx > 0.$$

It remains unknown whether the equation $L^*\mu = 0$ can have only finitely many (but more than one) linearly independent probability solutions in the case of smooth b and $A = \mathrm{I}$. For singular b such examples can be easily constructed on the real line.

4.3. Integrable solutions

In this section we consider bounded signed measures μ on $\mathbb{R}^d$ satisfying the equation $L^*\mu = 0$ with an elliptic operator $L = L_{A,b}$, where it is assumed as in (4.1.5) that the functions a^{ij} are in the Sobolev class $W^{d+,1}_{\mathrm{loc}}(\mathbb{R}^d)$, the functions b^i are in $L^{d+}_{\mathrm{loc}}(\mathbb{R}^d)$, and $\det A > 0$ for the continuous version of A. We know that under these conditions every solution of our equation is given by a continuous density $\varrho \in W^{d+,1}_{\mathrm{loc}}(\mathbb{R}^d)$; we call such densities *integrable solutions*. Hence this equation is equivalent to the equation

$$\partial_{x_i}\partial_{x_j}(a^{ij}\varrho) - \partial_{x_i}(b^i\varrho) = 0 \tag{4.3.1}$$

understood in the classical weak sense (in the case of smooth coefficients the density ϱ turns out to be smooth as well and equation (4.3.1) is fulfilled in the usual sense).

For example, for the unit matrix $A = \mathrm{I}$ we obtain the operator $Lf = \Delta f + \langle b, \nabla f\rangle$ and equation (4.3.1) takes the form

$$\Delta\varrho - \operatorname{div}(\varrho b) = 0.$$

Unlike § 4.1, we do not assume the existence of a probability solution, and the principal results here give conditions under which there are no signed solutions.

4.3.1. Lemma. *There is a bounded linear operator*

$$R\colon\, L^\infty(\mathbb{R}^d) \to L^\infty(\mathbb{R}^d) \cap W^{d+,2}_{\mathrm{loc}}(\mathbb{R}^d)$$

with the following properties:
(i) $(I - L)Rf = f$ *for all* $f \in L^\infty(\mathbb{R}^d)$,
(ii) $0 \leqslant Rf \leqslant 1$ *if* $0 \leqslant f \leqslant 1$ *(such an operator is called sub-Markov)*,
(iii) *there exists a measurable function* $G\colon \mathbb{R}^d \times \mathbb{R}^d \to [0, +\infty)$ *for which*

$$Rf(x) = \int_{\mathbb{R}^d} f(y)G(x, y)\, dy, \qquad \int_{\mathbb{R}^d} G(x, y)\, dy \leqslant 1.$$

In addition, the operator

$$R^*g(y) = \int_{\mathbb{R}^d} g(x)G(x, y)\, dx$$

is bounded in $L^1(\mathbb{R}^d)$ *and has* R *for its adjoint, moreover,* $R^*g \geqslant 0$ *if* $g \geqslant 0$. *Finally, if* $a^{ij} \in W^{p,1}_{\mathrm{loc}}(\mathbb{R}^d)$, $b^i \in L^p_{\mathrm{loc}}(\mathbb{R}^d)$, *where* $p > d$, *then the operator* R *takes values in* $L^\infty(\mathbb{R}^d) \cap W^{p,2}_{\mathrm{loc}}(\mathbb{R}^d)$.

PROOF. Let $f \in L^\infty(\mathbb{R}^d)$. The function Rf is constructed as a limit of functions g_n defined on the balls U_n of radius n centered at the origin that are solutions to the Dirichlet problem $(I - L)g_n = f$ in U_n with zero boundary conditions. It is known that such solutions exist and are unique in the class $W^{p_n,2}(U_n)$, where $p_n > d$, $a^{ij}|_{U_n} \in W^{p_n,1}(U_n)$, $b^i \in L^{p_n}(U_n)$. If $0 \leqslant f \leqslant 1$, then by the maximum principle we have $0 \leqslant g_n \leqslant 1$ and $g_{n+1} \geqslant g_n$ on U_n, whence the pointwise convergence of g_n follows. This yields convergence of g_n to a limit Rf also for functions f of variable sign. It is clear that (ii) is fulfilled. On every fixed ball U_n the norms in $W^{p_n,2}(U_n)$ of the functions g_k with $k > n$ are uniformly bounded, hence $Rf \in W^{d+,2}_{\mathrm{loc}}(\mathbb{R}^d)$ (and also $Rf \in W^{p,2}_{\mathrm{loc}}(\mathbb{R}^d)$ if $p_n = p > d$) and $(I - L)Rf = f$ in U_n. In addition, there exist the Green's functions $G_n(x, y) \geqslant 0$, which give integral representations of the solutions g_n in U_n. It is readily seen that $G_{n+1} \geqslant G_n$ on $U_n \times U_n$ and

$$\int_{U_n} G_n(x, y)\, dy \leqslant 1,$$

which enables us to obtain G as the limit of G_n and proves (iii). It is clear that $\|R^*g\|_{L^1} \leqslant \|g\|_{L^1}$, $R^*g \geqslant 0$ if $g \geqslant 0$, and for any $f \in L^\infty(\mathbb{R}^d)$ the integral of $f(y)R^*g(y)$ equals the integral of $g(x)Rf(x)$. □

We shall say that the measure μ is *subinvariant* for an operator R on $L^\infty(\mu)$ if

$$\int Tf\, d\mu \leqslant \int f\, d\mu \quad \forall f \geqslant 0.$$

If we have the equality above, then μ is called *invariant*. Such operators will be very important in Chapter 5, but they arise already here.

4.3.2. Lemma. *Let μ be a probability measure on some space Ω and let $R\colon L^\infty(\mu) \to L^\infty(\mu)$ be a bounded linear operator such that whenever $0 \leqslant f \leqslant 1$ we have $0 \leqslant Rf \leqslant 1$ and $\|Rf\|_1 \leqslant \|f\|_1$, i.e., R is a sub-Markov operator for which the measure μ is subinvariant. Then R extends to a bounded operator in each $L^p(\mu)$, the adjoint R^* to R on $L^1(\mu)$ is also sub-Markov, has μ as a subinvariant measure and extends to a bounded operator on all $L^p(\mu)$, $1 \leqslant p \leqslant +\infty$, and its adjoint on $L^1(\mu)$ is R. Finally, if $g \in L^1(\mu)$ and $R^*g = g$, then $R^*|g| = |g|$.*

PROOF. It follows from our hypotheses that R extends by continuity to an operator on $L^1(\mu)$ with norm at most 1, since

$$\|Rf\|_1 \leqslant \|Rf^+\|_1 + \|Rf^-\|_1 \leqslant \|f^+\|_1 + \|f^-\|_1 = \|f\|_1,$$

which by the interpolation theorem yields an extension to all $L^p(\mu)$, see Berg, Löfström [**108**, Theorem 1.1.1]. For every $q \in (1, +\infty]$, this gives a bounded operator $R^*\colon L^q(\mu) \to L^q(\mu)$. If $g \geqslant 0$, then $R^*g \geqslant 0$, since for any nonnegative $f \in L^\infty(\mu)$ the integral of fR^*g equals the nonnegative integral of gRf. In addition, the integral of R^*g equals the integral of $gR1$ and is estimated by the integral of g. Therefore, R^* extends also to $L^1(\mu)$. The condition $R^*1 \leqslant 1$, equivalent to

$$\int_\Omega R^*1f\,d\mu = \int_\Omega Rf\,d\mu \leqslant \int_\Omega f\,d\mu \qquad \text{whenever } f \in L^\infty(\mu),\ f \geqslant 0,$$

follows by the subinvariance of μ with respect to R. It is clear that R on $L^\infty(\mu)$ serves as the adjoint to R^* on $L^1(\mu)$.

Let us prove the last assertion of the lemma. Since $g^+ \geqslant g$, we have the inequalities $R^*g^+ \geqslant R^*g = g$ and $R^*g^+ \geqslant 0$, whence $R^*g^+ \geqslant g^+$, so

$$0 \leqslant \int_\Omega (R^*g^+ - g^+)\,dx = \int_\Omega g^+R1\,dx - \int_\Omega g^+\,dx \leqslant 0,$$

since $R1 \leqslant 1$. Hence $R^*g^+ = g^+$ and $R^*|g| = |g|$. □

4.3.3. Theorem. *Suppose that the set $(L-I)(C_0^\infty)$ is dense in $L^1(\sigma)$ for every absolutely continuous probability measure σ. Suppose in addition that the coefficient b is locally bounded. If a nonzero bounded measure μ satisfies the equation $L^*\mu = 0$, then it is either positive or negative, i.e., either $\mu = |\mu|$ or $\mu = -|\mu|$. Moreover, every solution of this equation is proportional to $|\mu|$ (so that a probability solution is unique). If it is given in advance that there exists a probability solution, then this is true even without local boundedness of b.*

PROOF. The measure μ has a continuous density g satisfying the identity

$$\int g(I-L)\varphi\,dx = \int \varphi g\,dx \quad \forall\, \varphi \in C_0^\infty,$$

which in the case of a locally bounded b is equivalent to the equality

$$\int g\psi\,dx = \int R\psi g\,dx = \int \psi R^*g\,dx \tag{4.3.2}$$

for all ψ of the form $(I-L)\varphi$, $\varphi \in C_0^\infty$, since $T(I-L)\varphi = \varphi$. Indeed, the function $\psi = (I-L)\varphi$ is bounded and φ is a solution to the Dirichlet problem $(I-L)\varphi = \psi$ on all balls containing the support of φ, but such a solution is unique. Now (4.3.2) yields the equality $g = R^*g$, since otherwise we obtain a contradiction to the assumption that the set of functions ψ of the indicated form is dense in $L^1(\sigma)$ for the measure $\sigma = |g - R^*g|\,dx$. Then, by the lemma above, $R^*|g| = |g|$, hence the previous relations yield the equality $L^*|\mu| = 0$. Therefore, the function $|g|$ is everywhere positive, that is, either $g > 0$ or $g < 0$ everywhere.

We now drop the condition of local boundedness of b, but shall assume that there is a probability measure μ_0 for which $L^*\mu_0 = 0$. The continuous density ϱ of this measure is positive. It will be proved in § 5.1 that the assumption that the range of $L - I$ is dense gives existence of a sub-Markov operator T on $L^1(\mu_0)$ (and also on all $L^p(\mu_0)$) extending $(I - L)^{-1}$ and that the operator T^* adjoint to T on $L^2(\mu_0)$ is sub-Markov as well; this operator also acts on all $L^p(\mu_0)$. Any measure μ satisfying the equation $L^*\mu = 0$ has a continuous density h with respect to μ_0. Hence $h = T^*h$, which, as above, yields that $|h| = T^*|h|$, i.e.,

$$\int T\psi\, d|\mu| = \int \psi\, d|\mu| \quad \forall \psi \in C_0^\infty,$$

which gives the identity

$$\int \varphi\, d|\mu| = \int (I - L)\varphi\, d|\mu| \quad \forall \varphi \in C_0^\infty,$$

that is, the equality $L^*|\mu| = 0$. Therefore, the continuous density $|h\varrho|$ of the measure $|\mu|$ is positive, hence either $h > 0$ or $h < 0$. Finally, we have the uniqueness of a probability solution, since the difference of two probability solutions has the zero integral and cannot be positive or negative. □

It is important to note that in the hypotheses of the theorem it is not sufficient that the set $(L - I)(C_0^\infty)$ be dense in $L^1(\sigma)$ for some fixed probability measure σ with a positive density, even satisfying the equation $L^*\sigma = 0$.

4.3.4. Example. As in Example 4.1.3, let $A(x) = 1$,

$$\varrho(x) = (1 + 4x^4)^{-1}e^{-x^4}, \quad b(x) = \varrho'(x)/\varrho(x).$$

Since $d = 1$, the measure $\mu_0 = \varrho\, dx$ is a unique probability solution of the corresponding equation $L^*\mu = 0$ (equivalent to the equation $(u' - bu)' = 0$ for densities), which also has a signed solution with density $u(x) = x(1 + 4x^4)^{-1}$. We have

$$xb(x) = -4x^4 - 16x^4/(1 + 4x^4) \to -\infty \quad \text{as } |x| \to +\infty,$$

whence it follows (see Example 5.3.4) that the set $(L - I)(C_0^\infty)$ is dense in $L^1(\mu_0)$.

On the other hand, it is left as Exercise 4.5.17 to show that for the coefficient $b(x) = -2x - 6e^{x^2}$ the measure $\mu = \pi^{-1/2}e^{-x^2}\, dx$ is a probability solution, the range of $L - I$ is not dense in $L^1(\mu)$, and it is directly verified that there are no signed solutions and μ is a unique probability solution. Thus, the condition indicated in our theorem is not necessary for the absence of signed solutions in the presence of a probability solution.

Let us give an effectively verified sufficient conditions in terms of Lyapunov functions in order that the range of the operator $L - I$ be dense.

4.3.5. Lemma. *Let $r \in \mathbb{R}$ and let μ be a Borel locally bounded measure on $\mathbb{R}^d$ satisfying the equation $L^*\mu = r\mu$ with coefficients satisfying the conditions indicated at the beginning of this section. Then the measure $|\mu|$ satisfies the Kato-type inequality $L^*|\mu| \geqslant r|\mu|$, understood as the integral inequality*

$$\int_{\mathbb{R}^d} L\psi\, d|\mu| \geqslant r \int_{\mathbb{R}^d} \psi\, d|\mu|$$

for every nonnegative function $\psi \in C_0^\infty$.

PROOF. The case $d = 1$ is verified directly taking into account that μ is given by a locally absolutely continuous density (it suffices to consider intervals on which the density does not change sign). Let $d > 1$. We know that $\mu = \varrho\, dx$, where $\varrho \in W^{d+,1}_{\text{loc}}(\mathbb{R}^d)$ has a continuous version, which we shall deal with. Let $b_0 = b - \beta_\mu$. The equation $L^*\mu = r\mu$ is equivalent to the integral identity

$$\int_{\mathbb{R}^d} \langle A\nabla\varrho - b_0\varrho, \nabla\varphi\rangle\, dx = -r\int_{\mathbb{R}^d} \varphi\varrho\, dx$$

for every function $\varphi \in C_0^\infty(\mathbb{R}^d)$. It is clear that this identity is fulfilled also for all functions φ in the class $W^{2,1}(\mathbb{R}^d)$ with compact support. Let $\delta > 0$. Set $f_\delta(t) = \operatorname{sign} t$ if $|t| > \delta$ and $f_\delta(t) = t/\delta$ if $|t| \leqslant \delta$. We observe that the function f is Lipschitzian. Let $\psi \in C_0^\infty$ and $\psi \geqslant 0$. Plugging $\varphi = f_\delta(\varrho)\psi$ in the integral identity, we obtain

$$\int_{\mathbb{R}^d} \langle A\nabla\varrho - b_0\varrho, \nabla\psi\rangle f_\delta(\varrho)\, dx + \int_{\mathbb{R}^d} \langle A\nabla\varrho - b_0\varrho, \nabla\varrho\rangle f'_\delta(\varrho)\psi\, dx$$
$$= -r\int_{\mathbb{R}^d} \psi f_\delta(\varrho)\varrho\, dx.$$

Let $Q = \sqrt{A}$. It easy to derive from the Cauchy–Bunyakovsky inequality that

$$\langle A\nabla\varrho - b_0\varrho, \nabla\varrho\rangle \geqslant |Q\nabla\varrho|^2 - |Q^{-1}b_0\varrho|\,|Q\nabla\varrho| \geqslant \frac{1}{2}|Q\nabla\varrho|^2 - \frac{1}{2}|Q^{-1}b_0\varrho|^2.$$

We observe that $f' \geqslant 0$. Hence, dropping the term

$$\frac{1}{2}\int_{\mathbb{R}^d} |Q\nabla\varrho|^2 f'_\delta(\varrho)\psi\, dx \geqslant 0,$$

we arrive at the inequality

$$\int_{\mathbb{R}^d} \langle A\nabla\varrho - b_0\varrho, \nabla\psi\rangle f_\delta(\varrho)\, dx \leqslant \frac{1}{2}\int_{\mathbb{R}^d} |Q^{-1}b_0|^2|\varrho|^2 f'_\delta(\varrho)\psi\, dx - r\int_{\mathbb{R}^d} \psi f_\delta(\varrho)\varrho\, dx.$$

Let us consider the first integral on the right side. Note that $f'(t) = \delta^{-1}$ if $|t| < \delta$ and $f'(t) = 0$ if $|t| > \delta$. If the support of ψ belongs to a ball U, then

$$\int_{\mathbb{R}^d} |Q^{-1}b_0|^2|\varrho|^2 f'_\delta(\varrho)\psi\, dx \leqslant \delta\|\psi\|_{L^\infty(U)}\int_U |Q^{-1}b_0|^2\, dx.$$

Therefore, the equality

$$\lim_{\delta\to 0}\int_{\mathbb{R}^d} |Q^{-1}b_0|^2|\varrho|^2 f'_\delta(\varrho)\psi\, dx = 0$$

holds. Since $f_\delta(t) \to \operatorname{sign} t$ as $\delta \to 0$, we obtain the estimate

$$\int_{\mathbb{R}^d} \langle A\nabla|\varrho| - b_0|\varrho|, \nabla\psi\rangle\, dx \leqslant -r\int_{\mathbb{R}^d} \psi|\varrho|\, dx.$$

Integrating by parts we obtain the inequality

$$\int_{\mathbb{R}^d} L\psi|\varrho|\, dx \geqslant r\int_{\mathbb{R}^d} \psi|\varrho|\, dx,$$

as required. □

4.3.6. Proposition. (i) *Let ν be a Borel probability measure on $\mathbb{R}^d$ and let the field b be locally integrable with respect to ν. Suppose that there exists a positive function V of class $C^2(\mathbb{R}^d)$ such that $V(x) \to +\infty$ as $|x| \to +\infty$ and for some number $C > 0$ and all $x \in \mathbb{R}^d$ we have*

$$LV(x) \geqslant -CV(x), \quad |\sqrt{A(x)}\nabla V(x)| \leqslant CV(x). \tag{4.3.3}$$

Then the set $(L - I)(C_0^\infty)$ is dense in $L^1(\nu)$.

Without loss of generality one can also require a formally stronger, but actually equivalent (with $\ln V$ in place of V and another constant C) condition

$$LV(x) \geqslant -C, \quad |\sqrt{A(x)}\nabla V(x)| \leqslant C. \tag{4.3.4}$$

(ii) *The set $(L - I)\big(C_0^\infty(\mathbb{R}^d)\big)$ is dense in $L^1(\nu)$ also under the following conditions:* $\dfrac{a^{ij}}{1+|x|^2}, \dfrac{b^i}{1+|x|} \in L^1(\nu)$.

PROOF. (i) We can pass from (4.3.3) to (4.3.4) by the change $W = \ln V$, since $\nabla W = V^{-1}\nabla V$,

$$LW = V^{-1}LV - V^{-2}\langle A\nabla V, \nabla V\rangle,$$

which under (4.3.3) yields (4.3.4) with the number $C + C^2$ in place of C.

Suppose that the range of $L - I$ is not dense in $L^1(\nu)$. Then there exists a bounded Borel function g such that $\|g\|_{L^\infty(\nu)} > 0$ and

$$\int_{\mathbb{R}^d} (L\varphi - \varphi) g \, d\nu = 0$$

for every function $\varphi \in C_0^\infty$. Therefore, the measure $g\nu$ satisfies the equation $L^*(g\nu) = g\nu$. Set $\sigma = g\nu$. By the previous lemma $L^*|\sigma| \geqslant |\sigma|$, i.e., for every nonnegative function $\psi \in C_0^\infty$ we have the inequality

$$\int_{\mathbb{R}^d} L\psi \, d|\sigma| \geqslant \int_{\mathbb{R}^d} \psi \, d|\sigma|. \tag{4.3.5}$$

Set $\psi_N(x) = \zeta\big(N^{-1}V(x)\big)$, where a nonnegative function ζ is such that $\zeta(x) = 1$ if $|x| \leqslant 1$, $\zeta(x) = 0$ if $|x| > 2$, $\zeta' \leqslant 0$ if $x > 0$, and there is a number $M > 0$ such that

$$|\zeta(x)| \leqslant M, \ |\zeta'(x)| \leqslant M, \ |\zeta''(x)| \leqslant M$$

for all x. Using the hypothesis of the proposition we find that

$$\begin{aligned} L\psi_N &= N^{-1}\zeta'(N^{-1}V)LV + N^{-2}\zeta''(N^{-1}V)|\sqrt{A}\nabla V|^2 \\ &\leqslant CMN^{-1} + C^2MN^{-2}. \end{aligned}$$

Thus,

$$\int_{\mathbb{R}^d} \psi_N \, d|\sigma| \leqslant CM\big(N^{-1} + CN^{-2}\big)|\sigma|(\mathbb{R}^d).$$

Letting $N \to \infty$ we obtain $|\sigma|(\mathbb{R}^d) = 0$, Therefore, $\sigma \equiv 0$, which is a contradiction. Justification of (ii) is even simpler: (4.3.5) is applied with $\psi_N(x) = \psi_0(x/N)$, where $\psi_0 \in C_0^\infty(\mathbb{R}^d)$, $\psi_0(x) = 1$ if $|x| \leqslant 1$. □

4.3.7. Corollary. *If b is locally bounded in the previous proposition and (4.3.3) holds, then its conclusion is true for any Borel probability measure ν, hence the equation $L^*\mu = 0$ has no signed solutions.*

This result should be compared with Theorem 4.1.9 with a similar conclusion, where b is not assumed to be locally bounded, but it is assumed that there is a probability solution. In addition, let us draw the reader's attention once again to what has been said after Theorem 4.1.9.

Let us give another sufficient condition that employs a nonlinear inequality with a Lyapunov function.

4.3.8. Proposition. *Suppose that there exist a function $V \in C^2(\mathbb{R}^d)$ and a locally integrable function β on $\mathbb{R}$ such that $\lim\limits_{|x|\to+\infty} V(x) = +\infty$, $\nabla V \neq 0$ outside some ball and for some $M > 0$ the following conditions are fulfilled:*

$$LV(x) \geqslant |\sqrt{A}(x)\nabla V(x)|^2 \beta\big(V(x)\big), \quad |\sqrt{A(x)}\nabla V(x)| \leqslant M,$$

$$\int_0^{+\infty} m(y) \int_{y_0}^{y} \frac{1}{m(s)}\, ds\, dy = +\infty, \quad \textit{where} \quad m(y) = \exp\Big(\int_{y_0}^{y} \beta(r)\, dr\Big).$$

Then the set $(L - I)(C_0^\infty)$ is dense in $L^1(\sigma)$ for every absolutely continuous probability measure σ on $\mathbb{R}^d$.

Proof. If the range of $L - I$ is not dense in $L^1(\sigma)$, where the measure σ has a density $f \in L^1(\mathbb{R}^d)$, then there exists a bounded measurable function h_0 such that the integrable function $h = h_0 f$ satisfies the equation $\operatorname{div}(A\nabla h - b_0 h) = h$ in the weak sense, where $b_0^i = b^i - \sum_j \partial_{x_j} a^{ij}$. Then h is locally Sobolev and $\operatorname{div}(A\nabla|h| - b_0|h|) \geqslant |h|$ in the weak sense. Set $g = |h|$. It is known (see Bogachev [**126**, Proposition 5.8.34] or Evans, Gariepy [**344**, § 3.4, Proposition 3]) that

$$\frac{d}{dt}\int_{\{V<t\}} g\, dx = \int_{\{V=t\}} \frac{g}{|\nabla V|}\, dS$$

if $t \geqslant t_0$, where t_0 is taken in such a way that $\nabla V(x) \neq 0$ if $|x| \geqslant t_0$. Set

$$G(t) = \int_{\{V<t\}} \langle A\nabla V, \nabla V\rangle g\, dx.$$

Then G is bounded: $0 \leqslant G(t) \leqslant M^2 \|g\|_{L^1}$, in addition,

$$G'(t) = \int_{\{V=t\}} g\langle A\nabla V, \nabla V|\nabla V|^{-1}\rangle\, dS = \int_{\{V<t\}} \operatorname{div}(gA\nabla V)\, dx,$$

$$G''(t) = \int_{\{V=t\}} \frac{\operatorname{div}(gA\nabla V)}{|\nabla V|}\, dS = \int_{\{V=t\}} \frac{\langle A\nabla g, \nabla V\rangle}{|\nabla V|}\, dS + \int_{\{V=t\}} \frac{g\operatorname{div}(A\nabla V)}{|\nabla V|}\, dS.$$

Since

$$\int_{\{V=t\}} \frac{\langle A\nabla g, \nabla V\rangle}{|\nabla V|}\, dS \geqslant \int_{\{V=t\}} g\frac{\langle b_0, \nabla V\rangle}{|\nabla V|}\, dS + \int_{\{V<t\}} g\, dx,$$

we have

$$G''(t) \geqslant \int_{\{V=t\}} g|\nabla V|^{-1} LV\, dS + \int_{\{V<t\}} g\, dx.$$

Taking into account the estimates in our condition, we obtain the inequality

$$G''(t) \geqslant \beta(t)G'(t) + M^{-1}G(t).$$

Since $\beta = m'/m$, one has $(G'/m)' \geqslant M^{-1}m^{-1}G$. If $G(t_1) > 0$, where $t_1 > t_0$, then

$$G'(t)/m(t) \geqslant M^{-1}G(t_1)\int_{t_1}^{t} m^{-1}(s)ds,$$

since $G'/m \geqslant 0$. Hence G cannot be bounded, which is a contradiction. So g has support in the ball of radius t_0, whence $h = 0$ (as a solution with zero boundary condition). Thus, $h_0 = 0$ σ-a.e. □

This proposition yields the result in Wu, Zhang [**943**] on uniqueness in the presence of the estimate $\langle b(x), x\rangle \geqslant \beta_0(|x|)|x|$, provided that

$$\int_1^\infty m_0(y)\int_1^y \frac{1}{m_0(s)}\,ds\,dy = +\infty, \quad m_0(y) = y^{d-1}\exp\biggl(\int_1^y \beta_0(s)\,ds\biggr).$$

To this end we take $\beta(s) = \beta_0(s) + (d-1)/s$ in the proposition.

Let us give an example showing that uniqueness cannot be always checked by using a radial estimate on $\langle x, b(x)\rangle$.

4.3.9. Example. Let $V(x) = \sum_{i=1}^d \ln\bigl(\ln(2+x_i^2)\bigr)$. In order to satisfy the condition $LV(x) \geqslant -C$ it suffices to have the following estimates:

$$\sum_{i=1}^d \frac{2x_i b^i(x)}{(2+x_i^2)\ln(2+x_i^2)} \geqslant -C_1$$

for some number $C_1 > 0$ and every $x \in \mathbb{R}^d$. Let $d = 2$. Then

$$V(x,y) = \ln\ln(2+x^2) + \ln\ln(2+y^2).$$

Set

$$b^1(x,y) = y(2+x^2)\ln(2+x^2) - x\ln(2+x^2),$$
$$b^2(x,y) = -x(2+y^2)\ln(2+y^2) - y\ln(2+y^2).$$

Then

$$\frac{2xb^1(x,y)}{(2+x^2)\ln(2+x^2)} + \frac{2yb^2(x,y)}{(2+y^2)\ln(2+y^2)} = -\frac{2x^2}{2+x^2} - \frac{2y^2}{2+y^2} \geqslant -2.$$

Therefore, the space of integrable solutions of the equation $L^*\mu = 0$ is at most one-dimensional. We observe that

$$\begin{aligned} &b^1(x,y)x + b^2(x,y)y \\ &= xy\bigl[(2+x^2)\ln(2+x^2) - (2+y^2)\ln(2+y^2)\bigr] - x^2\ln(2+x^2) - y^2\ln(2+y^2). \end{aligned}$$

It is clear that the expression $b^1(x,y)x + b^2(x,y)y$ cannot be estimated from below even by $-C(\sqrt{x^2+y^2})^3 - C$. Therefore, sufficient conditions of the form

$$\frac{b^1(x,y)x + b^2(x,y)y}{\sqrt{x^2+y^2}} \geqslant \beta(\sqrt{x^2+y^2}),$$

expressed in terms of a function β, do not work here. In this example, for V we have $LV(x,y) \leqslant -1$ outside some ball and $|\nabla V| \leqslant C$. Therefore, there exists a unique probability solution of the equation $L^*\mu = 0$, which is also a unique integrable solution.

We are unaware of examples where the equation $L^*\mu = 0$ has signed solutions in the class of bounded measures, but has no probability solutions. This is impossible for $d = 1$ (Exercise 4.5.22).

4.4. Equations with a potential

We now turn to the equation $L^*_{A,b,c}(\varrho\,dx) = 0$ already considered in § 2.4 and also in § 2.5(i), which we shall write as

$$L^*\varrho = 0,$$

with the operator $L = L_{A,b,c}$ with a potential term:

$$L_{A,b,c}u = a^{ij}\partial_{x_i}\partial_{x_j}u + b^i\partial_{x_i}u + cu.$$

We assume in this section that $A(x) = \big(a^{ij}(x)\big)_{i,j\leqslant d}$ is a positive definite matrix for all x, $a^{ij} \in W^{p,1}_{\mathrm{loc}}(\mathbb{R}^d)$, where the number $p > d$ is fixed, and we deal with the continuous versions of these functions, $b^i \in L^p_{\mathrm{loc}}(\mathbb{R}^d)$, $c \in L^{pd/(p+d)}_{\mathrm{loc}}(\mathbb{R}^d)$. We shall discuss the uniqueness of positive (not necessarily integrable) solutions and the uniqueness of probability solutions. We recall that in the case $c = 0$ for uniqueness of a probability solution it suffices to have an estimate $LV \leqslant CV$, where C is a positive number and the function $V \in C^2(\mathbb{R}^d)$ is such that $\lim\limits_{|x|\to\infty} V(x) = +\infty$. Here we consider a more general situation.

4.4.1. Theorem. *Suppose that for some positive solution ϱ of the equation $L^*\varrho = 0$ we have $c\varrho \in L^1(\mathbb{R}^d)$ and*

$$\int_{\mathbb{R}^d} c\varrho\,dx \geqslant 0.$$

Let $V \in C^2(\mathbb{R}^d)$ and $\lim\limits_{|x|\to\infty} V(x) = +\infty$. If

$$LV(x) \leqslant 0 \quad \text{and} \quad c(x) \leqslant 0 \qquad \textit{as } |x| > R$$

for some R, then ϱ is a unique positive solutions to the equation $L^*\varrho = 0$ up to multiplication by a number.

PROOF. Let $v\varrho$ and ϱ be two positive solutions of our equation. As in the reasoning in § 4.1, for any functions $\varphi \in C_0^\infty(\mathbb{R}^d)$ and $f \in C^2\big([0, +\infty)\big)$ we have the equality

$$\int_{\mathbb{R}^d} f''(v)|\sqrt{A}\nabla v|^2\varrho\varphi\,dx = \int_{\mathbb{R}^d} f(v)\varrho L\varphi\,dx. \tag{4.4.1}$$

Let $\lambda > 0$ and $f(v) = e^{-\lambda v} - 1$. Let $\zeta \in C^2(\mathbb{R})$, $\zeta(0) = 1$, $\zeta(t) = 0$ when $t \geqslant 1$, $\zeta' \leqslant 0$ and $\zeta'' \geqslant 0$. Set $\varphi_N(x) = \zeta\big(V(x)/N\big)$. Substituting such functions f and φ_N in equality (4.4.1) and taking into account the inequalities $f \leqslant 0$, $\zeta'_N \leqslant 0$ and $\zeta''_N \geqslant 0$, we obtain that

$$\begin{aligned}\lambda^2 \int_{\mathbb{R}^d} & e^{-\lambda v}|\sqrt{A}\nabla v|^2\varrho\varphi_N\,dx \\ &= \int_{\mathbb{R}^d} (e^{-\lambda v} - 1)\varrho\big(N^{-1}\zeta'_N LV + N^{-2}\zeta''_N|\sqrt{A}\nabla V|^2 + c\varphi_N\big)\,dx\end{aligned}$$

is estimated by the following sum of integrals:

$$\int_{|x|<R} (e^{-\lambda v} - 1)\varrho\big(N^{-1}\zeta'_N LV + N^{-2}\zeta''_N|\sqrt{A}\nabla V|^2\big)\,dx + \int_{\mathbb{R}^d} (e^{-\lambda v} - 1)c\varphi_N\varrho\,dx.$$

Letting $N \to \infty$ we have

$$\int_{\mathbb{R}^d} e^{-\lambda v} c^- \varrho\, dx + \lambda^2 \int_{\mathbb{R}^d} e^{-\lambda v} |\sqrt{A}\nabla v|^2 \varrho\, dx \leqslant \int_{|x|\leqslant R} e^{-\lambda v} c^+ \varrho\, dx.$$

Let $m = \min_{|x|\leqslant R} v(x)$. Multiplying the last inequality by $e^{\lambda m}$, we obtain

$$\lambda^2 \int_{\mathbb{R}^d} e^{-\lambda(v-m)} |\sqrt{A}\nabla v|^2 \varrho\, dx \leqslant \int_{|x|\leqslant R} e^{-\lambda(v-m)} c^+ \varrho\, dx.$$

We observe that $v(x) - m \geqslant 0$ when $|x| \geqslant R$. Indeed, if the open set $U = \{v < m\}$ is not empty, then the equality $\nabla v = 0$ must hold almost everywhere on it, since otherwise letting $\lambda \to +\infty$ we can make the left side of the previous equality as large as we wish, although the right side remains bounded by the integral of $|c|\varrho$. Then in every component of U the function v equals a constant $v_0 < m$, hence, $v = v_0$ also on the boundary of the component, which is impossible (if $v(x_0) = v_0$, then we have $v < m$ in a neighborhood of x_0). Hence U has no boundary, which contradicts the fact that there is a point x_m with $v(x_m) = m$. Therefore, it is proven that $v(x) \geqslant m$ for all x. Hence v attains its minimum in the interior of the ball $\{x\colon |x| \leqslant 2R\}$.

Finally, we observe that v satisfies the equation

$$\partial_{x_i}(a^{ij}\varrho\partial_{x_j} v) + (\partial_{x_j} a^{ij}\varrho + a^{ij}\partial_{x_j}\varrho - b^i\varrho)\partial_{x_i} v = 0,$$

the left side of which is $L^*(v\varrho) - vL^*\varrho$. Hence v satisfies the strong maximum principle. Therefore, v is constant (see the end of § 1.2). □

4.4.2. Remark. (i) If it is known in advance that ϱ is a probability solution (in the theorem only the integrability of $c\varrho$ is given), then the inequality $LV(x) \leqslant 0$ for $|x| > R$ required in the hypothesis can be replaced with the considerably weaker inequality $LV(x) \leqslant C_1 + C_2 V(x)$ for $|x| > R$. Indeed, for every $\gamma \in (0,1)$ and every $N > \gamma^{-1}$ we have the obvious estimate

$$\int_{\mathbb{R}^d} (e^{-\lambda v} - 1)\varrho N^{-1}\zeta_N' LV\, dx \leqslant N^{-1} \int_{|x|<R} |LV|\varrho\, dx$$
$$+ C_3\gamma \int_{\{R<|x|\}\cap\{V(x)<N\gamma\}} \varrho\, dx + C_4 \int_{\gamma N < V(x) < N} \varrho\, dx,$$

where C_3 and C_4 depend only on C_1 and C_2. Therefore,

$$\limsup_{N\to\infty} \int_{\mathbb{R}^d} (e^{-\lambda v} - 1)\varrho N^{-1}\zeta_N' LV\, dx \leqslant C_3\gamma \int_{|x|>R} \varrho\, dx.$$

Letting $\gamma \to 0$ we obtain

$$\limsup_{N\to\infty} \int_{\mathbb{R}^d} (e^{-\lambda v} - 1)\varrho N^{-1}\zeta_N' LV\, dx \leqslant 0.$$

The remaining reasoning repeats the proof of the theorem.

(ii) If $c = 0$, then the proven assertion takes an especially simple form: the inequality $LV(x) \leqslant 0$ for $|x| > R$ implies uniqueness of a positive solution, and the weaker inequality $LV(x) \leqslant C_1 + C_2 V(x)$ for $|x| > R$ yields that the given probability solution is a unique (up to proportionality) positive solution.

(iii) The condition $\displaystyle\int_{\mathbb{R}^d} c\varrho\, dx \geqslant 0$ for the solution ϱ of the equation $L^*\varrho = 0$ is fulfilled (even in the form of the exact equality) if the functions $c\varrho$, $(1+|x|)^{-2}a^{ij}\varrho$

and $(1+|x|)^{-1}b^i\varrho$ belong to $L^1(\mathbb{R}^d)$. In particular, this is true if $\varrho \in L^1(\mathbb{R}^d)$ and the functions a^{ij} and b^i are Lipschitzian.

Indeed, taking a function $\varphi \in C_0^\infty(\mathbb{R}^d)$ equal to 1 on the unit ball and 0 outside the doubled ball we obtain that for all $N \geqslant 1$ the integral of $\varrho L\varphi_N$ vanishes, where $\varphi_N(x) = \varphi(x/N)$. Here $\varrho(x)L\varphi_N(x)$ is the sum of the three summands: $N^{-2}a^{ij}\partial_{x_i}\partial_{x_j}\varphi(x/N)\varrho(x)$, $N^{-1}b^i\partial_{x_i}\varphi(x/N)\varrho(x)$ and $\varphi(x/N)c(x)\varrho(x)$. As $N \to \infty$, the integrals of the first two summands tend to zero (the integral of an integrable function over the set $\{N \leqslant |x| \leqslant 2N\}$ tends to zero) and the integral of the last summand tends to the integral of $c\varrho$.

(iv) One cannot drop the condition $c(x) \leqslant 0$ for $|x| > R$: let us take functions $b(x) = -x$, $c(x) = x^2/4 - 3/2$ on $\mathbb{R}$; then the corresponding equation has two different probability solutions $\varrho_1(x) = c_1 e^{x-x^2/4}$ and $\varrho_2(x) = c_2 e^{-x-x^2/4}$, where c_1 and c_2 are constants.

4.4.3. Remark. Let $A = \mathrm{I}$. It follows from the version of Liouville's theorem obtained in Nazarov, Ural'tseva [**736**] that the uniqueness of a positive solution ϱ holds if the function $|b - 2\nabla\varrho/\varrho|^d$ has uniformly bounded integrals over some sequence of sets of the form $\{R_k \leqslant |x| \leqslant 2R_k\}$ with $R_k \to +\infty$ (say, if we have $|b(x) - 2\nabla\varrho/\varrho(x)| \leqslant 1/|x|$ in these sets). Indeed, the function v satisfies the elliptic equation $\Delta v + \langle 2\nabla\varrho/\varrho - b, \nabla v\rangle = 0$, so the result of [**736**] applies.

4.5. Complements, comments and exercises

4.5(i). Equations on manifolds

Similar problems are studied also on Riemannian manifolds (see Bogachev, Röckner, Wang [**179**], [**180**], [**181**], Bogachev, Wang, Röckner [**186**]). The proofs of Theorem 4.1.11 and Theorem 4.1.12(ii) given above remain valid for connected complete Riemannian manifolds, because they possess sequences of Lipschitzian functions ζ_k with compact support such that $0 \leqslant \zeta_k(x) \leqslant 1$, $\sup_x |\nabla\zeta_k(x)| \to 0$ and $\zeta_k(x) \to 1$. However, there are connected complete Riemannian manifolds for which the analog of Theorem 4.1.12(i) is false. For example, there are connected complete Riemannian manifolds on which there exist nonconstant positive Harmonic functions ϱ integrable with respect to the Riemannian volume (see Grigoryan [**423**], Li, Schoen [**605**]); then the measure with density ϱ is bounded and satisfies the equation $L_{\mathrm{I},0}^*\mu = 0$, i.e., $b = 0$, so here the analog of Theorem 4.1.12(i) is not true. A condition under which it is true is given in Theorem 4.5.2.

Let us mention two results from [**179**] on the equation $L_Z^*\mu = 0$ with the operator $L_Z\varphi = \Delta\varphi + \langle Z, \nabla\varphi\rangle$, where Δ is the Laplace–Beltrami operator on a Riemannian manifold M, i.e., $\Delta\varphi = \operatorname{div}\nabla\varphi$, moreover, we consider a gradient vector field $Z = \nabla U$. We shall use the notation and assumptions from § 2.5(ii).

4.5.1. Theorem. *Suppose that $Z = \nabla U$, where $U \in W_{\mathrm{loc}}^{\alpha,1}(M) \cap L_{\mathrm{loc}}^\infty(\lambda_M)$ for some $\alpha > d$. Assume that there exists a nonnegative function $V \in C^2(M)$ such that the sets $\{V \leqslant c\}$ are compact for every $c < \sup_M V$ and cover M and there exists a compact set K such that $LV \leqslant -1$ outside K. Then $\exp U \in L^1(\lambda_M)$.*

If M is connected, then the normalized measure $\mu = C \exp U \, \lambda_M$ is a unique probability measure satisfying the equation $L_Z^* \mu = 0$.

4.5.2. Theorem. *Let M be connected with the Ricci curvature bounded from below and satisfy the following condition: $\inf_{x \in M} \lambda_M\bigl(B(x,r)\bigr) > 0 \ \forall\, r > 0$, where $B(x,r)$ is the geodesic ball. Suppose that μ is a Borel probability measure on M such that $L_Z^* \mu = 0$, where $|Z| \in L^2(\mu)$. Assume that $Z = \nabla U$, where $U \in W^{1,1}_{\mathrm{loc}}(\lambda_M)$. Then $\exp U \in L^1(\lambda_M)$ and $\mu = \|\exp U\|^{-1}_{L^1(\lambda_M)} \exp U \, \lambda_M$.*

4.5.3. Remark. (i) Let M be a connected complete Riemannian manifold with the Ricci curvature bounded from below and positive injectivity radius and let $Z = 0$. Then we arrive at the well-known fact (see Grigor'yan [**424**, § 13] for more general results) that every positive integrable harmonic function h on M is constant. Indeed, we consider $\mu = h \, \lambda_M$ and obtain $\nabla h / h = 0$ as the orthogonal projection of zero.

4.5(ii). Uniqueness of solutions of the Dirichlet problem for equations with lower order terms

Here we present the results of the paper Zhikov [**962**], which are close in the spirit of ideas to the results in § 4.2. They are concerned with uniqueness of solutions of the Dirichlet problem on a bounded domain for an operator with an unbounded drift.

Let $\Omega \subset \mathbb{R}^d$ be a bounded domain with a Lipschitzian boundary and let a be a divergence free vector field, i.e., $|a| \in L^1(\Omega)$ and

$$\int_\Omega \langle a, \nabla \varphi \rangle \, dx = 0 \quad \forall\, \varphi \in C_0^\infty(\Omega).$$

We consider the Dirichlet problem

$$u \in W_0^{2,1}(\Omega), \quad -\mathrm{div}(\nabla u + a u) = f \in W^{2,-1}(\Omega). \tag{4.5.1}$$

A solution is understood in the sense of the integral identity

$$\int_\Omega \langle \nabla u + a u, \nabla \varphi \rangle \, dx = 0 \quad \forall\, \varphi \in C_0^\infty(\Omega).$$

Suppose first that $|a| \in L^\infty(\Omega)$. Then there exists a solution u of the Cauchy problem (4.5.1), moreover, we have the equality

$$\int_\Omega |\nabla u|^2 \, dx = \langle f, u \rangle, \tag{4.5.2}$$

which is obtained by multiplication of the equation by u and integration by parts using the equality

$$\int_\Omega \langle a, \nabla u \rangle u \, dx = 0.$$

Since this estimate does not depend on the vector field a, it can be used for constructing solutions also in the case of unbounded a.

Suppose now that $|a| \in L^p(\Omega)$, where $p = 2d/(d-2)$ for $d > 2$ and $p = 1 + \delta$, $\delta > 0$ for $d = 2$. Then there exists a sequence $\{a_N\}$ with the following properties:

$$|a_N| \in L^\infty(\Omega), \quad \mathrm{div} a_N = 0, \quad \lim_{N \to \infty} \|a_N - a\|_{L^p(\Omega)} = 0.$$

For the corresponding solution u_N equality (4.5.2) is fulfilled. Therefore, the sequence $\{u_N\}$ is bounded in $W_0^{2,1}(\Omega)$ and passing to a subsequence we may assume that $\{u_N\}$ converges weakly to some function $u \in W_0^{2,1}(\Omega)$. Then $u_N \to u$ weakly in $L^q(\Omega)$, where $q = p/(p-1)$, and we can pass to the limit in the integral identity defining problem (4.5.1). Thus, the function u is a solution to the Dirichlet problem (4.5.1). Such solutions are naturally called approximate.

Note that for an approximate solution u we have the inequality

$$\int_\Omega |\nabla u|^2\,dx \leqslant \langle f, u\rangle.$$

Let D denote the set of all $v \in W_0^{2,1}(\Omega)$ for which the estimate

$$\left|\int_\Omega \langle av, \nabla\varphi\rangle\,dx\right| \leqslant c(v)\int_\Omega |\nabla\varphi|^2\,dx$$

holds for every function $\varphi \in C_0^\infty(\Omega)$.

Set

$$[v,\varphi] = \int_\Omega \langle av, \nabla\varphi\rangle\,dx, \quad v \in D, \quad \varphi \in C_0^\infty(\Omega).$$

If $\varphi \in W_0^{2,1}(\Omega)$, then we define $[v,\varphi]$ by continuity. Clearly, any solution of the Dirichlet problem (4.5.1) belongs to the set D and the integral identity takes the form

$$\int_\Omega \langle \nabla u, \nabla\varphi\rangle\,dx + [u,\varphi] = \langle f, u\rangle.$$

In addition, for an approximate solution we have $[u,u] \geqslant 0$.

Thus, a solution of problem (4.5.1) exists. It is proved in Zhikov [**962**] that if $|a| \in L^2(\Omega)$, then the solution is unique. In the same paper an example of nonuniqueness is constructed. For its construction the following simple observation is useful.

4.5.4. Proposition. *If there exists an element $v \in D$ for which $[v,v] < 0$, then the solution is not unique.*

PROOF. Let us consider the Dirichlet problem (4.5.1) with the right-hand side $f = -\mathrm{div}(\nabla v + av)$. Clearly, $f \in W^{2,-1}(\Omega)$ by the definition of the class D. One solution to this problem is v, but there exists also an approximate solution u for which $[u,u] \geqslant 0$. Hence u does not coincide with v. □

Let us now give an example of a and v such that $[v,v] < 0$. Let $d = 3$ and let $\Omega = \{|x| < 1\}$ be the open unit ball. Set

$$a(x) = x|x|^{-3}a_0(x|x|^{-1}), \quad v(x) = (1-|x|^4)v_0(x|x|^{-1}),$$

where a_0 and v_0 are smooth functions on the unit sphere. If a_0 and v_0 satisfy the conditions

$$\int_{|x|=1} a_0\,dS = 0, \quad \int_{|x|=1} v_0 a_0\,dS_x = 0, \quad \int_{|x|=1} v_0^2 a_0\,dS_x = -2,$$

then we obtain the desired vector field a and function v.

Note that even in this example it is not known whether the approximate solution is unique. In addition, there are no examples of nonuniqueness of approximate solutions.

4.5(iii). Weighted Sobolev classes

In § 4.1 and § 4.2 in the proof of uniqueness and in examples of nonuniqueness we studied a degenerate elliptic equation $\mathrm{div}\big(\varrho A\nabla v-(b-\beta_\mu)\varrho v\big)=0$. Zhikov [**961**] investigated similar degenerate equations

$$-\mathrm{div}(\varrho A\nabla v)+v\varrho=f\varrho, \tag{4.5.3}$$

where $A=A^*$, $\lambda^{-1}\cdot \mathrm{I}\leqslant A\leqslant \lambda\cdot \mathrm{I}$ and $f\in C_0^\infty(\mathbb{R}^d)$. Such equations are closely connected to Sobolev spaces with the weight ϱ.

Let $\varrho\geqslant 0$ and $\varrho,\varrho^{-1}\in L^1_{\mathrm{loc}}(\mathbb{R}^d)$. Set

$$\|u\|_{2,1,\varrho}=\biggl(\int_{\mathbb{R}^d}\bigl[|u|^2+|\nabla u|^2\bigr]\varrho\,dx\biggr)^{1/2}.$$

We define the space $H(\varrho)$ as the completion of the class $C_0^\infty(\mathbb{R}^d)$ with respect to the norm $\|u\|_{2,1,\varrho}$; in § 1.1 it was denoted by the symbol $W^{2,1}(\mu)$ with $\mu=\varrho\,dx$. Let $W(\varrho)$ denote the set of all functions $u\in W^{1,1}_{\mathrm{loc}}(\mathbb{R}^d)$ with $\|u\|_{2,1,\varrho}<\infty$. For simplicity, we write H and W for $H(\varrho)$ and $W(\varrho)$, respectively.

Clearly, H and W are Hilbert spaces and $H\subset W$, since for any ball U

$$\|\nabla u\|_{L^1(U)}\leqslant \|\nabla u\|_{L^2(\mu)}\|1/\varrho\|_{L^1(U)}$$

by the Cauchy–Bunyakovsky inequality. By definition, C_0^∞ is dense in H. The spaces H and W do not always coincide (see examples in [**961**, § 5]).

Let K be an intermediate space between H and W (it can coincide with one of them). A function $v\in K$ is called a K-solution if

$$\int_{\mathbb{R}^d}\bigl(\langle A\nabla u,\nabla\varphi\rangle+u\varphi\bigr)\varrho\,dx=\int_{\mathbb{R}^d}f\varphi\varrho\,dx$$

for every $\varphi\in K$. The existence and uniqueness of a K-solution follows by the Riezs theorem. A natural question arises: does the solution depend on our choice of K?

4.5.5. Proposition. *Suppose that the classes H and W do not coincide. Then there exists a function $f\in C_0^\infty(\mathbb{R}^d)$ such that the H-solution does not coincide with the W-solution.*

PROOF. Let us consider a more general equation

$$\mathrm{div}(\varrho A\nabla v)+\varrho v=F,$$

where F is a continuous functional on W. For any K-solution v we have the estimate

$$\|v\|_{2,1,\varrho}\leqslant\lambda\|F\|_{K^*},\quad \|F\|_{K^*}=\sup_{\varphi\in K,\|\varphi\|_{2,1,\varrho}}F(\varphi). \tag{4.5.4}$$

Since H and W do not coincide, there exists a continuous nonzero functional F_0 such that $F_0|_H=0$. The corresponding H-solution is zero, but the W-solution v is not zero. We now observe that the functionals of the form

$$\varphi\mapsto\int_{\mathbb{R}^d}f(x)\varphi(x)\,dx,$$

where $f\in C_0^\infty(\mathbb{R}^d)$, are dense in the dual space W^*. Let a sequence of functions f_k in $C_0^\infty(\mathbb{R}^d)$ be such that the corresponding functionals converge to $\mathcal{F}_0$ in W^*. Then by estimate (4.5.4) the corresponding H-solutions u_k converge in norm $\|\cdot\|_{2,1,\varrho}$ to zero, but the W-solutions v_k converge in norm $\|\cdot\|_{2,1,\varrho}$ to the nonzero function v. Therefore, for sufficiently large k the solutions u_k do not coincide with v_k. □

We draw the reader's attention to the fact that for every K-solution u we have the energy identity

$$\int_{\mathbb{R}^d} \bigl(\langle A\nabla u, \nabla u\rangle + u^2\bigr)\varrho\,dx = \int_{\mathbb{R}^d} fu\varrho\,dx.$$

We shall say that $v \in W$ is a weak solution of equation (4.5.3) if

$$\int_{\mathbb{R}^d} \bigl(\langle A\nabla u, \nabla\varphi\rangle + u\varphi\bigr)\varrho\,dx = \int_{\mathbb{R}^d} f\varphi\varrho\,dx$$

for every $\varphi \in C_0^\infty(\mathbb{R}^d)$. Any K-solution is a weak solution, but the converse is false. In addition, for weak solutions we do not always have the energy identity.

4.5.6. Proposition. *Suppose that H does not coincide with W. Then, for some function $f \in C_0^\infty(\mathbb{R}^d)$, equation (4.5.3) has a weak solution which for no K is a K-solution.*

PROOF. Let us take f such that the H-solution u does not coincide with the W-solution v. Then the function $w = (u+v)/2$ is a weak solution, but is not a K-solution, since it does not satisfy the energy identity by the strict convexity of the function on the space W taking w to the integral of $\bigl(\langle A\nabla w, \nabla w\rangle + w^2\bigr)\varrho$. □

A weak solution will be a K-solution if it satisfies the energy identity. Indeed, we take take for K the smallest subspace containing this weak solution and the whole set $C_0^\infty(\mathbb{R}^d)$.

Let us consider the functional

$$E(u) = \int_{\mathbb{R}^d} \Bigl[|\sqrt{A}\nabla u|^2 + |u|^2 - 2fu\Bigr]\varrho\,dx = \|u\|_{K,\varrho}^2 - 2\langle f, u\rangle_{L^2(\varrho\,dx)}.$$

Let us consider the variational problem $J(K) = \min_{u\in K} E(u)$. If the H-solution does not coincide with the W-solution, the so-called Lavrentiev phenomenon arises:

$$J(H) > J(W).$$

Indeed, a solution to the variational problem for the space K is its K-solution. Let us explain why a solution to the variational problem exists. It is clear that the functional $u \mapsto E(u)$ is bounded from below. Let $u_n \in K$ be functions such that $\lim\limits_{n\to\infty} E(u_n) = \inf_K E$. Then

$$\|u_n - u_m\|_{K,\varrho}^2 = 2\Bigl[\|u_n\|_{K,\varrho}^2 + \|u_m\|_{K,\varrho}^2 - 2\Bigl\|\frac{u_n+u_m}{2}\Bigr\|_{K,\varrho}^2\Bigr]$$
$$= 2\Bigl[E(u_n) + E(u_m) - 2E\Bigl(\frac{u_n+u_m}{2}\Bigr)\Bigr] \leqslant 2\Bigl[E(u_n) - \inf_K E + E(u_m) - \inf_K E\Bigr] \to 0.$$

Therefore, the sequence $\{u_n\}$ is fundamental and converges to some function $u \in K$ at which $\inf_K E$ is attained. This phenomenon was discovered by M. Lavrentiev [**592**], where an example of a functional

$$\int_0^1 f(x, u(x), u'(x))\,dx$$

was constructed such that its minimum on the space of absolutely continuous functions on $[0,1]$ with boundary conditions $u(0) = 0$, $u(1) = 1$ is strictly smaller than its minimum on the space of Lipschitz functions with the same boundary conditions. Lavrentiev's phenomenon is discussed in Zhikov [**960**] and Foss, Hrusa, Mizel [**382**].

Let us give sufficient conditions for the equality $H = W$. One such condition is the membership of the density ϱ in the Mackenhoupt class A_2, i.e., the inequality

$$\sup_{U\subset\mathbb{R}^d}\left(\frac{1}{|U|}\int_U \varrho\,dx\right)\left(\frac{1}{|U|}\int_U \varrho^{-1}\,dx\right) < \infty, \quad \text{where } U \text{ is a ball.}$$

For a long time De Giorgi's conjecture remained open: the condition

$$\exists\, t > 0: \quad \exp(t\varrho),\ \exp(t\varrho^{-1}) \in L^1_{\mathrm{loc}}(\mathbb{R}^d)$$

ensures the equality $H = W$. This old conjecture has recently been proved by Zhikov [**963**], who obtained an even more general sufficient condition:

$$\sup_n n^{-2}\left(\int_U \varrho^n\,dx\right)^{1/n}\left(\int_U \varrho^{-n}\,dx\right)^{1/n} < \infty$$

for every ball $U \subset \mathbb{R}^d$. We prove this result below.

Suppose that $\varrho \geqslant 0$ and ϱ^{-1} are integrable on the unit ball $B \subset \mathbb{R}^d$. Then the weighted Sobolev space

$$W_0(\varrho) = \{u \in W_0^{1,1}(B)\colon\ |\nabla u|^2\varrho \in L^1(B)\}$$

is complete with respect to the norm defined by

$$\|u\|^2_{W_0} := \int_B |\nabla u|^2\varrho\,dx,$$

because, as already noted above, we have

$$\int_B |\nabla u|\,dx \leqslant \left(\int_B |\nabla u|^2\varrho\,dx\right)^{1/2}\left(\int_B \varrho^{-1}\,dx\right)^{1/2}.$$

The norm in $W_0(\varrho)$ is generated by the inner product

$$(u,v)_{W_0} = \int_B \langle\nabla u, \nabla v\rangle\,\varrho\,dx.$$

Let $H_0(\varrho)$ be the closure of $C_0^\infty(B)$ in $W_0(\varrho)$.

Let us note that this norm in $W_0(\varrho)$ does not involve the integral of $u^2\varrho$, unlike the norm in $W(\varrho)$. The analog of $W(\varrho)$ in the present situation if defined by

$$\widetilde{W}_0(\varrho) = \{u \in W_0^{1,1}(B)\colon\ (|\nabla u|^2 + |u|^2)\varrho \in L^1(B)\}$$

and is equipped with the inner product

$$(u,v)_{\widetilde{W}_0} = \int_B [uv + \langle\nabla u, \nabla v\rangle]\,\varrho\,dx.$$

This norm is also complete. Both norms are equivalent if (and only if) the measure $\mu = \varrho\,dx$ satisfies the Poincaré inequality, i.e., $\|u\|_{L^2(\mu)}$ is estimated by $C\|\nabla u\|_{L^2(\mu)}$ for all $u \in W_0(\varrho)$. This inequality holds if there is $p \in \big(2d/(d+2), 2\big)$ such that $\varrho \in L^{q/(q-2)}(B)$ with $q = dp/(d-p)$ and $\varrho^{-1} \in L^{2/(2-p)}(B)$. Indeed, then $q > 2$, hence by Hölder's inequality and Sobolev's embedding $W_0^{p,1}(B) \subset L^q(B)$ we have

$$\begin{aligned}\int_B u^2\varrho\,dx &\leqslant \|u\|_q^2\|\varrho\|_{q/(q-2)} \leqslant C_p\|\nabla u\|_p^2\|\varrho\|_{q/(q-2)}\\ &\leqslant C_p\|\varrho\|_{q/(q-2)}\|\varrho^{-1}\|_{2/(2-p)}\int_B |\nabla u|^2\varrho\,dx.\end{aligned}$$

The closure of $C_0^\infty(B)$ in $\widetilde{W}_0(\varrho)$ is denoted by $\widetilde{H}_0(\varrho)$.

The classes $H_0(\varrho)$ and $W_0(\varrho)$ do not coincide precisely when there is a nonzero function $u \in W_0(\varrho)$ orthogonal to $H_0(\varrho)$, i.e.,

$$\int_B \langle \nabla u, \nabla \varphi \rangle \, \varrho \, dx = 0 \quad \forall \varphi \in C_0^\infty(B), \tag{4.5.5}$$

which is equivalent to the equality

$$\operatorname{div}(\varrho \nabla u) = 0. \tag{4.5.6}$$

Therefore, $H_0(\varrho) = W_0(\varrho)$ provided that there is no nonzero solution to the above equation in the class $W_0^{1,1}(B)$.

Similarly, if $\widetilde{W}_0(\varrho) \neq \widetilde{H}_0(\varrho)$, then there is a nonzero solution $u \in W_0^{1,1}(B)$ to the equation

$$\operatorname{div}(\varrho \nabla u) - u = 0. \tag{4.5.7}$$

The absence of such solutions yields that $\widetilde{W}_0(\varrho) = \widetilde{H}_0(\varrho)$.

If we formally multiply (4.5.6) or (4.5.7) by u and integrate by parts, then we obtain that $\|u\|_{W_0} = 0$ or $\|u\|_{\widetilde{W}_0} = 0$, respectively. However, the aforementioned examples of weights with $W \neq H$ show that this procedure is not always correct. Hence some restrictions on ϱ are needed.

We shall now prove a beautiful result due to Zhikov [**963**] giving a broad sufficient condition for the equality $W_0(\varrho) = H_0(\varrho)$ (which under this condition is also the equality $\widetilde{W}_0(\varrho) = \widetilde{H}_0(\varrho)$).

Recall that the Hardy–Littlewood maximal function for a locally integrable function f on $\mathbb{R}^d$ is defined by

$$M(f)(x) = \sup_D \frac{1}{|D|} \int_D |f(y)| \, dy,$$

where sup is taken over all balls D centered at x. It is known (see Stein [**877**, Chapter 1, § 1]) that for any $f \in L^p(\mathbb{R}^d)$ with $p > 1$

$$\|M(f)\|_p \leqslant M_p \|f\|_p, \quad M_p = 2\bigl(5^d/(p-1)\bigr)^{1/p}, \tag{4.5.8}$$

where $\|f\|_p = \|f\|_{L^p(B)}$. If $p \in [3/2, 2]$, then $M_p \leqslant 5^{d+1}$.

4.5.7. Theorem. *Suppose that there is $L < \infty$ such that*

$$\limsup_{n\to\infty} n^{-2} \Bigl(\int_B \varrho^n \, dx\Bigr)^{1/n} \Bigl(\int_B \varrho^{-n} \, dx\Bigr)^{1/n} \leqslant L. \tag{4.5.9}$$

Then $W_0(\varrho) = H_0(\varrho) = \widetilde{W}_0(\varrho) = \widetilde{H}_0(\varrho)$.

PROOF. (i) We start with a trivial observation: it suffices to prove the result for a weight ϱ_0 such that $k_1\varrho \leqslant \varrho_0 \leqslant k_2\varrho$ with positive constants k_1 and k_2. Next, we can change ϱ in order to obtain L in (4.5.9) as small as we need. Indeed, let $\alpha \in (0,1)$. We can assume that $\varrho^{-1}([\alpha, 1/\alpha]) = \varnothing$ by the above observation. Moreover, we can assume that both $B_0 := \{\varrho < \alpha\}$ and $B_1 := \{\varrho > 1/\alpha\}$ have positive measures: if, say, $|B_1| = 0$, we redefine ϱ on a subset of small positive measure in $\{\varrho > 0\}$ by the constant $2/\alpha$. Assume that $\|\varrho\|_n^n \|\varrho^{-1}\|_n^n \leqslant L^n n^{2n}$ for all sufficiently large n. Letting $\varrho_t(x) = t\varrho(x)$ if $x \in B_1$, $\varrho_t(x) = t^{-1}\varrho(x)$ if $x \in B_0$, we obtain that

$$\begin{aligned}\|\varrho_t\|_n^n\|\varrho_t^{-1}\|_n^n &= t^{2n}\int_{B_1}\varrho^n\,dx\int_{B_0}\varrho^{-n}\,dx + t^{-2n}\int_{B_0}\varrho^n\,dx\int_{B_1}\varrho^{-n}\,dx\\ &\quad+\int_{B_1}\varrho^n\,dx\int_{B_1}\varrho^{-n}\,dx+\int_{B_0}\varrho^n\,dx\int_{B_0}\varrho^{-n}\,dx\\ &\leqslant t^{2n}L^n n^{2n}+t^{-2n}|B|+\alpha^n\big(\|\varrho^n+\varrho^{-n}\|_1\big).\end{aligned}$$

The last term is dominated by $C\alpha^n\|\varrho\|_n^n\|\varrho^{-1}\|_n^n$ with some constant C, since we have $|B_0|>0$, $|B_1|>0$. Therefore, for large n the quantity $n^{-2}\|\varrho_t\|_n\|\varrho_t^{-1}\|_n$ is dominated by $t^2L+t^{-2}n^{-2}+\alpha L<3\alpha L$ if $t^2=\alpha$. We shall assume that ϱ is such that the set $\{1/\varrho\geqslant 2\}$ has positive measure and L in (4.5.9) is smaller than a certain constant depending only on d that will appear at the end of the proof.

(ii) Let now $u\in W_0(\varrho)$ be a solution to equation (4.5.6). We can assume that $u\in W_0^{1,1}(\mathbb{R}^d)$ by setting $u=0$ outside of B. Suppose that $\|u\|_{W_0}=1$. Set

$$g(x)=\max\big(M(|\nabla u|)(x),|u(x)|/r(x)\big),$$

where $r(x)=1-|x|$ is the distance from x to the boundary of B. By Hölder's inequality, for each $\varepsilon\in(0,1/2)$ we have

$$\|\nabla u\|_{2-\varepsilon}\leqslant\|u\|_{W_0}\|\varrho^{-1}\|_{(2-\varepsilon)/\varepsilon}^{2-\varepsilon/2}=\|\varrho^{-1}\|_{(2-\varepsilon)/\varepsilon}^{2-\varepsilon/2}.$$

By (4.5.8) we have $\|M(\nabla u)\|_{2-\varepsilon}\leqslant 5^{d+1}\|\nabla u\|_{2-\varepsilon}$. Now the Hardy inequality (see Exercise 4.5.24) yields that

$$\|u/r\|_{2-\varepsilon}\leqslant 3\|\nabla u\|_{2-\varepsilon},\quad \varepsilon\in(0,1/2).$$

Therefore,

$$\|g\|_{2-\varepsilon}\leqslant 2\cdot 5^{d+1}\|u\|_{W_0}\|\varrho^{-1}\|_{(2-\varepsilon)/\varepsilon}^{1/2}. \tag{4.5.10}$$

(iii) Let us consider the set $S_\lambda=\{x\colon g(x)<\lambda\}$. It is known (see Evans, Gariepy [**344**, §6.6.3, proof of Theorem 2, p. 255], where g is defined without u/r) that u is Lipschitz on S_λ with constant $c_d\lambda$, where $c_d>1$ depends only on d. But since $u\leqslant\lambda r$ on S_λ and $u=0$ outside B, it is $c_d\lambda$-Lipschitz on $S_\lambda\cup(\mathbb{R}^d\backslash B)$. Hence u can be extended from $S_\lambda\cup(\mathbb{R}^d\backslash B)$ to a $c_d\lambda$-Lipschitz function u_λ on $\mathbb{R}^d$ (see [**344**, §3.1]). Hence $u_\lambda\in W_0^{1,1}(\mathbb{R}^d)$, $\nabla u_\lambda=\nabla u$ a.e. on S_λ, and $|\nabla u_\lambda|\leqslant c_d\lambda$ a.e. The specific form of g introduced above is needed for these properties. Note that (4.5.5) remains valid for $\varphi=u_\lambda$, since we can find a sequence of functions $\varphi_j\in C_0^\infty(B)$ such that $\nabla\varphi_j(x)\to\nabla u_\lambda(x)$ a.e. and $|\nabla\varphi_j(x)|\leqslant c_d\lambda$. Therefore, we obtain that

$$\int_{S_\lambda}|\nabla u|^2\varrho\,dx=-\int_{B\backslash S_\lambda}\langle\nabla u,\nabla u_\lambda\rangle\varrho\,dx\leqslant c_d\lambda\int_{B\backslash S_\lambda}|\nabla u|\varrho\,dx. \tag{4.5.11}$$

(iv) We need an auxiliary estimate. Let $f\in C_b^1[0,+\infty)$ be a positive decreasing function with $\lim\limits_{\lambda\to+\infty}f(\lambda)=0$. Let us prove that

$$\int_B f\big(g(x)\big)|\nabla u(x)|^2\varrho(x)\,dx\leqslant c_d\int_B\Phi\big(g(x)\big)|\nabla u(x)|\varrho(x)\,dx, \tag{4.5.12}$$

where Φ is a C^1-function given by $\Phi'(\lambda)=-\lambda f'(\lambda)$, $\Phi(0)\geqslant 0$, provided that Φ' is bounded and $\Phi(\lambda)\leqslant C\lambda^{1-q}$ for some $q>0$. Indeed, we observe both sides are finite, since f is bounded, $g\in L^p(B)$ for all $p<2$, $\varrho\in L^n(B)$ for all n. Let us consider the measure σ given by density $|\nabla u|^2\varrho$ with respect to Lebesgue measure.

The image of this measure under the function g has the distribution function

$$F(\lambda) = \sigma\big(x \in B\colon\ g(x) < \lambda\big) = \sigma(S_\lambda) = \int_{S_\lambda} |\nabla u|^2 \varrho\, dx.$$

The left-hand side of (4.5.12) equals

$$\int_{\mathbb{R}} f(\lambda)\, dF(\lambda) = -f(0)F(0) - \int_0^\infty f'(\lambda)F(\lambda)\, d\lambda \leqslant -\int_0^\infty f'(\lambda)F(\lambda)\, d\lambda$$

by the integration by parts formula. Multiplying both sides of (4.5.11) by $-f' \geqslant 0$ and integrating over $(0, +\infty)$ we obtain

$$-\int_0^\infty f'(\lambda) \int_{S_\lambda} |\nabla u|^2 \varrho\, dx\, d\lambda \leqslant -c_d \int_0^\infty \lambda f'(\lambda) \int_{B\backslash S_\lambda} |\nabla u| \varrho\, dx\, d\lambda$$
$$= c_d \int_0^\infty \Phi'(\lambda) \int_{B\backslash S_\lambda} |\nabla u| \varrho\, dx\, d\lambda.$$

By the previous estimate the left-hand side of this inequality dominates the left-hand side of (4.5.12), since the inner integral in it is exactly $F(\lambda)$. The right-hand side of the latter inequality equals (again by the integration by parts formula)

$$-c_d\Phi(0)G(\infty) + c_d \int_0^\infty \Phi(\lambda)\, dG(\lambda) \leqslant c_d \int_0^\infty \Phi(\lambda)\, dG(\lambda), \quad G(\lambda) := \int_{S_\lambda} |\nabla u| \varrho\, dx,$$

which leads to the right-hand side of (4.5.12) by the change of variables formula (this time for the image of the measure $|\nabla u|\varrho\, dx$). Therefore, (4.5.12) is proven.

(v) Using (4.5.12) and the Cauchy–Bunyakovsky inequality, we obtain that

$$\int_B f(g(x)) |\nabla u(x)|^2 \varrho(x)\, dx \leqslant c_d \|u\|_{W_0} \Big(\int_B \Phi\big(g(x)\big)^2 \varrho(x)\, dx \Big)^{1/2}.$$

Let $f(\lambda) = (\lambda + \delta)^{-\varepsilon}$, where $\delta > 0$. Then $\Phi(\lambda) = \varepsilon(1-\varepsilon)^{-1}(\lambda+\delta)^{1-\varepsilon} + \delta(\lambda+\delta)^{-\varepsilon}$. Letting $\delta \to 0$ in the previous estimate, we arrive at the inequality

$$\int_B g^{-\varepsilon} |\nabla u|^2 \varrho\, dx \leqslant 2c_d\varepsilon \Big(\int_B g^{2(1-\varepsilon)} \varrho\, dx \Big)^{1/2}.$$

On the other hand, by Hölder's inequality and (4.5.10), we have

$$\int_B g^{2(1-\varepsilon)} \varrho\, dx \leqslant \|g^{2(1-\varepsilon)}\|_{(2-\varepsilon)/(2(1-\varepsilon))} \|\varrho\|_{(2-\varepsilon)/\varepsilon}$$
$$= \|g\|_{2-\varepsilon}^{2(1-\varepsilon)} \|\varrho\|_{(2-\varepsilon)/\varepsilon} \leqslant C_d^{2(1-\varepsilon)} \|\varrho\|_{(2-\varepsilon)/\varepsilon} \|\varrho^{-1}\|_{(2-\varepsilon)/\varepsilon}^{1-\varepsilon},$$

where $C_d = 3 \cdot 5^{d+1}$. Therefore, the previous estimate yields that

$$\int_B g^{-\varepsilon} |\nabla u|^2 \varrho\, dx \leqslant 2c_d\varepsilon C_d^{1-\varepsilon} \|\varrho\|_{(2-\varepsilon)/\varepsilon}^{1/2} \|\varrho^{-1}\|_{(2-\varepsilon)/\varepsilon}^{1/2-\varepsilon/2}.$$

We finally observe that

$$\lim_{\varepsilon\to 0} \int_B g^{-\varepsilon} |\nabla u|^2 \varrho\, dx = \|u\|_{W_0}^2 = 1$$

and that, letting $\varepsilon = 2/(n+1)$ with $n \in \mathbb{N}$, we have

$$\|\varrho\|_{(2-\varepsilon)/\varepsilon} \|\varrho^{-1}\|_{(2-\varepsilon)/\varepsilon}^{1-\varepsilon} = \|\varrho\|_n \|\varrho^{-1}\|_n^{1-\varepsilon} \leqslant \|\varrho\|_n \|\varrho^{-1}\|_n$$

for all n large enough, since $\varrho^{-1} \geqslant 2$ on a positive measure set, as we have agreed above. Since $n^{-2}\|\varrho\|_n \|\varrho^{-1}\|_n \leqslant L$ for all n large enough, we have $1 \leqslant 4c_d C_d L^{1/2}$, which leads to a contradiction (showing that we have $\|u\|_{W_0} = 0$), once we ensure that $4c_dC_dL^{1/2} < 1$, which is possible, as explained in (i). Finally, we recall that, under condition (4.5.9), we have $W_0(\varrho) = \widetilde{W}_0(\varrho)$ with equivalence of norms. □

4.5(iv). Bounds on the total variation and Kantorovich distances between solutions

Here we give upper bounds for the total variation and the quadratic Kantorovich distance W_2 between stationary distributions of two diffusion processes or, more generally, between probability solutions to a stationary Kolmogorov equation. These estimates are applied to nonlinear stationary Fokker–Planck–Kolmogorov equations for probability measures. Proofs can be found in Bogachev, Kirillov, Shaposhnikov [**147**]. These proofs employ the methods of proving the uniqueness developed above in this chapter. Also the typical assumptions in the results of this subsection resemble those needed in the uniqueness theorems. Certainly, the presented bounds yield uniqueness.

Let $\|\cdot\|$ denote the total variation norm on bounded measures. Using the Hahn–Jordan decomposition $\mu = \mu^+ - \mu^-$ of a signed measure μ on a space Ω, we have $\|\mu\| = \mu^+(\Omega) + \mu^-(\Omega)$.

Let us recall that, given two probability measures μ_1 and μ_2 on $\mathbb{R}^d$ such that $\mu_1 = w\cdot\mu_2$, the entropy $H(\mu_1|\mu_2)$ denoted also by $\operatorname{Ent}_{\mu_2} w$, is defined by

$$H(\mu_1|\mu_2) := \operatorname{Ent}_{\mu_2} w := \int w\ln w\, d\mu_2,$$

provided that $w\ln w \in L^1(\mu_2)$. If μ_1 and μ_2 are given by positive densities ϱ_1 and ϱ_2 such that $\varrho_1\ln(\varrho_1/\varrho_2) \in L^1(\mathbb{R}^d)$, then $H(\mu_1|\mu_2)$ is the integral of $\varrho_1\ln(\varrho_1/\varrho_2)$.

Given two probability measures μ_1 and μ_2 on $\mathbb{R}^d$ (or on a general metric space), the Kantorovich distance $W_p(\mu_1,\mu_2)$ of order $p\in[1,+\infty)$ is defined as the infimum of the integrals

$$\Bigl(\iint |x-y|^p\,\pi(dxdy)\Bigr)^{1/p}$$

over all probability measures π on $\mathbb{R}^d\times\mathbb{R}^d$ with projections μ_1 and μ_2 on the factors. For $p=1$ this gives the classical Kantorovich distance (sometimes mistakenly called the Wasserstein distance) and for $p=2$ we obtain the so-called quadratic Kantorovich distance W_2; see Bogachev, Kolesnikov [**148**] or Villani [**918**].

Given a locally bounded Borel vector field b on $\mathbb{R}^d$, we can consider the operator $L_b = L_{\mathrm{I},b}$ and the corresponding elliptic equation

$$L_b^*\mu = 0. \tag{4.5.13}$$

Suppose now that we are given two Borel probability measures μ and ν satisfying the equations $L_{b_\mu}^*\mu = 0$ and $L_{b_\nu}^*\nu = 0$ with certain locally bounded Borel vector fields b_μ and b_ν. We would like to estimate the quadratic Kantorovich distance between these measures via a suitable distance between b_μ and b_ν. Since the coefficients b_μ and b_ν are locally bounded, the measures μ and ν have continuous positive densities ϱ_μ and ϱ_ν with respect to Lebesgue measure, therefore, we can assume that

$$\nu = v\cdot\mu,$$

where $v = \varrho_\nu/\varrho_\mu$ is a continuous positive function.

The main result of this subsection gives the following estimate under broad conditions on b_μ and b_ν:

$$\int_{\mathbb{R}^d} \frac{|\nabla v|^2}{v}\, d\mu \leqslant \int_{\mathbb{R}^d} |b_\mu - b_\nu|^2\, d\nu. \tag{4.5.14}$$

This estimate resembles the inequality

$$\int_{\mathbb{R}^d} \frac{|\nabla \varrho_\nu|^2}{\varrho_\nu^2} d\nu \leqslant \int_{\mathbb{R}^d} |b_\nu|^2 d\nu,$$

established in Section 3.1 and meaning that ϱ_ν has finite Fisher's information number. The latter inequality can be formally obtained from the former one if for μ we take Lebesgue measure with $b_\mu = 0$.

Suppose now that the probability measures μ and ν have finite second moments, i.e., $|x| \in L^2(\mu + \nu)$. In the manifold case, the corresponding condition is that all Lipschitzian functions belong to $L^2(\mu+\nu)$. If it is given in addition that the measure μ satisfies the logarithmic Sobolev inequality

$$\mathrm{Ent}_\mu(f^2) := \int_{\mathbb{R}^d} f^2 \ln f^2\, d\mu - \int_{\mathbb{R}^d} f^2\, d\mu \ln \int_{\mathbb{R}^d} f^2\, d\mu \leqslant C \int_{\mathbb{R}^d} |\nabla f|^2\, d\mu$$

for all $f \in W^{2,1}(\mu)$ (equivalently, for all $f \in C_b^1(\mathbb{R}^d)$), then, as is well-known (see Otto, Villani [**758**], Villani [**918**, Section 9.3], [**919**, Chapters 21, 22] or Bogachev, Kolesnikov [**148**, Theorem 3.3.2]), the probability measure $\nu = v \cdot \mu$ satisfies the so-called transport inequality

$$W_2^2(\mu, v\mu) \leqslant C\mathrm{Ent}_\mu v = C \int_{\mathbb{R}^d} v \ln v\, d\mu, \tag{4.5.15}$$

hence, applying the logarithmic Sobolev inequality with $f = \sqrt{v}$, we arrive at the estimate

$$W_2^2(\mu, \nu) \leqslant 4^{-1} C^2 \int_{\mathbb{R}^d} |b_\mu - b_\nu|^2\, d\nu. \tag{4.5.16}$$

There is a vast literature on the logarithmic Sobolev inequality, see references in the comments to Chapter 5.

If in place of (4.5.15) we apply the Pinsker–Kullback–Csiszár inequality (see Bogachev [**125**, Theorem 2.12.24]) for the total variation distance

$$\|\mu - \nu\|^2 \leqslant 2\mathrm{Ent}_\mu v = 2H(\nu|\mu) \tag{4.5.17}$$

or its weighted generalizations obtained in Bolley, Villani [**194**], then we arrive at the estimate

$$\|\mu - \nu\|^2 \leqslant \frac{C}{2} \int_{\mathbb{R}^d} |b_\mu - b_\nu|^2\, d\nu. \tag{4.5.18}$$

It is known that entropy bounds yield many other bounds, see, e.g., Gozlan [**419**], Villani [**918**], [**919**] and Wang [**931**], where many additional references can be found.

Applications of estimate (4.5.16) to nonlinear stationary Fokker–Planck–Kolmogorov equations will be given below.

Let $W^{2,1}(\mu)$ denote the weighted Sobolev class with respect to the measure μ obtained by completing the class $C_0^\infty(\mathbb{R}^d)$ of smooth functions with compact support with respect to the weighted Sobolev norm given by the equality

$$\|\varphi\|_{2,1,\mu}^2 = \int_{\mathbb{R}^d} |\varphi|^2 \, d\mu + \int_{\mathbb{R}^d} |\nabla\varphi|^2 \, d\mu.$$

4.5.8. Theorem. *Let μ and ν be two probability solutions to equation* (4.5.13) *with locally bounded Borel coefficients b_μ and b_ν, respectively. Suppose that*

$$|b_\mu - b_\nu| \in L^2(\nu)$$

and that at least one of the following two conditions is fulfilled:

(i) $(1+|x|)^{-1}|b_\mu(x)| \in L^1(\nu)$,

(ii) *there exists a function $V \in C^2(\mathbb{R}^d)$ such that $L_{b_\mu}V(x) \leqslant MV(x)$ for all x and some $M > 0$ and*

$$\lim_{|x|\to\infty} V(x) = +\infty, \qquad \frac{\langle b_\mu - b_\nu, \nabla V\rangle}{1+V} \in L^1(\nu).$$

Then, the density $v := d\nu/d\mu$ satisfies the estimate

$$\int_{\mathbb{R}^d} \frac{|\nabla v|^2}{v} \, d\mu \leqslant \int_{\mathbb{R}^d} |b_\mu - b_\nu|^2 \, d\nu,$$

which yields, in particular, the inclusion $\sqrt{v} \in W^{2,1}(\mu)$. In case (ii), *a similar assertion is true for a smooth Riemannian manifold in place of $\mathbb{R}^d$, provided that the condition that $\lim\limits_{|x|\to\infty} V(x) = +\infty$ is replaced by the requirement that the sets $\{V \leqslant R\}$ are compact.*

Note that condition (i) is satisfied if we have the bound $|b_\mu(x)| \leqslant C_1 + C_2|x|$ or if $|b_\mu(x)| \leqslant C_1 + C_2|x|^k$ and the function $|x|^{k-1}$ is ν-integrable. If ν has all moments, then any polynomial bound on $|b_\mu|$ is sufficient.

4.5.9. Corollary. *Suppose that in addition to the hypotheses of Theorem* 4.5.8 *it is given that the solutions μ and ν have second moments and the measure μ satisfies the logarithmic Sobolev inequality with constant C. Then*

$$W_2(\mu,\nu)^2 \leqslant \frac{C^2}{4} \int_{\mathbb{R}^d} |b_\mu - b_\nu|^2 \, d\nu, \qquad \|\mu - \nu\|^2 \leqslant \frac{C}{2} \int_{\mathbb{R}^d} |b_\mu - b_\nu|^2 \, d\nu.$$

In case (ii) *of the theorem, these estimates hold on a Riemannian manifold.*

Let us consider an important partial case illustrating the last corollary.

Let b be a Borel locally bounded vector field on $\mathbb{R}^d$ satisfying the following condition:

(H$_\kappa$) $\langle b(x) - b(y), x - y\rangle \leqslant -\kappa|x-y|^2$ for some $\kappa > 0$ and all $x, y \in \mathbb{R}^d$.

It is known (see Theorem 5.6.36 or Röckner, Wang [**816**]) that condition (H$_\kappa$) ensures that every probability solution μ to the equation $L_{b_\mu}^*\mu = 0$ satisfies the logarithmic Sobolev inequality with constant $2/\kappa$.

Moreover, in this case $\langle b(x), x\rangle \to -\infty$ as $|x| \to \infty$ and condition (ii) of Theorem 4.5.8 is fulfilled with $V(x) = |x|^2$. In addition, the equation $L_b^*\mu = 0$ has the unique probability solution μ and $|x|^2 \in L^1(\mu)$.

Thus, we arrive at the following assertion.

4.5.10. Corollary. *Let b_μ and b_ν be locally bounded Borel vector fields satisfying condition* (H_κ) *and let μ and ν be the corresponding probability solutions to equation* (4.5.13) *such that $|b_\mu - b_\nu| \in L^2(\mu+\nu)$. Then, for every number $\gamma \in [0,1]$, one has the estimate*

$$W_2(\mu,\nu) \leqslant \kappa^{-1}\|b_\mu - b_\nu\|_{L^2(\mu_\gamma)},$$

where $\mu_\gamma = \gamma\mu + (1-\gamma)\nu$.

4.5.11. Remark. If (H_κ) is fulfilled just for b_μ and $|b_\mu - b_\nu| \in L^2(\nu)$, then

$$W_2(\mu,\nu) \leqslant \kappa^{-1}\|b_\mu - b_\nu\|_{L^2(\nu)}.$$

We now apply Corollary 4.5.10 to a nonlinear stationary Fokker–Planck–Kolmogorov equation discussed in § 2.5(iv).

Let $\mathcal{P}_2(\mathbb{R}^d)$ denote the space of all Borel probability measures on $\mathbb{R}^d$ with finite second moment. Suppose that for every measure $\mu \in \mathcal{P}_2(\mathbb{R}^d)$ we are given a locally bounded Borel vector field $b(\cdot,\mu)$ on $\mathbb{R}^d$. Set

$$L_\mu u(x) = \Delta u(x) + \langle b(x,\mu), \nabla u(x)\rangle, \quad u \in C^2(\mathbb{R}^d).$$

Then, as in § 2.5(iv), we can consider the nonlinear equation

$$L_\mu^*\mu = 0. \tag{4.5.19}$$

4.5.12. Corollary. *Suppose that for every measure $\mu \in \mathcal{P}_2(\mathbb{R}^d)$ the vector field $b(\cdot,\mu)$ satisfies condition* (H_κ) *with a common (for all measures μ) constant $\kappa > 0$. Assume also that there exists a number $C > 0$ such that*

$$|b(x,\mu) - b(x,\nu)| \leqslant CW_2(\mu,\nu) \quad \textit{for all } x \in \mathbb{R}^d,\ \mu,\nu \in \mathcal{P}_2(\mathbb{R}^d).$$

If $C < \kappa$, then there exists a unique solution to equation (4.5.19) *in $\mathcal{P}_2(\mathbb{R}^d)$.*

4.5.13. Corollary. *Suppose that for every measure $\mu \in \mathcal{P}_2(\mathbb{R}^d)$ the vector field $b(x,\mu)$ satisfies condition* (H_κ) *with a common (for all μ) constant $\kappa > 0$. Suppose that there exist a number $C > 0$ and a function $\Psi > 0$ with $\lim\limits_{|x|\to\infty} |x|^{-2}\Psi(x) = +\infty$ such that for every measure μ there is a Lyapunov function V_μ for which*

$$L_\mu V_\mu \leqslant C - \Psi.$$

Suppose also that if a sequence $\{\mu_n\}$ of measures in the set

$$K = \Bigl\{\mu \in \mathcal{P}_2(\mathbb{R}^d)\colon \int_{\mathbb{R}^d} \Psi\, d\mu \leqslant C\Bigr\}$$

converges to a measure μ with respect to the metric W_2, then the fields $b(\cdot,\mu_n)$ converge to $b(\cdot,\mu)$ in $L^2(\mu)$. Then equation (4.5.19) *has a solution in $\mathcal{P}_2(\mathbb{R}^d)$.*

4.5.14. Corollary. *Suppose that for every measure $\mu \in \mathcal{P}_2(\mathbb{R}^d)$ the vector field $b(\cdot,\mu)$ satisfies condition* (H_κ) *with a constant $\kappa(\mu) > 0$. Assume also that there exists a positive Borel function Q on $\mathbb{R}^d$ such that*

$$|b(x,\mu) - b(x,\sigma)| \leqslant Q(x)W_2(\mu,\sigma)$$

for all $x \in \mathbb{R}^d$, $\mu,\sigma \in \mathcal{P}_2(\mathbb{R}^d)$. If $\mu,\sigma \in \mathcal{P}_2(\mathbb{R}^d)$ satisfy equation (4.5.19) *and $\|Q\|_{L^2(\sigma)} < \kappa(\mu)$, then $\mu = \sigma$.*

Let us consider some examples illustrating these corollaries.

4.5.15. Example. Let

$$b(x,\mu) = -kx - \nabla U(x) - \int_{\mathbb{R}^d} \nabla W(x-y)\,\mu(dy),$$

where U and W are continuously differentiable convex functions such that

$$|\nabla W(z_1) - \nabla W(z_2)| \leqslant C|z_1 - z_2|.$$

It is clear that b satisfies the hypotheses of Corollary 4.5.12. If $k > C$, then the corresponding nonlinear equation possesses a unique probability solution.

Let us now give an example of nonuniqueness that exhibits a substantial difference between the nonlinear and linear cases.

4.5.16. Example. Let $d = 1$ and $b(x,\mu) = -x + B(\mu)$, where the vector $B(\mu)$ is the mean of the measure μ, i.e.,

$$B(\mu) = \int_{\mathbb{R}^1} x\,\mu(dx).$$

Any Gaussian measure μ_a with density

$$\varrho_a(x) = (2\pi)^{-1/2}\exp\bigl(-|x-a|^2/2\bigr)$$

is a solution to our nonlinear equation $L_\mu^*\mu = 0$. Indeed, $B(\mu_a) = a$ and

$$b(x,\mu_a) = \frac{\nabla\varrho_a(x)}{\varrho_a(x)} = -(x-a).$$

We observe that condition (H_κ) for such b is fulfilled with $\kappa = 1$, but not with $\kappa > 1$. Let us verify the second condition in Corollary 4.5.12:

$$|b(x,\mu) - b(x,\sigma)| \leqslant \iint_{\mathbb{R}^{2d}} |x-y|\,\pi(dxdy) \leqslant \biggl(\iint_{\mathbb{R}^{2d}} |x-y|^2\,\pi(dxdy)\biggr)^{1/2},$$

where π is an arbitrary probability measure on $\mathbb{R}^d \times \mathbb{R}^d$ having projections μ and σ on the factors. Since π is arbitrary, we arrive at the required estimate with constant $C = 1$. On the other hand, if $\kappa > 1$, then, by Corollary 4.5.12, the equation with the drift $b(x,\mu) = -\kappa x + B(\mu)$ has a unique solution given by density

$$\kappa^{1/2}(2\pi)^{-1/2}\exp\bigl(-\kappa|x|^2/2\bigr).$$

Thus, the condition $C < \kappa$ in Corollary 4.5.12 is sharp.

Comments

The problem of uniqueness of solutions to stationary Fokker–Planck equations was posed by A.N. Kolmogorov and for a long time it was studied as the uniqueness problem for stationary distributions of diffusions, since under standard assumptions about the coefficients these problems are equivalent. However, in the next chapter we shall see that even for smooth coefficients there is no such equivalence without additional conditions, moreover, the uniqueness of the stationary probability distribution of the diffusion holds under assumptions considerably more general than those ensuring the uniqueness of the probability solution of the corresponding elliptic equations. The first example of nonuniqueness in the case of the unit diffusion coefficient and a smooth drift on the whole space was constructed in the papers Bogachev, Röckner, Stannat [**178**], [**177**]. The phenomenon of nonuniqueness was thoroughly investigated by S.V. Shaposhnikov [**847**], [**845**], who obtained the results presented in § 4.2.

In § 4.3–4.5 we present the results of the papers Bogachev, Kirillov, Shaposhnikov [**143**]–[**146**], Bogachev, Röckner, Shaposhnikov [**171**], and also Bogachev, Röckner, Shaposhnikov [**169**]. Apart the open question about possible dimensions of the simplex of all probability solutions of the equation on the whole space with unit diffusion matrix and a smooth drift, it is of interest to describe the set of all solutions in the case of nonuniqueness. It is not clear whether all solutions can be obtained in the form of limits of solutions of boundary value problems on increasing balls with constant boundary values (or in some other unified manner). So far uniqueness in the case of coefficients of low regularity has not been studied in depth.

The problems for gradient drifts discussed in § 4.5 partly go back to Schrödinger [**837**] and Kolmogorov [**529**], although, certainly, the gradient case is very common in physical applications. There will be remarks about this case in the next chapter in § 5.6(i) and in the comments.

The discussed uniqueness problems and examples of nonuniqueness differ from analogous problems and examples for bounded domains (see Zhikov [**962**], Nadirashvili [**730**], Safonov [**832**]); however, in spite of all obvious distinctions, some similarities appear.

Exercises

4.5.17. Let $b(x) = -2x - 6e^{x^2}$, $x \in \mathbb{R}$ and let $L = L_{1,b}$. Show that the measure $\mu = \pi^{-1/2}e^{-x^2}\,dx$ is a probability solution to the equation $L^*\mu = 0$, the range of $L - I$ is not dense in $L^1(\mu)$, but there are no signed solutions and μ is the only probability solution.

HINT: Let

$$u(x) = \int_{-\infty}^{x} e^{-s^2}\,ds;$$

then $L_{1,\widehat{b}}u \geqslant u$, where $\widehat{b}(x) = -2x + 6e^{x^2}$; by Exercise 5.6.49 the measure μ is not invariant for the semigroup $\{T_t^\mu\}_{t\geqslant 0}$, hence the range of $L - I$ is not dense.

4.5.18. Let b be a locally bounded Borel vector field on $\mathbb{R}^d$ and $b(-x) = -b(x)$. Suppose that the equation $L_{I,b}^*\mu = 0$ has a unique probability solution μ. Prove that μ possesses a centrally symmetric density.

HINT: Show that the image of the measure μ under the mapping $x \mapsto -x$ satisfies the same equation.

4.5.19. Let b be a locally bounded Borel vector field on $\mathbb{R}^d$ such that for every orthogonal operator U we have $Ub(x) = b(Ux)$. Suppose that the equation $L_{I,b}^*\mu = 0$ has a unique probability solution μ. Prove that μ possesses a spherically symmetric density.

Apply this assertion to $b(x) = f(|x|)x$, where f is a locally bounded Borel function such that $f(|x|)|x|^2 \to -\infty$ as $|x| \to \infty$.

HINT: Show that for every orthogonal operator U the measure $\mu\circ U^{-1}$ satisfies the same equation.

4.5.20. Define $\text{ext}\mathcal{M}_{\text{ell}}^{A,b}$ as the set of all measures $\mu \in \mathcal{M}_{\text{ell}}^{A,b}$ on a domain $\Omega \subset \mathbb{R}^d$ which cannot be written as a nontrivial convex combination of two other elements in $\mathcal{M}_{\text{ell}}^{A,b}$. Let $\mu \in \mathcal{M}_{\text{ell}}^{A,b}$. Then the following assertions are equivalent:

(i) $\mu \in \text{ext}\mathcal{M}_{\text{ell}}^{A,b}$; (ii) if $\varrho \in L^\infty(\mu)$ and $\varrho \cdot \mu \in \mathcal{M}_{\text{ell}}^{A,b}$, then $\varrho = 1$ μ-a.e.

HINT: Suppose that (i) is fulfilled and a function $\varrho \in L^\infty(\mu)$ is such that $\varrho \cdot \mu \in \mathcal{M}_{\text{ell}}^{A,b}$. Then for $M := \|\varrho\|_\infty + 1$ we have

$$\mu_1 := \frac{M-\varrho}{M-1} \cdot \mu \in \mathcal{M}_{\text{ell}}^{A,b}, \ \mu = \frac{M-1}{M}\mu_1 + \frac{1}{M}(\varrho \cdot \mu).$$

Hence $\varrho = 1$ μ-a.e. Let (ii) be fulfilled and the measure $\mu_1, \mu_2 \in \mathcal{M}_{\text{ell}}^{A,b}$ and numbers $\alpha \in (0,1)$ be such that $\mu = \alpha\mu_1 + (1-\alpha)\mu_2$. Then both measures μ_1 and μ_2 are absolutely continuous with respect to μ and have bounded Radon–Nikodym densities. Hence $\mu_1 = \mu = \mu_2$.

4.5.21. Let in the previous exercise our usual conditions $a^{ij} \in C(\Omega) \cap W_{\text{loc}}^{d+,1}(\Omega)$, $\det A(x) \neq 0$, $b^i \in L_{\text{loc}}^{d+}(\Omega)$ be fulfilled. Let $\mu \in \text{ext}\mathcal{M}_{\text{ell}}^{A,b}$. Prove that for all $\nu \in \mathcal{M}_{\text{ell}}^{A,b} \setminus \{\mu\}$ the function $d\nu/d\mu$ is unbounded.

4.5.22. (Bogachev, Kirillov, Shaposhnikov [**144**]) Let $b \in L_{\text{loc}}^1(\mathbb{R}^1)$. Prove that the existence of a nonzero solution $\varrho \in L^1(\mathbb{R}^1)$ of the equation $\varrho'' - (b\varrho)' = 0$ implies the existence of a probability solution.

HINT: We have $\varrho' - b\varrho = C$, where C is a constant, whence

$$\varrho(x) = \biggl(\varrho(0) + C\int_0^x \frac{dt}{f(t)}\biggr) f(x), \quad f(x) = \exp\biggl(\int_0^x b(s)\,ds\biggr).$$

If there are two linearly independent integrable solutions, then f must be a positive integrable solution. Suppose that the function ϱ is signed and there is no integrable solution linearly independent with it. One can assume that $C = -1$ and the function f is not integrable at $+\infty$. Introducing the function $g(x)$ that is the integral of $1/f(t)$ over $[0,x]$, we obtain $\varrho(x) = \bigl(\varrho(0) - g(x)\bigr)/g'(x)$, where $g'(x) = 1/f(x) > 0$. Since $g(x)$ increases as $x \to +\infty$ and $1/g'(x)$ is not integrable at $+\infty$, we see that $g(x)$ increases to $\varrho(0)$, i.e., the function ϱ must be positive.

4.5.23. Let $f \in L^p(0,+\infty)$, $p \in (1,+\infty)$, and

$$F(x) = \frac{1}{x}\int_0^x f(y)\,dy.$$

Prove the following Hardy inequality:

$$\int_0^\infty |F(x)|^p\,dx \leqslant \frac{p^p}{(p-1)^p}\int_0^{+\infty} |f(x)|^p\,dx.$$

HINT: For bounded $f \geqslant 0$ with bounded support, let $f_1(x) = xF(x)$ and, integrating by parts, write the left-hand side of the Hardy inequality as

$$\int_0^\infty f_1(x)^p x^{-p}\,dx = \frac{p}{p-1}\int_0^\infty f_1(x)^{p-1} f(x) x^{1-p}\,dx = \frac{p}{p-1}\int_0^\infty F(x)^{p-1} f(x)\,dx,$$

which by Hölder's inequality is dominated by $p(p-1)^{-1}\|f\|_p\|F\|_p^{p-1}$.

4.5.24. (i) Let $f \in L^p[0,1]$, $p \in (1,+\infty)$, and

$$F(x) = \int_0^x f(y)\,dy.$$

Prove the following Hardy inequality:

$$\int_0^1 \frac{|F(x)|^p}{x^p}\,dx \leqslant \frac{p^p}{(p-1)^p}\int_0^1 |f(x)|^p\,dx.$$

(ii) Suppose that $f \in W_0^{p,1}(B)$, where B is the unit ball in $\mathbb{R}^d$ and $p > 1$. Prove the following Hardy inequality:

$$\int_B \frac{|f(x)|^p}{(1-|x|)^p}\,dx \leqslant \frac{p^p}{(p-1)^p}\int_B |\nabla f(x)|^p\,dx.$$

HINT: (i) For smooth f vanishing near zero observe that $F(x)/x$ equals the integral of $f(xt)$ over $[0,1]$, use the change of variable $s = xt$ and the previous exercise; (ii) Consider $f \in C_0^\infty(B)$, use spherical coordinates and (i).

4.5.25. Show that the construction of Example 4.2.1 can be performed in such a way that, for any given number C, one can obtain two probability densities ϱ_1 and ϱ_2 on $\mathbb{R}^2$ satisfying the same stationary equation with $A = \mathrm{I}$ and smooth b such that $\varrho_1(0) > C\varrho_2(0)$.

HINT: Observe that the function v Example 4.2.1 is not separated from zero, hence the ratio of any two solutions to the same equation cannot be bounded by a universal constant, which leads to the same affect at the origin, because a translation does not change the Laplacian.

4.5.26. Let μ and ν be two probability measures on $\mathbb{R}$ satisfying the equations $\mu'' - (b\mu)' = 0$ and $\nu'' - (c\nu)' = 0$, where $b, c \in C^\infty(\mathbb{R})$. Then the measure $\sigma = \mu\otimes\nu$ on $\mathbb{R}^2$ satisfies the equation $\Delta\sigma - \operatorname{div}(B\sigma) = 0$, where $B(x,y) = \bigl(b(x), c(y)\bigr)$, see Exercise 1.8.8. Construct smooth functions b and c such that the corresponding equation on $\mathbb{R}^2$ with $B(x,y) = \bigl(b(x), c(y)\bigr)$ has infinitely many probability solutions.

HINT: Let ϱ_1 and ϱ_2 be two smooth positive probability densities on $\mathbb{R}$; set $b = \varrho_1'/\varrho_1$ and $c = (\varrho_2' - 1)/\varrho_2$ and apply the method described in § 4.2 with the constant vector field $a = (0, -1)$.

4.5.27. In the situation of the previous exercise show that if $b = \varrho_1'/\varrho_1$ and $c = (\varrho_2' + 1)/\varrho_2$, then the solution $\mu\otimes\nu$, where $\mu = \varrho_1\,dx$ and $\nu = \varrho_2\,dy$, is the only probability solution of the corresponding equation.

CHAPTER 5

Associated Semigroups

Here we discuss semigroups associated with solutions to our elliptic equations for measures, along with relations between such equations and the proper invariance of measures with respect to their associated semigroups. It turns out that the latter is strongly connected with uniqueness of solutions to the elliptic equation.

The principal results of this chapter (see Theorem 5.2.2 and Theorem 5.3.1) say that (under rather broad local assumptions about the coefficients A and b) every probability solution μ to the equation $L^*_{A,b}\mu = 0$ produces some special strongly continuous semigroup of operators T^μ_t on $L^1(\mu)$ whose generator $L^\mu_{A,b}$ extends $L_{A,b}$ on C_0^∞ and has the property that $(I - L^\mu_{A,b})^{-1} f$ for $f \in C_0^\infty$ is a limit of solutions u_n to the equations $(I - L_{A,b})u_n = f$ on increasing balls B_n with zero boundary condition. The measure μ is subinvariant for this semigroup, but not always invariant. Its invariance is equivalent to the density of $(I - L_{A,b})(C_0^\infty)$ in $L^1(\mu)$ and in this case μ is a unique solution to the equation (but the converse is not true). A sufficient condition for invariance is the existence of a Lyapunov function V (a C^2-function V such that the sets $\{V \leqslant R\}$ are compact) with $L_{A,b}V \leqslant CV$. For $A = \mathrm{I}$ it is enough to have a bound $|b(x)| \leqslant c + c|x| \ln |x|$ (but $\ln |x|$ cannot be replaced by $(\ln |x|)^{1+\varepsilon}$). On the other hand, Proposition 5.3.9 shows that under broad assumptions (although not always) the existence of a Lyapunov function follows from the existence of a probability solution (which then necessarily will be unique).

5.1. Generalities on semigroups

Let us recall some general concepts related to operator semigroups. Let E be a Banach space. A family $\{T_t\}_{t\geqslant 0}$ of bounded linear operators on E is called an *operator semigroup* if

$$T_0 = I, \quad T_{t+s} = T_t T_s, \quad t, s \geqslant 0.$$

Such a family is called a *strongly continuous operator semigroup* (another term: C_0-semigroup) if

$$\lim_{t\to 0} \|T_t x - x\| = 0 \quad \forall x \in E.$$

If $\|T_t\| \leqslant 1$, then the semigroup is called *contractive.*

Let us consider the linear subspace

$$D(L) := \{x\colon \lim_{t\to 0} t^{-1}(T_t x - x) \text{ exists in } E\}.$$

It is known that for any strongly continuous operator semigroup the linear space $D(L)$ is dense in E. The linear operator

$$Lx := \lim_{t\to 0} t^{-1}(T_t x - x), \quad x \in D(L),$$

is called the *generator* of $\{T_t\}_{t\geqslant 0}$ and $D(L)$ is called its domain.

If $T_t = \exp(tA)$, where A is a bounded linear operator on E, then $D(L) = E$ and $L = A$. However, for typical operator semigroups arising in applications the domain $D(L)$ of generator is smaller than E and L is not continuous on it. Moreover, usually it is difficult or even impossible to explicitly find the domain of generator. Let us consider an interesting example where this is possible.

5.1.1. Example. Let E be a separable Hilbert space with an orthonormal basis $\{e_n\}$ and let

$$T_t x = \sum_{n=1}^{\infty} e^{-nt}(x, e_n)e_n.$$

Then

$$D(L) = \Bigl\{x \in H\colon \sum_{n=1}^{\infty} n^2|(x, e_n)|^2 < \infty\Bigr\}, \quad Lx = -\sum_{n=1}^{\infty} n(x, e_n)e_n.$$

In particular, this is true if $\{T_t\}_{t\geqslant 0}$ is the Ornstein–Uhlenbeck semigroup on the space $H = L^2(\gamma)$, where γ is the standard Gaussian measure on $\mathbb{R}^d$, given by

$$T_t f(x) = \int_{\mathbb{R}^d} f\Bigl(e^{-t}x - \sqrt{1 - e^{-2t}}\, y\Bigr)\, \gamma(dy), \tag{5.1.1}$$

$\{e_n\}$ is the basis formed by the Hermite polynomials H_n. The fact that in case $d = 1$ we have $T_t H_n = e^{-nt}H_n$, and consequently $T_t H_{n_1,\ldots,n_d} = e^{-(n_1+\cdots n_d)t}H_{n_1,\ldots,n_d}$, $H_{n_1,\ldots,n_d}(x_1,\ldots,x_d) = H_{n_1}(x_1)\cdots H_{n_d}(x_d)$ on $\mathbb{R}^d$, needs a proof (Exercise 5.6.42). The generator of the Ornstein–Uhlenbeck semigroup coincides with the Ornstein–Uhlenbeck operator on $C_0^\infty(\mathbb{R}^d)$ (Exercise 5.6.57). The calculation of $D(L)$ is straightforward, since $\|T_t x - x\|^2 = \sum_{n=1}^\infty (1 - e^{-nt})^2 |(x, e_n)|^2$. Note that the situation changes if we want to find the domain of generator of the Ornstein–Uhlenbeck semigroup on $L^p(\gamma)$ with $p \neq 2$.

The *heat semigroup* $\{P_t\}_{t\geqslant 0}$ on $L^p(\mathbb{R}^d)$, $p \in [1, +\infty]$, is defined by

$$P_t f(x) = \int_{\mathbb{R}^d} f(x - \sqrt{t}y)\, \gamma(dy).$$

The generator of $\{P_t\}_{t\geqslant 0}$ on $L^2(\mathbb{R}^d)$ is $\Delta/2$ on $W^{2,2}(\mathbb{R}^d)$ (Exercise 5.6.44).

It is important that not every explicitly defined operator semigroup is strongly continuous.

5.1.2. Example. Let us consider the same Ornstein–Uhlenbeck semigroup, but this time on the space $C_b(\mathbb{R})$ of bounded continuous functions with the norm $\|f\|_\infty = \sup_x |f(x)|$, which is mapped by the operators T_t into itself by the Lebesgue dominated convergence theorem. It is clear that the operators T_t have unit norm, but there is no continuity at zero: it is easy to find a function $f \in C_b(\mathbb{R})$ for which it is not true that $\|T_t f - f\|_\infty \to 0$ as $t \to 0$. For example, if a function f is Lipschitzian with constant 1, then $|T_t f(x) - f(e^{-t}x)| \leqslant \sqrt{1 - e^{-2t}}$, so it suffices to take such a function with the property that $\sup_x |f(x) - f(e^{-t}x)| \not\to 0$ as $t \to 0$. It is clear that the Ornstein–Uhlenbeck semigroup will not be strongly continuous also on the the space of bounded Borel (or measurable) functions with the same norm. The semigroup on $L^\infty(\mathbb{R})$ defined by the formula $T_t f(x) = f(x+t)$ is not strongly continuous, although it is strongly continuous on $L^1(\mathbb{R})$. See also Exercise 5.6.44.

This example is typical in that increasing the space, on which the semigroup acts, from bounded functions to functions integrable with respect to an invariant measure we improve the properties of the semigroups.

5.1.3. Remark. It is known (see Hille, Phillips [**454**, Theorem 10.5.5]) that in separable spaces the strong continuity of an operator semigroup with uniformly bounded norms follows by its weak measurability (which means the measurability of the functions $t \mapsto l(T_t x)$ for all $x \in E$, $l \in E^*$) combined with the property that the linear span of the elements $T_t x$, where $t > 0$, $x \in E$, is dense in E. For example, if the space $L^2(\mu)$ is separable (which is the case in all our examples), then for the strong continuity of a semigroup $\{T_t\}_{t\geqslant 0}$ of operators on $L^2(\mu)$ with uniformly bounded norms it suffices to have the measurability of the functions $t \mapsto (T_t f_i, f_j)$ for all elements of an everywhere dense sequence $\{f_i\}$ and the absence of nonzero elements g for which $(g, T_t f_i) = 0$ for all $t > 0$ and $i \in \mathbb{N}$. In this case the dual semigroup $\{T_t^*\}_{t\geqslant 0}$ will be also strongly continuous. It is also sufficient to have a strongly continuous extension of this semigroup to $L^1(\mu)$, see Exercise 5.6.43.

We recall that the closure $(\overline{L}, D(\overline{L}))$ of a linear operator (L, D) in a Banach space E is defined as follows: the graph of $\overline{L}$ on $D(\overline{L})$ is the closure of the graph of L on D, provided that this closure is the graph of a single-valued mapping.

Let $(X, \mathcal{B})$ be a measurable space and let $B_b(X)$ be the space of all bounded $\mathcal{B}$-measurable functions on X equipped with the sup-norm. If $\{T_t\}_{t\geqslant 0}$ is a semigroup of bounded linear operators on the space $B_b(X)$, then a bounded measure μ on $\mathcal{B}$ (possibly signed) is called *invariant* for $\{T_t\}_{t\geqslant 0}$ (or $\{T_t\}_{t\geqslant 0}$-invariant) if

$$\int_X T_t f \, d\mu = \int_X f \, d\mu \quad \forall f \in B_b(X). \tag{5.1.2}$$

In the case of a nonnegative measure μ similarly one defines the invariance for a semigroup on $L^1(\mu)$ or on $L^\infty(\mu)$, and then the semigroup $\{T_t\}_{\geqslant 0}$ extends naturally from $B_b(X)$ to $L^1(\mu)$, and the measure μ will be invariant also for the extension.

Such semigroups on $B_b(X)$ with nonnegative invariant measures in many cases turn out to be strongly continuous on the space $L^1(\mu)$ and not on $B_b(X)$, e.g., when they are defined by transition probabilities of random processes (some sufficient conditions are given below). Let L be the corresponding generator with domain $D(L) \subset L^1(\mu)$. Then (5.1.2) is equivalent to the equality

$$\int_X Lf \, d\mu = 0 \quad \forall f \in D(L).$$

It will be essential below that for the equivalence it is not sufficient to have this equality for all functions f in a dense set in $L^1(\mu)$, it is important to have the identity on all of $D(L)$.

Similarly one defines invariance in the case where X is a topological space, μ is a Borel measure on X, and $\{T_t\}_{t\geqslant 0}$ is a semigroup of bounded linear operators on the space $C_b(X)$ of bounded continuous functions. An obvious modification of the latter concept arises if the operators T_t are just operators from $C_b(X)$ to $L^\infty(\mu)$, not necessarily forming a semigroup.

A measure $\mu \geqslant 0$ on X is called *subinvariant* for an operator T on $L^\infty(\mu)$ (or on $B_b(X)$) if

$$\int_X Tf\,d\mu \leqslant \int_X f\,d\mu \quad \text{whenever } f \in L^\infty(\mu),\ f \geqslant 0 \text{ (or } f \in B_b(X),\ f \geqslant 0).$$

If for all bounded measurable functions f and g one has

$$\int_X T_t f(x)\, g(x)\, \mu(dx) = \int_X f(x)\, T_t g(x)\, \mu(dx),$$

then the semigroup is called *symmetric* and the measure μ is called *symmetric invariant.* It is clear that this case is characterized by the property that the generator L of the semigroup is symmetric in $L^2(\mu)$. If $\{T_t\}_{t\geqslant 0}$ is the transition semigroup of a Markov process, then this process is called *μ-symmetric.* A process is called *symmetrizable* if there is a measure μ making the process μ-symmetric.

A bounded operator T on $L^\infty(\mu)$ (or on $L^p(\mu)$) is called *sub-Markovian* if

$$0 \leqslant Tf \leqslant 1 \quad \text{whenever} \quad f \in L^\infty(\mu) \text{ and } 0 \leqslant f \leqslant 1.$$

If, in addition, $T1 = 1$, then T is called *Markovian.*

Note that if the measure μ is invariant for a Markov operator T in $L^\infty(\mu)$, then it is also invariant for the operator T^*, which arises naturally on $L^1(\mu)$ and on $L^\infty(\mu)$ (see Lemma 4.3.2, where a more general fact is established), since

$$\int_X T^* f\,d\mu = \int_X fT1\,d\mu = \int_X f\,d\mu, \quad f \in L^\infty(\mu).$$

If a bounded operator T on $L^1(\mu)$ is nonnegative in the sense that $Tf \geqslant 0$ whenever $f \geqslant 0$, then its sub-Markov property is equivalent to the condition $T1 \leqslant 1$, since in that case we have $Tf \leqslant T1 \leqslant 1$ whenever $f \leqslant 1$. The indicated condition (for nonnegative T) is equivalent to the property that the restriction of T to $L^\infty(\mu)$ is a contraction. Hence a sub-Markov operator T is a contraction on $L^\infty(\mu)$, and if the measure μ is subinvariant for it, it will be also a contraction on $L^1(\mu)$.

If we are given a contraction T on $L^1(\mu)$ such that T is a nonnegative operator, then its sub-Markov property is equivalent to the invariance of the measure μ for T^*, since the estimate $T1 \leqslant 1$ a.e. is equivalent to the property that for each bounded nonnegative measurable function g the integral of $gT1$ does not exceed the integral of the function g.

A *Markov semigroup* is a semigroup consisting of Markov operators on the space of bounded functions or on the space L^p. No strong continuity is required! For example, the Ornstein–Uhlenbeck semigroup on $C_b(\mathbb{R})$ is Markov, but not strongly continuous.

In general, the operators T_t^μ of the associated semigroup constructed in the next section are sub-Markovian, not Markovian. In addition, μ is merely subinvariant for them.

We have seen in Lemma 4.3.2 that if T is a sub-Markovian operator on $L^\infty(\mu)$ and μ is subinvariant for T, then T uniquely extends to a bounded operator on each $L^p(\mu)$. In addition, the absolute value of a function invariant for T^* is also invariant. We shall also need two other close assertions.

5.1.4. Lemma. (i) *If T is a bounded linear operator on $L^1(\mu)$ such that $T\bigl(L^\infty(\mu)\bigr) \subset L^\infty(\mu)$, then $T\bigl(L^p(\mu)\bigr) \subset L^p(\mu)$ and $T^*\bigl(L^p(\mu)\bigr) \subset L^p(\mu)$ for all numbers $p \in [1, +\infty)$.*

(ii) *If* $\{T_t\}_{t\geqslant 0}$ *is a semigroup of sub-Markovian operators on* $L^\infty(\mu)$ *such that* μ *is separable and subinvariant for each* T_t *and the integral of* gT_tf *is continuous in* t *on* $[0,+\infty)$ *for all* $g\in L^\infty(\mu)$ *and all* f *from a set dense in* $L^1(\mu)$, *then* $\{T_t\}_{t\geqslant 0}$ *uniquely extends to a strongly continuous semigroup on each* $L^p(\mu)$, $1\leqslant p<\infty$.

(iii) *If* $(X,\mathcal{B})$ *is a measurable space,* T *is a sub-Markov operator on* $B_b(X)$ *and a measure* ν *on* $\mathcal{B}$ *is invariant for* T, *then the measures* $\nu^+,\nu^-,|\nu|$ *are also invariant. The same is true if this operator is defined on* $L^\infty(\mu)$, *but the measure* ν *is absolutely continuous with respect to* μ.

PROOF. Assertion (i) follows by the interpolation theorem as in Lemma 4.3.2, and (iii) follows by Lemma 4.3.2 and the classical fact mentioned above that in separable spaces the strong continuity of operator semigroups follows by the weak measurability (which obviously holds in our case after extending to $L^1(\mu)$) combined with the property that the linear span of the elements T_tf, where $t>0$, is dense in $L^1(\mu)$ (which also obviously follows from our assumptions).

For the proof of (iii) take the Hahn–Jordan decomposition $X=X^+\cup X^-$, $\nu=\nu^+-\nu^-$ and observe that

$$\int_X TI_{X^+}\,d\nu^+ - \int_X TI_{X^+}\,d\nu^- = \int_X I_{X^+}\,d\nu = \int_X I_{X^+}\,d\nu^+,$$

whence taking into account the estimates $I_{X^+}TI_{X^+}\leqslant I_{X^+}$ and $TI_{X^+}\geqslant 0$ we obtain that $I_{X^-}TI_{X^+}=0$ a.e. Hence for every function $f\in B_b(X)$ with $0\leqslant f\leqslant 1$ we have $I_{X^-}T(fI_{X^+})=0$ a.e. Similarly, $I_{X^+}T(fI_{X^-})=0$ a.e. Therefore,

$$\int_X f\,d\nu^+ = \int_X fI_{X^+}\,d\nu \int_X T(fI_{X^+})\,d\nu = \int_X T(fI_{X^+})\,d\nu^+,$$

which proves the T-invariance of the measure ν^+, hence also of ν^-. □

5.1.5. Proposition. *Let* T *be a sub-Markovian operator on* $L^1(\mu)$, *where* μ *is a probability measure on* Ω *that is subinvariant with respect to* T. *Let* ν *be a probability measure on* Ω *equivalent to* μ. *If* ν *is invariant for* T, *then* μ *is invariant as well.*

PROOF. Let $\varrho:=d\nu/d\mu$ and let T^* be the adjoint of T. Then we obtain $T^*1\leqslant 1$ μ-a.e., since

$$\int_\Omega fT^*1\,d\mu = \int_\Omega Tf\,d\mu \leqslant \int_\Omega f\,d\mu \quad \forall\, f\in L^\infty(\mu),\ f\geqslant 0.$$

By using that $T1\leqslant 1$ μ-a.e. and $\varrho>0$ μ-a.e., we obtain from the equality

$$\int_\Omega T1\,\varrho\,d\mu = \int_\Omega \varrho\,d\mu$$

that $T1=1$ μ-a.e. It now follows that $T^*1=1$ μ-a.e., since

$$\int_\Omega T^*1\,d\mu = \int_\Omega T1\,d\mu = 1.$$

Hence μ is also invariant. □

5.1.6. Remark. Let T be a sub-Markov operator on $L^\infty(\mu)$ with a subinvariant probability measure μ on a space X. As we know, T extends to a bounded operator on $L^1(\mu)$, the dual to which is sub-Markov on $L^\infty(\mu)$.

(i) The invariance of μ for T is equivalent to the equality $T^*1=1$, and the invariance of μ for T^* is equivalent to the equality $T1=1$. Under our assumption of subinvariance of μ these two equalities are equivalent (however, without this assumption they are not equivalent, see Exercise 5.6.66). Indeed, if $T1 = 1$, then also $T^*1 = 1$, since otherwise by the inequality $T^*1 \leqslant 1$ (due to subinvariance of μ) the integral of T^*1 is strictly less than 1, but this integral equals 1, because $T1 = 1$. The same reasoning shows that $T1 = 1$ once $T^*1 = 1$. In particular, for a Markov semigroup, the invariant measure μ is also invariant for T^*.

(ii) Suppose now that μ is just subinvariant for T, but not invariant, hence is not invariant for T^*. Then $1 \notin \operatorname{Ker}(T - I)$. If in this case the kernel of $T - I$ in $L^1(\mu)$ is nontrivial, then by Lemma 4.3.2 it contains a nonzero nonnegative function $\varrho \in L^1(\mu)$. Let $\Omega = \{\varrho > 0\}$. We observe that the subspace in $L^1(\mu)$ consisting of all functions vanishing outside Ω is invariant for T, $TI_\Omega = I_\Omega$, $T^*I_\Omega = I_\Omega$, and the measure $\mu|_\Omega$ is invariant for the natural restriction of T to this subspace (which can be identified with $L^1(\mu|_\Omega)$.

Indeed, let $\Omega_n = \{\varrho \geqslant 1/n\}$. Then $TI_{\Omega_n} \leqslant nT\varrho = n\varrho$, hence $TI_{\Omega_n} = 0$ outside Ω. Letting $n \to \infty$, we obtain that $TI_\Omega = 0$ outside Ω. By the sub-Markov property, $Tf = 0$ outside Ω if $f \geqslant 0$ is bounded and $f = 0$ outside Ω. Then this is also true for any $f \in L^1(\mu)$ vanishing outside Ω. We have

$$\int_X \varrho T^* I_\Omega \, d\mu = \int_\Omega T\varrho = \int_X \varrho = 1$$

by the equality $T\varrho = \varrho$. Since the left integral is actually taken over Ω, the estimate $T^*I_\Omega \leqslant 1$ yields that $T^*I_\Omega = 1$ on Ω. Hence $T^*I_\Omega \geqslant I_\Omega$. However, we actually have the equality $T^*I_\Omega = I_\Omega$ a.e., since otherwise

$$\int_X T^* I_\Omega \, d\mu > \mu(\Omega),$$

but the left-hand side equals the integral of $I_\Omega T1 \leqslant I_\Omega$. Therefore, $I_\Omega T^* I_\Omega = I_\Omega$. Integrating this equality, we obtain that the integral of TI_Ω over Ω equals $\mu(\Omega)$. By the estimate $TI_\Omega \leqslant 1$, this is only possible if $TI_\Omega = 1$ on Ω. Then $TI_\Omega = I_\Omega$, because $TI_\Omega = 0$ outside Ω, as shown above. The equality $T^*I_\Omega = I_\Omega$ implies that the measure $\mu|_\Omega$ is invariant for the natural restriction of T to $L^1(\mu|_\Omega)$. Let us observe that $T1|_\Omega = 1$.

5.1.7. Definition. A linear operator L on a dense domain $\mathcal{D}$ in a Banach space X is called *dissipative* if, for every $u \in \mathcal{D}$, there exists $l_u \in X^*$ such that

$$\|l_u\|_{X^*} = \|u\|_X, \ l_u(u) = \|u\|_X^2, \ l_u(Lu) \leqslant 0.$$

A dissipative operator L is called *essentially m-dissipative* if it satisfies the following additional condition:

$$\overline{(L - \lambda I)(\mathcal{D})} = X \quad \forall \lambda > 0, \tag{5.1.3}$$

where $\overline{E}$ denotes the closure of E.

In fact, it suffices that (5.1.3) be true for *some* $\lambda > 0$.

If $E = L^p(\mu)$, where μ is a nonnegative measure and $p \in [1, +\infty)$, then a linear operator L on a dense domain $\mathcal{D}$ is dissipative precisely when

$$\int \operatorname{sign} u \, |u|^{p-1} Lu \, d\mu \leqslant 0 \quad \forall u \in \mathcal{D},$$

where $\operatorname{sign} u = 1$ if $u > 0$, $\operatorname{sign} u = 0$ if $u = 0$, $\operatorname{sign} u = -1$ if $u < 0$. The proof is delegated to Exercise 5.6.45. By using this criterion one readily proves (Exercise 5.6.46) the following sufficient condition.

5.1.8. Example. Let $\lambda \geqslant 0$ and let L be a densely defined operator on the space $L^p(\mu)$. Then $L - \lambda I$ is dissipative provided that

$$\int \psi(u) Lu \, d\mu \leqslant \lambda \int |u| \, d\mu \quad \forall\, u \in \mathcal{D}.$$

for each smooth increasing function ψ on the real line such that $\psi(0) = 0$ and $|\psi(s)| \leqslant |s|^{p-1}$ for all s.

5.1.9. Example. Let μ be a nonnegative measure on an open set Ω in $\mathbb{R}^d$ satisfying the equation $L^*_{A,b}\mu = 0$, where the functions a^{ij}, b^i are locally μ-integrable. Then $L_{A,b}$ with domain $C_0^\infty(\Omega)$ is dissipative on $L^p(\mu)$, $1 \leqslant p < \infty$. The same is true for the divergence form operator $\mathcal{L}_{A,b}$ in the manifold case (if in local charts the coefficients a^{ij} are locally Sobolev and their gradients are locally μ-integrable).

Indeed, given a smooth increasing function ψ on $\mathbb{R}$ with $\psi(0) = 0$ and letting

$$\Psi(t) = \int_0^t \psi(s) \, ds,$$

for every $u \in C_0^\infty(\Omega)$, we see that the function $\Psi(u)$ is in $C_0^\infty(\Omega)$. Hence the integral of $L\Psi(u)$ against μ vanishes. Since

$$L\Psi(u) = \psi(u) Lu + \psi'(u)\langle A\nabla u, \nabla u\rangle$$

and the second term is nonnegative, the integral of $\psi(u)Lu$ against μ is nonpositive.

5.1.10. Lemma. *A dissipative operator L is closable and its closure $\overline{L}$ is also dissipative.*

The proof of this lemma is left as Exercise 5.6.47.

We shall use the following Lumer–Phillips theorem (see Pazy [**767**, Chapter I, Theorem 4.3]).

5.1.11. Theorem. *A dissipative operator on a Banach space is essentially m-dissipative precisely when its closure coincides with the generator of a contractive semigroup.*

A useful concept related to semigroups is that of resolvent. Given a closed densely defined operator L on a Banach space E (real or complex), we consider the set $\varrho(L)$ of all numbers λ in the corresponding field such that there is a bounded inverse operator $R(\lambda, L) := (\lambda I - L)^{-1}$. Then we have Hilbert's identity

$$R(\lambda, L) - R(\mu, L) = (\mu - \lambda) R(\lambda, L) R(\mu, L)$$

for all $\lambda, \mu \in \varrho(L)$, which is verified directly. For the proof of the following result, see Pazy [**767**, Section 1.9].

5.1.12. Theorem. *Suppose that there is an unbounded set Λ of numbers and for each $\lambda \in \Lambda$ a bounded operator $R(\lambda)$ is given such that*

$$R(\lambda) - R(\mu) = (\mu - \lambda) R(\lambda) R(\mu), \quad \lambda, \mu \in \Lambda. \tag{5.1.4}$$

Assume also that there exist a sequence $\lambda_n \in \Lambda$ and a dense set $E_0 \subset E$ such that

$$\lim_{n\to\infty} |\lambda_n| = \infty, \ \sup_n \|R(\lambda_n)\| < \infty, \ \lim_{n\to\infty} \|\lambda_n R(\lambda_n)x - x\| = 0.$$

Then there exists a unique closed densely defined operator L such that $R(\lambda) = R(\lambda, L)$ for all $\lambda \in \Lambda$.

5.1.13. Corollary. *If for each $\lambda > 0$ there exists a bounded operator $R(\lambda)$ satisfying* (5.1.4), $\|\lambda R(\lambda)\| \leqslant 1$ *and* $\lim\limits_{\lambda\to\infty} \|\lambda R(\lambda)x - x\| = 0$ *on a dense set, then there exists a unique contractive C_0-semigroup $\{T_t\}_{t\geqslant 0}$ with generator L such that $R(\lambda) = R(\lambda, L)$ for all $\lambda > 0$.*

Note also the following useful relations between a contracting semigroup $\{T_t\}_{t\geqslant 0}$ and the resolvent $(I - \lambda L)^{-1}$ of its generator L:

$$(I - \lambda L)^{-1}x = \int_0^\infty e^{-\lambda t} T_t x\, dt, \quad \lambda > 0, \tag{5.1.5}$$

$$T_t x = \lim_{n\to\infty} \bigl((I - tn^{-1}L)^{-1}\bigr)^n x. \tag{5.1.6}$$

We are interested in the case where the semigroup $\{T_t\}_{t\geqslant 0}$ is in a certain sense generated by the elliptic operator $L_{A,b}$ for which there exists a probability measure μ satisfying the equation $L^*_{A,b}\mu = 0$. However, one should be very careful with a possible interpretation of the term "generated". We shall use instead the term "associated" and this will simply mean the following: the semigroup $\{T_t\}_{t\geqslant 0}$ on the space of bounded Borel functions will be called associated with $L_{A,b}$ and μ if it extends to a strongly continuous semigroup on $L^1(\mu)$ whose generator coincides with $L_{A,b}$ on the class of smooth compactly supported functions. Typically, such a semigroup is not continuous on B_b or C_b. Moreover, in general it is not unique (we shall see such examples; however, a certain special associated semigroup will be singled out), and this is closely connected with the nonuniqueness of solutions to the equation $L^*_{A,b}\mu = 0$ in the class of probability measures. On the other hand, we shall find conditions that guarantee the uniqueness of associated semigroups, and this uniqueness turns out to be equivalent to the proper invariance of μ with respect to its special associated semigroup in the case where the drift coefficient b is locally Lebesgue integrable to a power greater than the dimension of the space; moreover, under these conditions (but not always), the semigroup turns out to be indeed generated by $L_{A,b}$ in the classical sense that the closure of $(L_{A,b}, C_0^\infty)$ in $L^1(\mu)$ is the generator of the semigroup. By the Lumer–Phillips theorem formulated above, this is the case precisely when the range of $L_{A,b} - I$ is dense in $L^1(\mu)$; diverse sufficient conditions for that will be presented.

5.2. Associated semigroups

We now turn to the investigation of semigroups arising in the space $L^1(\mu)$ for a probability solution μ of the equation $L^*_{A,b}\mu = 0$ and associated with the operator $L_{A,b}$ in the sense that the generator of the semigroup extends the given operator. As we shall see, such a semigroup is not unique, but there is a certain canonical semigroup. Here there is a subtle point: the closure of the operator $L_{A,b}$ is not always the generator of a semigroup (which is the reason for a possible nonuniqueness of semigroups), so it is not always possible to speak about "generated semigroups" in the usual sense. However, there are effectively verified conditions (in terms of the operator coefficients) under which the closure of $L_{A,b}$ generates a semigroup. We first consider operators satisfying our standard local assumptions and then turn to the case where $A = \mathrm{I}$ and a drift b is such that $|b| \in L^2(\mu)$, but without additional local assumptions.

Let us fix an open subset Ω in $\mathbb{R}^d$ (or in a Riemannian manifold M of dimension d), which may coincide with the whole space.

Throughout this section we assume that A and b satisfy the following conditions:

(5.2.1) $$a^{ij} \in C(\Omega) \cap W^{d+,1}_{\text{loc}}(\Omega), \ \det A(x) > 0, \quad b^i \in L^{d+}_{\text{loc}}(\Omega).$$

Recall that $d+$ means the existence of a respective exponent $p_U > d$ for every ball $U \subset \Omega$ with closure in Ω, see p. 4.

We shall consider a measure $\mu \geqslant 0$ on Ω that satisfies the equation $L^*_{A,b}\mu = 0$; in some results we deal with the equation $\mathcal{L}^*_{A,b}\mu = 0$. Let ϱ be the density of μ. Let

$$\beta_\mu := \frac{\nabla \varrho}{\varrho}, \quad \beta_{\mu,A} := (\beta^i_{\mu,A})^d_{i=1}, \quad \beta^i_{\mu,A} := \partial_{x_j} a^{ij} + a^{ij}\frac{\partial_{x_j}\varrho}{\varrho},$$

where $\nabla\varrho(x)/\varrho(x) := 0$ if $\varrho(x) = 0$. Then $\beta^i_{\mu,A} \in L^p_{\text{loc}}(\Omega)$. One can write

$$L_{A,b}\varphi = L_{A,\beta_{\mu,A}} + \langle b - \beta_{\mu,A}, \nabla\varphi\rangle, \quad \varphi \in C_0^\infty(\Omega).$$

In the divergence form case, the mapping $A\beta_\mu$ will be employed instead of $\beta_{\mu,A}$. The operator $L_{A,\beta_{\mu,A}}$ is symmetric on $L^2(\Omega,\mu)$, i.e.,

(5.2.2) $$\int_\Omega L_{A,\beta_{\mu,A}}\varphi\,\psi\,d\mu = \int_\Omega \varphi\, L_{A,\beta_{\mu,A}}\psi\,d\mu \quad \forall \varphi,\psi \in C_0^\infty(\Omega).$$

Indeed, by the integration by parts formula, both sides of (5.2.2) are equal to

$$-\int_\Omega \langle A\nabla\varphi, \nabla\psi\rangle\,d\mu.$$

Similarly, for divergence form operators one has

$$\int_\Omega \psi\mathcal{L}_{A,A\beta_\mu}\varphi\,d\mu = \int_\Omega \varphi\,\mathcal{L}_{A,A\beta_\mu}\psi\,d\mu.$$

Note that if $A = \mathrm{I}$, then $b_{\mu,\mathrm{I}} = \beta_\mu$ and

$$L_{\mathrm{I},b}\varphi = \mathcal{L}_{\mathrm{I},b}\varphi = \Delta\varphi + \langle b, \nabla\varphi\rangle.$$

If μ is a nonnegative measure on Ω with a density $\varrho \in W^{1,1}_{\text{loc}}(\Omega)$, A and b satisfy conditions (5.2.1), and β_μ and $\beta_{\mu,A}$ are defined as above, then the equality $L^*_{A,b}\mu = 0$ is equivalent to the identity

(5.2.3) $$\int_\Omega \langle b - \beta_{\mu,A}, \nabla\varphi\rangle\,d\mu = 0 \quad \forall\varphi \in C_0^\infty(\Omega),$$

or shortly,

(5.2.4) $$\mathrm{div}_\mu(b - \beta_{\mu,A}) = 0.$$

The subindex μ indicates that divergence is taken with respect to μ, not with respect to Lebesgue measure. Indeed, it suffices to note that by the integration by parts formula, one has

$$\int_\Omega [\partial_{x_j}a^{ij}\partial_{x_i}\varphi\,\varrho + a^{ij}\partial_{x_i}\varphi\partial_{x_j}\varrho]\,dx = -\int_\Omega a^{ij}\partial_{x_j}\partial_{x_i}\varphi\,\varrho\,dx.$$

The divergence form analogue of (5.2.3) is

$$\int_\Omega \langle b - A\beta_\mu, \nabla\varphi\rangle\,d\mu = 0 \quad \forall\varphi \in C_0^\infty(\Omega),$$

or $\mathrm{div}_\mu\,(b - A\beta_\mu) = 0$.

The vector field

(5.2.5) $$\widehat{b} := 2\beta_{\mu,A} - b$$

is called the *dual drift* (in the theory of diffusion processes, this is the drift of the time-reversed process). In the divergence form case, the dual drift is defined by

$$\widetilde{b} := 2A\beta_\mu - b.$$

Thus,

$$b + \widehat{b} = 2\beta_{\mu,A}, \quad b + \widetilde{b} = 2A\beta_\mu.$$

Clearly, one has

$$L_{A,\widehat{b}} = L_{A,\beta_{\mu,A}} - (b^i - \beta^i_{\mu,A})\partial_{x_i} \quad \text{on } C_0^\infty(\Omega)$$

and obviously $L_{A,\widehat{b}}$ is a formal adjoint to $L_{A,b}$, i.e., one has

$$\int_\Omega \psi\, L_{A,b}\varphi\, d\mu = \int_\Omega \varphi\, L_{A,\widehat{b}}\psi\, d\mu, \quad \varphi, \psi \in C_0^\infty(\Omega).$$

Similarly,

$$\int_\Omega \psi\, \mathcal{L}_{A,b}\varphi\, d\mu = \int_\Omega \varphi\, \mathcal{L}_{A,\widetilde{b}}\psi\, d\mu, \quad \varphi, \psi \in C_0^\infty(\Omega).$$

We observe that by (5.2.2) and (5.2.4) one has

$$\mu \in \mathcal{M}_{\rm ell}^{A,\beta_{\mu,A}} \cap \mathcal{M}_{\rm ell}^{A,\widehat{b}}.$$

5.2.1. Lemma. *Let μ be a positive measure on a bounded domain U satisfying the equation $L^*_{A,b}\mu = 0$ in U and let $u \geqslant 0$ be in $W^{p,2}(U) \cap W_0^{p,1}(U)$, $p > d$. Then*

$$\int_U L_{A,b}u\, d\mu \leqslant 0.$$

PROOF. Let $u_\varepsilon := (u+\varepsilon)^{1+\varepsilon} - \varepsilon^{1+\varepsilon}$, $0 < \varepsilon < 1$. By Exercise 1.8.18 we have the inclusion $u_\varepsilon \in W_0^{p,2}(U)$. Hence the integral of $L_{A,b}u_\varepsilon$ over U against μ vanishes due to the fact that u_ε is a limit in $W_0^{p,2}(U)$ of a sequence of functions from $C_0^\infty(U)$ and $L^*_{A,b}\mu = 0$ in U. Since

$$\begin{aligned} L_{A,b}u_\varepsilon &= (1+\varepsilon)(u+\varepsilon)^\varepsilon L_{A,b}u + \varepsilon(1+\varepsilon)(u+\varepsilon)^{\varepsilon-1}\langle A\nabla u, \nabla u\rangle \\ &\geqslant (1+\varepsilon)(u+\varepsilon)^\varepsilon L_{A,b}u, \end{aligned}$$

the integral of $(u+\varepsilon)^\varepsilon L_{A,b}u$ over U against μ is nonpositive. Letting $\varepsilon \to 0$ and using that $(u+\varepsilon)^\varepsilon \to 1$ as $\varepsilon \to 0$ and $(u+\varepsilon)^\varepsilon \leqslant \sup_{x\in U} u(x) + 1$, we obtain the desired estimate. □

5.2.2. Theorem. (i) *Let $\Omega \subset \mathbb{R}^d$ be open and let $\mu \in \mathcal{M}_{\rm ell}^{A,b}$, where condition* (5.2.1) *holds. Then there exists a closed extension $\bigl(L^\mu_{A,b}, D(L^\mu_{A,b})\bigr)$ of $\bigl(L_{A,b}, C_0^\infty(\Omega)\bigr)$ that generates a sub-Markovian contractive C_0-semigroup $\{T^\mu_t\}_{t\geqslant 0}$ on $L^1(\mu)$ with the following properties*:

(a) *for every bounded measurable function f with compact support, the function $(I - L^\mu_{A,b})^{-1} f$ is the limit in $L^1(\mu)$ of the functions u_n that are unique solutions to the Dirichlet problems $(I - L_{A,b})u_n = f$ with zero boundary conditions on domains $B_n \subset \Omega$ with compact closures $\overline{B_n} \subset B_{n+1}$ and smooth boundaries ∂B_n such that $\Omega = \bigcup_{n=1}^\infty B_n$; furthermore, the measure μ is subinvariant for $\{T^\mu_t\}_{t\geqslant 0}$;*

(b) *the above mentioned semigroups are conjugated, i.e.,*

$$\int_\Omega g\, T^\mu_t f\, d\mu = \int_\Omega f\, \widehat{T}^\mu_t g\, d\mu, \quad f, g \in L^\infty(\mu); \tag{5.2.6}$$

the same is true for the corresponding resolvents $\{R^\mu_\alpha\}_{\alpha>0}$ and $\{\widehat{R}^\mu_\alpha\}_{\alpha>0}$.

(ii) *The same is true if* Ω *is an open subset of a complete Riemannian manifold* M *and the operator* $L_{A,b}$ *is replaced by the operator* $\mathcal{L}_{A,b}$ *in the form*

$$\mathcal{L}_{A,b}\varphi = \mathrm{div}(A\nabla\varphi) + \langle b, \nabla\varphi\rangle, \quad \varphi \in C_0^\infty(\Omega),$$

where b *is a vector field on* M *and* $A(x)$ *is a positive operator on* T_xM *such that the hypotheses of* (i) *are fulfilled in local coordinates.*

(iii) *Under the hypotheses of* (i) *or* (ii), *the semigroup* $\{T_t^\mu\}_{t\geqslant 0}$ *has the following property: for every* $\psi \in C_0^\infty(\Omega)$ *and every* $t \geqslant 0$, *the function* $T_t^\mu\psi$ *has a continuous modification* $\widetilde{T_t^\mu\psi}$ *such that, for every compact set* $K \subset \Omega$, *one has the equality* $\lim\limits_{t\to 0}\widetilde{T_t^\mu\psi}(x) = \psi(x)$ *uniformly with respect to* $x \in K$.

PROOF. (i) For notational simplicity we shall assume that the local order $p > d$ from our assumptions about the coefficients does not depend on B_n. Given a bounded nonnegative measurable function f, there is a unique function u_n in the class $W^{p,2}(B_n) \cap W_0^{p,1}(B_n)$ such that $(I - L_{A,b})u_n = f$ in B_n. By the maximum principle $u_n \geqslant 0$. Similarly, $u_n \leqslant 1$ if $f \leqslant 1$. In addition,

$$\int_{B_n} f\,d\mu = \int_{B_n} (I - L_{A,b})u_n\,d\mu \geqslant \int_{B_n} u_n\,d\mu, \tag{5.2.7}$$

since the integral of $L_{A,b}u_n$ over B_n with respect to μ is nonpositive by Lemma 5.2.1. Letting $R_nf := u_n$ and extending R_n by linearity to all bounded measurable functions we obtain a linear operator

$$R_n\colon L^\infty(\mathbb{R}^d) \to L^1(\mu|_{B_n})$$

with the following properties:

$$R_nf \geqslant 0 \text{ for all } f \geqslant 0 \quad\text{and}\quad \|R_nf\|_{L^1(\mu|B_n)} \leqslant \|f\|_{L^1(\mu|B_n)}.$$

The latter follows by inequality (5.2.7) since, writing $f = f^+ - f^-$, we obtain that $|R_nf| = |R_nf^+ - R_nf^-| \leqslant R_nf^+ + R_nf^-$. Let us also observe that $u_{n+1} \geqslant u_n$ in B_n if $f \geqslant 0$, which follows by the maximum principle, since $u_{n+1} \geqslant 0$ on ∂B_n. Hence the sequence $\{u_n(x)\}$ converges whenever $0 \leqslant f \leqslant 1$. It follows that a limit $Rf(x) = \lim\limits_{n\to\infty} R_nf(x)$ exists for every $f \in L^\infty(\mathbb{R}^d)$. Moreover, we obtain (see Chapter 1) that the sequence R_nf converges locally uniformly and locally in $W^{p,2}$, so that $Rf \in W^{p,2}_{\mathrm{loc}}(\mathbb{R}^d)$. It is clear that

$$(I - L_{A,b})Rf = f, \quad \|Rf\|_{L^1(\mu)} \leqslant \|f\|_{L^1(\mu)}.$$

In addition, $0 \leqslant Rf \leqslant 1$ whenever $0 \leqslant f \leqslant 1$. Now the operator R can be extended to a contraction on $L^1(\mu)$; actually, this could be done at once dealing with functions $f \in L^1(\mu)$ from the very beginning. The same construction applies to the operators $\lambda I - L_{A,b}$, $\lambda > 0$, which yields sub-Markovian operators $R(\lambda)$ on $L^1(\mu)$ that satisfy the hypotheses of Corollary 5.1.13. Namely, whenever $\lambda, \mu > 0$ we have

$$R(\lambda) - R(\mu) = (\mu - \lambda)R(\lambda)R(\mu), \quad \|\lambda R(\lambda)\| \leqslant 1,$$
$$\lim_{\lambda\to\infty} \|\lambda R(\lambda)f - f\|_{L^1(\mu)} = 0 \quad \forall f \in C_0^\infty(\mathbb{R}^n).$$

The latter holds since $\lambda R(\lambda)f - f = L_{A,b}R(\lambda)f = R(\lambda)L_{A,b}f$ and

$$\|R(\lambda)L_{A,b}f\|_{L^1(\mu)} \leqslant \lambda^{-1}\|L_{A,b}f\|_{L^1(\mu)}$$

due to the fact that $R(\lambda)f$ is the limit of $(\lambda I - L_{A,b})^{-1}f$, where the inverse is taken in $L^1(\mu|_{B_n})$ and commutes with $L_{A,b}$ on $C_0^\infty(B_n)$. Therefore, there is a unique contractive C_0-semigroup $\{T_t\}_{t\geqslant 0}$ on $L^1(\mu)$ with generator A such that $R(\lambda) = (\lambda I - A)^{-1}$ for all $\lambda > 0$. Then the domain of A coincides with the range of $R(\lambda)$, which contains $C_0^\infty(\mathbb{R}^d)$, because

$$R(\lambda)(\lambda I - L_{A,b})\varphi = (\lambda I - L_{A,b})R(\lambda)\varphi = \varphi \quad \forall \varphi \in C_0^\infty(\mathbb{R}^d),$$

as noted above. Therefore, $A\varphi = L_{A,b}\varphi$ for such φ.

(ii) It has been explained in § 1.4 that the operator $L_{A,b}$ in (i) can be written in the divergence form, i.e.,

$$L_{A,b}\varphi = \mathcal{L}_{A,b_0}\varphi = \operatorname{div}(A\nabla\varphi) + \langle b_0, \nabla\varphi\rangle, \quad b_0^j = -\partial_{x_i}a^{ij} + b^j,$$

hence a similar assertion is true for divergence form operators on manifolds.

(iii). Let U be a closed ball in Ω. Let $f_t(x) = T_t^\mu L_{A,b}\psi(x)$. For every $t > 0$, the measure $\nu_t = T_t^\mu\psi\,\mu = T_t^\mu\psi\varrho\,dx$ on Ω satisfies the equation $L_{A,b_0}^*\nu_t = f_t\,\mu$. Indeed, for any $\varphi \in C_0^\infty(\Omega)$, it follows by (5.2.6) that

$$\int L_{A,b_0}\varphi\,d\nu_t = \int \varphi T_t^\mu L_{A,b}\psi\,d\mu = \int \varphi f_t\,d\mu.$$

Note that, since $\varrho \leqslant C$ on U with some C, we have

$$\|f_t\varrho\|_{L^p(U)} \leqslant C^{(p-1)/p}\|f_t\|_{L^p(\mu)} \leqslant C^{(p-1)/p}\|L_{A,b}\psi\|_{L^p(\mu)}.$$

In addition, $\sup_t \|T_t^\mu\psi\varrho\|_{L^1(U)} < \infty$. Then $\nu_t = \varrho_t\,dx$, where $\varrho_t \in W^{p,1}_{loc}(\Omega)$ and $\sup_t \|\varrho_t\|_{W^{p,1}(U)} < \infty$ (see § 1.6). Therefore, the continuous versions of ϱ_t are Hölder continuous on U uniformly in $t > 0$. Hence the same is true for the functions $u(x,t) = \varrho_t(x)/\varrho(x)$. If $t_n \to 0$, then $u(x,t_n) \to \psi(x)$ in measure, since $T_{t_n}^\mu\psi \to \psi$ in $L^1(\Omega,\mu)$ and $T_t^\mu\psi = u(\,\cdot\,,t)$ μ-a.e. for each t. Setting $\widetilde{T_t^\mu\psi}(x) = u(t,x)$, it remains to note that if a sequence of functions is uniformly Hölder continuous and uniformly bounded on any compact K and converges in measure, then it converges uniformly on K (Exercise 1.8.17). □

5.2.3. Remark. The semigroup $\{T_t^\mu\}_{t\geqslant 0}$ does not depend on our choice of $\{B_n\}$, since the operators $(I - L_{A,b}^\mu)^{-1}$ have the minimality property: if $f \geqslant 0$, then $u = (I - L_{A,b}^\mu)^{-1}f \leqslant w$ for any nonnegative solution w of the equation $w - L_{A,b}w = f$. This follows by the maximum principle, since u is a limit of solutions to Dirichlet boundary value problems. Hence $T_t^\mu f$ for $f \in C_0^\infty(\Omega)$ does not depend on $\mu \in \mathcal{M}_{\rm ell}^{A,b}$, but only on A and b (this is not true for all $f \in L^1(\mu)$!).

The semigroup $\{T_t^\mu\}_{t\geqslant 0}$ constructed in this theorem will play a very important role below; in general, it is not the only semigroup associated with $L_{A,b}$ in the sense explained at the end of the previous section, but it is unique among associated semigroups with property (b). Below the super-index μ at T_t^μ will indicate this concrete semigroup. Such a semigroup exists also under weaker assumptions about A and b (see Stannat [**872**]). We emphasize that in (5.2.6) we are concerned with our canonical semigroups, not with arbitrary associated ones.

5.2.4. Remark. (i) Equality (5.2.6) implies that μ is $\{T_t^\mu\}_{t\geqslant 0}$-invariant if and only if for every $t \geqslant 0$ one has $\widehat{T}_t^\mu 1 = 1$. Hence, since both semigroups are sub-Markovian, this is the case if and only if $T_t^\mu 1 = 1$ for all $t \geqslant 0$, which in turn is equivalent to the invariance of μ for $\{\widehat{T}_t^\mu\}_{t\geqslant 0}$. The equality $T_t^\mu 1 = 1$ is equivalent to the inclusion $1 \in D(L_{A,b}^\mu)$ (or $1 \in D(L_{A,\widehat{b}}^\mu)$), which is not always fulfilled in spite of the pointwise equality $L_{A,b}1 = L_{A,\widehat{b}}1 = 0$.

(ii) For a symmetric operator $L_{A,b}$ (or $\mathcal{L}_{A,b}$) the closed extension mentioned in the theorem is its Friedrichs extension (Exercise 5.6.48). The condition of symmetry of the semigroup $\{T_t^\mu\}_{t\geqslant 0}$ is seen from (5.2.5); see also Proposition 1.4.8.

We do not know whether a sub-Markovian strongly continuous semigroup whose generator extends $L_{A,b}$ is unique. As we shall see below, this is true under certain additional assumptions.

We have seen in Example 5.1.9 that if a measure $\mu \geqslant 0$ satisfies the equation $L_{A,b}^*\mu = 0$, then the operator $L_{A,b}$ is dissipative on the domain $C_0^\infty(\Omega)$ in the space $L^1(\mu)$. In particular, it is closable. The same is true for the divergence form operator $\mathcal{L}_{A,b}$ in the case of a manifold. Hence the operator $L_{A,b}$ is essentially m-dissipative on the domain $C_0^\infty(\Omega)$ in $L^1(\mu)$ if and only if the set $(L_{A,b}-\lambda I)\bigl(C_0^\infty(\Omega)\bigr)$ is dense in $L^1(\mu)$ for some (and then for all) $\lambda > 0$. In this case the operator $\bigl(L_{A,b}, C_0^\infty(\Omega)\bigr)$ is called $L^1(\mu)$-unique; for sufficient conditions for this are given below.

Let $\bigl(\overline{L}_{A,b}^\mu, D(\overline{L}_{A,b}^\mu)\bigr)$ denote the closure of the operator $\bigl(L_{A,b}, C_0^\infty(\Omega)\bigr)$ in the space $L^1(\mu)$. The closure $\bigl(\overline{\mathcal{L}}_{A,b}^\mu, D(\overline{\mathcal{L}}_{A,b}^\mu)\bigr)$ of the operator $\mathcal{L}_{A,b}$ is defined analogously.

5.2.5. Proposition. *Let $\mu \in \mathcal{M}_{\rm ell}^{A,b}$, where A and b satisfy* (5.2.1). *The following assertions are equivalent:*

(i) $\bigl(\overline{L}_{A,b}^\mu, D(\overline{L}_{A,b}^\mu)\bigr)$ *generates a C_0-semigroup $\{T_t\}_{t\geqslant 0}$, i.e., a strongly continuous semigroup of bounded operators T_t on $L^1(\mu)$;*

(ii) *for some (hence for all) $\lambda > 0$, the set*

$$\bigl(L_{A,b} - \lambda I\bigr)\bigl(C_0^\infty(\Omega)\bigr)$$

is dense in $L^1(\mu)$, i.e., $\bigl(L_{A,b}, C_0^\infty(\Omega)\bigr)$ is essentially m-dissipative on $L^1(\mu)$;

(iii) *there exists exactly one C_0-semigroup on $L^1(\mu)$ which has a generator extending the operator $\bigl(L_{A,b}, C_0^\infty(\Omega)\bigr)$.*

If some (hence each) of the assertions (i)–(iii) *is true, then the semigroups $\{T_t^\mu\}_{t\geqslant 0}$ and $\{\widehat{T}_t^\mu\}_{t\geqslant 0}$ are Markovian and μ is invariant for them. Finally, the same is true for the operator $\mathcal{L}_{A,b}$ in divergence form.*

Proof. The equivalence of (i) and (ii) is a consequence of the Lumer–Phillips theorem (Theorem 5.1.11). The implication "(i)⇒(iii)" is trivial, and "(iii)⇒(i)" is due to W. Arendt [**60**, A-II, Theorem 1.33].

The invariance of μ with respect to $\{T_t^\mu\}_{t\geqslant 0}$ readily follows from (i)–(iii), since the equality

$$\int_\Omega L_{A,b}\varphi\, d\mu = 0 \quad \forall \varphi \in C_0^\infty(\Omega)$$

implies in case (i) by the definition of the closure that

$$\int_\Omega \overline{L}_{A,b}^\mu \varphi\, d\mu = 0 \quad \forall \varphi \in D(\overline{L}_{A,b}^\mu), \tag{5.2.8}$$

whence we obtain

$$\int_\Omega \overline{L}_{A,b}^\mu T_t^\mu \varphi\, d\mu = 0 \quad \forall \varphi \in C_0^\infty(\Omega),$$

i.e., the integral of $T_t^\mu\varphi$ with respect to μ is constant. The remaining statements follow from Remark 5.2.4. □

5.2.6. Remark. For a bounded domain Ω (with smooth boundary), assertion (iii) (hence also (i) and (ii)) in Proposition 5.2.5 is not valid even if $A = \mathrm{I}$ and $b \equiv 0$. So we shall be mainly interested in the case when $\Omega = \mathbb{R}^d$. As we shall see below, assertions (i)–(iii) are equivalent to the invariance of the measure μ with respect to the semigroup $\{T_t^\mu\}_{t\geqslant 0}$.

We shall see below that the operator $L_{A,b}$ on $C_0^\infty(\Omega)$ can have different closed extensions (even generating strongly continuous semigroups). Hence its closure can fail to generate a strongly continuous semigroup (i.e., it may not be a generator). This occurs even if $A = \mathrm{I}$ and b is infinitely differentiable.

The next theorem gives useful information about the domain of generator of $\{T_t^\mu\}_{t\geqslant 0}$ on $L^p(\mu)$. We recall that, for every $r \in [1,\infty)$, the restriction of $\{T_t^\mu\}_{t\geqslant 0}$ to $L^r(\mu)$ is a strongly continuous semigroup on $L^r(\mu)$. Its generator will be denoted by $\bigl(L_{A,b}^{\mu,r}, D(L_{A,b}^{\mu,r})\bigr)$. It is not difficult to verify that

$$D(L_{A,b}^{\mu,r}) = \Bigl\{ f \in D(L_{A,b}^{\mu}) \cap L^r(\mu)\colon\ L_{A,b}^{\mu} f \in L^r(\mu) \Bigr\}.$$

5.2.7. Theorem. (i) *In the situation of Theorem* 5.2.2, *one has*

$$\begin{aligned} \bigl(L_{A,b}^{\mu,p}, D(L_{A,b}^{\mu,p})\bigr) &\subset \Bigl\{ f \in L^p(\mu) \cap W^{p,2}_{\mathrm{loc}}(\Omega)\colon\ L_{A,b} f \in L^p(\mu) \Bigr\}, \\ L_{A,b}^{\mu,p} f &= L_{A,b} f \quad \textit{for all } f \in D(L_{A,b}^{\mu,p}). \end{aligned} \tag{5.2.9}$$

The same is true for any extension $\bigl(L, D(L)\bigr)$ *of* $\bigl(L_{A,b}, C_0^\infty(\Omega)\bigr)$ *with the following property: it is the generator of a strongly continuous sub-Markovian semigroup* $\{T_t\}_{t\geqslant 0}$ *on* $L^1(\mu)$ *such that the adjoint semigroup* $\{T_t'\}_{t\geqslant 0}$ *on* $L^{p'}(\mu)$ *(which is defined after extending* $\{T_t\}_{t\geqslant 0}$ *to* $L^p(\mu)$ *as explained above) has a generator which coincides with* $L_{A,\widehat{b}}$ *on* $C_0^\infty(\Omega)$. *If one has an equality in* (5.2.9), *then* μ *is invariant for* $\{T_t^\mu\}_{t\geqslant 0}$ *(equivalently, the equality* $T_t^\mu 1 = 1$ *holds).*

(ii) *If the operator* $\bigl(L_{A,\widehat{b}}, C_0^\infty(\Omega)\bigr)$ *is essentially m-dissipative on* $L^{p'}(\mu)$, *i.e., the set* $(L_{A,\widehat{b}} - \lambda I)\bigl(C_0^\infty(\Omega)\bigr)$ *is dense in* $L^{p'}(\mu)$ *for some* $\lambda > 0$, *then one has an equality in* (5.2.9).

Finally, the same assertions are true for $\mathcal{L}_{A,b}$.

Proof. (i) We shall deal with the divergence form operator $\mathcal{L}_{A,b}$. Let $\{S_t\}_{t\geqslant 0}$ denote the sub-Markovian semigroup on the space $L^p(\mu)$ which is the adjoint semigroup to $\{T_t^\mu\}_{t\geqslant 0}$ on $L^{p'}(\mu)$. Let $f \in \bigl(L_{A,b}^{\mu,p}, D(L_{A,b}^{\mu,p})\bigr)$. Then $g := L_{A,b}^{\mu,p} f \in L^p(\mu)$. For any $\varphi \in C_0^\infty(\Omega)$, one has

$$\int_\Omega \varphi L_{A,b}^{\mu,p} f\, d\mu = \frac{d}{dt}\int_\Omega \varphi T_t^\mu f\, d\mu\Big|_{t=0} = \frac{d}{dt}\int_\Omega S_t\varphi\, f\, d\mu\Big|_{t=0} = \int_\Omega L_{A,\widetilde{b}}\varphi\, f\, d\mu,$$

since $\frac{d}{dt}S_t\varphi = S_t L_{A,\widetilde{b}}\varphi$ in $L^{p'}(\mu)$. Indeed, it suffices to note that the mapping $t \mapsto S_t\mathcal{L}_{A,\widetilde{b}}\varphi$ is continuous with values in $L^{p'}(\mu)$ and that, for every $\psi \in C_0^\infty(\mu)$, one has

$$\frac{d}{dt}\int_\Omega \psi S_t\varphi\, d\mu = \int_\Omega \psi S_t \mathcal{L}_{A,\widetilde{b}}\varphi\, d\mu,$$

since the generator of $(S_t)_{t\geqslant 0}$ on $L^{p'}(\mu)$ extends $\bigl(\mathcal{L}_{A,\widetilde{b}}, C_0^\infty(\Omega)\bigr)$. Thus, we have proved that $(\mathcal{L}_{A,\widetilde{b}})^*(f\cdot\mu) = g\cdot\mu$. Clearly,

$$|\widehat{b}| \in L^1_{\mathrm{loc}}(\Omega, |f|\cdot\mu) \quad \text{and } g\cdot\mu = g_0\, dx \text{ with } g_0 \in L^p_{\mathrm{loc}}(\Omega, dx).$$

Therefore, the measure $f \cdot \mu$ has a density $\zeta \in W^{p,1}_{\rm loc}(\Omega)$. Hence $f \in W^{p,1}_{\rm loc}(\Omega)$. Next we prove that f satisfies the elliptic equation $\mathcal{L}_{A,b} f = g$ in the sense of the classical weak solutions, i.e., for every $\varphi \in C_0^\infty(\Omega)$, one has

$$\int_\Omega \Bigl[-\langle A\nabla\varphi, \nabla f\rangle + \varphi\langle b, \nabla f\rangle - \varphi g\Bigr]\, dx = 0. \tag{5.2.10}$$

Then we obtain that $f \in W^{p,2}_{\rm loc}(\Omega)$, see Ladyzhenskaya, Ural'tseva [**577**, Ch. III, § 10, § 11] and Chicco [**240**]. Let us write the desired equality (5.2.10) in the form

$$\int_\Omega \varphi \mathcal{L}^\mu_{A,b} f\, dx = \int_\Omega \Bigl[-\langle A\nabla f, \nabla\varphi\rangle + \varphi\langle b, \nabla f\rangle\Bigr]\, dx. \tag{5.2.11}$$

Let us write φ in the form $\varphi = \psi\varrho$, $\psi \in W_0^{p,1}(\Omega)$. Then (1.7.12) becomes

$$\int_\Omega \psi \mathcal{L}^\mu_{A,b} f\, \varrho\, dx = \int_\Omega \Bigl[-\langle A\nabla f, \nabla\psi\rangle - \langle A\nabla f, \nabla\varrho/\varrho\rangle\psi + \psi\langle b, \nabla f\rangle\Bigr]\, \varrho\, dx. \tag{5.2.12}$$

It suffices to verify this equality for all $\psi \in C_0^\infty(\Omega)$. Then our claim reduces to the equality

$$\int_\Omega \mathcal{L}_{A,\widetilde{b}}\psi\, f\, \varrho\, dx = \int_\Omega \Bigl[-\langle A\nabla f, \nabla\psi\rangle - \langle A\nabla f, \nabla\varrho/\varrho\rangle\psi + \psi\langle b, \nabla f\rangle\Bigr]\, \varrho\, dx. \tag{5.2.13}$$

By the integration by parts formula, its left-hand side equals

$$\int_\Omega \Bigl[-\langle A\nabla\psi, \nabla f + f\nabla\varrho/\varrho\rangle - \langle\widetilde{b}, \nabla\psi\rangle f\Bigr]\, \varrho\, dx.$$

Therefore, (5.2.13) is equivalent to the equality

$$\int_\Omega \Bigl[-\langle A\nabla\psi, f\nabla\varrho/\varrho\rangle - \langle\widetilde{b}, \nabla\psi\rangle f\Bigr]\, \varrho\, dx = \int_\Omega \Bigl[-\langle A\nabla f, \nabla\varrho/\varrho\rangle\psi + \psi\langle b, \nabla f\rangle\Bigr]\, \varrho\, dx.$$

Substituting $\widetilde{b} = 2A\nabla\varrho/\varrho - b$, we arrive at the verification of the equality

$$\int_\Omega \Bigl[\langle A\nabla\psi, \nabla\varrho/\varrho\rangle f + \langle A\nabla f, \nabla\varrho/\varrho\rangle\psi - \langle b, \nabla\psi\rangle f - \langle b, \nabla f\rangle\psi\Bigr]\, \varrho\, dx = 0,$$

which can be written as

$$\int_\Omega \langle A\nabla\varrho/\varrho - b, \nabla(\psi f)\rangle\, \varrho\, dx = 0,$$

hence is true by (5.2.3) (it is clear that (5.2.3) remains valid also for every function $\varphi \in C_0^1(\Omega)$). We observe that we have used only those properties of $\mathcal{L}^\mu_{A,b}$ that are mentioned in the second assertions in (i).

If we have an equality in (5.2.9) for $\mathcal{L}_{A,b}$, then $1 \in D(\mathcal{L}^{\mu,p}_{A,b})$. Then, according to what has already been proved, $\mathcal{L}^{\mu,p}_{A,b}1 = \mathcal{L}^{\mu,p}_{A,b}1 = 0$. Hence $T^\mu_t 1 = 1$, which, as we know, is equivalent to the invariance of μ.

(ii) Suppose now that the operator $\bigl(L_{A,\widehat{b}}, C_0^\infty(\Omega)\bigr)$ is essentially m-dissipative on $L^{p'}(\mu)$. Let $f \in W^{p,2}_{\rm loc}(\Omega)$ be such that $f, L_{A,b}f \in L^p(\mu)$. For any $\psi \in D(L^{\mu,p'}_{A,\widehat{b}})$, one has the equality

$$\int_\Omega L^\mu_{A,\widehat{b}}\psi\, f\, d\mu = \int_\Omega \psi L_{A,b} f\, d\mu.$$

Indeed, for $\psi \in C_0^\infty(\Omega)$ this follows by the integration by parts. In the general case, by our hypothesis, there exist functions $\psi_k \in C_0^\infty(\Omega)$ such that $\psi_k \to \psi$ and

$L_{A,\widehat{b}}\psi_k \to L_{A,\widehat{b}}\psi$ in $L^{p'}(\mu)$. Therefore, since $S_t\varphi \in D(L^{\mu,p'}_{A,\widehat{b}})$ for any $\varphi \in C_0^\infty(\Omega)$, we obtain that

$$\frac{d}{dt}\int_\Omega \varphi T_t^\mu f\,d\mu = \int_\Omega L_{A,\widehat{b}} S_t\varphi\, f\,d\mu = \int_\Omega S_t\varphi L_{A,b} f\,d\mu = \int_\Omega \varphi T_t^\mu L_{A,b} f\,d\mu.$$

Since the mapping $t \mapsto T_t^\mu L_{A,b}f$ to $L^p(\mu)$ is continuous, we conclude that it is the derivative of $T_t^\mu f$. In particular, f is in the domain of $L^{\mu,p}_{A,b}$ and we have $L^{\mu,p}_{A,b}f = L_{A,b}f$ μ-a.e. □

Let us prove a useful technical result about the local second Sobolev class for the resolvent of our special semigroup.

5.2.8. Lemma. *Let a probability measure μ satisfy $L^*_{A,b}\mu = 0$ with our standard local assumptions about A and b. Suppose that the set $(I - L_{A,b})\big(C_0^\infty(\mathbb{R}^d)\big)$ is dense in $L^1(\mu)$. If $v \in L^1(\mu)$, $\alpha u - L^{\mu,1}_{A,b}u = v$ and $v \in L^p(U_0)$ on some ball U_0, where $p > d$, then $u \in W^{p,2}(U)$ for every ball U with closure in the interior of U_0, and $L^{\mu,1}_{A,b}u$ coincides with Lu given by the usual differential expression.*

Proof. We can assume that the closure of U is in the interior of U_1. By our construction, the function u on U_2 is the limit in $L^1(\mu)$ of the functions u_n. By the positivity and continuity of the density ϱ the same is true for the norm of $L^1(U_2)$. Hence on the smaller ball U_1 the sequence $\{u_n\}$ is norm bounded in the Sobolev space $W^{d,1}(U_1)$ (see § 1.7), which according to the L^p-theory (see Theorem 1.2.1) gives the norm boundedness in $W^{p,2}(U)$. □

The reader should be warned that Theorem 5.2.2 does not assert that a sub-Markovian semigroup whose generator extends $L_{A,b}$ is unique: we do not know whether this is true under the indicated hypotheses, and without the sub-Markovian property (i.e., with the only requirement of strong continuity) this is not true in general, as we shall see. This is why we always specify the semigroup $\{T_t^\mu\}_{t\geqslant 0}$.

In addition, the measure μ may be only subinvariant for the semigroup and not invariant. We already know that these two phenomena are closely related: Proposition 5.2.5 shows that the uniqueness of associated strongly continuous semigroups is equivalent to the essential m-dissipativity of $L_{A,b}$.

Let us prove another result about the generated semigroup, in which under considerably stronger global conditions we substantially weaken the condition of the local integrability of the drift. This result will be very useful in the infinite-dimensional case, since its assumptions do not depend on dimension.

5.2.9. Theorem. *Let a Borel probability measure μ on $\mathbb{R}^d$ satisfy the equation $L^*_{\mathrm{I},b}\mu = 0$ and $|b| \in L^2(\mu)$. Then the operator $\big(L_{\mathrm{I},b}, C_0^\infty(\mathbb{R}^d)\big)$ is essentially m-dissipative in $L^1(\mu)$ and its closure generates a unique strongly continuous semigroup $\{T_t\}_{t\geqslant 0}$ on $L^1(\mu)$, moreover, this semigroup is Markov and the measure μ is invariant for it.*

Proof. We check explicitly that the range of $I - L_{\mathrm{I},b}$ on $C_0^\infty(\mathbb{R}^d)$ is dense in $L^1(\mu)$. It suffices to show that for every function φ in $C_0^\infty(\mathbb{R}^d)$ there is a sequence of functions $\psi_n \in C_0^\infty(\mathbb{R}^d)$ for which $(I - L_{\mathrm{I},b})\psi_n \to \varphi$ in $L^1(\mu)$. We first find such functions ψ_n in the class $C_b^\infty(\mathbb{R}^d)$. Once this is done, for any fixed function ψ_n we can take a sequence of functions $\eta_k \in C_0^\infty(\mathbb{R}^d)$ with the following properties:

$0 \leqslant \eta_k \leqslant 1$, $\eta_k(x) = 1$ if $|x| \leqslant k$, $\eta_k(x) = 0$ when $|x| \geqslant k+1$, $\sup_{k,x} |\nabla\eta_k(x)| < \infty$, $\sup_{k,x} |\Delta\eta_k(x)| < \infty$. Then

$$(I - L_{\mathrm{I},b})(\eta_k\psi_n) = \eta_k(I - L_{\mathrm{I},b})\psi_n - \psi_n\Delta\eta_k - 2\langle\nabla\eta_k, \nabla\psi_n\rangle + \psi_n\langle b, \nabla\eta_k\rangle,$$

where the last three terms tend to zero in $L^1(\mu)$ by the Lebesgue dominated convergence theorem, since ψ_n and $|\nabla\psi_n|$ are bounded, $|b| \in L^1(\mu)$. In addition, the first term tends to $(I - L_{\mathrm{I},b})\psi_n$ in $L^1(\mu)$.

In order to find functions ψ_n in $C_b^\infty(\mathbb{R}^n)$ we take a sequence of smooth mappings b_n with compact supports such that $|b - b_n| \to 0$ in $L^2(\mu)$. It is known (see Krylov [**552**, Theorem 4.3.2 and Theorem 2.9.2]) that for each n the equation

$$\psi_n - L_{\mathrm{I},b_n}\psi_n = \varphi$$

has a unique solution in the class $C_b^\infty(\mathbb{R}^d)$ and $\sup_x |\psi_n(x)| \leqslant \sup_x |\varphi(x)|$, and if $\varphi \geqslant 0$, then also $\psi_n \geqslant 0$. We now need the fact established in Theorem 3.1.2 that μ has a locally Sobolev density ϱ with $|\nabla\varrho/\varrho| \in L^2(\mu)$. So, multiplying the equation by ψ_n and integrating with respect to μ, we obtain the equality

$$\int_{\mathbb{R}^d} \psi_n^2\, d\mu - \int_{\mathbb{R}^d} \psi_n L_{\mathrm{I},b_n}\psi_n\, d\mu = \int_{\mathbb{R}^d} \psi_n\varphi\, d\mu$$

where we can integrate by parts in the term with $\Delta\psi_n$, which yields that

$$\begin{aligned}\int_{\mathbb{R}^d} \psi_n^2\, d\mu &+ \int_{\mathbb{R}^d} |\nabla\psi_n|^2\, d\mu \\ &= \int_{\mathbb{R}^d} \psi_n\varphi\, d\mu + \int_{\mathbb{R}^d} \psi_n\langle b_n, \nabla\psi_n\rangle\, d\mu - \int_{\mathbb{R}^d} \psi_n\langle\nabla\psi_n, \nabla\varrho\rangle\, dx \\ &= \int_{\mathbb{R}^d} \psi_n\varphi\, d\mu + \int_{\mathbb{R}^d} \psi_n\langle b_n - \nabla\varrho/\varrho, \nabla\psi_n\rangle\, d\mu.\end{aligned}$$

We can assume that $|\varphi| \leqslant 1$. Then the right-hand side is estimated by

$$1 + 2\int_{\mathbb{R}^d} |b_n - \nabla\varrho/\varrho|^2\, d\mu + 2^{-1}\int_{\mathbb{R}^d} |\nabla\psi_n|^2\, d\mu,$$

whence we obtain

$$\int_{\mathbb{R}^d} \psi_n^2\, d\mu + \int_{\mathbb{R}^d} |\nabla\psi_n|^2\, d\mu \leqslant 2 + 4\int_{\mathbb{R}^d} |b_n - \nabla\varrho/\varrho|^2\, d\mu,$$

which is estimated by some number C independent of n, since $|b_n - b| \to 0$ in $L^2(\mu)$. Therefore,

$$\varphi - (I - L_{\mathrm{I},b})\psi_n = (I - L_{\mathrm{I},b_n})\psi_n - (I - L_{\mathrm{I},b})\psi_n = \langle b - b_n, \nabla\psi_n\rangle,$$

which tends to zero in $L^1(\mu)$, because the sequence $\{|\nabla\psi_n|\}$ is bounded in $L^2(\mu)$. Thus, by the Lummer–Phillips theorem the closure of $L_{\mathrm{I},b}$ is the generator of a contraction semigroup $\{T_t\}_{t\geqslant 0}$ in $L^1(\mu)$. The measure μ is invariant with respect to $\{T_t\}$, which is proved by the same reasoning as in Proposition 5.2.5. The operator $L_{\mathrm{I},\widehat{b}}$ with the dual drift $\widehat{b} = 2\nabla\varrho/\varrho - b$ also generates a unique contraction semigroup $\{\widehat{T}_t\}_{t\geqslant 0}$ in $L^1(\mu)$. Note that the operators $(\lambda - \overline{L}_{\mathrm{I},b})^{-1}$ are sub-Markov, since whenever $0 \leqslant \varphi \leqslant 1$ we have $0 \leqslant \psi_n \leqslant 1$, as explained above, which yields that $\psi_n \to (\lambda - \overline{L}_{\mathrm{I},b})^{-1}\varphi$ and $0 \leqslant (\lambda - \overline{L}_{\mathrm{I},b})^{-1}\varphi \leqslant 1$. Hence the semigroup $\{\widehat{T}_t\}_{t\geqslant 0}$ is also sub-Markov (see equality (5.1.6)). Moreover, it is Markov, since $T_t 1 = 1$ by the estimate $T_t 1 \leqslant 1$ and the equality of the integrals of both functions (by the invariance of μ with respect to T_t). □

Inessential changes in the proof enable us to extend this theorem to the case of a nonconstant diffusion matrix A which is globally Lipschitzian and bounded along with its inverse (see Exercise 5.6.50).

5.2.10. Proposition. *Suppose that μ_1 and μ_2 are two probability measures on $\mathbb{R}^d$ satisfying the equation $L^*\mu = 0$ with $L = L_{I,b}$ such that μ_2 is absolutely continuous with respect to μ_1 and $|b|$ is μ_i-square integrable for both measures. Let $\{T_t^{\mu_i}\}_{t\geqslant 0}$ denote the corresponding strongly continuous semigroups on $L^1(\mu_i)$ whose generators $(L^{\mu_i}, D(L^{\mu_i}))$ extend L, $i = 1, 2$, which exist by the previous theorem. Then $T_t^{\mu_1} f = T_t^{\mu_2} f$ μ_1-a.e. for every $t > 0$ and every bounded Borel measurable function f on $\mathbb{R}^d$.*

PROOF. It suffices to consider $f \in C_0^\infty(\mathbb{R}^d)$ and prove the equality

$$(\lambda - L^{\mu_1})^{-1} f = (\lambda - L^{\mu_2})^{-1} f \quad \mu_2\text{-a.e. for } \lambda > 0.$$

It is seen from the proof of the previous theorem that $(\lambda - L^{\mu_i})^{-1} f$ can be obtained as a limit μ_i-a.e. of a sequence of solutions to the same equations, provided that we pick smooth b_n such that $|b - b_n| \to 0$ in $L^2(\mu_1 + \mu_2)$. By our assumption convergence μ_1-a.e. implies convergence μ_2-a.e., which yields our assertion. □

We observe that the conclusion (as well as the hypothesis) is not symmetric with respect to the given measures: the semigroups are not always equal μ_1-a.e. For example, if μ_1 on the straight line has density $\varrho(x) = x^2 g(x)$, where g is the standard Gaussian density, $b(x) = \varrho'(x)/\varrho(x)$, μ_2 has the zero density on $(-\infty, 0]$ and the density 2ϱ on $(0, +\infty)$, then the function $T_t^{\mu_2} f(x)$ can be made equal to $f(x)$ whenever $x < 0$ for all f, which will give a version of the semigroup that is μ_1-different from $T_t^{\mu_1} f(x)$.

5.3. Invariance and m-dissipativity

Let us introduce the following subset of $\mathcal{M}_{\text{ell}}^{A,b}$:

$$\mathcal{M}_{\text{ell,md}}^{A,b} := \Bigl\{\mu \in \mathcal{M}_{\text{ell}}^{A,b}\colon \ \overline{(L_{A,b} - I)\bigl(C_0^\infty(\Omega)\bigr)} = L^1(\mu)\Bigr\}. \tag{5.3.1}$$

The same notation is used for the operator $\mathcal{L}_{A,b}$.

In Proposition 4.3.6 a sufficient condition for the essential m-dissipativity of $L_{A,b}$ was given. The next result gives a convenient technical characterization of the essential m-dissipativity of $L_{A,b}$ and shows that it is equivalent to the invariance of μ with respect to the associated semigroup $\{T_t^\mu\}_{t\geqslant 0}$ from Theorem 5.2.2.

5.3.1. Theorem. *Suppose that* (5.2.1) *holds. Let $\mu \in \mathcal{M}_{\text{ell}}^{A,b}$. Then the following assertions are equivalent:*

(i) $\mu \in \mathcal{M}_{\text{ell,md}}^{A,b}$;

(ii) *μ is invariant for $\{T_t^\mu\}_{t\geqslant 0}$;*

(iii) *there exist $\chi_n \in W_{\text{loc}}^{2,1}(\mu)$ and a number $\alpha > 0$ such that $(1-\chi_n)^+ \in L^\infty(\mu)$ and $(1-\chi_n)^+ = 0$ outside some compact sets, $\lim\limits_{n\to\infty} \chi_n(x) = 0$ μ-a.e., and for $\eta = 1$ or $\eta = -1$ one has*

$$\int \langle A\nabla\chi_n, \nabla\varphi\rangle\, d\mu + \alpha \int \chi_n \varphi\, d\mu + \eta \int \langle b - \beta_{\mu,A}, \nabla\chi_n\rangle\, \varphi\, d\mu \geqslant 0 \tag{5.3.2}$$

for all nonnegative $\varphi \in C_0^\infty(\Omega)$ *and all* $n \in \mathbb{N}$; *the corresponding condition in the case of* $\mathcal{L}_{A,b}$ *takes the form*

$$\int \langle A\nabla\chi_n, \nabla\varphi\rangle\, d\mu + \alpha \int \chi_n\varphi\, d\mu + \eta \int \langle b - A\beta_\mu, \nabla\chi_n\rangle\, \varphi\, d\mu \geqslant 0;$$

(iv) $\mu \in \mathcal{M}^{A,\widehat{b}}_{\mathrm{ell,md}}$ (*respectively,* $\mu \in \mathcal{M}^{A,\widetilde{b}}_{\mathrm{ell,md}}$ *in the case of* $\mathcal{L}_{A,b}$).

PROOF. We consider the case of $\mathcal{L}_{A,b}$. The case of $L_{A,b}$ is similar or can be derived from the case of $\mathcal{L}_{A,b}$ with b suitably changed. Let ϱ be the density of μ. Suppose that (i) holds. Then, by Proposition 5.2.5, the measure μ is invariant with respect to $\{T_t^\mu\}_{t\geqslant 0}$, i.e., we have (ii). Let (ii) be fulfilled. We shall assume for simplicity that $\alpha = 1$. Let B_n be increasing domains with compact closures and smooth boundaries covering Ω (e.g., the centered balls of radii n in the case of $\mathbb{R}^d$). It is known (see, e.g., Chicco [**240**], where domains in $\mathbb{R}^d$ are considered, but the same method applies to domains with compact closures in Riemannian manifolds) that there exists a continuous function $u_n \in W^{p,2}(B_n) \cap W_0^{p,1}(B_n)$ such that

$$(I - \mathcal{L}_{A,b})u_n = 1 \quad \text{in } B_n.$$

We extend u_n by zero outside B_n. Clearly, the extension, denoted again by u_n, belongs to $W_0^{p,1}(\Omega)$. Note that by the Sobolev embedding theorem u_n and ∇u_n are Hölder continuous on B_n and admit continuous extensions to the closure of B_n. Let $e(x)$ be the unit outward normal vector at the point $x \in \partial B_n$. Since $u_n \geqslant 0$ in B_n by the maximum principle and $u_n = 0$ on ∂B_n, we have

$$\langle \nabla u_n(x), \eta\rangle \leqslant 0 \text{ if } x \in \partial B_n \text{ and } \eta \text{ is a vector with } \langle \eta, e(x)\rangle \geqslant 0, \tag{5.3.3}$$

where $\nabla u_n(x)$ on ∂B_n is defined as the result of the continuous continuation of ∇u_n from inside. Let π_{B_n} denote the usual surface measure on ∂B_n. For every function $\varphi \in C_0^\infty(\Omega)$, we obtain

$$\begin{aligned}
&\int_{B_n} \langle A\nabla u_n, \nabla\varphi\rangle\, d\mu + \int_{B_n} u_n\varphi\, d\mu + \int_{B_n} \langle A\beta_\mu - b, \nabla u_n\rangle\varphi\, d\mu \\
&= -\int_{B_n} \mathcal{L}_{A,b}u_n\, \varphi\, d\mu + \int_{B_n} u_n\varphi\, d\mu + \int_{\partial B_n} \langle A\nabla u_n, e\rangle\varphi\varrho\, d\pi_{B_n} \\
&= \int_{B_n} \varphi\, d\mu + \int_{\partial B_n} \langle A\nabla u_n, e\rangle\varphi\varrho\, d\pi_{B_n}.
\end{aligned}$$

Now set $\chi_n = 1 - u_n$. Then $\nabla\chi_n = -\nabla u_n$ and $\nabla\chi_n = 0$ outside B_n. Therefore,

$$\begin{aligned}
&\int_\Omega \langle A\nabla\chi_n, \nabla\varphi\rangle\, d\mu + \int_\Omega \chi_n\varphi\, d\mu + \int_\Omega \langle A\beta_\mu - b, \nabla\chi_n\rangle\varphi\, d\mu \\
&= \int_{\Omega\backslash B_n} \varphi\, d\mu - \int_{\partial B_n} \langle A\nabla u_n, e\rangle\varphi\varrho\, d\pi_{B_n}.
\end{aligned}$$

The right-hand side is nonnegative for nonnegative φ according to (5.3.3), since $\langle A(x)e(x), e(x)\rangle > 0$. Let us show that $\lim\limits_{n\to\infty} \chi_n(x) = 0$ μ-a.e. We shall prove that μ-a.e. one has the equality $\lim\limits_{n\to\infty} u_n(x) = 1$. By the construction of the semigroup $\{T_t^\mu\}_{t\geqslant 0}$ one has $u_n \to (I - \mathcal{L}^\mu_{A,b})^{-1}1$ in $L^1(\mu)$. Passing to a subsequence, we may assume that we also have convergence a.e. (actually, one can show that the whole sequence $\{u_n\}$ is increasing, hence a.e. convergent). But since μ is invariant, we have $1 \in D(\mathcal{L}^{\mu,r}_{A,b})$ for all $r \in [1,\infty)$ and $(I - \mathcal{L}^{\mu,r}_{A,b})1 = 1$. Hence $\lim\limits_{n\to\infty} u_n(x) = 1$ μ-a.e. Thus, (iii) holds with $\eta = -1$.

Now let condition (iii) be satisfied with $\eta = -1$. We shall show that our condition (i) holds. Suppose that $h \in L^\infty(\mu)$ is such that

$$\int_\Omega (I - \mathcal{L}_{A,b})\varphi\, h\, d\mu = 0 \quad \forall \varphi \in C_0^\infty(\Omega).$$

Let us show that $h = 0$. If this is not true, we may assume that $\|h\|_\infty = 1$. The above equality means that the measure $\nu := h\cdot\mu$ satisfies the equation $\mathcal{L}^*_{A,b}\nu - \nu = 0$. We know that the measure ν has a density $f \in W^{p,1}_{\mathrm{loc}}(\Omega)$. Hence $h \in W^{p,1}_{\mathrm{loc}}(\Omega)$. It follows by the integration by parts formula that for any $\varphi \in W_0^{2,1}(\Omega)$ one has

$$\begin{aligned}\int_\Omega \langle A\nabla\varphi, \nabla h\rangle\, d\mu + \int_\Omega \varphi h\, d\mu - \int_\Omega \langle b - A\beta_\mu, \nabla\varphi\rangle h\, d\mu& \\ = \int_\Omega (I - \mathcal{L}_{A,b})\varphi\, h\, d\mu = 0.& \qquad (5.3.4)\end{aligned}$$

Let us set $v_n = \chi_n - h$. Then $v_n \in W^{2,1}_{\mathrm{loc}}(\Omega)$. In addition, since $\|h\|_\infty = 1$ and $(\chi_n - 1)^- \in W_0^{2,1}(\Omega)$, we obtain that $v_n^- := \min(v_n, 0) \in W_0^{2,1}(\Omega)$. Letting $\varphi = v_n^-$ in equality (5.3.4) and taking into account that $v_n v_n^- = -|v_n^-|^2$ and $\langle A\nabla v_n, v_n^-\rangle = -\langle A\nabla v_n^-, v_n^-\rangle$ a.e., we obtain by our hypothesis that

$$\begin{aligned} 0 &\leqslant \int_\Omega \langle A\nabla \chi_n, \nabla v_n^-\rangle\, d\mu + \int_\Omega \chi_n v_n^-\, d\mu - \int_\Omega \langle b - A\beta_\mu, \nabla v_n^-\rangle \chi_n\, d\mu \\ &= \int_\Omega \langle A\nabla v_n, \nabla v_n^-\rangle\, d\mu + \int_\Omega v_n v_n^-\, d\mu - \int_\Omega \langle b - A\beta_\mu, \nabla v_n^-\rangle v_n\, d\mu \\ &= -\int_\Omega \langle A\nabla v_n^-, \nabla v_n^-\rangle\, d\mu - \int_\Omega |v_n^-|^2\, d\mu. \end{aligned}$$

Therefore, $v_n^- = 0$ a.e., that is, $h \leqslant \chi_n$ a.e. The same reasoning applies to the function $-h$, hence $|h| \leqslant \chi_n$ a.e., which shows that $h = 0$ a.e. Thus, (i) holds.

If condition (iii) is satisfied with $\eta = 1$, then the same condition is satisfied with $\eta = -1$ for the dual drift $\widetilde{b}$. Therefore, according to what has already been proved, the measure μ is invariant for the semigroup $\{S_t^\mu\}_{t\geqslant 0}$ associated with $L_{A,\widetilde{b}}$. As explained in Remark 5.2.4, $\{S_t^\mu\}_{t\geqslant 0}$ coincides with the extension to $L^1(\mu)$ of the adjoint semigroup for $\{T_t^\mu\}_{t\geqslant 0}$ (the adjoint semigroup is originally defined on $L^\infty(\mu)$) and that μ is invariant for $\{T_t^\mu\}_{t\geqslant 0}$. As we have seen, this yields condition (iii) for $\eta = -1$. It follows from this reasoning that (iv) is equivalent to (i)–(iii), where (iii) can hold either with $\eta = -1$ or with $\eta = 1$. □

5.3.2. Remark. (i) The operator $L_{A,b}$ is not essentially m-dissipative on domain $C_0^\infty(\Omega)$ in $L^p(\mu)$ for some $p \in [1,\infty)$ precisely when there exists a nonzero function $h \in L^{p'}(\mu)$ with $p' = p/(p-1)$ such that the measure $h\cdot\mu$ satisfies the equation $(L_{A,b} - 1)^*(h\cdot\mu) = 0$. It follows from the proof of Theorem 5.3.1 that (under our conditions on A and b, of course) this is equivalent to the following: there exists a nonzero function $h \in L^{p'}(\mu) \cap W^{p,2}_{\mathrm{loc}}(\Omega)$ such that $L_{A,\widehat{b}}h = h$ a.e.

(ii) The proof employs some properties of our special semigroup $\{T_t^\mu\}_{t\geqslant 0}$, and we do not know whether in assertion (ii) one can use any associated semigroup.

(iii) It is seen from the proof that assertion (iii) can be replaced by the stronger one that for *each* $\alpha > 0$ there are such functions χ_n.

Let us give a simple sufficient condition for (iii) (hence also for (i) and (ii)) in terms of Lyapunov functions.

5.3.3. Theorem. *Let (5.2.1) be fulfilled and let Ω be connected. Suppose that there exist an unbounded compact function $V \in C^2(\Omega)$, a number $\alpha > 0$, and a compact set K such that*

$$L_{A,b}V(x) \leqslant \alpha V(x) \quad \text{for a.e. } x \in \Omega\backslash K. \tag{5.3.5}$$

Then $\mu \in \mathcal{M}^{A,b}_{\rm ell,md}$ and μ is invariant for $\{T^\mu_t\}_{t\geqslant 0}$. The analogous result holds for the operator $\mathcal{L}_{A,b}$.

PROOF. Note first that we may assume that

$$V \geqslant 0 \quad \text{and} \quad \mathcal{L}_{A,b}V \leqslant \alpha V \quad \text{a.e.} \tag{5.3.6}$$

Indeed, by our assumptions, there exists a compact set D such that $V(x) \geqslant 0$ if $x \notin D$ and $\mathcal{L}_{A,b}V(x) \leqslant \alpha V(x)$ a.e. outside D. Let $z = \max_D V$. It is clear that $z \geqslant 0$. Let $D^* := \{V \leqslant z+1\}$. We can find a C^2-function θ on $\mathbb{R}$ such that $\theta(s) = z + 1/2$ if $s \leqslant z$, $\theta(s) = s$ if $s \geqslant z+1$, $0 \leqslant \theta' \leqslant 1$ and $\theta'' \geqslant 0$. Set

$$N := \alpha(z+1) + \sup \theta''(s) \sup_{x\in D^*} \langle A(x)\nabla V(x), \nabla V(x)\rangle.$$

Let us consider the function

$$W := \theta \circ V + N.$$

Then $W \in C^2$, $W \geqslant 0$ and W is compact and unbounded (i.e., $W(x) \to +\infty$ as $|x| \to +\infty$ in the case $\Omega = \mathbb{R}^d$). In addition, we have

$$\mathcal{L}_{A,b}W = \theta'(V)\mathcal{L}_{A,b}V + \theta''(V)\langle A\nabla V, \nabla V\rangle.$$

If $x \in D$, then $\theta\bigl(V(x)\bigr) = z+1/2$ and $\mathcal{L}_{A,b}W(x) = 0$. If $x \notin D^*$, then $V(x) > z+1$, hence $W(x) = V(x) + N$. Therefore,

$$\mathcal{L}_{A,b}W(x) = \mathcal{L}_{A,b}V(x) \leqslant \alpha V(x) \leqslant \alpha W(x) \text{ a.e. outside } D^*.$$

Finally, a.e. in $D^*\backslash D$ we have

$$\mathcal{L}_{A,b}W(x) = \theta'\bigl(V(x)\bigr)\mathcal{L}_{A,b}V(x) + \theta''\bigl(V(x)\bigr)\langle A(x)\nabla V(x), \nabla V(x)\rangle \leqslant N,$$

since $0 \leqslant \theta' \leqslant 1$ and $\mathcal{L}_{A,b}V(x) \leqslant \alpha V(x) \leqslant \alpha(z+1)$ if $x \in D^*\backslash D$.

Now let us set $\chi_n := n^{-1}V$. Clearly, the function $(\chi_n - 1)^-$ belongs to $W^{2,1}_0(\Omega)$ and $\lim\limits_{n\to\infty} \chi_n(x) = 0$. Finally, if $\mu \in \mathcal{M}^{A,b}_{\rm ell}$, then, for an arbitrary nonnegative function $\varphi \in C_0^\infty(\Omega)$, one has by the integration by parts formula

$$\begin{aligned}\int \langle A\nabla\chi_n, \nabla\varphi\rangle\, d\mu + \alpha \int \chi_n\varphi\, d\mu - \int \langle b - A\beta_\mu, \nabla\chi_n\rangle\, \varphi\, d\mu \\ = \int (\alpha\chi_n - \mathcal{L}_{A,b}\chi_n)\varphi\, d\mu = n^{-1}\int (\alpha V - \mathcal{L}_{A,b}V)\varphi\, d\mu \geqslant 0,\end{aligned}$$

which completes the proof by virtue of Theorem 5.3.1. The same reasoning applies to the operator $L_{A,b}$. □

For example, taking the function $V(x) = \ln(|x|^2+1)$ on $\mathbb{R}^d$ and noting that

$$L_{A,b}V(x) = \frac{2{\rm tr}A(x)}{|x|^2+1} - \frac{4\langle A(x)x, x\rangle}{(|x|^2+1)^2} + \frac{2\langle b(x), x\rangle}{|x|^2+1},$$

we arrive at the following result.

5.3.4. Example. Let (5.2.1) be fulfilled. Suppose that $\Omega = \mathbb{R}^d$ and that there exists a number $C > 0$ such that a.e. outside some ball one has

$$-\frac{2}{1+|x|^2}\langle A(x)x, x\rangle + \mathrm{tr}A(x) + \langle b(x), x\rangle \leqslant C|x|^2 \ln|x|.$$

Then $\mu \in \mathcal{M}^{A,b}_{\mathrm{ell,md}}$. In particular, if A is uniformly bounded, then it suffices to have

$$\langle b(x), x\rangle \leqslant C|x|^2 \ln|x| \quad \text{a.e. outside some ball.}$$

For example, the estimate

$$|b(x)| \leqslant c + c|x|\ln|x|$$

is sufficient. However, an estimate $|b(x)| \leqslant c + c|x|\big(\ln(2+|x|)\big)^r$ with $r > 1$ is not enough: in Example 4.2.1 we take a smooth function f with $f'(s) = |s|^{-1}(\ln s)^{-r}$ outside $[-2, 2]$. Here we can take any positive integrable function f', which will give an example on $\mathbb{R}^d$, $d > 1$, such that $|b(x)| \leqslant C/f'(|x|)$, but $\mu \notin \mathcal{M}^{A,b}_{\mathrm{ell,md}}$ due to nonuniqueness.

5.3.5. Remark. (i) In Exercise 5.6.49 it is suggested to show that if there exist a bounded function $u \in C^2(\mathbb{R}^d)$ and a number $\alpha > 0$ such that $u > 0$ and $L_{A,b}u \geqslant \alpha u$, then μ is not invariant for $\{T^\mu_t\}_{t\geqslant 0}$.

(ii) Let A satisfy (5.2.1) and let $b^i \in W^{d+,1}_{\mathrm{loc}}(\Omega)$. Let $V \in C^3(\Omega)$ be an unbounded compact function and let $\theta(x) = |L_{A,b}V(x)| + 1$. Then the scaled operator $L_{A/\theta, b/\theta} = \theta^{-1}L_{A,b}$ satisfies condition (5.2.1) and condition (5.3.5), since $\theta^{-1}L_{A,b}V \leqslant 1$. This shows that condition (5.3.5) can always be obtained by a proper scaling of a given operator (with reasonable coefficients). In addition, we see that there exists at most one probability density $f \in L^1(\mu)$ such that $f\cdot\mu \in \mathcal{M}^{A,b}_{\mathrm{ell}}$ and $\theta f \in L^1(\mu)$. Indeed, $(\theta^{-1}L_{A,b})^*(\theta f\cdot\mu) = 0$.

Our next goal is to study relations between infinitesimal invariance and invariance. As we shall see, these two concepts are different, but under some additional assumptions they coincide. First we mention a useful technical result which shows that any reasonable invariant measure ν of the semigroup $\{T^\mu_t\}_{t\geqslant 0}$ on $L^1(\mu)$ associated (as explained in Theorem 5.2.2 above) with a probability measure μ satisfying the equation $L^*_{A,b}\mu = 0$ also satisfies this equation.

5.3.6. Proposition. *Let Ω be an open set in $\mathbb{R}^d$. Assume that* (5.2.1) *is fulfilled and $\mu \in \mathcal{M}^{A,b}_{\mathrm{ell}}$. Let $\{T^\mu_t\}_{t\geqslant 0}$ be the associated semigroup specified in Theorem* 5.2.2. *Suppose that ν is a probability measure on Ω such that $\nu \ll \mu$ and*

$$\int_\Omega T^\mu_t f\,d\nu = \int_\Omega f\,d\nu \quad \forall\, f \in C^\infty_0(\Omega),\ \forall\, t > 0.$$

Assume, in addition, that $b^i \in L^q_{\mathrm{loc}}(\nu)$ for some $q > 1$. Then $\nu \in \mathcal{M}^{A,b}_{\mathrm{ell}}$.

In particular, if $|b|$ is locally bounded, then any absolutely continuous probability measure ν that is invariant for $\{T^\mu_t\}_{t\geqslant 0}$ satisfies our elliptic equation, hence has a positive continuous density and is equivalent to μ.

Proof. Let $f \in L^\infty(\mu)$. Note that $T^\mu_t f \in L^\infty(\nu)$, since $\nu \ll \mu$. By using that T^μ_t is sub-Markovian, it is readily verified by the monotone convergence theorem that

$$\int_\Omega T^\mu_t f\,d\nu = \int_\Omega f\,d\nu$$

for all nonnegative ν-integrable functions f, hence for all ν-integrable f. Then, using again that $\{T_t^\mu\}_{t\geqslant 0}$ is sub-Markovian, for all $t>0$ we obtain

$$\int_\Omega |T_t^\mu f|^q\,d\nu \leqslant \int_\Omega T_t^\mu |f|^q\,d\nu = \int_\Omega |f|^q\,d\nu$$

for all $f\in L^q(\nu)$. Therefore, for all $\varphi\in C_0^\infty(\Omega)$, the assumption that $b^i\in L^q_{\rm loc}(\Omega,\nu)$ yields that

$$\int_\Omega |T_t^\mu \mathcal{L}_{A,b}\varphi|^q\,d\nu \leqslant \int_\Omega |\mathcal{L}_{A,b}\varphi|^q\,d\nu < \infty.$$

Hence we have the uniform ν-integrability of the family $\{T_t^\mu \mathcal{L}_{A,b}\varphi \,|\, t\geqslant 0\}$. Therefore, the mapping $t\to T_t^\mu \mathcal{L}_{A,b}\varphi$ is continuous as a map to $L^1(\nu)$ by the Lebesgue–Vitali theorem, since it is continuous as a map to $L^1(\mu)$. Hence the $L^1(\nu)$-valued Riemann integrals

$$\int_0^t T_s^\mu \mathcal{L}_{A,b}\varphi\,ds,\quad t>0, \tag{5.3.7}$$

exist and it follows that in the space $L^1(\nu)$ one has

$$\frac{d}{dt}\int_0^t T_s^\mu \mathcal{L}_{A,b}\varphi\,ds = T_t^\mu \mathcal{L}_{A,b}\varphi,\quad t>0.$$

On the other hand, since $\nu\ll\mu$, the integral in (5.3.7) is equal to the corresponding $L^1(\mu)$-integral, therefore, it coincides with $T_t^\mu\varphi-\varphi$. Hence

$$\frac{d}{dt}T_t^\mu\varphi = T_t^\mu \mathcal{L}_{A,b}\varphi \quad \text{in } L^1(\nu),$$

and therefore, by the invariance of ν with respect to T_t^μ, one has

$$\begin{aligned}\int_\Omega \mathcal{L}_{A,b}\varphi\,d\nu &= \int_\Omega \frac{d}{dt}\bigl(T_t^\mu \mathcal{L}_{A,b}\varphi\bigr)_{t=0}\,d\nu\\ &= \frac{d}{dt}\Bigl(\int_\Omega T_t^\mu \mathcal{L}_{A,b}\varphi\,d\nu\Bigr)_{t=0} = \frac{d}{dt}\Bigl(\int_\Omega \mathcal{L}_{A,b}\varphi\,d\nu\Bigr)_{t=0} = 0,\end{aligned}$$

which completes the proof. □

This proposition is not true without the assumption that $\nu\ll\mu$. Indeed, let us fix $x_0\in\Omega$ and consider versions of $T_t^\mu f$, where $f\in C_0^\infty(\Omega)$, such that $T_t^\mu f(x_0)=f(x_0)$. Then Dirac's measure at x_0 is invariant for T_t^μ. We do not know, however, how essential the assumption that $b^i\in L^q(\nu)$ is. It will be explained below (Theorem 5.4.5) that in the case when (5.2.1) holds with $p>d+2$ in place of $d+$, there exist uniquely defined subprobability kernels $K_t(\,\cdot\,,\,\cdot\,)$ such that K_tf is a version of $T_t^\mu f$ for each $f\in L^1(\mu)$ and the assertion of this proposition is true for any invariant measure ν of the semigroup $\{K_t\}_{t\geqslant 0}$.

The next technical result gives a sufficient condition for a solution of the elliptic equation to be invariant for our special associated semigroup. However, as we shall see below, such a case does not really arise unless $\nu=\mu$.

5.3.7. Proposition. *Let $\mu\in\mathcal{M}^{A,b}_{\rm ell,md}$ and let $\{T_t^\mu\}_{t\geqslant 0}$ be the corresponding semigroup specified in Theorem 5.2.2. Suppose that ν is a bounded Borel measure on Ω (possibly signed) such that $L^*_{A,b}\nu=0$ and the function $\varrho:=d\nu/d\mu$ is bounded. Then ν is $\{T_t^\mu\}_{t\geqslant 0}$-invariant. The same is true in the case of $\mathcal{L}_{A,b}$.*

PROOF. We consider again the case of $\mathcal{L}_{A,b}$. Let $\varphi \in C_0^\infty(\Omega)$. Then, since ϱ is bounded and since by our assumption the operator $(\overline{\mathcal{L}}^\mu_{A,b}, D(\overline{\mathcal{L}}^\mu_{A,b}))$ is the generator of the semigroup $\{T_t^\mu\}_{t\geqslant 0}$ on $L^1(\mu)$, we have

$$\int_\Omega (T_t^\mu \varphi - \varphi)\, d\nu = \int_0^t \int_\Omega \overline{\mathcal{L}}^\mu_{A,b}(T_s^\mu \varphi)\, d\nu\, ds \quad \forall t > 0. \tag{5.3.8}$$

In order to conclude that the right-hand side of this equality is zero, it suffices to verify that, for every fixed $s \geqslant 0$, one has

$$\int_\Omega \overline{\mathcal{L}}^\mu_{A,b}(T_s^\mu \varphi)\, d\nu = 0. \tag{5.3.9}$$

Indeed, since $T_s^\mu \varphi$ belongs to the domain of $\overline{\mathcal{L}}^\mu_{A,b}$, by the definition of the closure, there exist functions $\varphi_k \in C_0^\infty(\Omega)$ such that

$$\lim_{k\to\infty} \mathcal{L}_{A,b}\varphi_k = \overline{\mathcal{L}}^\mu_{A,b} T_s^\mu \varphi$$

in $L^1(\mu)$, hence in $L^1(\nu)$. Since we have

$$\int_\Omega \mathcal{L}_{A,b}\varphi_k\, d\nu = 0$$

by our hypothesis, we obtain (5.3.9). Thus, ν is invariant for $\{T_t^\mu\}_{t\geqslant 0}$. □

5.3.8. Remark. It follows by Proposition 5.1.5 that the semigroup $\{T_t^\mu\}_{t\geqslant 0}$ cannot have invariant measures with positive densities if μ itself is not invariant. When applied to $\{T_t^\mu\}_{t\geqslant 0}$ in the case of a locally bounded function $|b|$, this fact (along with Proposition 5.3.6 and subinvariance of μ with respect to $\{T_t^\mu\}_{t\geqslant 0}$ taken into account) shows that if μ is not invariant for $\{T_t^\mu\}_{t\geqslant 0}$, then no measure equivalent to μ can be invariant for $\{T_t^\mu\}_{t\geqslant 0}$. Below we return to this question.

We now give conditions under which the existence of a Lyapunov function becomes also necessary for the existence of a probability solution (which is not always the case).

5.3.9. Proposition. *Let μ be a probability solution of the equation $L^*_{A,b}\mu = 0$ on $\mathbb{R}^d$, $a^{ij} \in C(\mathbb{R}^d) \cap W^{d+,1}_{\mathrm{loc}}(\mathbb{R}^d)$, $\det A > 0$, $b^i \in L^{d+}_{\mathrm{loc}}(\mathbb{R}^d)$, and*

$$\frac{\operatorname{tr} A(x)}{1+|x|^2},\ \frac{|b(x)|}{1+|x|} \in L^1(\mu).$$

Then there is a Lyapunov function $V \in W^{d+,2}_{\mathrm{loc}}(\mathbb{R}^d)$ such that $V(x) \to +\infty$ and $L_{A,b}V(x) \to -\infty$ as $|x| \to \infty$.

PROOF. We take a positive Borel function $h \in L^1(\mu)$ with $\|h\|_{L^1(\mu)} = 1$ such that $h \in L^{d+}_{\mathrm{loc}}(\mathbb{R}^d)$, $\lim\limits_{|x|\to+\infty} h(x) = +\infty$ and

$$\frac{|b(x)|}{|x|h(x)} + \frac{\operatorname{tr} A(x)}{|x|^2 h(x)} \to 0 \quad \text{as } |x| \to \infty.$$

For example, whenever $k \leqslant |x| < k+1$, we can set

$$h(x) = C_k + \frac{C_k \operatorname{tr} A(x)}{1+|x|^2} + \frac{C_k |b(x)|}{1+|x|},$$

where $C_k \to +\infty$ are chosen in such a way that the series of the integrals of h over the sets $\{k \leqslant |x| < k+1\}$ with respect to the measure μ converges; such numbers

C_k exists by the hypothesis of the theorem. Let $L = L_{A,b}$ and L^μ be the closure of L in $L^1(\mu)$, which exists and is the generator of $\{T_t^\mu\}_{t\geqslant 0}$ by Proposition 4.3.6(ii) and Proposition 5.2.5. By Proposition 4.3.6 the range of $I - L$ is dense in $L^1(\mu)$. By Lemma 5.2.8, for every $k \in \mathbb{N}$, the equation $k^{-1}u_k - Lu_k = h$ has a unique positive solution u_k in $L^1(\mu) \cap W^{d+,2}_{\text{loc}}(\mathbb{R}^d)$. By the embedding theorem u_k has a continuous gradient. We observe that the integral of $Lu_k = L^\mu u_k$ with respect to the measure μ vanishes, which follows from the construction of L^μ as the closure and the fact that the integral of Lu with respect to the measure μ vanishes for all functions $u \in C_0^\infty(\mathbb{R}^d)$. Hence $\|u_k\|_{L^1(\mu)} = k$.

We also observe that $u_{k+1} \geqslant u_k$ and $u_1(x) \to +\infty$ as $|x| \to \infty$. The first assertion follows from the fact that

$$(k+1)^{-1}(u_{k+1} - u_k) - L(u_{k+1} - u_k) = \bigl(k(k+1)\bigr)^{-1}u_k \geqslant 0.$$

For the proof of the second assertion we take $C > 0$ and pick $R > 0$ such that

$$h(x) > C + \frac{12C\,\mathrm{tr}\,A(x)}{|x|^2} + \frac{6C|b(x)|}{|x|} \tag{5.3.10}$$

as $|x| \geqslant R$. Then, for a fixed point x_0, for the function

$$G(x) = C\bigl(1 - 4|x - x_0|^2/|x_0|^2\bigr)$$

we have

$$-LG(x) = \frac{8C\,\mathrm{tr}\,A(x)}{|x_0|^2} + \frac{8C\langle b(x), x - x_0\rangle}{|x_0|}.$$

In addition,

$$G(x) = 0 \quad \text{if } |x - x_0| = |x_0|/2.$$

If $|x_0| > 2R$ and $|x - x_0| \leqslant |x_0|/2$, then, taking into account the trivial estimate $|x| \leqslant 3|x_0|/2$, which holds in this case, by (5.3.10) we obtain that

$$G(x) - LG(x) > h(x).$$

Hence by the maximum principle

$$G(x) \leqslant u_1(x) \quad \text{whenever } |x - x_0| \leqslant |x_0|/2.$$

In particular, $u_1(x_0) \geqslant C$. The same is true for all u_k, since in place of L we can take kL.

We now show that the minima of all u_k are attained in some ball. Let x_k be a point of minimum of u_k, which exists by the previous step. The equality $\|u_k\|_{L^1(\mu)} = k$ yields that $u_k(x_k) \leqslant k$, since otherwise $u_k(x) > k$ for all x. There is a closed ball U outside which we have $h \geqslant 3$. Suppose that x_k is outside U. Since $\nabla u_k(x_k) = 0$ and $|b(x)| \leqslant Ch(x)$ for some C in a sufficiently small neighborhood W of the point x_k, by the continuity of ∇u_k we can choose the neighborhood W so small that it will be contained in the complement of U and the estimate $|\langle b, \nabla u_k\rangle| \leqslant h(x)/3$ will hold on it. Then on W we have the estimate

$$\mathrm{tr}\,(AD^2u) = Lu_k - \langle b, \nabla u_k\rangle = -h + u_k/k - \langle b, \nabla u_k\rangle \leqslant -1.$$

By nondegeneracy of A and the strong maximum principle we conclude that the function u_k is constant in W, which is impossible, since this would give $D^2u_k = 0$ in W. Thus, $x_k \in U$.

Finally, let us consider the functions $v_k(x) = u_k(x) - u_k(x_k) \geqslant 0$ satisfying the equations $v_k/k - Lv_k = h - u_k(x_k)/k$. Passing to a subsequence, we may assume that $u_k(x_k)/k \to c \leqslant 1$. Since $v_k(x_k) = 0$, the sequence $\{v_k\}$ is norm bounded in

$W^{p,2}(U)$ on every ball U, where $p = p(U) > d$. We may assume that $\{v_k\}$ converges locally uniformly to a function $V \in W^{d+,2}_{\rm loc}(\mathbb{R}^d)$ satisfying the equation $LV = c - h$. As above, by using the function G, we verify that $V(x) \to +\infty$ as $|x| \to \infty$. Indeed, $L(V - G)(x) < 0$ if $|x - x_0| \leqslant |x_0|/2$, $|x_0| > 2R$, where we take the same R as above and $C > c$. Then $V(x_0) > G(x_0) = C$. □

5.3.10. Example. Let μ be a probability solution of the equation $L^*_{A,b}\mu = 0$, where $a^{ij} \in C(\mathbb{R}^d) \cap W^{d+,1}_{\rm loc}(\mathbb{R}^d)$, $\det A > 0$. If

$$\sqrt{|a^{ij}(x)|} + |b^i(x)| \leqslant C_1 + C_2|x|,$$

then there is a Lyapunov function $V \in W^{d+,2}_{\rm loc}(\mathbb{R}^d)$ for which $V(x) \to +\infty$ and $LV(x) \to -\infty$ as $|x| \to \infty$.

We do not know whether the conclusion of the proposition remains true if the integrability assumptions are replaced with the assumption that $(I - L)\bigl(C_0^\infty(\mathbb{R}^d)\bigr)$ is dense in $L^1(\mu)$, which follows from the indicated integrability conditions, and also follows from the existence of a Lyapunov function, so that this is the minimal requirement sufficient for the validity of the proposition.

5.4. Invariance and uniqueness

In this section we return to the investigation of the uniqueness of probability solutions to our elliptic equations, which turns out to be closely connected with the invariance of the measure on a domain Ω in $\mathbb{R}^d$ or in a manifold with respect to the associated semigroup.

5.4.1. Theorem. *Assume that* (5.2.1) *is fulfilled and that* Ω *is connected. Then*

$$\mathcal{M}^{A,b}_{\rm ell,md} \neq \varnothing \implies \#\mathcal{M}^{A,b}_{\rm ell} = 1.$$

Proof. Let $\mu = \varrho_\mu\,dx \in \mathcal{M}^{A,b}_{\rm ell,md}$ and $\nu = \varrho_\nu\,dx \in \mathcal{M}^{A,b}_{\rm ell}$, where ϱ_μ and ϱ_ν are the strictly positive, continuous versions of the respective Radon–Nikodym derivatives from the class $W^{p,1}_{\rm loc}(\Omega)$, where for notational convenience we assume that $p > d$ is common for the whole Ω. Such densities exist according to the results in Chapter 1. Let us apply Theorem 5.3.1 to μ. Then, there exist functions χ_n, where $n \in \mathbb{N}$, with the properties indicated in assertion (iii) of the theorem cited.

Now we shall show that assertion (ii) of Theorem 5.3.1 also holds for ν in place of μ with the same functions χ_n. Since both densities are strictly positive continuous, we have $\varrho_\nu/\varrho_\mu \in W^{p,2}_{\rm loc}(\Omega)$. Suppose first that $\Omega \subset \mathbb{R}^d$. Then, $\varphi\varrho_\nu/\varrho_\mu \in W^{2,1}_0(\Omega)$ for every $\varphi \in C_0^\infty(\Omega)$. Therefore, if $\varphi \geqslant 0$, we obtain by (5.3.2) that

$$\begin{aligned}
&\int a^{ij}\,\partial_{x_i}\chi_n\,\partial_{x_j}\varphi\,d\nu + \alpha\int \chi_n\varphi\,d\nu + \int \bigl(\beta^i_{\nu,A} - b^i\bigr)\,\partial_{x_i}\chi_n\,\varphi\,d\nu \\
&= \int a^{ij}\partial_{x_i}\chi_n\,\partial_{x_j}\Bigl(\varphi\frac{\varrho_\nu}{\varrho_\mu}\Bigr)\,d\mu + \alpha\int \chi_n\varphi\frac{\varrho_\nu}{\varrho_\mu}\,d\mu - \int a^{ij}\partial_{x_i}\chi_n\,\partial_{x_j}\Bigl(\frac{\varrho_\nu}{\varrho_\mu}\Bigr)\varphi\,d\mu \\
&+ \int \bigl(\beta^i_{\mu,A} - b^i\bigr)\,\partial_{x_i}\chi_n\Bigl(\varphi\frac{\varrho_\nu}{\varrho_\mu}\Bigr)\,d\mu + \int a^{ij}\Bigl(\frac{\partial_{x_j}\varrho_\nu}{\varrho_\nu} - \frac{\partial_{x_j}\varrho_\mu}{\varrho_\mu}\Bigr)\frac{\varrho_\nu}{\varrho_\mu}\,\partial_{x_i}\chi_n\,\varphi\,d\mu \geqslant 0,
\end{aligned}$$

since the last summand on the right cancels the third one due to the equality

$$\partial_{x_j}\Bigl(\frac{\varrho_\nu}{\varrho_\mu}\Bigr) = \Bigl(\frac{\partial_{x_j}\varrho_\nu}{\varrho_\nu} - \frac{\partial_{x_j}\varrho_\mu}{\varrho_\mu}\Bigr)\frac{\varrho_\nu}{\varrho_\mu}.$$

The reasoning is similar in the manifold case. Namely, one has

$$\begin{aligned}
&\int \langle A\nabla\chi_n, \nabla\varphi\rangle\, d\nu + \alpha \int \chi_n\varphi\, d\nu + \int \langle A\beta_\nu - b, \nabla\chi_n\rangle\varphi\, d\nu \\
&= \int \Big\langle A\nabla\chi_n, \nabla\Big(\varphi\frac{\varrho_\nu}{\varrho_\mu}\Big)\Big\rangle\, d\mu + \alpha\int \chi_n\varphi\frac{\varrho_\nu}{\varrho_\mu}\, d\mu - \int \Big\langle A\nabla\chi_n, \nabla\Big(\frac{\varrho_\nu}{\varrho_\mu}\Big)\Big\rangle\varphi\, d\mu \\
&+ \int \langle A\beta_\mu - b, \nabla\chi_n\rangle\varphi\frac{\varrho_\nu}{\varrho_\mu}\, d\mu + \int \Big\langle A\Big(\frac{\nabla\varrho_\nu}{\varrho_\nu} - \frac{\nabla\varrho_\mu}{\varrho_\mu}\Big)\frac{\varrho_\nu}{\varrho_\mu}, \nabla\chi_n\Big\rangle\varphi\, d\mu \\
&= \int \Big\langle A\nabla\chi_n, \nabla\Big(\varphi\frac{\varrho_\nu}{\varrho_\mu}\Big)\Big\rangle\, d\mu + \alpha\int \chi_n\varphi\frac{\varrho_\nu}{\varrho_\mu}\, d\mu + \int \langle A\beta_\mu - b, \nabla\chi_n\rangle\varphi\frac{\varrho_\nu}{\varrho_\mu}\, d\mu \geqslant 0.
\end{aligned}$$

By Theorem 5.3.1 we have $\nu \in \mathcal{M}^{A,b}_{\mathrm{ell,md}}$. Then we have the same for the measure $\mu_0 := (\mu+\nu)/2$, which also satisfies the elliptic equation. By Proposition 5.3.7, both measures μ and ν (which have bounded densities with respect to μ_0) are invariant for the semigroup $\{T^{\mu_0}_t\}_{t\geqslant 0}$ associated with μ_0. Hence the measure $|\mu-\nu|$ is also invariant for $\{T^{\mu_0}_t\}_{t\geqslant 0}$ (see, e.g., Proposition 2.2 in [**18**]). Since this measure has the density $|\varrho_\mu - \varrho_\nu|$, which is continuous, Proposition 5.3.6 yields that $L^*_{A,b}|\mu-\nu| = 0$. Hence $|\varrho_\mu - \varrho_\nu| > 0$ (unless $\mu = \nu$), which is impossible, since the continuous function $\varrho_\mu - \varrho_\nu$ has zero integral and must vanish somewhere. □

The importance of this result is seen from the fact that the uniqueness of solutions is ensured by information about *some* solution.

Certainly, it can happen that $\mathcal{M}^{A,b}_{\mathrm{ell}} = \varnothing$ (for example, if $\Omega = \mathbb{R}^d$, $A = \mathrm{I}$, $b \equiv 0$). But even in the case where $\Omega = \mathbb{R}^d$, $A = \mathrm{I}$ and b is infinitely differentiable, it can also happen that $\#\mathcal{M}^{A,b}_{\mathrm{ell}} > 1$ and so $\mathcal{M}^{A,b}_{\mathrm{ell,md}}$ is empty (see Example 4.2.1).

5.4.2. Remark. The inverse implication is false if $\Omega = \mathbb{R}^1$, $A(x)=1$ and $b(x) = -2x - 6e^{x^2}$. Here $\mathcal{M}^{A,b}_{\mathrm{ell}} = \{\mu\}$, where μ is given by density $\pi^{-1/2}e^{-x^2}$, but by Exercise 4.5.17 the operator $\big(L_{A,b}, C_0^\infty(\mathbb{R}^1)\big)$ is not essentially m-dissipative on the space $L^1(\mu)$.

Now Theorem 5.3.1 yields the following fact.

5.4.3. Corollary. *Under the assumptions of the previous theorem, if there exists a measure* $\mu \in \mathcal{M}^{A,b}_{\mathrm{ell}}$ *invariant with respect to the corresponding semigroup* $\{T^\mu_t\}_{t\geqslant 0}$, *then* $\#\mathcal{M}^{A,b}_{\mathrm{ell}} = 1$.

5.4.4. Remark. Suppose that the domain Ω is connected. Then, under the assumptions of Theorem 5.4.1, $\#\mathcal{M}^{A,b}_{\mathrm{ell}} \leqslant 1$ if and only if for any $\mu, \nu \in \mathcal{M}^{A,b}_{\mathrm{ell}}$, one has $|\mu-\nu| \in \mathcal{M}^{A,b}_{\mathrm{ell}}$. Indeed, the indicated condition yields that $\mu = \nu$, since otherwise $|\mu-\nu|$ must have a strictly positive continuous density. The converse is trivial.

This leads us to the following question: if a signed measure μ satisfies the equation $L^*_{A,b}\mu = 0$, then does $|\mu|$ also satisfy the same equation?

In general, this is not true even if $A = \mathrm{I}$ and b is smooth. Indeed, as explained above, in the situation where every probability measure satisfying the equation $L^*_{A,b}\mu = 0$ possesses a positive continuous density, which is the case if $a^{ij} \in W^{p,1}_{\mathrm{loc}}(\mathbb{R}^d)$, $|b| \in L^p_{\mathrm{loc}}(\mathbb{R}^d)$ with $p > d$, and A is nondegenerate, the nonuniqueness of solutions to this equation in $\mathcal{P}(\mathbb{R}^d)$ always yields signed solutions whose absolute values are not solutions. Of course, this is not surprising for locally integrable solutions. For example, the absolute value of a harmonic function may not be harmonic, but for globally integrable solutions this phenomenon is more interesting. It is worth noting here that if μ is an invariant measure for a semigroup $\{T_t\}_{t\geqslant 0}$ whose generator extends $(L_{A,b}, C_0^\infty)$, then $|\mu|$ is also an invariant measure, which again exhibits some difference between invariant measures for semigroups and infinitesimally invariant measures. As we shall see below, under our typical assumptions about A and b involving Lyapunov functions, the equation $L_{A,b}^*\mu = 0$ has one solution in $\mathcal{P}(\mathbb{R}^d)$. However, we do not know whether in such a case the space of all solutions in the class of bounded signed measures is one-dimensional.

We know from the previous chapter that it can really happen for $\Omega = \mathbb{R}^d$ that $\#\mathcal{M}_{\mathrm{ell}}^{A,b} > 1$. The above results give conditions on A and b under which $\#\mathcal{M}_{\mathrm{ell}}^{A,b} = 1$. Other results will be given below. However, the following question (already touched upon in Remark 5.3.8) arises:

Can the semigroup $\{T_t^\mu\}_{t\geqslant 0}$ have invariant measures if μ is not invariant or can it have invariant measures distinct from μ in the case when μ is invariant?

The following result answers this question in the case $p > d+2$. The same is likely to hold with our standard assumption that $p > d$.

5.4.5. Theorem. *Suppose that $\mu \in \mathcal{M}_{\mathrm{ell}}^{A,b}$ and that (5.2.1) is fulfilled with some $p > d+2$ in place of $d+$, i.e., $a^{ij} \in C(\Omega)\cap W_{\mathrm{loc}}^{p,1}(\Omega)$, $\det A > 0$, $b^i \in L_{\mathrm{loc}}^p(\Omega)$. Then, there exist subprobability kernels $K_t(\,\cdot\,, dy)$, $t > 0$, on Ω such that*

$$K_t(x, dy) = p_{A,b}(t, x, y)\, dy,$$

where $p_{A,b}(t, x, y)$ is a locally Hölder continuous nonnegative function defined on the set $(0, +\infty)\times\Omega\times\Omega$, and for every $f \in L^1(\Omega, \mu)$, the function

$$x \mapsto K_t f(x) := \int_\Omega f(y)\, p_{A,b}(t, x, y)\, dy$$

is a μ-version of $T_t^\mu f$ such that the function $(t, x) \mapsto K_t f(x)$ is continuous on the set $(0, +\infty)\times\Omega$. The function $p_{A,b}$ is positive if Ω is connected.

In addition, if Ω is connected and ν is a bounded Borel measure on Ω that is invariant for $\{K_t\}_{t\geqslant 0}$, i.e.,

$$\nu = K_t^*\nu(dy) := \int_\Omega K_t(x, dy)\,\nu(dx) \quad \forall\, t \geqslant 0,$$

then $\nu = c\mu$ for some constant c. In particular, if $\nu \neq 0$, then μ is also invariant. Hence $\{K_t\}_{t\geqslant 0}$ cannot have invariant probability measures different from μ.

The proof will be given in Chapter 6 on the basis of the results about parabolic equations (see Theorem 6.4.7). Note that p can be replaced by $(d+2)+$.

It is important to note that the obtained canonical version of T_t^μ enables us to speak of Borel invariant measures for $\{T_t^\mu\}$; without such a version this concept will be correct only for measures absolutely continuous with respect to μ, since the semigroup is given originally on $L^1(\mu)$.

5.4.6. Remark. By this theorem, the semigroup $\{T_t^\mu\}_{t\geqslant 0}$ is strong Feller: it takes bounded Borel functions to continuous functions (see Definition 5.6.9). In addition, it is stochastically continuous in the sense that $\lim_{t\to 0} T_t^\mu I_{U(z,r)}(x) = 1$ for every $z \in \Omega$ and $r > 0$ such that $U(z, r) \subset \Omega$. This follows from the last assertion in Theorem 5.2.2 and the estimate $T_t^\mu I_{U(z,r)} \geqslant T_t^\mu \psi$, where $\psi \in C_0^\infty(\Omega)$ is such that $0 \leqslant \psi \leqslant 1$, $\psi(z) = 1$ and $\psi = 0$ outside $U(z,r)$. If Ω is connected, we have the equivalence of all measures

$$B \mapsto T_t^\mu I_B(x) = K_t^*\delta_x(B) = \int_B p_{A,b}(t,x,y)\,dy.$$

Therefore, if μ is invariant for $\{T_t^\mu\}_{t\geqslant 0}$ (which is not automatically fulfilled in our situation!), then, by Doob's theorem (see Theorem 5.6.26 below), for every Borel set $B \subset \Omega$ we obtain

$$\lim_{t\to\infty} T_t^\mu I_B(x) = \mu(B) \quad \forall x \in \Omega,$$

where the jointly continuous version of $T_t^\mu I_B(x)$ is considered. Certainly, this yields that $\lim\limits_{t\to\infty} \|T_t^\mu f - \overline{f}\|_{L^p(\mu)} = 0$ for all $f \in L^p(\mu)$, where $\overline{f}$ is the integral of f with respect to μ. Moreover, it follows from this that $K_t f(x) \to \overline{f}$ as $t \to \infty$ uniformly on compact sets for all $f \in L^1(\mu)$, since the uniform Hölder continuity on a ball and convergence in measure give uniform convergence on this ball (Exercise 1.8.17).

Thus, one of the main results of the last two sections is the following alternative under local assumptions (A1) and (A2), in case when the equation $L_{A,b}^*\mu = 0$ has a solution μ in the class of probability measures: either μ is a unique probability measure satisfying this equation and its special associated semigroup $\{T_t^\mu\}_{t\geqslant 0}$ is a unique associated semigroup, and if μ is invariant for $\{T_t^\mu\}_{t\geqslant 0}$ (which is not implied automatically by the uniqueness of μ even on the real line, see Remark 5.4.2), then there are no other invariant probability measures for $\{T_t^\mu\}_{t\geqslant 0}$; or no probability measure μ satisfying this equation is invariant for its semigroup $\{T_t^\mu\}_{t\geqslant 0}$, and then for every such measure μ necessarily there are different associated semigroups in $L^1(\mu)$. An important qualitative conclusion is that, under reasonable local assumptions about A and b, one has uniqueness of invariant probability measures for our special associated semigroups, but not uniqueness for the elliptic equation, and that better smoothness of the coefficients does not help to obtain uniqueness of solutions for the elliptic equation.

5.5. Examples of uniqueness

We now consider some examples in which the uniqueness of solutions to elliptic equations and invariance with respect to semigroups is effectively verified in terms of the coefficients of the equations.

5.5.1. Example. Let the hypotheses of Theorem 5.3.3 be satisfied, i.e., let (5.2.1) be fulfilled and let Ω be connected. Suppose that there exist an unbounded compact function $V \in C^2(\Omega)$, a number $\alpha > 0$, and a compact set K such that

$$L_{A,b}V(x) \leqslant \alpha V(x) \quad \text{for a.e. } x \in \Omega\backslash K.$$

Then $\#\mathcal{M}_{\rm ell}^{A,b} \leqslant 1$. The analogous result holds for the operator $\mathcal{L}_{A,b}$. This follows by Theorem 5.4.1. For $\Omega = \mathbb{R}^d$ this follows also by Theorem 4.1.6.

As in the case with Theorem 5.3.3, we arrive at the following result.

5.5.2. Example. Let (5.2.1) be fulfilled. Suppose that $\Omega = \mathbb{R}^d$ and that there exists a number $C > 0$ such that a.e. outside some ball one has

$$-\frac{2}{1+|x|^2}\langle A(x)x, x\rangle + \mathrm{tr}A(x) + \langle b(x), x\rangle \leqslant C|x|^2 \ln|x|.$$

Then $\#\mathcal{M}_{\rm ell}^{A,b} \leqslant 1$. In particular, if A is uniformly bounded, then it suffices to have the estimate

$$\langle b(x), x\rangle \leqslant C|x|^2 \ln |x| \quad \text{a.e. outside some ball.}$$

For example, the estimate

$$|b(x)| \leqslant c + c|x| \ln |x|$$

is sufficient. However, an estimate $|b(x)| \leqslant c + c|x|\bigl(\ln(2+|x|)\bigr)^r$ with $r > 1$ is not enough: in Example 4.2.1 we take a smooth function f with $f'(s) = |s|^{-1}(\ln s)^{-r}$ outside $[-2, 2]$. Here one can take any positive integrable function f', which will give an example of nonuniqueness with the bound $|b(x)| \leqslant C/f'(|x|)$.

In the next example, continuing Theorem 4.1.6(ii), in order to conclude that μ is a unique element in $\mathcal{M}_{\rm ell}^{A,b}$ some information about μ itself is used.

5.5.3. Example. Suppose that conditions (5.2.1) hold. Assume also that $\Omega = \mathbb{R}^d$, $\mu \in \mathcal{M}_{\rm ell}^{A,b}$, and

$$\lim_{j\to\infty} \int_{j\leqslant|x|\leqslant j+1} \bigl[|a^{ik}| + |b - \beta_{\mu,A}|\bigr]\, d\mu = 0$$

for all i, k. Then $\#\mathcal{M}_{\rm ell}^{A,b} = 1$. In particular, it suffices to have the integrability of a^{ik} and $|b - \beta_{\mu,A}|$ with respect to μ.

In the case of the operator $\mathcal{L}_{A,b}$ on a connected complete Riemannian manifold M, a sufficient condition for μ to be a unique measure in $\mathcal{M}_{\rm ell}^{A,b}$ is the relation

$$\lim_{j\to\infty} \int_{B_{j+1}\backslash B_j} \bigl[\|A\| + |b - A\beta_\mu|\bigr]\, d\mu = 0.$$

where B_j is the geodesic ball of radius j centered at a fixed point $o \in M$. In particular, it suffices to have the integrability of $\|A\|$ and $|b - A\beta_\mu|$ with respect to the measure μ.

PROOF. We shall verify only the second assertion, since the first one is completely analogous. According to Theorem 5.4.1, it suffices to show that the set $(\mathcal{L}_{A,b} - I)(C_0^\infty(M))$ is dense in $L^1(\mu)$. Let g be a measurable function with $|g| \leqslant 1$ and

$$\int_M (\mathcal{L}_{A,b}\varphi - \varphi) g\, d\mu = 0$$

for all $\varphi \in C_0^\infty(M)$. We know that the function g belongs to $W_{\rm loc}^{d+,1}(M)$. The measure μ has a continuous strictly positive density of class $W_{\rm loc}^{d+,1}(M)$. Therefore, the above equality can be written in the form

$$\int_M \langle A\nabla\varphi, \nabla g\rangle\, d\mu + \int_M \langle A\nabla\varphi, \beta_\mu\rangle g\, d\mu + \int_M \varphi g\, d\mu = \int_M \langle b, \nabla\varphi\rangle g\, d\mu. \tag{5.5.1}$$

It is clear that equality (5.5.1) remains valid for all $\varphi \in W_0^{d+,1}(M)$. By the completeness of M, there exists a sequence of Lipschitzian functions ψ_j on M with the following properties: $0 \leqslant \psi_j \leqslant 1$, $\psi_j = 1$ on B_j, $\psi_j = 0$ outside B_{j+1}, $|\nabla\psi_j| \leqslant C$ for all j and some number C. By virtue of equality (5.5.1) applied to $\varphi = \psi_j^2 g$ we obtain with the aid of obvious transformations that

$$\begin{aligned}
&\int_M \psi_j^2 g^2\,d\mu + \int_M \langle A\nabla(\psi_j g), \nabla(\psi_j g)\rangle\,d\mu \\
&\quad = \int_M \psi_j^2 g^2\,d\mu + \int_M \langle A\nabla(\psi_j^2 g), \nabla g\rangle\,d\mu + \int_M \langle A\nabla\psi_j, \nabla\psi_j\rangle g^2\,d\mu \\
&\quad = \int_M \langle A\nabla\psi_j, \nabla\psi_j\rangle g^2\,d\mu + \int_M \langle b, \nabla(\psi_j^2 g)\rangle g\,d\mu - \int_M \langle A\nabla(\psi_j^2 g), \beta_\mu\rangle g\,d\mu \\
&\quad = \int_M \langle A\nabla\psi_j, \nabla\psi_j\rangle g^2\,d\mu + \int_M \langle b - A\beta_\mu, \nabla\psi_j\rangle \psi_j g^2\,d\mu \\
&\qquad + \int_M \langle b - A\beta_\mu, \nabla(\psi_j g)\rangle \psi_j g\,d\mu \\
&\quad = \int_M \langle A\nabla\psi_j, \nabla\psi_j\rangle g^2\,d\mu + \int_M \langle b - A\beta_\mu, \nabla\psi_j\rangle \psi_j g^2\,d\mu \\
&\quad \leqslant \int_{B_{j+1}\backslash B_j} \bigl[C^2\|A\| + C|b - A\beta_\mu|\bigr]\,d\mu,
\end{aligned}$$

since

$$\int_M \langle b - A\beta_\mu, \nabla(\psi_j g)\rangle \psi_j g\,d\mu = \frac{1}{2}\int_M \langle b - A\beta_\mu, \nabla(\psi_j g)^2\rangle\,d\mu = 0$$

by the equality $\mathcal{L}_{A,b}^*\mu = 0$. Hence the integrals of $\psi_j^2 g^2$ converge to zero and we have $g = 0$. For a compact manifold the reasoning is simpler, since we take at once the function $\varphi = g$. □

In the next example, it is not necessary to know β_μ.

5.5.4. Example. Let $\Omega = \mathbb{R}^d$ and $\mu \in \mathcal{M}_{\mathrm{ell}}^{A,b}$. Assume that A and A^{-1} are uniformly bounded, A is globally Lipschitzian, and that $|b| \in L_{\mathrm{loc}}^{d+}(\mathbb{R}^d)$. Suppose also that $|b| \in L^2(\mu)$. Then $\#\mathcal{M}_{\mathrm{ell}}^{A,b} = 1$ by Proposition 4.3.6.

5.5.5. Remark. In the last example, one cannot omit the assumption that $|b| \in L_{\mathrm{loc}}^{d+}(\Omega, dx)$, as is clear from Example 1.4.4. If in that example we take $\mu = \varrho\,dx$ with $\varrho(x) = Cx^2\exp(-x^2)$, where C is a normalizing constant, then μ satisfies the equation $L_{1,b}^*\mu = 0$ with $b(x) = 2x^{-1} - 2x$. The probability measure with density g such that $g(x) = f(x)/2$ if $x < 0$ and $g(x) = 3f(x)/2$ if $x \geqslant 0$ is another solution. For both solutions the operator $L_{1,b}$ is even symmetric and the coefficient b is quadratically integrable with respect to both measures. One can verify (Exercise 5.6.61) that these two measures are not merely solutions to the same elliptic equation, but are also invariant for the same Markovian semigroup associated to $L_{1,b}$ (in the case under consideration such a semigroup is unique).

The results presented above have the following character: either it is asserted that $\#\mathcal{M}_{\mathrm{ell}}^{A,b} \leqslant 1$ or it is asserted that $\#\mathcal{M}_{\mathrm{ell}}^{A,b} = 1$, but the existence of some solution is part of the hypotheses. However, the existence results presented in § 2.4 in terms of Lyapunov functions yield also the uniqueness of solutions, as is clear from the results of this section.

5.5.6. Example. Let $\Omega = \mathbb{R}^d$, $A = \mathrm{I}$, $b^i \in L_{\mathrm{loc}}^{d+}(\mathbb{R}^d)$ and

$$\limsup_{|x|\to\infty} |x|^{\gamma-1}\bigl[2(\gamma - 1) + d + \langle b(x), x\rangle\bigr] = -\infty$$

for some $\gamma \geqslant 1$. Then $\#\mathcal{M}_{\mathrm{ell}}^{A,b} = 1$.

The presented uniqueness results assume the Sobolev differentiability of a^{ij}. The case of a general nondegenerate measurable diffusion coefficient has not been studied. For example, the uniqueness of solutions to the equation $L^*_{A,b}\mu = 0$ with uniformly bounded Borel A, A^{-1}, b has not been studied. It is worth noting that if a^{ij} and b^i are locally Hölder continuous, $\det A > 0$, and a probability measure μ is invariant for the semigroup $\{T_t^\mu\}_{t\geqslant 0}$ associated with $L_{A,b}$, then it has a density locally uniformly separated from zero and hence is a unique invariant measure (see, e.g., Lorenzi, Bertoldi [**642**, § 8.1]).

In § 5.6(i) below we briefly discuss the so-called symmetric case, that is, the case $b = \beta_{A,\mu}$ (or $b = A\beta_\mu$ for operators in divergence form). If $A = \mathrm{I}$, then b is just $\nabla\varrho/\varrho$.

5.6. Complements, comments, and exercises

5.6(i). Essential self-adjointness of Dirichlet operators

Let μ be a probability measure on $\mathbb{R}^d$ with a density $\varrho \in W^{1,1}_{\mathrm{loc}}(\mathbb{R}^d)$. We set $\beta = \nabla\varrho/\varrho$, where $\beta = 0$ on the set $\{\varrho = 0\}$. The operator

$$Lf = \Delta f + \langle f, \beta\rangle,$$

called the Dirichlet operator, is symmetric on the domain $C_0^\infty(\mathbb{R}^d)$. Here we briefly discuss its essential self-adjointness, which attracted considerable attention in the literature. The operator $L = L_{\mathrm{I},b}$ turns out to be symmetric nonpositive. Its essential self-adjointness on $C_0^\infty(\Omega)$ is equivalent to the density of the range of $L - I$ in $L^2(\mu)$, which is stronger than the L^1-uniqueness (the density of its range in $L^1(\mu)$). Various results can be found in Bogachev, Krylov, Röckner [**150**], [**226**], Eberle [**320**], [**321**], Kolb [**525**], Liskevich [**620**], Liskevich, Semenov [**623**], Liskevich, Us [**627**], Röckner, Zhang [**819**], [**820**], Semenov [**841**], Stannat [**872**], Takeda [**891**], [**892**], Wielens [**939**], and in the references therein. Let us mention some principal results with the simplest formulations. The essential self-adjointness of $L = L_{\mathrm{I},b}$ on $C_0^\infty(\mathbb{R}^d)$ has been proved under certain global or local conditions. Here is a typical example.

5.6.1. Theorem. *Let the density $\varrho \geqslant 0$ be such that $\sqrt{\varrho} \in W^{2,1}_{\mathrm{loc}}(\mathbb{R}^d)$. For the essential self-adjointness of $L_{\mathrm{I},b}$ on domain $C_0^\infty(\mathbb{R}^d)$ in $L^2(\varrho\, dx)$ it suffices to have at least one of the following conditions*:

(i) (Liskevich, Semenov [**623**]) $|\nabla\varrho|/\varrho \in L^4(\varrho\, dx)$;

(ii) (Bogachev, Krylov, Röckner [**150**]) $|\nabla\varrho|/\varrho \in L^p_{\mathrm{loc}}(\mathbb{R}^d)$, *where* $p > d$.

It follows from the proof in [**150**] that in place of $p > d$ it suffices to have its own $p > d$ for each ball. Moreover, if $d > 3$, then yet a weaker local condition is sufficient, as shown in Liskevich [**620**]; for example, in terms of L^p_{loc} it suffices to have $p = d$. Further improvements were obtained in Kolb [**525**], where the following result was established.

5.6.2. Theorem. *Let $\varrho \geqslant 0$, $\sqrt{\varrho} \in W^{2,1}_{\rm loc}(\mathbb{R}^d)$ and $\nabla\varrho/\varrho \in L^4_{\rm loc}(\varrho\,dx)$. Suppose additionally that for every ball U_0 there is $\varepsilon > 0$ such that*

$$\sup_{U \subset U_0} \left(|U|^{-1}\int_U \varrho(x)^{1+\varepsilon}\,dx\right)\left(|U|^{-1}\int_U \varrho(x)^{-1-\varepsilon}\,dx\right) < \infty,$$

where sup is taken over all balls $U \subset U_0$ and $|U|$ is the volume of U. Then the operator $L_{{\rm I},b}$ with $b = \nabla\varrho/\varrho$ is essentially self-adjoint on $C_0^\infty(\mathbb{R}^d) \subset L^2(\varrho\,dx)$.

It is clear that if $|\nabla\varrho|/\varrho \in L^p_{\rm loc}(\mathbb{R}^d)$, where $p > d > 3$, then the hypotheses of this theorem are satisfied, since ϱ has a positive continuous density of class $W^{p,1}_{\rm loc}(\mathbb{R}^d)$.

It is still unknown whether the single condition $\nabla\varrho/\varrho \in L^4_{\rm loc}(\varrho\,dx)$, independent of the dimension, is sufficient (but such a global condition is sufficient by the previous theorem).

In the case of a nonconstant diffusion coefficient and a Riemannian manifold of dimension d with the Riemannian volume λ, it was shown in Bogachev, Röckner [**162**] that, in the symmetric case, the following conditions are sufficient for the essential self-adjointness of $L_{A,b}$ on $C_0^\infty(M)$: the matrix A is uniformly nondegenerate and uniformly bounded and in local coordinates is given by continuous functions $a^{ij} \in W^{p,1}_{\rm loc}$, where $p > d$, $|b| \in L^p_{\rm loc}(\lambda)$, provided that M is complete with respect to the metric generated by A^{-1}. It was also shown in [**162**] that under the same assumptions the stronger L^p-uniqueness holds for the symmetric operator $L_{A,b}$, i.e., $(I - L_{A,b})\big(C_0^\infty(M)\big)$ is dense in $L^p(\mu)$.

Suppose now that in the symmetric case we have $A = {\rm I}$. Let b be a fixed Borel measurable dx-version of the mapping $\nabla\varrho_0/\varrho_0$ for some probability density $\varrho_0 \in W^{1,1}_{\rm loc}(\mathbb{R}^d)$, where $\nabla\varrho_0/\varrho_0 := 0$ on the set $\{\varrho_0 = 0\}$, i.e., we take a dx-version of the logarithmic gradient β^{μ_0} of the measure $\mu_0 := \varrho_0\,dx$. Then $\mu_0 := \varrho_0\,dx \in \mathcal{M}^{{\rm I},b}_{\rm ell}$. Our precise specification of the character of a version is due to the fact that the measure μ_0 need not be equivalent to Lebesgue measure, so a μ_0-version of b may fail to be a dx-version. Clearly, $|b| \in L^1_{\rm loc}(\mu_0)$. In this case we have the following result (see Albeverio, Bogachev, Röckner [**18**]) with a rather weak local condition on b, but it requires the global condition (5.6.1).

5.6.3. Theorem. *Let b be of the indicated form and let $|b| \in L^2_{\rm loc}(\mu_0)$.*

(i) *Let $\mu \in \mathcal{M}^{{\rm I},b}_{\rm ell}$ with $|b| \in L^2_{\rm loc}(\mu)$ be such that the density ϱ of the measure μ belongs to $W^{1,1}_{\rm loc}(\mathbb{R}^d)$ and satisfies the condition*

$$|b - \nabla\varrho/\varrho| \in L^2(\mu). \tag{5.6.1}$$

Then $\big(L, C_0^\infty(\mathbb{R}^n)\big)$ is symmetric on $L^2(\mu)$ and

$$\frac{\nabla\varrho}{\varrho} = b \quad \mu\text{-a.e.}$$

(ii) *If $|b| \in L^1_{\rm loc}(U)$ for some connected open set $U \subset \mathbb{R}^d$ whose complement has Lebesgue measure zero, then μ_0 is the only measure $\mu \in \mathcal{M}^{{\rm I},b}_{\rm ell}$ such that $|b| \in L^2_{\rm loc}(\mu)$ and $\varrho = d\mu/dx$ satisfies* (5.6.1).

Assertion (i) gives conditions under which infinitesimally invariant measures are symmetrizing. However, if a probability measure μ solves the equation $L^*_{{\rm I},b} = 0$, where $b = \nabla V$ for some $V \in C^\infty(\mathbb{R}^d)$, then it is not always true that b coincides with the logarithmic gradient of μ and then $\mu = c\exp V\,dx$ (see the example before

Theorem 4.1.11). We have also seen in § 4.4 that the answer is positive if we have $|\nabla V| \in L^1(\mu)$ or if $\exp V \in L^1(\mathbb{R}^d)$.

5.6(ii). Markov and Feller semigroups

Here we present some additional material about certain special semigroups related to this chapter, but in a broader context than is really needed for the main purposes of this chapter. Let X be a completely regular topological space, let $B_b(X)$ be the space of bounded Borel functions on X, and let $\{P_t\}_{t\geqslant 0}$ be a semigroup of nonnegative operators on $B_b(X)$ (for example, Markov or sub-Markov) with the property that the adjoint semigroup takes Radon measures to Radon measures, i.e., for every Radon measure σ on X there exists a Radon measure $P_t^*\sigma$ satisfying the equality

$$\int_X P_t f\, d\sigma = \int_X f\, d(P_t^*\sigma), \quad f \in B_b(X).$$

In other words, the functional $P_t^*\sigma$ is represented by a Radon measure (in the general case it is represented by a just additive set function). In particular, for Dirac's measure δ_x at the point x we have $P_t^*\delta_x(B) = P_t I_B(x)$. Note that

$$P_t^*\sigma(B) = \int_X I_B\, d(P_t^*\sigma) = \int_X P_t I_B\, d\sigma = \int_X P_t^*\delta_x(B)\, \sigma(dx) \tag{5.6.2}$$

for all $B \in \mathcal{B}(X)$. This condition is related to the property of integral representation. Certainly, it can be introduced without the Radon property of measures in the case of an arbitrary measurable space $(X, \mathcal{B})$ without any topology.

We recall that an operator $T\colon L^p(\mu) \to L^r(\nu)$, where μ and ν are bounded nonnegative measures on spaces X and Y and $p, r \in [1, +\infty]$, is called *integral* if

$$Tf(y) = \int_X \mathcal{K}(x,y) f(x)\, \mu(dx)$$

for some measurable function $\mathcal{K}$ on $X \times Y$, called an *integral kernel*, for which for every function $f \in L^p(\mu)$ the function $\mathcal{K}(x,y)f(x)$ is integrable in x for ν-almost every y. For example, it is known that every Hilbert–Schmidt operator on $L^2(\mu)$ is integral. The next theorem of Bukhvalov gives a description of integral operators (see Bukhvalov et al. [**211**] or Kantorovich, Akilov [**498**, Chapter XI, § 1]).

5.6.4. Theorem. *A bounded operator $T\colon L^p(\mu) \to L^r(\nu)$ is integral precisely when $Tf_n(y) \to 0$ a.e. for each sequence $\{f_n\} \subset L^p(\mu)$ converging to zero in measure μ and majorized in absolute value by a function of class $L^p(\mu)$ (if $p < \infty$, this is equivalent to norm convergence and existence of a majorant in L^p).*

Every bounded operator from $L^1(\mu)$ to $L^\infty(\nu)$ is integral (Exercise 5.6.55). We prove a close fact. We recall that the topological support $\operatorname{supp}\mu$ of a Radon measure $\mu \geqslant 0$ on X is the smallest closed set of measure $\mu(X)$, see Bogachev [**125**, Proposition 7.2.9].

5.6.5. Proposition. *Let μ be a Radon probability measure on a completely regular space X and let T be a bounded operator on $L^1(\mu)$ with its range in $C(X)$. Then T is an integral operator.*

Proof. Let us take a compact set S in X of positive measure. The restriction of μ to S has a compact topological support K. Then $\mu(K \cap U) > 0$ for every open set U intersecting K. The operator $T\colon L^1(\mu) \to C(K)$ taking f to the restriction of Tf to K is continuous by the closed graph theorem. Indeed, if $f_n \to f$ in $L^1(\mu)$ and $Tf_n \to g$ uniformly on K for a function $g \in C(K)$, then $Tf_n|_K \to Tf|_K$ a.e. on K, whence $Tf|_K = g$ a.e. on K, so $Tf|_K(x) = g(x)$ for all $x \in K$ by the continuity of both functions on K and the absence of nonempty relatively open subsets of K of positive measure. Therefore, if a sequence $\{f_n\}$ converges to zero in $L^1(\mu)$, we have $\sup_{x\in K} |Tf_n(x)| \to 0$. Taking compact sets S_j with $\mu(S_j) \to 1$, we obtain compact sets K_j that are the supports of the restrictions of μ to S_j, on which one has the pointwise convergence $Tf_n(x) \to 0$, moreover, the union of K_j has full measure. By Bukhvalov's theorem T is an integral operator. □

Certainly, the condition that T takes $L^1(\mu)$ to $C(X)$ can be stronger than even the strong Feller property (Exercise 5.6.53).

5.6.6. Remark. In relation to the proof of this proposition the question arises: is it true that for every Radon probability measure μ on a completely regular space X every point s in the topological support of μ can be included in a compact set K of positive measure such that s will belong also to the topological support of the restriction of μ to K? In the general case this is false, as the following example shows. We recall that the Stone–Čech compactification of the space X is a compact space such that X is homeomorphic to its dense subset and every bounded continuous function on X extends to a continuous function on βX. A Stone–Čech compactification exists for every completely regular space, but it is very complicated already for the set of natural numbers (in particular, it is not metrizable).

Let us now take for X the part of the Stone–Čech compactification $\beta\mathbb{N}$ of the set of natural numbers consisting of $\mathbb{N}$ and one more point $s \in \beta\mathbb{N}\backslash\mathbb{N}$ and equip it with the measure $\mu = \sum_{n=1}^{\infty} 2^{-n}\delta(n)$, where $\delta(n)$ is Dirac's measure at n. It is known that every compact K in this space is finite (see Engelking [**335**, Theorem 3.6.14]). Since $\mu(s) = 0$, the point s does not belong to the support of the measure $\mu|_K$. However, for all complete metric spaces the answer is positive (Exercise 5.6.54).

5.6.7. Proposition. *If $T\colon L^1(\mu) \to L^1(\mu)$ is a bounded nonnegative integral operator defined by some $\mathcal{B}\otimes\mathcal{B}$-measurable integral kernel $\mathcal{K}$ by the formula*

$$Tf(x) = \int_X \mathcal{K}(x,y)f(y)\,\mu(dy)$$

and $Tf \in B_b(X)$ for all $f \in B_b(X)$, then for every nonnegative measure ν on $\mathcal{B}$ the functional $T^*\nu$ on $B_b(X)$ is given by the nonnegative measure

$$T^*\nu = g\cdot\mu, \quad g(y) = \int_X \mathcal{K}(x,y)\,\nu(dx).$$

In the case of a Radon measure ν the measure $T^*\nu$ is also Radon.

PROOF. The function

$$\varphi(x) = T1(x) = \int_X \mathcal{K}(x,y)\,\mu(dy)$$

is $\mathcal{B}$-measurable and bounded. We observe that the function g is integrable with respect to the measure μ, since the integral of $\mathcal{K}$ against the measure $\mu\otimes\nu$ is finite by Fubini's theorem (if we integrate first in μ, then we obtain a bounded function φ). It is readily verified that the integral of a function $f \in B_b(X)$ with respect to the measure $g\cdot\mu$ equals the integral of the function Tf with respect to the measure ν. □

If we use a somewhat weaker condition $T(L^\infty(\mu))\subset L^\infty(\mu)$, then the indicated formula defines $T^*\nu$ for every measure ν absolutely continuous with respect to μ. Here the following nuance appears: we defined the operator T^* on measures by using a fixed Borel version of the integral kernel $\mathcal{K}$, but this version can be replaced without changing the operator T on $L^1(\mu)$, although the action on measures will change (say, on Dirac measures we can obtain some other values). For absolutely continuous measures this change of a version has no effect. The reason is that $B_b(X)$ consists of individual functions, but not of the equivalence classes as $L^1(\mu)$.

5.6.8. Remark. If a sub-Markov integral operator T on the space $L^1(\mu)$, where the measure μ is subinvariant for T, is given by a measurable integral kernel $\mathcal{K}\geqslant 0$, then the natural extension of the operator T^* to $L^1(\mu)$ is given by the integral kernel $\mathcal{K}^*(x,y)=\mathcal{K}(y,x)$. This can be easily seen from Fubini's theorem and the fact that, for any integrable nonnegative functions f and g, the function $\mathcal{K}(x,y)f(x)g(y)$ must be integrable.

Let us give definitions of a number of frequently used properties of semigroups in functional spaces. Let $\{P_t\}_{t\geqslant 0}$ be a semigroup on $B_b(X)$, where X is a completely regular space with its Borel σ-algebra $\mathcal{B}=\mathcal{B}(X)$; cases (ii) and (v) of the following definition are applicable also to arbitrary measurable spaces $(X,\mathcal{B})$.

5.6.9. Definition. (i) The semigroup $\{P_t\}_{t\geqslant 0}$ is called Feller at a time $t_0>0$ if $P_{t_0}(C_b(X))\subset C_b(X)$, and strong Feller at $t_0>0$ if $P_{t_0}(B_b(X))\subset C_b(X)$. If this true for all $t_0>0$, then $\{P_t\}_{t\geqslant 0}$ is called Feller (respectively, strong Feller).

(ii) The semigroup $\{P_t\}_{t\geqslant 0}$ is called regular at a time $t_0>0$ if for every $x\in X$ the element $P_{t_0}^*\delta_x$ is given by a measure and all measures $P_{t_0}^*\delta_x$ are equivalent. A semigroup regular at every positive time is called regular.

(iii) The semigroup $\{P_t\}_{t\geqslant 0}$ is called topologically irreducible at a time $t_0>0$ if $P_{t_0}I_U(x)>0$ for every nonempty open set $U\subset X$ for all $x\in U$. If this is true for all $t_0>0$, then the semigroup is called topologically irreducible.

(iv) The semigroup $\{P_t\}_{t\geqslant 0}$ is called stochastically continuous if

$$\lim_{t\to 0} P_t I_U(x)=1$$

for every nonempty open set $U\subset X$ for all $x\in U$.

(v) The semigroup $\{P_t\}_{t\geqslant 0}$ is called irreducible if there is no measurable set $Y\subset X$ such that $0<\mu(Y)<\mu(X)$ and the set of functions vanishing outside Y is taken into itself under the action of the semigroup. Similarly we define the irreducibility of a semigroup on $L^p(\mu)$.

5.6.10. Lemma. *Suppose that a semigroup $\{P_t\}_{t\geqslant 0}$ is strong Feller and has an invariant Radon probability measure μ with topological support X. If there exists a positive function $f\in C_b(X)$ invariant with respect to $\{P_t\}_{t\geqslant 0}$ (say, if the semigroup is Markov) and X is connected, then the extension of $\{P_t\}_{t\geqslant 0}$ to $L^1(\mu)$ is irreducible.*

PROOF. Suppose that there is a measurable set $Y\subset X$ with $0<\mu(Y)<\mu(X)$ for which the subspace of functions in $L^1(\mu)$ vanishing outside Y is invariant with respect to the operators P_t extended to $L^1(\mu)$. Then, whenever $t>0$, the functions $P_t(fI_Y)$ are continuous by the strong Feller property and nonnegative, vanish outside Y, and $P_t(fI_Y)\leqslant P_tf=f$. Hence $P_t(fI_Y)\leqslant fI_Y$. Therefore, $P_t(fI_Y)=fI_Y$, since the integrals of both functions are equal by the invariance of the measure. Since $f>0$, the function I_Y is continuous, which is impossible for connected X. $\square$

5.6.11. Proposition. (i) *If the semigroup* $\{P_t\}_{t\geqslant 0}$ *is regular at a time* $t_0 > 0$, *then all measures* $P_t^*\sigma$ *are equivalent for all* $t \geqslant t_0$ *for any probability measure* σ *on* $\mathcal{B}$.

(ii) *If the semigroup* $\{P_t\}_{t\geqslant 0}$ *is strong Feller at a time* $t_0 > 0$ *and irreducible at a time* $s_0 > 0$, *then it is regular at the time* $t_0 + s_0$.

PROOF. (i) If a set $B \in \mathcal{B}$ is such that for all x the equality $P_{t_0}^*\delta_x(B) = 0$ holds, then the integral of $P_{t_0}I_B(x)$ against any probability measure σ vanishes, and this integral is precisely $P_{t_0}^*\sigma(B)$ according to (5.6.2). Conversely, if $P_{t_0}^*\sigma(B) = 0$, then for some x the equality $P_{t_0}I_B(x) = 0$ holds, hence it is true for all x. Thus, $P_{t_0}^*\delta_x \sim P_{t_0}^*\sigma$. Then for $t = t_0 + s$ we obtain $P_t^*\delta_x = P_{t_0}^*P_s^*\delta_x \sim P_{t_0}^*\delta_x$.

(ii) Suppose that for some $x_0 \in X$ and $B \in \mathcal{B}(X)$ we have $P_{t_0+s_0}^*\delta_{x_0}(B) > 0$. Hence the integral of the function $P_{t_0}I_B$ against the measure $P_{s_0}^*\delta_{x_0}$ is positive, but this function is nonnegative and continuous by assumption, which gives an open set U on which it is positive. Now for every $x \in X$ we have

$$P_{t_0+s_0}^*\delta_x(B) = \int_X P_{t_0}^* I_B \, d(P_{s_0}^*\delta_{x_0}) \geqslant \int_U P_{t_0}^* I_B \, d(P_{s_0}^*\delta_{x_0}) > 0,$$

since $P_{s_0}^*\delta_{x_0}(U) = P_{s_0}I_U(x_0) > 0$ by the irreducibility at the time s_0. Thus, all measures $P_{t_0+s_0}^*\delta_x$ are equivalent. □

5.6.12. Remark. In the discussion of many questions arising below it will be useful, for a given collection of functions ξ_s on a probability space $(X, \mathcal{B}, \mu)$ indexed by a parameter s from a measurable space $(S, \mathcal{S})$, to choose a version ξ_s such that the function $(x, s) \mapsto \xi_s(x)$ be measurable in both variables. In the language of random processes this means a choice of a measurable modification. Among diverse sufficient conditions for that the following one will be useful below: the measure μ is separable, the functions ξ_s are integrable and for every set $B \in \mathcal{B}$ the integral of ξ_s over B is a $\mathcal{S}$-measurable function of s. A justification is left as Exercise 1.8.14.

5.6.13. Lemma. *Let a Markov semigroup* $\{P_t\}_{t\geqslant 0}$ *on* $B_b(X)$ *with a separable invariant probability measure* μ *be regular at a time* τ. *Then the operators* P_t^* *on* $L^1(\mu)$ *are integral whenever* $t \geqslant \tau$.

PROOF. By condition the measures $P_t^*\delta_x$ for $t \geqslant \tau_0$ are equivalent to the measure $\mu = P_\tau^*\mu$ and are given by densities $y \mapsto \mathcal{K}_t(x, y)$ with respect to μ. According to the remark above applied to $\xi_x(y) = \mathcal{K}_t(x, y)$, we can choose the function $\mathcal{K}_t$ measurable on $X \times X$. For all $f \in B_b(X)$ we have

$$P_t f(x) = \int_X f(y)\, P_t^*\delta_x(dy) = \int_X f(y)\, \mathcal{K}_t(x, y)\, \mu(dy).$$

By the continuity of P_t in $L^1(\mu)$ this equality is true almost everywhere for every integrable function. By Lemma 5.6.8 the adjoint operators are integral as well. □

5.6.14. Proposition. *A Markov semigroup* $\{P_t\}_{t\geqslant 0}$ *is stochastically continuous precisely when*

$$\lim_{t\to 0} P_t f(x) = f(x) \quad \textit{for all } f \in C_b(X) \textit{ and } x \in X.$$

If X *is a metric space, then it suffices to have this equality for all* $f \in C_b(X)$ *with bounded support, and for* $X = \mathbb{R}^d$ *it is enough to have it for all* $f \in C_0^\infty(\mathbb{R}^d)$.

PROOF. Let a Markov semigroup $\{P_t\}_{t\geqslant 0}$ be stochastically continuous, let $f\in C_b(X)$, and let $x\in X$. We show that $\lim\limits_{t\to 0} P_tf(x)=f(x)$. We can assume that $|f|\leqslant 1$. Let $\varepsilon>0$. Let us find a neighborhood U of the point x such that $|f(y)-1|<\varepsilon$ whenever $y\in U$. Then

$$|P_tf(x)-f(x)|=\left|\int_X (f(y)-f(x))\,P_t^*\delta_x(dy)\right|\leqslant \varepsilon+2P_t^*\delta_x(X\backslash U)$$
$$=\varepsilon+2P_tI_{X\backslash U}(x).$$

We now pick $\delta>0$ such that $|1-P_tI_U(x)|<\varepsilon$ whenever $t<\delta$. Since $P_t1=1$, we have $P_tI_{X\backslash U}(x)<\varepsilon$, which gives $|P_tf(x)-f(x)|\leqslant 3\varepsilon$ when $t<\delta$.

Conversely, let the condition mentioned in the proposition be fulfilled, let a set $U\subset X$ be open, and let $x\in U$. We can take a function $f\in C_b(X)$ for which $0\leqslant f\leqslant 1$, $f(x)=1$, and $f=0$ outside U. Since $f\leqslant(1-\varepsilon)I_U$, we have $P_tf(x)\leqslant(1-\varepsilon)P_tI_U(x)\leqslant 1-\varepsilon$. It remains to take $\delta>0$ such that $|P_tf(x)-1|\leqslant\varepsilon$ whenever $t\leqslant\delta$. It is clear that for a metric space we can take a ball for U, so it suffices to consider functions f with bounded support. Furthermore, in $\mathbb{R}^d$ it is enough to consider only $f\in C_0^\infty(\mathbb{R}^d)$ by uniform approximation. □

5.6.15. Corollary. *Suppose that a Markov semigroup $\{P_t\}_{t\geqslant 0}$ is stochastically continuous and has an invariant Radon probability measure. Then its extension to the space $L^p(\mu)$ for $p\in[1,+\infty)$ is strongly continuous.*

PROOF. If $f\in C_b(X)$, then we have $\lim\limits_{t\to 0}\|P_tf-f\|_{L^1(\mu)}=0$ by the above proposition and the Lebesgue dominated convergence theorem. For a general function $f\in L^p(\mu)$ and every $\varepsilon>0$, there is a function $f_\varepsilon\in C_b(X)$ with $\|f-f_\varepsilon\|_{L^1(\mu)}\leqslant\varepsilon$, whence we have $\|P_tf-P_tf_\varepsilon\|_{L^p(\mu)}\leqslant\varepsilon$, since $\|T_t\|=1$ on $L^p(\mu)$, hence we obtain that $\|P_tf-f\|_{L^p(\mu)}\leqslant 3\varepsilon$ if $\|P_tf_\varepsilon-f_\varepsilon\|_{L^p(\mu)}\leqslant\varepsilon$. □

We observe that if our special semigroup $\{T_t^\mu\}_{t\geqslant 0}$ is Markov (which is not always the case), then it will be stochastically continuous under our usual conditions on A and b. This follows by the previous proposition and Theorem 5.2.2.

5.6(iii). Invariant measures of Markov semigroups

Here we introduce a number of concepts and prove some facts related to invariant measures of general Markov semigroups acting on spaces of functions on general measurable spaces or on completely regular topological spaces in those cases where continuous functions are involved. We first give a useful general fact from the theory of semigroups established in Greiner [**422**], but in the formulation from Arendt [**61**], where the proof can be found. Let $(X,\mathcal{B},\mu)$ be a probability space. As above, let $B_b(X)$ denote the Banach space of bounded $\mathcal{B}$-measurable functions on X with the sup-norm.

5.6.16. Theorem. *Let $\{T_t\}_{t\geqslant 0}$ be a contractive C_0-semigroup of nonnegative operators on $L^p(\mu)$, $1\leqslant p<\infty$, such that for some $\tau>0$ the operator T_τ is integral. Suppose also that this semigroup is irreducible in the sense that there is no set Ω with $0<\mu(\Omega)<1$ for which the subspace of functions in $L^p(\mu)$ vanishing outside Ω is invariant with respect to this semigroup and that the semigroup generator L has a nonzero kernel. Then, there exist two almost everywhere positive functions $\varrho\in\operatorname{Ker}L$ and $\psi\in\operatorname{Ker}L^*$, where L^* is the generator of the dual semigroup on the* *space* $L^{p/(p-1)}(\mu)$*, such that the following equality holds*:

$$\lim_{t\to\infty} \|T_t f - \langle f,\psi\rangle \varrho\|_{L^p} = 0, \quad \langle f,\psi\rangle = \int f\psi\, d\mu, \quad f \in L^p(\mu). \tag{5.6.3}$$

The condition that the kernel is nontrivial is obviously satisfied if $T_t 1 = 1$. The situation is of interest where there are no nonconstant invariant functions of the semigroup, i.e., functions f with $T_t f = f$ for all $t > 0$. Let $(X, \mathcal{B}, \mu)$ be a probability space and let $\{T_t\}_{t\geqslant 0}$ be a Markov semigroup on $L^1(\mu)$ such that μ is its invariant probability measure.

5.6.17. Definition. The measure μ is called ergodic if there are no nonconstant invariant functions for $\{T_t\}_{t\geqslant 0}$.

5.6.18. Proposition. *The following conditions are equivalent*:

(i) *the measure* μ *is ergodic*; (ii) *there are no nonconstant bounded measurable functions invariant for* $\{T_t\}_{t\geqslant 0}$; (iii) *if* $A \in \mathcal{B}$ *is such that* $T_t I_A = I_A$ *for all* t*, then either* $\mu(A) = 0$ *or* $\mu(A) = 1$.

PROOF. We only need to show that (iii) implies (i). Let $f \in L^1(\mu)$ and $T_t f = f$ for all t. Then the measure $f\cdot\mu$ is invariant with respect to the operators T_t^* in $L^\infty(\mu)$, since

$$\int_X T_t^* g f\, d\mu = \int_X g T_t f\, d\mu = \int_X g f\, d\mu, \quad g \in L^\infty(\mu).$$

By Lemma 5.1.4(iii) the measure $f^+\cdot\mu$ is invariant with respect to T_t^*, whence we have $T_t f^+ = f^+$, $T_t f^- = f^-$. From what we have proved we obtain also the following: for every function g invariant with respect to T_t and each number c, the functions $(g-c)^+$ and $(g-c)^-$ are invariant. Hence so are the functions $\max\bigl(a, \min(g, b)\bigr)$ for all numbers a, b. It follows that for every number c the indicator of the set $B = \{f > c\}$ is an invariant function. Indeed, the indicator of B is the limit in $L^1(\mu)$ of the invariant functions $\min\bigl(k\max(f-c, 0), 1\bigr)$ as $k \to \infty$. Condition (iii) says that each of these sets can have measure either 0 or 1. This is only possible for a function that almost everywhere coincides with a constant. □

5.6.19. Remark. (i) It is seen from the proof that for a single Markov operator T on $L^1(\mu)$ with an invariant probability measure μ the existence of a nonconstant function $f \in L^1(\mu)$ with $Tf = f$ implies the existence of a measurable set B with $0 < \mu(B) < 1$ and $TI_B = I_B$.

(ii) For a Markov semigroup $\{T_t\}_{t\geqslant 0}$ with an invariant measure μ ergodicity is equivalent to irreducibility. Indeed, if there is a nontrivial measurable set Ω for which the space of functions vanishing outside Ω is invariant, then $T_t I_\Omega = 0$ outside Ω, whence it follows that $T_t I_\Omega = I_\Omega$, since $T_t I_\Omega \leqslant 1$ and the integral of $T_t I_\Omega$ equals $\mu(\Omega)$. If, conversely, the equality $T_t I_\Omega = I_\Omega$ is given, then for every function f with $0 \leqslant f \leqslant I_\Omega$ we have $0 \leqslant T_t f \leqslant I_\Omega$, whence we obtain that $T_t(I_\Omega f) \leqslant I_\Omega$ for every bounded nonnegative function f, hence also for all functions $f \in L^1(\mu)$.

5.6.20. Example. We know that under the assumptions that $a^{ij} \in W^{p}_{loc}(\mathbb{R}^d)$, A is invertible, $b^i \in L^p_{loc}(\mathbb{R}^d)$, $p > d+2$, and assuming that there is a probability solution μ of the equation $L^*_{A,b}\mu = 0$ on $\mathbb{R}^d$, the associated semigroup $\{T_t^\mu\}_{t\geqslant 0}$ cannot have invariant probability measures distinct from μ, and if μ is invariant, then it is the unique invariant probability measure also for $\{(T_t^\mu)^*\}_{t\geqslant 0}$, whence we obtain the ergodicity of μ with respect to both semigroups. However, the condition that $b^i \in L^p_{\rm loc}(\mathbb{R}^d)$ cannot be replaced with the condition $b^i \in L^p_{\rm loc}(\mu)$, since already in the one-dimensional case there is no ergodicity in the presence of zeros of the density of μ.

Arendt [**61**] proved also the following result (he actually established a somewhat more general fact).

5.6.21. Theorem. *Let $\{T_t\}_{t\geqslant 0}$ be a strongly continuous Markov semigroup on $L^1(\mu)$ for which the measure μ is invariant and ergodic and such that for some $\tau > 0$ the operator T_τ is compact. Then there exist numbers $\varepsilon > 0$ and $M > 0$ such that $\|T_t - I\| \leqslant Me^{-\varepsilon t}$.*

Unfortunately, the condition of compactness of T_t in $L^1(\mu)$ is very restrictive, for example, it is not fulfilled for such a nice semigroup as the Ornstein–Uhlenbeck semigroup (Exercise 5.6.60).

For operator semigroups the following ergodic theorem holds, which we formulate in the form suitable for the questions discussed here (see Dunford, Schwartz [**316**, Chapter VIII, Section 7] and Yosida [**946**, Chapter XIII]).

5.6.22. Theorem. *Let $\{T_t\}_{t\geqslant 0}$ be a sub-Markov semigroup on $L^1(\mu)$ such that the probability measure μ is subinvariant. Then, for every $\tau > 0$ for all $f \in L^1(\mu)$, the averages*

$$S_n f = \frac{1}{n}(T_\tau f + \cdots + T_{n\tau} f)$$

converge in $L^1(\mu)$ and μ-a.e. to the function $\overline{f}$ in the subspace $\mathrm{Ker}(T_\tau - I)$ that is the conditional expectation of f with respect to the σ-algebra generated by this subspace.

If for all $f \in L^1(\mu)$ the mapping $t \mapsto T_t f$ is measurable with values in $L^1(\mu)$, then the averages

$$S_t f = \frac{1}{t}\int_0^t T_s f\, ds$$

converge in $L^1(\mu)$ and μ-a.e. to the function $\widetilde{f}$ in the subspace $\bigcap_{t>0}\mathrm{Ker}(T_t - I)$ that is the conditional expectation of f with respect to the σ-algebra generated by this subspace. In case of a Markov semigroup and an ergodic invariant measure convergence takes place to the integral of f.

Even for an ergodic measure the two aforementioned limits may differ. For example, the semigroup $T_t f(u) = f(e^{i2\pi t}u)$ on $L^1(S)$ on the unit circle with the linear Lebesgue measure is ergodic, but $T_n f = f$ for all n. However, the following fact is true (the proof given below was suggested by V.V. Ryzhikov).

5.6.23. Proposition. *Let $\{T_t\}_{t\geqslant 0}$ be a strongly continuous Markov semigroup on $L^1(\mu)$ such that the probability measure μ is invariant, ergodic and separable. Then there exists $\tau > 0$ such that the discrete system $T_{n\tau}$, $n = 0, 1\ldots$, is ergodic, i.e., the intersection of the kernels of $T_{n\tau} - I$ is trivial.*

Proof. First we consider our semigroup on $L^2(\mu)$ and take the linear span L of the kernels $T_t - I$, $t > 0$. Let H be the closure of L. We observe that $T_s(L) = L$ for all $s \geqslant 0$. Indeed, if $T_t f = f$ for some $t > 0$, then we have $T_s f \in L$, since $T_t T_s f = T_s T_t f = T_s f$ and for $k > s/t$ we have $f = T_{kt} f = T_s(T_{kt-s} f)$, where $T_{kt-s} f \in L$, since $T_t T_{kt-s} f = T_{kt-s} T_t f = T_{kt-s} f$. In addition, $\|T_s g\|_2 = \|g\|_2$ for all $g \in L$. Indeed, let $g = f_1 + \cdots + f_n$, $\|g\|_2 = 1$, where $f_i = T_{t_i} f_i$, $t_i > 0$. By the

Markov property of T_t we have $\|T_s g\|_2 \leqslant \|g\|_2$, and if $\|T_s g\|_2 \leqslant q\|g\|_2$, where $q < 1$, then $\|T_t g\|_2 \leqslant q\|g\|_2$ for all $t \geqslant s$. We show that there exists t large enough such that $\|T_t g\|_2 > q\|g\|_2$. Since $g = T_{k_1 t_1} f_1 + \cdots + T_{k_n t_n} f_n$ for all $k_1, \ldots, k_n \in \mathbb{N}$, for a given $\varepsilon > 0$ we can pick natural numbers k_i such that the points $k_1 t_1, \ldots, k_n t_n$ will belong to some interval $[t, t+\varepsilon]$. Then $g = T_t[T_{\varepsilon_1} f_1 + \cdots + T_{\varepsilon_n} f_n]$, where $\|g - T_{\varepsilon_1} f_1 + \cdots + T_{\varepsilon_n} f_n\|_2 < \varepsilon$ for ε sufficiently small, which gives the desired t.

Therefore, the operators T_t map isometrically H onto H. The complexification of T_t (denoted by the same symbol) is a unitary operator on the complexification of the space H, denoted by $H^{\mathbb{C}}$. Letting $T_{-t} := T_t^{-1}$, we obtain a strongly continuous group of unitary operators on $H^{\mathbb{C}}$, which by the Stone theorem (see Reed, Simon [**802**, Theorem VIII.8]) has the form $T_t = \exp(itA)$ with some self-adjoint operator A on $H^{\mathbb{C}}$. It remains to verify that for some $\tau > 0$ the intersection of the kernels of $T_{n\tau} - I$ in $H^{\mathbb{C}}$ is zero; then it will be trivial on all of $L^2(\mu)$. To this end, we recall that the operator A is unitary isomorphic to the operator A_φ of multiplication by a measurable real function φ defined by $A_\varphi f = \varphi f$ on $L^2(\nu)$ for some bounded Borel measure ν on the real line, so there exist only a finite or countable set of values c_i for which $\nu(\varphi^{-1}(c_i)) > 0$. Here $c_i \neq 0$, since the intersection of the kernels of all $T_t - I$ is trivial. Hence the operator of multiplication by $\exp(it\varphi) - 1$ has no kernel unless t has the form $t = 2\pi k/c_i$, $k \in \mathbb{Z}$. Thus, every $\tau > 0$ in the complement of $\{2\pi k/c_i\}$ is suitable. Finally, with this choice of τ the intersection of the kernels of $T_{n\tau} - I$ is trivial also in $L^1(\mu)$, which follows by Remark 5.6.19. □

Let us also observe that if the measure μ is separable (which is always fulfilled in the examples considered in this book), then for the measurability of the mapping $t \mapsto T_t f$ it suffices to have its scalar measurability, i.e., the measurability of the integrals of the functions $I_A T_t f$, $A \in \mathcal{B}$, for which the measurability of the functions $(x,t) \mapsto T_t f(x)$ for $f \in B_b(X)$ is sufficient (in the case of a semigroup on $L^\infty(\mu)$ one has to require that for every t there is a version of $T_t f$ such that the obtained function of two variables will be measurable).

Certainly, not every invariant measure of a Markov semigroup generated by an elliptic operator is ergodic. A counterexample can be easily constructed on the union of two disjoint intervals. Though, one has ergodicity for the semigroups constructed in this chapter on spaces of functions on all of $\mathbb{R}^d$ with invariant measures. The general situation is this.

5.6.24. Proposition. *Let $(X, \mathcal{B})$ be a measurable space with a countably generated σ-algebra $\mathcal{B}$ and let $\{T_t\}_{t\geqslant 0}$ be a Markov semigroup on the space $B_b(X)$ such that the set $\mathcal{P}_{\rm inv}$ of all probability measures on $\mathcal{B}$ invariant with respect to it is not empty. Suppose that for every function $f \in B_b(X)$ and every measure $\mu \in \mathcal{P}_{\rm inv}$ the function $(x,t) \mapsto T_t f(x)$ is measurable with respect to the measure $\mu \otimes dt$. Then*

(i) *different ergodic measures on $\mathcal{P}_{\rm inv}$ are mutually singular; in addition, if a measure $\mu \in \mathcal{P}_{\rm inv}$ is ergodic and a measure $\nu \in \mathcal{P}_{\rm inv}$ is absolutely continuous with respect to μ, then $\nu = \mu$;*

(ii) *a measure in $\mathcal{P}_{\rm inv}$ is ergodic precisely when it is an extreme point of $\mathcal{P}_{\rm inv}$, i.e., it cannot be represented as a nontrivial convex combination of two measures in $\mathcal{P}_{\rm inv}$. Therefore, if $\mathcal{P}_{\rm inv}$ consists of a unique measure, then it is ergodic.*

PROOF. (i) Suppose that two measures μ_1 and μ_2 in $\mathcal{P}_{\rm inv}$ are ergodic and there exists a set $B \in \mathcal{B}$ for which $\mu_1(B) > \mu_2(B)$. By the ergodic theorem (its hypotheses are fulfilled in this case), as $n \to \infty$, the functions

$$f_n(x) = \frac{1}{n}\int_0^n T_s I_B(x)\,ds$$

converge to $\mu_1(B)$ for μ_1-almost all x and converge to $\mu_2(B)$ for μ_2-almost all x. By our assumption these functions are measurable with respect to the measure $\mu = (\mu_1+\mu_2)/2 \in \mathcal{P}_{\rm inv}$, hence the corresponding sets of convergence X_1 and X_2 are measurable with respect to both measures and disjoint, which means the mutual singularity of these measures. For the proof of the second assertion we also use the fact that the functions $f_n(x)$ converge to $\mu(B)$ for μ-a.e. x for every $B \in \mathcal{B}$, hence they converge to the same limit ν-almost everywhere. The integral of f_n with respect to the invariant measure ν equals the integral of I_B, i.e., $\nu(B)$.

(ii) If $\mu \in \mathcal{P}_{\rm inv}$ is ergodic and $\mu = \alpha\mu_1 + (1-\alpha)\mu_2$, where $\mu_1, \mu_2 \in \mathcal{P}_{\rm inv}$ and $\alpha \in (0,1)$, then $\mu_1 \ll \mu$, $\mu_2 \ll \mu$, hence $\mu_1 = \mu_2 = \mu$ by (i).

Conversely, let $\mu \in \mathcal{P}_{\rm inv}$ be not ergodic. Then there exists a measurable set B with $T_t I_B = I_B$ and $0 < \mu(B) < 1$. We show that the measures $\mu_1 = \mu|_B/\mu(B)$ and $\mu_2 = \mu|_{X\backslash B}/\mu(X\backslash B)$ are invariant as well, which will give a contradiction, since $\mu = \mu(B)\mu_1 + \mu(X\backslash B)\mu_2$. It suffices to consider μ_1 and show that $T_t^* I_B = I_B$. Since $0 \leqslant T_t^* I_B \leqslant 1$ and the integral of $T_t^* I_B$ against the measure μ equals $\mu(B)$ by the equality $T_t 1 = 1$, it remains to verify that $T_t^* I_B = 0$ a.e. outside B, i.e., $I_{X\backslash B} T_t^* I_B = 0$ a.e., which is true, because this function is nonnegative and has a zero integral by the equality $T_t I_{X\backslash B} = I_{X\backslash B}$. □

5.6.25. Remark. In the situation of the proven proposition the measure μ is ergodic also for the semigroup $\{T_t^*\}_{t\geqslant 0}$, provided that it is ergodic for $\{T_t\}_{t\geqslant 0}$. Indeed, we know that the semigroup $\{T_t^*\}_{t\geqslant 0}$ is also Markov and that μ is invariant for it. If $B \in \mathcal{B}$ is such that $T_t^* I_B = I_B$ and $\mu(B) > 0$, then the measure $\mu|_B/\mu(B)$ is invariant with respect to $\{T_t\}_{t\geqslant 0}$, hence coincides with μ by assertion (i).

In many applications the convex set $\mathcal{P}_{\rm inv}$ turns out to be compact in some locally convex topology and hence by the Krein–Milman theorem it has extreme points, which gives ergodic measures. For example, in typical situations the set $\mathcal{P}_{\rm inv}$ is compact in the weak topology (say, if it uniformly tight and the semigroup is Feller).

Let us prove an important theorem due to Doob (see Doob [**310**] or Da Prato, Zabczyk [**281**, § 4.2]) in a more general form obtained in Gerlach, Nittka [**405**].

5.6.26. Theorem. *Let $\{P_t\}_{t\geqslant 0}$ be a Markov semigroup on $B_b(X)$ that is regular at some time. If it possesses a separable invariant probability measure μ for which the natural extension of the semigroup to $L^1(\mu)$ is strongly continuous (which is fulfilled for stochastically continuous semigroups), then this measure is a unique invariant probability measure, $\mu \sim P_t^*\delta_x$ for all $x \in X$ and $t > t_0$, and*

$$\lim_{t\to\infty} \|P_t^*\delta_x - \mu\| = 0.$$

Moreover, $\lim_{t\to\infty} \|P_t^*\nu - \mu\| = 0$ for every probability measure ν on $\mathcal{B}$.

PROOF. According to Proposition 5.6.11, the measures $P_t^*\delta_x$ are equivalent to the measure $P_t^* = \mu$ for all $t \geqslant \tau$ for some $\tau > 0$. Hence they have densities ϱ_t with respect to μ. The semigroup $\{P_t^*\}$ is also strongly continuous Markov in all $L^p(\mu)$. It is verified directly that $\varrho_{t+\tau} = P_t^*\varrho_\tau$. We now wish to apply Theorem 5.6.16 to the semigroup $\{P_t^*\}_{t\geqslant 0}$ in $L^1(\mu)$. Since $P_t^* 1 = 1$, the kernel of its generator contains 1. As we have already noted above, regularity implies that $\{P_t^*\}_{t\geqslant 0}$ is integral. Finally, the measure μ is ergodic for $\{P_t^*\}_{t\geqslant 0}$, since if $P_t^* I_B = I_B$ for some $B \in \mathcal{B}$ and $\mu(B) > 0$, then the probability measure $\nu = I_B \cdot \mu/\mu(B)$ is invariant for $\{P_t\}_{t\geqslant 0}$, whence we obtain $\nu = P_\tau^* \nu \sim \mu$, which is only possible if $\mu(B) = 1$. □

5.6(iv). Existence of invariant measures

There are diverse sufficient conditions for the existence of invariant probability measures for a Markov semigroup $\{T_t\}$ on $C_b(X)$ for a completely regular topological space X. General results that do not take into account the specific features of the semigroup are not sensitive to our choice of the space. The simplest case: X is compact. In this case the operators T_t^* form a commuting family of continuous affine operators on the space of Radon measures on X identified with the dual to $C_b(X)$. By the Banach–Alaoglu theorem the closed unit ball in the space of measures is compact in the weak topology (i.e., the $*$-weak topology of the dual space). Therefore, the set of probability measures is compact as well and is invariant with respect to the operators T_t^*. By the Markov–Kakutani theorem (see Dunford, Schwartz [**316**, Theorem V.10.6]) we obtain a common fixed point of the mappings T_t^*, which will be the desired invariant measure.

In the general case we may attempt to apply the proven fact to the Stone–Čech compactification βX of the space X (see Remark 5.6.6 above). The operators T_t extend naturally to $C(\beta X)$, so that according to what has been said above we obtain a Radon probability measure ν on βX invariant with respect to the extended operators. However, this does not always give invariant measures on the original space X (which need not exist at all). Let us give a sufficient condition which enables us to use such a measure on βX to produce an invariant measure on X. Let $\mathcal{P}_{\mathcal{B}a}(X)$ be the set of Baire probability measures on X, i.e., measures on the Baire σ-algebra $\mathcal{B}a(X)$ generated by continuous functions.

5.6.27. Lemma. *Let $T\colon C_b(X) \to C_b(X)$ be a Markov operator such that T^* takes Baire measures to Baire measures. Suppose that the extension of T to $C(\beta X)$ has an invariant Radon probability measure such that X is measurable with respect to it and has positive measure. Then T possesses an invariant Baire probability measure on X. If a family of Markov operators on $C_b(X)$ is such that their extensions to $C(\beta X)$ possess a common invariant Radon probability measure that is positive on X, then on X there is a Baire probability measure invariant for this family.*

PROOF. The extension of T to $C(\beta X)$ will be denoted also by T; let ν be the invariant measure mentioned in the hypotheses. The functional

$$\psi(f) = \int_{\beta X} \widehat{f}\, d\nu$$

on $C_b(X)$, where $\widehat{f}$ is the extension of f to βX, is nonnegative on nonnegative functions and can be written as $\psi = \psi_\sigma + \psi_a$, where the functionals ψ_σ and ψ_a are also nonnegative, ψ_σ is represented as the integral over a nonnegative Baire measure μ, and the functional ψ_a corresponds to a purely additive measure, i.e., it does not admit a subtraction of a countably additive component (see Bogachev [**125**, Theorem 7.10.8], Knowles [**523**]). Thus, ψ_σ is the maximal countably additive component. Then $\psi = \psi \circ T = \psi_\sigma \circ T + \psi_a \circ T$. We observe that the functional $\psi_\sigma \circ T$

is represented by the nonnegative Baire measure $T^*\mu$. It follows by the maximality of μ that $\mu \geqslant T^*\mu$. Hence $\mu = T^*\mu$, since $\mu(X) = T^*\mu(X)$ by the equality $T1 = 1$. We show that $\mu(X) > 0$. By assumption the measure $\nu|_X$ is well-defined and is not zero. Whenever $f \geqslant 0$, we have

$$\psi(f) = \int_X f\,d\nu + \int_{\beta X\backslash X} \widehat{f}\,d\nu \geqslant \int_X f\,d\nu,$$

hence $\mu \geqslant \nu|_X$ by the maximality of μ. The second assertion of the lemma is clear from the proof. □

We observe that if T is a Markov operator on $C_b(X)$, then, for every Radon measure ν on X, the functional $T^*\nu$ is represented by a measure on the compact βX (but not always by its restriction to X), so for every compact set K in X the value $T^*\nu(K)$ is well-defined, in particular, the numbers $T^*\delta_x(K)$ are defined for $x \in X$.

In the formulation of the next theorem we implicitly use the following facts: if X is a completely regular Souslin space, a function $f(x,t)$ is continuous in x and Borel measurable in t, then it is measurable with respect to the σ-algebra generated by Souslin sets in $X\times(0,+\infty)$ (see Bogachev [**125**, Exercise 6.10.39, § 8.6]), and if σ_t is a Radon probability measure on X for each t such that the integral of every function in $C_b(X)$ against σ_t is a Borel function of t, then every function $t \mapsto \sigma_t(B)$ for any Borel set B in X will be Borel measurable, which follows by the monotone class theorem and the coincidence of the Borel and Baire σ-algebras in X (see [**125**, Theorem 2.12.9 and Theorem 6.7.7]).

5.6.28. Theorem. *Let X be a completely regular Souslin space and let $\{T_t\}$ be a Markov semigroup on $C_b(X)$ such that, for all $f \in C_b(X)$ and $x \in X$, the function $t \mapsto T_t f(x)$ is Borel measurable, $T_t^*\delta_x$ is a Radon measure on X and there exist a compact $K \subset X$ and a point $z \in X$ for which*

$$\limsup_{t\to\infty} \frac{1}{t}\int_0^t T_s^*\delta_z(K)\,ds = \alpha > 0.$$

Then there is a Radon probability measure on X invariant with respect to this semigroup.

Proof. It suffices to find a Radon probability measure ν on βX invariant with respect to the extensions of the operators T_t on $C(\beta X)$ and positive on X. Let us consider the Radon probability measures ν_t on βX defined by the formula

$$\int_{\beta X} f\,d\nu_t := \frac{1}{t}\int_0^t T_s f(z)\,ds, \quad f \in C(\beta X).$$

In fact, this means that

$$\nu_t = \frac{1}{t}\int_0^t T_s^*\delta_z\,ds,$$

but the integral form simplifies the measurability issues mentioned above. It follows from our assumptions that there exists a strictly increasing sequence $t_n \to \infty$ for which

$$\nu_{t_n}(K) \geqslant \frac{\alpha}{2}.$$

The sequence of measures ν_{t_n} has a weak limit point ν. It is clear that $\nu(K) \geqslant \alpha/2$. We show that the measure ν is invariant with respect to the extensions of T_t.

Indeed, let $f \in C_b(\beta X)$, $0 \leqslant f \leqslant 1$. We verify that the integrals of f and $T_t f$ with respect to the measure ν are equal. Let us show that for every $\varepsilon > 0$ the absolute value of their difference does not exceed ε. There is $n > 4t/\varepsilon$ such that

$$\biggl|\int_{\beta X} f\,d(\nu_{t_n} - \nu)\biggr| + \biggl|\int_{\beta X} T_t f\,d(\nu_{t_n} - \nu)\biggr| \leqslant \frac{\varepsilon}{2}.$$

We have

$$\begin{aligned}\frac{1}{t_n}\int_0^{t_n} T_s T_t f\,ds &= \frac{1}{t_n}\int_0^{t_n} T_{s+t} f\,ds = \frac{1}{t_n}\int_t^{t_n+t} T_s f\,ds \\ &= \frac{1}{t_n}\int_0^{t_n} T_s f\,ds + \frac{1}{t_n}\int_{t_n}^{t_n+t} T_s f\,ds - \frac{1}{t_n}\int_0^{t} T_s f\,ds,\end{aligned}$$

whence

$$\biggl|\int_{\beta X} T_t f\,d\nu_{t_n} - \int_{\beta X} f\,d\nu_{t_n}\biggr| \leqslant \frac{2t}{t_n} \leqslant \frac{\varepsilon}{2},$$

which gives the required estimate. □

5.6.29. Remark. The condition of the theorem is also necessary (moreover, for every $\alpha < 1$ and with a point z from any given set with measure sufficiently close to 1) for the existence of a Radon probability measure invariant for $\{T_t\}_{t\geqslant 0}$. Indeed, if such a measure μ exists, then for every $\gamma \in (1/2, 1)$ there exists a compact set K with $\mu(K) > \gamma$. Since by (5.6.2) we have

$$\mu(K) = T_s^*\mu(K) = \int_X T_s^*\delta_x(K)\,\mu(dx)$$

for all s, the integral of the right-hand side over $[0,t]$ divided by t is greater than γ. Hence the integrals over X with respect to the measure μ of the functions

$$\varphi_n(x) = \frac{1}{n}\int_0^n T_s^*\delta_x(K)\,ds$$

are greater than γ, while the integrals over K are greater than $2\gamma - 1$, since we have $0 \leqslant \varphi_n \leqslant 1$. The bounded functions $\psi_n(x) = \sup_{k\geqslant n}\varphi_n(x) \geqslant \varphi_n(x)$ decrease pointwise and their integrals over K with respect to the measure μ are also greater than $2\gamma - 1$. Therefore, the integral over K of the function $\lim\limits_{n\to\infty}\psi_n(x)$ is not less than $2\gamma - 1$. Hence there exists a point $z \in K$ for which $\lim\limits_{n\to\infty}\psi_n(z) \geqslant \alpha$, which gives an increasing sequence of indices n_k such that $\lim\limits_{k\to\infty}\varphi_{n_k}(z) \geqslant 2\gamma - 1$. In this reasoning we have not used the compactness of K. Moreover, in place of n we could take any numbers increasing to infinity.

The previous reasoning can be modified to obtain the following fact. Let X be a Souslin completely regular space.

5.6.30. Theorem. *If a Markov semigroup $\{T_t\}_{t\geqslant 0}$ has an invariant Radon probability measure μ on X, then for every sequence $\tau_n \to +\infty$ there exist a point $z \in X$ and an increasing sequence $\{t_n\}$ such that the sequence of measures*

$$\nu_{t_n} = \frac{1}{t_n}\int_0^{t_n} T_s^*\delta_z\,ds$$

is uniformly tight and the sequence $\{\nu_{t_n}\}$ contains a weakly convergent subsequence, moreover, for a complete separable metric space X this gives weak compactness of the closure of the set of measures ν_t, $t \geqslant 1$.

PROOF. We take increasing compact sets K_n with $\mu(K_n) > 1 - 2^{-n}$ and use them to construct functions φ_n as above. The integral of φ_n over X is greater than $1 - 2^{-n}$, so $\lim\limits_{n\to\infty} \psi_n(x) = 1$ a.e. Finally, we find $z \in X$ and $n_k \to \infty$ with $\varphi_{n_k}(z) \to 1$. For a space X with metrizable compacta, which is the case for every Souslin space, this gives a weakly convergent subsequence in $\{\nu_{n_k}\}$, and for a complete separable metric space also weak compactness of the closure of the set of measures ν_t, $t \geqslant 1$ (see Bogachev [**125**, § 8.6]). □

If $\{T_t\}_{t\geqslant 0}$ is generated by a homogeneous Markov process $\{\xi_{t,x}\}$ in X with transition probabilities $P(x, t, \cdot\,)$, i.e., $T_t f(x) = \mathbb{E} f(\xi_{t,x})$, then $T_t^*\delta_x$ is exactly $P(x, t, \cdot\,)$, hence the condition indicated in the theorem can be written as

$$\limsup_{t\to\infty} \frac{1}{t}\int_0^t P(z, s, K)\, ds = \alpha > 0.$$

This condition is definitely fulfilled if there is a sequence $t_n \to +\infty$ for which the sequence of measures

$$\nu_{t_n}(\,\cdot\,) = \frac{1}{t_n}\int_0^{t_n} P(z, s, \cdot\,)\, ds$$

is uniformly tight. As we have seen, the latter stronger condition for a broad class of spaces follows from the weaker condition in the theorem. If, in addition, we are given the uniqueness of the invariant measure of the semigroup, then the measures ν_t converge weakly to this invariant measure as $t \to +\infty$. Under the assumptions of Theorem 5.6.26 one also has convergence in variation.

5.6(v). Compactness of semigroups

In many problems, the compactness of the operators of the semigroup $\{T_t\}_{t\geqslant 0}$ and the corresponding resolvents is useful (for example, it ensures discrete spectra), however, it does not always hold. Here we give a number of sufficient conditions for the compactness of Markov or sub-Markov semigroups in functional spaces. It turns out that there is a significant difference between the following three situations: compactness in C_b, compactness in L^1 and compactness in L^p with $p > 1$. The latter is rather typical and occurs quite often under broad assumptions, while the first two ones normally require more special conditions. A typical example: the Ornstein–Uhlenbeck semigroup in $L^1(\gamma)$, where γ is the standard Gaussian measure on the real line. As we shall see below, here there is no compactness in $L^1(\gamma)$ and in $C_b(\mathbb{R})$, but we have it in $L^p(\gamma)$ if $p > 1$.

The simplest sufficient condition for the compactness of the operators T_t defined by integral kernels $\mathcal{K}_t$ in $L^p(\mu)$ for some probability measure μ is the condition $\mathcal{K}_t \in L^q(\mu\otimes\mu)$, where $q = p/(p-1)$. However, this simple by appearance condition it difficult to verify (see Exercise 5.6.56), and for $p = 1$ it is seldom fulfilled for diffusion semigroups, since it requires the uniform boundedness of $\mathcal{K}_t(x, y)$ in x, y.

5.6.31. Theorem. *Let $\{T_t\}_{t\geqslant 0}$ be a strong Feller Markov semigroup on $B_b(X)$, where X is completely regular, $T_t^*\delta_x$ is a Radon measure for every $x \in X$ and for every compact K in X the restriction operators $T_t f|_K\colon B_b(X) \to C(K)$ are compact for $t > 0$. If for every $t > 0$ the family of measures $T_t^*\delta_x$, $x \in X$, is uniformly tight, then the operators T_t are compact for $t > 0$. If either there is an invariant Radon probability measure μ on X with topological support X or X is a countable union of compacts, then the converse is also true.*

PROOF. Suppose that the indicated family of measures is uniformly tight. We show that for any fixed $t > 0$ the set of functions $T_t f$, where $|f| \leqslant 1$, is completely bounded. It follows from our assumptions that for every compact K in X the operator $S_t f(x) = T_{t/2}(I_K T_{t/2} f)(x)$ in $C_b(X)$ is compact. Indeed, if we are given a bounded sequence $\{f_n\}$ in $C_b(X)$, then there exists its subsequence $\{n_i\}$ for which the sequence of functions $I_K T_{t/2} f_{n_i}$ converges uniformly, then the functions $T_{t/2}(I_K T_{t/2} f_{n_i})$ converge uniformly too. Let $\varepsilon > 0$. We choose a compact K such that $T^*_{t/2}\delta_x(X \backslash K) \leqslant \varepsilon$ for all x. Then

$$\begin{aligned} T_t f(x) &= \int_X T_{t/2} f(y)\, T^*_{t/2}\delta_x(dy) = \int_K T_{t/2} f(y)\, T^*_{t/2}\delta_x(dy) + r(x,t,f) \\ &= S_t f(x) + r(x,t,f), \end{aligned}$$

where $|r(x,t,f)| \leqslant \varepsilon \|f\|_\infty$. Hence $\|T_t - S_t\| \leqslant \varepsilon$, whence the compactness of T_t follows.

Conversely, if the operators T_t are compact for $t > 0$ and there is an invariant measure μ, then, taking increasing compacts K_n with $\mu(K_n) \to \infty$, we obtain that the sequence of functions $f_n = T_t I_{K_n}$ uniformly converges to some continuous function g with $0 \leqslant g \leqslant 1$. This follows by the estimate $0 \leqslant f_n \leqslant f_{n+1} \leqslant 1$ and the property that every sequence in $\{f_n\}$ contains a uniformly convergent subsequence. In addition, by the invariance of the measure μ, the integrals of the functions f_n with respect to the measure μ are increasing to 1, whence it follows that $g = 1$. Finally, $T^*_t\delta_x(K_n) \geqslant T_t I_{K_n}(x)$, which proves the uniform tightness. If X is a countable union of compacts K_n, then they can be assumed increasing, hence the same reasoning applies: for every x we have $T^*_t\delta_x(K_n) \to 1$, since $T^*_t\delta_x$ is a probability measure by the equality $T_t 1 = 1$. □

One of the most important conditions of this theorem, the compactness of the restrictions $T_t f|_K$, is fulfilled for the semigroups considered in this chapter on account of Theorem 5.4.5, where in the case of a bounded function f the corresponding Hölder norms of $T_t f$ on compacta are uniformly bounded also with respect to f with $|f| \leqslant 1$. However, the uniform tightness of transition probabilities is not so common (say, it fails in the case of the Ornstein–Uhlenbeck semigroup). Let us give an effectively verified sufficient condition in the case of the semigroup associated with a solution of an elliptic equation.

5.6.32. Example. Let a probability measure μ on $\mathbb{R}^d$ satisfy the equation $L^*_{A,b}\mu = 0$, where the functions a^{ij} are continuous and belong to $W^{p,1}_{\text{loc}}(\mathbb{R}^d)$, the matrix $A(x)$ is positive definite, $b^i \in L^p_{\text{loc}}(\mathbb{R}^d)$ with $p > d+2$, and there are a positive function $W \in C^2(\mathbb{R}^d)$ and a positive continuous increasing function G on the half-line $[0, +\infty)$ such that

$$\lim_{|x|\to\infty} W(x) = +\infty, \qquad \int_1^{+\infty} \frac{ds}{sG(s)} < +\infty$$

and for some number $C > 0$ and almost all x we have the inequality

$$L_{A,b}W(x) \leqslant C - W(x)G\bigl(W(x)\bigr).$$

Then, for the generated semigroup $\{T_t\}_{t\geqslant 0}$, the measures $T^*_t\delta_x$ are uniformly tight for any fixed $t > 0$, and the operators T_t with $t > 0$ are compact in $L^1(\mu)$.

This follows by case (ii) in Corollary 7.1.2, which gives a uniform estimate of the integrals of the function W with respect to the measures $T^*_t\delta_x$ for any fixed $t > 0$.

We recall that the indicated conditions give the existence and uniqueness of a probability solution of our equation, the uniqueness of the associated semigroup and the invariance of μ with respect to this semigroup.

Let us give yet another sufficient condition for the compactness of the operators of a semigroup.

5.6.33. Theorem. *If $\{T_t\}_{t\geqslant 0}$ is a Markov semigroup on $L^\infty(\mu)$ with an invariant probability measure μ and for any $t>0$ the operators T_t are compact in $L^\infty(\mu)$, then the operators $T_t\colon L^p(\mu)\to L^p(\mu)$ are compact for all $t>0$ and $p>1$.*

PROOF. Let $t>0$. For every $\varepsilon>0$ we find a continuous projection operator Π_ε on $L^1(\mu)$ mapping $L^1(\mu)$ onto a finite-dimensional subspace in $L^\infty(\mu)$ such that for the operator norm we obtain

$$\|\Pi_\varepsilon\|_{\mathcal{L}(L^\infty(\mu))}=\|\Pi_\varepsilon\|_{\mathcal{L}(L^1(\mu))}=1,\quad \|\Pi_\varepsilon T_t-T_t\|_{\mathcal{L}(L^\infty(\mu))}\leqslant\varepsilon.$$

To this end, in the image of the unit ball of $L^\infty(\mu)$ under T_t we find a finite $\varepsilon/4$-net $g_1,\dots,g_n$ and approximate g_i up to $\varepsilon/4$ by a function f_i with finitely many values. Hence we obtain an $\varepsilon/2$-net consisting of functions f_i that are linear combinations of the indicators of disjoint measurable sets $B_1,\dots,B_k$ with union X. Now we can take the projection operators

$$\Pi_\varepsilon f=\sum_{j=1}^k \mu(B_j)^{-1}\int_{B_j} f\,d\mu.$$

If $\|f\|_\infty\leqslant 1$, then $\|\Pi_\varepsilon T_tf-T_tf\|_\infty\leqslant\|\Pi_\varepsilon T_tf-f_j\|_\infty+\|f_j-T_tf\|_\infty\leqslant\varepsilon$, where f_j is such that $\|T_tf-f_j\|_\infty\leqslant\varepsilon/2$ (since $\Pi_\varepsilon f_j=f_j$, this gives also the estimate $\|\Pi_\varepsilon T_tf-f_j\|_\infty=\|\Pi_\varepsilon T_tf-\Pi_\varepsilon f_j\|_\infty\leqslant\varepsilon/2$). By the Riesz–Thorin interpolation theorem (see Bergh, Löfström [**108**, Theorem 1.1.1]) for the operator $\Pi_\varepsilon T_t-T_t$ on $L^p(\mu)$ with a norm at most 2 on $L^1(\mu)$ and a norm at most ε on $L^\infty(\mu)$ we have $\|\Pi_\varepsilon T_t-T_t\|_{\mathcal{L}(L^p(\mu))}\leqslant 2^{2-\theta}\varepsilon^\theta$, $1-\theta=1/p$. This proves the compactness of T_t on the space $L^p(\mu)$. □

5.6.34. Corollary. *Let $\{T_t\}_{t\geqslant 0}$ be a Markov semigroup on $C_b(X)$, where X is completely regular. If the operators T_t with $t>0$ are compact, then there is an invariant Baire probability measure μ (not always unique), and the operators $T_t\colon L^p(\mu)\to L^p(\mu)$ are compact for $t>0$ and $p>1$.*

PROOF. The existence of an invariant measure $\mu\in\mathcal{P}_{\mathcal{B}a}(X)$ is ensured by the aforementioned Markov–Kakutani theorem: the operators T_t^* in $C_b(X)^*$ are compact, hence $T_t^*\bigl(\mathcal{P}_{\mathcal{B}a}(X)\bigr)$ has a convex and norm compact closure V_t. Moreover, $V_t\subset\mathcal{P}_{\mathcal{B}a}(X)$, since if $f_n\in C_b(X)$, $f_n(x)\downarrow 0$, then $\sup_x|T_tf_n(x)|\to 0$, whence $T_t^*\nu(f_n)\to 0$ as $\nu\in\mathcal{P}_{\mathcal{B}a}(X)$, because $T_t^*\nu(f_n)$ is the integral of T_tf_n against ν. Thus, $T_t^*\nu$ is represented by a measure (see Bogachev [**125**, Theorem 7.10.1]). Furthermore, $V_t\subset V_s$ if $t>s$, hence $V=\bigcap_{t>0}V_t\neq\varnothing$, $T_t^*(V)\subset V$. Our assumption gives the compactness of the operators $T_t\colon L^\infty(\mu)\to L^\infty(\mu)$, $t>0$ (Exercise 5.6.65). □

Let us give another sufficient condition for the compactness of T_t.

5.6.35. Theorem. *Let $\{T_t\}_{t\geqslant 0}$ be a Markov semigroup on $L^p(\mu)$, where μ is an invariant probability measure, $p\in(1,+\infty)$ and there are measurable sets K_n such that $\mu(K_n)\to 1$ and for all $t>0$ the operators $f\mapsto I_{K_n}T_tf$ from $L^p(\mu)$*

to $L^1(\mu|_{K_n})$ are compact. Suppose that for every $t > 0$ there exist $q(t) > p$ and $C(t,p) > 0$ such that

$$\|T_t f\|_{q(t)} \leqslant C(t,p)\|f\|_p \quad \forall f \in L^p(\mu). \tag{5.6.4}$$

Then the operators T_t are compact on $L^p(\mu)$ for $t > 0$.

PROOF. Condition (5.6.4) for a fixed $t > 0$ gives the uniform integrability of the functions $T_t f$ with f in a ball in $L^p(\mu)$. Hence along with the compactness of the operators $S_{t,n} f = I_{K_n} T_t f$ from $L^p(\mu)$ to $L^1(\mu|_{K_n})$ it implies the compactness of these operators also with values in $L^p(\mu|_{K_n})$. We have $\|T_t - S_{t,n}\|_{\mathcal{L}(L^p(\mu))} \to 0$, since

$$\|I_{X\backslash K_n} T_t f\|_p \leqslant \|I_{X\backslash K_n}\|_r^{1/p} \|T_t f\|_{q(t)} \leqslant C(t,p)\mu(X\backslash K_n)^{\frac{pq(t)}{q(t)-p}} \|f\|_p$$

by Hölder's inequality with the exponents $q(t)/p$ and $r = q(t)/(q(t) - p)$ and inequality (5.6.4). □

Estimate (5.6.4) is called the *hypercontractivity* of the semigroup. This property is closely related to the *logarithmic Sobolev inequality*

$$\int f^2 \ln f^2 \, d\mu \leqslant r \int |\nabla f|^2 \, d\mu + \beta(r), \quad f \in C_0^\infty(\mathbb{R}^d), \ \|f\|_{L^2(\mu)} = 1. \tag{5.6.5}$$

The logarithmic Sobolev inequality, as the usual Sobolev inequality (which increases the order of integrability), also ensures the uniform integrability. Let us give a sufficient condition for the hypercontractivity of the semigroup $\{T_t^\mu\}_{t\geqslant 0}$ associated with a probability solution μ of the elliptic equation $L_{\mathrm{I},b}^* \mu = 0$ on $\mathbb{R}^d$.

5.6.36. Theorem. *Suppose that a Borel vector field b is locally bounded and*

$$\langle b(x) - b(y), x - y\rangle \leqslant -\kappa |x-y|^2, \quad x, y \in \mathbb{R}^d$$

for some $\kappa > 0$. Then, there exists a probability measure μ on $\mathbb{R}^d$ satisfying the equation $L^*\mu = 0$, where $L = L_{\mathrm{I},b}$, the associated semigroup $\{T_t^\mu\}_{t\geqslant 0}$ is hypercontractive on $L^p(\mu)$ for $p \in [1, +\infty)$, inequality (5.6.4) is fulfilled with $C(t,p) = 1$, and one also has

$$\int_{\mathbb{R}^d} |f|^p \ln |f| \, d\mu \leqslant \|f\|_{L^p(\mu)}^p \ln \|f\|_{L^p(\mu)} + \frac{p}{2\kappa} \int_{\mathbb{R}^d} |f|^{p-2} |\nabla f|^2 \, d\mu \tag{5.6.6}$$

for all $f \in W^{p,1}(\mu)$.

PROOF. We recall a standard method of proving hypercontractivity on the basis of (5.6.6). It suffices to prove (5.6.4) for nonnegative $f \in C_b^\infty(\mathbb{R}^d)$, moreover, for f of the form $f = f_0 + c$, where $c > 0$, $f_0 \in C_0^\infty(\mathbb{R}^d)$, $f_0 \geqslant 0$. Then the function $G\colon t \mapsto \|T_t f\|_{q(t)}$, where $q(t) = 1 + (p-1)e^{2\kappa t}$, is differentiable in t on $[0, \infty)$ (observe that $T_t f \geqslant c$). We show that $G'(t) \leqslant 0$. This will yield $G(t) \leqslant G(0)$, whence the desired estimate follows.

We first observe that the logarithmic Sobolev inequality for $f^{r/2}$ in place of f and formula (3.7.15), where we take functions $\psi \in C_b^\infty(\mathbb{R})$ such that $\psi(s) = s^{r-1}$ on the range of f, give the estimate

$$\begin{aligned} \int_{\mathbb{R}^d} f^r \ln f \, d\mu - \frac{1}{r} \int_{\mathbb{R}^d} f^r \, d\mu \Bigl(\ln \int_{\mathbb{R}^d} f^r \, d\mu\Bigr) \\ \leqslant \frac{r}{2\kappa} \int_{\mathbb{R}^d} f^{r-2} |\nabla f|^2 \, d\mu = -\frac{r}{2\kappa(r-1)} \int_{\mathbb{R}^d} f^{r-1} Lf \, d\mu. \end{aligned} \tag{5.6.7}$$

It is important that this estimate remains valid for all $f \in W^{p,1}(\mu)$. Set

$$F(t) = \int_{\mathbb{R}^d} (T_t f)^{q(t)}\, d\mu.$$

Then $G(t) = F(t)^{1/q(t)}$, hence

$$G'(t) = G(t)\Big[-\frac{q'(t)}{q(t)^2}\ln F(t) + \frac{F'(t)}{q(t)F(t)}\Big].$$

Since $q'(t) = 2\kappa q(t) - 2\kappa > 0$, it suffices to verify that

$$-\frac{1}{q(t)}F(t)\ln F(t) + \frac{F'(t)}{q'(t)} \leqslant 0. \tag{5.6.8}$$

Observing that

$$F'(t) = \int_{\mathbb{R}^d} (T_t f)^{q(t)}\Big[q'(t)\ln T_t f + q(t)\frac{LT_t f}{T_t f}\Big]\, d\mu,$$

we see that (5.6.8) is equivalent to the inequality

$$-\frac{F(t)\ln F(t)}{q(t)} + \int_{\mathbb{R}^d} (T_t f)^{q(t)} \ln T_t f\, d\mu + \frac{q(t)}{q'(t)}\int_{\mathbb{R}^d} (T_t f)^{q(t)-1}\, LT_t f\, d\mu \leqslant 0,$$

which is exactly (5.6.7) for $r = q(t)$ and $T_t f$ in place of f, since $q'(t) = 2\kappa q(t) - 2\kappa$. It is possible to use (5.6.7) for the bounded function $T_t f$, since the function $|\nabla T_t f|$ is bounded due to Theorem 5.6.41 proven below.

We now turn to the logarithmic Sobolev inequality. Its justification also employs T_t. Again we first take f of the form $f = f_0 + c$, $f_0 \in C_b^\infty(\mathbb{R}^d)$, $f_0 \geqslant 0$, $c > 0$. Set $\varphi = f^2$. Then $\nabla f = \frac{1}{2}\nabla\varphi/\sqrt{\varphi}$, and the desired inequality takes the form

$$\int_{\mathbb{R}^d} \varphi\ln\varphi\, d\mu - \int_{\mathbb{R}^d} \varphi\, d\mu\, \ln\Big(\int_{\mathbb{R}^d} \varphi\, d\mu\Big) \leqslant \frac{1}{2\kappa}\int_{\mathbb{R}^d} \frac{1}{\varphi}|\nabla\varphi|^2\, d\mu. \tag{5.6.9}$$

Note that $T_t\varphi \geqslant c^2$. Since by the ergodicity of μ (Example 5.6.20) we have

$$T_t\varphi\ln T_t\varphi \to \int_{\mathbb{R}^d} \varphi\, d\mu\, \ln\int_{\mathbb{R}^d} \varphi\, d\mu \quad \text{if } t\to\infty,$$

the left-hand side of (5.6.9) is represented in the form

$$-\int_0^\infty \Big(\frac{d}{dt}\int_{\mathbb{R}^d} T_t\varphi\ln T_t\varphi\, d\mu\Big)\, dt,$$

which by the semigroup property can be written as

$$-\int_0^\infty \Big(\int_{\mathbb{R}^d} (LT_t\varphi)\, \ln T_t\varphi\, d\mu\Big)\, dt - \int_0^\infty \Big(\int_{\mathbb{R}^d} T_t\varphi\frac{1}{T_t\varphi}\frac{d}{dt}T_t\varphi\, d\mu\Big)\, dt.$$

Since the second term here vanishes by the invariance of μ, on the left in (5.6.9) we have

$$-\int_0^\infty \Big(\int_{\mathbb{R}^d} (LT_t\varphi)\, \ln T_t\varphi\, d\mu\Big)\, dt.$$

Applying (3.7.15) once again, we rewrite this as

$$\int_0^\infty \Big(\int_{\mathbb{R}^d} \langle\nabla T_t\varphi, \nabla(\ln T_t\varphi)\rangle\, d\mu\Big)\, dt = \int_0^\infty \Big(\int_{\mathbb{R}^d} \frac{1}{T_t\varphi}|\nabla T_t\varphi|^2\, d\mu\Big)\, dt.$$

Set

$$F(t) = \int_{\mathbb{R}^d} \frac{1}{T_t\varphi}|\nabla T_t\varphi|^2\, d\mu.$$

Using the estimate $|\nabla T_t\varphi| \leqslant e^{-\kappa t}T_t|\nabla\varphi|$ from Theorem 5.6.41 below and the Cauchy–Bunyakovsky inequality giving the estimate

$$\frac{1}{T_t(\varphi)}\big[T_t|\nabla\varphi|\big]^2 \leqslant T_t\Big(\frac{1}{\varphi}|\nabla\varphi|^2\Big),$$

we obtain

$$e^{-2\kappa t}\int_{\mathbb{R}^d}\frac{1}{T_t\varphi}(T_t|\nabla\varphi|)^2\,d\mu \leqslant e^{-2\kappa t}\int_{\mathbb{R}^d}T_t\Big(\frac{1}{\varphi}|\nabla\varphi|^2\Big)\,d\mu$$
$$= e^{-2\kappa t}\int_{\mathbb{R}^d}\frac{1}{\varphi}|\nabla\varphi|^2\,d\mu,$$

which gives (5.6.9) after integration in t over $(0,\infty)$.

Let now $f\in W^{2,1}(\mu)$ and $f\geqslant 0$. Then the logarithmic Sobolev inequality follows by the Fatou theorem, since there is a sequence $\{\varphi_k\}\subset C_b^\infty(\mathbb{R}^k)$ of functions of the indicated form for which $\varphi_k\geqslant 1/k$ and $\varphi_k\to f$ in $W^{2,1}(\mu)$ and almost everywhere. In order to apply the Fatou theorem we note also that $t^2\ln t > -1$ if $t>0$ and $\varphi_k(x)^2\ln\varphi_k(x)\to 0$, if $\varphi_k(x)\to 0$. Finally, for every function $f\in W^{2,1}(\mu)$, it remains to observe that $|f|\in W^{2,1}(\mu)$ and $|\nabla|f||=|\nabla f|$ almost everywhere. □

The indicated condition is obviously fulfilled for $b(x)=-x$, which corresponds to the standard Gaussian measure, and also for $b(x) = -x+v(x)$, where v is Lipschitzian with the Lipschitz constant less than 1.

In the paper Röckner, Wang [**816**], the semigroup $\{T_t\}_{t\geqslant 0}$ generated by the operator $Lf = \Delta f + \langle b,\nabla f\rangle$ on a manifold was studied in the space $L^1(\mu)$ under the assumption that μ is an invariant probability measure for the semigroup. The goal was to study the properties of hypercontractivity, ultracontractivity, i.e., $T_t\big(L^2(\mu)\big)\subset L^\infty(\mu)$ for $t>0$, and also supercontractivity, i.e., $T_t\big(L^2(\mu)\big)\subset L^4(\mu)$ for $t>0$. Let us mention the main results of this work conformably to $\mathbb{R}^d$. Let $\|T\|_{p\to q}$ denote the norm of the operator $T\colon\ L^p(\mu)\to L^q(\mu)$.

5.6.37. Theorem. *Let $b\in C^1(\mathbb{R}^d,\mathbb{R}^d)$.* (i) *If* (5.6.5) *holds, then*

$$\|T_t\|_{p\to q}\leqslant \exp\big[4\beta(r)(1/p-1/q)\big]$$

for $t>0$, $q>p>1$, $\exp(4t/r)>(q-1)/(p-1)$.

(ii) *Let*

$$\langle Db(x)h,h\rangle \leqslant K|h|^2, \quad \text{where } K\geqslant 0. \tag{5.6.10}$$

If there are numbers $C,t>0$ and $q>p>1$ for which $\|T_t\|_{p\to q}\leqslant C$, then (5.6.5) *is true for*

$$r = 2[\exp(2Kt)-1]p(q-1)\big(K(q-p)\big)^{-1}, \quad \beta(r) = (q-p)^{-1}pq\ln C,$$

where $r = 4tp(q-1)(q-p)^{-1}$ if $K=0$. In addition, the supercontractivity of the semigroup $\{T_t\}_{t\geqslant 0}$ is equivalent to (5.6.5).

(iii) *Under condition* (5.6.10) *the ultracontractivity of $\{T_t\}_{g\geqslant 0}$ is equivalent to the inclusion $T_t\exp(\lambda|x|^2)\in L^\infty(\mu)$ for all $\lambda,t>0$, and its supercontractivity is equivalent to the inclusion $\exp(\lambda|x|^2)\in L^1(\mu)$ for all $\lambda>0$. If there exists $\lambda>K/2$ such that $\exp(\lambda|x|^2)\in L^1(\mu)$, then $\|T_t\|_{2\to 4}<\infty$ for some $t>0$, hence the hypercontractivity holds.*

The hypercontractivity of diffusion semigroups on finite-dimensional spaces, the Poincaré inequality, the logarithmic Sobolev inequality, and other related inequalities for them are studied in Bakry [**82**], Bakry, Bolley, Gentil [**83**], Bakry, Cattiaux, Guillin [**84**], Bakry, Emery [**85**], Bakry, Gentil, Ledoux [**86**], Bolley, Gentil [**188**], Cattiaux, Guillin, Roberto [**228**], Cattiaux, Guillin, Wang, Wu [**229**], Deuschel, Stroock [**299**], Fujita [**390**], Hino [**456**], Gross [**426**], Ledoux [**594**], Lin, Huang [**616**], Matthes, Jüngel, Toscani [**679**], Royer [**827**], Varopoulos [**910**], Varopoulos, Saloff-Coste, Coulhon [**911**], and Wang [**928**], where one can find numerous additional references.

5.6(vi). Estimates of diffusion semigroups

Let now X be a complete separable metric space with a positive σ-finite measure μ and let $\{P_t\}_{t\geqslant 0}$ be a strongly continuous symmetric Markov semigroup on $L^2(\mu)$. Let L denote the generator of this semigroup and let $\mathcal{E}$ be the corresponding Dirichlet form defined by the equality

$$\mathcal{E}(f,f) := -(Lf,f)_2 = -(\sqrt{-L}f,\sqrt{-L}f)_2.$$

The right-hand side defines a form on an even broader domain than the domain of definition of L, namely, on the domain of definition of $\sqrt{-L}$. The following interesting result was proved in Carlen, Kusuoka, Stroock [**219**].

5.6.38. Theorem. *Let $\nu>0$ and $\delta>0$. If for some K one has the estimate*

$$\|f\|_2^{2+4/\nu} \leqslant K\bigl[\mathcal{E}(f,f)+\delta\|f\|_2^2\bigr]\,\|f\|_1^{4/\nu}, \quad f\in L^2(\mu), \tag{5.6.11}$$

then there exists $M=M(K,\nu)$ such that

$$\|P_t\|_{\mathcal{L}(L^1,L^\infty)} \leqslant Me^{\delta t}t^{-\nu/2}, \quad t>0. \tag{5.6.12}$$

Conversely, if (5.6.12) *holds with some number M, then also* (5.6.11) *holds with some number $K=K(M,\nu)$.*

In Hino, Ramírez [**457**], asymptotics of the transition semigroup for small times were studied. Let $\{T_t\}_{t\geqslant 0}$ be a Markov semigroup on $L^2(\mu)$ with an invariant probability measure μ on $(\Omega,\mathcal{B})$ with generator L possessing the following properties: if $f\in D(\sqrt{-L})$, then $\min(f,1)\in D(\sqrt{-L})$ and $\mathcal{E}\bigl(\min(f,1),\min(f,1)\bigr)\leqslant\mathcal{E}(f,f)$, and also $\mathcal{E}(f,g)=0$ if $f,g\in D(\sqrt{-L})$ and $fg=0$. Let

$$P_t(A,B)=\int_A T_tI_B\,d\mu, \quad A,B\in\mathcal{B}.$$

Set $\mathcal{E}(f,g)=(\sqrt{-L}f,\sqrt{-L}g)_{L^2}$ and

$$D:=\bigl\{f\in\mathcal{D}(\sqrt{-L})\cap L^\infty(\mu)\colon \\ 2\mathcal{E}(fh,h)-\mathcal{E}(f^2,h)\leqslant\|h\|_{L^1}\ \forall\, h\in\mathcal{D}(\sqrt{-L})\cap L^\infty(\mu)\bigr\}.$$

Then the following equality holds:

$$\lim_{t\to 0} t\ln P_t(A,B) = -\frac{1}{2}d(A,B)^2,$$

where $d(A,B):=\sup_{f\in D}\{\operatorname{essinf}_{x\in B}f(x)-\operatorname{essinf}_{y\in A}f(y)\}$.

Let us mention two results on estimates of the semigroups associated with elliptic operators. The next result is obtained in Metafune, Pallara, Rhandi [**695**].

5.6.39. Theorem. *Let $a_{ij} \in C_b^2(\mathbb{R}^d)$ and $b^i \in C^2(\mathbb{R}^d)$. Suppose that A and A^{-1} are uniformly bounded and there exist numbers $\varepsilon, c_1, c_2, c > 0$, $\beta > 2$ such that*

$$\limsup_{|x|\to\infty} |x|^{-\beta}\langle b(x), x\rangle \leqslant -c, \quad |b(x)| + |\nabla b^i(x)| + \|D^2 b^i(x)\| \leqslant c_1 \exp\bigl(c_2|x|^{\beta-\varepsilon}\bigr).$$

Then the semigroup $\{T_t^\mu\}_{t\geqslant 0}$ on $L^1(\mu)$, where μ is the probability solution of the equation $L_{A,b}^*\mu = 0$, which exists and is unique under the indicated conditions, is defined by probability densities $p(y,t,x)$, i.e.,

$$T_t^\mu f(x) = \int_{\mathbb{R}^d} p(y,t,x)\, dy,$$

for which the estimate

$$p(y,t,x) + |\nabla_x p(y,t,x)| + |D_x^2 p(y,t,x)| + |\partial_t p(y,t,x)| \leqslant \kappa \exp\Bigl(\alpha t^{-\frac{\beta}{\beta-2}} - \gamma|y|^\beta\Bigr)$$

holds with some $\alpha, \kappa, \gamma > 0$.

In Arendt, Metafune, Pallara [**63**], the following result is obtained. Let

$$Lu = \sum_{i=1}^d \partial_{x_i}(a^{ij}\partial_{x_j}u) - \langle b, \nabla u\rangle + \operatorname{div}(uc) - Vu,$$

where $a^{ij} \in C_b^1(\mathbb{R}^d)$, b, $c \in C^1(\mathbb{R}^d, \mathbb{R}^d)$, $V \in L^\infty_{\rm loc}(\mathbb{R}^s)$. It is shown in [**63**] that on $L^2(\mathbb{R}^d)$ there exists the minimal sub-Markov C_0-semigroup $\{T_t\}_{t\geqslant 0}$ the generator of which extends L (the minimality means the estimate $T_t f \leqslant S_t f$ for $f \geqslant 0$ for every other sub-Markov C_0-semigroup $\{S_t\}_{t\geqslant 0}$ whose generator extends L). Note that in the situation of the previous theorem $T_t = T_t^\mu$.

5.6.40. Theorem. *Let $V \geqslant 0$, $\operatorname{div} b \leqslant \beta V$, $\operatorname{div} c \leqslant \beta V$, $0 < \beta < 1$, and also $|b| + |c| \leqslant k_1 V^\alpha + k_2$. If $1/2 \leqslant \alpha < 1$, then T_t is given by an integral kernel k_t from the class $L^\infty(\mathbb{R}^d \times \mathbb{R}^d)$ for which*

$$k_t(x,y) \leqslant c_1 e^{\omega t} t^{-d/2} \exp\Bigl(-\frac{c_2|x-y|^{1/(1-\alpha)}}{t}\Bigr)^{-\alpha}.$$

Unlike the previous theorem, here an invariant measure need not exist. See also Metafune, Ouhabaz, Pallara [**690**], Metafune, Spina [**703**] and § 7.5.

5.6(vii). Gradient estimates in the case of dissipative drifts

Here we present the results of the papers Bogachev, Da Prato, Röckner, Sobol [**140**], [**141**]. Let us consider the operator $L = L_{A,b}$, where A is a constant positive definite matrix (independent of x) and b satisfies our standard condition that $\|b\|_{L^p(U)} < \infty$ for every ball U in $\mathbb{R}^d$ with some $p = p(U) > d$, $p \geqslant 2$. Assume also that b is *dissipative* in the following sense: for every $h \in \mathbb{R}^d$ there exists a set $N_h \subset \mathbb{R}^d$ of measure zero such that

$$\langle b(x+h) - b(x), h\rangle \leqslant 0 \quad \text{for all } x \in \mathbb{R}^d \setminus N_h. \tag{5.6.13}$$

Finally, suppose that there exists a C^2-function V (a Lyapunov function) such that $V(x) \to +\infty$ and $LV(x) \to -\infty$ as $|x| \to \infty$.

It follows from this that there exists a unique probability measure μ on $\mathbb{R}^d$ with a strictly positive continuous density $\varrho \in W^{p,1}_{\rm loc}(\mathbb{R}^d)$ satisfying the equation $L^*\mu = 0$. We know that the closure $\overline{L}$ of the operator L on domain $C_0^\infty(\mathbb{R}^d)$ in $L^1(\mu)$ generates a Markov semigroup $\{T_t\}_{t\geqslant 0}$ for which μ is invariant. Let $D(\overline{L})$ denote the domain of definition of $\overline{L}$ in $L^1(\mu)$ and let $\{R_\lambda\}_{\lambda>0}$ denote the corresponding resolvent, i.e., $R_\lambda = (\lambda - \overline{L})^{-1}$. The restrictions of T_t and R_λ to $L^2(\mu)$ are contractions of $L^2(\mu)$. In particular, if $v \in D(\overline{L})$ and $\lambda v - \overline{L}v = g \in L^2(\mu)$, then $v \in L^2(\mu)$. Moreover, $v \in W^{2,2}_{loc}(\mathbb{R}^d)$, one even has $v \in W^{p,2}_{loc}(\mathbb{R}^d)$, and $\overline{L}v = Lv$ a.e., since a.e.

$$\lambda v - Lv = g. \tag{5.6.14}$$

Let $\widehat{L}$ be the elliptic operator with the same second order part as L, but with the dual drift $\widehat{b} = 2A\nabla\varrho/\varrho - b$. By the integration by parts formula

$$\int_{\mathbb{R}^d} \psi L\varphi\, d\mu = \int_{\mathbb{R}^d} \varphi\widehat{L}\psi\, d\mu \quad \text{for all } \psi, \varphi \in C_0^\infty(\mathbb{R}^d).$$

In addition, for every $\lambda > 0$ the ranges of $\lambda - L$ and $\lambda - \widehat{L}$ on $C_0^\infty(\mathbb{R}^d)$ are dense in $L^1(\mu)$ due to existence of a Lyapunov function.

The operator $\widehat{L}$ also generates a Markov semigroup on $L^1(\mu)$ with respect to which the measure μ is invariant. The corresponding resolvents will be denoted by $\widehat{R}_\lambda$.

5.6.41. Theorem. *For every Lipschitzian function $f \in L^1(\mu)$ and all $t, \lambda > 0$, the functions $T_t f$ and $R_\lambda f$ have Lipschitz version and*

$$|\nabla T_t f(x)| \leqslant T_t|\nabla f|(x) \quad \textit{and} \quad |\nabla R_\lambda f(x)| \leqslant R_\lambda|\nabla f|(x) \tag{5.6.15}$$

for the corresponding continuous versions. In addition,

$$\sup_{x,t} |\nabla T_t f(x)| \leqslant \sup_x |\nabla f(x)|, \quad \sup_x |\nabla R_\lambda f(x)| \leqslant \frac{1}{\lambda}\sup_x |\nabla f(x)|. \tag{5.6.16}$$

If we replace (5.6.13) *with the stronger estimate*

$$\langle b(x+h) - b(x), h\rangle \leqslant -\kappa|h|^2 \quad \textit{for all } x \in \mathbb{R}^d \setminus N_h \tag{5.6.17}$$

with some number $\kappa > 0$, then we obtain

$$|\nabla T_t f(x)| \leqslant e^{-\kappa t} T_t|\nabla f|(x) \quad \textit{and} \quad |\nabla R_\lambda f(x)| \leqslant R_{\lambda+\kappa}|\nabla f|(x). \tag{5.6.18}$$

PROOF. Suppose first that the mapping b is infinitely differentiable, Lipschitz and strongly dissipative, i.e., for some $\alpha > 0$ we have

$$\langle b(x+h) - b(x), h\rangle \leqslant -\alpha|h|^2 \quad \text{for all } x, h \in \mathbb{R}^d.$$

We show that for every $\lambda > 0$ and every smooth bounded Lipschitzian function f one has the pointwise estimate

$$|\nabla R_\lambda f(x)| \leqslant R_\lambda|\nabla f|(x). \tag{5.6.19}$$

It will give us the desired estimate for $|\nabla T_t f(x)|$. Let us prove the following: if $\lambda > 0$ and $v \in L^2(\mu)$ satisfies the inequality $(\lambda - L)v \leqslant 0$ in the weak sense, i.e., the integral of $v(\lambda - \widehat{L})\varphi$ with respect to μ is nonpositive for all nonnegative functions $\varphi \in C_0^\infty(\mathbb{R}^d)$, then $v \leqslant 0$. Indeed, letting $V(x) = |x|^2$, we obtain

$$\begin{aligned} LV(x) &= 2\mathrm{tr}\, A + 2\langle b(x), x\rangle \leqslant 2\mathrm{tr}\, A - 2\alpha(x,x) + 2\langle b(0), x\rangle \\ &\leqslant 2\mathrm{tr}\, A + \alpha^{-1}|b(0)|^2 - \alpha|x|^2. \end{aligned}$$

According to the results in § 2.3, the measure μ has all moments, hence $|b| \in L^2(\mu)$. By Theorem 3.1.2 we obtain $|\nabla\varrho/\varrho| \in L^2(\mu)$. Let $\zeta_0 \in C_0^\infty(\mathbb{R}^d)$ be such that $0 \leqslant \zeta_0 \leqslant 1$ and $\zeta_0(x) = 1$ if $|x| \leqslant 1$. Let $\zeta_k(x) = \zeta_0(x/k)$, $k \in \mathbb{N}$. Then

$0 \leqslant \zeta_k \leqslant 1$, the functions $|\nabla\zeta_k|$ are bounded uniformly in k, $\zeta_k \to 1$ pointwise and $|\nabla\zeta_k|, L\zeta_k, \widehat{L}\zeta_k$ converge to zero in $L^2(\mu)$ as $k \to \infty$.

Let $\eta \in C_0^\infty(\mathbb{R}^d)$, $\eta \geqslant 0$ and $u := \widehat{R}_\lambda \eta$. Then the function u is bounded and nonnegative by the Markov property, is smooth by the elliptic regularity, since $\widehat{b}$ is smooth. We show that $|\nabla u| \in L^2(\mu)$. We have

$$\int_{\mathbb{R}^d} u(\lambda u - \widehat{L}u)\zeta\, d\mu = \int_{\mathbb{R}^d} \zeta u \eta\, d\mu, \quad \zeta \in C_0^\infty(\mathbb{R}^d).$$

Since $u\widehat{L}u = \frac{1}{2}\widehat{L}(u^2) - \langle A\nabla u, \nabla u\rangle$, we have

$$\int_{\mathbb{R}^d} u^2\Big(\lambda - \frac{1}{2}L\Big)\zeta\, d\mu + \int_{\mathbb{R}^d} \langle A\nabla u, \nabla u\rangle \zeta\, d\mu = \int_{\mathbb{R}^d} u\eta\zeta\, d\mu.$$

Taking ζ_k as indicated above, we obtain that $|\nabla u| \in L^2(\mu)$.

Now let $\varphi_k := \zeta_k u$. Then $\varphi_k \in C_0^\infty(\mathbb{R}^d)$, $\varphi_k \geqslant 0$ and

$$(\lambda - \widehat{L})\varphi_k = \zeta_k \eta + u\,\widehat{L}\zeta_k + 2\langle A\nabla\zeta_k, \nabla u\rangle \to \eta \quad \text{in } L^2(\mu) \text{ as } k \to \infty$$

by the dominated convergence theorem, since u is bounded and $|\nabla u| \in L^2(\mu)$. Hence

$$\int_{\mathbb{R}^d} v\eta\, d\mu = \lim_{k\to\infty} \int_{\mathbb{R}^d} v(\lambda - \widehat{L})\varphi_k\, d\mu \leqslant 0,$$

whence we conclude that $v \leqslant 0$, since η was an arbitrary nonnegative smooth function with compact support. We now establish (5.6.19). Let $u = R_\lambda f$. As above, we have $u \in C^\infty(\mathbb{R}^d)$ and $|\nabla u| \in L^2(\mu)$. We observe that

$$(\lambda - \overline{L})R_\lambda|\nabla f| = (\lambda - L)R_\lambda|\nabla f| = |\nabla f|,$$

since the function $|\nabla f|$ is bounded and Lipschitzian and $R_\lambda|\nabla f| \in W^{p,2}_{\text{loc}}(\mathbb{R}^d)$. Hence it suffices to show that $v := |\nabla u|$ satisfies the inequality

$$\int_{\mathbb{R}^d} v(\lambda\varphi - \widehat{L}\varphi)\, d\mu \leqslant \int_{\mathbb{R}^d} |\nabla f|\varphi\, d\mu \tag{5.6.20}$$

for every nonnegative function $\varphi \in C_0^\infty(\mathbb{R}^d)$ and apply the previous step. Let $\varepsilon > 0$ and $v_\varepsilon := \big(|\nabla u|^2 + \varepsilon\big)^{1/2}$. It is readily seen that v_ε satisfies the equation $\lambda v_\varepsilon - Lv_\varepsilon - w_\varepsilon v_\varepsilon = f_\varepsilon$, where

$$w_\varepsilon := \langle Db\nabla u, \nabla u\rangle v_\varepsilon^{-2} \leqslant 0,$$

$$f_\varepsilon := \lambda\varepsilon v_\varepsilon^{-1} + \langle \nabla u, \nabla f\rangle v_\varepsilon^{-1} - v_\varepsilon^{-1}\big[\mathrm{tr}(D^2uAD^2u) - v_\varepsilon^{-2}\langle \nabla u, D^2uAD^2u\nabla u\rangle\big].$$

Let $\xi := v_\varepsilon^{-1}\nabla u$ and $S := D^2u$. Observing that $|\xi| \leqslant 1$ and the matrix SAS is symmetric and nonnegative definite, we obtain the estimates $\langle \xi, \nabla f\rangle \leqslant |\nabla f|$ and $\mathrm{tr}(SAS) - \langle \xi, SAS\xi\rangle \geqslant 0$. Hence $(\lambda - L)v_\varepsilon \leqslant \lambda\varepsilon v_\varepsilon^{-1} + |\nabla f|$ pointwise, then also in the weak sense. Letting $\varepsilon \to 0$, we arrive at (5.6.20). The first estimate in (5.6.16) in the case under consideration follows from the already proven facts, and for obtaining the second estimate we have to employ also the estimate $\|R_\lambda f\|_\infty \leqslant \lambda^{-1}\|f\|_\infty$, following from the fact that $\lambda - L = \lambda(1 - \lambda^{-1}L)$, where the operator $\lambda^{-1}L$ satisfies the same conditions as L.

Let us proceed to the general case. We recall that if a sequence of functions on $\mathbb{R}^d$ is uniformly Lipschitzian with constant Λ and bounded at some point, then it contains a subsequence that converges uniformly on every ball to a function that is Lipschitzian with the same constant. Therefore, approximating f in $L^1(\mu)$ by a sequence of bounded smooth functions f_j with the estimate

$$\sup_x |\nabla f_j(x)| \leqslant \sup_x |\nabla f(x)|,$$

it suffices to prove (5.6.16) for smooth bounded f. Moreover, by Euler's formula

$$T_t f = \lim_{n\to\infty} \Bigl(\frac{t}{n} R_{\frac{t}{n}}\Bigr)^n f,$$

it suffices to establish our estimate for the resolvent. We first construct a suitable sequence of smooth strongly dissipative Lipschitz vector fields b_k such that $b_k^i \to b^i$ in $L^p(U)$ for every ball U as $k \to \infty$.

Let $\sigma_j(x) = j^{-d}\sigma(x/j)$, where σ is a smooth probability density with compact support. Set $\beta_j := b * \sigma_j$. Then β_j is a smooth dissipative vector field and $\beta_j^i \to b^i$ as $j \to \infty$ in $L^p(U)$ for every ball U. For $\alpha > 0$ the mapping $I - \alpha\beta_j$ is a homeomorphism of $\mathbb{R}^d$, and the inverse mapping $(I - \alpha\beta_j)^{-1}$ is Lipschitzian with constant α^{-1} (see Brezis [**207**]). Let us consider the Yosida approximations

$$F_\alpha(\beta_j) := \alpha^{-1}\bigl((I - \alpha\beta_j)^{-1} - I\bigr) = \beta_j \circ (I - \alpha\beta_j)^{-1}.$$

It is known (see Brezis [**206**, Chapter II]) that $|F_\alpha(\beta_j)(x)| \leqslant |\beta_j(x)|$, the mappings $F_\alpha(\beta_j)$ converge locally uniformly to β_j as $\alpha \to 0$, and

$$\bigl\langle F_\alpha(\beta_j)(x) - F_\alpha(\beta_j)(y), x - y\bigr\rangle \leqslant 0.$$

Thus, the sequence $b_k := F_{\frac{1}{k}}(b * \sigma_k) - \frac{1}{k} I$ is the desired one. For every $k \in \mathbb{N}$ let L_k be the elliptic operator with the same constant matrix A and the drift b_k in place of b. Let $\mu_k = \varrho_k\,dx$ be the corresponding invariant probability measure and $R_\lambda^{(k)}$ the associated family of resolvents on $L^1(\mu_k)$. Since b_k is smooth, Lipschitzian and strongly dissipative, the function $v_k := R_\lambda^{(k)} f$ is smooth, bounded and Lipschitzian, moreover

$$\sup_x |v_k(x)| \leqslant \lambda^{-1} \sup_x |f(x)| \quad \text{and} \quad \sup_x |\nabla v_k(x)| \leqslant \lambda^{-1} \sup_x |\nabla f(x)|$$

by the first step of the proof. In addition, for every ball $U \subset \mathbb{R}^d$ the functions v_k are uniformly bounded in the Sobolev space $W^{2,2}(U)$, since the mappings $|b_k|$ are bounded in $L^p(U)$ uniformly in k and the function f is bounded. This follows by the fact that for every solution $w \in W^{2,2}(U)$ of the elliptic equation

$$a^{ij}\partial_{x_i}\partial_{x_j} w + b^i \partial_{x_i} w - \lambda w = g$$

we have the estimate $\|w\|_{W^{2,2}(U)} \leqslant C\|w\|_{L^2(U)}$, where C is a constant depending on U, A and the quantity $\kappa := \|g\|_{L^2(U)} + \bigl\||b|\bigr\|_{L^p(U)}$ in such a way that as a function of κ it is locally bounded. Thus, the sequence $\{v_k\}$ contains a subsequence, again denoted by $\{v_k\}$, which converges locally uniformly to a bounded Lipschitzian function $v \in W^{2,2}_{\mathrm{loc}}(\mathbb{R}^d)$ satisfying the inequalities

$$\sup_x |v(x)| \leq \lambda^{-1} \sup_x |f(x)| \quad \text{and} \quad \sup_x |\nabla v(x)| \leq \lambda^{-1} \sup_x |\nabla f(x)|,$$

where the restrictions of the functions v_k to any ball U converge to $v|_U$ weakly in the space $W^{2,2}(U)$.

We now show that $v = R_\lambda f$. We observe that $\varrho_k \to \varrho$ uniformly on balls according to the results in § 1.6. Hence for every function $\varphi \in C_0^\infty(\mathbb{R}^d)$ with support in a ball U we have

$$\int_{\mathbb{R}^d} (\lambda v - Lv - f)\varphi\varrho\, dx = \lim_{k\to\infty} \int_{\mathbb{R}^d} (\lambda v_k - L_k v_k - f)\varphi\varrho_k\, dx = 0$$

by the weak convergence of v_k to v in $W^{2,2}(U)$ combined with convergence of b_k^i to b^i in $L^p(U)$. Therefore, by the integration by parts formula we have

$$\int_{\mathbb{R}^d} v(\lambda\varphi - \widehat{L}\varphi)\, d\mu = \int_{\mathbb{R}^d} f\varphi\, d\mu$$

for all $\varphi \in C_0^\infty(\mathbb{R}^d)$. The function $R_\lambda f$ is bounded and satisfies the same relation, hence it remains to recall that if a bounded function u satisfies the equality

$$\int_{\mathbb{R}^d} u(\lambda\varphi - \widehat{L}\varphi)\, d\mu = 0 \quad \forall\, \varphi \in C_0^\infty(\mathbb{R}^d),$$

then $u = 0$ a.e., since $(\lambda - \widehat{L})\bigl(C_0^\infty(\mathbb{R}^d)\bigr)$ is dense in $L^1(\mu)$.

We now turn to the pointwise estimate $|\nabla R_\lambda f(x)| \leqslant R_\lambda |\nabla f|(x)$. Suppose first that $f \in C_0^\infty(\mathbb{R}^d)$. The desired estimate holds for every $R_\lambda^{(k)}$ in place of R_λ. It has been shown above that $v = R_\lambda f$ is the weak limit of $v_k = R_\lambda^{(k)} f$ in $W^{2,2}(U)$ for every ball U. In addition, the functions $R_\lambda^{(k)}|\nabla f|$ converge weakly in $W^{2,2}(U)$ to the function $R_\lambda|\nabla f|$, which is clear from the same reasoning. Since the embedding of $W^{2,2}(U)$ to $W^{2,1}(U)$ is compact, we can assume, passing to a subsequence, that $\nabla R_\lambda^{(k)} f(x) \to \nabla R_\lambda f(x)$ and $R_\lambda^{(k)}|\nabla f|(x) \to R_\lambda|\nabla f|(x)$ almost everywhere in U. Hence we arrive at the desired estimate. If f is Lipschitzian and has bounded support, then we can find uniformly Lipschitzian functions $f_n \in C_0^\infty(\mathbb{R}^d)$ vanishing outside some ball such that $f_n \to f$ uniformly and $\nabla f_n \to \nabla f$ a.e. Then with the aid of the same reasoning as above we obtain that $R_\lambda|\nabla f_n| \to R_\lambda|\nabla f|$ and $\nabla R_\lambda f_n \to \nabla R_\lambda f$ in $L^2(U)$. Passing to an almost everywhere convergent subsequence, we obtain the pointwise inequality. Finally, in the case of a general Lipschitzian function $f \in L^1(\mu)$ one can find uniformly Lipschitzian functions ζ_n such that $0 \leqslant \zeta_n \leqslant 1$ and $\zeta_n(x) = 1$ whenever $|x| \leqslant n$. Let $f_n = f\zeta_n$. By the previous step we have

$$|\nabla R_\lambda f_n(x)| \leqslant R_\lambda |\nabla f_n|(x).$$

The functions f_n are uniformly Lipschitzian. Hence for every ball U the sequence of functions $R_\lambda f_n|_U$ is norm bounded in $W^{2,2}(U)$. In addition, the functions $R_\lambda|\nabla f_n|$ on U converge to $R_\lambda|\nabla f|$ in $L^2(U)$, since $|\nabla f_n| \to |\nabla f|$ in $L^2(\mu)$ by the Lebesgue dominated convergence theorem. Now the same reasoning as above completes the proof for R_λ. The estimates for the resolvents yield the desired estimates for T_t. The case $\kappa > 0$ is similar, with a minor difference in the estimate for the function w_ε. □

Note that (5.6.16) differs from (5.6.15) by the factor $1/\lambda$, which cannot be put in the pointwise estimate, as the following example shows: let L be the one-dimensional Ornstein–Uhlenbeck operator and let $f(x) = x$; then we have the equality $R_\lambda f(x) = x/(1+\lambda)$, $R_\lambda|f'|(x) = 1/\lambda$.

Comments

Some minimal background on operator semigroups necessary for our discussion can be found in the books Clément et als. [**251**] and Pazy [**767**].

Diffusion semigroups on L^p-spaces with respect to invariant measures, domains of definition of generators and related problems for elliptic operators are studied in Chill, Fašangovà, Metafune, Pallara [**241**], [**242**], Da Prato, Lunardi [**272**], Farkas, Lunardi [**351**], Kusuoka, Stroock [**574**], Lunardi, Metafune [**648**], Lunardi, Metafune, Pallara [**649**], Lorenzi, Bertoldi [**642**], Lorenzi, Lunardi [**643**], Lorenzi, Zamboni [**644**], Metafune, Pallara [**691**], Metafune, Pallara, Prüss, Schnaubelt [**693**], Metafune, Pallara, Vespri [**696**], Metafune, Pallara, Wacker [**697**], [**698**], Metafune, Priola [**699**], Metafune, Prüss, Schnaubelt, Rhandi [**701**], Metafune, Spina [**702**], Metafune, Spina, Tacelli [**704**], [**705**], Prüss, Rhandi, Schnaubelt [**793**], Röckner, Wang [**817**]. On estimates for diffusion semigroups, see also Azencott [**81**], Chen, Qian, Hu, Zheng [**238**], Liskevich, Sobol [**624**], Priola, Wang [**791**], and Wang [**926**], [**930**].

On the L^p-uniqueness of elliptic operators, see Albanese, Lorenzi, Mangino [**14**], Albanese, Mangino [**15**], [**16**], Wu, Zhang [**943**]. Apart the uniqueness of a continuous semigroup on $L^1(\mu)$ for an infinitesimally invariant measure of an elliptic operator whose generator extends the given elliptic operator, also of interest is the question of uniqueness of a Markov or sub-Markov semigroup with such a property (see Cattiaux, Fradon [**226**]). On diffusion semigroups on $L^1(\mathbb{R}^d)$ and $L^p(\mathbb{R}^d)$, see Davies [**283**], Liskevich, Sobol, Vogt [**626**]. Diverse aspects of the theory of elliptic operators possessing infinitesimally invariant measures and also the properties of the corresponding semigroups are considered in the papers Da Prato, Goldys [**270**], Da Prato, Vespri [**278**].

Operators and semigroups of Ornstein–Uhlenbeck type (which corresponds to a constant diffusion matrix and a linear drift) are studied in Bramanti, Cupini, Lanconelli, Priola [**203**], [**204**], Geissert, Lunardi [**404**], Hieber et al. [**453**], Lanconelli [**578**], Lunardi [**647**], Mauceri, Noselli [**680**], Metafune [**689**], Metafune, Pallara, Priola [**692**], and Metafune, Prüss, Rhandi, Schnaubelt [**700**].

In case $\Omega = \mathbb{R}^d$ Theorem 5.3.1 was proved in Stannat [**872**, Proposition 1.9] under somewhat more general assumptions about A, b and μ (the validity of those assumptions in our case follows by the results in § 1.6).

Theorem 5.4.5 complements an earlier result from Albeverio, Bogachev, Röckner [**18**] and under only local assumptions gives an affirmative answer to a question posed by S.R.S. Varadhan in [**909**].

Proposition 5.3.9 was proved in Bogachev, Röckner, Shaposhnikov [**174**]. An assertion close to Example 5.3.10(ii) can be found in Arapostathis, Borkar, Ghosh [**57**, Theorem 2.6.10], but without justification of compactness of V. On the converse to the condition with Lyapunov functions, see also Kushner [**567**].

In many applications, one has to deal with operators of the form $\Delta + b \cdot \nabla$ with drift coefficients b integrable with respect to some absolutely continuous measure $\varrho\, dx$. Such a drift can be very singular from the Lebesgue measure point of view. This is why in many theorems presented in this book we include conditions expressed in terms of the measure μ. Here the question arises about the existence of a diffusion (in a suitable sense) whose generator extends the given operator. Methods of constructing and studying such diffusions constitute the theory of Dirichlet forms. There is a vast literature devoted to this direction, see the books Fukushima [**391**],

Fukushima, Oshima, Takeda [**393**], Ma, Röckner [**653**] and the papers Cattiaux, Léonard [**230**]–[**232**], Fattler, Grothaus [**352**], [**353**], Fukushima [**392**], Kondratiev, Konstantinov, Röckner [**535**], Meyer, Zheng [**707**], Stannat [**871**], Wallstrom [**925**].

On existence of invariant measures of finite-dimensional diffusions, see Bhattacharya [**115**], Bhattacharya, Ramasubramanian [**116**], Khasminskii [**502**], Kushner [**569**], Miyahara [**717**], Wonham [**942**], Zakai [**950**]. General problems related to ergodicity of diffusions and convergence to invariant measures are considered in Arnold, Kliemann [**67**], Borovkov [**196**], Cohen de Lara [**253**], Ganidis, Roynette, Simonot [**398**], Kushner [**566**], Malyshkin [**657**], Meyn, Tweedie [**708**], Pardoux, Veretennikov [**764**], Pinsky, Pinsky [**778**], Pinsky [**780**], Shurenkov [**860**], Skorohod [**862**], Veretennikov [**915**], [**914**], Veretennikov, Klokov [**917**]. The case of a linear drift is studied in Dym [**318**], Snyders [**864**], and Zakai, Snyders [**951**]. It is shown in Zhang [**957**] that a diffusion with a uniformly elliptic continuous diffusion matrix and a linear drift B has a unique invariant probability measure precisely when the eigenvalues of the operator B have negative real parts.

In some special cases invariant measures of diffusions can be found explicitly, see, for example, Liberzon, Brockett [**609**].

The case of symmetric diffusions is rather special, see Ichihara [**470**], [**471**].

A criterion of invariance of an infinitesimally invariant measure in terms of martingale problems is obtained in Echeverria [**322**], see also Bhatt, Karandikar [**113**].

There is a rich literature on stationary distributions of general Markov processes, irreducibility of processes, uniqueness of stationary distributions and convergence to them of transition probabilities, and Markov operators and their generalizations. In the first place one should mention the classical corn-stone papers of Doeblin [**306**], Doob [**310**] and Harris [**441**], [**442**]; let us also mention the works Abourashchi, Veretennikov [**1**], Ahmed, Ding [**9**], Arnold, Kliemann [**68**], Arnold, Eizenberg, Wihstutz [**66**], Azéma, Kaplan-Duflo, Revuz [**80**], Baxendale [**98**], Beneš [**103**], [**104**], Bhatt, Karandikar [**114**], Borovkov [**196**], Brunel, Horowitz, Lin [**209**], Carmona, Klein [**220**], Chen [**236**], Delbrouck [**292**], Doob [**311**], Dubins, Freedman [**315**], Foguel [**369**]–[**376**], Getoor [**406**], Hairer, Mattingly [**435**], Hairer, Mattingly, Scheutzow [**437**], Hernandez-Lerma, Lasserre [**450**], [**451**], Horowitz [**462**]–[**465**], Isaac [**477**]–[**479**], Ito [**483**], Jain, Krylov [**489**], Khasminskii [**502**], Kliemann [**518**], Kolokoltsov [**531**], Kushner [**571**], Kushner, Yu [**572**], Lasota, Yorke [**585**], Lasserre [**589**]–[**591**], Lyons [**652**], Neveu [**743**], Nummelin [**752**], Pichór, Rudnicki [**775**], [**776**], Portenko, Skorohod, Shurenkov [**789**], Revuz [**805**], [**806**], Rosenblatt [**826**], Stettner [**879**], Stroock [**881**], Sunyach [**886**], Szarek [**887**]–[**889**], Zaharopol [**947**], [**948**]; see also the papers Angst [**54**], Arnold, Kliemann [**67**], Butkovsky [**213**], Cattiaux, Guillin [**227**], Davies [**285**], Kulik [**561**], Veretennikov, Klokov [**917**], and Zhang [**956**], concerned with the rate of convergence, where one can find additional references on diverse aspects of these problems. Many of these works deal with ergodic problems.

Among early works which are of historical interest one should mention the papers Kryloff, Bogoliouboff [**545**], Kakutani, Yosida [**497**], Maruyama, Tanaka [**671**], Nelson [**740**], and Itô, Nisio [**482**].

Ergodic properties of diffusions in relation to controlled diffusion processes are studied in the paper Arapostathis, Borkar [**55**], [**56**] Arapostathis, Borkar, Kumar [**58**], and in the recent book Arapostathis, Borkar, Ghosh [**57**].

Transition probabilities can be used for obtaining information about the drift of a diffusion, see Albeverio, Marinelli [**30**], Hwang, Hwang-Ma, Sheu [**469**].

The so-called time reversal for finite-dimensional and infinite-dimensional diffusions (i.e., the process ξ_{T-t} for a given diffusion ξ_t) is discussed in Durrett [**317**], Föllmer [**378**], Haussmann, Pardoux [**446**], Millet, Nualart, Sanz [**711**], [**712**], Nagasawa [**731**], [**732**]. A.N. Kolmogorov in his famous paper [**529**] studied the following problem. Let $\{\xi_t\}_{t\geqslant 0}$ be a diffusion in a finite-dimensional Riemannian manifold X (in Kolmogorov's case it was compact) with generator $L = (\Delta + b)/2$, where b is a smooth vector field on X. When is the process $\eta_t = \xi_{T-t}$ governed by the same equation, i.e., has the same generator L? The answer found by Kolmogorov says: if and only if b is the gradient of a function. Earlier this question was considered by E. Schrödinger [**837**] in the one-dimensional case. Kolmogorov in [**529**] only considered solutions to the Fokker–Planck equations as densities of transition probabilities; no stochastic integration was developed that time (the corresponding stochastic equation on $\mathbb{R}^d$ would have the form $d\xi_t = dw_t + 2^{-1}b(\xi_t)dt$). Moreover, mainly the case of compact X was considered, when an invariant probability measure always exists. In this case the property studied by Kolmogorov is equivalent to the symmetrizability of $\{\xi_t\}_{t\geqslant 0}$. Thus, Kolmogorov's result is a criterion of the symmetrizability of a diffusion in a compact manifold. In the noncompact case additional conditions are required to ensure the existence of invariant probability measures. The difference between the existence of invariant measures $\mu = \varrho\,dx$ and the symmetrizability of the diffusion $\{\xi_t\}_{t\geqslant 0}$ with generator $L = (\Delta + b)/2$ can be seen from Corollary 3.1.3, which extends Kolmogorov's theorem to general drifts in $L^2(\mu)$. According to this corollary, the drifts of symmetrizable diffusions in $\mathbb{R}^d$ are the logarithmic gradients of measures (the analogous fact is true also for manifolds, see Bogachev, Röckner, Wang [**179**]). If $|b| \in L^p_{\rm loc}(\mathbb{R}^d)$ for some $p > d$, then we obtain the exact analog, since the measure μ has a positive density ϱ and $b = \nabla \ln \varrho$, i.e., b is indeed the gradient of a function.

It is of interest to study the behavior of the invariant measure of the diffusion with the diffusion coefficient εA as $\varepsilon \to 0$, see Nevelson [**741**], Stettner [**878**].

In this book we mention (with a few exceptions) only those properties of Fokker–Planck–Kolmogorov equations or associated diffusions in manifolds which do not use any specific features of manifolds; however, for many questions geometric properties of manifolds (especially related to curvatures) play an important role, and there is a vast literature on these questions, see Arendt, Biegert, Elst [**62**], Bogachev, Wang, Röckner [**185**], [**186**], Grigor'yan [**424**], Molchanov [**721**]. In addition, there are many works concerned with semigroups generated by diffusions in manifolds, see Elworthy [**331**], Elworthy, Le Jan, Li [**332**], [**333**], Elworthy, Li [**334**], Qian [**795**], Wang [**927**].

On invariant measures of diffusions in domains, see Bass, Burdzy, Chen, Hairer [**92**], Gong [**418**], Harrison, Williams [**443**], O'Connell, Ortmann [**753**], Pardoux, Williams [**765**], Pinsky [**779**], [**780**], Trutnau [**906**], and Williams [**941**].

Exercises

5.6.42. Let $\{T_t\}_{t\geqslant 0}$ be the Ornstein–Uhlenbeck semigroup defined by formula (5.1.1) on the space $L^2(\gamma)$, where γ is the standard Gaussian measure on the real line. The Hermite–Chebyshev polynomials H_n arise as the result of orthogonalization of the functions x^n, $n = 0, 1, \ldots$, in $L^2(\gamma)$ and are defined by the formula

$$H_0(x) = 1, \quad H_n(x) = (-1)^n (n!)^{-1/2} \exp(x^2/2) \frac{d^n}{dx^n} \exp(-x^2/2).$$

Prove the equality $T_t H_n = e^{-nt} H_n$.

HINT: Observe that $T_t H_n$ is a polynomial of degree n, use induction on n and the symmetry of T_t in $L^2(\gamma)$.

5.6.43. Suppose that on a separable space $L^2(\mu)$ we are given a semigroup $\{T_t\}_{t\geqslant 0}$ of operators with uniformly bounded norms obtained by restricting a strongly continuous semigroup on $L^1(\mu)$. Show that this semigroup on $L^2(\mu)$ is also strongly continuous.

Show also that if a semigroup on $L^1(\mu)$ with uniformly bounded norms possesses a strongly continuous restriction to $L^2(\mu)$, then it is strongly continuous also on $L^1(\mu)$.

HINT: Use Remark 5.1.3 and the fact that if $f \in L^2(\mu)$ and $T_t f \to f$ as $t \to 0$ in $L^1(\mu)$, then by the uniform boundedness of the norms of T_t on $L^2(\mu)$ there is an increasing sequence $n_k \to \infty$ for which the averages $(T_{1/n_1} f + \cdots + T_{1/n_k} f)/k$ converge in $L^2(\mu)$; due to convergence in $L^1(\mu)$ to f the limit in $L^2(\mu)$ will be the same, hence f is the limit of the elements $T_{1/n_k}\big((T_{1/n_1 - 1/n_k} f + \cdots + f)/k\big)$.

5.6.44. (i) Show that the heat semigroup $\{P_t\}_{t\geqslant 0}$ (see Example 5.1.1) is strongly continuous on the space $L^p(\mathbb{R}^d)$ with $p < \infty$ and is not strongly continuous on $L^\infty(\mathbb{R}^d)$ and on $C_b(\mathbb{R}^d)$.

(ii) Show that the generator of the heat semigroup $\{P_t\}_{t\geqslant 0}$ on $L^2(\mathbb{R}^d)$ is $\Delta/2$ with domain $W^{2,2}(\mathbb{R}^d)$.

HINT: (i) Check that $\|P_t\| \leqslant 1$ on $L^p(\mathbb{R}^d)$ and use that $C_0^\infty(\mathbb{R}^d)$ is dense in $L^p(\mathbb{R}^d)$ with $p < \infty$; the failure of the strong continuity on $L^\infty(\mathbb{R}^d)$ is seen from the fact that $P_t f$ is continuous for each $f \in L^\infty(\mathbb{R}^d)$; to disprove the strong continuity on $C_b(\mathbb{R}^d)$ take a function $f \in C_b(\mathbb{R})$ such that $f(n) = 0$, $f(x) = 1$ on $[n + 1/n, n + 2/n]$ for $n > 4$, and extend f in an appropriate way to the whole real line; (ii) Use that the Fourier transform is an isometry on L^2 and that $P_t f$ is the convolution of f and a Gaussian density, hence the Fourier transform of $P_t f$ is the product of the Fourier transform of f and an explicitly given function.

5.6.45. Let $X = L^p(\mu)$, where μ is a nonnegative measure on a measurable space and $p \in [1, +\infty)$. Prove that a linear operator L on a dense domain $\mathcal{D}$ is dissipative precisely when

$$\int \operatorname{sign} u\, |u|^{p-1} Lu \, d\mu \leqslant 0 \quad \forall\, u \in \mathcal{D},$$

where $\operatorname{sign} u = 1$ if $u > 0$, $\operatorname{sign} u = 0$ if $u = 0$, $\operatorname{sign} u = -1$ if $u < 0$.

5.6.46. Justify Example 5.1.8.

5.6.47. Prove Lemma 5.1.10.

5.6.48. Justify Remark 5.2.4 by showing that for a symmetric operator $L_{A,b}$ the closed extension obtained in Theorem 5.2.2 is its Friedrichs extension. The latter is obtained by the closure of the quadratic form $(f, f) - (L_{A,b} f, f)$, which gives a self-adjoint operator $\mathfrak{R} \geqslant 0$ with $(\mathfrak{R} f, \mathfrak{R} f) = -(L_{A,b} f, f)$, then $-\mathfrak{R}^2$ is the Friedrichs extension of the operator $L_{A,b}$; see Yosida [**945**, Chapter XI, § 7].

5.6.49. Let μ be a probability measure on $\mathbb{R}^d$ satisfying the equation $L_{A,b}^*\mu = 0$ under condition (5.2.1). (i) Show that if there are a bounded nonzero function u in $C^2(\mathbb{R}^d)$ and a number $\alpha > 0$ such that $u \geqslant 0$ and $L_{A,\widehat{b}}u \geqslant \alpha u$, where $\widehat{b} = 2\beta_{\mu,A} - b$, then μ is not invariant for the semigroup $\{T_t^\mu\}_{t\geqslant 0}$.

(ii) Show that the same is true under the existence of a bounded nonzero function $w \in C^2(\mathbb{R}^d)$ and a number $\alpha > 0$ for which $w \geqslant 0$ and $L_{A,b}w \geqslant \alpha w$.

HINT: (i) If μ is invariant, then by Theorem 5.3.1 and Remark 5.3.2(iii) there are functions $\chi_n \in W^{2,1}_{\text{loc}}(\mu)$ such that $(1-\chi_n)^+ \in L^\infty(\mu)$ and $(1-\chi_n)^+ = 0$ outside some compact sets, $\lim\limits_{n\to\infty} \chi_n(x) = 0$ μ-a.e., and

$$\int \langle A\nabla\chi_n, \nabla\varphi\rangle\, d\mu + \alpha \int \chi_n \varphi\, d\mu + \int \langle b - \beta_{\mu,A}, \nabla\chi_n\rangle\, \varphi\, d\mu \geqslant 0$$

for all nonnegative $\varphi \in C_0^\infty(\mathbb{R}^d)$ and all $n \in \mathbb{N}$. We can assume that $u \leqslant 1$. Substituting in place of φ the function $v = (u - \chi_n)^+$ and taking into account the inequality

$$\int \langle A\nabla u, \nabla v\rangle\, d\mu + \int \langle b - \beta_{\mu,A}, \nabla u\rangle\, v\, d\mu \leqslant -\alpha \int uv\, d\mu,$$

we obtain that the integral of v^2 with respect to the measure μ is nonpositive. Therefore, $0 \leqslant u \leqslant \chi_n$. Since $\chi_n \to 0$, we have $u = 0$, which contradicts our assumption. Note that a proof of (i) is also given in Stannat [**872**, Remark 1.11(ii)]. For justification of (ii) we observe that by (i) the measure μ will not be invariant for the semigroup associated with $L_{A,\widehat{b}}$, which excludes the invariance with respect to $\{T_t^\mu\}_{t\geqslant 0}$.

5.6.50. Prove the following converse of Theorem 5.2.9. Let a Borel probability measure μ on $\mathbb{R}^d$ satisfy the equation $L_{A,b}^*\mu = 0$, where the mapping A is bounded and Lipschitzian on $\mathbb{R}^d$, A^{-1} is bounded as well, $|b| \in L^2(\mu)$. Then the operator $\bigl(L_{A,b}, C_0^\infty(\mathbb{R}^d)\bigr)$ is essentially m-dissipative in the space $L^1(\mu)$, its closure generates a unique C_0-semigroup $\{T_t\}$ in $L^1(\mu)$, this semigroup is Markov, and the measure μ is invariant with respect to it.

5.6.51. Give an example of a Feller, but not strong Feller Markov semigroup.
HINT: Consider $T_t f(x) = e^{-t} f(x)$.

5.6.52. (Lasota, Szarek [**584**]) Let X be a complete separable metric space and let $\{T_t\}_{t\geqslant 0}$ be a Feller Markov semigroup on $C_b(X)$ such that for every bounded Lipschitzian function f the family of functions $\{T_t f\}_{t\geqslant 0}$ is equicontinuous. Suppose that there exists a compact set $K \subset X$ such that for every open neighborhood U of K there is a point $x \in X$ with the property that

$$\limsup_{t\to\infty} \frac{1}{t} \int_0^t T_s^*\delta_x(U)\, ds > 0.$$

Prove that this semigroup possesses an invariant probability measure.

5.6.53. Give an example of a nonnegative operator T from the space $L^1[0,1]$ to $L^1[0,1]$ such that its image does not belong to $C[0,1]$, although T takes $L^\infty[0,1]$ to $C[0,1]$.

HINT: Take the kernel $K(t,s) = (1-s)^{-1/2} I_{[0,t]}(s)$, $t, s \in [0,1]$.

5.6.54. Prove that for every Borel probability measure μ on a complete separable metric space X that is positive on all nonempty open sets, given a point $x \in X$, there is a compact K of positive measure such that x belongs to the topological support of the restriction of μ to K, i.e., every ball of positive radius centered at x intersects K in a set of positive measure.

HINT: Reduce the assertion to the case of a measure without atoms, verify it for Lebesgue measure on the set $\mathcal{R}$ of irrational numbers in $(0, 1)$, then apply the Oxtoby theorem (see Bogachev [**125**, Theorem 9.6.3]), according to which in X there is a subset Y of full measure such that there exists a homeomorphism between Y and $\mathcal{R}$ taking $\mu|_Y$ to Lebesgue measure, moreover, make sure that Y can be chosen such that it contains the point x.

5.6.55. Prove that every bounded operator $T\colon L^1(\mu) \to L^\infty(\nu)$ is integral and is represented by a kernel $\mathcal{K}$ with $\|\mathcal{K}\|_\infty \leqslant \|T\|$.

HINT (see Bukhvalov et al. [**211**, Lemma 4.2.5]): Consider the space $L^1(\mu\otimes\nu)$ and on all elements of the form $u = c_1 I_{A_1}(x) I_{B_1}(y) + \cdots + c_n I_{A_n}(x) I_{B_n}(y)$, where A_i, B_i are measurable, $A_i \cap A_j = \varnothing$ if $i \neq j$, define a functional F by the formula

$$Fu = \sum_{i=1}^{n} c_i \int_{B_i} T I_{A_i}\, d\nu,$$

verify its boundedness observing that one has the equality

$$\|u\| = |c_1|\mu(A_1)\nu(B_1) + \cdots + |c_n|\mu(A_n)\nu(B_n),$$

extend F to the whole space; this gives a function $\mathcal{K} \in \mathcal{L}^\infty(\mu\otimes\nu)$ for which Fu is the integral of $\mathcal{K}u$ with respect to $\mu\otimes\nu$; verify that T is represented by the kernel $\mathcal{K}$ on all indicators, hence on all functions.

5.6.56. Let γ be the standard Gaussian measure on $\mathbb{R}^d$ and let $\{T_t\}_{t\geqslant 0}$ be the Ornstein–Uhlenbeck semigroup on $L^1(\gamma)$ (see (5.1.1)). Write $T_t f$ in the form

$$T_t f(x) = \int_{\mathbb{R}^d} \mathcal{K}_t(x, y) f(y)\, \gamma(dy)$$

with explicit kernels $\mathcal{K}_t$ and investigate when $\mathcal{K}_t \in L^r(\gamma\otimes\gamma)$ for given numbers $t > 0$ and $r > 1$.

5.6.57. Prove that the generator L of the Ornstein–Uhlenbeck semigroup on $L^2(\gamma)$, where γ is the standard Gaussian measure on $\mathbb{R}^d$, on the class $C_0^\infty(\mathbb{R}^d)$ coincides with the Ornstein–Uhlenbeck operator introduced in Chapter 1 (see p. 16).

HINT: Differentiate the expression for $T_t f(x)$ at zero and apply the integration by parts formula.

5.6.58. Prove that the domain of definition of the generator of the Ornstein–Uhlenbeck semigroup in $L^2(\gamma)$ is strictly smaller than the domain of definition of its generator in $L^1(\gamma)$. Moreover, show that the inclusion remains strict even if we take the intersection of the latter domain with $L^2(\gamma)$.

5.6.59. Show that the operators T_t of the Ornstein–Uhlenbeck semigroup have eigenfunctions different from polynomials.

5.6.60. Let γ be the standard Gaussian measure on the real line and let $\{T_t\}_{t\geqslant 0}$ be the Ornstein–Uhlenbeck semigroup on $L^1(\gamma)$. Show that the operators T_t are not compact. Show also that for $t > 0$ the compactness holds in $L^p(\gamma)$ for $p \in (1, +\infty)$.

HINT: For proving noncompactness in $L^1(\gamma)$ consider the functions $C_n I_{[n,n+1]}$, where $C_n = \gamma([n, n+1])^{-1}$.

5.6.61. Justify Remark 5.5.5.

5.6.62. Suppose that a probability measure μ is subinvariant with respect to a sub-Markov strongly continuous semigroup $\{T_t\}_{t\geqslant 0}$ on $L^1(\mu)$ with generator L. Show that μ is invariant with respect to $\{T_t\}_{t\geqslant 0}$ precisely when it is invariant with respect to the operator $(I-L)^{-1}$.

HINT: Write $(I-L)^{-1}$ as the integral of $e^{-t}T_t$ over $(0,\infty)$.

5.6.63. Let $T\colon L^1(\mu)\to L^1(\mu)$ be a continuous linear operator, where μ is a probability measure, and let $T\bigl(L^\infty(\mu)\bigr)\subset L^\infty(\mu)$. Prove that $T\bigl(L^p(\mu)\bigr)\subset L^p(\mu)$ for all $p\geqslant 1$ and that T is continuous in $L^p(\mu)$.

HINT: See Bergh, Löfström [**108**, Theorem 1.1.1].

5.6.64. Let $\{T_t\}_{t\geqslant 0}$ be a Markov semigroup on $B_b(X)$, where X is completely regular, and let $T_t^*\delta_x$ be Radon measures for all $x\in X$, $t\geqslant 0$. Prove that if the family of measures $\{T_\tau^*\delta_x\}_{x\in X}$ is uniformly tight for some $\tau>0$, then the larger family of measures $\{T_t^*\delta_x\}_{x\in X, t\geqslant\tau}$ is uniformly tight as well.

HINT: Write $T_{\tau+\varepsilon}^*\delta_x(B)$ as the integral of $T_\tau^*\delta_y(B)$ with respect to $T_\varepsilon^*\delta_x$.

5.6.65. Let μ be a Radon probability measure on a completely regular space X and let $T\colon L^1(\mu)\to L^1(\mu)$ be a bounded linear operator such that T maps $C_b(X)$ to $C_b(X)$ as a compact operator. Show that T maps the closed unit ball U of $L^\infty(\mu)$ into a compact set in $C_b(X)$.

HINT: Show that $T(U)$ is contained in the closure of the image of the closed unit ball in $C_b(X)$ by observing that, given $f\in U$, there is a sequence of functions $f_n\in U\cap C_b(X)$ converging to f in $L^1(\mu)$, which yields that there is a subsequence $\{f_{n_i}\}$ such that $\{Tf_{n_i}\}$ converges in $C_b(X)$.

5.6.66. Let μ be Lebesgue measure on $[0,1]$ and $Tf(x)=f(x/2)$. Show that T is a sub-Markov operator on $L^\infty(\mu)$, but μ is not subinvariant for it. Show that $T^*g = 2gI_{[0,1/2]}$ and that T^* is not sub-Markov.

CHAPTER 6

Parabolic Fokker–Planck–Kolmogorov Equations

In this chapter we turn to evolution Fokker–Planck–Kolmogorov equations, which are parabolic equations for measures (such equations can be called weak parabolic equations for measures), and obtain analogs of the results of Chapter 1. In the parabolic case, in addition to second order parabolic equations on domains in $\mathbb{R}^d\times\mathbb{R}^1$, which are exact analogs of elliptic equations and which can be actually regarded as degenerate elliptic equations of a very special type, one can also consider the Cauchy problem, i.e., a parabolic equation with an initial data. This circumstance causes considerable differences as compared to the elliptic case.

The main local regularity results are Theorem 6.3.1 and Corollary 6.3.2 saying that any probability solution to the parabolic equation $\partial_t\mu = L^*_{A,b}\mu$ has the property that $(\det A)^{1/(d+1)}\mu$ is always absolutely continuous, so that μ has a density if $\det A > 0$, and the results of § 6.4 and § 6.5, according to which, under broad local assumptions about A and b (in particular, A is locally Sobolev), the solution density belongs to a certain local Sobolev class. The main existence results are Theorem 6.6.2, which ensures a subprobability solution to the Cauchy problem with any subprobability initial condition in case of locally bounded coefficients, and Theorem 6.6.4, in which A is more regular and $|b|$ is locally Lebesgue integrable to some power greater than $d+2$.

6.1. Basic problems

Suppose we are given an open set $\Omega_T = \Omega\times(0,T)$ in $\mathbb{R}^d\times\mathbb{R}^1$, where $\Omega\subset\mathbb{R}^d$ is an open set and $T>0$, and Borel functions a^{ij}, b^i, c on Ω_T, where $i,j=1,\dots,d$. In this and the subsequent chapters we discuss the Fokker–Planck–Kolmogorov equation of the form

$$\partial_t\mu = \partial_{x_i}\partial_{x_j}(a^{ij}\mu) - \partial_{x_i}(b^i\mu) + c\mu.$$

Let

$$L_{A,b,c}\varphi(x,t) = a^{ij}(x,t)\partial_{x_i}\partial_{x_j}\varphi(x,t) + b^i(x,t)\partial_{x_i}\varphi(x,t) + c(x,t)\varphi(x,t),$$

which can be written in a more compact way as

$$L_{A,b,c}\varphi = \operatorname{tr}(AD^2_x\varphi) + \langle b,\nabla_x\varphi\rangle + c\varphi,$$

where $A=(a^{ij})_{i,j\leqslant d}$, $b=(b^1,\dots,b^d)$, the matrices $A(x,t)$ are symmetric nonnegative, and D^2_x and ∇_x denote the second and first order derivatives in x.

Then the Fokker–Planck–Kolmogorov equation can be compactly written as

$$\partial_t\mu = L^*_{A,b,c}\mu. \tag{6.1.1}$$

Let $C_0^\infty(\Omega_T)$ be the class of infinitely differentiable functions with compact support in Ω_T and let $C^{2,1}(\Omega_T)$ be the class of functions on Ω_T with continuous derivatives up to the second order in x and a continuous derivative in t.

6.1.1. Definition. A locally bounded Borel measure μ on the domain Ω_T satisfies equation (6.1.1) if a^{ij}, b^i, $c \in L^1_{\rm loc}(\mu)$ and for every function $\varphi \in C_0^\infty(\Omega_T)$ we have

$$\int_{\Omega_T} [\partial_t\varphi + L_{A,b,c}\varphi]\, d\mu = 0. \tag{6.1.2}$$

We can also consider the Cauchy problem for (6.1.1) corresponding to an initial data. It is more convenient to do this in the special case where μ can be written as

$$\mu(dx\, dt) = \mu_t(dx)\, dt$$

with some family of measures μ_t on Ω such that, for every Borel set B with compact closure in Ω, the function $t \mapsto \mu_t(B)$ is Lebesgue measurable and the function $t \mapsto |\mu_t|(B)$ is integrable on compact intervals in $(0, T)$. Then, for every bounded Borel function f with compact support in Ω_T, the function

$$t \mapsto \int_\Omega f(x,t)\, \mu_t(dx)$$

is also Lebesgue measurable and the previous equality means that

$$\int_{\Omega_T} f\, d\mu = \int_0^T\int_\Omega f(x,t)\, \mu_t(dx)\, dt.$$

Such measures μ_t exist under broad assumptions, in particular, they exist if μ is absolutely continuous with respect to Lebesgue measure (a necessary and sufficient condition for their existence is the absolute continuity of the projection of $|\mu|$ on $(0, T)$). We shall write $\mu = (\mu_t)_{t\in(0,T)}$ or $\mu = \mu_t\, dt$.

If almost every μ_t is a probability measure, then $\mu = (\mu_t)_{t\in(0,T)}$ will be called a *probability solution* (although it is not a probability measure on Ω_T if $T \neq 1$). Similarly, $\mu = (\mu_t)_{t\in(0,T)}$ is a *subprobability solution* if a.e. μ_t is a subprobability measure (i.e., nonnegative of total mass at most 1).

We shall say that a locally finite measure ν on Ω is the initial condition for $\mu = (\mu_t)_{t\in(0,T)}$ and write $\mu|_{t=0} = \nu$ if for every function $f \in C_0^\infty(\Omega)$ there is a set of full measure $J_f \subset (0, T)$ such that

$$\int_\Omega f(x)\, \nu(dx) = \lim_{t\to 0, t\in J_f} \int_\Omega f(x)\, \mu_t(dx). \tag{6.1.3}$$

In general, this condition is weaker than weak convergence of μ_t to ν as $t \to 0$. Note that if the function $t \mapsto \displaystyle\int_\Omega f(x)\, \mu_t(dx)$ is continuous on the interval $(0, T)$, then we can take $J_f = (0, T)$.

Below we need certain equivalent definitions of solutions to the Cauchy problem (6.1.1) and (6.1.3), collected in the following proposition.

6.1.2. Proposition. *Suppose that we are given a locally finite Borel measure $\mu = (\mu_t)_{t\in(0,T)}$ on Ω_T and a^{ij}, b^i, $c \in L^1_{\rm loc}(\mu)$. Then the following assertions are equivalent:*

(i) *the measure μ is a solution to the Cauchy problem* (6.1.1) *and* (6.1.3);

(ii) *for every function* $\varphi \in C_0^\infty(\Omega)$ *there exists a set of full measure* $J_\varphi \subset (0,T)$ *such that for all* $t \in J_\varphi$ *we have*

$$\int_\Omega \varphi\, d\mu_t - \int_\Omega \varphi\, d\nu = \lim_{\tau\to 0+} \int_\tau^t \int_\Omega L_{A,b,c}\varphi\, d\mu_s\, ds;$$

(iii) *for every function* φ *of class* $C^{2,1}(\Omega_T)\cap C\big(\Omega\times[0,T)\big)$ *such that* $\varphi(x,t)=0$ *if* $x\notin U$ *for some ball* U *with closure in* Ω, *there is a full measure set* $J_\varphi \subset (0,T)$ *such that for all* $t\in J_\varphi$ *we have*

$$\int_\Omega \varphi(x,t)\,\mu_t(dx) = \int_\Omega \varphi(x,0)\,\nu(dx) + \lim_{\tau\to 0+}\int_\tau^t\int_\Omega \big[\partial_s\varphi(x,s) + L_{A,b,c}\varphi(x,s)\big]\,\mu_s(dx)\,ds.$$

PROOF. It is clear that (iii) implies (ii). Let us show that (ii) implies (i). Let us fix $\varphi\in C_0^\infty(\Omega)$ and $t_0\in J_\varphi$. Since the function

$$g(\tau) = \int_\tau^{t_0}\int_\Omega L_{A,b,c}\varphi\, d\mu_s\, ds$$

has a finite limit as $\tau\to 0+$, by the Cauchy criterion for every $\varepsilon>0$ there is $\delta>0$ such that for all $\tau,\tau'\in(0,\delta)$ we have the estimate

$$\left|\int_\tau^{\tau'}\int_\Omega L_{A,b,c}\varphi\, d\mu_s\, ds\right| < \varepsilon.$$

Therefore, for all $t\in J_\varphi\cap(0,\delta)$ we have

$$\left|\int_\Omega \varphi\, d\mu_t - \int_\Omega \varphi\, d\nu\right| < \varepsilon.$$

Thus, the measure μ satisfies (6.1.3). Let us verify (6.1.2). We observe that there is version of the function $t\mapsto \int_\Omega \varphi\, d\mu_t$ absolutely continuous on closed intervals in $(0,T)$ such that

$$\frac{d}{dt}\int_\Omega \varphi\, d\mu_t = \int_\Omega L_{A,b,c}\varphi\, d\mu_t$$

for almost all points $t\in(0,T)$. Multiplying this equality by an arbitrary function $\eta\in C_0^\infty\big((0,T)\big)$ and integrating in t, we obtain

$$\int_0^T\int_\Omega [\partial_t(\eta\varphi) + L_{A,b,c}(\eta\varphi)]\, d\mu = 0.$$

Thus, (6.1.2) holds for finite linear combinations of functions of the form $\eta(t)\varphi(x)$, and using such combinations one can uniformly approximate an arbitrary function in $C_0^\infty(\Omega_T)$ along with its derivatives up to the second order. It remains to prove that (i) implies (iii). Again, by approximation, it suffices to prove (iii) for functions φ of the form $\varphi(x,t)=\eta(t)f(x)$, where $f\in C_0^\infty(\Omega)$ and $\eta\in C^\infty(\mathbb{R})$. Suppose first that $\eta\in C_0^\infty\big((0,T)\big)$. Then $\partial_t\varphi(x,t)=\eta'(t)f(x)$, $L_{A,b,c}\varphi(x,t)=\eta(t)L_{A,b,c}f(x,t)$. Integrating by parts in (6.1.2) in t in the first term, we obtain that the function

$$u(t) = \int_\Omega L_{A,b,c}f(x,t)\,\mu_t(dx)$$

on $(0,T)$ is the generalized derivative of the function

$$v(t) = \int_\Omega f(x)\,\mu_t(dx).$$

As above, we take the absolutely continuous version of v, for which $v'(t) = u(t)$ a.e. on $(0,T)$. This yields the equality in (iii) for any $\eta \in C^\infty(\mathbb{R})$ by using the Newton–Leibniz formula for ηv on $[\tau, t]$ and applying (6.1.3). □

Note that in (ii) it would be enough to take also $\tau \to 0$ in the set J_φ, since it is dense in $(0,T)$.

6.1.3. Remark. If $L_{A,b,c}\varphi \in L^1(U\times[0,T],\mu)$ in (ii), then

$$\lim_{\tau\to 0,\tau\in J_\varphi} \int_\tau^t\!\int_\Omega L_{A,b,c}\varphi\,d\mu_s\,ds = \int_0^t\!\int_\Omega L_{A,b,c}\varphi\,d\mu_s\,ds.$$

A similar assertion is true in (iii).

6.1.4. Remark. The following equality, which is a corollary of (iii), is useful: for any function $\varphi \in C^{2,1}(\Omega_T)$ such that $\varphi(t,x) = 0$ whenever $x \notin U$ for some ball U with closure in Ω, there is a set of full measure $J_\varphi \subset (0,T)$ such that for all $t, \tau \in J_\varphi$ we have

$$\int_\Omega \varphi(x,t)\,\mu_t(dx) = \int_\Omega \varphi(x,\tau)\,\nu(dx) + \int_\tau^t\!\int_\Omega \bigl[\partial_t\varphi(x,s) + L_{A,b,c}\varphi(x,s)\bigr]\,\mu_s(dx)\,ds.$$

Indeed, it suffices to redefine $\varphi(x,t)$ for points $t \in [0,\tau)$ with τ from a countable set dense in the initial set J_φ to obtain a function of class $C^{2,1}(\Omega_T)\cap C\bigl(\Omega\times[0,T)\bigr)$.

We shall also consider equations with divergence form operators

$$\mathcal{L}_{A,b,c}u = \partial_{x_i}(a^{ij}\partial_{x_j}u) + b^i\partial_{x_i}u + cu.$$

A locally finite Borel measure μ on Ω_T is said to satisfy the weak parabolic equation $\partial_t\mu = \mathcal{L}^*_{A,b,c}\mu$ if the functions a^{ij}, b^i, c are integrable on every compact set in the domain Ω_T with respect to μ and for every function $\varphi \in C_0^\infty(\Omega_T)$ we have the equality

$$\int_{\Omega_T} \mathcal{L}_{A,b,c}\varphi\,d\mu = 0, \tag{6.1.4}$$

which is understood in one of the following two ways.

(I) For every $t \in (0,T)$ and every ball U with $\overline{U} \subset \Omega$ the functions $x \mapsto a^{ij}(x,t)$ belong to the class $W^{1,1}(U)$, the functions a^{ij}, $\partial_{x_i}a^{ij}$ and b^i are Borel measurable and locally integrable with respect to $|\mu|$, and we have the equality

$$\int_{\Omega_T} [\partial_t\varphi + a^{ij}\partial_{x_i}\partial_{x_j}\varphi + \partial_{x_i}a^{ij}\partial_{x_j}\varphi + b^j\partial_{x_j}\varphi + c\varphi]\,d\mu = 0. \tag{6.1.5}$$

(II) For every closed interval $J \subset (0,T)$ and every ball U with $\overline{U} \subset \Omega$ the restriction of the measure μ to $U\times J$ has a density ϱ such that for every $t \in J$ the function $x \mapsto \varrho(x,t)$ belongs to the class $W^{1,1}(U)$, the functions $a^{ij}\partial_{x_i}\varrho$, $b^i\varrho$, $c\varrho$ are locally Lebesgue integrable and we have

$$\int_{\Omega_T} [\partial_t\varphi\varrho - a^{ij}\partial_{x_i}\varrho\partial_{x_j}\varphi + b^i\partial_{x_i}\varphi\varrho + c\varphi\varrho]\,dx\,dt = 0. \tag{6.1.6}$$

Throughout we always indicate which case is regarded by referring to (6.1.5) in Case (I) and to (6.1.6) in Case (II). It should be noted that an alternative assumption in Case (II) that μ has a locally Sobolev density is fulfilled automatically if in Case (I) the coefficients A and b satisfy certain additional assumptions, which will be indicated in the subsequent sections of this chapter.

For this reason in some results we consider solutions a priori having locally Sobolev densities and make no assumptions about the regularity of A; however, in other results dealing with applications to transition probabilities we require a suitable local Sobolev regularity of A in order to guarantee the Sobolev regularity of all solutions of the considered equation.

The principal problems for the Fokker–Planck–Kolmogorov equation (6.1.1) and the Cauchy problem (6.1.1) and (6.1.3) have many similarities with the corresponding problems for stationary equations and can be described as follows. Local regularity: existence of densities of solutions with respect to Lebesgue measure, smoothness of solutions, the absence of zeros of densities of nonnegative solutions. Global regularity: upper and lower estimates of solutions. Existence and uniqueness of solutions to the Cauchy problem.

In this and the subsequent chapters we discuss all these problems, not only in the linear case, but also for nonlinear equations.

We recall that the interest in the considered Fokker–Planck–Kolmogorov equations is explained to a large degree by the fact that such equations are satisfied by transition probabilities of diffusion processes and also by the integral kernels of the semigroup with generator $L_{A,b,c}$.

6.2. Background material

In this section we formulate certain auxiliary results on parabolic Sobolev spaces and solutions to parabolic equations for functions. To be consistent with the parabolic case, the Sobolev classes $W^{p,k}$ on domains in $\mathbb{R}^d$ will be denoted by $H^{p,k}$, as we did in Chapter 1 for fractional Sobolev classes.

Let J be an interval (open or closed) and let U be an open set in $\mathbb{R}^d$. Let $\mathbb{H}^{p,s}(U,J)$ denote the space of all measurable functions u on the set $U\times J$ such that $u(\,\cdot\,,t)\in H^{p,s}(U)$ for almost all t and the norm

$$\|u\|_{\mathbb{H}^{p,s}(U,J)} = \biggl(\int_J \|u(\,\cdot\,,t)\|^p_{H^{p,s}(U)}\,dt\biggr)^{1/p}$$

is finite. The space $\mathbb{H}^{p,s}_0(U,J)$ is defined similarly, but with $H^{p,s}_0(U)$ in place of $H^{p,s}(U)$, and $\mathbb{H}^{p',-s}(U,J)$ denotes its dual. In connection with parabolic equations, it is useful to introduce also the following spaces. Let $\mathcal{H}^{p,1}(U,J)$ be the space of all functions $u\in\mathbb{H}^{p,1}(U,J)$ with $\partial_t u\in\mathbb{H}^{p,-1}(U,J)$ and finite norm

$$\|u\|_{\mathcal{H}^{p,1}(U,J)} = \|\partial_t u\|_{\mathbb{H}^{p,-1}(U,J)} + \|u\|_{\mathbb{H}^{p,1}(U,J)}.$$

We denote by U_R an open ball of radius $R>0$ centered at some point in $\mathbb{R}^d$; in the case where the center is important we use the notation $U(a,R)$, as we did in the elliptic case (these symbols will not be used for sets in $\mathbb{R}^d\times\mathbb{R}^1$).

Let $H^{p,2;1}_0(U_R,[0,T])$ denote the closure of the space of smooth functions u on the cylinder $U_{R,T} := U_R\times(0,T)$ that vanish on $(\partial U_R\times[0,T])\bigcup(U_R\times\{0\})$, i.e., have zero limits on this part of the boundary, with respect to the norm

$$\|u\|_{H_0^{p,2;1}(U_R,[0,T])} = \|u\|_{L^p(U_R\times[0,T])} + \|\partial_t u\|_{L^p(U_R\times[0,T])} + \|\nabla_x u\|_{L^p(U_R\times[0,T])} + \sum_{i,j=1}^{d} \|\partial_{x_i}\partial_{x_j} u\|_{L^p(U_R\times[0,T])}.$$

Let $C^\gamma(U_R)$ be the space of γ-Hölder functions on U_R.

The next lemma is a simple generalization of the Sobolev inequality (see Chapter 1).

6.2.1. Lemma. *Let J be a compact interval in $(0,T)$ and let a function u of class $\mathbb{H}^{2,1}(\mathbb{R}^d,(0,T))$ be such that the mapping $x\mapsto u(x,t)$ has compact support for almost all $t\in J$. Then there is a number $C>0$, depending only on dimension d, such that*

$$\|u\|_{L^{2(d+2)/d}(\mathbb{R}^d\times J)} \leqslant C\bigl(\sup_{t\in J}\|u(\,\cdot\,,t)\|_{L^2(\mathbb{R}^d)} + \|\nabla u\|_{L^2(\mathbb{R}^d\times J)}\bigr).$$

PROOF. By Hölder's inequality

$$\|u(\,\cdot\,,t)\|^{2(d+2)/d}_{L^{2(d+2)/d}(\mathbb{R}^d)} \leqslant c(d)\|u(\,\cdot\,,t)\|^2_{L^{2d/(d-2)}(\mathbb{R}^d)}\|u(\,\cdot\,,t)\|^{2/d}_{L^2(\mathbb{R}^d)}.$$

The Sobolev inequality yields

$$\|u(\,\cdot\,,t)\|^{2(d+2)/d}_{L^{2(d+2)/d}(\mathbb{R}^d)} \leqslant c(d)\|\nabla_x u(\,\cdot\,,t)\|^2_{L^2(\mathbb{R}^d)}\|u(\,\cdot\,,t)\|^{2/d}_{L^2(\mathbb{R}^d)}.$$

Integrating this inequality in t and applying Young's inequality, we arrive at the desired estimate. □

Recall also the parabolic embedding theorem for the spaces $\mathcal{H}^{p,1}$. Its proof can be found in Krylov [**554**, Corollary 7.6] or in Krylov [**553**, Theorem 7.2] (where the restriction $p>2$ was needed only in the case of stochastic Sobolev spaces), and also in Besov, Il'in, Nikolskiĭ [**112**, Chapter III], Metafune, Pallara, Rhandi [**695**].

6.2.2. Theorem. (i) *Let $q>p>1$ and $(d+2)(1/p-1/q)<1$. Then there is a number $N(d,p,q,R,T)$ such that for every $u\in\mathcal{H}^{p,1}(U_R,[0,T])$ we have*

$$\|u\|_{L^q(U_{R,T})} \leqslant N(d,p,q,R,T)\|u\|_{\mathcal{H}^{p,1}(U_R,[0,T])}. \tag{6.2.1}$$

(ii) *Let $p>d+2$, $\alpha>0$ and $\beta>0$, where $\beta>\alpha>1/p$ and $p(1-2\beta)>d$. If $u\in\mathcal{H}^{p,1}(U_R,[0,T])$, then $u\in C^{\alpha-1/p}\bigl([0,T],H^{p,1-2\beta}(U_R)\bigr)$, where the right-hand side is the space of $(\alpha-1/p)$-Hölder mappings from $[0,T]$ to $H^{p,1-2\beta}(U_R)$. The latter spaces is contained in*

$$C^{\alpha-1/p}\bigl([0,T],C^\gamma(U_R)\bigr)$$

for some $\gamma>0$, and there is a number $N(d,p,\alpha,\beta,R,T)$ such that

$$\|u\|_{C^{\alpha-1/p}\left([0,T],C^\gamma(U_R)\right)} \leqslant N(d,p,\alpha,\beta,R,T)\|u\|_{\mathcal{H}^{p,1}(U_R,[0,T])}.$$

As before, set

$$\Omega_T=\Omega\times(0,T), \quad \text{where } T>0.$$

Let $A(\,\cdot\,,\cdot\,)=\bigl(a^{ij}(\,\cdot\,,\cdot\,)\bigr)_{i,j=1}^d$ be a Borel mapping on Ω_T with values in the space of nonnegative symmetric operators on $\mathbb{R}^d$.

Let $U_R=U_R(x_0)$ be a ball with some center x_0 and compact closure in Ω.

Assume that the mapping A is uniformly continuous in x uniformly in t, i.e., for all indices i, j, one has

$$\lim_{\delta\to 0}\sup_t \sup_{|x-y|\leqslant\delta} |a^{ij}(x,t)-a^{ij}(y,t)| = 0. \tag{6.2.2}$$

Note that a^{ij} has a modification with this property if

$$\sup_t \|a^{ij}(\,\cdot\,,t)\|_{W^{p,1}(U_R)} < \infty, \quad \text{where } p > d. \tag{6.2.3}$$

We shall also assume that for some numbers $c_1, c_2 > 0$ we have

$$c_1\cdot \mathrm{I} \leqslant A(x,t) \leqslant c_2\cdot \mathrm{I} \quad \text{for all } (x,t)\in U_{R,T}.$$

For this operator A, let us consider the following boundary value problem in the cylinder $U_{R,T} = U_R\times(0,T)$:

$$\partial_t u + a^{ij}\partial_{x_i}\partial_{x_j} u = f, \quad u(x,0)=0, \ u|_{\partial U_R\times[0,T]} = 0. \tag{6.2.4}$$

We shall need the following estimate for the solutions of (6.2.4).

6.2.3. Lemma. *There exist two numbers $R(d,c_1,c_2), N_0(d,c_1,c_2) > 0$, depending only on d, c_1, c_2, and the modulus of continuity of A in x, such that, for every $R < R(d,c_1,c_2)$ and every function $f\in C_0^\infty(U_{R,T})$, there exists a function $u\in H_0^{p,2;1}(U_R,[0,T])$ that is a solution of* (6.2.4) *and satisfies the estimate*

$$\|u\|_{H_0^{p,2;1}(B_R,[0,T])} \leqslant N_0(d,c_1,c_2)\|f\|_{L^p(U_{R,T})}. \tag{6.2.5}$$

In addition, $R(d,c_1,c_2)$ and $N_0(d,c_1,c_2)$ can be found as locally bounded functions of the variables $c_1, c_2 > 0$.

PROOF. It is known (see, e.g., Lieberman [**611**, Chapter IV], Krylov [**552**, Chapter 10], Ladyz'enskaya, Solonnikov, Ural'tseva [**576**, Chapter IV]) that if A is Hölder continuous with respect to (x,t), then u has two derivatives in x and one derivative in t that are continuous on the closure of $U_{R,T}$. Moreover, according to Knerr [**522**] or Lieberman [**610**] (see [**610**, Corollary 15.4, Theorem 15.6]), if A is Hölder continuous in x uniformly with respect to t, then, for every compact set $K\subset U_R$, the second order derivatives in x and the first order derivative in t are continuous in $x\in K$ uniformly with respect to t.

Estimate (6.2.5) for all $R>0$ is obtained in [**576**, Chapter IV, Theorem 9.2] in the case where A is continuous in (x,t). Let us explain how to prove (6.2.5) and the local existence result under condition (6.2.2). In fact, the existence result is obtained in [**610**, Section 16], but we need also (6.2.5). To this end, note that (6.2.5) holds if A is independent of x (see, e.g., Krylov [**553**, Theorem 4.10]). Suppose first that there is a solution u. Let $a_0^{ij}(t) = a^{ij}(t,x_0)$ and $b^{ij} = a_0^{ij} - a^{ij}$. Let us rewrite the equation as

$$\partial_t u + a_0^{ij}\partial_{x_i}\partial_{x_j}u = f + b^{ij}\partial_{x_i}\partial_{x_j}u.$$

By (6.2.2), we can choose $R>0$ such that

$$\|b^{ij}\partial_{x_i}\partial_{x_j}u\|_{L^p(U_{R,T})} \leqslant \frac{1}{2N_0(d,c_1,c_2)}\|u\|_{H_0^{p,2;1}(U_R,[0,T])}.$$

Then we obtain (6.2.5) with $2N_0(d,c_1,c_2)$ in place of $N_0(d,c_1,c_2)$. This a priori estimate implies the existence result. Indeed, we can find a sequence of Hölder continuous mappings A_k such that $A_k(x,t)\to A(x,t)$ almost everywhere, $c_1\mathrm{I}\leqslant A_k\leqslant c_2\mathrm{I}$, and (6.2.2) is fulfilled uniformly in k. Then the sequence of the corresponding solutions u_k is bounded in the space $H_0^{p,2;1}(U_R,[0,T])$, hence a subsequence of $\{u_k\}$

converges weakly in this space to some u, which is a solution corresponding to A (note that, for every $f \in C_0^\infty(U_{R,T})$, this is the solution constructed in [**610**, Section 16], i.e., the functions $\partial_{x_i}\partial_{x_j}u$ and $\partial_t u$ are locally Hölder continuous in x uniformly with respect to t). □

6.2.4. Corollary. *Now let $p > d+2$. Then in the situation of the previous lemma there is a constant $N(d, c_1, c_2)$, depending only on d, c_1, c_2 and the modulus of continuity of A, such that u has a modification which is continuously differentiable in x and*

$$\sup_{U_{R,T}} |\nabla_x u| \leqslant N(d, c_1, c_2)\|f\|_{L^p(U_{R,T})}. \tag{6.2.6}$$

PROOF. The assertion follows immediately by (6.2.5) combined with the embedding theorem (see Besov, Il'in, Nikol'skii [**112**, Chapter III, Theorem 10.4] or Krylov [**553**, Theorem 7.2]). □

We now consider the following more general equation:

$$\partial_t u + a^{ij}\partial_{x_i}\partial_{x_j}u = \partial_{x_i} f^i, \quad f_i \in L^q(U_{R,T}), \tag{6.2.7}$$

where $a^{ij} \in \mathbb{H}^{p,1}(U_R, [0,T])$. Recall that $u \in \mathbb{H}_0^{q,1}(U_R, [0,T])$, where $q \geqslant 1$, is called a solution of (6.2.7) with zero boundary condition

$$u(x, 0) = 0, \quad u|_{\partial U_R \times [0,T]} = 0$$

if for all $\varphi \in C_0^\infty(U_{R,T})$ we have

$$\int \Bigl[u\partial_t\varphi + a^{ij}\partial_{x_i}u\partial_{x_j}\varphi + u\partial_{x_i}a^{ij}\partial_{x_j}\varphi\Bigr]\,dx\,dt = \int f^i\partial_{x_i}\varphi\,dx\,dt.$$

If $f^i \in C_0^\infty(U_{R,T})$ and $u \in H^{q,2;1}(U_R, [0,T])$, then (6.2.7) holds almost everywhere.

6.2.5. Lemma. *Suppose that (6.2.3) is fulfilled and $q \geqslant p'$. Then there exist numbers $R_0(d, c_1, c_2), M(d, c_1, c_2) > 0$, depending only on d, c_1, c_2 and the quantities from (6.2.3), such that for every $R < R_0$ we have the estimate*

$$\|u\|_{\mathcal{H}^{q,1}(U_R,[0,T])} \leqslant M\|\partial_{x_i} f^i\|_{\mathbb{H}^{q,-1}(U_R,[0,T])} \leqslant M\|(f^i)\|_{L^q(U_{R,T})}$$

for every $u \in H_0^{q,1}(U_R, [0,T])$ satisfying (6.2.7) in the sense indicated above.

PROOF. We first obtain an estimate on the norm $\|u\|_{\mathbb{H}^{q,1}(U_R,[0,T])}$. To this end, set $a_0^{ij}(t) = a^{ij}(x_0, t)$. Let us write our equation in the form

$$\partial_t u + a_0^{ij}\partial_{x_i}\partial_{x_j}u = \partial_{x_i} f^i + \partial_{x_i}(b^{ij}\partial_{x_j}u) + c^i\partial_{x_i}u,$$

$$b^{ij}(x,t) = a_0^{ij}(t) - a^{ij}(x,t), \quad c^i(x,t) = \partial_{x_j}a^{ij}(x,t).$$

Let $c(x,t) = \bigl(c^i(x,t)\bigr)_{i=1}^d$. According to Krylov [**554**, Theorem 4.10], there exists a number N_0 independent of u and $R < 1$ such that

$$\|u\|_{\mathbb{H}^{q,1}(U_R,[0,T])} \leqslant N_0\|\partial_{x_i} f^i + \partial_{x_i}(b^{ij}\partial_{x_j}u) + c^i\partial_{x_i}u\|_{\mathbb{H}^{q,-1}(U_R,[0,T])}.$$

It should be noted that although the cited theorem was proved for $\mathbb{R}^d$ in place of a ball, the same assertion is true also for U_R. Indeed, this can be derived from the case $\mathbb{R}^d$, since the extension of a function in $H_0^{q,1}(U_R)$ by zero outside U_R belongs to $H_0^{q,1}(\mathbb{R}^d)$ and has the same norm.

Let $q > d'$. By Lemma 1.1.7(i) for every fixed t we have

$$\|c^i\partial_{x_i}u(\,\cdot\,,t)\|_{H^{q,-1}(U_R)} \leqslant N\|c(\,\cdot\,,t)\|_{L^d(U_R)}\|\nabla_x u(\,\cdot\,,t)\|_{L^q(U_R)},$$

where N does not depend on t, u and R. By (6.2.3) we can choose $R > 0$ so small that

$$\sup_t \|c(\,\cdot\,,t)\|_{L^d(U_R)} \leqslant \frac{1}{4NN_0}.$$

Hence

$$\|c^i\partial_{x_i}u\|_{\mathbb{H}^{q,-1}(U_R,[0,T])} \leqslant \frac{1}{4N_0}\|\nabla_x u\|_{L^q(U_{R,T})} \leqslant \frac{1}{4N_0}\|u\|_{\mathbb{H}^{q,1}(U_R,[0,T])}.$$

In addition, setting $v = (b^{1j}\partial_{x_j}u,\ldots,b^{dj}\partial_{x_j}u)$, by (1.1.10) we have

$$\|\partial_{x_i}(b^{ij}\partial_{x_j}u)\|_{\mathbb{H}^{q,-1}(U_R,[0,T])} \leqslant \|v\|_{L^q(U_{R,T})} \leqslant \sup_{U_{R,T}} |b^{ij}|\,\|\nabla_x u\|_{L^q(U_{R,T})}.$$

Since A is uniformly continuous in x uniformly in t, we can make R so small that

$$\sup_t \sup_{x\in U_R} |b^{ij}(x,t)| \leqslant \frac{1}{4N_0}.$$

Then we obtain

$$\|u\|_{\mathbb{H}^{q,1}(U_R,[0,T])} \leqslant N_0\|\partial_{x_i}f^i\|_{\mathbb{H}^{q,-1}(U_R,[0,T])} + \frac{1}{2}\|u\|_{\mathbb{H}^{q,1}(U_R,[0,T])}.$$

The two remaining cases $q < d'$ and $q = d'$ are considered similarly by using assertions (ii) and (iii) in Lemma 1.1.7 for estimating $\|c^i\partial_{x_i}u\|_{\mathbb{H}^{q,-1}(U_R,[0,T])}$. Finally, we observe that

$$\|\partial_t u\|_{\mathbb{H}^{q,-1}} = \|\partial_{x_i}f^i - \partial_{x_j}(a^{ij}\partial_{x_i}u) + \partial_{x_j}a^{ij}\partial_{x_i}u\|_{\mathbb{H}^{q,-1}} \leqslant \|f\|_{\mathbb{H}^{q,-1}} + C\|u\|_{\mathbb{H}^{q,-1}}.$$

Thus, we have obtained the first inequality in the lemma, the second one follows by inequality (1.1.10). □

6.2.6. Corollary. *In the situation of Lemma 6.2.5 for every collection of functions $f^1,\ldots,f^d \in L^q(U_{R,T})$, there exists $u \in \mathbb{H}_0^{q,1}(U_R,[0,T])$ satisfying equation (6.2.7) in the sense indicated above.*

PROOF. It suffices to approximate a^{ij} and f^i by smooth functions a_k^{ij} and f_k^i in $\mathbb{H}^{p,1}(U_R,[0,T])$ and $L^q(U_{R,T})$, respectively, in such a way that condition (6.2.3) will be satisfied uniformly in k and A_k will be uniformly nondegenerate. Then by Lemma 6.2.5 the corresponding solutions u_k will be uniformly bounded in $\mathbb{H}_0^{q,1}(U_R,[0,T])$, hence they will have a weak limit point $u \in \mathbb{H}_0^{q,1}(U_R,[0,T])$, which will be a solution corresponding to a^{ij} and f^i. □

Let $\Omega \subset \mathbb{R}^d$ be a bounded domain and $\Omega_T = \Omega\times(0,T)$. We consider the parabolic equation in divergence form

$$\partial_t u = \partial_{x_i}\big(a^{ij}\partial_{x_j}u - b^i u\big) + \beta^i\partial_{x_i}u + cu, \tag{6.2.8}$$

which is understood in the sense of the integral identity

$$\int_{\Omega_T}\big[-u\partial_t\varphi + \langle A\nabla u - bu, \nabla\varphi\rangle + \langle\beta,\nabla u\rangle\varphi + cu\varphi\big]\,dx\,dt = 0$$

for all $\varphi \in C_0^\infty(\Omega_T)$.

For positive numbers p and q, let $L^{p,q}(\Omega_T)$ denote the space of measurable functions f on Ω_T which have finite norm

$$\|f\|_{p,q} = \biggl(\int_0^T\Bigl(\int_\Omega |f(x,t)|^p\,dx\Bigr)^{q/p}dt\biggr)^{1/q}.$$

Let $R(r)$ denote an open cube in $\mathbb{R}^d$ with edge length r.

Set also $Q(r) = R(r)\times(\tau - r^2, \tau)$, where $\tau \in (0, T)$.

We shall need the following assertions proved in the paper Aronson, Serrin [**74**, Theorems 2, 3, 4] (see also the books Ladyz'enskaya, Solonnikov, Ural'tseva [**576**], Trudinger [**902**]).

6.2.7. Theorem. *Let $\lambda \cdot \mathrm{I} \leqslant A \leqslant \lambda^{-1} \cdot \mathrm{I}$, $b, \beta \in L^{p,q}(\Omega_T)$, where $p > 2$ and $d/(2p) + 1/q < 1/2$, and let $c\in L^{m,l}(\Omega_T)$, where $m>1$, $d/(2p) + 1/l < 1$. Suppose also that $u \in \mathbb{H}^{2,1}(\Omega_T)$ is a solution of equation (6.2.8). Let $Q(3r) \subset \Omega_T$ for some number $r > 0$. Then the following assertions are true.*

(i) (*Boundedness of solutions.*) *We have the estimate*

$$\sup_{Q(r)} |u(x,t)| \leqslant Cr^{-(d+2)/2}\|u\|_{L^{2,2}(Q(3r))}.$$

(ii) (*Harnack's inequality.*) *If u is a nonnegative function, then*

$$\sup_{Q^*(r)} u(x,t) \leqslant C \inf_{Q(r)} u(x,t),$$

where $Q^*(r) = R(r)\times(\tau - 8r^2, \tau - 7r^2)$.

(iii) (*Hölder's continuity.*) *For all $(x,t), (y,s)$ in $Q(r)$, we have*

$$|u(x,t) - u(y,s)| \leqslant Cr^{-\alpha}(|x-y| + \sqrt{|t-s|})^\alpha \sup_{Q(3r)} |u| \quad \textit{for some } \alpha \in (0,1).$$

The constants C, α depend only on d, λ, $\|b\|_{p,q}$, $\|\beta\|_{p,q}$, $\|c\|_{m,l}$.

In Chapter 8 we return to Harnack's inequality for solutions of divergence form parabolic equations and discuss in more detail the case of locally bounded coefficients.

As in the elliptic case, in the theory of parabolic equations an important role is played by various forms of the maximum principle.

The next assertion is proved in [**552**, Theorem 8.1.4 and Corollary 8.1.5].

6.2.8. Proposition. *Let $u \in C^{2,1}\bigl(\mathbb{R}^d\times(0,T)\bigr) \cap C_b\bigl(\mathbb{R}^d\times[0,T)\bigr)$ satisfy the inequality*

$$\partial_t u \leqslant a^{ij}\partial_{x_i}\partial_{x_j}u + b^i\partial_{x_i}u + cu,$$

where a^{ij}, b^i, c, f are bounded functions on $\mathbb{R}^d\times(0,T)$, $c \leqslant 0$. If $u(\,\cdot\,,0) \leqslant 0$, then we have $u \leqslant 0$ in $\mathbb{R}^d\times(0,T)$. If u satisfies the equation, then

$$\sup_{\mathbb{R}^d\times(0,T)} |u(x,t)| \leqslant \sup_{\mathbb{R}^d} |u(x,0)|.$$

Finally, let us mention one more important result from Krylov [**550**, §3 in Chapter III] or [**548**, Theorem 2], which generalizes to parabolic equations the maximum principle of A.D. Aleksandrov.

6.2.9. Proposition. *There are two numbers $C(R,d)$, $\kappa(R,d) > 0$ such that for every nonnegative function $f \in C_0^\infty(U_{R,T})$ there exists a bounded function $z \leqslant 0$ on $U_{4R}(x_0)\times\mathbb{R}^1$ the following properties: (i) the function $x \mapsto z(x,t)$ is convex on $U_{4R}(x_0)$ for every fixed $t \in \mathbb{R}^1$, (ii) for every smooth probability density g with support in the unit ball in $\mathbb{R}^{d+1}$, setting $g_\varepsilon(y,t) := \varepsilon^{-d-1}g(y/\varepsilon, t/\varepsilon)$ and $\psi_\varepsilon := \psi * g_\varepsilon$*

for an integrable function ψ, *we have*

$$C(R,d)\big(\det(\alpha^{ij})\big)^{1/(d+1)} f_\varepsilon \leqslant \partial_t z_\varepsilon + \alpha^{ij}\partial_{x_i}\partial_{x_j} z_\varepsilon,$$

(6.2.9)

$$|z(x,t)| \leqslant \kappa(R,d)\|f\|_{L^{d+1}(U_{R,T})}, \quad (x,t)\in U_{2R}(x_0)\times\mathbb{R}^1$$

for every $\varepsilon\in(0,R)$ *and every nonnegative definite symmetric matrix* (α^{ij}).

Note that in [**548**, Theorem 2] there is the minus sign in front of $\partial_t z_\varepsilon$. However, the same is true also with plus, since we can replace t with $-t$, which does not effect the other terms.

6.3. Existence of densities

Let us investigate the conditions for the existence of densities similar to the elliptic case (see § 1.5). Let $\Omega_T=\Omega\times(0,T)$, where $\Omega\subset\mathbb{R}^d$ is an open set and $T>0$ is a fixed number.

6.3.1. Theorem. *Let* μ *be a locally finite Borel measure on* Ω_T *such that* $a^{ij}\in L^1_{\rm loc}(\Omega_T,\mu)$ *and*

$$\int_{\Omega_T}\Big[\partial_t\varphi + a^{ij}\partial_{x_i}\partial_{x_j}\varphi\Big]\,d\mu \leqslant C\big(\sup_{\Omega_T}|\varphi| + \sup_{\Omega_T}|\nabla_x\varphi|\big) \tag{6.3.1}$$

for all nonnegative $\varphi\in C_0^\infty(\Omega_T)$. *Then the following assertions are true.*

(i) *If* $\mu\geqslant 0$, *then* $(\det A)^{1/(d+1)}\mu=\varrho\,dx\,dt$, *where* $\varrho\in L^{(d+1)'}_{\rm loc}(\Omega_T)$.

(ii) *If, on every compact set in* Ω_T, *the mapping* A *is uniformly bounded, uniformly nondegenerate, and Hölder continuous in* x *uniformly with respect to* t, *then* $\mu=\varrho\,dx\,dt$, *where* $\varrho\in L^r_{\rm loc}(\Omega_T)$ *for every* $r\in[1,(d+2)')$.

PROOF. In order to prove statement (i) we shall modify the reasoning used in the elliptic case as follows. By choosing an appropriate cut-off function ζ with compact support in D, where $D=U\times J$, J is a subinterval with compact closure in $(0,T)$ and $U=U_R(x_0)$ is a ball with compact closure in Ω, and letting $\nu=\zeta\mu$, we arrive at the estimate

$$\int_D\Big[\partial_t\varphi + a^{ij}\partial_{x_i}\partial_{x_j}\varphi\Big]\,d\nu \leqslant N\sup_D|\nabla_x\varphi| + N\sup_D|\varphi| \tag{6.3.2}$$

for all φ that are continuously differentiable in t and twice continuously differentiable in x on $\overline{D}$.

By Proposition 6.2.9, for every nonnegative function f of class $C_0^\infty(D)$, there is a bounded function $z\leqslant 0$ on the cylinder $U_{4R}(x_0)\times\mathbb{R}^1$ such that the function $x\mapsto z(t,x)$ is convex on $U_{4R}(x_0)$ for every fixed $t\in J$ and the following two properties hold:

$$C(R,d)\big(\det(\alpha^{ij})\big)^{1/(d+1)} f_\varepsilon \leqslant \partial_t z_\varepsilon + \alpha^{ij}\partial_{x_i}\partial_{x_j} z_\varepsilon,$$

(6.3.3)

$$|z(x,t)| \leqslant \kappa(R,d)\|f\|_{L^{d+1}(D)}, \quad (x,t)\in U_{2R}(x_0)\times J$$

for $\varepsilon\in(0,R)$ and every nonnegative definite symmetric matrix (α^{ij}), where z_ε is defined in Proposition 6.2.9.

Clearly, for all sufficiently small ε, we obtain

$$|z_\varepsilon(x,t)| \leqslant 2\kappa(R,d)\|f\|_{L^{d+1}(D)}.$$

We observe that z_ε is nonpositive and convex on $U_{3R}(x_0)$ for $\varepsilon < R$. Therefore, as in the elliptic case, one has

$$\sup_{x\in U_R(x_0)} |\nabla_x z_\varepsilon(x,t)| \leqslant 2R^{-1} \sup_{x\in U_{2R}(x_0)} |z_\varepsilon(x,t)| \leqslant 2R^{-1}2\kappa(R,d)\|f\|_{L^{d+1}(D)}.$$

Hence, from (6.2.9) applied to $\alpha^{ij} = a^{ij}(x,t)$, for all sufficiently small ε, taking into account that ν is nonnegative we obtain

$$\begin{aligned}\int_D |\det A|^{1/(d+1)} f_\varepsilon\, d\nu &\leqslant \alpha^{-1}\int_D \Big[\partial_t z_\varepsilon + a^{ij}\partial_{x_i}\partial_{x_j} z_\varepsilon\Big]\, d\nu \\ &\leqslant \alpha^{-1}N\Big(\sup_D|z_\varepsilon| + \sup_D|\nabla_x z_\varepsilon|\Big) \leqslant \alpha^{-1}N(1+2R^{-1})\sup_D|z_\varepsilon| \\ &\leqslant 2\alpha^{-1}k(R,d)N(1+2R^{-1})\|f\|_{L^{d+1}(D)}.\end{aligned}$$

Letting ε go to zero, we obtain the same estimate for the function f, which yields that $\sqrt[d+1]{\det A}\,\nu = \varrho\, dxdt$ with $\varrho \in L^{(d+1)'}(D)$.

Assertion (ii) is also proved similarly to the elliptic case. Let $p > d+2$. As above, we arrive at the estimate (6.3.2). Let $f \in C_0^\infty(D)$. We shall first assume that A is Hölder continuous in both variables. This case is simpler, but exhibits the main idea of the general case. Let u be the solution of the equation $\partial_t u + a^{ij}\partial_{x_i}\partial_{x_j} u = f$ on D with zero initial and boundary data. We recall that u is twice continuously differentiable in x and continuously differentiable in t. Let $R > 0$ be small enough. By (6.2.6), there exists a constant N_0 independent of f such that

$$\sup_D |\nabla_x u| \leqslant N_0\|f\|_{L^p(D)}. \tag{6.3.4}$$

Therefore, estimating $\sup_D |u|$ by $R\sup_D |\nabla_x u|$, we obtain

$$\begin{aligned}\int_D f\, d\nu &= \int_D \Big[\partial_t u + a^{ij}\partial_{x_i}\partial_{x_j} u\Big]\, d\nu \\ &\leqslant N(1+R)\sup_D|\nabla_x u| \leqslant N_0N(1+R)\|f\|_{L^p(D)},\end{aligned}$$

whence it follows that $\nu = \varrho_\nu\, dx\, dt$ with $\varrho_\nu \in L^{p'}(D)$.

Now let us explain how to handle the general case where A is Hölder continuous in x uniformly with respect to t. We can extend a^{ij} from D to $\mathbb{R}^{d+1}$ such that

$$\lambda_1 \cdot \mathrm{I} \leqslant A \leqslant \lambda_2 \cdot \mathrm{I} \quad \text{for some } \lambda_1, \lambda_2 > 0$$

and (6.2.2) is fulfilled on $\mathbb{R}^{d+1}$. As noted above, for every $f \in C_0^\infty(D)$, there exists a function $\varphi \in H_0^{p,2;1}(U_{3R}(x_0),[0,T])$ satisfying $\partial_t\varphi + a^{ij}\partial_{x_i}\partial_{x_j}\varphi = f$ in $U_{3R}(x_0)\times(0,T)$ such that its second order derivatives in x and the first order derivative in t are uniformly bounded and continuous in $x \in U_{2R}(x_0)$ uniformly with respect to the variable $t \in [0,T]$. Moreover, if $R > 0$ is sufficiently small, then, by (6.2.6), we have

$$\begin{aligned}\|\varphi\|_{H_0^{p,2;1}(U_R(x_0),[0,T])} &\leqslant \|\varphi\|_{H_0^{p,2;1}(U_{3R}(x_0),[0,T])} \\ &\leqslant N_0(d,c_1,c_2)\|f\|_{L^p(U_{3R}(x_0)\times(0,T))} = N_0(d,c_1,c_2)\|f\|_{L^p(D)}.\end{aligned}$$

Let us verify (6.3.2) for such φ.

Step 1. Let $\nu(dxdt) = \sigma_t(dx)\eta(dt)$ be a factorization of ν such that, for *every* point $t \in \mathbb{R}^1$, σ_t is a non zero measure on $U_R(x_0)\times\{t\}$. First, we prove that η is absolutely continuous.

By substituting $\varphi(x,t) = f(t)h(x)$ in (6.3.2), for any smooth compactly supported functions f and h, we find that

$$\left|\int_{\mathbb{R}} f'\psi\,\eta(dt)\right| \le N \sup|f|,$$

where N is independent of f and

$$\psi(t) = \int_{\mathbb{R}^d} h(x)\,\sigma_t(dx).$$

It follows that the measure $\psi\eta$ is absolutely continuous and the measure η is absolutely continuous on the set $\Gamma(h) := \Bigl\{t\colon\ |\psi(t)| \neq 0\Bigr\}$. Since the space $C(\overline{U})$ is separable and σ_t is nonzero, we can find $h_k \in C_0^\infty(\mathbb{R}^d)$ such that $\mathbb{R}^1 = \cup_i \Gamma(h_i)$, which proves our claim.

Step 2. We observe that (6.3.2) also holds for any φ such that its first and second order derivatives in x and the first order derivative in t are bounded and continuous in x uniformly with respect to t. Indeed, let $\varphi_\varepsilon = \varphi * g_\varepsilon$, where g_ε is the same as above in Proposition 6.2.9 with $g(y,t) = g_1(y)g_2(t)$, where g_1 and g_2 are smooth probability densities. Then the functions $\partial_t\varphi_\varepsilon$, $\partial_{x_i}\varphi_\varepsilon$, and $\partial_{x_i}\partial_{x_j}\varphi_\varepsilon$ are bounded uniformly in ε and converge ν-a.e. to the corresponding derivatives of φ. To see this, let us observe that if a function ψ is bounded and continuous in x uniformly with respect to t, then $\psi_\varepsilon \to \psi$ ν-a.e. Indeed,

$$\lim_{\varepsilon\to 0}\sup_x\left|\int \psi(x+\varepsilon y, t+\varepsilon s)g_1(y)g_2(s)\,dy\,ds - \int \psi(x, t+\varepsilon s)g_2(s)\,ds\right| = 0.$$

On the other hand, for every fixed x and almost all t, one has

$$\int \psi(x, t+\varepsilon s)g_2(s)\,ds \to \psi(x,t).$$

Since η is absolutely continuous, we obtain almost sure convergence with respect to ν, which yields (6.3.2) by the dominated convergence theorem. Now the rest of the proof is the same as in the special case considered above. □

6.3.2. Corollary. *Let μ be a locally finite Borel measure on Ω_T such that a^{ij}, b^i, $c \in L^1_{\mathrm{loc}}(\Omega_T,\mu)$, and*

$$\int_{\Omega_T}\Bigl[\partial_t\varphi + a^{ij}\partial_{x_i}\partial_{x_j}\varphi + b^i\partial_{x_i}\varphi + c\varphi\Bigr]\,d\mu = 0 \quad \forall\,\varphi \in C_0^\infty(\Omega_T). \tag{6.3.5}$$

Then assertions (i) *and* (ii) *of Theorem 6.3.1 are true. In addition, in case* (ii), *if $J = [T_0, T_1] \subset (0,T)$, U is a ball with compact closure in Ω, and W is a neighborhood of $\overline{U}\times J$ with compact closure in Ω_T, then, for each $r < (d+2)'$, one has*

$$\|\varrho\|_{L^r(U\times J)} \leqslant C(d,r,A,W)\Bigl(|\mu|(W) + \|c\|_{L^1(W,\mu)} + \|b\|_{L^1(W,\mu)}\Bigr),$$

where $C(d,r,A,W)$ depends only on d, r, the Hölder norms of a^{ij} with respect to x on W, $\inf\limits_W \det A$, $\sup\limits_{i,j}\sup\limits_W |a^{ij}|$, and the distance from $U\times J$ to ∂W. An analogous assertion is true in case (i).

Proof. Let us consider the measure $\nu = \zeta\,\mu$, where $\zeta \in C_0^\infty(W)$ is such that $0 \leqslant \zeta \leqslant 1$ and $\zeta|_{U\times J} = 1$. Then ν satisfies (6.3.2) with W in place of D and some constant $N = N_0\Bigl(|\mu|(W) + \|c\|_{L^1(W,\mu)} + \|b\|_{L^1(W,\mu)}\Bigr)$, where N_0 depends on

$\sup_{i,j}\sup_{W}\bigl[|\partial_t\zeta|+|\partial_{x_i}\zeta|+|\partial_{x_i}\partial_{x_j}\zeta|+|a^{ij}|\bigr]$. Now the assertion is readily seen from the proof of the above theorem. □

6.3.3. Remark. Assume that in the situation of Corollary 6.3.2, one has, in addition, that $|b|+|c|\in L^p_{loc}(\Omega_T)$, where $p>r'$. Then

$$\|\varrho\|_{L^r(U\times J)}\leqslant C(d,r,A,W)\Bigl(|\mu|(W)+(\|c\|_{L^p(W)}+\|b\|_{L^p(W)})\|\varrho\|_{L^{p'}(W)}\Bigr).$$

6.3.4. Remark. (i) If there exists a diffusion process $\xi=(\xi_t^{s,x})$ governed by the stochastic differential equation

$$d\xi_t^{s,x}=\sqrt{2A(\xi_t^{s,x},t)}dW_t+b(\xi_t^{s,x},t)dt,\quad \xi_s^{s,x}=x,$$

then the above results apply to the transition probabilities $P(s,x;t,dy)$ of the diffusion ξ. Namely, for any fixed (x,s), the measure $\mu=P(s,x;t,dy)\,dt$ satisfies (6.3.5) with $c=0$. Hence the measure $P(s,x;t,dy)$ is absolutely continuous for almost every t. This fact is well known for locally bounded b (see Krylov [**549**, Chapter II, § 2], Stroock, Varadhan [**884**, Chapters 7, 9]). However, the measure $P(s,x;t,dy)$ can fail to be absolutely continuous for all t. In Fabes, Kenig [**346**], an example is constructed such that $b=0$ and $A(x,t)$ is uniformly continuous, uniformly bounded, and uniformly positive, but, for some fixed t, the measures $P(s,x;t,dy)$ are purely singular with respect to Lebesgue measure for all $s\in(0,t)$ and all x (a similar example is constructed in Safonov [**831**]).

(ii) It is worth noting that Portenko [**788**] employed analogous assumptions about A and b in his study of generalized diffusion processes. In particular, it is shown in [**788**, Chapter II] that if A is uniformly bounded, uniformly positive, and uniformly Hölder continuous, and b is a measurable vector field on $\mathbb{R}^d$ such that $|b|\in L^p(\mathbb{R}^d)$ for some $p>d+2$, then there exists a continuous function $G(s,x,t,y)$, $0\leqslant s<t\leqslant T$, $x,y\in\mathbb{R}^d$, that is the transition probability density for a continuous Markov process $\bigl(x(t),\mathcal{M}_t^s,P_{s,x}\bigr)$, where $P_{s,x}$ is a probability measure generated on the σ-field $\mathcal{M}_t^s$ on the space Ω of continuous paths $x(\,\cdot\,)\colon[0,+\infty)\to\mathbb{R}^d$ by the evaluation mappings $x(\,\cdot\,)\mapsto x(u)$ with $u\in[s,t]$, $P_{s,x}\bigl(x(s)=x\bigr)=1$, and $P_{s,x}$-almost surely

$$x(t)-x(s)=\int_s^t b\bigl(x(\tau),\tau\bigr)\,d\tau+\int_s^t\sqrt{A\bigl(x(\tau),\tau\bigr)}\,dw_s(\tau)$$

with a certain Wiener process $\bigl(w_s(t),\mathcal{M}_t^s,P_{s,x}\bigr)$.

(iii) We also note that the parabolic equations considered in Corollary 6.3.2 have been used in Cattiaux, Léonard [**230**] for the study of certain flows of probability measures. The above results yield the absolute continuity of such measures.

6.4. Local regularity

Here we obtain parabolic analogs of the results in the elliptic case on Sobolev regularity of solution densities, i.e., their membership in certain parabolic Sobolev classes, which will be the usual Sobolev regularity in the space variable. As above, $U_{R,T}=U_R\times(0,T)$, U_R is an open ball in $\mathbb{R}^d$ of radius R.

6.4.1. Theorem. *Let $d\geqslant 2$, $p>d$, $q\in[p',+\infty)$. Let A and A^{-1} be uniformly bounded and let (6.2.3) hold. Suppose that μ is a finite measure on $U_{R,T}$ such that,* *for some $N > 0$ one has*

$$\left|\int_{U_{R,T}} \left[\partial_t\varphi + a^{ij}\partial_{x_i}\partial_{x_j}\varphi\right] d\mu\right| \leqslant N\|\nabla_x\varphi\|_{L^q(U_{R,T})} \quad \forall\varphi\in C_0^\infty(U_{R,T}). \tag{6.4.1}$$

Then $\mu \in \mathbb{H}^{q',1}(U_{R'}, [t_0, t_1])$ and $\mu \in \mathcal{H}^{q',1}(U_{R'}, [t_0, t_1])$ for every $R' < R$ and $[t_0, t_1] \subset (0, T)$.

PROOF. The reasoning is similar to the elliptic case. Passing to smaller R and a time interval $[t_0, t_1] \subset (0, T)$, replacing the measure μ by the measure $\zeta\,\mu$, where $\zeta \in C_0^\infty\big(U_{R'} \times (t_0, t_1)\big)$, and arguing as in the proof of Theorem 6.3.1 (see Step 2 in the proof of assertion (ii)), we may assume that (6.4.1) is fulfilled for all functions φ such that $\partial_t\varphi$ and $\partial_{x_i}\partial_{x_j}\varphi$ are uniformly bounded and continuous in x uniformly with respect to t. Let $f_j \in C_0^\infty\big(U_{R'} \times (t_0, t_1)\big)$, $j = 1, \ldots, d$. Let us solve the equation

$$\partial_t\varphi + a^{ij}\partial_{x_i}\partial_{x_j}\varphi = \partial_{x_i} f^i$$

with zero boundary conditions on $U_{R_0,T}$. By our hypotheses, φ has two derivatives in x and one derivative in t that are bounded and continuous in $x \in U_{R'}$ uniformly with respect to $t \in [t_0, t_1]$. Letting $\psi = \partial_{x_i} f^i$, we have, by Lemma 6.2.5,

$$\begin{aligned}\left|\int \psi\, d\mu\right| &\leqslant N\|\nabla_x\varphi\|_{L^q(U_{R,T})} \leqslant NM\|\psi\|_{\mathbb{H}^{q,-1}(U_R,[0,T])}\\ &= NM\|\psi\|_{\mathbb{H}^{q,-1}(U_{R'},[t_0,t_1])}.\end{aligned}$$

Hence $\mu \in \mathbb{H}_0^{q',1}(U_{R'}, [t_0, t_1])$. Moreover, we have by the above estimate

$$-\partial_t\mu + \partial_{x_i}\partial_{x_j}(a^{ij}\mu) \in \mathbb{H}^{q',-1}(U_{R'}, [t_0, t_1]).$$

Therefore, in order to show that we have $\partial_t\mu \in \mathbb{H}^{q',-1}(U_{R'}, [t_0, t_1])$, it remains to check that $\partial_{x_i}\partial_{x_j}(a^{ij}\mu) \in \mathbb{H}^{q',-1}(U_{R'}, [t_0, t_1])$. For this it suffices to prove that $a^{ij}\mu \in \mathbb{H}_0^{q',1}(U_{R'}, [t_0, t_1])$. This is proved exactly as in the proof of Lemma 1.2.5. Let us recall that both inclusions have been established, of course, for $\zeta\mu$ in place of μ, so that for the initial measure μ we obtain the inclusions announced in the theorem for yet smaller R' and $[t_0, t_1]$. □

6.4.2. Theorem. *Let the maps A and A^{-1} be locally bounded on $U_{R,T}$ and let (6.2.3) hold, where now $p > d + 2$. Let μ be a finite Borel measure on $U_{R,T}$ identified with its density μ such that $\mu \in L^r(U_{R,T})$ with some $r > p'$. Let*

$$\beta \in L^p(U_{R,T}) + L^p(U_{R,T}, |\mu|), \quad \gamma \in L^{p/2}(U_{R,T}) + L^{p/2}(U_{R,T}, |\mu|).$$

Suppose that for all $\varphi \in C_0^\infty(U_{R,T})$ one has

$$\left|\int \left[\partial_t\varphi + a^{ij}\partial_{x_i}\partial_{x_j}\varphi\right] d\mu\right| \leqslant \int \big(|\varphi|\,|\gamma| + |\nabla_x\varphi|\,|\beta|\big)|\mu|\, dx. \tag{6.4.2}$$

Then μ has a density that is locally Hölder continuous on $U_R\times(0, T)$ and belongs to the classes $\mathbb{H}^{p,1}(U_{R'}, [T_0, T_1])$ and $\mathcal{H}^{p,1}(U_{R'}, [T_0, T_1])$ for all radii $R' < R$ and any compact interval $[T_0, T_1] \subset (0, T)$.

PROOF. Fix $R' < R$ and a closed interval $[T_0, T_1] \subset (0, T)$. Since $\mu \in L^r(U_{R,T})$ with some $r > p(p-1)^{-1} = p'$, it follows by inequality (1.1.1) that $|\beta|\,\mu \in L^s(U_{R,T})$ with $s = pr/(p+r) < p$ and $|\gamma|\,|\mu| \in L^{\widetilde{s}}(U_{R,T})$, where $\widetilde{s} = \big(pr/(p + 2r)\big) \wedge 1$. Note that the inequality $\widetilde{s} > 1$ is equivalent to the inequality $r > p/(p-2)$, which in turn is equivalent to the inequality $s' < p$. Note also that by the parabolic embedding theorem 6.2.2 we have

$$\|\varphi\|_{L^{\tilde{s}}(U_{R,T})} \leqslant N\|\varphi\|_{\mathcal{H}^{s',1}(U_R,[0,T])}.$$

Hence inequality (6.4.1) holds with $q = s' > p'$. Then, by Theorem 6.4.1, we have $\mu \in \mathbb{H}^{s,1}(U_{R_1}, [t_0, t_1])$ and $\mu \in \mathcal{H}^{s,1}(U_{R_1}, [t_0, t_1])$ if $[t_0, t_1] \subset (0, T)$ and $R_1 < R$. In particular, we can take any $t_0 \in (0, T_0)$, $t_1 \in (T_1, T)$, and $R_1 \in (R', R)$.

According to (6.2.1), we obtain that $\mu \in L^{r_1}(U_{R_1} \times [t_0, t_1])$ for each r_1 with

$$\frac{1}{r_1} > \frac{1}{s} - \frac{1}{d+2}.$$

This is equivalent to the following relation:

$$\frac{1}{r_1} > \frac{p+r}{pr} - \frac{1}{d+2} = \frac{(d+2)(p+r) - pr}{pr(d+2)},$$

which can be written as

$$r_1 < r\frac{p(d+2)}{(d+2)(p+r) - pr},$$

provided $(d+2)(p+r) - pr > 0$ (otherwise one can take for r_1 any number $r_1 > 1$). We observe that one can choose r_1 sufficiently close to the right-hand side of the above relation in such a way that

$$\frac{r_1}{r} = \frac{p(d+2)}{(d+2)(p+r) - pr} = \frac{p}{p + r - \frac{pr}{d+2}}$$
$$= \frac{p}{p - r(\frac{p}{d+2} - 1)} \geqslant \frac{p}{p - \frac{p}{d+2} + 1} > 1.$$

Therefore, repeating the above procedure finitely many times we arrive at the situation with $(d+2)(p+r) - pr \leqslant 0$, when $\mu \in L^r(U_{R_0} \times [\tau_0, \tau_1])$ for all $r \in (1, +\infty)$, where $\tau_0 < T_0$, $\tau_1 > T_1$, $R_0 > R'$. Hence $\mu \in \mathbb{H}^{s,1}(U_{R_0}, [\tau_0, \tau_1])$ and $\mu \in \mathcal{H}^{s,1}(U_{R_0}, [\tau_0, \tau_1])$ for each $s < p$. Let us choose $s > d + 2$, $\lambda > 0$ and $\kappa > 0$ such that $\kappa > \lambda > 1/s$ and $s(1 - 2\kappa) > d$, which is possible, since $p > d + 2$. By assertion (ii) in the embedding theorem 6.2.2 we conclude that $\mu \in C^{\lambda - 1/s}([\tau_0, \tau_1], C^\delta(U_{R_0}))$ for some $\delta > 0$. Thus, μ has a Hölder continuous density. Then (6.4.1) holds with $q = p'$, whence our statement follows by Theorem 6.4.1. □

For solutions of equations we obtain the following assertion.

6.4.3. Corollary. *Let $p > d + 2$. Suppose that A and A^{-1} are locally bounded on Ω_T and (6.2.3) holds with $p > d + 2$ for every ball U_R with closure in Ω. Let $b^i \in L^p_{\rm loc}(\Omega_T)$ and $c \in L^{p/2}_{\rm loc}(\Omega_T)$. Assume that μ is a locally finite (possibly signed) Borel measure on Ω_T such that b^i, $c \in L^1_{\rm loc}(\Omega_T, \mu)$ and*

$$\int_{\Omega_T} \Big[\partial_t\varphi + a^{ij}\partial_{x_i}\partial_{x_j}\varphi + b^i\partial_{x_i}\varphi + c\varphi\Big]\, d\mu = 0 \quad \forall\, \varphi \in C_0^\infty(\Omega_T). \tag{6.4.3}$$

Then μ has a locally Hölder continuous density that belongs to the spaces $\mathbb{H}^{p,1}(U, J)$ and $\mathcal{H}^{p,1}(U, J)$ for every interval J and open set U such that $U \times J$ has compact closure in Ω_T.

PROOF. By Theorem 6.3.1, the measure μ has a density in $L^r_{\rm loc}(\Omega_T)$ with any order $r < (d+2)'$, in particular, with some $r > p'$. Hence Theorem 6.4.2 applies with $\beta = |b|$ and $\gamma = |c|$. □

It is worth noting that if μ is nonnegative, then, by Harnack's inequality (Theorem 6.2.7), the continuous version of its density is strictly positive in every component of Ω_T in which it is not identically zero. This will be considered in more detail in Chapter 8, where lower estimates are discussed.

6.4.4. Corollary. *Assume that, for all $\varphi \in C_0^\infty(\Omega_T)$, one has*

$$\int_{\Omega_T} \Bigl[\partial_t\varphi + a^{ij}\partial_{x_i}\partial_{x_j}\varphi + b^i\partial_{x_i}\varphi + c\varphi\Bigr]\,d\mu = \int_{\Omega_T} f^i\partial_{x_i}\varphi\,dx\,dt, \tag{6.4.4}$$

where $f^i \in L^p_{\rm loc}(\Omega_T)$ and a^{ij}, b^i, c satisfy the same hypotheses as in Corollary 6.4.3. Then the assertion of Corollary 6.4.3 with condition (6.4.3) replaced by (6.4.4) is still true.

PROOF. As in the elliptic case, our assertion reduces to the case where we have identity (6.4.3). Indeed, for sufficiently small $R > 0$ and any $[T_0, T_1] \subset (0, T)$, there exists $u \in \mathbb{H}_0^{p,1}(U_R, [T_0, T_1])$ satisfying the equation

$$-\partial_t u + \partial_{x_i}(a^{ij}\partial_{x_j}u + \partial_{x_j}a^{ij}u - b^i u) + cu = -\partial_{x_i}f^i.$$

This fact is proved in Ladyz'enskaya, Solonnikov, Ural'tseva [**576**, Ch. III] for the space $\mathbb{H}_0^{2,1}(U_R, [T_0, T_1])$, but the same is true for $\mathbb{H}_0^{p,1}(U_R, [T_0, T_1])$, which follows from Lemma 6.2.5. Then the measure $\mu - u dx dt$ satisfies (6.4.3) with $U_R \times (T_0, T_1)$ in place of Ω_T. □

A closer look at the above proof leads to the following result.

6.4.5. Corollary. *Let μ satisfy the hypotheses of Corollary 6.4.3, let U_{R_0} be an open ball with closure in Ω, and let $[t_1, t_2] \subset (0, T)$. Then, for every closed interval $[\tau_1, \tau_2] \subset (t_1, t_2)$ and any $R < R_0$, there exists a constant N depending on*

$$t_1,\ t_2,\ \tau_1,\ \tau_2,\ R_0,\ R,\ \|c\|_{L^{p/2}(U_{R_0}\times[t_1,t_2])},$$

$$\inf_{U_{R_0}\times[t_1,t_2]} \det A,\ \sup_{t\in[t_1,t_2]} \|a^{ij}(\,\cdot\,,t)\|_{W^{p,1}(U_{R_0})},\ \|b^i\|_{L^p(U_{R_0}\times[t_1,t_2])},$$

such that N is a locally bounded function of the indicated quantities and one has

$$\|\mu\|_{\mathbb{H}^{p,1}(U_R,[\tau_1,\tau_2])} \leqslant N\|\mu\|_{L^{p'}(U_{R_0}\times[t_1,t_2])}.$$

Moreover, one can choose N so that if μ is nonnegative, then

$$\|\mu\|_{\mathbb{H}^{p,1}(U_R,[\tau_1,\tau_2])} \leqslant N\|\mu\|_{L^1(U_{R_0}\times[t_1,t_2])}. \tag{6.4.5}$$

PROOF. By Remark 6.3.3 we obtain an a priori estimate for the L^r-norm of any solution with any $r \in \bigl(p', (d+2)'\bigr)$. Now the first claim follows by the reasoning in the proof of Theorem 6.4.2. Finally, estimate (6.4.5) for nonnegative solutions follows from Harnack's inequality. Indeed, suppose that $s_1 < s_2 < s_3 < s_4$ and $W_1 = U_{R_1}\times[s_1, s_2]$, $W_2 = U_{R_1}\times[s_3, s_4]$ have compact closures in $U_R\times(t_1, t_2)$. Letting ϱ be the continuous version of the density of a nonnegative solution μ, according to Harnack's inequality (see Theorem 6.2.7), we have

$$\max_{W_1} \varrho \leqslant C \min_{W_2} \varrho,$$

where C depends (as a locally bounded function) on $\|c\|_{L^{p/2}(U_{R_0}\times[t_1,t_2])}$, s_i, t_i, R_0, R, R_1, $\|b^i\|_{L^p(U_{R_0}\times[t_1,t_2])}$, $\sup_{t\in[t_1,t_2]}\|a^{ij}(\,\cdot\,,t)\|_{H^{p,1}(U_{R_0})}$, and $\inf_{U_{R_0}\times[t_1,t_2]}\det A$. Therefore, we have

$$\max_{W_1}\varrho \leqslant C|W_2|^{-1}\int_{U_{R_0}\times[t_1,t_2]}\varrho(x,t)\,dx\,dt,$$

where $|W_2|$ is Lebesgue measure of W_2. Now (6.4.5) follows by the first claim (we take s_i such that $[\tau_1,\tau_2]\subset(s_1,s_2)$). □

6.4.6. Remark. The assumption that $p>d+2$ was used in the proof of Theorem 6.4.2 in order to improve the initial integrability of a solution and also to guarantee its Hölder continuity (by an appropriate embedding theorem). It is clear from the first step of that proof (or from Theorem 6.4.1) that if we assume only that $p>d$, then we obtain that $\mu\in\mathbb{H}^{s,1}(B_{R'},[T_0,T_1])$ and $\mu\in\mathcal{H}^{s,1}(B_{R'},[T_0,T_1])$ with $s=pr/(p+r)>d$, provided that we have $r>pd/(p-d)$.

Let us return to justification of Theorem 5.4.5. The next result shows that $K_t^*\delta_x = p_{A,b}(t,x,y)dy$ for $t>0$, where $p_{A,b}$ is the function from Theorem 5.4.5, and the Fokker–Planck–Kolmogorov equation is satisfied.

6.4.7. Theorem. *Suppose that $\mu\in\mathcal{M}_{\rm ell}^{A,b}$, where A is continuous, $\det A>0$, $a^{ij}\in W^{p,1}_{\rm loc}(\Omega)$, $b^i\in L^p_{\rm loc}(\Omega)$, and $p>d+2$. Then there exist subprobability kernels $K_t(\,\cdot\,,dy)$, $t>0$, on Ω such that*

$$K_t(x,dy)=p_{A,b}(t,x,y)\,dy,$$

where $p_{A,b}(t,x,y)$ are locally Hölder continuous nonnegative functions defined on $(0,+\infty)\times\Omega\times\Omega$, for each function $f\in L^1(\mu)$ the function

$$x\mapsto K_tf(x):=\int_\Omega f(y)\,p_{A,b}(t,x,y)\,dy$$

serves as a μ-version of $T_t^\mu f$ such that the function $(t,x)\mapsto K_tf(x)$ is continuous on $(0,+\infty)\times\Omega$, and on each compact set in Ω it is Hölder continuous in x uniformly in $t\geqslant\tau$ for every fixed $\tau>0$. If Ω is connected, then $p_{A,b}>0$.

If we are given a bounded Borel measure μ_0 on Ω and

$$K_t^*\mu_0(dy):=\int_\Omega K_t(x,dy)\,\mu_0(dx)=\int_\Omega p_{A,b}(t,x,y)\,\mu_0(dx)\,dy,$$

then the measure $K_t^*\mu_0(dy)\,dt$ satisfies the equation $\partial_t\mu=L_{A,b}^*\mu$ for all $T>0$. In particular, $K_t^*\mu_0$ has a positive continuous density in $W^{p,1}_{\rm loc}(\Omega)$.

PROOF. *Step 1.* Consider the operator L_{A,b_0} on $C_0^\infty(\Omega)$ with the first-order term $b_0=(b_0^1,\ldots,b_0^d)$ given by

$$b_0^j=2\partial_{x_i}a^{ij}+2a^{ij}\frac{\partial_{x_i}\varrho}{\varrho}-b^j,$$

where ϱ is the continuous version of the density of μ, which is in $W^{p,1}_{\rm loc}(\Omega)$ and is strictly positive in every connected component of Ω where it is not identically zero. We may assume that ϱ does not vanish in Ω. By our assumption, $b_0^j\in L^p_{\rm loc}(\Omega)$. We know from Chapter 5 that

$$L_{A,b_0}^*\mu=0$$

and that the operator $\big(L_{A,b_0}, C_0^\infty(\Omega)\big)$ has a closed extension $\big(\widehat{L}_{A,b_0}, D(\widehat{L}_{A,b_0})\big)$ generating a sub-Markovian C_0-semigroup $\{T_t'\}_{t\geqslant 0}$ on $L^1(\mu)$. Moreover,

$$\int_\Omega T_t f\, g\, d\mu = \int f\, T_t' g\, d\mu, \quad f, g \in L^\infty(\mu),\ t \geqslant 0. \tag{6.4.6}$$

Step 2. Let us show that for all $g \in L^1(\mu)$, $t > 0$, there exists a μ-version $\widetilde{T_t g}$ of $T_t g$ such that

$$(x,t) \mapsto \widetilde{T_t g}(x) \text{ is continuous on } \Omega\times(0,+\infty).$$

First, by the continuity of $t \mapsto T_t g$ in $L^1(\mu)$, for each fixed $t \geqslant 0$, one can choose a version of $T_t g$ (denoted by the same symbol) such that the function $(x,t) \mapsto T_t g(x)$ is jointly measurable (see Exercise 1.8.15).

Assume that $g \in L^\infty(\mu)$. Define a measure ν on $\Omega\times(0,+\infty)$ by

$$\nu(dx\, dt) := T_t g(x)\, \mu(dx)\, dt. \tag{6.4.7}$$

Then

$$\int_{\Omega_T} \big(\partial_t \varphi + a^{ij}\partial_{x_i}\partial_{x_j}\varphi + b_0^i \partial_{x_i}\varphi\big)\, d\nu = 0, \quad \varphi \in C_0^\infty(\Omega_T). \tag{6.4.8}$$

Indeed, since $\{T_t\}_{t\geqslant 0}$ is continuous on $L^{p'}(\mu)$, it suffices to verify (6.4.8) for functions $g \in C_0^\infty(\Omega)$. Moreover, it suffices to do this for φ of the form $\varphi = \varphi_1\varphi_2$, where $\varphi_1 \in C_0^\infty(\Omega)$, $\varphi_2 \in C_0^\infty\big((0,T)\big)$. It follows by (6.4.6) that

$$\int_\Omega L_{A,b_0}\varphi_1\, T_t g\, d\mu = \int_\Omega \varphi_1 T_t L_{A,b} g\, d\mu.$$

Integrating by parts with respect to t, we obtain

$$\begin{aligned}\int_{\Omega_T} &\big(\partial_t \varphi + a^{ij}\partial_{x_i}\partial_{x_j}\varphi + b_0^i \partial_{x_i}\varphi\big)\, d\nu \\ &= \int_0^T\!\!\int_\Omega \varphi_1(x)\varphi_2(t) T_t L_{A,b} g(x)\, \mu(dx)\, dt - \int_0^T \varphi_2(t)\frac{d}{dt}\int_\Omega \varphi_1(x) T_t g(x)\, \mu(dx)\, dt.\end{aligned}$$

Since

$$\frac{d}{dt}\int_\Omega \varphi_1(x) T_t g(x)\, \mu(dx) = \int_\Omega \varphi_1(x) T_t L_{A,b} g(x)\, \mu(dx),$$

we obtain (6.4.8). Because of the continuity and strict positivity of the density ϱ, equality (6.4.8) and Corollary 6.4.3 imply that there exists a continuous function $u\colon \Omega\times(0,+\infty) \to \mathbb{R}$ such that for dt-a.e. $t \in (0,+\infty)$ we have that $u(\,\cdot\,,t)$ is a version of $T_t g$. However, we observe that the latter is true for every $t > 0$. Indeed, for any $t > 0$, there exists a sequence $t_n \to t$ and a Borel set $S_t \subset \Omega$ such that $\mu(S_t) = 1$ and $T_{t_n} g(x) = u(x, t_n)$ for all $x \in S_t$. It remains to note that the right-hand side of this equality converges to $u(x,t)$ for each x and that $T_{t_n} g \to T_t g$ in the norm of $L^1(\mu)$, hence in measure. Therefore, we can set $\widetilde{T_t g}(x) = u(x,t)$.

Now let $g \in L^1(\mu)$. We may assume that $g \geqslant 0$. Let $g_k = \min(g,k)$. According to Corollary 6.4.5 (see (6.4.5)), equality (6.4.8) implies that, for every ball U_R with compact closure in Ω and every $[T_1,T_2] \subset (0,+\infty)$, the sequence of functions $(x,t) \mapsto \widetilde{T_t g_k}(x)$ is bounded in $\mathbb{H}^{p,1}(U_R,[T_1,T_2])$. For every fixed t, the functions $T_t g_k$ converge to $T_t g$ in $L^1(\mu)$. Hence the function $(x,t) \mapsto \widetilde{T_t g}(x)$ is in $\mathbb{H}^{p,1}(U_R,[T_1,T_2])$. Moreover, the functions $\widetilde{T_t g_k}(x)$ converge to $\widetilde{T_t g}(x)$ in

$L^p(U_R\times[T_1,T_2])$. Therefore, since the measure $T_t g_k\,\mu(dx)\,dt$ satisfies (6.4.8), the measure $T_t g\,\mu(dx)\,dt$ also does. As above, this yields our claim by Corollary 6.4.3.

We observe that if a sequence $\{g_j\}$ converges in $L^1(\mu)$, then, for every compact set $Q\subset\Omega$, the continuous functions $\widetilde{T_t g_j}$ converge uniformly on Q for every $t>0$. This follows by the closed graph theorem and the above reasoning (note that the positive and negative parts of g_j converge in $L^1(\mu)$), since the natural embedding $C(Q)\to L^1(Q,\mu)$ is continuous (certainly, this is also seen from the above reasoning). In fact, this shows that there exists a number $C(Q,t)$ such that

$$\sup_{x\in Q}|\widetilde{T_t g}(x)|\leqslant C(Q,t)\|g\|_{L^1(\mu)}\quad\forall\, g\in L^1(\mu). \tag{6.4.9}$$

Moreover, there are numbers $\alpha>0$ and $C_1(Q,T)$ such that

$$\sup_{x,y\in Q}\frac{|\widetilde{T_t g}(x)-\widetilde{T_t g}(y)|}{|x-y|^\alpha}\leqslant C_1(Q,t)\|g\|_{L^1(\mu)}\quad\forall\, g\in L^1(\mu). \tag{6.4.10}$$

which is proved by applying the same reasoning to the Banach space of α-Hölder functions.

Step 3. By estimate (6.4.9) we obtain that there exist subprobability measures $K_t(x,dy)$ such that, for every $x\in\Omega$ and $t>0$, one has

$$K_t(x,dy)=p_t(x,y)\mu(dy),$$

where $y\mapsto p_t(x,y)$ is a bounded μ-measurable function, and

$$K_t g(x):=\int_\Omega g(y)\,K_t(x,dy)=\widetilde{T_t g}(x) \tag{6.4.11}$$

for all $g\in L^1(\mu)$, $x\in\Omega$, and $t>0$.

Let $K_t'(\,\cdot\,,dy)$ be defined analogously to $K_t(\,\cdot\,,dy)$ above, but with T_t' in place of T_t. Then $K_t'(x,dy)=p_t'(x,y)\mu(dy)$, where $p_t'(x,\,\cdot\,)\in L^\infty(\mu)$.

Step 4. Now we apply (6.4.6) and Fubini's theorem to (6.4.8), which yields that for all $g\in L^\infty(\mu)$ and all $\varphi\in C_0^\infty(\Omega_T)$ we have

$$\int_\Omega g(y)\int_{\Omega_T}(\partial_t\varphi+a^{ij}\partial_{x_i}\partial_{x_j}\varphi+b_0^i\partial_{x_i}\varphi)K_t'(y,dx)\,dt\,\mu(dy)=0. \tag{6.4.12}$$

Interchanging the roles of b_0 and b as well as T_t and T_t' in our reasoning, we also have for all $g\in L^\infty(\mu)$ and all $\varphi\in C_0^\infty(\Omega_T)$ that

$$\int_\Omega g(y)\int_{\Omega_T}(\partial_t\varphi+a^{ij}\partial_{x_i}\partial_{x_j}\varphi+b^i\partial_{x_i}\varphi)K_t(y,dx)\,dt\,\mu(dy)=0. \tag{6.4.13}$$

We observe that the function

$$\Phi\colon\ (y,t)\mapsto\int_\Omega(\partial_t\varphi+a^{ij}\partial_{x_i}\partial_{x_j}\varphi+b^i\partial_{x_i}\varphi)K_t(y,dx)$$

is continuous on $\Omega\times(0,+\infty)$. Indeed, if we denote the integrand by $\xi(t,x)$, then $|\xi(t,x)-\xi(s,x)|\leqslant\theta(x)|t-s|$, where θ is μ-integrable. Now our claim readily follows from the joint continuity of the function $(x,t)\mapsto\widetilde{T_t g}(x)$ for any $g\in L^1(\mu)$. Integrating Φ with respect to t yields a continuous function. Since (6.4.13) holds for any g, for all $y\in\Omega$ one has

$$\int_{\Omega_T}(\partial_t\varphi+a^{ij}\partial_{x_i}\partial_{x_j}\varphi+b^i\partial_{x_i}\varphi)K_t(y,dx)\,dt=0.$$

This means that, for every $y \in \Omega$, the measure $\nu = K_t(y, dx)\,dt$ satisfies equality (6.4.8) with b^i in place of b_0^i. Therefore, there exists a locally Hölder continuous function $(x,t) \mapsto p_t(y,x)$ (that is a version of the function $p_t(y,x)$ defined above) on $\Omega\times(0,+\infty)$ such that

$$K_t(y, dx) = p_t(y,x)\,\mu(dx), \quad y \in \Omega, t > 0.$$

According to Corollary 6.4.5 (estimate (6.4.5)) and Theorem 6.2.2, the functions $(x,t) \mapsto p_t(y,x)$, $y \in \Omega$, have locally uniformly bounded Hölder and supremum norms. In addition, (6.4.10) implies that, for every $t > 0$, we have

$$\sup_{x,y\in Q}\sup_{z\in\Omega} \frac{|p_t(x,z) - p_t(y,z)|}{|x-y|^\alpha} \leqslant C_1(Q,t).$$

Note also that (6.4.9) yields that

$$\sup_{x\in Q}\sup_{z\in\Omega} p_t(x,z) \leqslant C(Q,t). \tag{6.4.14}$$

It is worth noting that, for a fixed compact set Q, we can choose quantities $C(Q,t)$ and $C_1(Q,t)$ such that they are bounded in t in every compact interval in $(0,+\infty)$. By the same reasoning we obtain that one can choose densities $p_t'(y,x)$ of the measures $K_t'(y,dx)$, $t > 0$, $y \in \Omega$, with respect to μ such that $p_t'(y,x)$ is locally Hölder continuous in (t,x) and y. It is readily seen from identity (6.4.6) that $p_t(x,y) = p_t'(y,x)$ for all $x,y \in \Omega$, $t > 0$. Setting $p(t,x,y) = p_t(x,y)\varrho(y)$ for $x,y \in \Omega$, $t > 0$, we complete the proof of the first part of the theorem.

Step 5. Since, as shown in the previous step, the measure $K_t(y,dx)\,dt$ satisfies (6.4.8) with b^i in place of b_0^i for all $y \in \Omega$, by integrating over y with respect to μ_0 we obtain that the same holds for the measure $K_t^*\mu_0\,dt$. Now the last part of the assertion follows by Corollary 6.4.3. In order to apply this corollary, we have to verify that $|b|^p$ is locally integrable with respect to the measure $K_t^*\mu_0\,dt$. This measure has the density

$$f(x,t) = \int_\Omega p_t(y,x)\,\mu_0(dy)$$

with respect to $\mu\,dt$. It suffices to note that $f(x,t)$ is bounded on compact sets in $\Omega\times(0,+\infty)$. Indeed, the function $p_t'(x,y)$ satisfies an estimate of the type (6.4.14). By the equality $p_t'(x,y) = p_t(y,x)$ we obtain that $\sup_{x\in Q}\sup_{y\in\Omega} p_t(y,x) \leqslant C(Q,t)$ for every compact set $Q \subset \Omega$, where $C(Q,t)$ is locally bounded in t. □

6.4.8. Remark. It follows from the presented proof that for every compact set $E \subset \Omega\times(0,+\infty)$, there exists a constant $C(E)$ such that

$$\sup_{(x,t)\in E}\sup_{y\in\Omega} p_t(x,y) \leqslant C(E),$$

where $p_t(x,y)$ is the jointly continuous version of the Radon–Nikodym density of the measure $K_t(x,dy)$ with respect to the measure $\mu = \varrho\,dx$ from Theorem 5.4.5, i.e., $p_{A,b}(t,x,y) = p_t(x,y)\varrho(y)$. Hence the continuous density of the measure $K_t^*\mu_0(dy)$ is given by

$$\int_\Omega p_t(x,y)\,\mu_0(dx)\,\varrho(y).$$

We observe that the subinvariance of μ with respect to T_t^μ can be written as the estimate $K_t^*\mu \leqslant \mu$.

6.5. Local estimates

Let us consider in more detail the situation where the coefficients b and c are integrable with respect to $|\mu|$, but are not necessarily integrable with respect to Lebesgue measure. Applying Moser's iteration techniques, we refine the results on regularity obtained above.

6.5.1. Proposition. *Let $p > d+2$, let A and A^{-1} be locally bounded on Ω_T and let (6.2.3) be fulfilled with some $p > d+2$ for every ball U_R with closure in Ω. Suppose that μ is a locally finite (possibly signed) Borel measure on Ω_T such that $b^i \in L^p_{\rm loc}(\mu)$, $c \in L^{p/2}_{\rm loc}(\mu)$, and*

$$\int_{\Omega_T} [\partial_t\varphi + a^{ij}\partial_{x_i}\partial_{x_j}\varphi + \partial_{x_i}a^{ij}\partial_{x_j}\varphi + b^i\partial_{x_i}\varphi + c\varphi]\, d\mu = 0 \quad \forall\varphi\in C_0^\infty(\Omega_T).$$

Then μ possesses a locally Hölder continuous density which belongs to the spaces $\mathbb{H}^{p,1}(U,J)$ and $\mathcal{H}^{p,1}(U,J)$ for every interval J and every open ball U such that the closure of $U\times J$ is compact in Ω_T.

Proof. By Theorem 6.3.1 the measure μ has a density of class $L^r_{\rm loc}(\Omega_T)$ with every $r < (d+2)'$, in particular, with some $r > p'$. Hence Theorem 6.4.2 applies with $\beta = |b|$ and $\gamma = |c|$. □

Throughout this section we assume that A satisfies condition (6.2.3) with some $p > d+2$ and

$$0 < \lambda(U,J) := \inf\bigl\{\langle A(x,t)\xi,\xi\rangle\colon\ |\xi| = 1,\ (x,t)\in U\times J\bigr\}$$

for every ball U with closure in Ω and every interval $J\subset(0,T)$, and also that

$$b^i \in L^p_{\rm loc}(\mu), \quad c\in L^{p/2}_{\rm loc}(\mu).$$

Let a *nonnegative* locally finite Borel measure μ on Ω_T satisfy the equation

$$\partial_t\mu = L^*_{A,b,c}\mu.$$

Then, by Proposition 6.5.1, the measure μ possesses a Hölder continuous density $\varrho\in\mathcal{H}^{p,1}_{\rm loc}$ with respect to Lebesgue measure.

Set $B^i = b^i - \partial_{x_j}a^{ij}$. Then the equation $\partial_t\mu = L^*_{A,b,c}\mu$ for the measure can be written as the divergence form parabolic equation for the density

$$\partial_t\varrho = {\rm div}\bigl(A\nabla\varrho - B\varrho\bigr) + c\varrho, \tag{6.5.1}$$

understood, as before, in the sense of the integral identity

$$\int_0^T\!\!\int_\Omega [-\varrho\partial_t\varphi + \langle A\nabla\varrho,\nabla\varphi\rangle]\,dx\,dt = \int_0^T\!\!\int_\Omega [\langle B,\nabla\varphi\rangle\varrho + c\varrho\varphi]\,dx\,dt \tag{6.5.2}$$

for every function $\varphi\in C_0^\infty\bigl(\Omega\times(0,T)\bigr)$.

As above, we set $c^+(x,t) = \max\{c(x,t),0\}$.

The next lemma is the key step of the iteration techniques.

6.5.2. Lemma. *Let $m\geqslant 1$, let $U\subset\Omega$ be a ball, and let $[s_1,s_2]\subset(0,T)$. Then there exists a number $C(d)>0$, depending only on d, such that for every function*

$\psi \in C_0^\infty(\Omega\times(0,T))$ with support in $U\times(0,T)$ for which $\psi(x,s_1)=0$ for all x, we have the inequality

$$(6.5.3)\quad \biggl(\int_{s_1}^{s_2}\int_U |\varrho^m\psi|^{2(d+2)/d}\,dx\,dt\biggr)^{d/(d+2)}$$
$$\leqslant 32C(d)m^2(1+\lambda^{-1})\int_{s_1}^{s_2}\int_U \Bigl[|\psi|\,|\partial_t\psi| + \|A\|\,|\nabla\psi|^2 + |\sqrt{A^{-1}}B|^2\psi^2 + c^+\psi^2\Bigr]\varrho^{2m}\,dx\,dt,$$

where $\|A(x,t)\| = \max_{|\xi|=1}\langle A(x,t)\xi,\xi\rangle$, and the number $\lambda=\lambda(U,[s_1,s_2])$ is defined above.

PROOF. Let $f\in C^\infty([0,+\infty))$ be such that $f\geqslant 0$, $f'\geqslant 0$, $f''\geqslant 0$. Substituting the function $\varphi = f'(\varrho)\psi^2$ in (6.5.2) and applying the Cauchy–Bunyakovsky inequality, for every $t\in[s_1,s_2]$ we obtain the inequality

$$\int_\Omega f(\varrho(x,t))\psi^2(x)\,dx - \int_\Omega f(\varrho(x,s_1))\psi^2(x)\,dx + \frac{1}{3}\int_{s_1}^t\int_\Omega |\sqrt{A}\nabla\varrho|^2 f''(\varrho)\psi^2\,dx\,d\tau$$
$$\leqslant \int_{s_1}^t\int_\Omega \Bigl[2|\psi|\,|\partial_t\psi| f(\varrho) + 3|\sqrt{A}\nabla\psi|^2\frac{f'(\varrho)^2}{f''(\varrho)} + 3|\sqrt{A^{-1}}B|^2\varrho^2 f''(\varrho)\psi^2 + 2|\langle B,\nabla\psi\rangle|\psi\varrho f'(\varrho) + c^+\varrho f'(\varrho)\psi^2\Bigr]\,dx\,d\tau.$$

Indeed, it suffices to observe that

$$2\langle A\nabla\varrho,\nabla\psi\rangle\psi f'(\varrho) \leqslant 3^{-1}|\sqrt{A}\nabla\varrho|^2 f''(\varrho)\psi^2 + 3|\sqrt{A}\nabla\psi|^2\frac{f'(\varrho)^2}{f''(\varrho)},$$
$$\langle B,\nabla\varrho\rangle\varrho f''(\varrho)\psi^2 \leqslant 3^{-1}|\sqrt{A}\nabla\varrho|^2 f''(\varrho)\psi^2 + 3|\sqrt{A^{-1}}B|^2\varrho^2 f''(\varrho)\psi^2.$$

Set $f(\varrho)=\varrho^{2m}$. Recall that $\psi(x,s_1)=0$. We have

$$\sup_{t\in[s_1,s_2]}\int_\Omega \varrho^{2m}(x,t)\psi^2(x)\,dx + \frac{4m-2}{3m}\int_{s_1}^{s_2}\int_\Omega |\sqrt{A}\nabla(\varrho^m\psi)|^2\,dx\,d\tau$$
$$\leqslant 32m^2\int_{s_1}^{s_2}\int_\Omega \Bigl[|\psi|\,|\partial_t\psi| + |\sqrt{A}\nabla\psi|^2 + |\sqrt{A^{-1}}B|^2\psi^2 + c^+\psi^2\Bigr]\varrho\,dx\,d\tau.$$

Now our assertion follows by Lemma 6.2.1. □

6.5.3. Theorem. (L^p-estimates.) *Let $p\geqslant 2(d+2)/d$ and let U and U' be two balls in Ω with $\overline{U'}\subset U$. Let also $[s_1,s_2]\subset(0,T)$. Then, for every $s\in(s_1,s_2)$, there exists a number $C>0$, depending only on U, U', s, s_1, d and p, such that*

$$\|\varrho\|_{L^p(U'\times[s,s_2])} \leqslant C(1+\lambda^{-1})^\gamma\int_{s_1}^{s_2}\int_U \Bigl[1+\|A\|^\gamma + |c^+|^\gamma + |\sqrt{A^{-1}}B|^{2\gamma}\Bigr]\varrho\,dx\,dt,$$

where $\gamma=(d+2)/2p'$, $p'=p/(p-1)$ and the numbers $\lambda=\lambda(U,[s_1,s_2])$ and $\|A\|$ are defined above.

PROOF. Set $m = dp/2(d+2)$ and

$$\alpha = 1+\frac{4m}{(2m-1)d},\quad \alpha' = 1+\frac{(2m-1)d}{4m},\quad \delta = \frac{4}{d(2m-1)+4m}.$$

We observe that $m \geqslant 1$. Set $\psi(x,t) = \zeta(x)\eta(t)$, where $\zeta \in C_0^\infty(U)$, $\zeta(x) = 1$ if $x \in U'$, $0 \leqslant \psi \leqslant 1$, $\eta \in C_0^\infty\bigl((s_1, T)\bigr)$, $\eta(t) = 1$ if $t \in [s, s_2]$, $0 \leqslant \eta \leqslant 1$ and

$$|\partial_t \eta(t)| \leqslant K\eta^{1-\delta}(t), \quad |\nabla\zeta(x)| \leqslant K\zeta^{1-\delta}(x)$$

for some numbers $K > 0$ and all $(x,t) \in U\times[s_1, s_2]$. We observe that K depends only on U, U', s and s_1. Applying Lemma 6.5.2, we obtain

$$\biggl(\int_{s_1}^{s_2}\int_U |\varrho^m\psi|^{2(d+2)/d}\,dx\,dt\biggr)^{d/(d+2)} \leqslant 32C(d)m^2(1+\lambda^{-1})$$
$$\times \int_s^{s_2}\int_U \Bigl[|\psi|\,|\partial_t\psi| + \|A\|\,|\nabla\psi|^2 + |\sqrt{A^{-1}}B|^2\psi^2 + c^+\psi^2\Bigr]\varrho^{2m}\,dx\,dt.$$

By using Hölder's inequality with the exponents α and α', we estimate the integral on the right-hand side by the product of the quantities

$$K^2\biggl(\int_{s_1}^{s_2}\int_U \bigl(1+\|A\| + |\sqrt{A^{-1}}B|^2 + c^+\bigr)^{\alpha'}\varrho^{2m}\,dx\,dt\biggr)^{1/\alpha'},$$

$$\biggl(\int_{s_1}^{s_2}\int_U |\varrho^m\psi|^{2(d+2)/d}\,dx\,dt\biggr)^{1/\alpha}.$$

Applying the inequality $xy \leqslant \varepsilon x^\alpha + C(\alpha,\varepsilon)y^{\alpha'}$ with a sufficiently small $\varepsilon > 0$, we obtain the assertion of the theorem. □

6.5.4. Theorem. (L^∞-estimates.) *Let $\gamma > (d+2)/2$ and U and U' be two balls in Ω with $\overline{U'} \subset U$. Let also $[s_1, s_2] \subset (0,T)$. Then, for every $s \in (s_1, s_2)$, there exists a number $C > 0$, depending only on U, U', s, s_1, d, and γ, such that*

$$\|\varrho\|_{L^\infty(U'\times[s,s_2])} \leqslant C(1+\lambda^{-1})^\gamma \int_{s_1}^{s_2}\int_U \Bigl[1 + \|A\|^\gamma + |c^+|^\gamma + |\sqrt{A^{-1}}B|^{2\gamma}\Bigr]\varrho\,dx\,dt,$$

where the numbers $\lambda = \lambda(U, [s_1, s_2])$ and $\|A\|$ are defined above.

Proof. If $\varrho \equiv 0$ on $U\times[s_1, s_2]$, then our assertion is trivial. Let us consider the case where $\varrho \not\equiv 0$. Multiplying the solution ϱ by the number

$$(1+\lambda^{-1})^{-\gamma}\biggl(\int_{s_1}^{s_2}\int_U \Bigl[1 + \|A\|^\gamma + |c^+|^\gamma + |\sqrt{A^{-1}}B|^{2\gamma}\Bigr]\varrho\,dx\,dt\biggr)^{-1},$$

we can assume that

$$(1+\lambda^{-1})^\gamma \int_{s_1}^{s_2}\int_U \Bigl[1 + \|A\|^\gamma + |c^+|^\gamma + |\sqrt{A^{-1}}B|^{2\gamma}\Bigr]\varrho\,dx\,dt = 1.$$

In this case for the proof of the theorem it suffices to find a number C, depending only on U, U', s, s_1, s_2, d and γ, such that

$$\|\varrho\|_{L^\infty(U'\times[s,s_2])} \leqslant C.$$

Let $U = U(x_0, R)$, $U' = U(x_0, R')$ and $R' < R$. For every number $n \in \mathbb{N}$, we set $R_n = R' + (R - R')2^{-n}$, $s_n = s - (s - s_1)2^{-n}$, and also $U_n = U(x_0, R_n)$. Let us consider the system of decreasing cylinders

$$Q_n = U_n\times[s_n, s_2], \quad Q_0 = U\times[s_1, s_2].$$

We define functions $\psi_n \in C_0^\infty\bigl(\Omega\times(0,T)\bigr)$ as in the proof of Theorem 6.5.3, i.e., $\psi_n(x,t) = 1$ if $(x,t) \in Q_{n+1}$, $0 \leqslant \psi_n \leqslant 1$, the support of ψ_n belongs to $U_n\times(s_n, T)$ and $|\partial_t\psi_n(x,t)| + |\nabla\psi_n(x,t)| \leqslant K^n$ for all $(x,t) \in \Omega_T$ and some number $K > 1$

depending only on the numbers s, s_1, R, R'. Applying Lemma 6.5.2 and Hölder's inequality with the exponents γ and γ', we obtain

$$\Bigl(\int_{Q_n} |\varrho^m\psi_n|^{2(d+2)/d}\,dx\,dt\Bigr)^{d/(d+2)} \leqslant 32m^2C(d,s)K^{2n}\Bigl(\int_{Q_n} \varrho^{(2m-1)\gamma'+1}\,dx\,dt\Bigr)^{1/\gamma'}.$$

Set

$$p_{n+1} = \beta p_n + (\gamma'-1)\gamma'^{-1}, \quad p_1 = \gamma'+1, \quad \beta = (d+2)d^{-1}\gamma'^{-1}.$$

We observe that $\beta^{n-1}p_1 \leqslant p_n \leqslant \beta^{n-1}(p_1+1)$. Letting

$$m = p_{n+1}d/(2d+4),$$

we arrive at the inequality

$$\|\varrho\|_{L^{p_{n+1}}(Q_{n+1})} \leqslant C^{n\beta^{-n}}\|\varrho\|_{L^{p_n}(Q_n)}^{p_n/(p_n+\gamma'-1)},$$

where C depends only on K, d and γ. We observe that $\sum_n n\beta^{-n} < \infty$. In addition, Theorem 6.5.3 estimates the norm $\|\varrho\|_{L^{p_1}(Q_1)}$ by a constant depending only on the numbers p_1, d, s, s_1, U and U_1. Therefore, we obtain an estimate of the norms $\|\varrho\|_{L^{p_{n+1}}(Q_{n+1})}$ uniform in n, which gives the desired L^∞-estimate. □

6.5.5. Remark. (i) We observe that the constant C in the inequalities in Theorem 6.5.3 and Theorem 6.5.4 does not depend on the number s_2.

(ii) If $c \leqslant 0$, then all estimates obtained above remain true without the coefficient c in the right side.

The obtained estimates will be used in Chapter 7 in the study of global regularity of nonnegative solutions.

6.6. Existence of solutions to the Cauchy problem

Here we discuss sufficient conditions for the existence of solutions to the Cauchy problem for Fokker–Planck–Kolmogorov equations in the case of a locally nondegenerate diffusion matrix A. The key role in the proof of the existence theorem will be played by the local estimates proved in the preceding sections. Degenerate equations are discussed in § 6.7(ii).

Let $\mathbb{R}^d_T = \mathbb{R}^d \times (0,T)$ and let ν be a subprobability measure on $\mathbb{R}^d$. We study the Cauchy problem (6.1.1) and (6.1.3), namely,

$$\partial_t\mu = L^*_{A,b,c}\mu, \quad \mu|_{t=0} = \nu.$$

Recall that

$$L_{A,b,c}\varphi = a^{ij}\partial_{x_i}\partial_{x_j}\varphi + b^i\partial_{x_i}\varphi + c\varphi,$$

where a^{ij}, b^i, c are Borel functions on $\mathbb{R}^d_T$ and the matrix $A = (a^{ij})$ is symmetric and positive definite.

In this section we assume that $c \leqslant 0$. This condition can be replaced with the inequality $c \leqslant c_0$ for some constant c_0. Indeed, one condition reduces to the other by the multiplication of the solution by $e^{-c_0 t}$.

Let $\mathcal{M}_\nu$ be the set of measures μ defined by a family of nonnegative Borel measures $(\mu_t)_{0<t<T}$ on $\mathbb{R}^d$ such that μ is a solution to the Cauchy problem (6.1.1) and (6.1.3), $|c| \in L^1(\mathbb{R}^d_T, \mu)$ and for almost all $t \in (0,T)$ we have the inequality

$$\mu_t(\mathbb{R}^d) \leqslant \nu(\mathbb{R}^d) + \int_0^t\int_{\mathbb{R}^d} c(x,s)\,\mu_s(dx)\,ds. \tag{6.6.1}$$

The main goal of this section is to obtain sufficient conditions under which the set $\mathcal{M}_\nu$ is not empty.

We first consider the classical case of smooth bounded coefficients.

6.6.1. Theorem. *Suppose that there exist positive numbers m, M, Λ and $\alpha \in (0,1)$ such that $m \cdot \mathrm{I} \leqslant A(x,t) \leqslant M \cdot \mathrm{I}$ and*

$$|a^{ij}(x,t) - a^{ij}(y,s)| \leqslant \Lambda(|x-y|^\alpha + |t-s|^{\alpha/2}) \quad \forall\,(x,t) \in \mathbb{R}^d_T.$$

Suppose also that the functions

$$a^{ij},\ \partial_{x_k} a^{ij},\ \partial_{x_k}\partial_{x_m} a^{ij},\ b^i,\ \partial_{x_k} b^i,\ c$$

are bounded and continuous in $\mathbb{R}^d_T$ and Hölder continuous of order α in x uniformly with respect to t. Then there is a nonnegative function $G(x,t,y,s)$ such that

$$G(x,t,y,s) \leqslant C_1(t-s)^{-d/2} \exp\Bigl(-C_2|x-y|^2/4(t-s)\Bigr)$$

and for every subprobability density $\varrho_0 \in C_b(\mathbb{R}^d)$ the formula

$$\varrho(x,t) = \int_{\mathbb{R}^d} G(x,t,y,0)\varrho_0(y)\,dy$$

defines a solution $\mu = \varrho\,dx\,dt$ of the Cauchy problem (6.1.1) *and* (6.1.3) *of class $\mathcal{M}_\nu$, where $\nu = \varrho_0\,dx$, moreover, $\varrho \in C^{2,1}(\mathbb{R}^d_T) \cap C\bigl(\mathbb{R}^d \times [0,T)\bigr)$.*

PROOF. The main assertion in this theorem is a special case of the classical results (see Friedman [**386**, Theorems 11 and 12, Chapter I, Theorem 11, Chapter II]). We only explain why the measure μ with density ϱ satisfies condition (6.6.1), even with the exact equality. It is readily seen that $\varrho \in L^1(\mathbb{R}^d_T)$. Multiplying our equation (6.1.1) by the function $\psi \in C_0^\infty(\mathbb{R}^d)$ and integrating by parts, we obtain the equality

$$\int_{\mathbb{R}^d} \psi(x)\varrho(x,t)\,dx = \int_{\mathbb{R}^d} \psi(x)\varrho_0(x)\,dx + \int_0^t\!\int_{\mathbb{R}^d} \varrho L_{A,b,c}\psi\,dx\,ds.$$

Let $\zeta \in C_0^\infty(\mathbb{R}^d)$ be such that $\zeta(x) = 1$ if $|x| \leqslant 1$, $\zeta(x) = 0$ if $|x| > 2$, $|\zeta| \leqslant 1$ and ζ has two bounded derivatives. Let us substitute $\psi(x) = \zeta(x/N)$ and let N increase to infinity. Taking into account that the functions a^{ij}, b^i, c are globally bounded, we obtain the equality

$$\int_{\mathbb{R}^d} \varrho(x,t)\,dx = \int_{\mathbb{R}^d} \varrho_0(x)\,dx + \int_0^t\!\int_{\mathbb{R}^d} c(x,s)\varrho(x,s)\,dx\,ds,$$

which means that condition (6.6.1) holds for μ. □

In the following theorem existence is proved just under the assumption of local boundedness of coefficients.

6.6.2. Theorem. *Suppose that $c \leqslant 0$, for every ball U in $\mathbb{R}^d$ the functions a^{ij}, b^i and c are bounded on $U \times [0,T]$ and that there exist positive numbers $m(U)$ and $M(U)$ such that*

$$m(U) \cdot \mathrm{I} \leqslant A(x,t) \leqslant M(U) \cdot \mathrm{I} \quad \forall\,(x,t) \in U \times [0,T].$$

Then, for every probability measure ν, the set $\mathcal{M}_\nu$ is not empty.

PROOF. The main idea is this. We construct a sequence of solutions to the Cauchy problems with smooth coefficients and smooth initial conditions, approximating the data of the original problem. Using Corollary 6.3.2 we show that the constructed sequence is uniformly bounded in $L^{(d+2)'}$ on balls, hence it contains a weakly convergent subsequence. Finally, we prove that this subsequence converges to a solution of class $\mathcal{M}_\nu$. We split the proof in several steps.

1. Set $a^{ij}(x,t) = \delta^{ij}$, $b^i(x,t) = 0$ and $c(x,t) = 0$ for all points $t \notin [0,T]$. Let ω be an averaging kernel, i.e.,

$$\omega \in C_0^\infty(\mathbb{R}^{d+1}), \quad \omega \geqslant 0, \quad \int_{\mathbb{R}^{d+1}} \omega(x,t)\,dx\,dt = 1,$$

where $\omega(x,t) = 0$ if $|x| > 1$. Set

$$\omega_\varepsilon(x,t) = \varepsilon^{-d-1}\omega(x\varepsilon^{-1}, t\varepsilon^{-1}).$$

Let $a_n^{ij} = a^{ij} * \omega_{1/n} + n^{-1}\delta^{ij}$, $b_n^i = b^i * \omega_{1/n}$ and $c_n = c * \omega_{1/n}$. It is clear that for any fixed n the smooth functions a_n^{ij}, b_n^i and c_n are uniformly bounded along with all derivatives. In addition, we have $A_n(x,t) \geqslant n^{-1} \cdot \mathrm{I}$ for all $(x,t) \in \mathbb{R}^d_T$. Let U_n be the ball of radius n centered at the origin in $\mathbb{R}^d$ and let $U_{n,T}$ be the cylinder $U_n \times [0,T]$. For every $k \in \mathbb{N}$ and $p \geqslant 1$, the sequences of functions a_n^{ij}, b_n^i, c_n converge in $L^p(U_k \times [0,T])$ to the functions a^{ij}, b^i and c, respectively.

Let $\eta_n \in C_0^\infty(\mathbb{R}^d)$ be a sequence of nonnegative functions such that the measures $\nu_n = \eta_n\,dx$ are probabilities and converge weakly to the measure ν. Let us consider the following Cauchy problem on $\mathbb{R}^d \times [0,T)$:

$$\partial_t u_n = \partial_{x_i}\partial_{x_j}(a_n^{ij} u_n) - \partial_{x_i}(b_n^i u_n) + c_n u_n, \quad u_n|_{t=0} = \eta_n.$$

By Theorem 6.6.1 this Cauchy problem has a solution u_n of class $\mathcal{M}_{\nu_n}$.

2. We choose a convergent subsequence in $\{u_n\}$. We observe that

$$A_n(x,t) \geqslant m_{k+1} \cdot \mathrm{I} \quad \text{for all } (x,t) \in U_k \times [0,T],$$

where $m_{k+1} = m(U_{k+1})$ is the number from the hypotheses of the theorem corresponding to the set U_{k+1}. Moreover, for any fixed k and sufficiently large n we have

$$\|a_n^{ij}\|_{L^\infty(U_{k,T})} \leqslant \|a\|_{L^\infty(U_{k+1,T})} + 1,$$
$$\|b_n^i\|_{L^\infty(U_{k,T})} \leqslant \|b^i\|_{L^\infty(U_{k+1,T})}, \|c_n\|_{L^\infty(U_{k,T})} \leqslant \|c\|_{L^\infty(U_{k+1,T})}.$$

Recall that the measures $u_n(x,t)\,dx$ are subprobability measures, which follows immediately from condition (6.6.1). Then, by Corollary 6.3.2, for every $k > 2$ we have the estimate

$$\int_{U_k \times [Tk^{-1}, T(1-k^{-1})]} u_n^{(d+1)/d}\,dx\,dt \leqslant C_k,$$

where C_k depends on m_{k+1}, $\|a\|_{L^\infty(U_{k+1,T})}$, $\|b^i\|_{L^\infty(U_{k+1,T})}$, $\|c\|_{L^\infty(U_{k+1,T})}$ and does not depend on n.

Since the unit ball in the reflexive space $L^{(d+1)/d}$ is weakly compact, for every number $k > 2$ the sequence $\{u_n\}$ contains a subsequence weakly converging in the space $L^{(d+1)/d}(U_k \times [Tk^{-1}, T(1-k^{-1})])$. Applying the diagonal procedure, without loss of generality we can assume that $\{u_n\}$ converges weakly to a nonnegative function u belonging to $L^{(d+1)/d}(U_k \times [Tk^{-1}, T(1-k^{-1})])$ for every k.

3. As we have noted in the previous step, for every k the coefficients a_n^{ij}, b_n^i and c_n are bounded on $U_k \times [0,T]$ uniformly in n. Let $\psi \in C_0^\infty(\mathbb{R}^d)$. Then the support

of ψ belongs to U_k for some k and there exists a number $C(\psi)$ independent of n such that

$$\left|\int_{\mathbb{R}^d} \psi(x)u_n(x,t)\,dx - \int_{\mathbb{R}^d} \psi(x)u_n(x,s)\,dx\right| = \left|\int_s^t\int_{\mathbb{R}^d} L_{A_n,b_n,c_n}\psi(x,\tau)\,u_n(x,\tau)dx\,d\tau\right| \leqslant C(\psi)|t-s|$$

for all n and $s,t\in[0,T]$. Therefore, the functions

$$f_n(t) := \int_{\mathbb{R}^d} \psi(x)u_n(x,t)\,dx$$

are Lipschitzian with the constant $C(\psi)$, which does not depend on n. Thus, for every function ψ this is a uniformly bounded and equicontinuous family of functions. Hence by the Ascoli–Arzelà theorem every sequence in it contains a subsequence uniformly converging on $[0,T]$. We observe that in the space $L^{(d+1)/d}[0,T]$ this sequence converges to

$$f(t) := \int_{\mathbb{R}^d} \psi(x)u(x,t)\,dx,$$

and since the weak and uniform limits coincide almost everywhere, we conclude that every subsequence in $\{f_n\}$, so the whole sequence, converges uniformly to one and the same Lipschitzian function $\widetilde{f}$ that coincides with f on a set of full measure in $[0,T]$. Certainly, this set depends on our choice of ψ. We denote it by $\mathbb{T}(\psi)$.

4. We now show that the function u constructed in Step 2 is a solution to the Cauchy problem (6.1.1) and (6.1.3). Fix a function $\psi\in C_0^\infty(\mathbb{R}^d)$ with support in some U_k. By hypothesis we have $L_{A,b,c}\psi\in L^\infty(U_k\times[0,T])$. There is a constant C_k such that

$$\sup_n \|L_{A_n,b_n,c_n}\psi\|_{L^\infty(U_k\times[0,T])} \leqslant C_k.$$

In addition, $L_{A_n,b_n,c_n}\psi\to L_{A,b,c}\psi$ in $L^p(U_k\times[0,T])$ for every $p\geqslant 1$. Let us take t in the full measure set $\mathbb{T}(\psi)$ on which we have convergence

$$\int_{\mathbb{R}^d} \psi(x)u_n(x,t)\,dx \to \int_{\mathbb{R}^d} \psi(x)u(x,t)\,dx.$$

Let $0<s<t$. The following relations hold:

$$\left|\int_{\mathbb{R}^d} \psi(x)u_n(x,t)\,dx - \int_{\mathbb{R}^d} \psi(x)\eta_n(x)\,dx - \int_s^t\int_{\mathbb{R}^d} u_n L_{A_n,b_n,c_n}\psi\,dx\,d\tau\right| = \left|\int_{\mathbb{R}^d} \psi(x)u_n(x,s)\,dx - \int_{\mathbb{R}^d} \psi(x)\eta_n(x)\,dx\right| \leqslant C(\psi)s, \tag{6.6.2}$$

where $C(\psi)$ does not depend on n and s. We observe that

$$\lim_{n\to\infty}\int_s^t\int_{\mathbb{R}^d} u_n L_{A_n,b_n,c_n}\psi\,dx\,d\tau = \int_s^t\int_{\mathbb{R}^d} uL_{A,b,c}\psi\,dx\,d\tau.$$

Indeed,

$$\begin{aligned}\Bigl|\int_s^t\int_{\mathbb{R}^d} u_n L_{A_n,b_n,c_n}\psi\,dx\,d\tau - \int_s^t\int_{\mathbb{R}^d} uL_{A,b,c}\psi\,dx\,d\tau\Bigr| \\ \leqslant |L_{A_n,b_n,c_n}\psi - L_{A,b,c}\psi\|_{L^{d+1}(U_k\times[s,t])}\|u_n\|_{L^{(d+1)/d}(U_k\times[s,t])} \\ + \Bigl|\int_s^t\int_{\mathbb{R}^d} u_n L_{A,b,c}\psi\,dx\,d\tau - \int_s^t\int_{\mathbb{R}^d} uL_{A,b,c}\psi\,dx\,d\tau\Bigr|,\end{aligned}$$

where the first term on the right tends to zero by virtue of convergence of $L_{A_n,b_n,c_n}\psi$ to $L_{A,b,c}\psi$ and the uniform boundedness of the norms of u_n which was proved at Step 2, and the second term converges to zero by weak convergence of $\{u_n\}$ and the boundedness of $L_{A,b,c}\psi$. Thus, letting $n\to\infty$ in (6.6.2), we obtain

$$\Bigl|\int_{\mathbb{R}^d}\psi(x)u(x,t)\,dx - \int_{\mathbb{R}^d}\psi(x)\,d\nu - \int_s^t\int_{\mathbb{R}^d} uL_{A,b,c}\psi\,dx\,d\tau\Bigr| \leqslant C(\psi)s.$$

As $s\to 0$ we obtain

$$\int_{\mathbb{R}^d}\psi(x)u(x,t)\,dx = \int_{\mathbb{R}^d}\psi(x)\,\nu(dx) + \int_0^t\int_{\mathbb{R}^d} uL_{A,b,c}\psi\,dx\,d\tau.$$

Therefore, the function u is a nonnegative solution to the Cauchy problem (6.1.1) and (6.1.3).

5. We now show that the measure $u(x,t)dx\,dt$ is a solution in $\mathcal{M}_\nu$. Recall that $c_n\leqslant 0$ and $\eta_n\,dx$ is a probability measure. Since we have $u_n\in\mathcal{M}_\nu$, every function $\psi\in C_0^\infty(\mathbb{R}^d)$ with $0\leqslant\psi\leqslant 1$ satisfies the inequality

$$\int_{\mathbb{R}^d}\psi(x)u_n(x,t)\,dx - \int_0^t\int_{\mathbb{R}^d}\psi(x)c_n(x,s)u_n(x,s)\,dx\,ds \leqslant 1. \tag{6.6.3}$$

Let $\psi_N\in C_0^\infty(\mathbb{R}^d)$, $0\leqslant\psi_N\leqslant 1$ and $\psi_N(x)=1$ if $x\in U_N$. Let also t belong to the full measure set $\mathbb{T}=\bigcap_{N\in\mathbb{N}}\mathbb{T}(\psi_N)$, i.e., for all $N\in\mathbb{N}$ one has convergence

$$\int_{\mathbb{R}^d}\psi_N(x)u_n(x,t)\,dx \to \int_{\mathbb{R}^d}\psi_N(x)u(x,t)\,dx.$$

Substituting such ψ_N and t in inequality (6.6.3) and letting $n\to\infty$, we obtain the estimate

$$\int_{\mathbb{R}^d}\psi_N(x)u(x,t)\,dx - \int_0^t\int_{\mathbb{R}^d}\psi_N(x)c(x,s)u(x,s)\,dx\,ds \leqslant 1.$$

Finally, letting $N\to\infty$, we arrive at the inequality

$$\int_{\mathbb{R}^d} u(x,t)\,dx - \int_0^t\int_{\mathbb{R}^d} c(x,s)u(x,s)\,dx\,ds \leqslant 1 = \nu(\mathbb{R}^d)$$

by Fatou's theorem. □

6.6.3. Remark. According to Corollary 6.3.2, under the hypotheses of Theorem 6.6.2 any solution μ in $\mathcal{M}_\nu$ is given by a density ϱ with respect to Lebesgue measure, and $\varrho\in L^{(d+1)'}_{\mathrm{loc}}(\mathbb{R}^d_T)$.

The condition of local boundedness can be weakened under the assumption that the elements of the matrix A belong to the Sobolev class.

6.6.4. Theorem. *Let $p > d+2$. Suppose that $c \leqslant 0$ and for every ball $U \subset \mathbb{R}^d$ we have the following condition: $a^{ij}(\,\cdot\,,t) \in W^{p,1}(U)$,*

$$\sup_{t\in(0,T)} \|a^{ij}(\,\cdot\,,t)\|_{W^{1,p}(U)} < \infty, \quad A(x,t) \geqslant m(U)\cdot \mathrm{I}$$

for all points $(x,t) \in U\times[0,T]$ and some numbers $m(U) > 0$. Assume also that $b \in L^p(U\times[0,T])$, $c \in L^{p/2}(U\times[0,T])$. Then, for every probability measure ν, the set $\mathcal{M}_\nu$ is not empty.

PROOF. 1. As at the first step of the proof of the previous theorem, we find sequences of smooth bounded functions a_n^{ij}, b_n^i and c_n for which for every ball $U \subset \mathbb{R}^d$ we have the relations

$$\lim_{n\to\infty} \|a_n^{ij} - a^{ij}\|_{L^p(U\times[0,T])} = 0, \quad \lim_{n\to\infty} \|b_n^i - b^i\|_{L^p(U\times[0,T])} = 0,$$

$$\lim_{n\to\infty} \|c_n - c\|_{L^{p/2}(U\times[0,T])} = 0,$$

in particular, the norms $\|a_n^{ij}\|_{L^p(U\times[0,T])}$, $\|b^i\|_{L^p(U\times[0,T])}$, $\|c_n\|_{L^{p/2}(U\times[0,T])}$ are bounded uniformly in n. In addition, $c_n \leqslant 0$ and $A_n(x,t) \geqslant n^{-1}\cdot \mathrm{I}$ for all $(x,t) \in \Omega_T$. Let $\eta_n \in C_0^\infty(\mathbb{R}^d)$ be nonnegative functions such that the measures $\eta_n\,dx$ are probabilities and converge weakly to the measure ν. Let u_n be the smooth bounded solution of the Cauchy problem

$$\partial_t u_n = \partial_{x_i}\partial_{x_j}(a_n^{ij}u_n) - \partial_{x_i}(b_n^i u_n) + c_n u_n, \quad u_n|_{t=0} = \eta_n,$$

where $u_n\,dx\,dt \in \mathcal{M}_\nu$ (the solution exists by Theorem 6.6.2).

2. Choose in $\{u_n\}$ a convergent subsequence in the following way. Let U_k be the ball of radius k centered at the origin. According to Corollary 6.4.3, for every $k > 2$ one has the estimate on the Hölder norm

$$\|u_n\|_{C^\alpha(U_k\times[Tk^{-1},T(1-k^{-1})])} \leqslant C_k,$$

where $\alpha \in (0,1)$ and the number C_k do not depend on n. Applying the Ascoli–Arzelà theorem, the diagonal procedure and passing to a subsequence, we can assume that $\{u_n\}$ converges uniformly to some function u on $U_k\times[Tk^{-1},T(1-k^{-1})]$ for every k. Clearly, u is a nonnegative continuous function. We show that u satisfies (6.1.1). Let $\psi \in C_0^\infty(\mathbb{R}^d)$. Then the support of ψ belongs to U_k for some k. The uniform convergence of $\{u_n\}$ yields the equality

$$\lim_{n\to\infty} \int_{\mathbb{R}^d} \psi(x)u_n(x,t)\,dx = \int_{\mathbb{R}^d} \psi(x)u(x,t)\,dx$$

for all $t \in (0,T)$. Let now $0 < s < t < T$. We observe that

$$\begin{aligned} \biggl|\int_s^t\!\!\int_{\mathbb{R}^d} L_{A_n,b_n,c_n}\psi u_n\,dx\,d\tau - \int_s^t\!\!\int_{\mathbb{R}^d} L_{A,b,c}\psi u\,dx\,d\tau\biggr| \\ \leqslant \|L_{A_n,b_n,c_n}\psi - L_{A,b,c}\psi\|_{L^1(U_k\times[s,t])}\|u_n\|_{L^\infty(U_k\times[s,t])} \\ + \|L_{A,b,c}\psi\|_{L^1(U_k\times[s,t])}\|u - u_n\|_{L^\infty(U_k\times[s,t])}, \end{aligned}$$

where the first term in the right side tends to zero by convergence of the coefficients a_n^{ij}, b_n^i and c_n to a^{ij}, b^i and c, respectively, and the second term tends to zero by the uniform convergence of $\{u_n\}$ to the function u. Therefore, one has the equality

$$\lim_{n\to\infty} \int_s^t\!\!\int_{\mathbb{R}^d} L_{A_n,b_n,c_n}\psi u_n\,dx\,d\tau = \int_s^t\!\!\int_{\mathbb{R}^d} L_{A,b,c}\psi u\,dx\,d\tau.$$

Thus, letting $n \to \infty$, for all $s, t \in (0, T)$ we obtain

$$\int_{\mathbb{R}^d} \psi(x)u(x,t)\,dx = \int_{\mathbb{R}^d} \psi(x)u(x,s)\,dx + \int_s^t \int_{\mathbb{R}^d} L_{A,b,c}\psi u\,dx\,d\tau.$$

3. Let us justify passage to the limit as $s \to 0$.

Let $0 < \tau < T$ and $y \in C_0^\infty(\mathbb{R}^d)$. Let also $w_{n,\tau}$ be the solution of the adjoint problem

$$\partial_t w_{n,\tau} + a_n^{ij}\partial_{x_i}\partial_{x_j} w_{n,\tau} + b_n^i \partial_{x_i} w_{n,\tau} + c_n w_{n,\tau} = 0, \quad w_{n,\tau}|_{t=\tau} = y.$$

Let $\zeta \in C_0^\infty(\mathbb{R}^d)$ be such that $\zeta(x) = 1$ if $|x| \leqslant 1$, $\zeta(x) = 0$ if $|x| > 2$, $|\zeta| \leqslant 1$ and ζ has two bounded derivatives. Multiplying the adjoint equation by $\zeta_N u_n$ and integrating by parts, and then letting $N \to \infty$, we obtain the equality

$$\int_{\mathbb{R}^d} y(x)u_n(x,\tau)\,dx = \int_{\mathbb{R}^d} w_{n,\tau}(x,0)\eta_n(x)\,dx.$$

Note that, according to Ladyz'enskaya, Solonnikov, Ural'tseva [**576**, Chapter III, Theorem 10.1, p. 238], for every ball $U \subset \mathbb{R}^d$ one has the estimate

$$\|w_{n,\tau}(x,0) - y(x)\|_{L^\infty(U)} \leqslant C(U)\tau^\alpha,$$

where C and α do not depend on n. Thus,

$$\begin{aligned}
&\left|\int_{\mathbb{R}^d} y(x)u_n(x,\tau)\,dx - \int_{\mathbb{R}^d} y\,d\nu\right| \\
&\qquad \leqslant \int_{\mathbb{R}^d} |w_{n,\tau}(x,0) - y(x)|\eta_n(x)\,dx + \left|\int_{\mathbb{R}^d} y\eta_n\,dx - \int_{\mathbb{R}^d} y\,d\nu\right| \\
&\qquad\qquad \leqslant C\tau^\alpha + \left|\int_{\mathbb{R}^d} y\eta_n\,dx - \int_{\mathbb{R}^d} y\,d\nu\right|.
\end{aligned}$$

Letting $n \to \infty$, we obtain the inequality

$$\left|\int_{\mathbb{R}^d} y(x)u(x,\tau)\,dx - \int_{\mathbb{R}^d} y\,d\nu\right| \leqslant C\tau^\alpha,$$

which yields the required equality

$$\lim_{\tau \to 0} \int_{\mathbb{R}^d} y(x)u(x,\tau)\,dx = \int_{\mathbb{R}^d} y\,d\nu.$$

Finally, the inequality

$$\int_{\mathbb{R}^d} u(x,t)\,dx - \int_0^t \int_{\mathbb{R}^d} c(x,s)u(x,s)\,dx\,ds \leqslant 1 = \nu(\mathbb{R}^d)$$

is obtained in the same manner as in the proof of Theorem 6.6.2. □

6.6.5. Remark. Analogs of the last two theorems obviously remain true for signed bounded measures ν. By Corollary 6.4.3, in Theorem 6.6.4 any $\mu \in \mathcal{M}_\nu$ is given by a locally Hölder continuous density with respect to Lebesgue measure.

6.6.6. Corollary. *If under the hypotheses of Theorem* 6.6.2 *and Theorem* 6.6.4 *it is known additionally that there exists a function $V \in C^2(\mathbb{R}^d_T)$ such that*

$$\lim_{|x| \to +\infty} V(x) = +\infty, \quad L_{A,b,c}V(x,t) \leqslant C + CV(x)$$

for all $(x,t) \in \mathbb{R}^d_T$ *and some* $C > 0$, *then the set* $\mathcal{M}_\nu$ *not only is nonempty for any subprobability measure* ν, *but, in addition, for every solution* $\mu \in \mathcal{M}_\nu$ *the equality*

$$\mu_t(\mathbb{R}^d) = \nu(\mathbb{R}^d) + \int_0^t \int_{\mathbb{R}^d} c(x,s)\mu_s(dx)\,ds$$

is true for almost all $t \in (0,T)$. *In particular, for* $c = 0$ *and a probability measure* ν, *the measures* μ_t *are probability measures for almost all* $t \in (0,T)$.

PROOF. This assertion reduces to probability initial measures ν and follows from the a priori estimates with Lyapunov functions discussed in detail in § 7.1. □

6.7. Complements, comments, and exercises

(i) Gradient flows (272). (ii) Solvability of degenerate Fokker–Planck–Kolmogorov equations (275). (iii) Nonlinear parabolic equations for measures (278). Comments (283). Exercises (285).

6.7(i). Gradient flows

A gradient flow in $\mathbb{R}^d$ is the solution of the equation

$$\dot{x}(t) = -\nabla\Phi\big(x(t)\big), \quad x(0) = x_0.$$

Fix $\tau > 0$. Let x_n be a point of minimum of the function

$$x \mapsto \Phi(x) + (2\tau)^{-1}|x_{n-1} - x|^2,$$

where for $n = 0$ we take the initial value x_0. Set $x_\tau(t) = x_n$ if $t \in \big[n\tau, (n+1)\tau\big)$. We show that the functions x_τ converge to a solution x as $\tau \to 0$. Indeed, at the point of minimum x_n the gradient of the function is zero:

$$(x_n - x_{n-1})/\tau = -\nabla\Phi(x_n).$$

If we join the points x_n by segments, we obtain Euler's approximations. It is known that for smooth functions Φ Euler's approximations converge to a solution. The constructed piece-wise constant curve x_τ differs from the corresponding polygon line by a quantity of magnitude τ.

This simple scheme extends to metric spaces, in particular, to $\mathcal{P}_2(\mathbb{R}^d)$, the space of probability measures μ on $\mathbb{R}^d$ with finite second moment, i.e., $|x|^2 \in L^1(\mu)$. The space $\mathcal{P}_2(\mathbb{R}^d)$ is a complete metric space with respect to the Kantorovich metric

$$W_2(\mu,\nu) = \inf\biggl(\int_{\mathbb{R}^d\times\mathbb{R}^d} |x-y|^2\,d\pi\biggr)^{1/2},$$

where inf is taken over all probability measures π on $\mathbb{R}^d\times\mathbb{R}^d$ with projections μ and ν on the factors. If the measure μ is sufficiently regular, then the same metric can be characterized in a different way, namely: there exists a mapping $T\colon \mathbb{R}^d \mapsto \mathbb{R}^d$ taking the measure μ to the measure ν such that

$$W_2(\mu,\nu) = \biggl(\int_{\mathbb{R}^d} |T(x) - x|^2\,d\mu\biggr)^{1/2}.$$

The mapping T is called optimal for the measures μ and ν (here inf is attained at a measure π whose support is the graph of T). It is known that it is the gradient of a convex function.

We shall say that a curve $(\mu_t)_{t\in[\alpha,\beta]}$ in the space of measures $\mathcal{P}_2(\mathbb{R}^d)$ belongs to the class $AC^2\big([\alpha,\beta],\mathcal{P}_2(\mathbb{R}^d)\big)$ if there exists a function $m\in L^2[\alpha,\beta]$ such that

$$W_2(\mu_t,\mu_s)\leqslant\int_s^t m(\tau)\,d\tau\quad\text{whenever } \alpha<s\leqslant t<\beta.$$

Note that for, each curve $(\mu_t)_{t\in[\alpha,\beta]}\in AC^2\big([\alpha,\beta],\mathcal{P}_2(\mathbb{R}^d)\big)$, for almost all $t\in[\alpha,\beta]$ we can define the metric derivative

$$|\mu'|_t:=\lim_{s\to t}\frac{W_2(\mu_t,\mu_s)}{|t-s|},$$

where $|\mu'|_t\in L^2\big([\alpha,\beta]\big)$. We have $|\mu'|_t\leqslant m(t)$ a.e., and in place of the function m one can substitute $|\mu'|_t$.

It turns out (see Ambrosio, Gigli, Savaré [**45**, Theorem 8.3.1]), that for, every curve $(\mu_t)_{t\in[\alpha,\beta]}\in AC^2\big([\alpha,\beta],\mathcal{P}_2(\mathbb{R}^d)\big)$, there exists a vector field $v=(v_t)$, where $v_t\in L^2(\mu_t)$, such that

$$\partial_t\mu_t=\operatorname{div}(v_t\mu_t)$$

in the sense of generalized functions. Moreover, for almost all $t\in[\alpha,\beta]$, the vector v_t belongs to the closure of the set $\{\nabla\varphi\colon\ \varphi\in C_0^\infty(\mathbb{R}^d)\}$ in the $L^2(\mu_t)$-space of $\mathbb{R}^d$-valued functions and $\|v_t\|_{L^2(\mu_t)}=|\mu'|_t$.

Such a vector v_t is called a tangent vector to the curve (μ_t) at the time t. The closure of $\{\nabla\varphi\colon\ \varphi\in C_0^\infty(\mathbb{R}^d)\}$ in the space $L^2(\mu)$ is called the tangent spaces and denoted by $\operatorname{Tan}_\mu\mathcal{P}_2(\mathbb{R}^d)$.

Let $w_t\in L^2(\mu_t)$ and $\partial_t\mu_t=\operatorname{div}(w_t\mu_t)$. The vector v_t-w_t is orthogonal to the subspace $\operatorname{Tan}_{\mu_t}\mathcal{P}_2(\mathbb{R}^d)$, in particular, to the vector v_t itself. Therefore,

$$\|w_t\|^2_{L^2(\mu_t)}=\|v_t-w_t\|^2_{L^2(\mu_t)}+\|v_t\|^2_{L^2(\mu_t)}\geqslant\|v_t\|^2_{L^2(\mu_t)}.$$

Thus, the vectors v_t are the minimal selections in the norm of $L^2(\mu_t)$ among all vectors for which the continuity equation holds on the curve (μ_t). Finally, the vector field (v_t) is the only vector field for which the continuity equation holds on the curve (μ_t) and $\|v_t\|_{L^2(\mu_t)}=|\mu'|_t$.

A curve $(\mu_t)_{t\in[0,1]}$ is called a geodesic of constant velocity if

$$W_2(\mu_t,\mu_s)=|t-s|W_2(\mu_0,\mu_1).$$

For such a geodesic we can take the curve $(\mu_t)_{t\in[0,1]}$ defined as follows. Suppose that $\mu\in\mathcal{P}_2(\mathbb{R}^d\times\mathbb{R}^d)$ is the optimal plan for μ_1 and μ_0, i.e.,

$$W_2^2(\mu_0,\mu_1)=\int_{\mathbb{R}^d\times\mathbb{R}^d}|x-y|^2\,d\mu.$$

Let $Pr_x,Pr_y\colon\ \mathbb{R}^d\times\mathbb{R}^d\mapsto\mathbb{R}^d$ be projections: $Pr_x(x,y)=x$, $Pr_y(x,y)=y$. Set $Pr_t=tPr_x+(1-t)Pr_y$ and define our curve by $\mu_t=\mu\circ Pr_t^{-1}$.

A generalized geodesic joining μ_0 and μ_1 with base μ_2 is defined as the curve of measures $\mu_t=\mu\circ Pr_t^{-1}$, where $Pr_t=tPr_x+(1-t)Pr_y$ and μ is a probability measure on $\mathbb{R}^d_z\times\mathbb{R}^d_x\times\mathbb{R}^d_y$ such that its projection on $\mathbb{R}^d_z\times\mathbb{R}^d_x$ is the optimal plan for $W_2(\mu_2,\mu_0)$ and its projection on $\mathbb{R}^d_z\times\mathbb{R}^d_y$ is the optimal plan for $W_2(\mu_2,\mu_1)$.

Necessity in generalized geodesics is due to failure of 1-convexity for the mapping $\mu\mapsto W_2^2(\mu,\sigma)$, which becomes essential when working with gradient flows.

We now define convex functionals on $\mathcal{P}_2(\mathbb{R}^d)$. Recall that a function f on $\mathbb{R}^d$ is called λ-convex if the real function $f(x)-\lambda|x|^2/2$ is convex. Note that λ can assume any value, including zero (which corresponds to the usual convexity).

A functional $\Phi\colon \mathcal{P}_2(\mathbb{R}^d) \to (-\infty, +\infty]$ is called λ-convex (along generalized geodesics) if, for all μ_0, μ_1, μ_2 in the domain of definition of Φ, there is a a generalized geodesic μ_t joining μ_0 and μ_1 with base μ_2 and generated by the measure μ (i.e., $\mu_t = \mu\circ Pr_t^{-1}$) such that

$$\Phi(\mu_t) \leqslant t\Phi(\mu_0) + (1-t)\Phi(\mu_1) - \frac{\lambda}{2}t(1-t)\int_{\mathbb{R}^d_z\times\mathbb{R}^d_x\times\mathbb{R}^d_y} |x-y|^2\,d\mu$$

for all $t \in [0,1]$.

An important example of such a functional is

$$\Phi(\mu) = \int_{\mathbb{R}^d} F\bigl(\varrho(x)\bigr)\,dx + \int_{\mathbb{R}^d} V(x)\varrho(x)\,dx$$

for measures of the form $\mu = \varrho\,dx$ and $\Phi(\mu) = +\infty$ otherwise. Here V is some λ-convex function and the function $t \mapsto t^d F(t^{-d})$ is convex and increasing. For example, we can take for F the function $F(t) = t\ln t$ or $F(t) = (m-1)^{-1}t^m$ if $m \geqslant (d-1)/d$.

Let $\mathcal{D}(\Phi)$ denote the set of measures μ for which $\Phi(\mu) < \infty$. For $\mu \in \mathcal{D}(\Phi)$ we define the absolute value of the metric gradient by the formula

$$|\partial\Phi|(\mu) = \limsup_{\sigma\to\mu} \frac{\bigl(\Phi(\mu) - \Phi(\sigma)\bigr)_+}{W_2(\mu,\sigma)}.$$

If the measure μ is sufficiently regular (for example, has a density), then for every measure ν there exists an optimal mapping T of the measure μ to the measure ν. Let $\mu \in \mathcal{D}(|\partial\Phi|)$. We say that a vector field $\xi \in L^2(\mu)$ belongs to the subdifferential $\partial\Phi(\mu)$ if

$$\Phi(\nu) - \Phi(\mu) \geqslant \int_{\mathbb{R}^d} \langle \xi(x), T(x) - x\rangle\,\mu(dx) + o\bigl(W_2(\mu,\nu)\bigr).$$

The next theorem (see Ambrosio, Gigli, Savaré [**45**, Theorem 10.4.13]) gives an example of evaluation of a subdifferential.

6.7.1. Theorem. *Consider on $\mathcal{P}_2(\mathbb{R}^d)$ the functional*

$$\Phi(\mu) = \int_{\mathbb{R}^d} \varrho\ln\varrho\,dx + \int_{\mathbb{R}^d} V\varrho\,dx$$

if $\mu = \varrho\,dx$ and $\Phi(\mu) = \infty$ if μ has no density. Suppose that V is a lower semicontinuous λ-convex function and the set $\mathcal{D}(V)$ is not empty. Then the measure $\mu = \varrho\,dx$ belongs to $\mathcal{D}(|\partial\Phi|)$ precisely when $\varrho \in W^{1,1}_{\mathrm{loc}}(\mathbb{R}^d)$ and

$$\varrho\xi = \nabla\varrho + \varrho\nabla V$$

for some $\xi \in L^2(\mu)$. The vector ξ is the minimal selection in $\partial\Phi(\mu)$, i.e., has the minimal $L^2(\mu)$-norm among vectors in $\partial\Phi(\mu)$.

A curve $(\mu_t) \in AC^2\bigl([0,+\infty), \mathcal{P}_2(\mathbb{R}^d)\bigr)$ is called the gradient flow for the functional Φ if its tangent vector v_t belongs to the set $\partial\Phi(\mu_t)$ for each t. Note that it is proved in [**45**] that if a tangent vector v_t belongs to $\partial\Phi(\mu_t)$, then it is the minimal selection in $\partial\Phi(\mu_t)$.

Let us now consider the functional Φ from Theorem 6.7.1. Let a curve (μ_t), where $\mu_t = \varrho(x,t)\,dx$, be the gradient flow for Φ. By definition, there exists a vector v_t tangent to the curve (μ_t) for each t, i.e., $\partial_t\mu_t = \operatorname{div}(v_t\mu_t)$, and giving the minimal selection in $\partial\Phi(\mu_t)$. By Theorem 6.7.1 we obtain the equality

$$v_t(x) = \varrho(x,t)^{-1}\nabla\varrho(x,t) + \nabla V(x).$$

Therefore, we have

$$\partial_t\varrho = \Delta\varrho + \operatorname{div}(\varrho\nabla V).$$

Thus, the Fokker–Planck–Kolmogorov equation becomes an equation on the gradient flow for the functional Φ.

For constructing gradient flows the following method is used. Let $\tau > 0$ and let us set $M_\tau^0 := \mu_0$. Let M_τ^n be a point of minimum of the functional

$$\Phi(\mu) + \frac{1}{2\tau}W_2^2(\mu, M_\tau^{n-1}).$$

Set $\mu_t^\tau = M_n^\tau$ if $t \in [n\tau, (n+1)\tau)$.

The next theorem is a corollary of Theorem 11.2.1 in [**45**].

6.7.2. Theorem. *If Φ is a λ-convex functional, then the discrete broken lines (μ_t^τ) converge locally uniformly in the metric W_2 to a unique gradient flow for the functional Φ.*

For the functional Φ in Theorem 6.7.1, convergence of discrete broken lines to a solution of the Fokker–Planck–Kolmogorov equation was justified in the paper Otto, Jordan, Kinderlehrer [**495**] (see also Peletier, Renger, Veneroni [**770**]).

Let us consider a more general functional

$$\Phi(\mu) = \int_{\mathbb{R}^d} \varrho \ln \varrho\, dx + \int_{\mathbb{R}^d} V\varrho\, dx + \frac{1}{2}\int_{\mathbb{R}^d}\int_{\mathbb{R}^d} W(x-y)\varrho(x)\varrho(y)\, dx\, dy$$

if $\mu = \varrho\, dx$ and $\Phi(\mu) = \infty$ if μ has no density. The gradient flow (μ_t) corresponding to this functional satisfies the nonlinear Fokker–Planck–Kolmogorov equation

$$\partial_t\mu_t = \Delta\mu_t - \operatorname{div}\bigl(b(x,\mu_t)\mu_t\bigr),$$

where $b(x,\mu_t) = -\nabla V(x) - \displaystyle\int \nabla W(x-y)\,\mu_t(dy)$. A detailed presentation of the theory of gradient flows in spaces of probability measures is given in the book Ambrosio, Gigli, Savaré [**45**].

6.7(ii). Solvability of degenerate Fokker–Planck–Kolmogorov equations

We now consider the case of a degenerate diffusion matrix A. Let U_k be the ball of radius k centered at the origin.

6.7.3. Theorem. *Let a^{ij}, b^i, c be continuous in x, measurable in t and bounded in $U_k\times[0,T]$ for every k. Suppose that the matrix A is symmetric and nonnegative definite. Then, for every probability measure ν, the set $\mathcal{M}_\nu$ is nonempty.*

Proof. Let us use the known method of "vanishing viscosity". 1. Let us introduce the operator $L_\varepsilon := \varepsilon\Delta + L_{A,b,c}$ for $\varepsilon > 0$ and consider the Cauchy problem

$$\partial_t\mu_t = L_\varepsilon^*\mu, \quad \mu|_{t=0} = \nu. \tag{6.7.1}$$

It is clear that the hypotheses of Theorem 6.6.4 are fulfilled. Hence for every n the Cauchy problem (6.7.1) with $\varepsilon = 1/n$ has a solution μ^n given by a family of subprobability measures $(\mu_t^n)_{t\in(0,T)}$ and, taking into account that $\nu(\mathbb{R}^d) = 1$,

$$\mu_t^n(\mathbb{R}^d) \leqslant 1 + \int_0^t\int_{\mathbb{R}^d} c(x,s)\mu_s^n(dx)\,ds.$$

2. Next we choose a convergent subsequence of solutions in the following way. There is a subsequence of indices n_l such that the measures $\mu_t^{n_l}$ converge weakly on every compact set $\overline{U_k}$ for every $t\in[0,T]$. For the proof we take a countable dense set $S\in[0,T]$. Applying the Prohorov theorem and the diagonal procedure, one can subtract a subsequence of measures $\mu_t^{n_l}$ weakly converging on all balls for every $t\in S$. We show that the constructed subsequence converges weakly for every $t\in[0,T]$. Let $t\in[0,T]$, $s\in S$ and $\varphi\in C_0^\infty(\mathbb{R}^d)$. By the local boundedness of the coefficients of L we have

$$\biggl|\int_{\mathbb{R}^d}\varphi\,d\mu_t^{n_p} - \int_{\mathbb{R}^d}\varphi\,d\mu_t^{n_k}\biggr| \leqslant 2C(\varphi)\cdot|t-s| + \biggl|\int_{\mathbb{R}^d}\varphi\,d\mu_s^{n_p} - \int_{\mathbb{R}^d}\varphi\,d\mu_s^{n_k}\biggr|.$$

For a given $\varepsilon>0$ one can take s sufficiently close to t and make the first term less than $\varepsilon/2$. Since the sequence of measures $\mu_s^{n_l}$ converges, it is fundamental, so there exists a number N such that for all $p,k>N$ the second term will be less that $\varepsilon/2$. Hence it is shown that for every function $\varphi\in C_0^\infty(\mathbb{R}^d)$ the sequence of its integrals against the measures $\mu_t^{n_l}$ is fundamental. It follows from this that for each t the sequence of measures $\mu_t^{n_l}$ converges weakly to some subprobability measure μ_t on every ball (Exercise 6.7.22). We observe that for every continuous function f the mapping $t\mapsto \int_{U_k} f\,d\mu_t$ is Borel measurable on $[0,T]$ as the limit of measurable functions. Consider the class Φ of bounded Borel functions φ on U_k for which the mapping $t\mapsto\int_{U_k}\varphi\,d\mu_t$ is Borel measurable on $[0,\tau]$. The class Φ contains the algebra of bounded continuous functions on U_k and is closed with respect to uniform and monotone limits. By the monotone class theorem (see Bogachev [**125**, Theorem 2.12.9]) the set Φ contains all bounded Borel functions on U_k. In particular, the function $t\mapsto\mu_t(B)$ is Borel measurable on $[0,T]$ for every Borel set $B\subset U_k$. Since k was arbitrary, this is true for every Borel set in $\mathbb{R}^d$. Let μ be the measure defined by the family $(\mu_t)_{t\in[0,T]}$. Clearly, $\{\mu^{n_l}\}$ converges weakly to μ on $U_k\times[0,T]$ for each k.

3. Passing to the subsequence constructed above we conclude that the sequence of measures $\mu^n = (\mu_t^n)_{t\in(0,t)}$ is such that $\{\mu^n\}$ converges weakly to $\mu=(\mu_t)_{t\in(0,T)}$ and $\{\mu_t^n\}$ converges weakly to μ_t for each t. Fix $\varphi\in C_0^\infty(\mathbb{R}^d)$. For every n we have

$$\int_{\mathbb{R}^d}\varphi\,d\mu_t^n - \int_{\mathbb{R}^d}\varphi\,d\nu = \int_0^t\int_{\mathbb{R}^d} L_{1/n}\varphi\,d\mu_s^n\,ds. \tag{6.7.2}$$

Since

$$\biggl|\int_0^t\int_{\mathbb{R}^d}(L_{1/n}-L)\varphi\,d\mu_s^n\,ds\biggr| \leqslant \frac{T}{n}\sup_{\mathbb{R}^d}|\Delta\varphi|,$$

by weak convergence of μ^n to μ we conclude that

$$\lim_{n\to\infty}\int_0^t\int_{\mathbb{R}^d} L_{1/n}\varphi\,d\mu_s^n\,ds = \int_0^t\int_{\mathbb{R}^d} L\varphi\,d\mu_s\,ds,$$

and also for every t one has the equality

$$\lim_{n\to\infty}\int_{\mathbb{R}^d}\varphi\,d\mu_t^n=\int_{\mathbb{R}^d}\varphi\,d\mu_t.$$

Hence we can let $n\to\infty$ in (6.7.2) and obtain the equality

$$\int_{\mathbb{R}^d}\varphi\,d\mu_t-\int_{\mathbb{R}^d}\varphi\,d\nu=\int_0^t\int_{\mathbb{R}^d}L\varphi\,d\mu_s\,ds.$$

So μ is the desired solution to the Cauchy problem. Finally, the inequality

$$\int_{\mathbb{R}^d}u(x,t)\,dx-\int_0^t\int_{\mathbb{R}^d}c(x,s)u(x,s)\,dx\,ds\leqslant 1=\nu(\mathbb{R}^d)$$

is justified similarly to Step 5 in the proof of Theorem 6.6.2. □

If we assume that the coefficients and the initial value are more regular, then one can construct a solution given by a density even in the case of a degenerate diffusion matrix. Such results are obtained in the papers Figalli [**362**], Le Bris, Lions [**593**]. Let us prove a close result. Uniqueness will be discussed in Chapter 9.

6.7.4. Theorem. *Let $q>1$, $b,c\in L^q(U\times[0,T])$, $a^{ij}(\,\cdot\,,t)\in W^{q,1}(U)$, $\sup_{t\in[0,T]}\|a^{ij}(\,\cdot\,,t)\|_{W^{q,1}(U)}<\infty$ for every ball U. Let $p=q/(q-1)$. Suppose that*

$$\bigl(pc+(p-1)\mathrm{div}h\bigr)^+\in L^1\bigl([0,T],L^\infty(\mathbb{R}^d)\bigr),\quad \textit{where } h^i=\partial_{x_j}a^{ij}-b^i.$$

Let $\varrho_0\in L^p(\mathbb{R}^d)$. Then there exists a solution $\varrho\in L^\infty\bigl([0,T],L^p(\mathbb{R}^d)\bigr)$ of the Cauchy problem $\partial_t\varrho=L^*_{A,b,c}\varrho$, $\varrho|_{t=0}=\varrho_0$.

PROOF. We give a justification of the last assertion in the case of locally smooth coefficients. The general case reduces to this one by approximating the coefficients by smooth functions. Thus, the only difficulty is global unboundedness of the coefficients and degeneracy of the diffusion matrix. Let $p>1$. We can find functions $g_k\in C_0^\infty(\mathbb{R}^d)$ for which $\|\varrho_0-g_k\|_p\to 0$. Let $\psi\in C_0^\infty(\mathbb{R}^d)$, where $\psi(x)=1$ if $|x|<1$. Set $\psi_k(x)=\psi(x/k)$. For every k we consider the Cauchy problem $\partial_t u_k=L_k^*u_k$, $u_k|_{t=0}=g_k$, where

$$\begin{aligned}L_k=(\psi_k a^{ij}+k^{-1}\delta^{ij})\partial_{x_i}\partial_{x_j}+\psi_k b^i\partial_{x_i}\\ +p^{-1}(p-1)\bigl((2\partial_{x_j}a^{ij}-b^i)\partial_{x_i}\psi_k+a^{ij}\partial_{x_i}\partial_{x_j}\psi_k\bigr)+\psi_k c.\end{aligned}$$

There is a bounded smooth solution $u_k\in L^\infty\bigl([0,T],L^1(\mathbb{R}^d)\bigr)$. Then the inequality

$$\begin{aligned}\partial_t|u_k|^p\leqslant\partial_{x_i}\partial_{x_j}\bigl((\psi_k a^{ij}+k^{-1}\delta^{ij})|u_k|^p\bigr)\\ -\partial_{x_i}(\psi_k b|u_k|^p)+\psi_k(pc+(p-1)\mathrm{div}h)|u_k|^p\end{aligned}$$

and Gronwall's inequality yield that

$$\sup_{t\in[0,T]}\int_{\mathbb{R}^d}|u_k(x,t)|^p\,dx\leqslant M$$

with a constant M independent of k. It remains to choose a subsequence in $\{u_k\}$ that converges to a solution ϱ of class $L^\infty\bigl([0,T],L^p(\mathbb{R}^d)\bigr)$. □

6.7(iii). Nonlinear parabolic equations for measures

Here, following the papers Manita, Shaposhnikov [**666**], [**667**], we consider nonlinear parabolic Fokker–Planck–Kolmogorov equations similar to the elliptic equations discussed in § 2.5(iv) and the following Cauchy problem for them:

$$\partial_t \mu_t = \partial_{x_i}\partial_{x_j}\big(a^{ij}(x,t,\mu)\mu_t\big) - \partial_{x_i}\big(b^i(x,t,\mu)\mu_t\big), \quad \mu|_{t=0} = \nu, \tag{6.7.3}$$

where ν is a probability measure on $\mathbb{R}^d$.

Consider the operator $L_\mu = a^{ij}(x,t,\mu)\partial_{x_i}\partial_{x_j} + b^i(x,t,\mu)\partial_{x_i}$.

Let $\mathcal{M}(\mathbb{R}^d\times[0,\tau])$ denote the linear space of finite Borel measures on $\mathbb{R}^d\times[0,\tau]$. The measure $\mu \in \mathcal{M}(\mathbb{R}^d\times[0,\tau])$ given by a family $(\mu_t)_{t\in[0,\tau]}$ is called a probability solution to the Cauchy problem (6.7.3) if each μ_t is a probability measure, for $i,j \leqslant d$ we have Borel functions $(x,t)\mapsto a^{ij}(x,t,\mu)$, $(x,t)\mapsto b^i(x,t,\mu)$ on $\mathbb{R}^d\times[0,\tau]$, and for every closed ball $U \subset \mathbb{R}^d$ we have $a^{ij}, b^i \in L^1(U\times[0,\tau],\mu)$, and for all $t\in[0,\tau]$ and all $\varphi \in C_0^\infty(\mathbb{R}^d)$ the equality

$$\int_{\mathbb{R}^d} \varphi\, d\mu_t - \int \varphi\, d\nu = \int_0^t\!\!\int_{\mathbb{R}^d} L_\mu\varphi\, d\mu_s\, ds \tag{6.7.4}$$

holds. Typical examples of nonlinear parabolic equations for measures are equations with coefficients containing expressions of the form

$$\int_{\mathbb{R}^d} K(x,y,t)\,\mu_t(dy) \quad \text{or} \quad \int_0^t\!\!\int_{\mathbb{R}^d} K(x,y,s)\,\mu_s(dy)\,ds,$$

where the kernel K is growing at infinity.

Let $C^+[0,\tau]$ denote the space of nonnegative continuous functions on $[0,\tau]$.

Let τ_0 be a fixed positive number and let $V\geqslant 0$ be a Borel function on $\mathbb{R}^d$. For every function $\alpha\in C^+[0,\tau_0]$ and every $\tau\in(0,\tau_0]$ let $M_{\tau,\alpha}(V)$ denote the set of *nonnegative* measures μ in $\mathcal{M}(\mathbb{R}^d\times[0,\tau])$ given by a family of probability measures $(\mu_t)_{t\in[0,\tau]}$ such that for all $t\in[0,\tau]$ the following inequality holds:

$$\int_{\mathbb{R}^d} V\, d\mu_t \leqslant \alpha(t).$$

Note that the function α is chosen in the class $C^+[0,\tau_0]$, but not in $C^+[0,\tau]$ in order to be able to choose a function α and a number τ independently of each other. Let us formulate our conditions on the coefficients a^{ij} and b^i.

(H1) We are given a function $V\in C^2(\mathbb{R}^d)$ with

$$V(x) > 0, \qquad \lim_{|x|\to+\infty} V(x) = +\infty,$$

two mappings $\Lambda_1,\Lambda_2\colon C^+[0,\tau_0]\to C^+([0,\tau_0]$, and for every number $\tau\in(0,\tau_0]$ and every function $\alpha\in C^+[0,\tau_0]$, for every measure μ in the class $M_{\tau,\alpha}=M_{\tau,\alpha}(V)$ we are given functions $(x,t)\mapsto a^{ij}(x,t,\mu)$ and $(x,t)\mapsto b^i(x,t,\mu)$ such that

$$L_\mu V(x,t) \leqslant \Lambda_1(\alpha)(t) + \Lambda_2(\alpha)(t)V(x) \quad \forall\,(x,t)\in\mathbb{R}^d\times[0,\tau].$$

Such a function V will be called a Lyapunov function for L_μ. It is important for the sequel that V is a strictly positive function.

6.7.5. Example. Condition (H1) is obviously fulfilled with the function $V(x) = 1+|x|^2$ for the operator

$$L_\mu u(x,t) = \langle x, \nabla u(x)\rangle \int_{\mathbb{R}^d} |y|^2\,\mu_t(dy).$$

In this case one has

$$L_\mu V(x,t) = 2|x|^2 \int_{\mathbb{R}^d} |y|^2 \, \mu_t(dy) \leqslant 2\alpha(t)V(x)$$

and $L_\mu V \leqslant \Lambda_1 + \Lambda_2 V$, $\Lambda_1(\alpha) \equiv 0$ and $\Lambda_2(\alpha) = 2\alpha$.

We shall also need the continuity of the coefficients with respect to a suitable convergence of measures. However, weak convergence is not sufficient, since in typical examples the coefficients have the form of integrals of unbounded functions. We shall say that measures $\mu^n = (\mu_t^n)_{t\in[0,\tau]}$ in the set $M_{\tau,\alpha}$ converge V-weakly to a measure $\mu = (\mu_t)_{t\in[0,\tau]}$ in $M_{\tau,\alpha}$ if for all $t \in [0,\tau]$ one has the equality

$$\lim_{n\to\infty} \int_{\mathbb{R}^d} F \, d\mu_t^n = \int_{\mathbb{R}^d} F \, d\mu_t$$

for every continuous function F on $\mathbb{R}^d$ with $\lim\limits_{|x|\to\infty} F(x)/V(x) = 0$.

Let us introduce the following condition.

(H2) For all $\tau \in (0,\tau_0]$, $\alpha \in C^+[0,\tau_0]$, $\sigma \in M_{\tau,\alpha}$ and all $x \in \mathbb{R}^d$ the functions $t \mapsto a^{ij}(x,t,\sigma)$, $t \mapsto b^i(x,t,\sigma)$ are Borel measurable on $[0,\tau]$, the matrices $\bigl(a^{ij}(x)\bigr)_{i,j\leqslant d}$ are symmetric and nonnegative definite, for every closed ball $U \subset \mathbb{R}^d$ the functions $x \mapsto b^i(x,t,\sigma)$, $x \mapsto a^{ij}(x,t,\sigma)$ are bounded on U uniformly in $\sigma \in M_{\tau,\alpha}$ and $t \in [0,\tau]$ and continuous on U uniformly in $\sigma \in M_{\tau,\alpha}$ and $t \in [0,\tau]$. In addition, if a sequence $\mu^n \in M_{\tau,\alpha}$ converges V-weakly to $\mu \in M_{\tau,\alpha}$, then for all $(x,t) \in \mathbb{R}^d \times [0,\tau]$ we have

$$\lim_{n\to\infty} a^{ij}(x,t,\mu^n) = a^{ij}(x,t,\mu), \qquad \lim_{n\to\infty} b^i(x,t,\mu^n) = b^i(x,t,\mu).$$

6.7.6. Remark. (i) The continuity of the functions a^{ij} and b^i in x and measurability in t imply the joint measurability in the variables (x,t) with respect to the Borel σ-algebra $\mathcal{B}(\mathbb{R}^d \times [0,\tau])$, see Bogachev [**125**, Lemma 6.4.6, Exercise 6.10.39].

(ii) For every $t \in [0,\tau]$ the sequences $a^{ij}(x,t,\mu^n)$ and $b^i(x,t,\mu^n)$ converge uniformly in x on every ball $U \subset \mathbb{R}^d$, which follows from the pointwise convergence and equicontinuity.

Note that no continuity of the coefficients in t is assumed.

6.7.7. Example. Condition (H2) is fulfilled for

$$b(x,t,\mu) = \int_{\mathbb{R}^d} K(x,y) \, \mu_t(dy)$$

with a given Lyapunov function V and an integral kernel K that is a continuous vector field on $\mathbb{R}^d \times [0,\tau]$ such that

$$|K(x,y)| \leqslant C_1(x) + C_2(x)V^{1-\gamma}(y),$$

where $\gamma \in (0,1)$ is a number and $C_1(x), C_2(x)$ are continuous functions. Justification is left as Exercise 6.7.19.

Here is the main result in this subsection.

6.7.8. Theorem. *Let Conditions* (H1), (H2) *be fulfilled and let the initial condition ν be such that $V \in L^1(\nu)$. Then the following assertions are valid.*

(i) *There is $\tau \in (0,\tau_0]$ such that the Cauchy problem* (6.7.3) *on* $[0,\tau]$ *has a solution.*

(ii) *If the mappings* Λ_1 *and* Λ_2 *in Condition* (H1) *are constant, then the Cauchy problem* (6.7.3) *has a solution on the whole interval* $[0, \tau_0]$.

(iii) *If* $\Lambda_1(\alpha) = 0$ *and* $\Lambda_2(\alpha)(t) = G\bigl(\alpha(t)\bigr)$, *where* $G > 0$ *is a strictly increasing continuous function on* $[0, +\infty)$, *then the Cauchy problem* (6.7.3) *has a solution on every interval* $[0, \tau]$, *where* $\tau \in (0, \tau_0]$, $\tau < T$ *and*

$$T = \int_{u_0}^{+\infty} \frac{1}{uG(u)}\, du, \quad u_0 = \int_{\mathbb{R}^d} V\, d\nu.$$

(iv) *If* $\Lambda_1(\alpha)(t) = G\bigl(\alpha(t)\bigr)$ *and* $\Lambda_2(\alpha) = 0$, *where* $G > 0$ *is a strictly increasing continuous function on* $[0, +\infty)$, *then the Cauchy problem* (6.7.3) *has a solution on every interval* $[0, \tau]$, *where* $\tau \in (0, \tau_0]$, $\tau < T$ *and*

$$T = \int_{u_0}^{+\infty} \frac{1}{G(u)}\, du, \quad u_0 = \int_{\mathbb{R}^d} V\, d\nu.$$

In addition, in all these assertions the constructed solution μ *is given by a family of probability measures* $(\mu_t)_{t\in[0,\tau]}$ *such that*

$$\sup_{t\in[0,\tau]} \int_{\mathbb{R}^d} V\, d\mu_t < \infty.$$

For the proof, see Manita, Shaposhnikov [**667**].

Let us give an example. Consider the following Cauchy problem:

$$\partial_t \mu_t = \operatorname{div}_x\Bigl(\mu_t \int_{\mathbb{R}^d} |x-y|^{m-1}(x-y)\,\mu_t(dy)\Bigr), \quad \mu_0 = \nu. \tag{6.7.5}$$

6.7.9. Proposition. *Let* $m > 0$ *and let* ν *be a probability measure on* $\mathbb{R}^d$ *such that* $|x|^{m+1} \in L^1(\nu)$, *and if* $m < 1$, *then* $|x|^2 \in L^1(\nu)$. *Then, there exists a number* $\tau > 0$ *such that on* $[0, \tau]$ *the Cauchy problem* (6.7.5) *has a solution* μ *given by a family of probability measures* $(\mu_t)_{t\in[0,\tau]}$. *In addition,*

$$\sup_{t\in[0,\tau]} \int_{\mathbb{R}^d} |x|^{m+1}\,\mu_t(dx) < \infty.$$

If $0 < m < 1$, *then, for every number* $\tau > 0$, *there exists a solution of the Cauchy problem* (6.7.5) *on the interval* $[0, \tau]$.

PROOF. By condition, $a^{ij} = 0$ and

$$b(x, t, \mu) = -\int_{\mathbb{R}^d} |x-y|^{m-1}(x-y)\,\mu_t(dy).$$

Let $m \geqslant 1$, $V(x) = 1 + |x|^{m+1}$. Since the function $K(x, y) = -|x-y|^{m-1}(x-y)$ is continuous on the space $\mathbb{R}^d \times \mathbb{R}^d$ for $m > 0$ and $|K(x, y)| \leqslant 2^m|x|^m + 2^m V^{1-\gamma}(y)$, where $\gamma = 1/(m+1)$, we see from Example 6.7.7 that the coefficient b satisfies condition (H2). Let us verify condition (H1). We have

$$\begin{aligned} L_\mu V(x,t) &= \langle b(x,t,\mu), \nabla V(x)\rangle \\ &= -(m+1)|x|^{m-1} \int_{\mathbb{R}^d} |x-y|^{m-1}\langle x-y, x\rangle\,\mu_t(dy). \end{aligned}$$

We observe that

$$-\int_{\mathbb{R}^d} |x-y|^{m-1}\langle x-y, x\rangle\, \mu_t(dy)$$

$$= -\int_{\mathbb{R}^d} |x-y|^{m+1}\, \mu_t(dy) - \int_{\mathbb{R}^d} |x-y|^{m-1}\langle x-y, y\rangle\, \mu_t(dy),$$

which is estimated from above by

$$-\int_{\mathbb{R}^d} |x-y|^{m+1}\, \mu_t(dy) + \int_{\mathbb{R}^d} |x-y|^{m}|y|\, \mu_t(dy).$$

By Young's inequality

$$|x-y|^m|y| \leqslant \frac{m}{m+1}|x-y|^{m+1} + \frac{1}{m+1}|y|^{m+1}.$$

Therefore,

$$\langle b(x,t,\mu), \nabla V(x)\rangle \leqslant |x|^{m-1}\int_{\mathbb{R}^d} |y|^{m+1}\, \mu_t(dy).$$

By Young's inequality we obtain $L_\mu V(x,t) \leqslant \dfrac{m-1}{m+1}V(x) + \dfrac{2}{m+1}\alpha^{\frac{m+1}{2}}(t)$. Therefore, Condition (H1) is fulfilled and we can apply assertion (i) in Theorem 6.7.8. Finally, in the case $0 < m < 1$ take $V(x) = 1+|x|^2$. A similar reasoning yields the inequality $L_\mu V(x,t) \leqslant 2\alpha^{(m+1)/2}(t)$. We observe that $\Lambda_2(\alpha) = 0$, $\Lambda_1(\alpha) = 2\alpha^{(m+1)/2}$ and

$$\int_{u_0}^{\infty} \frac{du}{u^{(m+1)/2}} = +\infty, \quad u_0 = \int_{\mathbb{R}^d} (1+|x|^2)\, \nu(dx),$$

since $(m+1)/2 < 1$. Thus, we can apply assertion (iv) in Theorem 6.7.8. □

Let us consider a more general Cauchy problem:

$$\partial_t\mu_t = \operatorname{div}\Bigl(\mu_t\int_{\mathbb{R}^d} \nabla W(x-y)\, \mu_t(dy)\Bigr), \quad \mu_0 = \nu. \tag{6.7.6}$$

6.7.10. Proposition. *Let $W(x) = K(|x|)$, where $K \in C^2[0,+\infty)$, $K(0) = 0$, $K'(0) = 0$, $K'(u) > 0$, $K''(u) > 0$ if $u > 0$ and there exist numbers $C_1 > 0$, $C_2 > 0$ such that $K'(u+v) \leqslant C_1K(u) + C_2K(v)$ if $u, v > 0$. Suppose also that $K^m(|x|) \in L^1(\nu)$ for some $m > 1$. Then there exists a number $\tau > 0$ such that on the interval $[0,\tau]$ the Cauchy problem (6.7.6) has a solution μ given by a family of probability measures $(\mu_t)_{t\in[0,\tau]}$. In addition,*

$$\sup_{t\in[0,\tau]} \int K^m(|x|)\, d\mu_t < \infty.$$

The proof is given in Manita, Shaposhnikov [**667**].

Suppose that $\tau_0 > 0$, $b\colon \mathbb{R}^d\times[0,\tau_0] \to \mathbb{R}^d$ is a continuous mapping, $F \geqslant 0$ is a continuous function on $\mathbb{R}^d$, $G > 0$ is a strictly increasing continuous function on the half-line $[0,+\infty)$, $G(0) > 1$. Set

$$b(x,t,\mu) = b(x,t)G\Bigl(\int_{\mathbb{R}^d} F(y)\, \mu_t(dy)\Bigr).$$

Let us consider the Cauchy problem

$$\partial_t\mu_t = \Delta\mu_t + \operatorname{div}\bigl(\mu_t b(x,t,\mu)\bigr), \quad \mu_0 = \nu. \tag{6.7.7}$$

6.7.11. Proposition. *Suppose that there exists a function* $V \in C^2(\mathbb{R}^d)$ *such that* $V \geqslant 1$, $\lim\limits_{|x|\to+\infty} V(x) = +\infty$, $\lim\limits_{|x|\to+\infty} F(x)/V(x) = 0$ *and for some numbers* C_1 *and* C_2 *and all* $(x,t) \in \mathbb{R}^d \times [0,\tau_0]$ *we have the inequality*

$$|\Delta V(x)| + \langle b(x,t), \nabla V(x)\rangle \leqslant C_1 + C_2 V(x).$$

Let $V \in L^1(\nu)$. *Then there exists a number* $\tau \in (0,\tau_0]$ *such that on the interval* $[0,\tau]$ *the Cauchy problem* (6.7.7) *has a solution* μ *given by a family of probability measures* $(\mu_t)_{t\in[0,\tau]}$. *In addition,*

$$\sup_{t\in[0,\tau]} \int_{\mathbb{R}^d} V\, d\mu_t < \infty.$$

If it is given that

$$\int_{u_0}^{+\infty} \frac{du}{uG(u)} = +\infty, \quad u_0 = \int_{\mathbb{R}^d} V\, d\nu.$$

then a solution to the Cauchy problem exists on the whole interval $[0,\tau_0]$.

PROOF. Let $\mu \in M_{\tau,\alpha}$. We observe that

$$\begin{aligned} L_\mu V(x,t) &= \Delta V + \langle b(x,t), \nabla V(x)\rangle G\big(\alpha(t)\big) \\ &= \Delta V + \langle b(x,t), \nabla V(x)\rangle + \langle b(x,t), \nabla V(x)\rangle\big[G\big(\alpha(t)\big) - 1\big] \\ &\leqslant (C_1 + C_2 V)G\big(\alpha(t)\big). \end{aligned}$$

Hence our assertion follows by Theorem 6.7.8. □

Assume, for example, that $V(x) = \exp\big(M|x|^r\big)$, where $M > 0$ and $r \geqslant 2$. To satisfy the hypotheses in Proposition 6.7.11 it suffices that

$$\exp\big(M|x|^r\big) \in L^1(\nu), \quad |F(x)| \leqslant \exp\big(M'|x|^r\big), \quad M' < M$$

and that for some numbers c_1 and $c_2 > drM$ and all $(x,t) \in \mathbb{R}^d \times [0,\tau_0]$ the inequality $\langle b(x,t), x\rangle \leqslant c_1 - c_2|x|^r$ holds.

A solution to the Cauchy problem (6.7.3) on the whole interval $[0,\tau_0]$ is called a global solution, and a solution on $[0,\tau]$, where $\tau < \tau_0$, is called a local solution. Thus, in assertion (i) in Theorem 6.7.8 the existence of a local solution is asserted, while in (ii) we are concerned with a global one.

Let us give sufficient conditions for the absence of a global solution. The problem of "blow-up" of solutions of equations of the considered form is addressed in many works, see Mitidieri, Pokhozhaev [**714**], [**715**], Bertozzi, Carrillo, Laurent [**111**], Carrillo et al. [**221**], Li, Toscani [**604**].

Recall that if in Condition (H1) we have $\Lambda_1(\alpha) = 0$ and $\Lambda_2(\alpha)(t) = G\big(\alpha(t)\big)$, then under Condition (H2) assertion (iii) in Theorem 6.7.8 gives the existence of a solution on the whole interval $[0,\tau]$, where

$$\tau < \int_{u_0}^{+\infty} \frac{du}{uG(u)}.$$

A similar assertion is true in (iv). The next theorem from [**667**] shows that in a sense such estimates of the interval of existence of a solution are sharp.

6.7.12. Theorem. *Let* $V \in C^2(\mathbb{R}^d)$, $V \geqslant 0$, $\lim\limits_{|x|\to\infty} V(x) = +\infty$. *Suppose also that we are given a continuous positive increasing function* G *on* $[0,+\infty)$, a^{ij}, b^i *are defined on every set* $M_{\tau,\alpha}(V) \times \mathbb{R}^d \times [0,\tau]$ *and satisfy the inequality*

(i) $L_\mu V(x,t) \geqslant G\Bigl(\int_{\mathbb{R}^d} V\,d\mu_t\Bigr)V(x)$

or the inequality

(ii) $L_\mu V(x,t) \geqslant G\Bigl(\int_{\mathbb{R}^d} V\,d\mu_t\Bigr)$.

Let also $|\sqrt{A(x,t,\mu)}\nabla V(x)|^2 \leqslant C_1 + C_2 V(x)$ *for some constants* $C_1 > 0$ *and* $C_2 > 0$. *Suppose that* $u_0 = \|V\|_{L^1(\nu)} > 0$, *and in case* (i) *the function* $1/\bigl(uG(u)\bigr)$ *is integrable on* $[u_0, +\infty)$, *in case* (ii) *the function* $1/G(u)$ *is integrable. Then the Cauchy problem* (6.7.3) *on the interval* $[0,\tau]$ *with* $\tau \geqslant T$ *has no solution in the class of measures* $\mu = (\mu_t)_{t\in[0,T]}$ *for which each* μ_t *is a probability measure and*

$$\sup_{t\in[0,\tau]} \int V\,d\mu_t < \infty.$$

6.7.13. Example. Suppose that in the Cauchy problem (6.7.3) we have

$$a^{ij} = 0, \quad b(x,t,\mu) = x\int_{\mathbb{R}^d} |y|^2\,\mu_t(dy).$$

Suppose also that $|x|^p \in L^1(\nu)$ for some $p > 2$. Then a solution exists on the interval $[0,\tau]$ provided that

$$\tau^{-1} > 2\int_{\mathbb{R}^d} |x|^2\,\nu(dx).$$

If the initial condition is not Dirac's measure δ_0 at zero, then the condition in assertion (i) in Theorem 6.7.12 is fulfilled with the function $V(x) = |x|^2$, hence there is no global solution on the interval $[0,\tau]$ if $\tau^{-1} \leqslant 2\int |x|^2\,\nu(dx)$. It is readily seen that $\mu_t \equiv \delta_0$ is a solution to the Cauchy problem on every interval $[0,\tau]$ with the initial condition δ_0. Justification is left as Exercise 6.7.20.

Note that even in the linear case the continuity and boundedness of the coefficients is not sufficient for the uniqueness of solutions to the Cauchy problem. Indeed, let $A \equiv 0$ and $b(x) = x^{2/3}$. Then the Cauchy problem for the ordinary equation $\dot{x} = b(x)$, $x(0) = 0$ has two solutions: $x_1(t) = t^3/3$ and $x_2(t) = 0$. The measures $\delta_{x_1(t)}$ and $\delta_{x_2(t)}$ are different solutions to the corresponding Cauchy problem (6.7.3). In the nonlinear case the situation is even more complicated (see Exercise 6.7.21).

Comments

There is a vast literature devoted to the general theory of second order parabolic equations, including the books Èĭdel'man [**326**], Eidelman, Ivasyshen, Kochubei [**327**], Eidelman, Zhitarashu [**328**], Friedman [**386**], Hörmander [**461**], Kondrat'ev, Landis [**533**], Kresin, Maz'ya [**542**], Krylov [**552**], [**556**], Ladyz'enskaya, Solonnikov, Ural'tseva [**576**], Landis [**581**], Lieberman [**611**], Lions, Magenes [**618**], Lunardi [**646**], Maugeri, Palagachev, Softova [**681**], Oleĭnik, Radkevič [**757**], Watson [**935**], Wu, Yin, Wang [**944**], Zelenyak, Lavrentiev, Vishnevskii [**954**]. However, as in the elliptic case, the general theory is not sufficient to deal with many specific features of the Fokker–Planck–Kolmogorov equations.

Parabolic equations for bounded measures (linear, including Fokker–Planck–Kolmogorov equations, and also nonlinear) is a very important object in statistical physics and related fields, see Akhiezer, Peletminskii [**13**], Arsen'ev [**75**], Coffey,

Kalmykov [**252**], Frank [**384**], Galkin[**396**], Gardiner [**399**], Grasman, van Herwaarden [**420**], Isihara [**481**], Keizer [**500**], Klimontovich [**520**], Klyatskin [**521**], Liboff [**608**], Lifshitz, Pitaevskii [**615**], Malanin, Poloskov [**655**], McCauley [**686**], Pavliotis [**766**], Risken [**807**], Rytov [**829**], Schuss [**838**], Soize [**865**], Vlasov [**923**], and also Peeters, Strintzi [**768**], Mamontov [**658**].

In these books and in many papers the authors consider also cases of explicit solutions (which are not often, of course), see, for example, Araujo, Drigo Filho [**59**], Dechant, Lutz, Barkai, Kessler [**291**], Englefield [**336**], Lin, Ho [**617**], Lo [**635**], Makov, Hohlov [**654**], Meyer, Schröter [**706**], and Miyazawa [**719**].

In the proof of existence of densities of solutions to parabolic equations a very important role is played by the parabolic version of Aleksandrov's inequality obtained in Krylov [**548**].

In § 6.1–6.4 we follow the paper Bogachev, Krylov, Röckner [**152**], in § 6.5 we use the paper Shaposhnikov [**852**], § 6.6 is based on Manita, Shaposhnikov [**668**], and partly Bogachev, Da Prato, Röckner, Stannat [**142**], Bogachev, Da Prato, Röckner [**132**].

Equations with singular or rapidly increasing coefficients are considered also in Bogachev, Da Prato, Röckner [**129**], Bogachev, Krylov, Röckner [**151**], Kunze, Lorenzi, Lunardi [**564**].

Powerful methods of proving existence of probability solutions to parabolic Fokker–Planck–Kolmogorov equations are developed in the theory of diffusion processes, see the books Gihman, Skorohod [**407**], Krylov [**549**], Stroock, Varadhan [**884**], and the papers Baur, Grothaus [**96**], Baur, Grothaus, Stilgenbauer [**97**].

Asymptotic properties of solutions depending on a parameter are studied in Il'in, Khasminskii, Yin [**476**], Khasminskii, Yin [**503**] – [**507**].

Fokker–Planck–Kolmogorov equations on domains and manifolds are considered in Chupin [**249**], Li [**607**], Soloveitchik [**866**], Zhang [**958**], and also in a number of papers on degenerate equations mentioned in Chapter 9 (in several subsections in § 9.8).

Among works of historical interest, we note Kryloff, Bogoliouboff [**546**] (see also Bogolyubov [**187**]), Feller [**355**], and Yosida [**945**].

Nonlinear first and second order parabolic equations for probability measures belong to the most widely used equations in statistical mechanics and physics (see the books mentioned above, especially Frank [**384**]). Such equations are studied in many mathematical works, especially of the last years, for example, on transport and continuity equations see Ambrosio [**38**], [**39**], Belgacem, Jabin [**100**], Lions, Perthame [**619**], Maniglia [**663**], Ouédraogo, De Dieu Zabsonré [**761**], and Perthame [**771**], where additional references can be found. In a great number of these works mainly equations with coefficients of convolution type are considered, for example, Vlasov equations, see Bouchut [**121**], Dobrushin [**305**], Kozlov [**539**], [**540**], Vedenyapin [**912**], but also more general equations are studied, see Bogachev, Röckner, Shaposhnikov [**168**], Carrillo, McCann, Villani [**223**], Carrillo et al. [**221**], Carrillo, Toscani [**225**], Dreyer et al. [**314**], Funaki [**395**], Lorenz [**639**], [**638**], McKean [**687**], [**688**]. Certainly, here substantial intersections with the general theory of nonlinear equations arise, a very large area (see, e.g., Quittner, Souplet [**797**]), which do not touch.

Diverse results on nonlinear equations related to functional inequalities are obtained in Blanchet, Carlen, Carrillo [**119**].

Assertion (ii) in Theorem 6.7.8 was obtained in a more special case in Bogachev, Röckner, Shaposhnikov [**168**]. Assertion (iii) in the case $T = +\infty$ was obtained in Funaki [**395**] by probabilistic methods, in fact, the solvability of the corresponding martingale problem was proved. The main difference between Theorem 6.7.8 and other known results is the absence of any restrictions on the growth of the coefficients of the operator L_μ (only the existence of a Lyapunov function is important), and also a rather general dependence of the coefficients on the measure μ, which covers typical special cases considered in Carrillo et al. [**221**], Carrillo, McCann, Villani [**223**], DiPerna, Lions [**304**], Dobrushin [**305**], Kozlov [**539**], [**540**], Lorenz [**639**]. In addition, it does not assume the membership of the coefficients in the Sobolev class or in the class of functions of bounded variation in the space variable, but just their continuity. Note that in [**305**] the Vlasov equations are investigated in the space of finite measures with the classical Kantorovich L^1-metric, and the existence is proved by applying the theorem on fixed points for contracting mappings. In the situation of Theorem 6.7.8 that theorem does not work just because a solution, generally speaking, is not unique. In Carrillo, McCann, Villani [**223**], Carrillo et al. [**221**], Li, Toscani [**604**], Lorenz [**639**] nonlinear transport equations with coefficients of convolution type are also studied in spaces of measures with suitable Kantorovich metrics, and solutions of equations are considered as geodesics. However, this approach is heavily based on a specific form of coefficients. There is an extensive literature on relations between Fokker–Planck–Kolmogorov equations (especially nonlinear) and the Monge–Kantorovich problem, one aspect of which has already been noted in § 6.7(i); see the influential paper Otto, Jordan, Kinderlehrer [**495**] and Agueh [**8**], Peletier, Renger, Veneroni [**770**], and also surveys Ambrosio, Gigli [**44**], Bogachev, Kolesnikov [**148**], Villani [**918**], where further references can be found.

Convergence of solutions to a stationary one is investigated in Bolley, Gentil, Guillin [**190**], Bolley, Guillin, Malrieu [**192**], Bolley, Guillin, Villani [**193**], Carrillo et al. [**222**], Desvillettes, Villani [**298**], and Villani [**920**].

In Bogachev, Röckner, Shaposhnikov [**173**] the result of Proposition 6.2.9 is used to establish the existence of a density of a solution to a parabolic inequality for measures that contains an integro-differential part. This gives a condition for the existence of densities of transition probabilities of processes with jumps.

The book Jikov, Kozlov, Oleinik [**491**] has some common points with the themes in this chapter.

Exercises

6.7.14. A measure μ with density $\varrho(x,t)$ on $\mathbb{R}^d\times(0,1)$ satisfies the parabolic equation $\partial_t\mu = L^*_{A,b}\mu$. Write down the equation satisfied by the image of this measure under the mapping $(x,t)\mapsto \bigl((Fx),t\bigr)$ for a diffeomorphism $F\colon \mathbb{R}^d\to\mathbb{R}^d$, and also under the mapping $(x,t)\mapsto(x,1-t)$.

6.7.15. A probability measure μ on the strip $\mathbb{R}\times(-1,1)$ satisfies the equation $\partial_t\mu = L^*_{1,b}\mu$ on $\mathbb{R}\times(-1,0)$ and on $\mathbb{R}\times(0,1)$ with some continuous function b on the real line. Is it true that μ satisfies this equation on the domain $\mathbb{R}\times(-1,1)$?

6.7.16. Show that the assertion of Theorem 6.3.1 is sharp: give an example where the density of the measure $(\det A)^{1/(d+2)}\mu$ belongs to $L^{(d+1)'}$, but does not belong to any L^p with $p>(d+1)'$.

6.7.17. Give an example of a nonnegative function a and a probability measure ν on $\mathbb{R}^1$ such that the Cauchy problem $\partial_t\mu = \partial_x^2(a\mu)$, $\mu|_{t=0} = \nu$ has no solution given by a family of subprobability measures.

HINT: Consider $a(x) = 0$ if $x \neq 0$, $a(0) = 1$, and $\nu = \delta_0$; show that $\mu_t = e^{2t}\delta_0$ is the only solution.

6.7.18. Let a measure $\mu = (\mu_t) \geqslant 0$ on the strip $\mathbb{R}^d\times[0,T]$ satisfy the equation $\partial_t\mu = L_{A,b}^*\mu$, where $\lambda_1 \cdot \mathrm{I} \leqslant A(x,t) \leqslant \lambda_1 \cdot \mathrm{I}$, $|a^{ij}(x,t) - a^{i,j}(y,t)| \leqslant \Lambda|x-y|$, the function $|b|^p$ is integrable with respect to μ on bounded sets and $p > d+2$. Let $[\tau_1,\tau_2] \subset (0,T)$. Using the proof of Theorem 6.4.2 derive an estimate for the density ϱ of the measure μ of the form

$$\|\varrho\|_{\mathbb{H}^{p,1}(K\times[\tau_1,\tau_2])} \leqslant C\mu(2K\times(0,T)) + C\int_0^T\int_{2K} |b|^p\, d\mu,$$

where K is the unit cube and C depends only on $d, \lambda_1, \lambda_2, \Lambda, \tau_1, \tau_2$.

6.7.19. Justify Example 6.7.7.

HINT: See Manita, Shaposhnikov [**667**].

6.7.20. Justify Example 6.7.13.

6.7.21. Let $d = 1$, $A \equiv 0$,

$$b(x,t,\mu) = b(t,\mu) = G\Bigl(\int |y|^2\, \mu_t(dy)\Bigr).$$

Let $x(t)$ be the solution of the Cauchy problem $\dot{x} = G(x^2)$, $x(0) = 0$, where $G(u) = u^{1/3}$. Show that the Cauchy problem (6.7.3) has two different solutions.

6.7.22. Let μ_n be Borel subprobability measures on $\mathbb{R}^d$ such that for every function $\varphi \in C_0^\infty(\mathbb{R}^d)$ the integrals of φ with respect to the measures μ_n have a finite limit. Prove that there exists a subprobability measure μ such that on every ball the measures μ_n converge weakly to μ.

HINT: Observe that the limit measure is defined uniquely once it exists and that every subsequence in $\{\mu_n\}$ contains a further subsequence weakly converging on all balls.

6.7.23. Let ν be a probability measure on $\mathbb{R}^2$ satisfying the equation $L_{\mathrm{I},b}^*\nu = 0$ with smooth b such that ν is not its unique probability solution (see Chapter 5). Show that the Cauchy problem $\partial_t\mu = L_{\mathrm{I},b}^*\mu$ has a family of solutions $\mu(t,\eta)$ with initial conditions η defined for all bounded measures η given by bounded densities with respect to ν such that this family does not possess the semigroup property, which means that it is not always true that $\mu(t+\tau,\nu) = \mu(\tau,\mu(t,\nu))$, which is the case for the solution $K_t^*\nu$ constructed in Theorem 6.4.7.

HINT: As shown in Chapter 5, our assumption yields the existence of another (different from $\{T_t^\nu\}_{t\geqslant 0}$) strongly continuous semigroup $\{S_t\}_{t\geqslant 0}$ on $L^1(\nu)$ with generator extending $L_{\mathrm{I},b}$. Consider the family of measures on $\mathbb{R}^2$ defined by $\mu(t,\eta) = K_t^*\nu$ if $\eta = \nu$ and by $S_t^*\eta := S_t^*\varrho_\eta$ if $\eta = \varrho_\eta \cdot \nu \neq \nu$, $\varrho_\eta \in L^\infty(\nu)$, where S_t^* is defined on $L^\infty(\nu)$. Observe that $\mu(t,\eta)$ satisfies the Cauchy problem for $L_{\mathrm{I},b}^*$ with initial condition η and that there are $t > 0$ and a sequence of times $\tau_n \to 0$ for which $\mu(t+\tau_n,\nu) \neq \mu(\tau_n,\mu(t,\nu))$, because ν is not invariant for $\{T_t^\nu\}_{t\geqslant 0}$ and the measures $K_t^*\nu$ and $S_t^*\nu$ cannot coincide for all t. Note that $\mu(t,\eta)$ is not a probability solution.

CHAPTER 7

Global Parabolic Regularity and Upper Bounds

In this chapter we obtain upper bounds of densities of solutions to Fokker–Planck–Kolmogorov equations with possibly unbounded coefficients. These estimates generalize the classical Gaussian estimates in the case of bounded coefficients or coefficients of linear growth. The main idea enabling us to consider unbounded coefficients is a combination of local estimates from Chapter 6 with scaling and the method of Lyapunov functions.

In § 7.1 we obtain sufficient conditions for the integrability of a given function $V(x,t) \geqslant 0$ with respect to every probability solution to the Cauchy problem for the equation $\partial_t \mu = L^*_{A,b}\mu$ in terms of the coefficients A and b. The main results are presented in Theorem 7.1.1 and its corollary and also in several examples.

In § 7.2 and § 7.3 we deal with global upper bounds for the solution densities, in particular, given a function Φ satisfying certain conditions that are easy to verify in terms of A and b and certain integrability conditions, Corollary 7.2.5 provides an upper bound of the form $\varrho(x,t) \leqslant Ct^{-d/2}\Phi(x,t)^{-1}$.

7.1. A priori estimates with Lyapunov functions

Here we derive a priori estimates with Lyapunov functions, i.e., obtain sufficient conditions under which certain increasing at infinity functions V are integrable with respect to the solution μ to the Cauchy problem for the Fokker–Planck–Kolmogorov equation. In addition, we obtain estimates of the integrals

$$\int_{\mathbb{R}^d} V(x,t)\,\mu_t(dx).$$

At the end of this section we give some examples with exponential and power Lyapunov functions, in particular, we obtain estimates that in a sense do not depend on the initial condition.

We recall (see § 1.3) that if we are given a diffusion process $\{\xi_t\}$ with the generator L of the transition semigroup and a Lyapunov function V satisfying the inequality $\partial_t V + LV \leqslant CV$ for some number C, then Itô's formula and Gronwall's inequality yield the inequality

$$\mathbb{E}V(\xi_t,t) \leqslant e^{Ct}\mathbb{E}V(\xi_0,0).$$

Set $\mu_t(B) := P(\xi_t \in B)$. Then

$$\mathbb{E}V(\xi_t,t) = \int_{\mathbb{R}^d} V(x,t)\,\mu_t(dx).$$

We shall derive similar estimates for solutions without using probabilistic methods, but basing essentially only on the definition of a solution. In this case the restrictions on the coefficients are minimal.

We consider the Cauchy problem (6.1.1) and (6.1.3) on $\mathbb{R}^d_T = \mathbb{R}^d \times (0,T)$:

$$\partial_t \mu = L^*_{A,b,c}\mu, \quad \mu|_{t=0} = \nu,$$

where ν is a subprobability measure and $c \leqslant 0$. As above in Chapter 6, we shall study solutions μ of class $\mathcal{M}_\nu$, i.e., solutions μ defined by a family of nonnegative measures $(\mu_t)_{t\in(0,T)}$ such that $c \in L^1(\mu)$ and for a.e. $t \in [0,T]$ we have the inequality

$$\mu_t(\mathbb{R}^d) \leqslant \nu(\mathbb{R}^d) + \int_0^t \int_{\mathbb{R}^d} c(x,s)\,\mu_s(dx)\,ds. \tag{7.1.1}$$

In particular, μ_t is a subprobability measure on $\mathbb{R}^d$. No other conditions on the coefficients a^{ij}, b^i and c are imposed so far, except for their local integrability with respect to μ.

In the following theorem we derive an a priori estimate with a Lyapunov function.

7.1.1. Theorem. *Let $c \leqslant 0$ and let ν be a subprobability measure on $\mathbb{R}^d$. Suppose that we have a solution $\mu = (\mu_t)_{0<t<T} \in \mathcal{M}_\nu$ and there is a function V in the class $C^{2,1}(\mathbb{R}^d_T)\cap C\big(\mathbb{R}^d\times[0,T)\big)$ such that for every compact interval $[\alpha,\beta] \subset (0,T)$ we have*

$$\lim_{|x|\to+\infty} \min_{t\in[\alpha,\beta]} V(x,t) = +\infty$$

and for some $K, H \in L^1\big((0,T)\big)$, where $H \geqslant 0$, and for all $(x,t) \in \mathbb{R}^d_T$

$$\partial_t V(x,t) + L_{A,b,c}V(x,t) \leqslant K(t) + H(t)V(x,t).$$

Suppose also that $V(\,\cdot\,,0) \in L^1(\nu)$. Then for almost all $t \in (0,T)$ we have

$$\mu_t(\mathbb{R}^d) = \nu(\mathbb{R}^d) + \int_0^t \int_{\mathbb{R}^d} c(x,s)\,\mu_s(dx)\,ds,$$

$$\int_{\mathbb{R}^d} V(x,t)\,\mu_t(dx) \leqslant Q(t) + R(t)\int_{\mathbb{R}^d} V(x,0)\,\nu(dx),$$

where the functions R and Q are defined by the following formulas:

$$R(t) = \exp\biggl(\int_0^t H(s)\,ds\biggr), \quad Q(t) = R(t)\int_0^t \frac{K(s)}{R(s)}\,ds.$$

Proof. Let a function $\zeta_N \in C^2\big([0,+\infty)\big)$ be such that $0 \leqslant \zeta_N' \leqslant 1$, $\zeta_N'' \leqslant 0$, $\zeta_N(z) = z$ if $z \leqslant N-1$, $\zeta(z) = N$ if $z > N+1$. Substituting the function $\varphi = \zeta_N(V) - N$ into the equality in assertion (iii) in Proposition 6.1.2, we obtain

$$\begin{aligned}
\int_{\mathbb{R}^d} \zeta_N\big(V(x,t)\big)\,\mu_t(dx) &= \int_{\mathbb{R}^d} \zeta_N\big(V(x,s)\big)\,\mu_s(dx) \\
&\quad + N\biggl(\mu_t(\mathbb{R}^d) - \nu(\mathbb{R}^d) - \int_s^t\int_{\mathbb{R}^d} c(x,\tau)\,\mu_\tau(dx)\,d\tau\biggr) \\
&\quad + \int_s^t\int_{\mathbb{R}^d} \Big(\zeta_N'(V)(\partial_t V + L_{A,b,c}V) + \zeta_N''(V)|\sqrt{A}\nabla V|^2\Big)\,\mu_\tau(dx)\,d\tau \\
&\quad + \int_s^t\int_{\mathbb{R}^d} c\big(\zeta_N(V) - \zeta_N'(V)V\big)\,\mu_\tau(dx)\,d\tau.
\end{aligned}$$

We observe that $z\zeta_N'(z) \leqslant \zeta_N(z)$. Therefore, we have

$$\int_{\mathbb{R}^d} \zeta_N\big(V(x,t)\big)\,\mu_t(dx) \leqslant \int_{\mathbb{R}^d} \zeta_N\big(V(x,s)\big)\,\mu_s(dx) + N\Big(\mu_t(\mathbb{R}^d) - \nu(\mathbb{R}^d) - \int_s^t\int_{\mathbb{R}^d} c(x,\tau)\,\mu_\tau(dx)\,d\tau\Big) + \int_s^t \Big[K(\tau) + H(\tau)\int_{\mathbb{R}^d} \zeta_N\big(V(x,\tau)\big)\,\mu_\tau(dx)\Big]\,d\tau.$$

Letting $s \to 0$, we arrive at the inequality

$$\int_{\mathbb{R}^d} \zeta_N\big(V(x,t)\big)\,\mu_t(dx) \leqslant \int_{\mathbb{R}^d} \zeta_N\big(V(x,0)\big)\,\nu(dx) + N\Big(\mu_t(\mathbb{R}^d) - \nu(\mathbb{R}^d) - \int_0^t\int_{\mathbb{R}^d} c(x,\tau)\,\mu_\tau(dx)\,d\tau\Big) + \int_0^t \Big[K(\tau) + H(\tau)\int_{\mathbb{R}^d} \zeta_N(V(x,\tau))\,\mu_\tau(dx)\Big]\,d\tau. \tag{7.1.2}$$

Since one has the estimate

$$\mu_t(\mathbb{R}^d) \leqslant \nu(\mathbb{R}^d) + \int_0^t\int_{\mathbb{R}^d} c(x,\tau)\,\mu_\tau(dx)\,d\tau,$$

the last inequality can be rewritten in the following form:

$$\int_{\mathbb{R}^d} \zeta_N\big(V(x,t)\big)\,\mu_t(dx) \leqslant \int_{\mathbb{R}^d} \zeta_N\big(V(x,0)\big)\,\nu(dx) + \int_0^t \Big[K(\tau) + H(\tau)\int_{\mathbb{R}^d} \zeta_N\big(V(x,\tau)\big)\,\mu_\tau(dx)\Big]\,d\tau.$$

Applying Gronwall's inequality (see Exercise 7.5.3), we obtain

$$\int_{\mathbb{R}^d} \zeta_N\big(V(x,t)\big)\,\mu_t(dx) \leqslant Q(t) + R(t)\int_{\mathbb{R}^d} \zeta_N\big(V(x,0)\big)\,\nu(dx).$$

Letting $N \to \infty$, we obtain the desired estimate. We observe that if

$$\mu_t(\mathbb{R}^d) < \nu(\mathbb{R}^d) + \int_0^t\int_{\mathbb{R}^d} c(x,\tau)\,\mu_\tau(dx)\,d\tau,$$

then

$$\int_{\mathbb{R}^d} V(x,t)\,\mu_t(dx) - \int_{\mathbb{R}^d} V(x,0)\,\nu(dx) - \int_0^t \Big[K(\tau) + H(\tau)\int_{\mathbb{R}^d} V(x,\tau)\,\mu_\tau(dx)\Big]\,d\tau = -\infty,$$

which is impossible. Thus, the desired equality holds. □

7.1.2. Corollary. *Let $c \leqslant 0$ and let ν be a subprobability measure on $\mathbb{R}^d$. Suppose that $\mu = (\mu_t)_{0<t<T} \in \mathcal{M}_\nu$ and that a positive function $W \in C^2(\mathbb{R}^d)$ is given such that*

$$\lim_{|x|\to+\infty} W(x) = +\infty.$$

(i) *If for some number $C > 0$ and all $(x,t) \in \mathbb{R}^d_T$ we have*

$$L_{A,b,c}W(x,t) \leqslant C + CW(x),$$

then for almost all $t \in (0,T)$ *we have the inequality*

$$\int_{\mathbb{R}^d} W(x)\,\mu_t(dx) \leqslant \exp(Ct) - 1 + \exp(Ct)\int_{\mathbb{R}^d} W(x)\,\nu(dx).$$

(ii) *Let* G *be a positive continuous increasing function on* $[0,+\infty)$ *such that*

$$\int_1^{+\infty} \frac{ds}{sG(s)} < +\infty.$$

Let η *be an continuous function on* $[0,T)$ *defined by the equality*

$$t = \int_0^{\eta(t)} \frac{ds}{sG(s^{-\delta})}, \quad \delta \in (0,1).$$

If for some number $C>0$ *and all* $(x,t)\in\mathbb{R}^d_T$ *we have*

$$L_{A,b,c}W(x,t) \leqslant C - W(x)G\bigl(W(x)\bigr),$$

then for almost all $t \in (0,T)$ *the inequality*

$$\int_{\mathbb{R}^d} W(x)\,\mu_t(dx) \leqslant \frac{1}{(1-\delta)\eta^\delta(t)} + \frac{C}{\eta(t)}\int_0^t \eta(s)\,ds$$

holds.

(iii) *Let* G *and* η *be the functions in* (ii) *such that for some number* $C>0$ *and all* $(x,t)\in\Omega_T$ *we have*

$$L_{A,b,c}W(x,t) + \eta(t)\bigl|\sqrt{A(x,t)}\nabla W(x)\bigr|^2 \leqslant C - W(x)G\bigl(W(x)\bigr).$$

Then for almost all $t \in (0,T)$ *the inequality*

$$\int_{\mathbb{R}^d} \exp\bigl(\eta(t)W(x)\bigr)\,\mu_t(dx) \leqslant \exp\Bigl((1-\delta)^{-1}\eta^{1-\delta}(t) + C\int_0^t \eta(s)\,ds\Bigr)$$

holds.

PROOF. For the proof of (i) it suffices to apply Theorem 7.1.1 with functions $H(t)=K(t)=C$ and $V(x,t)=W(x)$.

Let us prove (ii). Set $V(x,t)=\eta(t)W(x)$. Then

$$\partial_t V(x,t) + L_{A,b,c}V(x,t) \leqslant \eta'(t)W(x) - \eta(t)W(x)G\bigl(W(x)\bigr) + C\eta(t).$$

We observe that for all $\alpha,\beta\geqslant 0$ the inequality

$$\alpha\beta \leqslant \alpha G^{-1}(\alpha) + \beta G(\beta)$$

holds, where G^{-1} is the inverse function to G. Using this inequality with numbers $\alpha=\eta'/\eta$ and $\beta=W$, we obtain

$$(\partial_t V + L_{A,b,c}V)(x,t) \leqslant \eta'(t)G^{-1}\bigl(\eta'(t)/\eta(t)\bigr) + C\eta(t) = \frac{\eta'(t)}{\eta^\delta(t)} + C\eta(t),$$

since $\eta'(t)=\eta(t)G\bigl(\eta^{-\delta}(t)\bigr)$ by hypothesis. Applying Theorem 7.1.1 with $H(t)=0$ and $K(t)=\eta'(t)/\eta^\delta(t)+C\eta(t)$, we obtain the desired fact.

Let us prove (iii). Let $V(x,t)=\exp\bigl(\eta(t)W(x)\bigr)$. Then we have

$$\begin{aligned}&\partial_t V(x,t) + L_{A,b,c}V(x,t)\\ &\qquad\leqslant \Bigl[\eta'(t)W(x) - \eta(t)W(x)G(W(x)) + C\eta(t)\Bigr]\exp\bigl(\eta(t)W(x)\bigr).\end{aligned}$$

Therefore, we arrive at the following estimate:

$$\partial_t V(x,t) + L_{A,b,c}V(x,t) \leqslant \Big[\frac{\eta'(t)}{\eta^\delta(t)} + C\eta(t)\Big]\exp(\eta(t)W(x)).$$

Applying Theorem 7.1.1 with functions $K(t)=0$ and

$$H(t) = \frac{\eta'(t)}{\eta^\delta(t)} + C\eta(t),$$

we arrive at our assertion. □

Let us consider several examples.

7.1.3. Example. Let $V(x,t) = |x|^r$, where $r \geqslant 2$. Then

$$\begin{aligned} L_{A,b,c}V(x,t) =& r|x|^{r-2}\mathrm{tr}\, A(x,t) + r(r-2)|x|^{r-4}\langle A(x,t)x,x\rangle \\ &+ r|x|^{r-2}\langle b(x,t),x\rangle + |x|^r c(x,t). \end{aligned}$$

Suppose that for some numbers $C_1>0$, $C_2>0$ and all $(x,t)\in\mathbb{R}^d\times(0,T)$ we have

$$r\,\mathrm{tr}\, A(x,t) + r(r-2)|x|^{-2}\langle A(x,t)x,x\rangle + r\langle b(x,t),x\rangle + |x|^2c(x,t) \leqslant C_1 + C_2|x|^2.$$

For example, this is fulfilled if A, c are bounded and $|b(x,t)| \leqslant C + C|x|$. Suppose also $|x|^r \in L^1(\nu)$. Then we obtain the estimate

$$\int_{\mathbb{R}^d} |x|^r\,\mu_t(dx) \leqslant e^{C_3t} - 1 + e^{C_3t}\int_{\mathbb{R}^d} |x|^r\,\nu(dx)$$

for almost all $t\in(0,T)$ and some number $C_3>0$.

7.1.4. Example. Let $V(x,t) = \exp(\alpha|x|^r)$, where $r\geqslant 2$. Then

$$\begin{aligned} L_{A,b,c}V(x,t) = \exp(\alpha|x|^r)\Big[&\alpha r|x|^{r-2}\mathrm{tr}\, A(x,t) \\ &+ \alpha r(r-2)|x|^{r-4}\langle A(x,t)x,x\rangle + \alpha^2r^2|x|^{2r-4}\langle A(x,t)x,x\rangle \\ &+ \alpha r|x|^{r-2}\langle b(x,t),x\rangle + c(x,t)\Big]. \end{aligned}$$

Suppose that for some number C_1 and all $(x,t)\in\mathbb{R}^d\times(0,T)$ we have the inequality

$$\begin{aligned} &\alpha r|x|^{r-2}\mathrm{tr}\, A(x,t) + \alpha r(r-2)|x|^{r-4}\langle A(x,t)x,x\rangle \\ &\quad + \alpha^2r^2|x|^{2r-4}\langle A(x,t)x,x\rangle + \alpha r|x|^{r-2}\langle b(x,t),x\rangle + c(x,t) \leqslant C_1. \end{aligned}$$

If $\exp(|x|^r)\in L^1(\nu)$, then

$$\int_{\mathbb{R}^d} \exp(\alpha|x|^r)\,\mu_t(dx) \leqslant e^{C_2t} - 1 + e^{C_2t}\int_{\mathbb{R}^d} \exp(\alpha|x|^r)\,\nu(dx).$$

for almost all $t\in(0,T)$ and some number $C_2>0$.

In the following three examples our estimates do not depend on the initial condition given by a subprobability measure ν.

7.1.5. Example. Suppose that $k>2$, $r\geqslant 2$ and that for all $(x,t)\in\mathbb{R}^d\times(0,T)$ and some numbers $C_1>0$ and $C_2>0$ we have

$$r\,\mathrm{tr}\, A(x,t) + r(r-2)|x|^{-2}\langle A(x,t)x,x\rangle + r\langle b(x,t),x\rangle + |x|^2c(x,t) \leqslant C_1 - C_2|x|^k.$$

Then $L_{A,b,c}|x|^r \leqslant C_3 - C_3|x|^{r+k-2}$ for some number $C_3>0$. Set $W(x)=|x|^r$ and $G(z) = C_3z^\sigma$, where $\sigma = (k-2)/r > 0$. Then

$$L_{A,b,c}W(x,t) \leqslant C_3 - WG(W(x))$$

and $\eta(t) = C_4 t^{1/(\delta\sigma)}$, where the number C_4 depends only on C_3, δ and σ. Applying Corollary 7.1.2, we obtain

$$\int_{\mathbb{R}^d} |x|^r \,\mu_t(dx) \leqslant \gamma t^{-r/(k-2)},$$

where γ depends on C_1, C_2, δ, σ.

7.1.6. Example. Suppose that $r > 2$, $k > r$ and that for some numbers $C_1 > 0$ and $C_2 > 0$ and all $(x,t) \in \mathbb{R}^d \times (0,T)$ we have

$$\alpha r|x|^{r-2}\mathrm{tr}\, A(x,t) + \alpha r(r-2)|x|^{r-4}\langle A(x,t)x, x\rangle + c(x,t) \\ + \alpha^2 r^2 |x|^{2r-4}\langle A(x,t)x,x\rangle + \alpha r|x|^{r-2}\langle b(x,t),x\rangle \leqslant C_1 - C_2|x|^k.$$

Then $L_{A,b,c}\exp(\alpha|x|^r) \leqslant C_3 - C_3|x|^k \exp(\alpha|x|^r)$, where $C_3 > 0$ is a constant. Set $W(x) = \exp(\alpha|x|^r)$ and $G(z) = C_3|\ln z|^\sigma$ if $z \geqslant 2$, where $\sigma = k/r > 1$. We obtain $L_{A,b,c}W(x,t) \leqslant C_3 - WG(W(x))$ and $\eta(t) = C_4\exp(-C_5 t^{1/(1-\sigma-1)})$, where the numbers $C_4 > 0$ and $C_5 > 0$ depend only on C_3, δ and σ. Applying Corollary 7.1.2, we arrive at the inequality

$$\int_{\mathbb{R}^d} \exp(\alpha|x|^r)\,\mu_t(dx) \leqslant \gamma_1 \exp\Bigl(\gamma_2 t^{-r/(k-r)}\Bigr),$$

where γ_1 and γ_2 depend only on C_1, C_2, δ and σ.

7.1.7. Example. Let $r > 2$, $k > 2$ and $\alpha > 0$. Suppose that for some numbers $C_1 > 0$ and $C_2 > 0$ and all $(x,t) \in \mathbb{R}^d \times (0,T)$ we have

$$\alpha r\,\mathrm{tr}\, A(x,t) + \alpha r(r-2)|x|^{-2}\langle A(x,t)x,x\rangle + \alpha r\langle b(x,t),x\rangle \\ + \alpha|x|^2 c(x,t) + \alpha^2 r^2|x|^{r-2}\langle A(x,t)x,x\rangle \leqslant C_1 - C_2|x|^k.$$

Then $\alpha L_{A,b,c}|x|^r + \alpha^2 r^2|x|^{2r-4}\langle A(x,t)x,x\rangle \leqslant C_3 - C_3|x|^{k+r-2}$. Letting

$$W(x) = \alpha|x|^r \quad \text{and} \quad G(z) = C_3\alpha^{-(1+\sigma)/\sigma} z^\sigma, \quad \text{where } \sigma = (k-2)/r > 0,$$

we obtain

$$L_{A,b,c}W(x,t) + \bigl|\sqrt{A(x,t)}\nabla W(x)\bigr|^2 \leqslant C_3 - WG(W(x)).$$

Applying Corollary 7.1.2 with $\delta \in (0,1)$, $\eta(t) = C_4 t^{1/(\delta\sigma)}$, where C_4 depends only on C_3, δ and σ, we obtain for all $\beta > r/(k-2)$ the estimate

$$\int_{\mathbb{R}^d} \exp(\alpha t^\beta|x|^r)\,\mu_t(dx) \leqslant \gamma_1\exp\Bigl(\gamma_2\bigl(t^{\beta-\frac{r}{k-2}} + t^{\beta+1}\bigr)\Bigr),$$

where positive numbers γ_1 and γ_2 depend only on C_1, C_2, r and β.

Note that the estimates in Examples 7.1.5, 7.1.6 and 7.1.7 do not depend on the initial condition. Applying them to the transition probabilities $P(0,y,t,dx)$ of the diffusion process with generator L, one can obtain estimates uniform in y. Estimates of such a form were first obtained for the kernels of the transition semigroup in the papers Metafune, Pallara, Rhandi [**695**] and Spina [**869**].

The hypotheses of Theorem 7.1.1 not only guarantee some estimates on the integral of the function V, but also imply for almost all $t \in (0,T)$ the equality

$$\mu_t(\mathbb{R}^d) = \nu(\mathbb{R}^d) + \int_0^t\int_{\mathbb{R}^d} c(x,s)\,\mu_s(dx)\,ds \tag{7.1.3}$$

for every measure $\mu = (\mu_t)_{0<t<T}$ of class $\mathcal{M}_\nu$. Since for this equality it suffices to have some Lyapunov function V, we shall now prove that transforming a given function V we can neglect the condition $V(\,\cdot\,,0) \in L^1(\nu)$, because it can be ensured for a new Lyapunov function.

7.1.8. Proposition. *Let $c \leqslant 0$ and let ν be a subprobability measure on $\mathbb{R}^d$. Suppose that $\mu = (\mu_t)_{0<t<T} \in \mathcal{M}_\nu$ and that there exists a nonnegative function V such that $V \in C^{2,1}(\mathbb{R}^d_T) \cap C(\mathbb{R}^d \times [0,T))$, for every compact interval $[\alpha,\beta] \in (0,T)$ we have*

$$\lim_{|x|\to\infty} \min_{t\in[\alpha,\beta]} V(x,t) = +\infty,$$

and for some nonnegative functions $K, H \in L^1\big((0,T)\big)$ we have

$$\partial_t V(x,t) + L_{A,b,c}V(x,t) \leqslant K(t) + H(t)V(x,t).$$

Then there exists a function $W \in C^{2,1}(\mathbb{R}^d_T) \cap C\big(\mathbb{R}^d \times [0,T)\big)$ such that $W \geqslant 0$ and for every compact interval $[\alpha,\beta] \in (0,T)$ we have

$$\lim_{k\to\infty} \min_{t\in[\alpha,\beta]} W(x,t) = +\infty,$$

$$\partial_t W(x,t) + L_{A,b,c}W(x,t) \leqslant K(t) + H(t) \quad \textit{and} \quad W(x,0) \in L^1(\nu).$$

PROOF. We construct a function $\theta \in C^2(\mathbb{R})$ with the following properties: $\theta \geqslant 0$, $\theta(0) = 0$, $\lim_{r\to\infty} \theta(r) = +\infty$, $0 \leqslant \theta'(r) \leq 1$, $\theta''(r) \leqslant 0$ and $\theta\big(V(\,\cdot\,,0)\big) \in L^1(\nu)$. For this it suffices to find a function θ with the indicated properties and integrable with respect to the measure $\sigma = \nu \circ W^{-1}(\,\cdot\,,0)$. We find an increasing sequence of numbers z_k such that $z_{k+1} - z_k \geqslant z_k - z_{k-1} \geqslant 1$ and $\sigma\big([z_k,\infty)\big) \leqslant 2^{-k}$. Let a function θ_0 be affine on each interval $[z_k, z_{k+1}]$ and $\theta_0\big(z_k\big) = k-1$. We have obtained a σ-integrable increasing concave function θ_0, which, however, does not belong to the class C^2. Let us take for θ the function

$$\theta(z) = \int_0^z g(s)\,ds, \quad g \in C^1(\mathbb{R}),$$

where $g(z) \geqslant 0$, $g'(z) \leqslant 0$ and if $z \in (z_k, z_{k+1} - k^{-1})$, then $g(z) = \theta_0'(z)$. Clearly, this is a suitable function. Furthermore, taking into account the concavity of θ and the sign of c, we obtain

$$\begin{aligned}\partial_t\theta(V) + L_{A,b,c}\theta(V) &= \theta'(V)\big(\partial_t V + L_{A,b,c}V\big)\\ &\quad + \theta''(V)\langle A\nabla V, \nabla V\rangle + c\big(\theta(V) - \theta'(V)V\big) \leqslant K + H\theta(V).\end{aligned}$$

The function $W := \ln\big(1 + \theta(V)\big)$ is the desired one. □

7.1.9. Remark. Arguing somewhat differently, one can obtain sufficient conditions for equality (7.1.3) of another type. Say, let a function V be such that $V \in C^2(\mathbb{R}^d)$ and $\lim_{|x|\to\infty} V(x) = +\infty$. Let $\zeta \in C^\infty\big([0,+\infty)\big)$, $\zeta(x) = 1$ if $x < 1$ and $\zeta(x) = 0$ if $x > 2$. Set $\varphi_N(x) = \zeta\big(V(x)/N\big)$. If $\mu \in \mathcal{M}_\nu$, then

$$\int_{\mathbb{R}^d} \varphi_N\,d\mu_t = \int_{\mathbb{R}^d} \varphi_N\,d\nu + \int_0^t\!\!\int_{\mathbb{R}^d} [L_0\varphi_N + c\varphi_N]\,d\mu_s\,ds,$$

where $L_0 = a^{ij}\partial_{x_i}\partial_{x_j} + b^i\partial_{x_i}$. We observe that $\varphi_N \to 1$ as $N \to \infty$. Thus, for equality (7.1.3) it suffices to have the equality

$$\lim_{N\to\infty} \int_0^t \int_{\mathbb{R}^d} L_0\varphi_N(x,s)\,\mu_s(dx)\,ds = 0,$$

which fulfills, for example, if

$$\lim_{N\to\infty} \int_0^T \int_{N\leqslant V\leqslant 2N} \Big[N^{-1}|L_0V| + N^{-2}|\sqrt{A}\nabla V|^2\Big]\,d\mu_s\,ds = 0.$$

If it is known that $a^{ij}(\,\cdot\,,t) \in W^{1,1}_{\rm loc}(\mathbb{R}^d)$, $\mu_t = \varrho(x,t)\,dx$, and $\varrho(\,\cdot\,,t) \in W^{1,1}_{\rm loc}(\mathbb{R}^d)$, then for equality (7.1.3) it is sufficient to have the equality

$$\lim_{N\to\infty} \int_0^T \int_{N\leqslant V\leqslant 2N} \Big[N^{-1}|(b-\beta_\mu)\nabla V| + N^{-2}|\sqrt{A}\nabla V|^2\Big]\,d\mu_s\,ds = 0,$$

where we set by definition $\beta^i_\mu = \sum_{j=1}^d \big(\partial_{x_j}a^{ij} + a^{ij}\varrho^{-1}\partial_{x_j}\varrho\big)$.

Observe also that the probabilistic sense of equality (7.1.3) is that the diffusion process corresponding to the operator $L_{A,b,c}$ does not reach infinity in finite time.

7.2. Global upper bounds

In this section we use the local estimates from Chapter 6 and scaling to deduce some global bounds. Here we do not assume that $c \geqslant 0$.

Suppose that for every ball $U \subset \mathbb{R}^d$ the matrix A satisfies the conditions

$$\sup_{t\in(0,T)} \|a^{ij}(\,\cdot\,,t)\|_{W^{p,1}(U)} < \infty, \quad \text{where } p > d+2,$$

and for every compact interval $J \subset (0,T)$ satisfies the condition

$$0 < \lambda_0(U,J) := \inf\big\{\langle A(x,t)\xi,\xi\rangle\colon\ |\xi| = 1,\ (x,t) \in U\times J\big\}.$$

Set

$$\|A(x,t)\| = \max_{\xi\colon\ |\xi|=1} \langle A(x,t)\xi,\xi\rangle.$$

Suppose that a *nonnegative* locally finite Borel measure μ on $\mathbb{R}^d_T$ satisfies the equation

$$\partial_t\mu = L^*_{A,b,c}\mu.$$

We shall assume that

$$b^i \in L^p_{\rm loc}(\mu), \quad c \in L^{p/2}_{\rm loc}(\mu).$$

Then, according to Proposition 6.5.1, the measure μ possesses a locally Hölder continuous density $\varrho \in \mathcal{H}^{p,1}_{\rm loc}$ with respect to Lebesgue measure.

Set $B^i = b^i - \partial_{x_j}a^{ij}$.

We recall that the equation $\partial_t\mu = L^*_{A,b,c}\mu$ can be written as a divergence form equation on the density ϱ of the measure μ:

$$\partial_t\varrho = \operatorname{div}\big(A\nabla\varrho - B\varrho\big) + c\varrho. \tag{7.2.1}$$

The next result is Theorem 6.5.4, the formulation of which is repeated here for convenience.

7.2.1. Theorem. *Let $\gamma > (d+2)/2$ and let U and U' be two balls in $\mathbb{R}^d$ with $\overline{U'} \subset U$. Let also $[s_1, s_2] \subset (0, T)$. Then, for every $s \in (s_1, s_2)$, there exists a number $C > 0$, depending only on U, U', s, s_1, d and γ, such that*

$$\|\varrho\|_{L^\infty(U'\times[s,s_2])} \leqslant C(1+\lambda_0^{-1})^\gamma \int_{s_1}^{s_2}\int_U \Big[1 + \|A\|^\gamma + |c^+|^\gamma + \big|\sqrt{A^{-1}}B\big|^{2\gamma}\Big]\varrho\, dx\, dt,$$

where the numbers $\lambda_0 = \lambda_0(U, [s_1, s_2])$, $\|A\|$ are defined above.

Using the simple observation that after scaling the equation keeps its form and the coefficients change in a simple way: are just multiplied by a distortion coefficient in the corresponding degree, we derive an estimate on all of $\mathbb{R}^d_T$.

We recall that $U(x, r)$ is the open ball in $\mathbb{R}^d$ of radius r centered at x.

7.2.2. Corollary. *Let $\gamma > (d+2)/2$, $\kappa > 0$ and $t_0 \in (0, T)$. Then there exists a number $C > 0$, depending only on κ, t_0, d, γ, such that for all $(x, t) \in \mathbb{R}^d \times (t_0/2, T)$ we have the estimate*

$$\varrho(x,t) \leqslant C\big(1+\lambda_0^{-1}(x,t)\big)^\gamma \int_{t_0/2}^t \int_{U(x,\kappa)} \Big[1 + \|A\|^\gamma + |c^+|^\gamma + \big|\sqrt{A^{-1}}B\big|^{2\gamma}\Big]\varrho\, dy\, d\tau,$$

where

$$\lambda_0(x,t) = \inf\big\{\langle A(y,\tau)\xi, \xi\rangle \colon |\xi| = 1,\ (y,\tau) \in U(x,\kappa)\times[t_0/2, t]\big\}.$$

In particular, if $\mu_t(dx) = \varrho(x,t)\,dx$ is a subprobability measure for almost all points $t \in (0, T)$, in addition,

$$\|A\|^\gamma,\ |c^+|^\gamma,\ |B|^{2\gamma} \in L^1\big(\mathbb{R}^d \times (t_0/2, T), \mu\big)$$

and the function $\|A\|^{-1}$ is uniformly bounded, then $\varrho \in L^\infty\big(\mathbb{R}^d \times (t_0, T)\big)$.

PROOF. We shift x to 0 and apply Theorem 7.2.1 for the sets $U = U(0, \kappa)$, $U' = U(0, \kappa/2)$ and $s_1 = t_0/2$, $s = t_0$, $s_2 = t$. □

7.2.3. Corollary. *Let $\gamma > (d+2)/2$ and $\Theta \in (0,1)$. Then there exists a number $C > 0$, depending only on γ, d, Θ, such that for all points $(x, t) \in \mathbb{R}^d_T$ one has the following estimate:*

$$\varrho(x,t) \leqslant C\big(1+\lambda_0^{-1}(x,t)\big)^\gamma t^{-(d+2)/2}$$
$$\times \int_{\Theta t}^t \int_{U(x,\sqrt{t})} \Big[1 + \|A\|^\gamma + t^{2\gamma}|c^+|^\gamma + t^{2\gamma}\big|\sqrt{A^{-1}}B\big|^{2\gamma}\Big]\varrho\, dy\, d\tau,$$

where

$$\lambda_0(x,t) = \inf\big\{\langle A(y,\tau)\xi, \xi\rangle \colon |\xi| = 1, \quad (y,\tau) \in U(x,\sqrt{t})\times[\Theta t, t]\big\}.$$

In particular, if $\mu_t(dx) = \varrho(x,t)\,dx$ is a subprobability measure for almost all points $t \in (0, T)$, the functions $\|A\|^\gamma$, $|c^+|^\gamma$, $|B|^{2\gamma}$ belong to $L^1(\Omega_T, \mu)$ and the function $\|A\|^{-1}$ is uniformly bounded, then there exists a number $\widetilde{C} > 0$ such that

$$\varrho(x,t) \leqslant \widetilde{C} t^{-d/2} \quad \text{for all } (x,t) \in \mathbb{R}^d_T.$$

PROOF. For obtaining the desired estimate at (x_0, t_0) we change coordinates $x \mapsto (x - x_0)/\sqrt{t_0}$, $t \mapsto t/t_0$ and apply Theorem 7.2.1 for the balls $U = U(0,1)$, $U' = U(0, 1/2)$ and the points $s_1 = \Theta$, $s = (1+\Theta)/2$, and $s_2 = 1$. □

7.2.4. Remark. Suppose that

$$m\cdot \mathrm{I}\leqslant A(x,t)\leqslant M\cdot \mathrm{I},\quad |a^{ij}(x,t)-a^{ij}(y,t)|\leqslant \lambda|x-y|$$

for all $(x,t)\in\mathbb{R}^d_T$ and some numbers $m, M, \lambda>0$. Then the estimate from Corollary 7.2.3 can be written as

$$\varrho(x,t)\leqslant Ct^{-(d+2)/2}\mu\bigl(U(x,\sqrt{t})\times[\Theta t,t]\bigr) + Ct^{-(d+2)/2+2\gamma}\int_{\Theta t}^{t}\int_{U(x,\sqrt{t})}\bigl[|b|^{2\gamma}+|c^+|^{\gamma}\bigr]\,d\mu.$$

We observe that $-(d+2)/2+2\gamma>0$. Therefore, for obtaining estimates as $t\to 0$ it suffices to study the behavior of the quantity

$$\mu\bigl(U(x,\sqrt{t})\times[\Theta t,t]\bigr).$$

This idea is developed in the next section.

7.2.5. Corollary. *Let $\Phi\in C^{2,1}(\mathbb{R}^d_T)$ and $\Phi>0$. Set*

$$c_0=c+\bigl(\partial_t\Phi+\operatorname{div}(A\nabla\Phi)+B\nabla\Phi\bigr)\Phi^{-1},\quad B_0=B+\Phi^{-1}A\nabla\Phi.$$

Let $\gamma>(d+2)/2$ and $\Theta\in(0,1)$. Then there is a number $C>0$, depending only on γ, d, Θ, such that for all $(x,t)\in\mathbb{R}^d\times(0,T)$ we have

$$\varrho(x,t)\leqslant C\Phi(x,t)^{-1}\bigl(1+\lambda_0^{-1}(x,t)\bigr)^{\gamma}t^{-(d+2)/2} \times\int_{\Theta t}^{t}\int_{U(x,\sqrt{t})}\Bigl(1+\|A\|^{\gamma}+t^{2\gamma}|c_0^+|^{\gamma}+t^{2\gamma}\bigl|\sqrt{A^{-1}}B_0\bigr|^{2\gamma}\Bigr)\Phi\varrho\,dy\,d\tau,$$

where λ_0 is defined in Corollary 7.2.2. In particular, if

$$\sup_{t\in(0,T)}\int_{\mathbb{R}^d}\Phi(x,t)\varrho(x,t)\,dx<\infty,$$

$\|A\|^{\gamma}\Phi$, $|c_0^+|^{\gamma}\Phi$, $|B_0|^{2\gamma}\Phi\in L^1(\mathbb{R}^d_T,\mu)$ and the function $\|A\|^{-1}$ is uniformly bounded, then there exists a number $\widetilde{C}>0$ such that

$$\varrho(x,t)\leqslant \widetilde{C}t^{-d/2}\Phi(x,t)^{-1}\quad\text{for all } (x,t)\in\mathbb{R}^d_T.$$

PROOF. It suffices to notice that the function $\Phi\varrho$ satisfies equation (7.2.1) with the new coefficients c_0 and B_0. □

Let us consider two typical examples of application of the obtained results. Suppose that $c\leqslant 0$ and $\mu_t(dx)=\varrho(x,t)\,dx$ is a subprobability solution to the Cauchy problem (6.1.1) and (6.1.2) of class $\mathcal{M}_\nu$, where the initial condition ν is a subprobability measure on $\mathbb{R}^d$. We shall obtain upper estimates for the density ϱ in different situations.

7.2.6. Example. Let $\alpha>0$, $r>2$ and $k>r$. Suppose that

$$\alpha r|x|^{r-2}\mathrm{tr}\,A(x,t)+\alpha r(r-2)|x|^{r-4}\langle A(x,t)x,x\rangle+c(x,t) +\alpha^2r^2|x|^{2r-4}\langle A(x,t)x,x\rangle+\alpha r|x|^{r-2}\langle b(x,t),x\rangle\leqslant C-C|x|^k$$

for some $C>0$ and all $(x,t)\in\mathbb{R}^d_T$. Assume that for all $(x,t)\in\mathbb{R}^d_T$ one has

$$C_1\exp\bigl(-\kappa_1|x|^{r-\delta}\bigr)\leqslant\|A(x,t)\|\leqslant C_2\exp\bigl(\kappa_2|x|^{r-\delta}\bigr),$$

$$|b^i(x,t)|+|\partial_{x_j}a^{ij}(x,t)|\leqslant C_3\exp(\kappa_3|x|^{r-\delta})$$

where C_1, C_2, C_3, κ_1, κ_2, κ_3 are positive numbers and $\delta \in (0,r)$. Let $\alpha' \in (0,\alpha)$. Then, the density ϱ satisfies the estimate

$$\varrho(x,t) \leqslant C_4 \exp(-\alpha'|x|^r) \exp\bigl(C_5 t^{-\frac{r}{k-r}}\bigr)$$

for all $(x,t) \in \mathbb{R}^d_T$ with some positive numbers C_4 and C_5. In particular, such an estimate holds in the following situation. Suppose that $d = d_1 + d_2 \geqslant 2$; we shall write $x = (x', x'')$, where $x' \in \mathbb{R}^{d_1}$, $x'' \in \mathbb{R}^{d_2}$. Let $r > 2$, $k > r$ and $\delta \in (0,1)$. Set

$$A(x,t) = e^{|x'|^{r-\delta} - |x''|^{r-\delta}} \cdot \mathrm{I}, \; b(x,t) = -x|x|^{r-2} e^{|x'|^{r-\delta} - |x''|^{r-\delta}}, \; c(x,t) = -|x|^k.$$

PROOF. According to Example 7.1.6 we have

$$\int_{\mathbb{R}^d} \exp\bigl(\alpha|x|^r\bigr)\mu_t(dx) \leqslant \gamma_1 \exp\Bigl(\gamma_2 t^{-\frac{r}{k-r}}\Bigr)$$

for a.e. $t \in (0,T)$ and some numbers γ_1 and γ_2. Set $\Phi(x) = \exp\bigl(\alpha'|x|^r\bigr)$. We observe that $\widetilde{c}^+ \leqslant \gamma_3$ for some number γ_3 and

$$\Bigl(1 + \|A(x,t)\|^\gamma + t^{2\gamma}\bigl|\sqrt{A^{-1}(x,t)}\widetilde{B}(x,t)\bigr|^{2\gamma}\Bigr)\Phi \leqslant \gamma_4 \exp\bigl(\alpha|x|^r\bigr)$$

for all $(x,t) \in \mathbb{R}^d \times (0,T)$. Now Corollary 7.2.5 applies. □

7.2.7. Example. Suppose we are given numbers $r > 2$, $k > 2$, $\gamma > d+2$, $\alpha > 0$ and $\beta > r/(k-2)$, and for all $(x,t) \in \mathbb{R}^d_T$ and some number $C > 0$ the following inequality is true (for example, it is true in the case where $A = \mathrm{I}$, $c = 0$, $\langle b(x,t), x\rangle \leqslant C - C|x|^k$):

$$\alpha r \operatorname{tr} A(x,t) + \alpha r(r-2)|x|^{-2}\langle A(x,t)x, x\rangle + \alpha r\langle b(x,t), x\rangle$$
$$+ |x|^2 c(x,t) + \alpha^2 r^2 |x|^{r-2}\langle A(x,t)x, x\rangle \leqslant C - C|x|^k.$$

Suppose also that for all $(x,t) \in \mathbb{R}^d_T$ we have the estimates

$$C_1\bigl(1 + |x|^{\frac{m}{\gamma}}\bigr)^{-1} \leqslant \|A(x,t)\| \leqslant C_2\bigl(1 + |x|^{\frac{m}{\gamma}}\bigr),$$

$$|b^i(x,t)|^{2\gamma} + |\partial_{x_j} a^{ij}(x,t)|^{2\gamma} \leqslant C_3(1 + |x|^m),$$

where C_1, C_2, C_3 are positive numbers and $m \geqslant \gamma \max\{r-1, r\beta^{-1}\}$. Let $\alpha' \in (0,\alpha)$. Then the density ϱ satisfies the estimate

$$\varrho(x,t) \leqslant C_4 t^{-\kappa} \exp\bigl(-\alpha' t^\beta |x|^r\bigr), \quad \kappa = 4m\beta/r + d/2 - 2\gamma$$

for all $(x,t) \in \mathbb{R}^d \times (0,T)$ and some number $C_4 > 0$.

PROOF. According to Example 7.1.7 we have

$$\int_{\mathbb{R}^d} \exp\bigl(\alpha t^\beta |x|^r\bigr)\, \mu_t(dx) \leqslant \gamma_1$$

for almost all $t \in (0,T)$ and some number γ_1. We observe that for all $p \geqslant 1$ and $\varepsilon > 0$ there holds the estimate $|x|^p \leqslant \gamma_2 t^{-\beta p/r} \exp\bigl(\varepsilon t^\beta |x|^r\bigr)$. It remains to apply Corollary 7.2.5 with the function $\Phi(x,t) = \exp\bigl(\alpha t^\beta |x|^r\bigr)$. □

7.3. Upper bounds for solutions to the Cauchy problem

The results presented in the previous section do not use information about the distribution at $t = 0$. Here we sharpen the estimates obtained above taking into account the initial distribution, including the case where the initial distribution has a locally bounded density. Moreover, it turns out that, for a sufficiently regular diffusion matrix, the drift coefficient has practically no influence on the behavior of the solution density as $t \to 0$.

In this section we assume that the matrix A satisfies the following two conditions:

$$
\begin{aligned}
&\text{(P1)} \quad \text{for some numbers } m, M > 0 \text{ we have} \\
&\qquad\quad m \cdot \mathrm{I} \leqslant A(x,t) \leqslant M \cdot \mathrm{I} \quad \text{for all } (x,t) \in \mathbb{R}^d_T, \\
&\text{(P2)} \quad \text{for some number } \lambda > 0 \text{ we have} \\
&\qquad\quad |a^{ij}(x,t) - a^{ij}(y,s)| \leqslant \lambda(|t-s|^{1/2} + |x-y|)
\end{aligned} \tag{7.3.1}
$$

for all $t, s \in [0, T]$ and $x, y \in \mathbb{R}^d$.

Set

$$
Z_{x,t}(y,s) = (2\pi s)^{-d/2} |\det A(x,t)|^{-1/2} \exp\Bigl[-\frac{\langle A^{-1}(x,t)y, y\rangle}{2s}\Bigr],
$$

where A^{-1} is the inverse matrix to the matrix $A = (a^{ij})$. Clearly,

$$
\int_{\mathbb{R}^d} Z_{x,t}(y,s)\, dy = 1.
$$

For a function $\varphi \in C_0^\infty(\mathbb{R}^d)$ we set

$$
Q(y,s) = \int_{\mathbb{R}^d} Z_{x,t}(y-z, \tau - s)\varphi(z)\, dz.
$$

We observe that $\partial_s Q(y,s) + a^{ij}(x,t)\partial_{y_i}\partial_{y_j} Q(y,s) = 0$ and $Q(y,\tau) = \varphi(y)$. In addition, we obviously have the estimate $|Q(y,s)| \leqslant \max_x |\varphi(x)|$.

7.3.1. Lemma. *Let $\varphi \geqslant 0$, $\varphi = 0$ outside the ball $U(y_0, R)$ and*

$$
M_1 = \sup_{x\in\mathbb{R}^d} |\nabla\varphi(x)|\varphi(x)^{(1-q)/q}, \quad M_2 = \sup_{x\in\mathbb{R}^d} \|D^2\varphi(x)\|\varphi(x)^{(1-q)/q}
$$

for some $q > 1$. Then

$$
|\nabla Q(y,s)| \leqslant M_1 Q(y,s)^{(q-1)/q}, \quad \|D^2 Q(y,s)\| \leqslant M_2 Q(y,s)^{(q-1)/q},
$$

$$
|y - y_0|\, \|D^2 Q(y,s)\| \leqslant \bigl(R + N\sqrt{\tau - s}\bigr) M_2 Q(y,s)^{(q-1)/q},
$$

where the number N depends only on q, m, M, d.

PROOF. Applying Hölder's inequality with the weight function $Z_{x,t}$, we obtain

$$
\begin{aligned}
|\nabla Q(y,s)| &\leqslant \int_{\mathbb{R}^d} Z_{x,t}(y-z, \tau-s)|\nabla\varphi(z)|\, dz \\
&\leqslant M_1 \int_{\mathbb{R}^d} Z_{x,t}(y-z,\tau-s)|\varphi(z)|^{(q-1)/q}\, dz \leqslant M_1 Q(y,s)^{(q-1)/q}.
\end{aligned}
$$

The second inequality is proved similarly. We proceed to the third estimate:

$$|y-y_0|\,\|D^2Q(y,s)\|\leqslant\int_{|z-y_0|<R}|y-y_0|Z_{x,t}(y-z,\tau-s)\|D^2\varphi(z)\|\,dz$$
$$\leqslant\int_{|z-y_0|<R}|z-y_0|Z_{x,t}(y-z,\tau-s)\|D^2\varphi(z)\|\,dz$$
$$+\int_{|z-y_0|<R}|y-z|Z_{x,t}(y-z,\tau-s)\|D^2\varphi(z)\|\,dz.$$

The first term is easily estimated by $RM_2Q(y,s)^{(q-1)/q}$, and the second term is estimated by the expression

$$M_2\sqrt{\tau-s}\biggl(\int_{\mathbb{R}^d}|z|^qZ_{x,t}(z,1)\,dy\biggr)^{1/q}Q(y,s)^{(q-1)/q},$$

which completes the proof. □

The next lemma plays the key role in this section.

7.3.2. Lemma. *Let ν be a subprobability measure on $\mathbb{R}^d$ and let a nonnegative Borel measure $\mu=(\mu_t)_{0<t<T}$, where each μ_t is a subprobability measure on $\mathbb{R}^d$, be a solution to the Cauchy problem* (6.1.1) *and* (6.1.3) *with initial condition ν. Suppose that the coefficients a^{ij} satisfy condition* (7.3.1) *and that for some number $p>d+2$ and every ball $U\subset\mathbb{R}^d$ we have*

$$b\in L^p\bigl(\mu,U\times(0,T)\bigr)\quad\text{and}\quad c\in L^{p/2}\bigl(\mu,U\times(0,T)\bigr).$$

Then, for every $\theta\in(0,1)$ and $\kappa>1$, there exists a number $C>0$ such that

$$\mu\bigl(U(x,\sqrt{t})\times[\theta t,t]\bigr)\leqslant C\sigma\bigl(U(x,\kappa\sqrt{t})\times[\theta t,t]\bigr)$$
$$+Ct^{p/2}\Bigl(1+\|b\|_{L^p(\mu,U(x,2)\times(0,T))}+\|c^+\|_{L^p(\mu,U(x,2)\times(0,T))}\Bigr)^p$$

for all $t\in(0,\theta)$ and $x\in\mathbb{R}^d$, where the measure σ is defined by the density

$$\int_{|y-z|<2}Z_{x,t}(z-y,s)\,\nu(dz)$$

with respect to Lebesgue measure on $\mathbb{R}^d\times(0,T)$.

PROOF. Let us fix x_0 and t_0. Let $\varphi\in C_0^\infty(\mathbb{R}^d)$ be such that $0\leqslant\varphi\leqslant1$, $\varphi(y)=1$ if $|y-x_0|\leqslant\sqrt{t_0}$ and $\varphi(y)=0$ if $|y-x_0|\geqslant\kappa\sqrt{t_0}$, and there exists a constant K such that

$$\sup_{y\in\mathbb{R}^d}|\nabla\varphi(y)|\varphi(y)^{(1-p)/p}\leqslant Kt_0^{-1/2},\quad\sup_{y\in\mathbb{R}^d}\|D^2\varphi(y)\|\varphi(y)^{(1-p)/p}\leqslant Kt_0^{-1}.$$

For the function φ defined above and an arbitrary number $t\in(\theta t_0,t_0)$ we define, as above, the function Q by the formula

$$Q(y,s)=\int_{\mathbb{R}^d}Z_{x_0,t_0}(y-z,t-s)\varphi(z)\,dz.$$

Let $\zeta\in C_0^\infty(\mathbb{R}^d)$ be a nonnegative function such that $\zeta\leqslant1$, $\zeta(y)=1$ if $|x_0-y|<1$ and $\zeta(y)=0$ if $|x_0-y|>2$. Moreover, suppose that there exists a constant K_1 such that

$$|\nabla\zeta(y)|\zeta(y)^{(1-p)/p}\leqslant K_1,\quad\|D^2\zeta(y)\|\zeta(y)^{(1-p)/p}\leqslant K_1.$$

Assertion (iii) in Proposition 6.1.2 with the function $Q(y,s)\zeta(y)$ gives

$$\int_{\mathbb{R}^d} Q(y,s)\zeta(y)\,d\mu_s - \int_{\mathbb{R}^d} Q(y,0)\zeta(y)\,\nu(dy)$$
$$= \lim_{\delta\to 0}\int_\delta^s\int_{\mathbb{R}^d}\Big[\big(a^{ij}(y,\tau)-a^{ij}(x,t)\big)\partial_{y_i}\partial_{y_j}Q(y,\tau)\zeta(y)$$
$$+ a^{ij}(y,\tau)\partial_{y_i}Q(y,\tau)\partial_{y_j}\zeta(y) + a^{ij}\partial_{y_i}\partial_{y_j}\zeta(y)Q(y,\tau)$$
$$+ b^i(y,\tau)\partial_{y_i}Q(y,\tau)\zeta(y) + b^i(y,\tau)\partial_{y_i}\zeta(y)Q(y,\tau) + c(y,\tau)\zeta(y)Q(y,\tau)\Big]\,\mu_\tau(dy)\,d\tau.$$

According to Condition (P1) in (7.3.1) we have

$$|a^{ij}(y,\tau)-a^{ij}(x,t)| \leqslant \lambda(|\tau-t|^{1/2}+|x-y|).$$

Let $s\in[\theta t_0,t]$. By Lemma 7.3.1 we estimate the right-hand side in this relation by

$$C_1 t_0^{-1/2}\Big(1+\|b\|_{L^p(\mu,U(x_0,2)\times(0,T))}+\|c^+\|_{L^p(\mu,U(x_0,2)\times(0,T))}\Big)$$
$$\times\bigg(\int_0^s\int_{\mathbb{R}^d} Q(y,\tau)\zeta(y)\,\mu_\tau(dy)\,d\tau\bigg)^{(p-1)/p},$$

where the number C_1 depends only on θ, m, M, K, K_1, d. Set

$$h(s)=\int_0^s\int_{\mathbb{R}^d} Q(y,\tau)\zeta(y)\,\mu_\tau(dy),\quad D_0=\int_{\mathbb{R}^d} Q(y,0)\zeta(y)\,\nu(dy),$$
$$D_1 = C_1 t^{-1/2}\big(1+\|b\|_{L^p(\mu,U(x_0,2)\times(0,T))}+\|c^+\|_{L^p(\mu,U(x_0,2)\times(0,T))}\big).$$

Then

$$h'(s)\leqslant D_0 + D_1 h(s)^{1-1/p},\quad h(0)=0.$$

Let us use the inequality $x^\alpha+y^\alpha\leqslant 2(x+y)^\alpha$ for all $x,y>0$ and $\alpha\in(0,1)$. Let $\alpha=1-1/p$. We obtain

$$h'(s)\leqslant 2\big(D_0^{1/\alpha}+D_1^{1/\alpha}h(s)\big)^\alpha.$$

Integrating, we obtain

$$\big(D_0^{1/\alpha}+D_1^{1/\alpha}h(s)\big)^{1-\alpha}\leqslant 2(1-\alpha)D_1^{1/\alpha}s + D_0^{(1-\alpha)/\alpha}.$$

The last inequality implies the estimate

$$h'(s)\leqslant C(p)D_0 + C(p)D_1^p s^{p-1}.$$

We recall that $Q(y,t)=\varphi(y)$. We obtain the inequality

$$\int_{\mathbb{R}^d}\varphi(y)\zeta(y)\,\mu_t(dy)\leqslant C(p)\int_{\mathbb{R}^d} Q(y,0)\zeta(y)\,\nu(dy) + C(p)(D_1)^p t^{p-1}.$$

Since $\zeta(y)\varphi(y)=1$ whenever $|y-x_0|\leqslant\sqrt{t_0}$, we have

$$\mu\big(U(x_0,\sqrt{t_0})\times[\theta t_0,t_0]\big)\leqslant\int_{\theta t_0}^{t_0}\int_{\mathbb{R}^d}\varphi(y)\,\mu_t(dy)\,dt$$
$$\leqslant C_2\bigg(\int_{\theta t_0}^{t_0}\int_{\mathbb{R}^d} Q(y,0)\zeta(y)\,\nu(dy)\,dt$$
$$+ t_0^{p/2}\big[1+\|b\|_{L^p(\mu,U(x_0,2)\times(0,T))}+\|c^+\|_{L^p(\mu,U(x_0,2)\times(0,T))}\big]^p\bigg).$$

For completing the proof we apply Fubini's theorem and the property that $\varphi(y) = 0$ if $|y - x_0| \geqslant \kappa\sqrt{t_0}$. Thus,

$$\begin{aligned}\int_{\mathbb{R}^d} Q(y,0)\zeta(y)\,\nu(dy) &= \int_{\mathbb{R}^d} \zeta(y)\int_{\mathbb{R}^d} Z_{x,t}(y-z,t)\varphi(z)\,dz\,\nu(dy)\\ &= \int_{\mathbb{R}^d}\int_{\mathbb{R}^d} Z_{x,t}(y-z,t)\zeta(y)\,\nu(dy)\varphi(z)\,dz\\ &\leqslant \int_{U(x,\kappa\sqrt{t_0})}\int_{|x_0-y|<2} Z_{x,t}(y-z,t)\,\nu(dy)\,dz.\end{aligned}$$

The lemma is proven. □

Combining now Lemma 7.3.2 and Corollary 7.2.3 (see also Remark 7.2.4), we arrive at the following result.

7.3.3. Theorem. *Let ν be a subprobability measure on $\mathbb{R}^d$ and let a nonnegative Borel measure $\mu = (\mu_t)_{0<t<T}$, where each μ_t is a subprobability measure on $\mathbb{R}^d$, be a solution to the Cauchy problem (6.1.1) and (6.1.3) with initial condition ν. Suppose also that the coefficients a^{ij} satisfy condition (7.3.1) and that for some number $p > d+2$ and every ball $U \subset \mathbb{R}^d$ we have*

$$b \in L^p\bigl(\mu, U\times(0,T)\bigr) \quad\text{and}\quad c \in L^{p/2}\bigl(\mu, U\times(0,T)\bigr).$$

Then the measure μ has a density ϱ with respect to Lebesgue measure and for every $\theta \in (0,1)$ and $\kappa > 1$ there exists a number $C = C(\theta,\kappa) > 0$ such that

$$\begin{aligned}\varrho(x,t) &\leqslant Ct^{-(d+2)/2}\sigma\bigl(U(x,\kappa\sqrt{t})\times[\theta t,t]\bigr)\\ &\quad + Ct^{(p-d-2)/2}\bigl(1 + \|b\|_{L^p(\mu,U(x,2)\times(0,T))} + \|c^+\|_{L^{p/2}(\mu,U(x,2)\times(0,T))}\bigr)^p\end{aligned}$$

for all $(x,t) \in \mathbb{R}^d_T$, where the measure σ is given by the density

$$\int_{|x-z|<2} Z_{x,t}(z-y,s)\,\nu(dz)$$

with respect to Lebesgue measure on $\mathbb{R}^d_T$.

7.3.4. Remark. The measure σ is a solution to the Cauchy problem

$$\begin{cases}\partial_s\sigma - a^{ij}(x,t)\partial_{y_i}\partial_{y_j}\sigma = 0,\\ \sigma|_{s=0} = \widetilde{\nu},\end{cases}$$

where $\widetilde{\nu} = I_{U(x,2)}\cdot\nu$. Thus, Theorem 7.3.3 yields that if, for every ball $U \subset \mathbb{R}^d$, we have $b \in L^p\bigl(\mu, U\times(0,T)\bigr)$, $c \in L^{p/2}\bigl(\mu, U\times(0,T)\bigr)$ with $p > d+2$, then for establishing upper estimates we have to study the solution of the Cauchy problem with a simpler equation without the coefficients b and c.

7.3.5. Remark. Assertion of Theorem 7.3.3 remains in force if the condition that each measure μ_t is subprobability is replaced with the following condition: $\mu_t \geqslant 0$ and $\sup_{t\in(0,T)}\mu_t(\mathbb{R}^d) < \infty$. Note that the number C in this case depends on $\sup_{t\in(0,T)}\mu_t(\mathbb{R}^d)$.

Let us give an example of direct application of Theorem 7.3.3.

7.3.6. Example. Let A satisfy conditions (7.3.1) and

$$b\in L^p\bigl(\mu,\mathbb{R}^d\times(0,T)\bigr),\ c\in L^{p/2}_{\mathrm{loc}}\bigl(\mu,\mathbb{R}^d\times(0,T)\bigr),\ c^+\in L^{p/2}\bigl(\mu,\mathbb{R}^d\times(0,T)\bigr).$$

Suppose that the initial distribution ν is Dirac's measure at the origin. Then the density of the measure σ coincides with the function

$$(2\pi s)^{-d/2}(\det A)^{-1/2}\exp\Bigl(-\frac{1}{2s}\langle A^{-1}y,y\rangle\Bigr).$$

Let $\theta\in(0,1)$ and $\kappa>1$. It is verified directly that

$$\sigma\bigl(U(x,\kappa\sqrt{t})\times[\theta t,t]\bigr)\leqslant C_1\int_{\theta t}^{t}\int_{|x-y|\leqslant\kappa\sqrt{t}}(2\pi s)^{-d/2}\exp\Bigl(-\frac{1}{2Ms}|y|^2\Bigr)\,dy\,ds.$$

If $|x-y|\leqslant\kappa\sqrt{t}$, then $|y|^2\geqslant 2^{-1}|x|^2-2\kappa^2t$. Hence

$$\int_{\theta t}^{t}\int_{|x-y|\leqslant\kappa\sqrt{t}}(2\pi s)^{-d/2}\exp\Bigl(-\frac{1}{2Ms}|y|^2\Bigr)\,dy\,ds\leqslant C_2t\exp\Bigl(-\frac{1}{2Mt}|x|^2\Bigr),$$

where C_2 depends only on d,κ,M,θ. Then the density ϱ is estimated by means of the previous theorem as follows:

$$\varrho(x,t)\leqslant Ct^{-d/2}\exp\Bigl(-\frac{1}{2Mt}|x|^2\Bigr)+Ct^{(p-d-2)/2}\bigl(1+\|b\|_{L^p(\mu)}+\|c^+\|_{L^{p/2}(\mu)}\bigr)^p$$

for all $t\in(0,T)$ and $x\in\mathbb{R}^d$.

If, in addition, there is a function $\Phi\in C^2(\mathbb{R}^d))$ such that $\Phi\geqslant C_0>0$ and for some $p>d+2$ we have

$$|b|^p\Phi,\ |A\nabla\Phi|^p\Phi^{1-p},\ |(L_{A,b,c}\Phi)^+|^{p/2}\Phi^{1-p/2}\in L^1(\mu,\mathbb{R}^d\times(0,T)),$$

$$\sup_{t\in(0,T)}\int_{\mathbb{R}^d}\Phi(x)\varrho(x,t)\,dx<\infty,$$

then there exists a number $C>0$ such that

$$\varrho(x,t)\leqslant C\Phi(x)^{-1}\Bigl[t^{(p-d-2)/2}+t^{-d/2}\exp\Bigl(-\frac{1}{2Mt}|x|^2\Bigr)\Bigr].$$

7.3.7. Corollary. *Suppose that the hypotheses of Theorem 7.3.3 are fulfilled and $\nu=\varrho_0\,dx$, where $\varrho_0\in L^\infty(U)$ for every ball U. Then the measure μ has a density ϱ with respect to Lebesgue measure such that for every ball U one has the inclusion $\varrho\in L^\infty(U\times[0,T])$.*

PROOF. The assertion follows immediately by Theorem 7.3.3 if we observe that

$$\int_{U(x,2)}Z_{x,t}(y-z,s)\,\nu(\,dz)\leqslant\|\varrho_0\|_{L^\infty(U(x,2))}.$$

Thus, the density of the measure σ is bounded, hence one has the estimate

$$\sigma\bigl(U(x,\kappa\sqrt{t})\times[\theta t,\theta^{-1}t]\bigr)\leqslant C(\theta,\kappa,d)\|\varrho_0\|_{L^\infty(U(x,2))}t^{(d+2)/2},$$

which implies our assertion. □

7.3.8. Corollary. *Suppose that the hypotheses of Theorem 7.3.3 are fulfilled, $\nu=\varrho_0\,dx$, $\varrho_0\in L^\infty(\mathbb{R}^d)$, $b\in L^p(\mu,\mathbb{R}^d_T)$, $c\in L^{p/2}_{\rm loc}(\mu,\mathbb{R}^d_T)$, $p>d+2$, and $c^+\in L^{p/2}(\mu,\mathbb{R}^d_T)$. Then the measure μ has a density ϱ with respect to Lebesgue measure and $\varrho\in L^\infty(\mathbb{R}^d_T)$, moreover, there exists a number $C>0$ such that*

$$\|\varrho\|_{L^\infty(\mathbb{R}^d_T)}\leqslant C\|\varrho_0\|_{L^\infty(\mathbb{R}^d)}+Ct^{(p-d-2)/2}\bigl(1+\|b\|_{L^p(\mu)}+\|c^+\|_{L^{p/2}(\mu)}\bigr)^p.$$

PROOF. We observe that

$$\int_{\mathbb{R}^d} Z_{x,t}(y-z,s)\,\nu(dz) \leqslant \|\varrho_0\|_{L^\infty(\mathbb{R}^d)}.$$

Therefore, we have the estimate

$$\sigma\big(U(x,\kappa\sqrt{t})\times[t/2,t]\big) \leqslant C(\kappa,d)\|\varrho_0\|_{L^\infty(\mathbb{R}^d)}t^{(d+2)/2}.$$

By Theorem 7.3.3 we obtain the desired estimate. □

7.3.9. Corollary. *Suppose that the hypotheses of Theorem 7.3.3 are fulfilled, $\nu=\varrho_0\,dx$, $p>d+2$ and that we are given a function $\Phi\in C^2(\mathbb{R}^d)$ with the following properties: $\Phi(x)\geqslant C_0>0$, $\Phi\varrho_0\in L^\infty(\mathbb{R}^d)\cap L^1(\mathbb{R}^d)$ and*

$$\Phi,\ |b|^p\Phi,\ |A\nabla\Phi|^p\Phi^{1-p},\ |(L_{A,b,c}\Phi)^+|^{p/2}\Phi^{1-p/2}\in L^1(\mu,\mathbb{R}^d_T).$$

Then there exists a number $C>0$ such that

$$\varrho(x,t)\leqslant C\Phi(x)^{-1},\quad (x,t)\in\mathbb{R}^d_T.$$

PROOF. Let us consider a new measure $\chi=\Phi\cdot\mu$, which satisfies the Cauchy problem with the initial condition $\Phi\cdot\nu$ and the parabolic operator $\widetilde{L}$ of the form

$$\widetilde{L}u=\partial_t u+a^{ij}\partial_{x_i}\partial_{x_j}u+(b^i+\Phi^{-1}a^{ij}\partial_{x_j}\Phi)\partial_{x_i}u+(\Phi^{-1}L_{A,b,c}\Phi)u.$$

It remains to apply Theorem 7.3.3 to the measure $\Phi\varrho\,dx\,dt$ and repeat the reasoning from the previous corollary. □

7.3.10. Example. Let $A=(a^{ij})$ satisfy conditions (7.3.1) and let $c=0$. Suppose that $\nu=\varrho_0\,dx$. Let $r\geqslant 2$, $\delta>0$ and $p>d+2$. Suppose also that

$$\langle b(x,t),x\rangle\leqslant c_1-c_2|x|^r,\quad |b(x,t)|\leqslant c_3\exp\big(c_4|x|^{r-\delta}\big),$$

where $c_1,c_2,c_3,c_4>0$, and for some $\alpha\in(0,c_2r^{-1}M^{-1})$ we have

$$\varrho_0\exp\big(\alpha|x|^r\big)\in L^\infty(\mathbb{R}^d)\cap L^1(\mathbb{R}^d).$$

According to Example 7.1.4, the inequality

$$\sup_{t\in(0,T)}\int_{\mathbb{R}^d}\exp\big(\alpha|x|^r\big)\,\mu_t(dx)<\infty$$

holds. Therefore, the function $\Phi(x)=\exp\big(\alpha|x|^r\big)$ satisfies the hypotheses in Corollary 7.3.9, so one has the estimate

$$\varrho(x,t)\leqslant C\exp(-\alpha|x|^r),\quad (x,t)\in\mathbb{R}^d_T.$$

Finally, let us prove yet another result which better describes the behavior of the solution as $t\to 0$.

7.3.11. Theorem. *Suppose that a measure μ given by a density ϱ with respect to Lebesgue measure is a solution to the Cauchy problem (6.1.1) and (6.1.3) with initial condition $\nu=\varrho_0\,dx$, where the density ϱ_0 is locally bounded, the functions a^{ij} satisfy Conditions (7.3.1), $b\in L^p\big(\mu,U\times(0,T)\big)$, $c\in L^{p/2}\big(\mu,U\times(0,T)\big)$ for every ball $U\subset\mathbb{R}^d$ and some $p>d+2$. Then, for every ball U, we have*

$$\lim_{t\to 0}\int_U|\varrho(x,t)-\varrho_0(x)|^2\,dx=0.$$

PROOF. We shall deal with the continuous version of the density ϱ. For every number $\varepsilon \in (0,1)$, set $\omega_\varepsilon(x) = \varepsilon^{-d}\omega(x\varepsilon^{-1})$, where $\omega \in C_0^\infty(\mathbb{R}^d)$ is a probability density with support in the unit ball. The function $\varrho_\varepsilon = \varrho * \omega_\varepsilon$ is smooth in the variable x and satisfies the equation

$$\partial_t \varrho_\varepsilon = (a^{ij}\varrho) * \partial_{x_i}\partial_{x_j}\omega_\varepsilon + (b^i\varrho) * \partial_{x_i}\omega_\varepsilon + (c\varrho) * \omega_\varepsilon. \tag{7.3.2}$$

Since a^{ij} are bounded and $\varrho, b^i\varrho, c\varrho \in L^1\bigl(U\times(0,T)\bigr)$ for any ball U, the function ϱ_ε is absolutely continuous in t on $(0,T)$. In addition, for any x, as $t\to 0$ one has the relation

$$\varrho_\varepsilon(x,t) = \int_{\mathbb{R}^d} \omega_\varepsilon(x-y)\varrho(y,t)\,dy \to \int_{\mathbb{R}^d} \omega_\varepsilon(x-y)\varrho_0(y)\,dy = \varrho_{0,\varepsilon}(x).$$

Note that $\varrho_\varepsilon \in L^\infty(\mathbb{R}^d_T)$ for any fixed ε, moreover,

$$\begin{aligned}(a^{ij}\varrho) * \partial_{x_i}\partial_{x_j}\omega_\varepsilon(x,t) &= \partial_{x_i}\bigl((a^{ij}\varrho) * \partial_{x_j}\omega_\varepsilon\bigr)(x,t)\\ &= \partial_{x_i}\bigl(a^{ij}(x,t)\partial_{x_j}\varrho_\varepsilon(x,t)\bigr) + \partial_{x_i}\bigl[\bigl(a^{ij}(y,t)-a^{ij}(x,t)\bigr)\varrho(y,t)\bigr] * \partial_{x_j}\omega_\varepsilon(x,t).\end{aligned}$$

Let us rewrite equation (7.3.2) in the form

$$\begin{aligned}\partial_t\varrho_\varepsilon(x,t) = \partial_{x_i}\bigl(a^{ij}(x,t)\partial_{x_j}\varrho_\varepsilon(x,t)\bigr) &+ \partial_{x_i}\bigl[\bigl(a^{ij}(y,t)-a^{ij}(x,t)\bigr)\varrho\bigr] * \partial_{x_j}\omega_\varepsilon(x,t)\\ &+ (b^i\varrho) * \partial_{x_i}\omega_\varepsilon(x,t) + (c\varrho) * \omega_\varepsilon(x,t).\end{aligned} \tag{7.3.3}$$

Let $\psi \in C_0^\infty(\mathbb{R}^d)$. Multiplying equality (7.3.3) by $\psi^2\varrho_\varepsilon$ and integrating by parts, we obtain

$$\begin{aligned}\int_{\mathbb{R}^d} \psi^2(x)\varrho_\varepsilon^2(x,t)\,dx - \int_{\mathbb{R}^d}\psi^2(x)\varrho_\varepsilon^2(x,0)\,dx&\\ + \int_0^t\!\!\int_{\mathbb{R}^d} \langle A\nabla\varrho_\varepsilon, \nabla\varrho_\varepsilon\rangle\psi^2\,dx\,d\tau &= I + J + K,\end{aligned} \tag{7.3.4}$$

where I, J and K stand for the following expressions, respectively:

$$\begin{aligned}I(t,\varepsilon) &= 2\int_0^t\!\!\int_{\mathbb{R}^d} \varrho_\varepsilon\psi a^{ij}\partial_{x_i}\psi\partial_{x_j}\varrho_\varepsilon\,dx\,d\tau,\\ J(t,\varepsilon) &= \int_0^t\!\!\int_{\mathbb{R}^d}\Bigl[-\bigl[\bigl(a^{ij}(y,t)-a^{ij}(x,\tau)\bigr)\varrho(y,t)\bigr] * \partial_{x_j}\omega_\varepsilon(x,t)\partial_{x_i}\varrho_\varepsilon\psi^2(x,t)\\ &\qquad - 2\bigl[\bigl(a^{ij}(y,t)-a^{ij}(x,\tau)\bigr)\varrho(y,t)\bigr] * \partial_{x_j}\omega_\varepsilon(x,t)\partial_{x_i}\psi(x,t)\varrho_\varepsilon(x,t)\Bigr]\,dx\,d\tau,\\ K(t,\varepsilon) &= \int_0^t\!\!\int_{\mathbb{R}^d}\bigl[(b^i\varrho) * \omega_\varepsilon\partial_{x_i}\varrho_\varepsilon\psi^2 + 2\psi(b^i\varrho) * \omega_\varepsilon\partial_{x_i}\psi\varrho_\varepsilon + (c\varrho) * \omega_\varepsilon\psi\varrho_\varepsilon\bigr]\,dx\,d\tau.\end{aligned}$$

We first prove that $|\nabla\varrho|\psi \in L^2\bigl(\mathbb{R}^d\times(0,T)\bigr)$. Suppose that the support of ψ is in the ball $U(0,R)$. By Corollary 7.3.7 there is a number $\Lambda>0$ such that $\varrho(x,t)\leqslant\Lambda$ whenever $t\in(0,T)$ and $x\in U(0,R+2)$. Therefore, $\varrho_\varepsilon(x,t)\leqslant\Lambda$ for all $t\in(0,T)$ and $x\in U(0,R)$. Using (7.3.1) and applying the Cauchy–Bunyakovsky inequality, we obtain the estimate

$$I(t,\varepsilon) \leqslant \delta\int_0^t\!\!\int_{\mathbb{R}^d} |\nabla\varrho_\varepsilon|^2\psi^2\,dx\,d\tau + C(\delta,\lambda,M,\Lambda)|U(0,R)|\sup_{x\in U(0,R)}|\nabla\psi(x)|^2.$$

Let us estimate K. Since $\|b\varrho\|_{L^2(U(0,R+2)\times(0,T))} \leqslant \sqrt{\Lambda}\|b\|_{L^2(\mu,U(0,R+2)\times(0,T))}$, we have

$$\int_0^t\int_{U(0,R)} |(b^i\varrho)*\omega_\varepsilon|^2\,dx\,d\tau \leqslant \Lambda\|b\|^2_{L^2(\mu,U(0,R+2)\times(0,T))}.$$

Then

$$\begin{aligned}K(t,\varepsilon) &\leqslant \delta\int_0^t\int_{\mathbb{R}^d} |\nabla\varrho_\varepsilon|^2\psi^2\,dx\,d\tau\\ &+ C\big(\|b\|^2_{L^2(\mu,U(0,R+2)\times(0,T))} + \|c\|_{L^1(\mu,U(0,R+2)\times(0,T))}\big)\sup_x\big[|\psi(x)|+|\nabla\psi(x)|\big]^2.\end{aligned}$$

Finally, let us estimate J. To this end we observe that

$$|(a^{ij}\varrho)*\partial_{x_j}\omega_\varepsilon(x,\tau) - a^{ij}(x,\tau)\varrho*\partial_{x_j}\omega_\varepsilon(x,\tau)| \leqslant \lambda\Lambda\int_{\mathbb{R}^d}|x|\omega(x)\,dx = \Lambda_1.$$

Acting in the same manner as above, we obtain

$$J(t,\varepsilon) \leqslant \delta\int_0^t\int_{\mathbb{R}^d} |\nabla\varrho_\varepsilon|^2\psi^2\,dx\,d\tau + C(\delta,M,\lambda,\Lambda,\Lambda_1)\sup_{x\in U(0,R)}\big[|\psi(x)|^2+|\nabla\psi(x)|\big].$$

According to (7.3.1), we have $\langle A\nabla\varrho,\nabla\varrho\rangle \geqslant m|\nabla\varrho|^2$. Taking $\delta = m/4$, we obtain the desired estimate for $|\nabla\varrho_\varepsilon|\psi$. It remains to let $\varepsilon\to 0$. Note that ψ was an arbitrary smooth function with compact support. Therefore, we have $|\nabla\varrho| \in L^2\big(U\times(0,T)\big)$ for every ball $U\subset\mathbb{R}^d$. Let us now show that in equality (7.3.4) one can pass to the limit as $\varepsilon\to 0$. To this end we first observe that for every ball $U\subset\mathbb{R}^d$ we have

$$\varrho,\ \nabla\varrho,\ a^{ij}\nabla\varrho,\ b^i\varrho,\ c\varrho \in L^2(U\times[0,\tau]).$$

In addition, $\varrho(\cdot,t)\in L^2(U)$ for all $t\in(0,T)$. We recall that for every function $f\in L^2(U)$ or $f\in L^2\big(U\times(0,T)\big)$ one has $\|f-f_\varepsilon\|_{L^2}\to 0$ as $\varepsilon\to 0$, where the L^2-norm is taken on U' or on $U'\times(0,\tau)$, $\overline{U'}\subset U$. Applying the last observations, one can pass to the limit in the terms I and K. In order to pass to the limit in the term J, one should observe that

$$\begin{aligned}&\big[\big(a^{ij}(y,t)-a^{ij}(x,\tau)\big)\varrho(y,t)\big]*\partial_{x_j}\omega_\varepsilon(x,t)\\ &\qquad= \big[\big(a^{ij}(y,t)-a^{ij}(x,\tau)\big)\partial_{y_i}\varrho(y,t)\big]*\omega_\varepsilon(x,t) + (\varrho\partial_{y_i}a^{ij})*\omega_\varepsilon(x,t).\end{aligned}$$

If in (7.3.4) we let $\varepsilon\to 0$, we obtain the equality

$$\int_{\mathbb{R}^d}\psi^2(x)\varrho^2(x,t)\,dx - \int_{\mathbb{R}^d}\psi^2(x)\varrho_0^2(x)\,dx + \int_0^t\int_{\mathbb{R}^d}\langle A\nabla\varrho,\nabla\varrho\rangle\psi^2\,dx\,d\tau = \widehat{I}+\widehat{K},$$

where $\widehat{I}$ and $\widehat{K}$ are defined by the formulas

$$\widehat{I}(t) = 2\int_0^t\int_{\mathbb{R}^d}\big[\varrho\psi a^{ij}\partial_{x_i}\psi\partial_{x_j}\varrho - \varrho^2\psi\partial_{x_i}a^{ij}\partial_{x_j}\psi\big]\,dx\,d\tau,$$

$$\widehat{K}(t) = \int_0^t\int_{\mathbb{R}^d}\big[(b^i\varrho)\partial_{x_i}\varrho\psi^2 + 2\psi(b^i\varrho)\partial_{x_i}\psi\varrho + c\psi\varrho\big]\,dx\,d\tau.$$

Thus,

$$\int_{\mathbb{R}^d}\psi^2(x)\varrho^2(x,t)\,dx - \int_{\mathbb{R}^d}\psi(x)^2\varrho_0^2(x)\,dx = C(t),\quad \lim_{t\to 0}C(t)=0.$$

Therefore, $\|\psi\varrho(\,\cdot\,,t)\|_{L^2}\to\|\psi\varrho_0\|_{L^2}$ and $\psi\varrho(\cdot,t)\to\psi\varrho_0$ weakly in $L^2(\mathbb{R}^d)$, which gives the assertion to be proven. □

7.3.12. Remark. Since in the hypotheses of the theorem the density ϱ is bounded on $U\times(0,T)$ for every ball $U\subset\mathbb{R}^d$, locally we have convergence to the initial condition in all L^p, i.e., for every ball U we have

$$\lim_{t\to 0}\int_U |\varrho(x,t)-\varrho_0(x)|^p\,dx=0.$$

For probability densities $\varrho(\,\cdot\,,t)$ and ϱ_0 this gives convergence of $\|\varrho(\,\cdot\,,t)-\varrho_0\|_{L^1(\mathbb{R}^d)}$ to zero as $t\to 0$ by the Vitaly–Scheffé theorem (see Bogachev [**125**, p. 135]).

7.4. Square integrability of logarithmic gradients

Here we establish square integrability of the logarithmic gradient of the solution μ to the Fokker–Planck–Kolmogorov equation, i.e., of the mapping $\nabla\varrho/\varrho$, where $\nabla\varrho(x,t):=\nabla_x\varrho(x,t)$, with respect to the measure μ itself. For $\varrho(\,\cdot\,,t)\in W^{1,1}_{\mathrm{loc}}$ we use the following convention:

$$\nabla\varrho(x,t)/\varrho(x,t):=0 \quad \text{whenever } \varrho(x,t)=0.$$

We recall that a probability measure ν on $\mathbb{R}^d$ has finite entropy if $\nu=\varrho_0\,dx$ and

$$\int_{\mathbb{R}^d}|\ln\varrho_0(x)|\varrho_0(x)\,dx<\infty.$$

Let $\Omega\subset\mathbb{R}^d$ and $J\subset[0,1]$. Let $L^{p,q}(\Omega\times J)$ denote the space of measurable functions u on $\Omega\times J$ with finite norm

$$\|u\|_{p,q}=\biggl(\int_J\|u(\,\cdot\,,t)\|^q_{L^p(\Omega)}\,dt\biggr)^{1/q}.$$

In the preceding sections we heavily used the integrability of the coefficients to power $p>d+2$ (or $p>(d+2)/2$) with respect to the solution μ. It turns out that under considerably weaker conditions one can also prove the Sobolev regularity of the solution, namely: the quadratic integrability of the drift b and the function $\ln\max(|x|,1)$ with respect to the solution μ implies that the measure μ has a density ϱ in $W^{1,1}(\mathbb{R}^d)$ and its logarithmic gradient μ is square integrable, provided that the initial condition ν has finite entropy; see also Theorem 7.4.3.

In this section we assume that the matrix A satisfies condition (P1) in (7.3.1) and the condition

$$|a^{ij}(x,t)-a^{ij}(y,t)|\leqslant\lambda|x-y|,\quad x,y\in\mathbb{R}^d,\ t\in(0,T) \tag{7.4.1}$$

for some number $\lambda>0$.

In addition, for simplification of our exposition we shall assume that $c=0$. The case where the coefficient c does not vanish is considered in Exercise 7.5.4 and Exercise 7.5.5. As above, let $\mathbb{R}^d_T=\mathbb{R}^d\times(0,T)$.

7.4.1. Theorem. *Suppose that a measure $\mu=(\mu_t)$ is a solution to the Cauchy problem* (6.1.1), (6.1.3), *each μ_t is a probability measure, condition* (P1) *in* (7.3.1) *and* (7.4.1) *are fulfilled, and that $|b|\in L^2(\mu,\mathbb{R}^d_T)$. Suppose also that the function $\Lambda(x):=\ln\max(|x|,1)$ belongs to $L^2(\mu,\mathbb{R}^d_T)$ (which is true if, for example, $\langle b(x,t),x\rangle\leqslant C_1|x|^2\Lambda(x)+C_2$ with some constants C_1 and C_2 and $\Lambda\in L^2(\nu)$, see* §7.1). *If the initial distribution $\nu=\varrho_0\,dx$ on $\mathbb{R}^d$ has finite entropy, then $\mu_t=\varrho(\,\cdot\,,t)\,dx$, where $\varrho(\,\cdot\,,t)\in W^{1,1}(\mathbb{R}^d)$, moreover, for every $\tau<T$ we have*

$$\int_0^\tau\int_{\mathbb{R}^d}\frac{|\nabla\varrho(x,t)|^2}{\varrho(x,t)}\,dx\,dt<\infty. \tag{7.4.2}$$

In particular, $\sqrt{\varrho} \in \mathbb{H}^{2,2}(\mathbb{R}^d \times [0,\tau])$, $\varrho \in L^{d/(d-2),1}(\mathbb{R}^d \times [0,\tau])$ *if* $d > 2$, *and* $\varrho \in L^{s,1}(\mathbb{R}^d \times [0,\tau])$ *for all* $s \in [1,\infty)$ *if* $d = 2$.

If the integrals $\int_{\mathbb{R}^d} \varrho(x,t)\Lambda(x)\,dx$ *remain bounded as* $t \to T$ *(which holds, for example, if we have* $\langle b(x,t), x\rangle \leqslant C_1|x|^2 + C_2$ *with some constants* C_1 *and* C_2 *and* $\Lambda \in L^1(\nu)$*), then (7.4.2) is true with* $\tau = T$.

PROOF. We use the following fact which is obvious from Lemma 3.1.1: given two nonnegative functions f_1 and f_2 in $L^1(\mathbb{R}^d)$, for every measurable function ψ with the property that $|\psi|^2 f_1 \in L^1(\mathbb{R}^d)$ we have

$$\int_{\mathbb{R}^d} \frac{|(\psi f_1) * f_2|^2}{f_1 * f_2}\,dx \leqslant \int_{\mathbb{R}^d} |\psi|^2 f_1\,dx \int_{\mathbb{R}^d} f_2\,dx, \tag{7.4.3}$$

where $|(\psi f_1) * f_2(x)|^2/(f_1 * f_2(x)) := 0$ if $f_1 * f_2(x) = 0$.

For any function $w \in C_0^\infty(\mathbb{R}^d)$ we set

$$\varrho * w(x,t) := \int_{\mathbb{R}^d} w(x-y)\varrho(y,t)\,dy, \quad x \in \mathbb{R}^d.$$

Here and below the convolutions are always taken with respect to the variable in $\mathbb{R}^d$. Since the measure $\mu = \varrho\,dx\,dt$ satisfies the equation $\partial_t \mu = L_{A,b}^* \mu$ and $|b| \in L^2(\mu)$, in the Sobolev sense we have the equality

$$\partial_t(\varrho * w) = (a^{ij}\varrho) * \partial_{x_i}\partial_{x_j} w - (b^i \varrho) * \partial_{x_i} w. \tag{7.4.4}$$

We shall deal with the version of $\varrho * w$ (denoted by the same symbol) defined by the formula

$$\varrho * w(x,t) := \varrho * w(x,0) + \int_0^t v(x,s)\,ds, \tag{7.4.5}$$

where v is the right side of (7.4.4). Since $|b| \in L^2(\mu)$ and the functions a^{ij} are bounded, we have $v \in L^1(\mathbb{R}^d \times [0,T])$. Hence the function $\varrho * w$ is absolutely continuous in t on $[0,T]$ and belongs to the class $C_b^\infty(\mathbb{R}^d)$ in x. For almost every t the indicated version coincides for all x with the original version defined by the convolution. It is also important below that this is true for $t = 0$. Since the original version does not exceed $\sup_x |w(x)|$, this is true also for the new version for almost all t, hence this is true pointwise by the continuity in t. It is readily seen from our assumptions and the inclusion $b \in L^2(\mu)$ that everything said above, including (7.4.4), remains true also for the functions $w_\varepsilon(x) = \varepsilon^{-d} g(x/\varepsilon)$, where g is the standard Gaussian density and $\varepsilon \in (0,T)$. Below we take for ε only numbers of the form $1/n$, $n \in \mathbb{N}$. Set

$$\varrho_\varepsilon := \varrho * w_\varepsilon, \quad f_\varepsilon(x,t) := \varrho_\varepsilon(x,t) + \varepsilon \max(1,|x|)^{-d-1},$$

where for $\varrho * w_\varepsilon$ we take the version indicated in (7.4.5). Since the function $\varrho\Lambda$ is integrable, there is τ as close to T as we wish such that

$$\int_{\mathbb{R}^d} \varrho(x,\tau)\Lambda(x)\,dx < \infty. \tag{7.4.6}$$

The number τ for which (7.4.6) holds can be taken in such a way that for every $\varepsilon = 1/n$ our version of $\varrho_\varepsilon(\tau,x)$ will coincide with the function $\varrho(\,\cdot\,,\tau) * w_\varepsilon(x)$ for all x. Then the easily verified inequality

$$\ln \max(|x+y|,1) \leqslant \ln \max(|x|,1) + |y|$$

gives the estimate

$$
\begin{aligned}
(7.4.7)\quad \int_{\mathbb{R}^d} f_\varepsilon(x,\tau)\Lambda(x)\,dx &\leqslant \int_{\mathbb{R}^d}\Big[\varrho_\varepsilon(x,\tau)\Lambda(x)+\varepsilon\max(|x|,1)^{-d-1}\Lambda(x)\Big]\,dx\\
&\leqslant \int_{\mathbb{R}^d}\varrho(x,\tau)\Lambda(x)\,dx+\int_{\mathbb{R}^d}|y|w_\varepsilon(y)\,dy+\varepsilon\int_{\mathbb{R}^d}\max(|x|,1)^{-d-1}\Lambda(x)\,dx\\
&\leqslant \int_{\mathbb{R}^d}\varrho(x,\tau)\Lambda(x)\,dx+\varepsilon M_1,
\end{aligned}
$$

where M_1 is a number independent of ε. By (7.4.4) we have

$$
\begin{aligned}
(7.4.8)\quad &\int_0^T\!\!\int_{\mathbb{R}^d}\partial_t(\varrho*w_\varepsilon)\ln f_\varepsilon\,dx\,dt\\
&\qquad=\int_0^T\!\!\int_{\mathbb{R}^d}\Big[(a^{ij}\varrho)*\partial_{x_i}\partial_{x_j}w_\varepsilon-(b^i\varrho)*\partial_{x_i}w_\varepsilon\Big]\ln f_\varepsilon\,dx\,dt,
\end{aligned}
$$

since $|\ln f_\varepsilon|\leqslant c_1+c_2\Lambda$ with some constants c_1 and c_2, and the functions

$$
\big((b^i\varrho)*\partial_{x_i}w_\varepsilon\big)\Lambda,\quad \big((a^{ij}\varrho)*\partial_{x_i}\partial_{x_j}w_\varepsilon\big)\Lambda
$$

are integrable on $\mathbb{R}^d\times(0,T)$. Indeed, since $\varrho*|\partial_{x_i}w_\varepsilon|>0$, we have

$$
\big((b^i\varrho)*\partial_{x_i}w_\varepsilon\big)\Lambda=\big((b^i\varrho)*\partial_{x_i}w_\varepsilon\big)\big(\varrho*|\partial_{x_i}w_\varepsilon|\big)^{-1/2}\big(\varrho*|\partial_{x_i}w_\varepsilon|\big)^{1/2}\Lambda.
$$

Then $\big((b^i\varrho)*\partial_{x_i}w_\varepsilon\big)\big(\varrho*|\partial_{x_i}w_\varepsilon|\big)^{-1/2}\in L^2\big(\mathbb{R}^d\times(0,T)\big)$ according to (7.4.3), since $|b|\in L^2(\mu)$. In addition, $\big(\varrho*|\partial_{x_i}w_\varepsilon|\big)^{1/2}\Lambda\in L^2\big(\mathbb{R}^d\times(0,T)\big)$ by the estimate

$$
|\ln\max(|x+y|,1)|^2\leqslant 4+2|\ln\max(|x|,1)|^2+2|\ln\max(|y|,1)|^2
$$

and the same calculations as in (7.4.7). Similarly we verify the integrability of $[(a^{ij}\varrho)*\partial_{x_i}\partial_{x_j}w_\varepsilon]\Lambda$ over $\mathbb{R}^d\times(0,T)$. We observe that one can integrate by parts in the right side of (7.4.8). Indeed,

$$
\int_0^T\!\!\int_{\mathbb{R}^d}\frac{|\nabla f_\varepsilon|^2}{f_\varepsilon}\,dx\,dt\leqslant 2\int_0^T\!\!\int_{\mathbb{R}^d}\frac{|\nabla\varrho_\varepsilon|^2}{\varrho_\varepsilon}\,dx\,dt+2\varepsilon(d+1)^2\int_{\{|x|\geqslant 1\}}|x|^{-d-3}\,dx,
$$

which is finite by (7.4.3), since $\nabla\varrho_\varepsilon=(\nabla w_\varepsilon)*\varrho$ and $|\nabla w_\varepsilon|^2/w_\varepsilon$ belongs to $L^1(\mathbb{R}^d)$. In addition, $[(b^i\varrho)*w_\varepsilon]\varrho_\varepsilon^{-1/2}\in L^2\big(\mathbb{R}^d\times(0,T)\big)$ also by (7.4.3) along with the inclusion $|b|\in L^2(\mu)$. Similarly we have

$$
[(a^{ij}\varrho)*\partial_{x_i}w_\varepsilon]\varrho_\varepsilon^{-1/2}\in L^2\big(\mathbb{R}^d\times(0,T)\big).
$$

Since $f_\varepsilon>\varrho_\varepsilon$, we obtain $\partial_{x_i}f_\varepsilon\big(\partial_{x_j}[(a^{ij}\varrho)*w_\varepsilon]-(b^i\varrho)*w_\varepsilon\big)f_\varepsilon^{-1}\in L^1(\mathbb{R}^d)$. Therefore,

$$
\begin{aligned}
(7.4.9)\quad &\int_0^T\!\!\int_{\mathbb{R}^d}\partial_t\varrho_\varepsilon\ln f_\varepsilon\,dx\,dt\\
&\qquad=-\int_0^T\!\!\int_{\mathbb{R}^d}\frac{\partial_{x_i}f_\varepsilon}{f_\varepsilon}\big(\partial_{x_j}[(a^{ij}\varrho)*w_\varepsilon]-(b^i\varrho)*w_\varepsilon\big)\,dx\,dt.
\end{aligned}
$$

The expression under the integral sign in the left side of this equality can be written as $\partial_t(f_\varepsilon\ln f_\varepsilon)-\partial_t\varrho_\varepsilon$. Since the integrals of $\varrho_\varepsilon(x,\tau)$ and $\varrho_\varepsilon(x,0)$ in x equal 1, the left side in (7.4.9) equals

$$
L_\varepsilon:=\int_{\mathbb{R}^d}[f_\varepsilon(x,\tau)\ln f_\varepsilon(x,\tau)-f_\varepsilon(x,0)\ln f_\varepsilon(x,0)]\,dx.
$$

Since $f_\varepsilon(\,\cdot\,,\tau)\ln f_\varepsilon(\,\cdot\,,\tau) \in L^1(\mathbb{R}^d)$ by (7.4.7) and the aforementioned estimate $|\ln f_\varepsilon| \leqslant c_1 + c_2\Lambda$, we have $f_\varepsilon(\,\cdot\,,0)\ln f_\varepsilon(\,\cdot\,,0) \in L^1(\mathbb{R}^d)$. We need a lower bound on L_ε. To this end, we observe that by convexity of the function $s \mapsto s\ln s$ on $(0,+\infty)$ and Jensen's inequality we have

$$\begin{aligned}\int_{\mathbb{R}^d} f_\varepsilon(x,0)\ln f_\varepsilon(x,0)\,dx &\leqslant \int_{\mathbb{R}^d} \varrho_\varepsilon(x,0)\ln\bigl(2\varrho_\varepsilon(x,0)\bigr)\,dx \\ &\quad + \int_{\mathbb{R}^d} \varepsilon\max(|x|,1)^{-d-1}\ln\bigl(2\varepsilon\max(|x|,1)^{-d-1}\bigr)\,dx \\ &\leqslant \ln 2 + \int_{\mathbb{R}^d} \varrho_\varepsilon(x,0)\ln\varrho_\varepsilon(x,0)\,dx + \varepsilon\ln 2\int_{\mathbb{R}^d}\max(|x|,1)^{-d-1}\,dx \\ &\leqslant \ln 2 + \int_{\mathbb{R}^d}\varrho_0(x)\ln\varrho_0(x)\,dx + \varepsilon\ln 2\int_{\mathbb{R}^d}\max(|x|,1)^{-d-1}\,dx =: M(\varepsilon).\end{aligned}$$

On the other hand, (7.4.7) and inequality $|z\ln z| \leqslant e^{-1}$ for $z \in (0,1)$ give

$$\begin{aligned}&\int_{\mathbb{R}^d} f_\varepsilon(x,\tau)\ln f_\varepsilon(x,\tau)\,dx \\ &\quad\geqslant -(d+1)\int_{\mathbb{R}^d} f_\varepsilon(x,\tau)\Lambda(x)\,dx - e^{-1}\int_{\mathbb{R}^d}\max\{|x|,1\}^{-d-1}\,dx \\ &\quad\geqslant -(d+1)\int_{\mathbb{R}^d}\varrho(x,\tau)\Lambda(x)\,dx - e^{-1}\int_{\mathbb{R}^d}\max\{|x|,1\}^{-d-1}\,dx \\ &\qquad\qquad - \varepsilon M_1(d+1) =: -K(\varepsilon).\end{aligned}$$

We observe that for every bounded Borel function a on the set $\mathbb{R}^d_T$ that is Lipschitzian in the second variable with the Lipschitz constant λ, for every index j, we have the equality

$$\begin{aligned}\partial_{x_j}[(a\varrho)*w_\varepsilon](x,t) &= a(x,t)\,\partial_{x_j}\varrho_\varepsilon(x,t) \\ &\quad + \int_{\mathbb{R}^d}\partial_{x_j}w_\varepsilon(x-y)[a(y,t)-a(x,t)]\,\varrho(y,t)\,dy,\end{aligned} \tag{7.4.10}$$

and also the following equality:

(7.4.11)

$$\begin{aligned}\left|\int_{\mathbb{R}^d}\partial_{x_j}w_\varepsilon(x-y)[a(y,t)-a(x,t)]\,\varrho(y,t)\,dy\right| &\leqslant \lambda\int_{\mathbb{R}^d}|\partial_{x_j}w_\varepsilon(x-y)|\,|y-x|\,\varrho(y,t)\,dy \\ &\leqslant \lambda\int_{\mathbb{R}^d}\varepsilon^{-d}\frac{|x-y|^2}{\varepsilon^2}\,g\Bigl(\frac{x-y}{\varepsilon}\Bigr)\,\varrho(y,t)\,dy = \lambda(\varrho*q_\varepsilon)(x,t),\end{aligned}$$

where $q_\varepsilon(x) := w_\varepsilon(x)|x/\varepsilon|^2$, $x \in \mathbb{R}^d$. Note that in our derivation of (7.4.10) and (7.4.11) we have not used the μ-integrability of Λ^2 and the existence of entropy for ν. Using (7.4.9) and (7.4.10), we obtain

$$\begin{aligned}&\int_0^\tau\!\!\int_{\mathbb{R}^d} a^{ij}\frac{\partial_{x_i}f_\varepsilon}{f_\varepsilon}\partial_{x_j}f_\varepsilon\,dx\,dt \\ &\quad= \int_0^\tau\!\!\int_{\mathbb{R}^d}\frac{\partial_{x_i}f_\varepsilon}{f_\varepsilon}\Bigl[(b^i\varrho)*w_\varepsilon + \varepsilon a^{ij}\partial_{x_j}\frac{1}{\max(|x|,1)^{d+1}}\Bigr]\,dx\,dt - L_\varepsilon \\ &\quad\quad - \int_0^\tau\!\!\int_{\mathbb{R}^d}\frac{\partial_{x_i}f_\varepsilon(x,t)}{f_\varepsilon(x,t)}\int_{\mathbb{R}^d}\partial_{x_j}w_\varepsilon(x-y)[a^{ij}(y,t)-a^{ij}(x,t)]\,\varrho(y,t)\,dy\,dx\,dt.\end{aligned}$$

The right side in this equality does not exceed

$$\left|\int_0^\tau\!\!\int_{\mathbb{R}^d}\frac{|\nabla f_\varepsilon|^2}{f_\varepsilon}\,dx\,dt\right|^{1/2}\left(\left|\int_0^\tau\!\!\int_{\mathbb{R}^d}\frac{\sum_{i=1}^d[(b^i\varrho)*w_\varepsilon]^2}{f_\varepsilon}\,dx\,dt\right|^{1/2}+\varepsilon MC_d\right)$$
$$+\,d^{3/2}\lambda\left|\int_0^\tau\!\!\int_{\mathbb{R}^d}\frac{|\nabla f_\varepsilon|^2}{f_\varepsilon}\,dx\,dt\int_0^\tau\!\!\int_{\mathbb{R}^d}\frac{(\varrho*q_\varepsilon)^2}{f_\varepsilon}\,dx\,dt\right|^{1/2}+M(\varepsilon)+K(\varepsilon),$$

where C_d is the integral of $(d+1)^2|x|^{-d-3}$ over the set $\{|x|\geqslant 1\}$ with respect to Lebesgue measure. By inequality (7.4.3) we have

$$\int_0^\tau\!\!\int_{\mathbb{R}^d}\frac{[(b^i\varrho)*w_\varepsilon]^2}{f_\varepsilon}\,dx\,dt\leqslant\int_0^\tau\!\!\int_{\mathbb{R}^d}|b^i|^2\,d\mu,\quad 1\leqslant i\leqslant d\,,$$
$$\int_0^\tau\!\!\int_{\mathbb{R}^d}\frac{(\varrho*q_\varepsilon)^2}{f_\varepsilon}\,dx\,dt\leqslant\gamma:=\int_{\mathbb{R}^d}|x|^4g(x)\,dx.$$

Since $A\geqslant m\cdot\mathrm{I}$, we arrive at the following estimate:

$$m\int_0^\tau\!\!\int_{\mathbb{R}^d}\frac{|\nabla f_\varepsilon|^2}{f_\varepsilon}\,dx\,dt\leqslant M(\varepsilon)+K(\varepsilon)$$
$$+\left(\int_0^\tau\!\!\int_{\mathbb{R}^d}\frac{|\nabla f_\varepsilon|^2}{f_\varepsilon}\,dx\,dt\right)^{1/2}\Bigl(\|b\|_{L^2(\mu)}+\varepsilon MC_d+\lambda d^{3/2}\sqrt{\gamma}\Bigr),$$

which by the inequality $c\sqrt{x}\leqslant mx/2+c^2/(2m)$ gives the estimate

$$\int_0^\tau\!\!\int_{\mathbb{R}^d}\frac{|\nabla f_\varepsilon|^2}{f_\varepsilon}\,dx\,dt\leqslant m^{-2}\Bigl(\|b\|_{2,\mu}+\varepsilon MC_d+\lambda d^{3/2}\sqrt{\gamma}\Bigr)^2+2m^{-1}\bigl(M(\varepsilon)+K(\varepsilon)\bigr).\tag{7.4.12}$$

The quantities $M(\varepsilon)$ and $K(\varepsilon)$ are uniformly bounded in ε. Letting $\varepsilon\to 0$, we obtain that $\sqrt{\varrho(\,\cdot\,,t)}\in W^{2,1}(\mathbb{R}^d)$ for almost all $t\in(0,\tau)$. Hence $\varrho(\,\cdot\,,t)\in W^{1,1}(\mathbb{R}^d)$ for almost all $t\in(0,\tau)$. In addition, the integral of $|\nabla\varrho|^2/\varrho$ does not exceed the right side of (7.4.12) with $\varepsilon=0$. Therefore, $\sqrt{\varrho}\in\mathbb{H}^{2,2}(\mathbb{R}^d\times[0,\tau])$. By the Sobolev embedding theorem we have the inclusion $\varrho\in L^{d/(d-2),1}(\mathbb{R}^d\times[0,\tau])$ if $d>2$, and $\varrho\in L^{s,1}(\mathbb{R}^d\times[0,\tau])$ for all $s\in[1,\infty)$ if $d=2$. The last assertion of the theorem is clear from our reasoning. □

The proof yields the estimate

$$\int_0^\tau\!\!\int_{\mathbb{R}^d}\frac{|\nabla\varrho|^2}{\varrho}\,dx\,dt\leqslant m^{-2}\Bigl(\|b\|_{L^2(\mu)}+\lambda d^{3/2}\sqrt{\gamma}\Bigr)^2+2\ln 2m^{-1}$$
$$+\,2m^{-1}\int_{\mathbb{R}^d}\varrho_0(x)\ln\varrho_0(x)\,dx+2m^{-1}(d+1)\int_{\mathbb{R}^d}\varrho(x,\tau)\Lambda(x)\,dx$$
$$+\,2m^{-1}e^{-1}\int_{\mathbb{R}^d}\max\{|x|,1\}^{-d-1}\,dx.\tag{7.4.13}$$

7.4.2. Remark. It is clear from the proof that for the entropy of $\varrho_\varepsilon(x,0)$ only an upper bound is needed, so in place of the integrability of the function $\varrho_0\ln\varrho_0$ it suffices to require the integrability of $\varrho_0\max(0,\ln\varrho_0)$ (then Jensen's inequality can be applied to the function $s\max(0,\ln s)$). The effect of this is that in estimate (7.4.13) in place of $\varrho_0\ln\varrho_0$ we will have $\varrho_0\max(0,\ln\varrho_0)$. However, the obtained estimates and (7.4.9) show that if we keep the other conditions, the entropy of ϱ_0

will be finite anyway. If we do not require the μ-integrability of Λ, the situation can change (see Exercise 7.5.7).

Let us give a condition for the membership of the density of a probability solution in the parabolic Sobolev class $\mathbb{H}^{p,1}(\mathbb{R}^d \times [\tau_1, \tau_2])$. Parabolic embedding theorems can be used to derive some further properties.

7.4.3. Theorem. *Suppose that condition* (P1) *in* (7.3.1) *and condition* (7.4.1) *are fulfilled,* $\mu \in \mathcal{P}(\mathbb{R}^d_T)$ *and* $\partial_t \mu = L^*_{A,b}\mu$, *where* $b \in L^p(\mathbb{R}^d_T, \mu)$ *and* $p > d+2$. *Then* $\mu = \varrho\, dx\, dt$ *and* $\varrho \in \mathbb{H}^{p,1}(\mathbb{R}^d \times [\tau_1, \tau_2])$ *for every* $[\tau_1, \tau_2] \subset (0, T)$.

PROOF. It follows from Theorem 6.4.2 and Exercise 6.7.18 that $\mu = \varrho\, dx\, dt$, $\varrho \in \mathbb{H}^{p,1}_{\rm loc}(\mathbb{R}^d \times (0,T))$ and

$$\|\varrho\|^p_{\mathbb{H}^{p,1}(K(x,1)\times[\tau_1,\tau_2])} \leqslant C\mu\bigl(K(x,2)\times(0,T)\bigr) + C\int_0^T\int_{K(x,2)} |b|^p\, d\mu,$$

where $K(x,r) = x + [-r, r]^d$ and C does not depend on x. It remains to cover the space $\mathbb{R}^d$ by cubes $K(x_j, 1)$ without common inner points. □

7.5. Complements, comments, and exercises

7.5(i). The Nash–Aronson estimates

Here we briefly discuss the classic Nash–Aronson estimates (see Nash [**734**], Aronson [**71**], and Fabes, Stroock [**349**]). Let us consider the elliptic operator $\mathcal{L}u(x) = \operatorname{div}\bigl(A(x)\nabla u(x)\bigr)$, where $A(x) = \bigl(a^{ij}(x)\bigr)_{1\leqslant ij\leqslant d}$ is a symmetric matrix, $a^{ij} \in C_b^\infty(\mathbb{R}^d)$, and a uniform bound $\lambda^{-1}\cdot \mathrm{I} \leqslant A(x) \leqslant \lambda\cdot\mathrm{I}$ holds for all $x \in \mathbb{R}^d$ and some number λ. Similar bounds hold in case of bounded lower order terms.

Let $u \in C_b\bigl(\mathbb{R}^d \times [0,T)\bigr) \bigcap C^2\bigl(\mathbb{R}^d \times (0,T)\bigr)$ be a solution to the Cauchy problem $\partial_t u = \mathcal{L}u$, $u|_{t=0} = f$, where $f \in C_0^\infty(\mathbb{R}^d)$. It is well-known that in this case the solution u can be represented by means of a Markov semigroup T_t in the form

$$u(x,t) = T_t f(x) = \int_{\mathbb{R}^d} K(t,x,y) f(y)\, dy,$$

where $K(t,x,y)$ is the integral kernel of the semigroup and $K(t,x,y) = K(t,y,x)$ is a smooth nonnegative function on $(0,T)\times\mathbb{R}^d\times\mathbb{R}^d$. Moreover, $K(t,x,y)$ as a function of (x,t) or (y,t) satisfies the equation $\partial_t u = \mathcal{L}u$ and

$$\int_{\mathbb{R}^d} K(t,x,y)\, dy = 1;$$

the semigroup property can be written as

$$K(t,x,y) = \int_{\mathbb{R}^d} K(t-s,z,y)K(s,x,z)\, dz, \quad 0 < s < t.$$

The latter equality enables one to derive upper bounds on K from integral estimates. Let us set

$$E(t) = \int_{\mathbb{R}^d} |K(t,x,y)|^2\, dx.$$

Then

$$E'(t) = \int_{\mathbb{R}^d} 2K\partial_t K\,dx = -2\int_{\mathbb{R}^d} \langle A\nabla K, \nabla K\rangle\,dx.$$

Using the condition $A(x) \geqslant \lambda^{-1}\cdot \mathrm{I}$, we arrive at the inequality

$$-E'(t) \geqslant 2\lambda^{-1}\int_{\mathbb{R}^d} |\nabla K(t,x,y)|^2\,dy.$$

The Sobolev inequality yields the estimate

$$\int_{\mathbb{R}^d} |\nabla K|^2\,dy \geqslant C(d)\Bigl(\int_{\mathbb{R}^d} K^{2d/(d-2)}\,dy\Bigr)^{(d-2)/d}.$$

By Hölder's inequality

$$\int_{\mathbb{R}^d} K^2\,dy \leqslant \Bigl(\int_{\mathbb{R}^d} K^{2d/(d-2)}\,dy\Bigr)^{(d-2)/(d+2)}\Bigl(\int_{\mathbb{R}^d} K\,dy\Bigr)^{2/(d+2)}.$$

Hence we arrive at the inequality $-E'(t) \geqslant 2C(d)\lambda^{-1}E(t)^{(d+2)/d}$, which yields the bound $E(t) \leqslant Ct^{-d/2}$. Therefore,

$$K(t,x,y)^2 = \Bigl(\int_{\mathbb{R}^d} K(t/2,z,y)K(t/2,x,z)\,dz\Bigr)^2 \leqslant E(t/2)^2 \leqslant C^2(t/2)^{-d}.$$

Thus, we arrive at the following result.

7.5.1. Theorem. *There is a number $C > 0$, depending only on d and λ, such that*

$$K(t,x,y) \leqslant Ct^{-d/2} \quad \textit{for all } t \in (0,T),\ x,y \in \mathbb{R}^d.$$

A similar reasoning leads to the bound

$$K(t,x,y) \leqslant C_1 t^{-d/2}\exp\bigl(-C_2|x-y|^2/t\bigr)$$

Generalization of these estimates are discussed below; see also § 5.6.

7.5(ii). A generalization of Nash–Aronson type estimates

Here we mention some generalizations of the Nash–Aronson estimates obtained in Zhikov [**964**]. Let us consider the Cauchy problem

$$\varrho\partial_t u - \operatorname{div}(\varrho A\nabla u) = 0, \quad u|_{t=0} = f. \tag{7.5.1}$$

Suppose that $\varrho \geqslant 0$ is a measurable function on $\mathbb{R}^d$, $\varrho, \varrho^{-1} \in L^1_{\mathrm{loc}}(\mathbb{R}^d)$, the functions a^{ij} are measurable, the matrices $A(x) = (a^{ij})_{i,j\leqslant d}$ are symmetric for all x and $\lambda\cdot\mathrm{I} \leqslant A(x) \leqslant \lambda^{-1}\cdot\mathrm{I}$ for some $\lambda > 0$. For $\mu = \varrho\,dx$ we introduce the weighted Sobolev class

$$W(\mu) = \Bigl\{u \in W^{1,1}_{\mathrm{loc}}(\mathbb{R}^d)\colon \int_{\mathbb{R}^d} [u^2 + |\nabla u|^2]\,d\mu < \infty\Bigr\},$$

which is complete with respect to the norm $\|u\|_{L^2(\mu)} + \|\nabla u\|_{L^2(\mu)}$ due to the condition that $\varrho^{-1} \in L^1_{\mathrm{loc}}(\mathbb{R}^d)$. Let $H(\mu)$ denote the closure of $C_0^\infty(\mathbb{R}^d)$ in $W(\mu)$. These classes have already been discussed in § 4.6(iii). The quadratic form

$$\int_{\mathbb{R}^d} \langle A\nabla u, \nabla u\rangle\,d\mu$$

with the domain of definition $H(\mu) \subset L^2(\mu)$ is closed, hence it generates a self-adjoint operator $\mathcal{L}f = -\operatorname{div}(A\nabla f)$ in the space $L^2(\mu)$. A solution to the Cauchy problem (7.5.1) is understood in the sense of semigroups as the equality $u = e^{t\mathcal{L}}f$. Suppose that the density ϱ satisfies the following condition: there are numbers α and β such that $\alpha > 1$, $\beta > 1$, $\alpha^{-1} + \beta^{-1} < 2d^{-1}$, and

$$\sup_{r\geqslant 1} r^{-d} \int_{|x|\leqslant r} (\varrho^\alpha + \varrho^{-\beta})\, dx = l < \infty.$$

If $\varrho, \varrho^{-1} \in L^\infty(\mathbb{R}^d)$, then $\alpha = \beta = \infty$. We determine the exponent γ from the relation $1/(\gamma\alpha') = 2/d - 1/\alpha - 1/\beta$, $\alpha' = \alpha/(\alpha-1)$. It is easy to show that $\gamma \geqslant d/2$ and $\gamma = d/2$ only if $\alpha = \beta = \infty$.

7.5.2. Theorem. *For the solution $u(x,t)$ to the Cauchy problem (7.5.1) we have the estimate*

$$|u(x,t)| \leqslant c_1 t^{-\gamma}(1+|x|+\sqrt{t})^{\gamma-d/2} e^{-\psi(x)} e^{2\lambda^{-1}\Lambda^2 t} \times \int_{\mathbb{R}^d} (1+|y|+\sqrt{t})^{\gamma-d/2} e^{\psi(y)} |f(y)|\, \mu(dy).$$

Here ψ is an arbitrary Lipschitz function, $\Lambda = \sup_x |\nabla\psi(x)| < \infty$. The constant c_1 depends only on the quantities d, λ, α, β, and l.

If $\psi_z(x) = (z,x)$, then $\Lambda = |z|$ and

$$\min_z e^{-\psi_z(x)} e^{2\lambda^{-1}\Lambda^2 t} = e^{-\lambda|x-y|^2/(8t)}.$$

One can show that the operator $e^{-t\mathcal{L}}$ is an integral operator and for its integral kernel $K(x,y,t)$ the following estimate holds:

$$K(x,t,y) \leqslant c_1 t^{-\gamma}\bigl(1+|x|+\sqrt{t}\bigr)^{\gamma-d/2}\bigl(1+|y|+\sqrt{t}\bigr)^{\gamma-d/2} e^{-\lambda|x-y|^2/(8t)}.$$

In the case of a nondegenerate density ϱ, where $\alpha = \beta = \infty$ and $\gamma = d/2$, we obtain the classical Nash–Aronson estimate

$$K(x,t,y) \leqslant c_1 t^{-d/2} e^{-\lambda|x-y|^2/(8t)}.$$

Theorem 7.5.2 generalizes to the case of a Borel measure μ on $\mathbb{R}^d$ satisfying the Sobolev inequality: for some $q > 1$

$$\biggl(\int_{|x|<r} |\varphi(x)|^{2q}\, \mu(dx)\biggr)^{1/q} \leqslant c_0 r^{d/q-d+2} \int_{|x|<r} |\nabla\varphi(x)|^2\, \mu(dx)$$

for all $\varphi \in C_0^\infty\bigl(U(0,r)\bigr)$ and $r \geqslant r_0 > 0$, where c_0 does not depend on r. If μ is Lebesgue measure and $q = d/(d-2)$, then this is the classical Sobolev inequality. Under this condition on μ, Theorem 7.5.2 is true with $\gamma = q/(q-1)$.

Comments

In our exposition in this chapter we used the papers Bogachev, Röckner, Shaposhnikov [**164**], Shaposhnikov [**850**]. Theorem 7.4.3 was proved in Shaposhnikov [**850**, Theorem 2.2, Corollary 2]. On logarithmic gradients, see also Dermoune, Filali [**296**]. There is an extensive literature on estimates of fundamental solutions to parabolic equations related to upper bounds, see Davies [**284**], Eidelman, Zhitarashu [**328**], Grigor'yan [**425**], Kusuoka [**573**], Porper, Eidelman [**785**], [**786**], and also § 5.6.

A survey about stabilization and behavior of solutions of parabolic equations for large times is given in Denisov [**294**] (see also [**295**] and [**328**, Chapter VII]), the case of Fokker–Planck–Kolmogorov equations was considered already in Il'in, Has'minskiĭ [**474**], [**475**]. Estimates of the rate of convergence of solutions of Fokker–Planck–Kolmogorov equations to stationary solutions are studied in many works, see Arnold, Carlen, Ju [**64**], Arnold, Markowich, Toscani, Unterreiter [**65**], Calogero [**216**], Carrillo, Toscani [**224**]; convergence in the Kantorovich metric is considered in Bolley, Gentil, Guillin [**189**], Natile, Peletier, Savaré [**735**]; in § 9.8(iii) we mention a result of the last paper in relation to the uniqueness problem. Angiuli, Lorenzi, Lunardi [**53**] study supercontractivity, ultraboundedness and ultracontractivity for nonautonomous parabolic equations. For global properties of solutions, see Angiuli, Lorenzi [**51**], [**52**], Bertoldi, Lorenzi [**110**], Kondratiev, Liskevich, Sobol, Us [**534**], Lorenzi [**641**], Mikami [**709**], [**710**], Semenov [**840**], Sheu [**853**].

Exercises

7.5.3. Let $f \geqslant 0$ be a locally integrable function on $[0, +\infty)$ such that

$$f(t) \leqslant C + \int_0^t [K(s) + H(s)f(s)]\, ds \quad \text{for a.e. } t,$$

where C is a constant, K and H are locally integrable functions, $H \geqslant 0$. Prove that, for almost all $t > 0$, there holds Gronwall's inequality

$$f(t) \leqslant CR(t) + R(t) \int_0^t \frac{K(s)}{R(s)}\, ds, \quad R(t) := \exp \int_0^t H(s)\, ds.$$

HINT: Denote by $g(t)$ the right side of the inequality in the formulation and observe that $g'(t) \leqslant K(t) + H(t)g(t)$ a.e., whence we obtain the estimate $g(t) \leqslant u(t)$ for the solution of the linear equation $u' = K + Hu$, $u(0) = C$, finally, solve this equation by the method of variation of the constant.

7.5.4. Prove Theorem 7.4.1 in the case of a nonzero potential c under the condition that $c \leqslant 0$ and $(1 + |x|^d)c \in L^1(\mu)$.

7.5.5. Prove Theorem 7.4.1 in the case of a nonzero potential c under the condition that $c \leqslant 0$ and $c \ln(|c| + 1), c \ln(1 + |x|) \in L^1(\mu)$.

7.5.6. Give an example of A, b, c such that for solutions $\mu = (\mu_t)$ of class $\mathcal{M}_\nu$ to the Cauchy problem for the equation $\partial_t \mu_t = L^*_{A,b,c}\mu$, where the initial condition ν is a probability measure, there are no estimates independent of ν (analogous to those in Example 7.1.5) for the integral of $|x|$ against μ_t.

HINT: Consider the heat equation for the Gaussian family

$$(2\pi t)^{-1/2} \exp\bigl(-(x - m)^2/(2t)\bigr), \quad m \in \mathbb{R}.$$

7.5.7. Let $d = 1$, $b = 0$ and $A = 1/2$. Show that, for every initial probability distribution μ_0, the solution to the Cauchy problem is given by the convolution $\mu_0 * g_t$, where $g_t(x) = (2\pi t)^{-1/2} \exp\bigl(-x^2/(2t)\bigr)$, and if μ_0 has a density ϱ_0 such that $|\varrho_0'|^2/\varrho_0 \in L^1(\mathbb{R}^1)$, but the function $\varrho_0 \ln \varrho_0$ is not integrable, then the solution $\varrho(t, x)$ has no entropy for any t, although the integrals of $|\partial_x \varrho(t, x)|^2 \varrho(t, x)^{-1}$ over $\mathbb{R}$ are uniformly bounded. With the aid of this example show also that for the validity of estimate (7.4.2) it is necessary to have certain conditions on the initial distribution, by taking for the measure μ_0 Dirac's measure at zero (then the function $|\partial_x \varrho|^2/\varrho$ is not integrable on all of $\mathbb{R}^1_T$).

CHAPTER 8

Parabolic Harnack Inequalities and Lower Bounds

In this chapter we discuss lower bounds for the densities of solutions to parabolic Fokker–Planck–Kolmogorov equations. Two principally different situations are considered: locally bounded drift coefficients and locally unbounded drift coefficients that need not be even integrable with respect to Lebesgue measure. In the first case we use Harnack's inequality, in which we study the dependence of the constant on the equation's coefficients (see Theorem 8.1.2 and Theorem 8.1.3). In the second case we replace the local boundedness of the drift coefficient with its integrability with respect to the solution.

The main results of the chapter are Theorem 8.2.1 and its Corollary 8.2.2, which give constructive uppers bound on the solution density, and also Corollary 8.3.7 and Corollary 8.3.9, which ensure positivity of densities of probability solutions.

8.1. Parabolic Harnack inequalities

Here we turn to Harnack inequalities for parabolic equations in divergence form.

Suppose that Ω is a bounded domain in $\mathbb{R}^d$, $\Omega_T = \Omega\times(0,T)$ and that we are given a measurable matrix-valued mapping $A = (a^{ij})_{1\leqslant i,j\leqslant d}$ on Ω_T such that there exist constants $\gamma \geqslant 0$ and $\alpha > 0$ for which

$$(8.1.1)\qquad \sum_{i,j} |a^{ij}(x,t)|^2 \leqslant \gamma^2 \quad\text{and}\quad A(x,t) \geqslant \alpha\cdot \mathrm{I} \quad\text{for all}\quad (x,t)\in\Omega_T.$$

In addition, let $b\colon\ \Omega_T \to \mathbb{R}^d$ be a measurable vector field with

$$(8.1.2)\qquad \sup_{(x,t)\in\Omega_T} |b(x,t)| \leqslant B < \infty.$$

Suppose that a *nonnegative* function $u \in \mathbb{H}^{2,1}(\Omega_T)$ satisfies the equation

$$(8.1.3)\qquad \partial_t u = \partial_{x_i}(a^{ij}\partial_{x_j}u - b^i u),$$

i.e., for every function $\varphi \in C_0^1(\Omega_T)$ one has the equality

$$\iint_{\Omega_T} \Bigl[-\partial_t\varphi u + \partial_{x_i}\varphi\bigl(a^{ij}\partial_{x_j}u - b^i u\bigr)\Bigr]\,dx\,dt = 0.$$

It is known from more general results in Chapter 6 that under our assumptions any solution u has a locally Hölder continuous version.

Let us fix a point $(\bar{x},\bar{t}) \in \Omega\times(0,T]$. Let $R(\bar{x},r)$ be the open cube of edge length r centered at the point $\bar{x}$. Set

$$Q(r) = R(\bar{x},r)\times(\bar{t}-r^2,\bar{t}),\quad Q^*(r) = R(\bar{x},r)\times(\bar{t}-8r^2,\bar{t}-7r^2).$$

Throughout $U(a,r) = \{x\colon\ |x-a| < r\}$. We recall a result from Chapter 6.

8.1.1. Theorem. *Let $Q(3r) \subset \Omega_T$. Then for the continuous version of the function u satisfying equation* (8.1.3) *we have*

$$\sup_{(x,t)\in Q^*(r)} u(x,t) \leqslant C \inf_{(x,t)\in Q(r)} u(x,t),$$

where the number $C = C(d,\alpha,\gamma,Br)$ depends only on d,α,γ, and Br.

As in the elliptic case, we are interested in a more precise form of dependence of C on the indicated parameters in (8.1.1), (8.1.2).

8.1.2. Theorem. *Let $Q(3r) \subset \Omega_T$. Then for the continuous version of the function u the following Harnack inequality holds:*

$$\sup_{(x,t)\in Q^*(r)} u(x,t) \leqslant C \inf_{(x,t)\in Q(r)} u(x,t),$$

where we set

$$C := C(d,\alpha,\gamma,B,r) := \exp\Bigl\{c(d)\bigl|1+\alpha^{-1}+(\alpha^{-1/2}+\alpha^{-1})(Br+\gamma)\bigr|^2\Bigr\}.$$

The proof can be found in Bogachev, Röckner, Shaposhnikov [**165**], where a method of Aronson, Serrin [**74**, Theorem 3] is employed. An analogous reasoning is used in the next result, which refines the obtained estimate with respect to its dependence on r. Its advantage as compared to the previous theorem is that now B appears in the estimate without the factor r.

8.1.3. Theorem. *Suppose that $U(z_0,\theta r) \subset \Omega$ for some $\theta > 1$ and $r > 0$. Then, whenever $0 < s < t < T$ and $x,y \in U(z_0,r)$, the continuous version of the function u satisfies the following inequality:*

$$u(y,s) \leqslant u(x,t)\exp\Bigl\{K\Bigl(\frac{|x-y|^2}{t-s} + (B+1)^2\frac{t-s}{\delta^2} + 1\Bigr)\Bigr\},$$

where $\delta = \min\{(\theta-1)r, \sqrt{s}, 1\}$ and

$$K := c(d)\Bigl|1+\alpha^{-1}+(\alpha^{-1}+\alpha^{-1/2})\gamma\Bigr|^2$$

with some number $c(d)$ depending only on d.

PROOF. Let $x \in U(z_0,r)$, $y \in U(z_0,r)$,

$$q_0 = \delta/\bigl(9d(B+1)\bigr).$$

Then $\delta \leqslant 1$, $q_0B \leqslant 1$ and for all $z \in U(z_0,r)$ we have

$$R(z,3q_0) \subset U(z_0,\theta r) \subset \Omega.$$

Indeed, for every $z_1 \in K(z,3q_0)$ we have

$$|z_1 - z_0| \leqslant |z_1 - z| + |z - z_0| \leqslant 3dq_0 + r \leqslant \theta r.$$

We observe that, whenever $s < \tau < t$, the inequality

$$\tau - 9q_0^2 \geqslant s - 9q_0^2 \geqslant 0$$

holds, i.e., $(\tau - 9q_0^2, \tau) \subset [0,T]$. Therefore,

$$Q(3q_0) = R(z,3q_0) \times \bigl(\tau - 9q_0^2, \tau\bigr) \subset \Omega_T.$$

Let $x_n = x + n(y-x)/N$, $t_n = s + n(t-s)/N$, where $n = 0,\dots,N$. Then $x = x_0$, $s = t_0$, $y = x_N$, $t = t_N$ and

$$|x_n - x_{n-1}| = |x-y|/N, \quad |t_n - t_{n-1}| = (t-s)/N.$$

Set

$$q=\frac{1}{2}\Big(\frac{4}{q_0^2}+\frac{56}{t-s}+\frac{64|x-y|^2}{(t-s)^2}\Big)^{-1/2}.$$

Then the following easily verified estimate holds:

$$q\leqslant \min\Big\{\frac{q_0}{2},\sqrt{\frac{t-s}{56}},\frac{t-s}{8|x-y|}\Big\}.$$

Let us set

$$N=\Big[\frac{t-s}{8q^2}\Big]+1,$$

where $[r]$ denotes the integer part of r. We observe that with this choice of N the following inequalities hold:

$$\frac{|x-y|}{q}\leqslant\frac{t-s}{8q^2}\leqslant N,\qquad \frac{t-s}{7q^2}-\frac{t-s}{8q^2}=\frac{t-s}{56q^2}\geqslant 1.$$

From this we obtain at once $|x_n-x_{n-1}|=|x-y|/N\leqslant q\leqslant q_0/2$. Since N is the smallest natural number greater than $(t-s)/8q^2$ and the above estimate on $t-s$ holds, we have $(t-s)/(8q^2)\leqslant N\leqslant(t-s)/(7q^2)$. Therefore,

$$7q^2\leqslant\frac{t-s}{N}\leqslant 8q^2.$$

Taking into account that $x_{n-1}\in R(x_n,q)$, $7q^2\leqslant t_n-t_{n-1}\leqslant 8q^2$ and

$$R(x_n,3q)\times(t_n-9q^2,t_n)\subset R(x_n,3q_0)\times(t_n-9q_0^2,t_n)\subset\Omega_T,$$

applying Theorem 8.1.2 with $r=q$, we obtain

$$\varrho(x_{n-1},t_{n-1})\leqslant\sup_{(z,\tau)\in Q^*(q)}\varrho(z,\tau)\leqslant C_0\inf_{(z,\tau)\in Q(q)}\varrho(z,\tau)\leqslant C_0\varrho(x_n,t_n),$$

where

$$Q^*(q)=R(x_n,q)\times(t_n-8q^2,t_n-7q^2),\ Q(q)=R(x_n,q)\times(t_n-q^2,t_n),$$

and C_0 depends only on d,m,M. Indeed, we have $qB\leqslant q_0B\leqslant 1$. We obtain the following recurrent relation:

$$u(x_{n-1},t_{n-1})\leqslant C_0u(x_n,t_n),\quad 0\leqslant n\leqslant N.$$

Therefore,

$$u(x,s)=u(x_0,t_0)\leqslant C_0^N u(x_N,t_N)=C_0^N u(y,t).$$

Substituting the value of N indicated above, we obtain that

$$u(x,s)\leqslant u(y,t)\exp\Big\{K_0\Big(\frac{t-s}{8q^2}+1\Big)\Big\},$$

where $K_0=\ln C_0$. We finally arrive at the estimate

$$\sup_{z\in U(z_0,r)}u(z,s)\leqslant\inf_{z\in U(z_0,r)}u(z,t)\exp\Big\{K\Big(\frac{|x-y|^2}{t-s}+(B+1)^2\frac{t-s}{\delta^2}+1\Big)\Big\},$$

where $K=K(d,m,M)$ and $\delta=\min\{(\theta-1)r,\sqrt{s},1\}$. □

A similar reasoning without investigation of dependence on B is used in the paper Aronson, Serrin [**74**].

8.2. Lower bounds for densities

We now proceed to lower bounds for densities of solutions to parabolic equations for measures. As we shall see, there are lower bounds similar to upper bounds discussed in Chapter 7; however, the difference between upper and lower bounds is more significant than in the elliptic case.

Let $A = (a^{ij})_{1\leqslant i,j\leqslant d}$ be a Borel measurable matrix-valued mapping defined on the set $\mathbb{R}^d\times(0,T)$ such that the matrix $A(x,t)$ is positive-definite for all (x,t) and let $b\colon \mathbb{R}^d\times(0,T)\to\mathbb{R}^d$ be a Borel measurable mapping.

As in the elliptic case, in the study of lower estimates it is more convenient to deal with divergence form operators $\mathcal{L}=\mathcal{L}_{A,b}$.

We recall that a Borel measure μ on $\mathbb{R}^d\times(0,T)$ satisfies the weak parabolic equation $\partial_t\mu=\mathcal{L}^*\mu$ if the functions a^{ij} and b^i are integrable on every compact set in $\mathbb{R}^d\times(0,T)$ with respect to μ and, for every function $\varphi\in C_0^\infty(\mathbb{R}^d\times(0,T))$, we have the equality

$$\int_{\mathbb{R}^d\times(0,T)}\mathcal{L}\varphi\,d\mu=0,$$

which is understood either as (6.1.5) or as (6.1.6) (see Chapter 6).

Let V be a continuous increasing function on the ray $[0,\infty)$ satisfying the condition $V(0)>0$.

8.2.1. Theorem. *Let $\sup_{t\in(0,T)}|b(x,t)|\leqslant V(|x|/\theta)$ for almost all $x\in\mathbb{R}^d$, where $\theta>1$. Let*

$$\alpha(r):=\sup_{t\in(0,T),|x|\leqslant r}\|A(x,t)^{-1}\|,\quad \gamma(r):=\sup_{t\in(0,T),|x|\leqslant r}\|A(x,t)\|.$$

Suppose that μ is a nonnegative measure with a density ϱ on $\mathbb{R}^d\times(0,T)$ such that $\varrho\in\mathbb{H}^{2,1}(U\times J)$ for every ball U in $\mathbb{R}^d$ and every compact interval J in $(0,T)$, and μ satisfies equation (6.1.4) in the sense of (6.1.6), i.e., we deal with Case (II). Then there exists a number $K=K(d)>0$ such that, whenever $0<s<t<T$, $x\in\mathbb{R}^d$, the continuous version of the function ϱ satisfies the inequality

$$\begin{aligned}\varrho(x,t)\geqslant\varrho(0,s)\exp\Bigl\{-K(d)\bigl|1+\alpha(\theta|x|)^{-1}+\bigl[\alpha(\theta|x|)^{-1}\\ +\alpha(\theta|x|)^{-1/2}\bigr]\gamma(\theta|x|)\bigr|^2\Bigl(1+\frac{t-s}{s}V(|x|)^2+\frac{1}{t-s}|x|^2\Bigr)\Bigr\}.\end{aligned}$$

In particular, if $\|A(x,t)\|\leqslant\gamma$ and $\|A(x,t)^{-1}\|\leqslant\alpha$, then there exists a positive number $K=K(d,\alpha,\gamma,\theta)$ such that the continuous version of the function ϱ satisfies the inequality

$$\varrho(x,t)\geqslant\varrho(0,s)\exp\Bigl\{-K\Bigl(1+\frac{t-s}{s}V(|x|)^2+\frac{1}{t-s}|x|^2\Bigr)\Bigr\},$$

where $0<s<t<T$, $x\in\mathbb{R}^d$.

PROOF. Take $B=V(|x|)$ in Theorem 8.1.3. □

8.2.2. Corollary. *If in the situation of Theorem 8.2.1 we have*

$$\|A(x,t)\|\leqslant\gamma\quad\text{and}\quad\|A(x,t)^{-1}\|\leqslant\alpha$$

and for a.e. $t\in(0,T)$ the function $x\mapsto\varrho(x,t)$ does not vanish identically, then for each compact interval $[\tau_1,\tau_2]$ in $(0,T)$ there is a number $K=K(d,\alpha,\gamma,\theta,\tau_1,\tau_2)\geqslant 0$

such that for all $t \in [\tau_1, \tau_2]$ *and* $x \in \mathbb{R}^d$ *we have the two-sided estimate*

$$\exp\Bigl(-K\bigl(1 + V(|x|)^2 + |x|^2\bigr)\Bigr) \leqslant \varrho(x,t) \leqslant \exp\Bigl(K\bigl(1 + V(|x|)^2 + |x|^2\bigr)\Bigr).$$

8.2.3. Example. Suppose that in the situation of Theorem 8.2.1 the matrices $A(x,t)$ and $A(x,t)^{-1}$ are uniformly bounded and that for some constants $c_1 > 0$ and $c_2 > 0$ the inequality

$$\sup_{t\in(0,T)} |b(x,t)| \leqslant c_1|x|^\beta + c_2$$

is fulfilled for almost all x. Then there exists a positive number K such that, whenever $s, t \in (0,T)$ and $s < t$, we have

$$\varrho(x,t) \geqslant \varrho(0,s) \exp\Bigl\{-K\Bigl(1 + \frac{t-s}{s}|x|^{2\beta} + \frac{1}{t-s}|x|^2\Bigr)\Bigr\}.$$

For example, if

$$L = \partial_t + \frac{1}{2}\Delta,$$

then the measure $(2\pi t)^{-1/2}e^{-|x|^2/2t}\,dx\,dt$ is a solution. For any given $\delta > 0$, our results yield numbers $K_1(\delta), K_2(\delta) > 0$ such that the lower bound

$$\varrho(x) \geqslant K_1(\delta)e^{-K_2(\delta)|x|^2/t}$$

holds in the strip $\mathbb{R}^d \times (\delta, T)$. Similarly, our lower estimates are sharp in the case of a linear drift coefficient, but they become less sharp in the case of quadratic growth of $|b|$; for example, if $\varrho(x,t) = C\exp(-|x|^3)$, then the function $\exp(-K|x|^4)$, but not $\exp(-|x|^3)$, appears in our lower bound.

Let us give conditions on the coefficients A and b ensuring two-sided exponential estimates of the density of the solution in the parabolic case.

8.2.4. Example. Suppose that the matrices $A(x,t)$ are symmetric and positive, $A(x,t)$ and $A(x,t)^{-1}$ are uniformly bounded, the functions $x \mapsto a^{ij}(x,t)$ are uniformly Lipschitzian with a common constant, and that for some $r > 1$, $\sigma \geqslant 0$, $K > 0$, and $K' > K$ we have

$$|b(x,t)| \leqslant C + C|x|^{r-1+\sigma}, \quad \varrho(x,0) \leqslant C\exp(-K'|x|^r),$$
$$\langle x, b(x,t)\rangle \leqslant c_1 - c_2|x|^r, \quad c_2 > 2rK\sup_{x,t}\|A(x,t)\|.$$

Suppose that a probability measure μ on $\mathbb{R}^d \times (0,1)$ satisfies equation (6.1.4) in the sense of (6.1.5), i.e., we deal with Case (I). Then μ has a continuous density ϱ such that, for every compact interval $[\tau_1, \tau_2] \subset (0,1)$, there exist numbers C_1, C_2, and K_0 for which

$$C_1\exp\Bigl(-K_0|x|^{2r+2\sigma-2} - K_0|x|^2\Bigr) \leqslant \varrho(x,t) \leqslant C_2\exp(-K|x|^r),$$

whenever $(x,t) \in \mathbb{R}^d \times [\tau_1, \tau_2]$. The upper bound follows by Example 7.2.6 and the lower bound follows from the results above. Unlike the elliptic case, here there is no coincidence of the powers of $|x|$ in the lower and upper bounds.

One more application of our results is concerned with the proof of the existence of finite entropy for any solution with respect to the space variable at any positive time t for an arbitrary initial distribution. The existence of finite entropy, which is useful in many respects, is necessary for applying those results of § 2.3 which give integrability of $|\nabla \varrho(x,t)|^2/\varrho(x,t)$.

Let W be a continuous increasing function on the half-line $[0,\infty)$ such that $W(0)>0$ and $\lim\limits_{r\to\infty} W(r)=+\infty$.

8.2.5. Proposition. *Suppose that the matrices $A(x,t)$ are positive definite and uniformly bounded along with the matrices $A(x,t)^{-1}$ and that, for some $\theta>1$, the following inequality holds:*

$$\sup_{t\in(0,T)} |b(x,t)|^2 \leqslant W(|x|/\theta), \quad x\in\mathbb{R}^d.$$

Let μ be a measure of the form $\mu=\mu_t\,dt$, where every μ_t is a probability measure on $\mathbb{R}^d$, and let μ have a density $\varrho\in\mathbb{H}^{2,1}_{\rm loc}(\mathbb{R}^d_T)$ and satisfy equation (6.1.4) *in the sense of* (6.1.6). *Suppose that*

$$\int_0^T\int_{\mathbb{R}^d} [|x|^2+W(|x|)]\,\mu_t(dx)\,dt<\infty.$$

Then, for every compact interval $[\tau_1,\tau_2]\subset(0,T)$, we have

$$\int_{\tau_1}^{\tau_2}\int_{\mathbb{R}^d} \varrho(x,s)|\ln\varrho(x,s)|\,dx\,ds<\infty.$$

PROOF. Let $[\tau_1,\tau_2]\subset(0,T)$. By Corollary 8.2.2 there is $K>0$ such that for all $t\in[\tau_1,\tau_2]$ and all x we have the inequality

$$\exp\bigl[-K\bigl(1+|x|^2+W(|x|)\bigr)\bigr]\leqslant\varrho(x,t)\leqslant\exp\bigl[K\bigl(1+|x|^2+W(|x|)\bigr)\bigr].$$

By Fubini's theorem the function $\bigl(|x|^2+W(|x|)\varrho(x,t)\bigr)$ is integrable over $\mathbb{R}^d$ for almost every $t\in(0,T)$. The integral of $\varrho(x,t)|\ln\varrho(x,t)|$ for such t is estimated by the integral of $K\bigl(|x|^2+W(|x|)\varrho(x,t)\bigr)$. □

8.2.6. Corollary. *Suppose that $a^{ij}\in\mathbb{H}^{2,1}_{\rm loc}\bigl(\mathbb{R}^d\times(0,T)\bigr)$, $A(x,t)$ is positive definite, and that $A(x,t)$ and $A(x,t)^{-1}$ are uniformly bounded. In addition, suppose that for some $\theta>1$ the following inequality holds:*

$$\sup_{t\in(0,T)} |b(x,t)|^2 \leqslant W(|x|/\theta), \quad x\in\mathbb{R}^d.$$

Let μ be a measure of the form $\mu=\mu_t\,dt$, where every μ_t is a probability measure. Assume that μ has a density $\varrho\in\mathbb{H}^{2,1}_{\rm loc}(\mathbb{R}^d_T)$ and satisfies equation (6.1.4) *in the sense of* (6.1.6) *and has the initial distribution μ_0 such that the function $W_0(x):=W(|x|)$ is integrable with respect to μ_0. Finally, let*

$$W(r)\geqslant c_1+c_2r^2 \quad\text{and}\quad \mathcal{L}W_0(x)\leqslant C$$

for some constants $c_1,c_2,C>0$. Then, for the continuous version ϱ of the density of μ and every compact interval $[\tau_1,\tau_2\subset(0,T)$, we have

$$\sup_{s\in(0,T)}\int_{\mathbb{R}^d}\varrho(x,s)|\ln\varrho(x,s)|\,dx<\infty.$$

PROOF. By Corollary 7.1.2(i) we have

$$\int_{\mathbb{R}^d} W(|x|)\varrho(x,t)\,dx\leqslant c^{Ct}+e^{Ct}\int_{\mathbb{R}^d} W(|x|)\,\nu(dx).$$

Now the reasoning from the previous corollary applies. □

We emphasize once again that if the functions $x \mapsto a^{ij}(x,t)$ are locally Lipschitz uniformly in $t \in (0,T)$, then in Theorem 8.2.1, Corollary 8.2.2, Example 8.2.3, Proposition 8.2.5, and Corollary 8.2.6 any solution μ of equation (6.1.4) in the sense of (6.1.5) is a solution in the sense of (6.1.6) as well.

8.2.7. Example. Suppose that in Corollary 8.2.6 it is known additionally that the functions $x \mapsto a^{ij}(x,t)$ are uniformly Lipschitzian with a common constant. Then, for every compact interval $[\tau_1, \tau_2] \subset (0,T)$, we have

$$\int_{\tau_1}^{\tau_2}\int_{\mathbb{R}^d} \frac{|\nabla\varrho(x,t)|^2}{\varrho(x,t)}\,dx\,dt < \infty.$$

This follows by the previous corollary and Theorem 7.4.1.

8.2.8. Example. Suppose that the matrices $A(x,t)$ and $A(x,t)^{-1}$ are uniformly bounded, the functions $x \mapsto a^{ij}(x,t)$ are uniformly Lipschitzian with a common constant and that there exist numbers $c, c_0, c_1, c_2, r > 0$ such that

$$|b(x,t)| \leqslant c_0 \exp(c|x|^r), \quad (b(x,t),x) \leqslant c_1 - c_2|x|^r, \quad c_2 > 2cr \sup_{x,t} \|A(x,t)\|.$$

Let μ be a probability measure on $\mathbb{R}^d{\times}(0,T)$ satisfying equation (6.1.4) in the sense of (6.1.5) with initial condition μ_0 such that the function $\exp(c|x|^r)$ is integrable with respect to μ_0. Then

$$\int_{\tau_1}^{\tau_2}\int_{\mathbb{R}^d} \frac{|\nabla\varrho(x,t)|^2}{\varrho(x,t)}\,dx\,dt < \infty$$

for every closed interval $[\tau_1, \tau_2] \subset (0,T)$. To prove this fact, it suffices to take the function $W_0(z) = \exp(M|z|^r)$, where $M > 2c$ is sufficiently close to $2c$.

8.3. Positivity of densities

We now turn to lower bounds without conditions on the growth of the drift coefficient. Until the end of this section we assume that $A(x,t) = \bigl(a^{ij}(x,t)\bigr)_{1\leqslant i,j\leqslant d}$ is a symmetric matrix that satisfies the following conditions:

(GP1) there is a constant $\lambda > 0$ such that for all $x, y \in \mathbb{R}^d$ and $t \in (0,T)$ one has

$$|a^{ij}(x,t) - a^{ij}(y,t)| \leqslant \lambda_0|x-y|,$$

(GP2) there are constants $m, M > 0$ such that for all $x, y \in \mathbb{R}^d$ and $t \in (0,T)$ one has

$$m\cdot \mathrm{I} \leqslant A(x,t) \leqslant M\cdot \mathrm{I}.$$

We shall study nonnegative solutions of the equation

$$\partial_t\mu = \mathcal{L}^*\mu \tag{8.3.1}$$

with an operator $\mathcal{L}$ in the divergence form

$$\mathcal{L}u(x,t) := \partial_{x_i}\bigl(a^{ij}(x,t)\partial_{x_j}u(x,t)\bigr) + b^i(x,t)\partial_{x_i}u(x,t).$$

The equation is understood in the sense of Definition (6.1.5).

Now we do not assume any integrability of b with respect to Lebesgue measure and only the inclusion $\Phi(|b|) \in L^1_{\mathrm{loc}}(\mu)$ is given, where Φ is an increasing positive function on $[0,+\infty)$.

According to Corollary 6.5.1, if in addition to conditions (GP1) and (GP2) we have $\Phi(z) = z^p$ c $p > d+2$, then the measure μ has a continuous density

ϱ belonging to the class $\mathbb{H}^{p,1}(U,J)$ for every ball U and every compact interval $J \subset (0,T)$. However, Harnack's inequality can fail even for stationary solutions.

As in Chapter 3, we are interested in the following questions.

Let $0 < \tau < T$ and $\sup_{(x,t)\in\mathbb{R}^d\times(0,\tau)} \varrho(x,t) > 0$. For which functions Φ does the condition $\Phi(|b|) \in L^1_{loc}\bigl(\mu, \mathbb{R}^d \times (0,T)\bigr)$ imply that the continuous version of ϱ on $\mathbb{R}^d \times [\tau, T)$ has no zeros? How can we estimate ϱ from below under the only condition $\Phi(|b|) \in L^1(\mu)$?

We note at once some distinctions from the stationary case. For stationary solutions, the condition $\exp(\delta|b|) \in L^1_{loc}(\mu)$ implies that the solution has a positive continuous density with respect to Lebesgue measure. In the nonstationary case such a condition is not sufficient anymore, as the example in Exercise 8.5.2 shows.

The principal results in this section are these.

Suppose that a function $f \in C^2\bigl([0,\infty)\bigr)$ and a nonnegative strictly increasing continuous function ψ mapping $[0,+\infty)$ onto $[0,+\infty)$ satisfy the following conditions:

(HP1) $f(z) > 0$, $f'(z) > 0$ and $f''(z) > 0$ if $z > 0$;

(HP2) there exist numbers $\gamma \in (0,1)$ and $z_0 \geqslant 0$ such that whenever $z > z_0$ the function $e^{-(1-\gamma)f(z)}$ is convex downwards (i.e., $(e^{-(1-\gamma)f})'' \geqslant 0$);

(HP3) $\psi^{-1}(z) \leqslant N\sqrt{f'(f^{-1}(z))}$ for all $z>0$ and some $N>0$.

If in addition to (GP1) and (GP2) we have the condition

(GP3) $|b|\exp\bigl(\psi(|b|)\bigr) \in L^p_{loc}\bigl(\mu, \mathbb{R}^d \times (0,T)\bigr)$, where $p > d+2$,

then the measure μ has a continuous density ϱ with respect to Lebesgue measure on the cylinder $\mathbb{R}^d \times (0,T)$, moreover, the density ϱ has no zeros in the cylinder $\mathbb{R}^d \times [\tau, T)$ if $\sup_{(x,t)\in\mathbb{R}^d\times(0,\tau)} \varrho(x,t) > 0$, where $\tau > 0$.

If in addition to (GP1) and (GP2) we have the condition

(GP3′) $|b|\exp\bigl(\psi(|b|)\bigr) \in L^p(\mu)$, where $p > d+2$,

and the density ϱ is positive on $\mathbb{R}^d\times(0,T)$ (which, for example, is the case where each $\mu_t = \varrho(x,t)\,dx$ is a probability measure, which follows by the previous assertion), then, for every compact interval $[\tau_0, \tau_1] \subset (0,T)$, there exist two numbers $c_1 > 0$ and $c_2 > 0$ such that

$$\varrho(x,t) \geqslant e^{-f(c_1|x|^2 + c_2)}, \quad x \in \mathbb{R}^d, t \in [\tau_0, \tau_1]. \tag{8.3.2}$$

For example, conditions (HP1), (HP2) and (HP3) are satisfied for the functions $f(z) = e^z$ and $\psi(z) = z^2$. On the ground of what has been said above we obtain that the condition $\exp(\delta|b|^2) \in L^1(\mu)$ implies that on every compact interval $[\tau_0, \tau_1]$ we have the estimate

$$\varrho(x,t) \geqslant \exp\bigl(-\gamma_1 \exp(\gamma_2|x|^2)\bigr)$$

for some numbers $\gamma_1, \gamma_2 > 0$ and all $(x,t) \in \mathbb{R}^d \times [\tau_0, \tau_1]$.

For the proof we need some auxiliary assertions. According to conditions (HP1) and (HP3), we have $f'(y) \to \infty$, hence there exists a number $y_0 > \max\{z_0, 1\}$ such that $f'(y) \geqslant 1$ and $\bigl(f'(y)\bigr)^{-1/2} e^{f(y)} \geqslant e^{\psi(0)}$ whenever $y > y_0$. Let $\omega_0 = e^{-f(\ln y_0)}$. Then $0 < \omega_0 < 1$. For $s \in (0, \omega_0)$ and $q \geqslant 0$ set

$$h_q(s) = -\frac{1}{sf'\bigl(f^{-1}(|\ln s|)\bigr)} e^{2qf^{-1}(|\ln s|)}.$$

Then $h_q(s) = H_q'(s)$, where

$$H_q(s) = \begin{cases} (2q)^{-1} e^{2qf^{-1}(|\ln s|)} & \text{if } q > 0, \\ f^{-1}(|\ln s|) & \text{if } q = 0. \end{cases}$$

Suppose that conditions (GP1), (GP2), and (GP3) are fulfilled and that a nonnegative measure μ is solution of equation (8.3.1). Let us fix a cube $Q \subset \mathbb{R}^d$ with edge length 2 and a compact interval $[T_0, T_1] \subset (0, T)$. According to (GP3), we have $|b| \in L^p(\mu, Q \times [T_0, T_1])$ with $p > d+2$. Therefore, the solution μ is given by a continuous density ϱ with respect to Lebesgue measure; in addition, we have $\varrho \in \mathbb{H}^{p,1}(Q \times [T_0, T_1])$. Let Λ be a positive number such that

$$\Lambda \leqslant \min\bigl\{\omega_0(2\|\varrho\|_{L^\infty(Q\times[T_0,T_1])})^{-1}, 1\bigr\}.$$

Set

$$\varrho_k = \Lambda\varrho + 1/k, \quad \xi_k = f^{-1}(|\ln \varrho_k|).$$

The next lemma is the key step in our reasoning. Its proof is delegated to the next section.

8.3.1. Lemma. *Let $\mu = \varrho\, dx\, dt$ be a nonnegative solution of equation* (8.3.1), *where the functions a^{ij}, b^i satisfy conditions* (GP1), (GP2) *and* (GP3) *and conditions* (HP1), (HP2) *and* (HP3) *are fulfilled. Let $s > 1$ and $s' = s/(s-1)$, where $2s' \leqslant p$, and let a function $\eta \in C_0^1\bigl(Q \times (T_0, T_1)\bigr)$ be given, where $T_0 < \tau_1 < \tau_2 < T_1$. Then the following inequality is fulfilled.*

If $q = 0$ and $\eta(t, x) = \eta_0(x)$ for $t \in [\tau_1, \tau_2]$, then

$$\int_Q [\xi_k(x,\tau_2)\eta_0^2(x) - \xi_k(x,\tau_1)\eta_0^2(x)]\, dx + \frac{m\gamma}{3}\int_{\tau_1}^{\tau_2}\int_Q |\nabla\xi_k|^2\eta_0^2\, dx\, dt \tag{8.3.3}$$
$$\leqslant N_1\Biggl[\int_{\tau_1}^{\tau_2}\int_Q |\nabla\eta|^2\, dx\, dt + (2-\gamma)\Bigl(\int_{\tau_1}^{\tau_2}\int_Q \eta^2\, dx\, dt\Bigr)^{1/s}\Biggr].$$

If $q > 0$, then

$$\frac{1}{2q}\int_Q \bigl[e^{2q\xi_k(x,\tau_2)}\eta^2(x,\tau_2) - e^{2q\xi_k(x,\tau_1)}\eta^2(x,\tau_1)\bigr]\, dx \tag{8.3.4}$$
$$+ \frac{m\gamma}{3q^2}\int_{\tau_1}^{\tau_2}\int_Q |\nabla e^{q\xi_k}|^2\eta^2\, dx\, dt \leqslant 2\int_{\tau_1}^{\tau_2}\int_\Omega |\eta|\,|\eta_t| e^{2q\xi_k}\, dx\, dt$$
$$+ N_1\Biggl[\int_{\tau_1}^{\tau_2}\int_Q |\nabla\eta|^2 e^{2q\xi_k}\, dx\, dt + (2+2q-\gamma)\Bigl(\int_{\tau_1}^{\tau_2}\int_Q e^{2sq\xi_k}\eta^2\, dx\, dt\Bigr)^{1/s}\Biggr]$$

and N_1 depends only on the following quantities: s, N, ω_0, m, M, γ, d,
$$\|\varrho\|_{L^\infty(Q\times[T_0,T_1])}, \quad \int_{T_0}^{T_1}\int_Q |b|^{2s'}\exp\bigl[2s'\psi(|b|)\bigr]\varrho\, dx\, dt.$$

8.3.2. Remark. If the nonnegative measure $\mu = \varrho\, dx\, dt$ is finite on $\mathbb{R}^d \times (0, T)$ and condition (GP3′) holds in place of condition (GP3), then, as shown in Chapter 7 (see Corollary 7.2.2), we have the inclusion $\varrho \in L^\infty(\mathbb{R}^d \times [T_0, T_1])$, and in the assertion of Lemma 8.3.3 the quantities

$$\|\varrho\|_{L^\infty(Q\times[T_0,T_1])}, \quad \int_{T_0}^{T_1}\int_Q |b|^p \exp\bigl[p\psi(|b|)\bigr]\varrho\, dx\, dt$$

can be replaced by the quantities

$$\|\varrho\|_{L^\infty(\mathbb{R}^d\times[T_0,T_1])}, \qquad \int_{T_0}^{T_1}\int_{\mathbb{R}^d} |b|^p \exp\bigl[p\psi(|b|)\bigr]\varrho\,dx\,dt.$$

In addition, for Λ one can take

$$\Lambda = \min\bigl\{\tau_0(2\|\varrho\|_{L^\infty(\mathbb{R}^d\times[T_0,T_1])})^{-1}, 1\bigr\}.$$

Let Ω be a bounded domain in $\mathbb{R}^d$ and $J\subset(0,T)$ a compact interval. Let $V^2(\Omega,J)$ denote the space of functions $u\in\mathbb{H}^{2,1}(\Omega,J)$ with finite norm

$$\|u\|_{V^2} = \|u\|_{\mathbb{H}^{2,1}(\Omega,J)} + \sup_{t\in J}\|u(\cdot,t)\|_{L^2(\Omega)}.$$

Our next lemma is a corollary of the embedding theorem from Chapter 6 (see Lemma 6.2.1).

8.3.3. Lemma. *Let $d>2$. Suppose that $v\in V^2(\Omega,J)$ and for almost all $t\in J$ the function $x\mapsto v(x,t)$ has compact support in Ω. Then*

$$\|v\|_{L^{2(d+2)/d}(\Omega\times J)} \leqslant C\|v\|_{V^2(\Omega,J)},$$

where C depends only on d and the diameter of the domain Ω.

For every cube $Q(r)=Q(y,r)=\prod_{i=1}^d[y_i-r,y_i+r]$ we define a function ψ_Q as follows:

$$\psi_Q(x) = \Pi_{i=1}^d\chi_i(x_i), \quad x=(x_1,\dots,x_d), \tag{8.3.5}$$

where $\chi_i\in C^1([y_i-2r,y_i+2r])$, $\chi_i(x_i)=1$ if $|x_i-y_i|\leqslant r$, $\chi_i(x_i)=0$ if $|x_i-y_i|\geqslant 2r$. Let also $0\leqslant\psi_Q\leqslant 1$ and $|\nabla\psi_Q|\leqslant cr^{-1}$ for some positive number c.

The next lemma is Trudinger's version of the classical Moser lemma. For a discussion and justification of this lemma, see § 8.5(i).

8.3.4. Lemma. *Suppose that $v\in V^2\bigl(Q(R)\times(T_0,T_1)\bigr)$ and that for every function $\psi\in C_0^1\bigl(Q(R)\bigr)$ the function*

$$t\mapsto\int_{Q(R)} v(x,t)\psi^2(x)\,dx$$

is absolutely continuous on compact intervals in (T_0,T_1). Assume also that there exist numbers $A_1,A_2>0$ such that for every cube $Q(r)\subset Q(R)$ and almost every $t\in(T_0,T_1)$ we have the inequality

$$\frac{d}{dt}\int_{Q(2r)} v(x,t)\psi_Q^2(x)\,dx + A_1\int_{Q(2r)}|\nabla v|^2\psi_Q^2\,dx \leqslant A_2|Q(r)|^{1-2/d}.$$

Then, for any numbers $s_1<s_2<t_1<t_2$ in $[T_0,T_1]$, there exist two numbers $\lambda>0$ and $\sigma>0$, depending only on s_1, s_2, t_1, t_2, d, A_1 and A_2, such that

$$\int_{s_1}^{s_2}\int_{Q(R)} e^{-\lambda v}\,dx\,dt\int_{t_1}^{t_2}\int_{Q(R)} e^{\lambda v}\,dx\,dt\leqslant\sigma.$$

Let $y\in\mathbb{R}^d$. Let us fix cubes $Q_0=Q(y,1)$, $Q=Q(y,1/2)$ and numbers $s_1<s_2<t_1<t_2$ such that $s_1,s_2,t_1,t_2\in[T_0,T_1]$, where $[T_0,T_1]\subset(0,T)$. Set $K^-:=Q\times[s_1,s_2]$ and $K^+:=Q\times[t_1,t_2]$.

8.3.5. Theorem. *Let a nonnegative measure $\mu = \varrho\,dx\,dt$ satisfy (8.3.1), where the functions a^{ij}, b^i satisfy conditions* (GP1), (GP2), (GP3) *and also conditions* (HP1), (HP2), (HP3) *hold. Then for some $\lambda > 0$ we have*

$$\sup_{K+} e^{f^{-1}(|\ln \varrho_k|)} \leqslant C\biggl(\int_{K-} e^{-\lambda f^{-1}(|\ln \varrho_k|)}\,dx\,dt\biggr)^{-1/\lambda},$$

where C depends only on the following quantities: s_1, s_2, t_1, t_2, p, N, ω_0, m, M, γ, d, $\|\varrho\|_{L^\infty(Q_0\times[T_0,T_1])}$, $\int_{T_0}^{T_1}\int_{Q_0} |b|^p \exp\bigl[p\psi(|b|)\bigr]\varrho\,dx\,dt$, but does not depend on k.

Proof. Let $n \geqslant 0$ and

$$Q_n := Q(2^{-1} + 2^{-n-1}), \quad \tau_n = t_1 - (t_1 - s_2)2^{-n-1}.$$

1. By Lemma 8.4.1 with the numbers $q = 0$ and $s = p/(p-2)$, whenever $T_0 < \kappa_1 < \kappa_2 < T_1$, one has the inequality

$$\int_{Q_0} [\xi_k(x,\kappa_2)\eta_0^2(x) - \xi_k(x,\kappa_1)|\eta_0^2(x)]\,dx + \frac{m\gamma}{3}\int_{\kappa_1}^{\kappa_2}\int_{Q_0} |\nabla\xi_k|^2\eta_0^2\,dx\,dt$$
$$\leqslant N_1(2-\gamma)\biggl[\int_{\kappa_1}^{\kappa_2}\int_{Q_0} |\nabla\eta|^2\,dx\,dt + \biggl(\int_{\kappa_1}^{\kappa_2}\int_{Q_0} \eta^2\,dx\,dt\biggr)^{(p-2)/p}\biggr] \tag{8.3.6}$$

for every function $\eta \in C_0^\infty(Q_0\times(T_0,T_1))$ such that $\eta(x,\tau) = \eta_0(x)$ for $\tau \in [\kappa_1, \kappa_2]$. According to equality (8.4.4) in the proof of our key lemma in the next section, the function

$$t \mapsto \int_{Q_0} \xi_k(x,t)\eta_0^2(x)\,dx$$

is absolutely continuous on every inner compact subinterval in $[T_0, T_1]$. For every cube $Q(r)$ with $Q(2r) \subset Q_0$ and each $\tau \in [\kappa_1, \kappa_2]$, let $\eta(x,\tau) = \eta_0(x) := \psi_Q(x)$, where the function ψ_Q is defined by (8.3.5). Since $s = p/(p-2) < d/(d-2)$, by estimate (8.3.6) there exist numbers A_1, A_2 such that

$$\frac{d}{dt}\int_{Q(2r)} \xi_k(x,t)\psi_Q^2(x)\,dx + A_1\int_{Q(2r)} |\nabla\xi_k(x,t)|^2\psi_Q^2(x)\,dx \leqslant A_2|Q(r)|^{1-2/d}.$$

Thus, the hypotheses of Lemma 8.3.4 are fulfilled. Hence there exist two numbers $\lambda, \sigma > 0$, depending only on the quantities indicated above, such that

$$\int_{s_1}^{s_2}\int_{Q_0} e^{-\lambda\xi_k}\,dx\,dt \int_{\tau_0}^{t_2}\int_{Q_0} e^{\lambda\xi_k}\,dx\,dt \leqslant \sigma. \tag{8.3.7}$$

2. Let $\eta \in C_0^1(Q_0\times(T_0,T_1))$. By Lemma 8.3.4 with $q > 0$ for all numbers κ_1, κ_2 in the interval (τ_0, t_2), where $\kappa_1 < \kappa_2$, we have

$$\frac{1}{2q}\int_{Q_0} \bigl[e^{2q\xi_k(x,\kappa_2)}\eta^2(x,\kappa_2) - e^{2q\xi_k(x,\kappa_1)}\eta^2(x,\kappa_1)\bigr]\,dx$$
$$+ \frac{m\gamma}{3q^2}\int_{\kappa_1}^{\kappa_2}\int_{Q_0} |\nabla e^{q\xi_k}|^2\eta^2\,dx\,dt \leqslant \frac{1}{q}\int_{\kappa_1}^{\kappa_2}\int_{Q_0} |\eta|\,|\eta_t|e^{2q\xi_k}\,dx\,dt$$
$$+ N_1(2+2q-\gamma)\biggl[\int_{\kappa_1}^{\kappa_2}\int_{Q_0} |\nabla\eta|^2 e^{2q\xi_k}\,dx\,dt + \biggl(\int_{\kappa_1}^{\kappa_2}\int_{Q_0} e^{2sq\xi_k}\eta^2\,dx\,dt\biggr)^{1/s}\biggr]. \tag{8.3.8}$$

Let η_n be a nonnegative continuously differentiable function on $\mathbb{R}^d\times\mathbb{R}^1$ such that $\eta_n(x,t)=1$ if $(x,t)\in Q_{n+1}\times(\tau_{n+1},t_2)$,

$$\eta_n(x,t)=0 \quad \text{if} \quad (x,t)\notin Q_n\times(\tau_n,T_1),$$

$|\eta(x,t)|\leqslant 1$, $|\partial_t\eta_n(x,t)|\leqslant c_1 2^n$, $|\nabla_x\eta_n(x,t)|\leqslant c_2 2^n$, where c_1,c_2 are some positive numbers. Applying Hölder's inequality with the exponents s and s', we deduce from (8.3.8) the following estimate:

$$\|\nabla(e^{q\xi_k}\eta_n)\|^2_{L^2(Q_{n+1}\times(\tau_{n+1},t_2))}\leqslant C_1 4^n(q+1)^3\|e^{q\xi_k}\|^2_{L^{2s}(Q_n\times(\tau_n,t_2))},$$

$$\sup_{t\in(\tau_{n+1},t_2)}\|e^{q\xi_k(\cdot,t)}\eta_n(\cdot,t)\|^2_{L^2(Q_{n+1})}\leqslant C_1 4^n(q+1)^2\|e^{q\xi_k}\|^2_{L^{2s}(Q_n\times(\tau_n,t_2))}.$$

By Lemma 8.3.3 there exists a constant $C_2>0$ such that

$$\|e^{q\xi_k}\|^2_{L^{2(d+2)/d}(Q_{n+1}\times(\tau_{n+1},t_2))}\leqslant C_2(q+1)^3 4^n\|e^{q\xi_k}\|^2_{L^{2s}(Q_n\times(\tau_n,t_2))}.$$

For $\alpha>0$ we set

$$F\big(\alpha,Q_n\times(\tau_n,t_2)\big)=\biggl(\int_{\tau_n}^{t_2}\int_{Q_n}e^{\alpha\xi_k}\,dx\,dt\biggr)^{1/\alpha}.$$

Let $s=p/(p-2)$. Since $p>d+2$, we have $\beta=(d+2)d^{-1}s^{-1}>1$. Set $q_0=\lambda/2$, $q_n=\beta^n\lambda/2$. Then $q_{n+1}=\beta q_n$ and $q_n\to\infty$. Substituting q_n in place of q, we find that

$$F\big(q_{n+1},Q_{n+1}\times(\tau_{n+1},t_2)\big)\leqslant C_3^{n\beta^{-n}}F\big(q_n,Q_n\times(\tau_n,t_2)\big).$$

Since $\sum_n n\beta^{-n}<\infty$, there exists a number C_4 such that

$$F\big(q_{n+1},Q_{n+1}\times(\tau_{n+1},t_2)\big)\leqslant C_4 F\big(\lambda,Q_0\times(\tau_0,t_2)\big),\ n\geqslant 0.$$

Taking into account estimate (8.3.7), we obtain the desired inequality. □

8.3.6. Remark. If the measure $\mu=\varrho\,dx\,dt$ is finite on $\mathbb{R}^d\times(0,T)$ and condition (GP3) is replaced with condition (GP3′), then in the assertion of the theorem one can replace the quantities

$$\|\varrho\|_{L^\infty(Q_0\times[T_0,T_1])},\qquad \int_{T_0}^{T_1}\int_{Q_0}|b|^p\exp\bigl[p\psi(|b|)\bigr]\varrho\,dx\,dt$$

with the quantities

$$\|\varrho\|_{L^\infty(\mathbb{R}^d\times[T_0,T_1])},\qquad \int_{T_0}^{T_1}\int_{\mathbb{R}^d}|b|^p\exp\bigl[p\psi(|b|)\bigr]\varrho\,dx\,dt.$$

In addition, for Λ one can take

$$\Lambda=\min\bigl\{\tau_0\bigl(2\|\varrho\|_{L^\infty(\mathbb{R}^d\times[T_0,T_1])}\bigr)^{-1},1\bigr\}.$$

8.3.7. Corollary. *Suppose that a nonnegative measure $\mu=\varrho\,dx\,dt$ satisfies equation (8.3.1), where a^{ij},b^i satisfy conditions (GP1), (GP2), (GP3) and also conditions (HP1), (HP2), (HP3) are fulfilled. Suppose that the continuous version of the density ϱ is fixed and for some $\tau>0$ we have*

$$\sup_{(x,t)\in\mathbb{R}^d\times(0,\tau)}\varrho(x,t)>0.$$

Then $\varrho>0$ in $\mathbb{R}^d\times[\tau,T)$. In particular, if $\mu_t=\varrho(x,t)\,dx$ is a probability measure for almost all $t\in(0,T)$, then $\varrho>0$ in $\mathbb{R}^d\times(0,T)$.

PROOF. Suppose that $0 < s_1 < s_2 < t_1 < t_2 < T$. Let us set

$$K_R^- = Q(0, R/2)\times[s_1, s_2], \quad K_R^+ = Q(0, R/2)\times[t_1, t_2].$$

According to the proven theorem, one has the inequality

$$\sup_{K_1^+} e^{f^{-1}(|\ln \varrho_k|)} \leqslant C\biggl(\int_{K_1^-} e^{-\lambda f^{-1}(|\ln \varrho_k|)}\,dx\,dt\biggr)^{-1/\lambda},$$

where the constant C does not depend on k. Let $\sup_{(x,t)\in K^-} \varrho(x,t) > 0$. Then

$$\int_{K_1^-} e^{-\lambda f^{-1}(|\ln \varrho_k|)}\,dx\,dt \geqslant \int_{K_1^-} e^{-\lambda f^{-1}(|\ln \varrho|)}\,dx\,dt > 0.$$

Therefore, $\sup_{K_1^+} e^{f^{-1}(|\ln \varrho_k|)} \leqslant C_1$, where

$$C_1 = C\biggl(\int_{K_1^-} e^{-\lambda f^{-1}(|\ln \Lambda\varrho|)}\,dx\,dt\biggr)^{-1/\lambda}.$$

Since the number C_1 does not depend on k, we arrive at the inequality

$$\varrho(x,t) \geqslant \Lambda^{-1}\exp\bigl(-f(\ln C_1)\bigr), \quad (x,t)\in K^+.$$

Thus, ϱ has no zeros in K^+ if $\sup_{(x,t)\in K^-} \varrho(x,t) > 0$. Changing coordinates $x = Ry$, $t = s$, we transfer this assertion to the case of K_R. Since by assumption $\sup_{(x,t)\in\mathbb{R}^d\times(0,\tau)} \varrho(x,t) > 0$, there exist $s_1, s_2 \in (0,\tau)$ and $R_0 > 0$ such that $\sup_{(x,t)\in K_R^-} \varrho(x,t) > 0$ for every radius $R > R_0$. Let us take $t_1 = \tau$ and $t_2 \in (\tau, T)$. Then, according to what has been proved above, the function ϱ has no zeros in K_R^+. Therefore, ϱ has no zeros in $\mathbb{R}^d\times[\tau, T)$. □

8.3.8. Example. Suppose that a nonnegative measure $\mu = \varrho\,dx\,dt$ satisfies equation (8.3.1), where a^{ij}, b^i satisfy conditions (GP1), (GP2), and (GP3). Suppose that for some $\delta > 0$ we have

$$\exp(\delta|b|^2) \in L^1_{\rm loc}(\mu).$$

Then, the continuous version of ϱ has no zeros in $\mathbb{R}^d\times[\tau, T)$ provided that we have $\sup_{(x,t)\in\mathbb{R}^d\times(0,\tau)} \varrho(x,t) > 0$. Indeed, it suffices to take the functions $f(z) = e^z$ and $\psi(z) = z^2$ and apply Corollary 8.3.7.

8.3.9. Corollary. *Suppose that a nonnegative measure $\mu = \varrho\,dx\,dt$ satisfies equation (8.3.1), where a^{ij}, b^i satisfy conditions (GP1), (GP2), (GP3′) and conditions (HP1), (HP2), (HP3) are fulfilled. Suppose that $\varrho > 0$ in $\mathbb{R}^d\times(0,T)$ for the continuous version (according to the previous corollary, this assumption holds if $\mu_t = \varrho(x,t)\,dx$ is a probability measure for a.e. $t\in(0,T)$). Finally, let us fix a closed interval $J = [\tau_1, \tau_2] \subset (T_0, T_1)$. Then, there exist positive numbers c_1 and c_2 such that*

$$\varrho(x,t) \geqslant \exp\bigl[-f\bigl(c_1|x|^2 + c_2\bigr)\bigr] \quad \forall\, x\in\mathbb{R}^d,\ t\in J.$$

PROOF. Let $T = 1$, $(x_0, t_0) \in \mathbb{R}^d\times(0,1)$, $0 < l < 1$,

$$v(x,t) = \varrho\bigl(x_0 + lx, t_0 + l^2(t - t_0)\bigr).$$

The measure $v(x,t)\,dx\,dt$ on $\mathbb{R}^d\times(0,1)$ satisfies equation (10.5.1) with the coefficients

$$\widehat{a}^{ij}(x,t) = a^{ij}\bigl(x_0 + lx, t_0 + l^2(t - t_0)\bigr), \quad \widehat{b}(x,t) = l\cdot b\bigl(x_0 + lx_0, t_0 + l^2(t - t_0)\bigr).$$

In order to apply the results obtained above it is important to have an estimate on the quantity

$$\int_0^1\int_{\mathbb{R}^d} |\widehat{b}|^p \exp\bigl[p\psi(|\widehat{b}|)\bigr] v\,dx\,dt.$$

Changing coordinates $y = x_0 + lx$ and $s = t_0 + l^2(t - t_0)$, we obtain

$$\int_0^1\int_{\mathbb{R}^d} |\widehat{b}|^p \exp\bigl[p\psi(|\widehat{b}|)\bigr] v\,dx\,dt = l^{p-d-2}\int_{t_0(1-l^2)}^{t_0(1-l^2)+l^2}\int_{\mathbb{R}^d} |b|^p \exp\bigl[p\psi(l|b|)\bigr]\varrho\,dy\,ds.$$

Taking into account that $p > d+2$, $0 < l < 1$ and ψ is an increasing function, we have

$$\int_0^1\int_{\mathbb{R}^d} |\widehat{b}|^p \exp\bigl[p\psi(|\widehat{b}|)\bigr] v\,dx\,dt \leqslant \int_0^1\int_{\mathbb{R}^d} |b|^p \exp\bigl[p\psi(|b|)\bigr]\varrho\,dy\,ds.$$

Set $w = e^{\lambda f^{-1}(|\ln \Lambda v|)}$. Then, by Theorem 8.4.1, for every cube $Q = Q(0, 1/2)$ we have

$$\sup_{[5/8,7/8]\times Q} w(x,t) \leqslant C\biggl(\int_{1/8}^{3/8}\int_Q \frac{1}{w(y,s)}\,dy\,ds\biggr)^{-1}.$$

Let h be a vector of length less than $1/2$. Replacing the cube $Q(0,1/2)$ by the cube $\widehat{Q} = Q(0,1/2)\cap Q(h,1/2)$, we increase the right-hand side of this inequality. Therefore,

$$\begin{aligned}\sup_{[5/8,7/8]\times Q} w(x,t) &\leqslant C\biggl(\int_{1/8}^{3/8}\int_{\widehat{Q}} \frac{1}{w(y,s)}\,dy\,ds\biggr)^{-1}\\ &\leqslant C \sup_{[1/8,3/8]\times\widehat{Q}} w(x,t) \leqslant C \sup_{[1/8,3/8]\times Q(h,1/2)} w(x,t).\end{aligned}$$

Set $u = e^{\lambda f^{-1}(|\ln \Lambda \varrho|)}$. Let $\tau_0 = t_0(1-l^2) + 3l^2/4$. Returning from v to ϱ, we arrive at the estimate

$$\sup_{S(\tau_0,x_0)} u(x,t) \leqslant C \sup_{S(\tau_0+l^2/2,x_0+lh)} u(x,t),$$

$$S(\tau_0,x_0) = [\tau_0 - l^2/8, \tau_0 + l^2/8]\times Q(x_0, l/2).$$

Let is fix $s \in (T_0, \tau_1)$. Set $R = \max\bigl\{1, 1/\sqrt{2(s-T_0)}\bigr\}$,

$$C_0 = \sup_{(x,t)\in U(0,R)\times(T_0,\tau_1)} u(x,t).$$

For $t \in (\tau_1, \tau_0)$ and $|x| > R$ set $N = 2\bigl[|x|^2/(t-s)\bigr]$, where $[\,\cdot\,]$ is the integer part. Let $h = -x/(2|x|)$, $l = 2|x|/N$. Set

$$x_{k+1} = x_k + lh,\ x_0 = x, \quad t_{k+1} = t_k - l^2/2,\ t_0 = t.$$

Then $x_N = x - Nlh = 0$ and $t_N = t - Nl^2/2 = t - 2|x|^2/N$, moreover, $T_0 < t_N < \tau_1$. Applying the inequalities obtained above, we find that

$$\sup_{S(t_k,x_k)} u(x,t) \leqslant C \sup_{S(t_{k+1},x_{k+1})} u(x,t), \quad 0 \leqslant k \leqslant N.$$

Therefore, $u(x,t) \leqslant C^N C_0$. Thus, there are numbers $\alpha_1 > 0$ and $\alpha_2 > 0$ such that $u(x,t) \leqslant e^{\alpha_1|x|^2+\alpha_2}$. We observe that $\varrho = \Lambda^{-1}e^{-f(\lambda^{-1}\ln u)}$, $\Lambda^{-1} \geqslant 1$ and f is a strictly increasing function. $\square$

8.3.10. Example. Suppose that conditions (GP1) and (GP2) are fulfilled and a measure $\mu = \varrho\, dx\, dt$ is a solution of equation (8.3.1), moreover, $\mu_t = \varrho(x,t)\, dx$ is a probability measure for a.e. $t \in (0,T)$. If for some $\delta > 0$ one has the inclusion $\exp(\delta|b|^2) \in L^1(\mu)$, then for every compact interval $[\tau_1, \tau_2]$ in $(0,T)$ there exist positive numbers γ_1 and γ_2 such that

$$\varrho(x,t) \geqslant \exp\bigl[-\gamma_1 \exp\bigl(\gamma_2|x|^2\bigr)\bigr] \quad \text{for all} \quad (x,t) \in \mathbb{R}^d \times [\tau_1, \tau_2].$$

Indeed, it is easy to see that for the functions $f(z) = e^z$ and $\psi(z) = \varepsilon z^2$ with $0 < \varepsilon < (2p)^{-1}\delta$ all conditions (HP1), (HP2), (HP3) are fulfilled.

8.3.11. Example. Suppose that conditions (GP1) and (GP2) are fulfilled and a measure $\mu = \varrho\, dx\, dt$ is a solution of equation (8.3.1), moreover, $\mu_t = \varrho(x,t)\, dt$ is a probability measure for a.e. $t \in (0,T)$. If for some $\delta > 0$ and $\beta > 1$ one has the inclusion $\exp\bigl(\delta|b|^{2\beta/(\beta-1)}\bigr) \in L^1(\mu)$, then for every compact interval $[\tau_1, \tau_2]$ there exist positive numbers γ_1 and γ_2 such that

$$\varrho(x,t) \geqslant \gamma_1 \exp\bigl(-\gamma_2|x|^{2\beta}\bigr) \quad \text{for all} \quad (x,t) \in \mathbb{R}^d \times [\tau_1, \tau_2].$$

This estimate holds, for example, if we take $f(z) = z^\beta$ and $\psi(z) = \varepsilon z^{2\beta/(\beta-1)}$, where $0 < \varepsilon < (2p)^{-1}\delta$.

A modification of conditions (HP1), (HP2), (HP3) in terms of ψ is given in Exercises 8.5.3, 8.5.4, and 8.5.5.

8.4. Proof of the key lemma

The definitions of the functions and quantities ω_0, h_q, Λ and ξ_k employed below are given on pages 322, 323. We observe that since ψ is an increasing function, for any numbers $\alpha \geqslant e^{\psi(0)}$, $\beta \geqslant 0$ the inequality

$$\alpha\beta \leqslant \alpha\psi^{-1}(\ln \alpha) + \beta e^{\psi(\beta)} \tag{8.4.1}$$

holds.

8.4.1. Lemma. *If* (HP1), (HP2), (HP3) *are fulfilled and* $s \in (0, \omega_0)$, *then*
(i) *we have*

$$\frac{\gamma e^{2qf^{-1}(|\ln s|)}}{s^2\bigl|f'\bigl(f^{-1}(|\ln s|)\bigr)\bigr|^2} \leqslant h_q'(s) \leqslant \frac{(2+2q-\gamma)e^{2qf^{-1}(|\ln s|)}}{s^2 f'\bigl(f^{-1}(|\ln s|)\bigr)};$$

(ii) *we have* $h_q^2(s) \leqslant \gamma^{-1} \exp\bigl(2qf^{-1}(|\ln s|)\bigr) h_q'(s)$.

PROOF. Direct calculations show that

$$\frac{d}{ds} h_q(s) = \frac{f'\bigl(f^{-1}(|\ln s|)\bigr) - \frac{f''(f^{-1}(|\ln s|))}{f'(f^{-1}(|\ln s|))} + 2q}{s^2\bigl[f'\bigl(f^{-1}(|\ln s|)\bigr)\bigr]^2}\, e^{2qf^{-1}(|\ln s|)}.$$

By condition (HP2) we have the estimate $f'(y) - f''(y)/f'(y) \geqslant \gamma f'(y) \geqslant \gamma$, which gives the left side of the inequality in (i). Again by condition (HP2) we obtain that $f''(y)/f'(y) \leqslant (1-\gamma)f'(y)$, which along with the inequality $f' \geqslant 1$ gives the right side of the inequality in (i). Estimate (ii) follows from (i). □

We observe that $\varrho_k < \omega_0$ for every natural number $k > 1/(2\omega_0)$. Hence, when we substitute ϱ_k in place of the variable τ in the function h_q, all assertions in Lemma 8.4.1 remain valid.

Now we can proceed to the proof of the key lemma.

PROOF. We can assume that $T=1$. For any function $\psi \in \mathbb{H}_0^{2,1}(Q,(0,1))$ the following inequality holds:

$$\int_0^1\int_Q \bigl[-\partial_t\psi\varrho + \langle A\nabla\varrho, \nabla\psi\rangle\bigr]\,dx\,dt = \int_0^1\int_Q \langle b,\nabla\psi\rangle\varrho\,dx\,dt. \tag{8.4.2}$$

Let $\omega_r(s) := r^{-1}\omega(sr^{-1})$, where $\omega \in C_0^1(\mathbb{R}^1)$ is a probability density, $\omega \leqslant 1$, $\operatorname{supp}\omega \subset [1/2, 3/4]$. Let also

$$\varrho_r(x,t) := \int_0^1 \omega_r(t-s)\varrho(x,s)\,ds, \quad t>r.$$

Then the function ϱ_r is continuously differentiable in t. According to (8.4.2), the function ϱ_r satisfies the integral identity

$$\int_0^1\int_Q \psi\partial_t\varrho_r + \langle (A\nabla\varrho)*\omega_r, \nabla\psi\rangle\,dx\,dt = \int_0^1\int_Q \langle (b\varrho)*\omega_r, \nabla\psi\rangle\,dx\,dt \tag{8.4.3}$$

for all functions $\psi \in \mathbb{H}_0^{2,1}(Q,(0,1))$ such that $\psi = 0$ for $t\in(0,r]$. We observe that the convolution $(b\varrho)*\omega_r$ is defined, since by Fubini's theorem we have the inclusion $b(x,\cdot)\varrho(x,\cdot) \in L^1[\delta, 1-\delta]$ for almost all x and all $\delta>0$.

Choose $\alpha_0>0$ such that $[T_0,T_1]\subset[\alpha_0, 1-\alpha_0]$ and $\eta(x,t)=0$ for $t\notin[\alpha_0,1-\alpha_0]$. We assume further that $r<\alpha_0$. Let us fix $k>0$. Let us introduce the following functions:

$$\varrho_{r,k} := \Lambda\varrho_r + k^{-1}, \quad \psi := I_{[\tau_1,\tau_2]}h_q(\varrho_{r,k})\eta^2.$$

Substituting such ψ in (8.4.3) and applying the Newton–Leibniz formula we obtain

$$\int_Q \bigl[H_q\bigl(\varrho_{r,k}(x,\tau_2)\bigr)\eta^2(x,\tau_2) - H_q\bigl(\varrho_{r,k}(x,\tau_1)\bigr)\eta^2(x,\tau_1)\bigr]\,dx$$
$$+\int_{\tau_1}^{\tau_2}\int_Q \langle (A\nabla\varrho)*\omega_r, \nabla\varrho_{r,k}\rangle h_q'(\varrho_{r,k})\eta^2\,dx\,dt = I(r,k)+J(r,k)+L(r,k)+N(r,k),$$

where I, J, L, and N are defined by the formulas

$$I(r,k) = -2\int_{\tau_1}^{\tau_2}\int_\Omega \langle (A\nabla\varrho)*\omega_r, \nabla\eta\rangle h_q(\varrho_{r,k})\eta\,dx\,dt,$$
$$J(r,k) = 2\int_{\tau_1}^{\tau_2}\int_\Omega \langle (b\varrho)*\omega_r, \nabla\eta\rangle h_q(\varrho_{r,k})\eta\,dx\,dt,$$
$$L(r,k) = \int_{\tau_1}^{\tau_2}\int_\Omega \langle (b\varrho)*\omega_r, \nabla\varrho_{r,k}\rangle h_q'(\varrho_{r,k})\eta^2\,dx\,dt,$$
$$N(r,k) = 2\int_{\tau_1}^{\tau_2}\int_\Omega H_q(\varrho_{r,k})\eta\partial_t\eta\,dx\,dt.$$

On the support of η the function $|H(\varrho_{r,k})|$ is uniformly bounded with respect to $r\leqslant\alpha_0$. Hence by the Lebesgue dominated convergence theorem, as $r\to 0$, we have

$$\int_Q H_q\bigl(\varrho_{r,k}(x,\tau_i)\bigr)\eta^2(x,\tau_i)\,dx \to \int_Q H_q\bigl(\varrho_k(x,\tau_i)\bigr)\eta^2(x,\tau_i)\,dx,\ i=1,2,$$

$$\int_{\tau_1}^{\tau_2}\int_Q H_q(\varrho_{r,k})\eta\partial_t\eta\,dx\,dt \to \int_{\tau_1}^{\tau_2}\int_Q H_q(\varrho_k)\eta\partial_t\eta\,dx\,dt.$$

We observe that $|h_q'(\varrho_{r,k})| \leqslant C(k)$ for some number $C(k)$ and all r in $(0,1)$. The functions $A\nabla\varrho$, $b\varrho$ and $\nabla\varrho_{r,k}$ belong to $L^2(Q\times[T_0,T_1])$ for all $r<\alpha_0$. Therefore, as $r\to 0$, we obtain

$$\|(A\nabla\varrho)*\omega_r - A\nabla\varrho\|_{L^2(Q\times[T_0,T_1])}\to 0,\ \|(b\varrho)*\omega_r - b\varrho\|_{L^2(Q\times[T_0,T_1])}\to 0,$$

$$\|\nabla\varrho_{r,k}-\nabla\varrho_{r,k}\|_{L^2(Q\times[T_0,T_1])}\to 0.$$

Hence the quantities

$$\int_{\tau_1}^{\tau_2}\int_Q \langle (A\nabla\varrho)*\omega_r, \nabla\varrho_{r,k}\rangle h_q'(\varrho_{r,k})\eta^2\,dx\,dt$$

and $I(r,k), J(r,k), L(r,k)$ converge as $r\to 0$. Therefore, we have

$$\int_Q\Big[H_q\big(\varrho_k(x,\tau_2)\big)\eta^2(x,\tau_2)-H_q\big(\varrho_k(x,\tau_1)\big)\eta^2(x,\tau_1)\Big]\,dx + \int_{\tau_1}^{\tau_2}\int_Q\langle A\nabla\varrho,\nabla\varrho\rangle h_q'(\varrho_k)\eta^2\,dx\,dt = I(k)+J(k)+L(k)+N(k), \tag{8.4.4}$$

where $I(k), J(k), L(k), N(k)$ are defined by the following formulas:

$$\begin{aligned}
I(k) &= -2\int_{\tau_1}^{\tau_2}\int_Q\langle A\nabla\varrho,\nabla\eta\rangle h_q(\varrho_k)\eta\,dx\,dt,\\
J(k) &= 2\int_{\tau_1}^{\tau_2}\int_Q\langle b\varrho,\nabla\eta\rangle h_q(\varrho_k)\eta\,dx\,dt,\\
L(k) &= \int_{\tau_1}^{\tau_2}\int_Q\langle b\varrho,\nabla\varrho\rangle h_q'(\varrho_k)\eta^2\,dx\,dt,\\
N(k) &= 2\int_{\tau_1}^{\tau_2}\int_Q H_q(\varrho_k)\eta\eta_t\,dx\,dt.
\end{aligned}$$

Let us estimate separately every term in the right-hand side of (8.4.4). Let us fix a number $\varepsilon>0$. Then

$$I(k)\leqslant\varepsilon\int_{\tau_1}^{\tau_2}\int_Q|\nabla\varrho|^2h_q'(\varrho_k)\eta^2\,dx\,dt+\varepsilon^{-1}M^2\int_{\tau_1}^{\tau_2}\int_Q|\nabla\eta|^2\frac{h_q^2(\varrho_k)}{h_q'(\varrho_k)}\,dx\,dt.$$

Inequality (ii) in Lemma 8.4.1 gives the estimate

$$I(k)\leqslant\varepsilon\int_{\tau_1}^{\tau_2}\int_Q|\nabla\varrho|^2h_q'(\varrho_k)\eta^2\,dx\,dt+\varepsilon^{-1}\gamma^{-1}M^2\int_{\tau_1}^{\tau_2}\int_Q e^{2q\xi_k}|\nabla\eta|^2\,dx.$$

Let us estimate $J(k)$. We have

$$J(k)\leqslant\int_{\tau_1}^{\tau_2}\int_Q|\nabla\eta|^2\frac{h_q^2(\varrho_k)}{h_q'(\varrho_k)}\,dx\,dt+\int_{\tau_1}^{\tau_2}\int_Q|b|^2\varrho^2h_q'(\varrho_k)\eta^2\,dx\,dt.$$

The first term is estimated as above. Let us consider the second term. Using estimate (i) in Lemma 8.4.1 and Hölder's inequality with the exponents s' and s, we obtain the following inequality:

$$\int_{\tau_1}^{\tau_2}\int_Q |b|^2\varrho^2 h_q'(\varrho_k)\eta^2\,dx\,dt$$
$$\leqslant (2+2q-\gamma)\Big(\int_{\tau_1}^{\tau_2}\int_Q |b|^{2s'}\varrho^{2s'}|f'(\xi_k)|^{-s'}e^{2s'f(\xi_k)}\eta^2\,dx\,dt\Big)^{1/s'}$$
$$\times\Big(\int_{\tau_1}^{\tau_2}\int_Q e^{2qs\xi_k}\eta^2\,dx\,dt\Big)^{1/s}.$$

We first estimate the first factor. According to inequality (8.4.1) and condition (H3), we have

$$\begin{aligned}\varrho|b|\,|f'(\xi_k)|^{-1/2}e^{f(\xi_k)} &\leqslant \varrho\big[|b|e^{\psi(|b|)}+|f'(\xi_k)|^{-1/2}e^{f(\xi_k)}\psi^{-1}(f(\xi_k))\big]\\ &\leqslant |b|e^{\psi(|b|)}\varrho+N/\Lambda.\end{aligned}$$

By the inequalities

$$(x+y)^{2s'}\leqslant 2^{2s'}(x^{2s'}+y^{2s'})$$

and $\eta^2\leqslant 1$ we obtain

$$\int_{\tau_1}^{\tau_2}\int_Q |b|^{2s'}\varrho^{2s'}|f'(\xi_k)|^{-s'}e^{2s'f(\xi_k)}\eta^2\,dx\,dt\leqslant (2N/\Lambda)^{2s'}|Q|(T_1-T_0)$$
$$+4^{s'}\|\varrho\|^{2s'-1}_{L^\infty(Q\times[T_0,T_1])}\int_{Q\times[T_0,T_1]}|b|^{2s'}e^{2s'\psi(|b|)}\varrho\,dx\,dt.$$

Therefore, there exists a number $C_1>0$, depending only on the quantities indicated in the formulation, such that

$$J\leqslant C_1(2+2q-\gamma)\Big[\int_{\tau_1}^{\tau_2}\int_Q|\nabla\eta|^2e^{2q\xi_k}\,dx\,dt+\Big(\int_{\tau_1}^{\tau_2}\int_Q e^{2sq\xi_k}\eta^2\,dx\,dt\Big)^{1/s}\Big].$$

Let us consider the term L. We have

$$L(k)\leqslant\varepsilon\int_{\tau_1}^{\tau_2}\int_Q|\nabla\varrho|^2h_q'(\varrho_k)\eta^2\,dx\,dt+4\varepsilon^{-1}\int_{\tau_1}^{\tau_2}\int_Q|b|^2\varrho^2h_q'(\varrho_k)\eta^2\,dx\,dt.$$

Estimating the second term as above, we find that

$$L(k)\leqslant\varepsilon\int_{\tau_1}^{\tau_2}\int_Q|\nabla\varrho|^2h_q'(\varrho_k)\eta^2\,dx\,dt$$
$$+4\varepsilon^{-1}(2+2q-\gamma)C_1\Big(\int_{\tau_1}^{\tau_2}\int_Q e^{2sq\xi_k}\eta^2\,dx\,dt\Big)^{1/s}.$$

We observe that by condition (GP2) the inequality

$$\int_{\tau_1}^{\tau_2}\int_\Omega\langle A\nabla\varrho,\nabla\varrho\rangle h_q'(\varrho_k)\eta^2\,dx\geqslant m\int_{\tau_1}^{\tau_2}\int_\Omega|\nabla\varrho|^2h_q'(\varrho_k)\eta^2\,dx.$$

holds. Applying inequality (i) in Lemma 8.4.1, we obtain

$$\int_{\tau_1}^{\tau_2}\int_Q\langle A\nabla\varrho,\nabla\varrho\rangle h_q'(\varrho_k)\eta^2\,dx\,dt\geqslant m\gamma\int_{\tau_1}^{\tau_2}\int_Q\frac{|\nabla\varrho|^2}{\varrho_k^2|f'(\xi_k)|^2}e^{2q\xi_k}\eta^2\,dx\,dt.$$

Summing the obtained inequalities and setting $\varepsilon := m\gamma/3$, we find that

$$\int_Q \bigl(H_q(\varrho_k)\eta^2\bigr)\,dx\Big|_{\tau_1}^{\tau_2} + 3^{-1}m\gamma\int_{\tau_1}^{\tau_2}\int_Q \frac{|\nabla\varrho|^2}{\varrho_k^2|f'(\xi_k)|^2}e^{2q\xi_k}\eta^2\,dx\,dt$$
$$\leqslant 2\int_{\tau_1}^{\tau_2}\int_\Omega H_q(\varrho_k)\eta\eta_t\,dx\,dt + N_1(2+2q-\gamma)\biggl[\int_{\tau_1}^{\tau_2}\int_Q |\nabla\eta|^2 e^{2q\xi_k}\,dx\,dt$$
$$+ (2+2q-\gamma)\biggl(\int_{\tau_1}^{\tau_2}\int_Q e^{2sq\xi_k}\eta^2\,dx\,dt\biggr)^{1/s}\biggr].$$

In the case where $q=0$ and $\eta(x,t)=\eta_0(x)$ for all $t\in[\tau_1,\tau_2]$, we have $\partial_t\eta=0$ on the interval $[\tau_1,\tau_2]$ and $N(k)=0$. □

8.5. Complements, comments, and exercises

8.5(i). Trudinger's version of Moser's lemma

An important step in applications of Moser's iteration techniques in the proof of Harnack's inequality and in the above reasoning justifying the positivity of solution densities is the following lemma. For every cube $Q(r)=Q(y,r)=\prod_{i=1}^d[y_i-r,y_i+r]$ the function ψ_Q was defined as follows:

$$\psi_Q(x)=\Pi_{i=1}^d\chi_i(x_i),\quad x=(x_1,\ldots,x_d),$$

where $\chi_i\in C^1([y_i-2r,y_i+2r])$, $\chi_i(x_i)=1$ if $|x_i-y_i|\leqslant r$, $\chi_i(x_i)=0$ if $|x_i-y_i|\geqslant 2r$. Let also $0\leqslant\psi_Q\leqslant 1$ and $|\nabla\psi_Q|\leqslant cr^{-1}$ for some positive number c.

8.5.1. Lemma. *Suppose that $v\in V^2\bigl(Q(R)\times(T_0,T_1)\bigr)$, $v\geqslant 0$ and for every function $\psi\in C_0^1\bigl(Q(R)\bigr)$ the function*

$$t\mapsto\int_{Q(R)} v(x,t)\psi^2(x)\,dx,\quad t\in(T_0,T_1)$$

is locally absolutely continuous. Suppose also that there exist numbers $A_1,A_2>0$ such that for every cube $Q(r)\subset Q(R)$ and a.e. $t\in(T_0,T_1)$ we have the inequality

$$\frac{d}{dt}\int_{Q(2r)} v(x,t)\psi_Q^2(x)\,dx + A_1\int_{Q(2r)}|\nabla v|^2\psi_Q^2\,dx\leqslant A_2|Q(r)|^{1-2/d}.$$

Then, for any numbers $s_1<s_2<t_1<t_2$ in $[T_0,T_1]$, there exist numbers $\lambda>0$ and $\sigma>0$, depending only on s_1, s_2, t_1, t_2, d, A_1 and A_2, such that

$$\int_{s_1}^{s_2}\int_{Q(R)} e^{-\lambda v}\,dx\,dt\int_{t_1}^{t_2}\int_{Q(R)} e^{\lambda v}\,dx\,dt\leqslant\sigma.$$

PROOF. It is convenient to split the proof in three steps.

1. The following Poincaré inequality holds:

$$\int_{Q(2r)}\psi_Q^2(x)\bigl(v(x,t)-V(t)\bigr)^2\,dx\leqslant\frac{C(d)}{\|\psi_Q\|^2_{L^2(Q(2r))}}\int_{Q(2r)}\psi_Q^2(x)|\nabla_x v(x,t)|^2\,dx,$$

where

$$V(t)=\frac{1}{\|\psi_Q\|^2_{L^2(Q(2r))}}\int_{Q(2r)}\psi_Q^2\,dx,\quad t\in(T_0,T_1).$$

The proof of this inequality is given in Lieberman [**611**, Lemma 6.12]. Now the hypothesis of the lemma can be written as

$$\frac{d}{dt}V(t) + A_3 \int_{Q(r)} \big(v(x,t) - V(t)\big)^2\, dx \leqslant A_4, \tag{8.5.1}$$

where the numbers A_3 and A_4 do not depend on r.

2. According to Aronson, Serrin [**74**, Lemma 7], inequality (8.5.1) yields the estimate

$$\begin{aligned} &\int_{\tau-r^2}^{\tau+r^2}\int_{Q(r)} \sqrt{\big(v(x,t) - V(\tau)\big)^+}\, dx\, dt \leqslant Cr^{d+2}, \\ &\int_{\tau-r^2}^{\tau+r^2}\int_{Q(r)} \sqrt{\big(V(\tau) - v(x,t)\big)^+}\, dx\, dt \leqslant Cr^{d+2} \end{aligned} \tag{8.5.2}$$

for some numbers C and all $r > 0$ and $\tau \in (T_0 + r^2, T_1 - r^2)$.

3. Now the conclusion follows by Moser's result, a simple proof of which is given in Fabes, Garofalo [**345**, Theorem 1]. Let

$$K^+ = Q(R)\times[t_1, t_2], \quad K^- = Q(R)\times[s_1, s_2].$$

According to this result, estimate (8.5.2) implies the following: there exist numbers k, m and M such that Lebesgue measures of the sets

$$\big\{(x,t) \in K^+ \colon \big(v(x,t) - M\big)^+ > \alpha\big\},\ \big\{(x,t) \in K^- \colon \big(M - v(x,t)\big)^+ > \alpha\big\}$$

do not exceed $ke^{-m\alpha/C}$, moreover k, m depend only on K^+, K^- and d. For all $\lambda < m/C$ we have

$$\int_{K^+} e^{\lambda v}\, dx\, dt \leqslant e^{-\lambda M}\int_{K^+} e^{\lambda(v-M)^+}\, dx\, dt \leqslant ke^{-\lambda M}\int_0^{+\infty} e^{(\lambda - m/C)\alpha}\, d\alpha,$$

$$\int_{K^-} e^{-\lambda v}\, dx\, dt \leqslant e^{\lambda M}\int_{K^-} e^{\lambda(M-v)^+}\, dx\, dt \leqslant ke^{\lambda M}\int_0^{+\infty} e^{(\lambda - m/C)\alpha}\, d\alpha.$$

Multiplying the obtained inequalities, we arrive at the desired estimate. □

Comments

Lower bounds for densities of solutions to parabolic equations for measures are presented here following the papers Bogachev, Röckner, Shaposhnikov [**166**], Shaposhnikov [**848**]. The important lemma of Moser is a generalization to the parabolic case of the known result of John and Nirenberg (see Lemma 3.4.1). It was obtained in the known paper of Moser [**725**] for divergence form equations without lower order terms. Later Moser's result was extended to general quasilinear divergence form parabolic equations in Aronson, Serrin [**74**], Ivanov [**484**] (where some comments on the history of the question are given), and Trudinger [**902**]. This lemma was used by Moser in [**725**] in his proof of Harnack's inequality, however, as acknowledged by Moser himself, that proof was quite intricate and a simplification was desirable. In his later paper [**726**], following ideas of Bombieri, he deduced Harnack's inequality without using this difficult lemma. Finally, Fabes and Garofalo [**345**] returned to Moser's lemma and found a relatively simple proof based on the ideas of Calderon's proof of the John–Nirenberg lemma. In the book Lieberman [**611**], an assertion similar to Lemma 8.5.1 is proved by using Moser's iteration techniques. Note also that in Ferretti, Safonov [**361**] the proof of Harnack's inequality is given with the aid of the maximum principle and a version of the growth lemma. A detailed survey on Harnack's inequality is given in Kassmann [**499**]. About the parabolic Harnack inequality, see also Barlow, Grigor'yan, Kumagai [**90**], Fabes, Stroock [**349**] (where some ideas from Nash [**734**] are used).

Concerning estimates of fundamental solutions to parabolic equations related to lower bounds, see Barlow, Grigor'yan, Kumagai [**90**], Eidelman, Zhitarashu [**328**], Gyrya, Saloff-Coste [**433**], Porper, Eidelman [**785**], [**786**], Davies [**284**], Grigor'-yan [**425**], Kusuoka [**573**], and Robinson [**810**].

Harnack's inequality holds also for many nonlinear equations, see, for example, Fornaro, Vespri [**381**].

Exercises

8.5.2. Let $d=1$, $Lf(x,t)=\partial_t f(x,t)+\partial_x^2 f(x,t)+b(x,t)\partial_x f(x,t)$,

$$b(x,t)=-\varepsilon(t-a)^{-2}\exp(\varepsilon x)+2\varepsilon^{-1}(t-a)^{-1},$$

where $\varepsilon, a\in(0,1)$. Show that the measure $\mu=\exp\bigl(-(t-a)^{-2}\exp(\varepsilon x)\bigr)\,dx\,dt$ satisfies the equation $\partial_t\mu=L^*\mu$ on $\mathbb{R}^1\times(0,1)$, but its density vanishes at $t=a$, although $\exp(\kappa|b|)$ is integrable with respect to the measure μ if $\varepsilon<\kappa^{-1}$.

8.5.3. Suppose that a nonnegative measure μ on $\mathbb{R}^d\times(0,T)$ has a continuous density $\varrho\in\mathbb{H}^{2,1}_{\mathrm{loc}}(\mathbb{R}^d,(0,T))$ and satisfy the inequality

$$\int_0^T\int_{\mathbb{R}^d}\bigl(\partial_t\varphi+\mathcal{L}_{A,b,c}\varphi\bigr)\,d\mu\leqslant 0\quad\text{if}\quad\varphi\in C_0^\infty\bigl(\mathbb{R}^d\times(0,T)\bigr),\ \varphi\geqslant 0.$$

Let conditions (GP1) and (GP2) be fulfilled. Assume also that $c\leqslant 0$ and

$$\exp\bigl(\delta|b|^2\bigr),\ \exp\bigl(\delta|c|\bigr)\in L^1_{\mathrm{loc}}(\mu)$$

for some $\delta>0$. Prove that if $\sup_{(x,t)\in\mathbb{R}^d\times[0,\tau)}\varrho(x,t)>0$, where $\tau\in(0,T)$, then the function ϱ is strictly positive on $\mathbb{R}^d\times[0,\tau)$.

HINT: See Bogachev, Röckner, Shaposhnikov [**166**].

8.5.4. Let $\mu=\varrho\,dx\,dt$ be a nonnegative solution of equation (8.3.1), where a^{ij} satisfy (GP1) and (GP2) on p. 321. Suppose that $\psi\colon[0,+\infty)\mapsto[0,+\infty)$ is a strictly increasing continuously differentiable function such that condition (GP3) on p. 322 is fulfilled and for some $\alpha>\psi(0)$ we have

(i) $\psi'(z)\geqslant(2+\varepsilon)z^{-1}$ for some $\varepsilon>0$ and all $z\geqslant\alpha$;

(ii) $\displaystyle\int_\alpha^{+\infty}\psi(z)z^{-3}\,dz=+\infty.$

Suppose that the continuous version of ϱ is taken and for some $\tau>0$ we have $\sup_{(x,t)\in\mathbb{R}^d\times(0,\tau)}\varrho(x,t)>0$. Prove that $\varrho>0$ in $\mathbb{R}^d\times[\tau,T)$.

HINT: Let

$$f^{-1}(z)=\int_\alpha^z\frac{1}{\bigl(\psi^{-1}(y)\bigr)^2}\,dy.$$

It suffices to verify that f satisfies conditions (HP1), (HP2), (HP3) on p. 322. Since

$$\int_\alpha^{+\infty}\frac{\psi(y)}{y^3}\,dy=+\infty,$$

according to (3.5.7), the function $f^{-1}(z)$ strictly increases to $+\infty$, hence we can define f as a strictly increasing twice continuously differentiable function on $[0,+\infty)$. Observe that

$$(f^{-1}(z))' = \frac{1}{\left(\psi^{-1}(z)\right)^2} > 0, \quad f'(y) = \left(\psi^{-1}(f(y))\right)^2.$$

Therefore, $\psi^{-1}(z) = \sqrt{f'(f^{-1}(z))}$ and condition (HP3) is fulfilled. It is clear that (HP1) holds. Let us check (HP2). Note that the inequality $\left(e^{-(1-\gamma)f}\right)'' \geqslant 0$ is equivalent to the inequality $(1-\gamma)(f')^2 - f'' \geqslant 0$. Since

$$f''(y) = \frac{2\psi^{-1}(f(y))f'(y)}{\psi'\left(\psi^{-1}(f(y))\right)},$$

it suffices to prove the inequality $(1-\gamma)\psi^{-1}(f(y))\psi'\left(\psi^{-1}(f(y))\right) - 2 \geqslant 0$. Let $z = \psi^{-1}(f(y))$. Then the last inequality takes the form $\psi'(z) \geqslant 2(1-\gamma)^{-1}z^{-1}$ and is the inequality from the formulation of the exercise with $\varepsilon = 2(1-\gamma)^{-1} - 2$.

8.5.5. Under the conditions of the previous exercise prove that if for almost all t the function $\varrho(x,t)$ is a probability density, then it satisfies estimate (8.3.2) with the function f defined by

$$f^{-1}(z) = \int_\alpha^z \frac{1}{\left(\psi^{-1}(y)\right)^2}\,dy.$$

8.5.6. Construct an example showing that in the situation of Exercise 8.5.3 the condition $\exp(\delta|c|) \in L^1(\mu)$ cannot be replaced with the condition $|c| \in L^p(\mu)$ for all $p \in (1,+\infty)$.

HINT: The conclusion fails already in the stationary case, which is seen for $d=1$, $\mu = e^{-x^{-2}}$, $A=1$, $b=0$ and $c(x) = -6/x^4$.

8.5.7. Show by example that in Harnack's inequality one cannot bound the maximum of a solution u on the cylinder $U \times [t_1, t_2]$ by the minimum of u on the same cylinder (unlike Theorem 8.1.2, where a shifted cylinder is used) with a constant independent of u. Show also that the time interval in the parabolic Harnack inequality must be separated from zero.

8.5.8. Show that a solution to the Cauchy problem for the Fokker–Planck–Kolmogorov equation with initial condition given by a probability density can have a signed density, so that in Harnack's inequality it is important to consider only nonnegative solutions.

HINT: Use Example 9.2.1 of two different probability solutions μ and η to the Cauchy problem with a common probability initial condition; in that example, $\mu_t = \nu\otimes\sigma_1 + (\nu - \nu_t)\otimes(\sigma_2 - \sigma_1)$, $\eta_t \equiv \nu\otimes\sigma_1$, the initial condition is $\nu\otimes\sigma_1$, where ν, σ_1 and σ_2 are certain probability measures on $\mathbb{R}^2$, $\nu_t \leqslant \nu$; observe that $2\mu - \eta$ is a solution with the same initial condition and use Exercise 4.5.25 to choose σ_1 and σ_2 in the justification of Example 9.2.1 such that this solution will be signed.

CHAPTER 9

Uniqueness of Solutions to Fokker–Planck–Kolmogorov Equations

This chapter is devoted to the investigation of the space of solutions to the Cauchy problem for the Fokker–Planck–Kolmogorov equation for measures. We study two classes of solutions: integrable (i.e., given by globally integrable densities with respect to Lebesgue measure) and probability (in the sense explained in Chapter 1). For every class of solutions we give sufficient conditions for uniqueness and construct examples showing that these conditions are rather sharp and also demonstrating some differences between the classes.

In § 9.2 we consider some examples of nonuniqueness in case of nice coefficients: $A = \mathrm{I}$ and b is infinitely differentiable. The main positive results are given in § 9.3 and § 9.4. Theorem 9.3.6 and Theorem 9.4.3 say that a probability solution μ to the Cauchy problem with the operator $L_{A,b}$ is unique under broad local assumptions about A and b if a^{ij} and b^i belong to $L^1(\mu)$ (or even if $a^{ij}(x)/(1+|x|^2)$ and $b^i(x)/(1+|x|)$ are μ-integrable). Theorem 9.4.6 and Theorem 9.4.8 give uniqueness in the presence of a Lyapunov function V such that $L_{A,b}V \leqslant C + CV$. Finally, Theorem 9.6.3 provides uniqueness of signed integrable solutions.

9.1. Setting of the problem

Suppose we are given a time-dependent elliptic operator of the form

$$Lu := L_{A,b,c}u = a^{ij}\partial_{x_i}\partial_{x_j}u + b^i\partial_{x_i}u + cu,$$

where a^{ij}, b^i and c are Borel functions on $\mathbb{R}^d_T = \mathbb{R}^d\times(0,T)$ for some $T>0$ such that $A(x,t) = \big(a^{ij}(x,t)\big)_{1\leqslant i,j\leqslant d}$ is a nonnegative-definite symmetric matrix for all (x,t).

As in Chapter 6, we shall consider Borel locally finite measures μ on $\mathbb{R}^d\times(0,T)$ given in the form $\mu = \mu_t\,dt$ by means of a family of measures $(\mu_t)_{t\in(0,T)}$ on $\mathbb{R}^d$ satisfying the Fokker–Planck–Kolmogorov equation introduced in Chapter 6

$$\partial_t\mu = L^*\mu$$

under the assumption that $a^{ij}, b^i, c \in L^1_{\rm loc}\big(|\mu|, \mathbb{R}^d\times(0,T)\big)$.

Throughout this chapter (excepting § 9.8(i)) we assume that $c \leqslant 0$ and the mapping A satisfies the following condition:

(H1) for each ball $U\subset\mathbb{R}^d$ there are numbers $\gamma=\gamma(U)>0$ and $M=M(U)>0$ such that

$$A(x,t)\geqslant\gamma\cdot\mathrm{I},\quad \|A(x,t)\|\leqslant M \quad \text{for all } (x,t)\in U\times(0,T).$$

In the case of nonnegative measures Condition (H1) ensures the existence of densities (see Theorem 6.3.1).

Let ν be a locally finite Borel measure on $\mathbb{R}^d$ (possibly signed). Then the uniqueness problem arises for solutions of the Cauchy problem

$$\partial_t\mu = L^*\mu, \quad \mu\big|_{t=0} = \nu. \tag{9.1.1}$$

We are interested in the two classes of solutions: subprobability and integrable.

A *subprobability solution* is a solution μ defined by a family of subprobability measures $(\mu_t)_{t\in(0,T)}$, i.e., $\mu_t \geqslant 0$ and $\mu_t(\mathbb{R}^d) \leqslant 1$. For a subprobability initial measure ν, let $\mathcal{SP}_\nu$ be the set of all subprobability solutions $\mu = \mu_t(dx)\,dt$ with the following properties: $c \in L^1\bigl(\mu, \mathbb{R}^d\times(0,T)\bigr)$, for every ball $U \subset \mathbb{R}^d$ we have

$$|b| \in L^2\bigl(\mu, U\times(0,T)\bigr), \tag{9.1.2}$$

and for almost every $t \in (0,T)$ we have

$$\mu_t(\mathbb{R}^d) \leqslant \nu(\mathbb{R}^d) + \int_0^t\!\!\int_{\mathbb{R}^d} c(x,s)\,\mu_s(dx)\,ds. \tag{9.1.3}$$

Condition (9.1.2) distinguishes $\mathcal{SP}_\nu$ and $\mathcal{M}_\nu$ (see p. 265). Since $c \leqslant 0$, we obviously have $\mathcal{SP}_0 = \{0\}$. If $\nu \neq 0$, we can assume that ν is a probability measure, dividing by $\nu(\mathbb{R}^d)$, since $\mu_t(\mathbb{R}^d) \leqslant \nu(\mathbb{R}^d)$.

If one has an equality in (9.1.3) and $c \equiv 0$, then $\mu_t(\mathbb{R}^d) = \nu(\mathbb{R}^d)$ for a.e. $t \in (0,T)$. The class of solutions with zero c, initial probability measure ν and measures μ_t that are probability measures for almost all t will be denoted by $\mathcal{P}_\nu$.

An *integrable solution* is a solution μ defined by a family of finite measures $(\mu_t)_{t\in(0,T)}$ such that $\sup_t \|\mu_t\| < \infty$. If the measure μ is given by a density ϱ with respect to Lebesgue measure on $\mathbb{R}^d\times(0,T)$, then the latter condition can be written as

$$\sup_{t\in(0,T)} \int_{\mathbb{R}^d} |\varrho(x,t)|\,dx < \infty.$$

The set of all integrable solutions $\mu = \mu_t(dx)\,dt$ such that for every ball $U \subset \mathbb{R}^d$ we have

$$|b| \in L^p\bigl(|\mu|, U\times(0,T)\bigr), \quad c \in L^{p/2}\bigl(|\mu|, U\times(0,T)\bigr), \quad \text{where } p > d+2, \tag{9.1.4}$$

is denoted by $\mathcal{I}_\nu$.

We emphasize that in the definitions of the classes $\mathcal{SP}_\nu$ and $\mathcal{I}_\nu$ we assume not only that the equation holds, but also (9.1.2) in the case of a subprobability solution and (9.1.4) in the case of an integrable solution. It is clear that locally bounded coefficients satisfy all our conditions. If the measure μ is given by a density $\varrho \in L^r\bigl(U\times(0,T)\bigr)$, then the condition on b can be replaced with the inclusion $b \in L^s\bigl(U\times(0,T)\bigr)$, where $2/r+1/s = 1$ in the case of the class $\mathcal{SP}_\nu$ and $p/r+1/s = 1$ in the case of the class $\mathcal{I}_\nu$. The condition $c \in L^{p/2}\bigl(|\mu|, U\times(0,T)\bigr)$ is replaced with $c \in L^s\bigl(U\times(0,T)\bigr)$, where $p/(2r) + 1/s = 1$.

The next assertion, which will be useful below, is a special case of Proposition 6.1.2.

9.1.1. Lemma. *Let $\mu = \mu_t(dx)\,dt$ be a solution to the Cauchy problem (9.1.1) such that $\sup_{t\in(0,T)} \|\mu_t\| < \infty$, $a^{ij}, b^i, c \in L^1\bigl(|\mu|, U\times(0,T)\bigr)$ for every ball $U \subset \mathbb{R}^d$. Let $\varphi \in C(\mathbb{R}^d\times[0,T]) \cap C^{2,1}(\mathbb{R}^d\times[0,T))$ be such that $\varphi(x,t) = 0$ if $|x| \geqslant R$ for some R. Then for almost all $t \in [0,T]$ we have the equality*

$$\int_{\mathbb{R}^d} \varphi(x,t)\,\mu_t(dx) = \int_{\mathbb{R}^d} \varphi(x,0)\,\nu(dx) + \int_0^t\!\!\int_{\mathbb{R}^d} [\partial_t\varphi + L\varphi]\,d\mu_s\,ds. \tag{9.1.5}$$

9.1.2. Remark. Let $\mathcal{J}$ be the set of all points t at which (9.1.5) holds. Then, whenever $\tau, t \in \mathcal{J}$, $\tau < t$, we have

$$\int_{\mathbb{R}^d} \varphi(x,t)\,\mu_t(dx) = \int_{\mathbb{R}^d} \varphi(x,\tau)\,\mu_\tau(dx) + \int_\tau^t \int_{\mathbb{R}^d} [\partial_t \varphi + L\varphi]\,d\mu_s\,ds.$$

Indeed, it suffices to subtract equality (9.1.5) for τ from the corresponding equality for t.

9.1.3. Remark. If $\varphi(\,\cdot\,,t) = \psi \in C_0^2(\mathbb{R}^d)$ for all $t \in [0,T]$, then one can write equality (9.1.5) as follows:

$$\int_{\mathbb{R}^d} \psi\,d\mu_t = \int_{\mathbb{R}^d} \psi\,d\nu + \int_0^t \int_{\mathbb{R}^d} L\psi\,d\mu_s\,ds \text{ for a.e. } t\in[0,T]. \tag{9.1.6}$$

Moreover, if J_ψ^μ is the set of all points $t \in [0,T]$ at which equality (9.1.6) holds, then J_ψ^μ is a full measure set in $[0,T]$ and the following map is continuous on J_ψ^μ:

$$t \mapsto \int_{\mathbb{R}^d} \psi(x)\,\mu_t(dx).$$

9.1.4. Remark. Let $T \in J^\mu_{\varphi(\,\cdot\,,T)}$. Then (9.1.5) holds with $t = T$. Indeed, the functions $\varphi(x,t)$ converge uniformly to $\varphi(x,T)$ as $t \to T$. Now let $\mathcal{J}$ be the set of all points $t \in [0,T]$ where equality (9.1.5) holds. Let $t_n \in J^\mu_{\varphi(\,\cdot\,,T)} \cap \mathcal{J}$ and $\lim\limits_{n\to\infty} t_n = T$. Then

$$\lim_{n\to\infty} \int_{\mathbb{R}^d} \varphi(x,t_n)\,\mu_{t_n}(dx) = \int_{\mathbb{R}^d} \varphi(x,T)\,\mu_T(dx)$$

and equality (9.1.5) holds for every t_n. Letting $n \to \infty$ we obtain equality (9.1.5) also for $t = T$.

The uniqueness problem for solutions to the Cauchy problem for parabolic equations is an old and classical object of investigation. Already A.N. Tychonoff in his classical paper [**907**] proved the uniqueness of solutions to the Cauchy problem for the heat equation in the class of functions growing not faster than $e^{C|x|^2}$, and in Widder [**938**] the uniqueness of the zero solution was established in the class of nonnegative functions. A.N. Tychonoff constructed also an example of a nonunique signed solution with zero initial condition growing not faster than $e^{C|x|^{2+\varepsilon}}$. Later similar results were obtained for a broader class of second order equations (see Friedman [**386**], Aronson [**70**], [**72**]). However, to solutions of the Fokker–Planck–Kolmogorov equation these results apply only in the case where it is known in advance that the densities of these solutions grow not too fast, which is a problem by itself. Hence it is important to have conditions for the uniqueness of integrable solutions without a priori assumptions about the growth of their densities. The first sufficient conditions of this sort were obtained in the papers Friedman [**385**] and Aronson, Besala [**73**]. In particular, Theorem 1 of the latter paper yields that a solution to the Cauchy problem (9.1.1) in the class of integrable solutions is unique if the functions $\partial_{x_i} a^{ij}$ and b^i are locally Lipschitzian in x uniformly in t and

$$|a^{ij}(x,t)| \leqslant K(1+|x|^2), \quad |b^i(x,t)| \leqslant K(1+|x|)$$

with some constant K. In particular, this is true if A is the unit matrix and the drift $b(x,t) = b(x)$ is Lipschitzian. We shall see below that already this last result is rather sharp if sufficient conditions are expressed in terms of the growth of coefficients.

There are several methods of proving uniqueness of a solution to the Cauchy problem. In the first place we mention the classical method going back to Holmgren and developing the reasoning the main idea of which can be illustrated by the following example: if $\mathfrak{A}$ is a bounded operator on a Hilbert space H, then its kernel is the orthogonal complement of the range of the adjoint operator $\mathfrak{A}^*$, i.e., for the proof of uniqueness of a solution of the equation $\mathfrak{A}x = f$ one has to verify that the range of $\mathfrak{A}^*$ is dense. Conformably to the Cauchy problem this approach requires rather high smoothness of the coefficients of the differential operator L. Thus, if a^{ij} and b^i are continuous in (x,t) and have continuous bounded derivatives in x up to the second order, then one can show (see Friedman [**386**] and Stroock, Varadhan [**884**]) that for every $t > 0$ and every function $\psi \in C_0^\infty(\mathbb{R}^d)$ there is a function $f \in C_b(\mathbb{R}^d \times [0,t]) \cap C_b^{2,1}(\mathbb{R}^d \times [0,t))$ satisfying the equation

$$\partial_s f(x,s) + Lf(x,s) = 0, \quad f(x,t) = \psi(x).$$

Using equality (9.1.5) and the fact that the measures μ_t are bounded, we obtain the equality

$$\int_{\mathbb{R}^d} \psi(x)\,\mu_t(dx) = \int_{\mathbb{R}^d} f(x,t)\,\nu(dx),$$

which immediately yields the uniqueness of solutions to our Cauchy problem. Indeed, for any two solutions μ^1 and μ^2 we have the equality

$$\int_{\mathbb{R}^d} \psi\,d\mu_t^1 = \int_{\mathbb{R}^d} \psi\,d\mu_t^2$$

for every function $\psi \in C_0^\infty(\mathbb{R}^d)$, whence $\mu^1 = \mu^2$. If the diffusion matrix is nondegenerate, then the condition on the coefficients can be weakened: it suffices to require the boundedness and Hölder continuity. This method of proving uniqueness for Fokker–Planck–Kolmogorov equations and for the corresponding martingale problem is presented in the book Stroock, Varadhan [**884**]. We aim at proving uniqueness under weaker local conditions and for globally unbounded coefficients.

Another approach to proving uniqueness of subprobability or integrable solutions to the Cauchy problem is based on the techniques of "renormalized solutions", see DiPerna, Lions [**302**], [**303**], [**304**], Le Bris, Lions [**593**], Figalli [**362**] and Theorem 9.8.1 in § 9.8(i).

The methods mentioned above enable us to obtain sufficient conditions for uniqueness of integrable solutions, hence also for uniqueness of probability solutions. However, as in the elliptic case, the two forms of uniqueness are not equivalent.

Let us consider the third method of proving uniqueness oriented towards probability solutions (we assume that $c=0$ and $\mu_t(\mathbb{R}^d)=1$) suggested in Bogachev, Da Prato, Röckner, Stannat [**142**] and using specific features of the probability case. Its main idea is this. Let $P = \partial_t + L$. Suppose that $\mathcal{K}_\nu$ is a *convex* subset in the set of probability solutions to the Cauchy problem with a probability initial condition ν such that for every measure $\mu \in \mathcal{K}_\nu$ we have the equality

$$\overline{P\bigl(C_0^\infty(\mathbb{R}^d \times [0,T))\bigr)} = L^1\bigl(\mu, \mathbb{R}^d \times (0,T)\bigr).$$

Then the set $\mathcal{K}_\nu$ contains at most one element. Indeed, if we have two measures $\eta, \sigma \in \mathcal{K}_\nu$, then $\mu = (\eta+\sigma)/2 \in \mathcal{K}_\nu$ and $\eta = \varrho_\eta \cdot \mu$, $\sigma = \varrho_\sigma \cdot \mu$, where the functions

ϱ_η and ϱ_σ are obviously uniformly bounded by 2. In addition, the following identity holds:

$$\int_0^T\int_{\mathbb{R}^d} Pu(\varrho_\eta-\varrho_\sigma)\,d\mu=0\quad \forall u\in C_0^\infty(\mathbb{R}^d\times[0,T)),$$

whence we obtain that $\varrho_\eta(x,t)-\varrho_\sigma(x,t)=0$ for almost all (x,t) with respect to μ.

In the paper Bogachev, Da Prato, Röckner, Stannat [**142**] examples of such convex sets $\mathcal{K}_\nu$ are constructed. In particular, if the diffusion matrix is unit and $b\in L^p(U\times(0,T))$ for every ball U and some $p>d+2$, then for the set $\mathcal{K}_\nu$ one can take the set of all probability solutions μ to the Cauchy problem such that $b\in L^1(\mu,U\times(0,T))$ for every ball $U\in\mathbb{R}^d$ and

$$b-\beta_\mu\in L^1(\mu,\mathbb{R}^d\times(0,T)),$$

where β_μ is the logarithmic gradient $\nabla_x\varrho(x,t)/\varrho(x,t)$ of the density ϱ of the measure μ. For verification of this condition the authors of [**142**] used the results on the integrability of the logarithmic gradient presented in Chapter 7, which assume that the initial condition has finite entropy. In this chapter we considerably reinforce this result by showing that it suffices to verify the condition indicated above just for one probability solution. Moreover, we give sufficient conditions in terms of Lyapunov functions and for the initial distribution we shall admit any probability measure.

Sufficient conditions for uniqueness in terms of a Lyapunov function $V\in C^2$ with $\lim\limits_{|x|\to\infty}V(x)=+\infty$, when for the initial condition an arbitrary probability measure (or a finite measure in case of integrable solutions) is admissible, have the following form. It turns out that for the uniqueness of subprobability solutions it suffices to have the estimate $LV\leqslant CV$, and for the uniqueness of integrable solutions the estimates $LV\geqslant -CV$ and $|\sqrt{A}\nabla V|\leqslant CV$ are sufficient. If the function V depends only on $|x|$ and the diffusion matrix is unit, then the conditions mentioned above can be informally reformulated as follows: for the uniqueness of subprobability solutions the function $\langle b(x),x\rangle$ must not go to $+\infty$ too fast, and for the uniqueness of integrable solutions it must not tend to fast to $-\infty$. It is interesting that these conditions coincide exactly with the sufficient conditions for the uniqueness of probability and integrable solutions of stationary equations (see Chapters 4 and 5).

Observe also that sufficient conditions for uniqueness in the class $\mathcal{SP}_\nu$ ensure an equality in (9.1.3). Uniqueness of subprobability solutions is studied below in § 9.3, 9.4, Uniqueness of integrable solutions is studied in § 9.5, 9.6.

In addition, in § 9.2 we construct examples of nonuniqueness of probability solutions to the Cauchy problem for the Fokker–Planck–Kolmogorov equation in the case of unit diffusion matrix, infinitely differentiable drift coefficient, $c=0$ and the initial condition given by a smooth probability density. Note that in this example the simplex of probability solutions to the Cauchy problem is infinite-dimensional (but we do not know whether it can be finite-dimensional of dimension bigger than one and whether there are examples of nonuniqueness of probability solutions in $\mathbb{R}^1$ and $\mathbb{R}^2$). In our construction of this example we rely on the examples of nonuniqueness of stationary solutions to the Fokker–Planck–Kolmogorov equation in Chapter 4. Moreover, in exercises we give examples that distinguish the classes of integrable, positive and subprobability solutions.

9.2. Examples of nonuniqueness

Here we construct examples of Cauchy problems with several probability solutions. Exercises 9.8.47 and 9.8.49 show that uniqueness in the classes of integrable, nonnegative, and probability solutions is indeed not the same thing.

Now we show that the set $\mathcal{P}_\nu$ can consist of several elements. It is well known that in the case where $A \equiv 0$ and the field b is just continuous, the Cauchy problem can have more than one solution. Say, let $b(x) = x^{2/3}$. Then the Cauchy problem for the ordinary equation $\dot{x} = b(x)$, $x(0) = 0$ has two distinct solutions: $x_1(t) = t^3/3$ and $x_2(t) = 0$. The measures $\delta_{x_1(t)}$ and $\delta_{x_2(t)}$ are two distinct solutions to the corresponding Cauchy problem (9.1.1). The question arises about nonuniqueness in the case where $A = \mathrm{I}$ and $b(x,t) = b(x)$, $b \in C^\infty(\mathbb{R}^d)$. It turns out that in this case there is also an example of nonuniqueness, the construction of which is based not on the local irregularity of the drift b, but on its sufficiently strong growth at infinity. Below we shall see more precisely how strong this growth must be.

Until the end of this section we assume that $T = 1$.

For constructing our example we need some auxiliary assertions about the stationary equation, following from the results of Chapters 4 and 5. Let $d \geqslant 2$. There is a vector field $B \in C^\infty(\mathbb{R}^d, \mathbb{R}^d)$ such that the stationary equation $L^*\mu = 0$ with the operator $Lf = \Delta f + \langle B, \nabla f\rangle$ has an infinite-dimensional simplex of probability solutions (see Examples 4.2.1, 4.2.13). We recall that all these solutions are given by positive smooth densities. Furthermore, by Theorems 5.2.2 and 6.4.7, for every probability solution μ of the equation $L^*\mu = 0$, there exists a sub-Markov contracting C_0-semigroup of operators $\{T_t^\mu\}_{t\geqslant 0}$ on $L^1(\mu)$ the generator of which coincides with L on $C_0^\infty(\mathbb{R}^d)$ such that the measure μ is subinvariant for it. Moreover, for every function $f \in C_0^\infty(\mathbb{R}^d)$ and for every $t \geqslant 0$, the function $T_t^\mu f$ possesses a continuous modification $\widetilde{T_t^\mu f}$ such that for every compact set $K \subset \mathbb{R}^d$ uniformly in $x \in K$ we have $\lim_{t\to 0} \widetilde{T_t^\mu f}(x) = f(x)$. There are subprobability kernels $K_t(\,\cdot\,, dy)$, where $t > 0$, such that $K_t(x, dy) = p(t,x,y)\,dy$, where $p(t,x,y)$ are locally Hölder continuous positive functions on $(0,+\infty)\times\mathbb{R}^d\times\mathbb{R}^d$ and for each $f \in L^1(\mu)$ the function

$$x \mapsto K_t f(x) := \int_{\mathbb{R}^d} f(y)p(t,x,y)\,dy$$

is a μ-version of $T^\mu f$ such that the function $(x,t) \mapsto K_t f(x)$ is continuous on the cylinder $\mathbb{R}^d\times(0,+\infty)$. Moreover, for every probability measure ν on $\mathbb{R}^d$ the measure $\sigma = \sigma_t(dy)\,dt$, where $\sigma_t = K_t^*\nu$, i.e.,

$$\sigma_t(dy) = K_t^*\nu(dy) := \int_{\mathbb{R}^d} K_t(x,dy)\,\nu(dx),$$

is a solution of the Cauchy problem (9.1.1) with initial condition ν. If $\nu = \mu$, then $\sigma_t \leqslant \mu$ for all t and $\sigma_t \not\equiv \mu$. Here only the last assertion requires an explanation: the equality $\sigma_t = \mu$ for all $t \geqslant 0$ is impossible, since otherwise the measure μ will be invariant for $\{T_t^\mu\}_{t\geqslant 0}$, which, as we know, yields the uniqueness of probability solutions of the stationary Fokker–Planck–Kolmogorov equation, while we are in the situation of several probability solutions.

Now we can start constructing an example of a Cauchy problem for the Fokker–Planck–Kolmogorov equation in $\mathbb{R}^4\times[0,1]$ with several probability solutions.

9.2.1. Example. (The Cauchy problem for the Fokker–Planck–Kolmogorov equation with $A = \mathrm{I}$ and $b \in C^\infty$ having several probability solutions.) The construction is divided in several steps.

Step I. Let

$$b(x, y) = \big(B(x), C(y)\big),$$

where $x, y \in \mathbb{R}^2$ and $B = (b^1, b^2)$ and $C = (c^1, c^2)$ are two vector fields of class $C^\infty(\mathbb{R}^2)$ such that the corresponding elliptic equations $L_1^*\nu = 0$ and $L_2^*\sigma = 0$ with respect to measures on $\mathbb{R}^2$, that is,

$$L_1 u = \Delta_x u + \langle B, \nabla_x u\rangle, \quad L_2 u = \Delta_y u + \langle C, \nabla_y u\rangle,$$

have at least two linearly independent probability solutions each. Set

$$Lu = L_1 u + L_2 u = \Delta_x u + \Delta_y u + \langle B, \nabla_x u\rangle + \langle C, \nabla_y u\rangle = \Delta u + \langle b, \nabla u\rangle.$$

Step II. Let ν be a probability solution of the equation $L_1^*\nu = 0$. Then the measure $\mu^\nu = \nu_t\,dt$, where $\nu_t = K_t^*\nu$, is a solution to the Cauchy problem with the operator L_1 and initial condition ν. Moreover, $\nu_t \leqslant \nu$ and $\nu_t \not\equiv \nu$. Set

$$\widehat{\nu}_t = \nu - \nu_t.$$

We observe that the measure $\widehat{\nu} = \widehat{\nu}_t\,dt$ is nonnegative and is a nonzero solution of the following Cauchy problem: $\partial_t\widehat{\nu} = L_1^*\widehat{\nu}$ and $\widehat{\nu}|_{t=0} = 0$.

Step III. Let σ_1 and σ_2 be two linearly independent probability solutions of the equation $L_2^*\sigma = 0$. Set

$$\mu_t = \nu\otimes\sigma_1 + \widehat{\nu}_t\otimes(\sigma_2 - \sigma_1).$$

We show that the measures $\nu\otimes\sigma_1\otimes dt$ and $\mu = \mu_t\,dt$ on $\mathbb{R}^4\times[0,1]$ are probability solutions to the Cauchy problem $\partial_t\mu = L^*\mu$, $\mu|_{t=0} = \nu\otimes\sigma_1$. Indeed,

$$L^*(\nu\otimes\sigma_1) = \sigma_1 L_1^*\nu + \nu L_2^*\sigma_1 = 0$$

and $\nu\otimes\sigma_1$ does not depend on t. Similarly,

$$L^*\big(\widehat{\nu}_t\otimes(\sigma_2 - \sigma_1)\big) = (\sigma_2 - \sigma_1)\otimes L_1^*\widehat{\nu}_t = (\sigma_2 - \sigma_1)\otimes\partial_t\widehat{\nu}_t = \partial_t\big(\widehat{\nu}_t\otimes(\sigma_2 - \sigma_1)\big).$$

Since $\nu - \widehat{\nu}_t = \nu_t \geqslant 0$, we have

$$\mu_t = \nu\otimes\sigma_1 + \widehat{\nu}_t\otimes(\sigma_2 - \sigma_1) = (\nu - \widehat{\nu}_t)\otimes\sigma_1 + \widehat{\nu}_t\otimes\sigma_2 \geqslant 0.$$

Taking into account that $\widehat{\nu}|_{t=0} = 0$, we obtain that $\mu|_{t=0} = \nu\otimes\sigma_1$. It remains to observe that for every t the measure μ_t is a probability measure, since we have the equality $(\sigma_2 - \sigma_1)(\mathbb{R}^2) = 0$.

We recall that vector fields B and C can be chosen such that the corresponding stationary Kolmogorov equations $L_1^*\nu = 0$ and $L_2^*\sigma = 0$ will have infinite-dimensional simplices of probability solutions. Therefore, changing in the example the measure σ_2, we obtain an infinite-dimensional simplex of probability solutions to the Cauchy problem.

Note that in the considered example nonuniqueness takes place in the state space $\mathbb{R}^4 = \mathbb{R}^2\times\mathbb{R}^2$. In Exercise 9.8.48 the reader is suggested to construct a modification for the space $\mathbb{R}^3$. It remains an open question whether examples of nonuniqueness with $A = \mathrm{I}$ and smooth b exist in $\mathbb{R}^1$ and $\mathbb{R}^2$.

In Exercise 9.8.47, one can find an example of a Cauchy problem which has a unique subprobability solution that is not a unique integrable solution or unique nonnegative solution.

9.3. The case of a diffusion matrix of class VMO

As we have already discussed above, in case of sufficiently smooth coefficients uniqueness can be proved by using Holmgren's principle. If the diffusion matrix is nondegenerate, then the assumptions about coefficients can be considerably weakened. It is well-known (see, for example, Stroock, Varadhan [**884**]) that in the nondegenerate case it suffices to have Hölder continuous coefficients. In this section we show that combining results of Chapter 6 and estimates of solutions of parabolic equations with VMO-coefficients from Krylov [**555**], one can obtain a considerably stronger result.

The next assertion is a special case of Theorem 6.3.1, which is mentioned here for the reader's convenience.

9.3.1. Proposition. *Let $Q=(q^{ij})_{i,j\leqslant d}$ be a Borel mapping on $\mathbb{R}^d\times(0,T)$ with values in the set of nonnegative-definite symmetric matrices and let μ be a locally finite nonnegative Borel measure on $\mathbb{R}^d\times(0,T)$ such that $q^{ij}\in L^1_{\rm loc}(\mu,\mathbb{R}^d\times(0,T))$ and for every nonnegative function φ of class $C_0^\infty(\mathbb{R}^d\times(0,T))$ we have*

$$\int_{\mathbb{R}^d\times(0,T)}[\partial_t\varphi+q^{ij}\partial_{x_i}\partial_{x_j}\varphi]\,d\mu\leqslant C\bigl(\sup_{\mathbb{R}^d\times(0,T)}|\varphi|+\sup_{\mathbb{R}^d\times(0,T)}|\nabla_x\varphi|\bigr).$$

Then the measure $(\det Q)^{1/(d+1)}\mu$ has a density in $L^{(d+1)'}_{\rm loc}(\mathbb{R}^d\times(0,T))$ with respect to Lebesgue measure on $\mathbb{R}^d\times(0,T)$.

9.3.2. Corollary. *If a locally finite nonnegative Borel measure μ on $\mathbb{R}^d\times(0,T)$ satisfies the equation $\partial_t\mu=L^*\mu$ and condition* (H1) *from* §9.1 *is fulfilled, then $\mu=\varrho\,dx\,dt$ and $\varrho\in L^{(d+1)'}_{\rm loc}(\mathbb{R}^d\times(0,T))$.*

The fact that, under rather general assumptions, nonnegative solutions are locally better integrable than just to power one is very useful.

Let g be a bounded function on $\mathbb{R}^{d+1}$. Set

$$O(g,R)=\sup_{(x,t)\in\mathbb{R}^{d+1}}\sup_{r\leqslant R} r^{-2}|U(x,r)|^{-2}$$
$$\times\int_t^{t+r^2}\iint_{y,z\in U(x,r)}|g(y,s)-g(z,s)|\,dy\,dz\,ds.$$

If $\lim\limits_{R\to 0}O(g,R)=0$, then we shall say that the function g belongs to the class $VMO_x(\mathbb{R}^{d+1})$. If $g\in VMO_x(\mathbb{R}^{d+1})$, then one can always assume that

$$O(g,R)\leqslant w(R)\quad\text{for all } R>0,$$

where w is a continuous function on $[0,+\infty)$ and $w(0)=0$.

Suppose that a function g is defined on $\mathbb{R}^d\times[0,T]$ and bounded on $U\times[0,T]$ for every ball U. We extend g by zero to the whole space $\mathbb{R}^{d+1}$. If for every function $\zeta\in C_0^\infty(\mathbb{R}^d)$ the function $g\zeta$ belongs to the class $VMO_x(\mathbb{R}^{d+1})$, then we say that g belongs to the class $VMO_{x,\rm loc}(\mathbb{R}^d\times[0,T])$.

9.3.3. Remark. Let $\omega\in C_0^\infty(\mathbb{R})$ be a probability density. Let us consider the functions

$$\omega_\varepsilon(x,t)=\varepsilon^{-d-1}\omega(|x|/\varepsilon)\omega(|t|/\varepsilon).$$

If $g \in VMO_x(\mathbb{R}^{d+1})$, then $g_\varepsilon = g * \omega_\varepsilon \in VMO_x(\mathbb{R}^{d+1})$ and $O(g_\varepsilon, R) \leqslant O(g, R)$. Indeed, it suffices to note that

$$r^{-2}|U(x,r)|^{-2} \int_t^{t+r^2} \iint_{y,z\in U(x,r)} |g_\varepsilon(y,s) - g_\varepsilon(z,s)|\, dy\, dz\, ds$$
$$\leqslant r^{-2}|U(x+\varepsilon\xi, r)|^{-2} \int_{\mathbb{R}^{d+1}} \omega(|\xi|)\omega(|\tau|)\, d\xi\, d\tau$$
$$\times \int_{t+\varepsilon\tau}^{t+\varepsilon\tau+r^2} \iint_{y,z\in U(x+\varepsilon\xi,r)} |g(y,s) - g(z,s)|\, dy\, dz\, ds,$$

where the right-hand side is estimated by the quantity $O(g,R)$. Observe also that $g_\varepsilon \to g$ in $L^p(U)$ on each ball $U \subset \mathbb{R}^{d+1}$ for every $p \geqslant 1$. Hence for every function $g \in VMO_x(\mathbb{R}^{d+1})$ one can find a sequence of smooth functions g_k converging to g in $L^p(U)$ on balls $U \subset \mathbb{R}^{d+1}$ for every $p \geqslant 1$ such that $O(g_k, R) \leqslant O(g, R)$.

9.3.4. Remark. Let functions $a^{ij} = a^{ji}$ be defined on $\mathbb{R}^{d+1}$ and let a matrix-valued mapping $A = (a^{ij})$ be such that $A(x,t) \geqslant \gamma \cdot \mathrm{I}$, $|A(x,t)| \leqslant M$ for some numbers $\gamma > 0$ and $M > 0$ for all $(x,t) \in \mathbb{R}^{d+1}$. Then these inequalities with the same numbers γ and M hold for the matrices $A_\varepsilon = (a_\varepsilon^{ij})$, where $a_\varepsilon^{ij} = a^{ij} * \omega_\varepsilon$ and the functions ω_ε are defined as in Remark 9.3.3 above. Indeed, it suffices to observe that

$$A_\varepsilon(x,t) = \int_{\mathbb{R}^{d+1}} A(x+\varepsilon\xi, t+\varepsilon\tau)\omega(|\xi|)\omega(|\tau|)\, d\xi\, d\tau.$$

Let $H^{p,2;1}\bigl(\mathbb{R}^d \times (-1,T)\bigr)$ be the space of all functions $u \in L^p\bigl(\mathbb{R}^d \times (-1,T)\bigr)$ having Sobolev derivatives $\partial_t u$, $\partial_{x_i} u$ and $\partial_{x_i}\partial_{x_j} u$ in the class $L^p\bigl(\mathbb{R}^d \times (-1,T)\bigr)$ and finite norm

$$\|u\|_{H^{p,2;1}(\mathbb{R}^d\times(-1,T))} = \|u\|_{L^p(\mathbb{R}^d\times(-1,T))} + \|\partial_{x_i} u\|_{L^p(\mathbb{R}^d\times(-1,T))}$$
$$+ \|\partial_{x_i}\partial_{x_j} u\|_{L^p(\mathbb{R}^d\times(-1,T))} + \|\partial_t u\|_{L^p(\mathbb{R}^d\times(-1,T))}.$$

The next result follows from Krylov [**555**]. For the reader's convenience we give a proof.

9.3.5. Proposition. *Suppose that* $q^{ij}, h^i, g \in C^\infty(\mathbb{R}^{d+1})$,

$$\sup_{x,t}\bigl(|q^{ij}(x,t)| + |h^i(x,t)| + |g(x,t)|\bigr) \leqslant M,$$

the matrix $Q = (q^{ij})$ *is symmetric and* $Q(x,t) \geqslant \kappa \cdot \mathrm{I}$ *for some number* $\kappa > 0$ *for all* $(x,t) \in \mathbb{R}^{d+1}$. *Suppose also that there exists a continuous function* w *on* $[0,+\infty)$ *such that* $w(0) = 0$ *and* $O(q^{ij}, R) \leqslant w(R)$ *for all* i, j. *Finally, assume that* $T > 0$ *and* $p > d+1$. *Then, for* $\psi \in C_0^\infty(\mathbb{R}^d)$, *the Cauchy problem*

$$\partial_t f + q^{ij}\partial_{x_i}\partial_{x_j} f + h^i \partial_{x_i} f + gf = 0, \quad f\big|_{t=T} = \psi \tag{9.3.1}$$

has a unique solution $f \in H^{p,2;1}\bigl(\mathbb{R}^d \times (-1,T)\bigr)$, *moreover, there is a number* $C > 0$, *depending only on the numbers* d, p, M, κ, T, *and the functions* ψ *and* w, *such that*

$$\|f\|_{H^{p,2;1}(\mathbb{R}^d\times(-1,T))} \leqslant C.$$

In addition, $f \in C^\infty\bigl(\mathbb{R}^d \times (-1,T)\bigr) \cap C_b(\mathbb{R}^d \times [-1,T])$.

PROOF. According to [**555**, Theorem 2.1], there is a unique solution of the Cauchy problem

$$\partial_t w + q^{ij}\partial_{x_i}\partial_{x_j} w + h^i\partial_{x_i} w + gw = -q^{ij}\partial_{x_i}\partial_{x_j}\psi - h^i\partial_{x_i}\psi - g\psi,$$

with boundary value $g\big|_{t=T} = 0$ in the class $H^{p,2;1}\big(\mathbb{R}^d\times(-1,T)\big)$. Moreover,

$$\|g\|_{H^{p,2;1}(\mathbb{R}^d\times(-1,T))} \leqslant C_1\|q^{ij}\partial_{x_i}\partial_{x_j}\psi + h^i\partial_{x_i}\psi + g\psi\|_{L^p(\mathbb{R}^d\times(-1,T))},$$

where C_1 depends only on d, p, M, κ, T, and w. The function $f = w + \psi$ is the unique solution of the Cauchy problem (9.3.1) in the class $H^{p,2;1}$. Since the matrix Q is nondegenerate and $q^{ij}, h^i, g \in C^\infty(\mathbb{R}^{d+1})$, we have $w \in C^\infty\big(\mathbb{R}^d\times(-1,T)\big)$ (see Ladyz'enskaya, Solonnikov, Ural'tseva [**576**, Theorem 12.2]). Therefore, we have $f \in C^\infty\big(\mathbb{R}^d\times(-1,T)\big)$. Let U be the unit ball in $\mathbb{R}^d$. Since $p > d+1$, the Sobolev embedding theorem gives the inclusion $f \in C^\alpha(U\times[-1,T])$ and the estimate

$$\|f\|_{C^\alpha(U\times[-1,T])} \leqslant C_2\|f\|_{H^{p,2;1}(\mathbb{R}^d\times[-1,T])},$$

where $\alpha = 1 - (d+1)/p$ and C_2 depends only on d and p. In particular, we have the inclusion $f \in C_b(\mathbb{R}^d\times[-1,T])$. □

We now prove the main result in this section. We recall that in this chapter we assume that $c \leqslant 0$.

9.3.6. Theorem. *Let $a^{ij} \in VMO_{x,\mathrm{loc}}(\mathbb{R}^d\times[0,T])$ and let the matrix $A = (a^{ij})$ satisfy condition* (H1) *in* §9.1. *Then the set*

$$\mathcal{SM}_\nu := \big\{\mu \in \mathcal{SP}_\nu\colon\ a^{ij}, b^i \in L^1\big(\mu, \mathbb{R}^d\times(0,T)\big)\big\}$$

contains at most one element.

PROOF. Let $\varphi_N(x) = \eta(x/N)$, where $\eta \in C_0^\infty(\mathbb{R}^d)$ is a nonnegative function such that $\eta(x) = 1$ if $|x| \leqslant 1$, $\eta(x) = 0$ if $|x| > 2$, $0 \leqslant \eta \leqslant 1$ and there is a number $K > 0$ such that for all x we have $|\nabla\eta(x)|^2\eta^{-1}(x) \leqslant K$. Let us fix $N \in \mathbb{N}$. Let U be an open ball in $\mathbb{R}^d$ containing the support of φ_N and let $\zeta \in C_0^\infty(\mathbb{R}^d)$ be a function such that $0 \leqslant \zeta \leqslant 1$ and $\zeta(x) = 1$ if $x \in U$. Set

$$\overline{A}(x,t) = \zeta(x)A(x,t) + \big(1 - \zeta(x)\big)\cdot \mathrm{I}.$$

Let $\overline{A}(x,t) = \mathrm{I}$ if $t < 0$ or $t > T$. Then $\overline{a}^{ij} \in VMO_x(\mathbb{R}^{d+1})$ and for some numbers $\kappa > 0$ and $M_1 > 0$ we have

$$\overline{A}(x,t) \geqslant \kappa\cdot\mathrm{I}, \quad |\overline{a}^{ij}(x,t)| \leqslant M_1$$

for all $(x,t) \in \mathbb{R}^{d+1}$. Moreover, we have $\overline{A}(x,t) = A(x,t)$ for all $(x,t) \in U\times[0,T]$.

Using Remark 9.3.3 and Remark 9.3.4 we can find a sequence of matrix-valued mappings $Q_n = (q_n^{ij})$ such that $q_n^{ij} \in C^\infty(\mathbb{R}^{d+1})$ and for all points $(x,t) \in \mathbb{R}^{d+1}$ there holds the inequality

$$Q_n(x,t) \geqslant \kappa\cdot\mathrm{I}, \quad |q_n^{ij}(x,t)| \leqslant M_1, \quad O(q_n^{ij}, R) \leqslant O(\overline{a^{ij}}, R),$$

moreover, $\lim\limits_{n\to\infty} \|\overline{A} - Q_n\|_{L^r(U\times[0,T])} = 0$, where $r = 2(d+1)$.

Let $\sigma^1 = \sigma_t^1(dx)\,dt$ and $\sigma^2 = \sigma_t^2(dx)\,dt$ belong to $\mathcal{SM}_\nu$. Then the measure $\sigma = (\sigma^1 + \sigma^2)/2$ also belongs to $\mathcal{SM}_\nu$. By assumption, $|b| \in L^2\big(\mu, U\times(0,T)\big)$. We can find a sequence of mappings $(h_m^i)_{1\leqslant i\leqslant d}\colon \mathbb{R}^{d+1} \to \mathbb{R}^d$, $m \in \mathbb{N}$, such that $h_m^i \in C^\infty(\mathbb{R}^{d+1})$, $|h_m(x,t)| \leqslant M_2(m)$ for all $(x,t) \in \mathbb{R}^{d+1}$ and

$$\lim_{m\to\infty} \int_0^T\!\!\int_U |b(x,t) - h_m(x,t)|^2\,\sigma_t(dx)\,dt = 0.$$

Let us also find a sequence of functions $g_m \in C^\infty(\mathbb{R}^{d+1})$ such that $g_m \leqslant 0$, $|g_m(x,t)| \leqslant M_3(m)$ for all $(x,t) \in \mathbb{R}^{d+1}$ and

$$\lim_{m\to\infty} \int_0^T \int_U |c(x,t) - g_m(x,t)|\,\sigma_t(dx)\,dt = 0.$$

Set

$$L_{n,m} = q_n^{ij}\partial_{x_i}\partial_{x_j} + h_m^i\partial_{x_i} + g_m.$$

Let $\psi \in C_0^\infty(\mathbb{R}^d)$ and $\max_x |\psi(x)| \leqslant 1$. Let $\mathcal{J}$ denote the set of all points $t \in [0,T]$ for which equality (9.1.6) holds for all functions $\psi\varphi_N$ and $\psi^2\varphi_N$ for both measures σ^1 and σ^2. It is clear that $\mathcal{J}$ is a full measure set in $[0,T]$, hence we can apply Remark 9.1.3 and Remark 9.1.4 to this set.

Let us fix a point $t \in \mathcal{J}$. According to Proposition 9.3.5, there exist solutions $f_{n,m} \in C^\infty(\mathbb{R}^d \times (-1,t)) \cap C_b(\mathbb{R}^d \times [-1,t])$ of the Cauchy problems

$$\partial_s f_{n,m} + L_{n,m} f_{n,m} = 0, \quad f_{n,m}\big|_{s=t} = \psi.$$

Let $\mathcal{I}$ be the set of all points $s \in [0,t]$ where equality (9.1.5) holds for all functions $f_{n,m}\varphi_N$ and $f_{n,m}^2\varphi_N$ for both measures σ^1 and σ^2. We observe that $t \in \mathcal{I}$ and $\mathcal{I}$ is a set of full measure in $[0,t]$. The measure $\mu = \sigma^1 - \sigma^2$ satisfies the Cauchy problem (9.1.1) with zero initial condition. Suppose that $\tau \in \mathcal{I}$ and $\tau < t$. Applying Lemma 9.1.1 to the function $f_{n,m}\varphi_N$, we obtain the equality

$$\begin{aligned}(9.3.2) \qquad &\int_{\mathbb{R}^d} \psi\varphi_N\,d\mu_t - \int_{\mathbb{R}^d} f_{n,m}(x,\tau)\varphi_N(x)\,\mu_\tau(dx) \\ &= \int_\tau^t \int_{\mathbb{R}^d} \bigl[\varphi_N(L - L_{n,m})f_{n,m} + \langle A\nabla f_{n,m}, \nabla\varphi_N\rangle + f_{n,m}L_{A,b}\varphi_N\bigr]\,d\mu_s\,ds.\end{aligned}$$

Let us estimate the quantity

$$\int_\tau^t \int_{\mathbb{R}^d} \varphi_N |\sqrt{A}\nabla f_{n,m}|^2\,d\sigma_s\,ds.$$

To this end we apply Lemma 9.1.1 to the function $f_{n,m}^2\varphi_N$ and the measure σ. We obtain

$$\begin{aligned}&\int_{\mathbb{R}^d} \psi^2\varphi_N(x)\,d\sigma_t - \int_{\mathbb{R}^d} f_{n,m}^2(x,\tau)\varphi_N(x)\,\sigma_\tau(dx) \\ &= 2\int_\tau^t \int_{\mathbb{R}^d} |\sqrt{A}\nabla f_{n,m}|^2\varphi_N\,d\sigma_s\,ds + \int_\tau^t \int_{\mathbb{R}^d} \bigl[2f_{n,m}\varphi_N(L - L_{n,m})f_{n,m} \\ &\quad + 2f_{n,m}\langle A\nabla f_{n,m}, \nabla\varphi_N\rangle + f_{n,m}^2 L_{A,b}\varphi_N\bigr]\,d\sigma_s\,ds.\end{aligned}$$

We observe that $0 \leqslant \varphi_N(x) \leqslant 1$, $|\psi(x)| \leqslant 1$ and $|f_{n,m}(x,s)| \leqslant 1$ by the maximum principle (we use that $g_m \leqslant 0$). Applying the inequality $2\alpha\beta \leqslant \alpha^2/2 + 2\beta^2$, we find that the following estimate holds:

$$\begin{aligned}&2\int_\tau^t \int_{\mathbb{R}^d} |\langle b - h_m, \nabla f_{n,m}\rangle|\,|f_{n,m}|\varphi_N\,d\sigma_s\,ds \\ &\leqslant 2\int_\tau^t \int_{\mathbb{R}^d} |A^{-1/2}(b - h_m)|^2\varphi_N\,d\sigma_s\,ds + \frac{1}{2}\int_\tau^t \int_{\mathbb{R}^d} |\sqrt{A}\nabla f_{n,m}|^2\varphi_N\,d\sigma_s\,ds.\end{aligned}$$

Similarly,

$$
\begin{aligned}
&2\int_\tau^t\int_{\mathbb{R}^d} |f_{n,m}|\,|\langle A\nabla f_{n,m}, \nabla\varphi_N\rangle|\,d\sigma_s\,ds\\
&\leqslant 2\int_\tau^t\int_{\mathbb{R}^d} \varphi_N^{-1}|\sqrt{A}\nabla\varphi_N|^2\,d\sigma_s\,ds + \frac{1}{2}\int_\tau^t\int_{\mathbb{R}^d}|\sqrt{A}\nabla f_{n,m}|^2\varphi_N\,d\sigma_s\,ds.
\end{aligned}
$$

Using the obtained inequalities and taking into account condition (H1), we obtain the estimate

$$
\int_\tau^t\int_{\mathbb{R}^d}\varphi_N|\sqrt{A}\nabla f_{n,m}|^2\,d\sigma_s\,ds \leqslant 1 + I + R_N,
$$

where I and R_N are defined by the following formulas:

$$
\begin{aligned}
I = 2\int_\tau^t\int_{\mathbb{R}^d} &|a^{ij} - q_n^{ij}|\,|\partial_{x_i}\partial_{x_j} f_{n,m}|\varphi_N\,d\sigma_s\,ds\\
&+ 2\kappa\int_\tau^t\int_{\mathbb{R}^d}|b^i - h_m^i|^2\varphi_N\,d\sigma_s\,ds + \int_\tau^t\int_{\mathbb{R}^d}|c - g_m|\varphi_N\,d\sigma_s\,ds,\\
R_N = \int_0^T\int_{\mathbb{R}^d}&\big[|L_{A,b}\varphi_N| + 2\varphi_N^{-1}|\sqrt{A}\nabla\varphi_N|^2\big]\,d\sigma_s\,ds.
\end{aligned}
$$

Returning to equality (9.3.2), applying the last estimate and taking into account that $|\mu|\leqslant 2\sigma$, we obtain the inequality

$$
\begin{aligned}
(9.3.3)\quad &\int_{\mathbb{R}^d}\psi\varphi_N\,d\mu_t - \int_{\mathbb{R}^d} f_{n,m}(x,\tau)\varphi_N(x)\,\mu_\tau(dx)\\
&\leqslant 2\int_\tau^t\int_{\mathbb{R}^d}|a^{ij} - q_n^{ij}|\,|\partial_{x_i}\partial_{x_j} f_{n,m}|\varphi_N\,d\sigma_s\,ds + \int_\tau^t\int_{\mathbb{R}^d}|c - g_m|\varphi_N\,d\sigma_s\,ds\\
&\quad + 2\kappa^{1/2}(1 + I + R_N)^{1/2}\left(\int_0^T\int_{\mathbb{R}^d}|b^i - h_m^i|^2\varphi_N\,d\sigma_s\,ds\right)^{1/2}\\
&\quad + 2(1 + I + R_N)^{1/2}\left(\int_0^T\int_{\mathbb{R}^d}\varphi^{-1}|\sqrt{A}\nabla\varphi_N|\,d\sigma_s\,ds\right)^{1/2}\\
&\qquad\qquad + 2\int_0^T\int_{\mathbb{R}^d}|L_{A,b}\varphi_N|\,d\sigma_s\,ds.
\end{aligned}
$$

By Corollary 9.3.2 the measure σ has a density $\varrho\in L^{(d+1)'}_{\rm loc}(\mathbb{R}^d\times(0,T))$. Let $p = 2(d+1)$. By Proposition 9.3.5 there is a number C depending only on p, d, κ, M_1, T, U, w, ψ, and h_m, but not depending on n, such that

$$
\max_{i,j}\|\partial_{x_i}\partial_{x_j} f_{n,m}\|_{L^p(U\times[0,T])}\leqslant C.
$$

We recall that $A = \overline{A}$ on U. Hölder's inequality yields the estimate

$$
\begin{aligned}
&\int_\tau^t\int_{\mathbb{R}^d}|a^{ij} - q_n^{ij}|\,|\partial_{x_i}\partial_{x_j} f_{n,m}|\varphi_N\,d\sigma_s\,ds\\
&\qquad\leqslant \|\overline{A} - Q_n\|_{L^r(U\times(0,T))}\|\partial_{x_i}\partial_{x_j} f_{n,m}\|_{L^p(U\times(0,T))}\|\varrho\|_{L^{(d+1)'}(U\times[\tau,t])}.
\end{aligned}
$$

Therefore, if τ and m are fixed, we have the equality

$$
\lim_{n\to\infty}\int_\tau^t\int_{\mathbb{R}^d}|a^{ij} - q_n^{ij}|\,|\partial_{x_i}\partial_{x_j} f_{n,m}|\varphi_N\,d\sigma_s\,ds = 0.
$$

According to Proposition 9.3.5 and the Sobolev embedding theorem, there exists a number C_1 independent of n such that

$$\|f_{n,m}\|_{C^\alpha(\overline{U}\times[0,t])} \leqslant C_1.$$

Hence on the compact set $\overline{U}\times[0,t]$ the family of functions $\{f_{n,m}\}_{n\geqslant 1}$ is uniformly bounded and equicontinuous, which enables us to pick a uniformly convergent subsequence. Keeping the same indices n, we can assume that $f_{n,m}\to f_m$ uniformly as $n\to\infty$ uniformly on $\overline{U}\times[0,t]$. Clearly, the function f_m is continuous on $\overline{U}\times[0,t]$. Keeping τ and m fixed, we let $n\to+\infty$ in (9.3.3). We obtain the inequality

$$\int_{\mathbb{R}^d}\psi\varphi_N\,d\mu_t-\int_{\mathbb{R}^d} f_m(x,\tau)\varphi_N(x)\,\mu_\tau(dx) \\ \leqslant I_m+2(1+I_m+R_N)^{1/2}I_m^{1/2}+2(1+I_m+R_N)^{1/2}R_N^{1/2}+2R_N, \tag{9.3.4}$$

where

$$I_m=2\kappa\int_0^T\!\!\int_{\mathbb{R}^d}|b^i-h^i_m|^2\varphi_N\,d\sigma_s\,ds+\int_0^T\!\!\int_{\mathbb{R}^d}|c-g_m|\varphi_N\,d\sigma_s\,ds.$$

Let us choose a sequence $\tau_k\in\mathcal{I}\cap(0,t)$ such that $\lim\limits_{k\to\infty}\tau_k=0$. Since the function f_m is continuous on $\overline{U}\times[0,t]$, the functions $f_m(x,\tau_k)$ converge to $f_m(x,0)$ uniformly in x as $k\to\infty$. In addition, since the initial condition is zero, we have

$$\lim_{k\to 0}\int_{\mathbb{R}^d} f_m(x,0)\varphi_N(x)\,\mu_{\tau_k}(dx)=0.$$

Therefore,

$$\lim_{k\to 0}\int_{\mathbb{R}^d} f_m(x,\tau_k)\varphi_N(x)\,\mu_{\tau_k}(dx)=0.$$

Replacing τ in inequality (9.3.4) with the number τ_k and letting first $k\to\infty$ and then $m\to\infty$, we arrive at the inequality

$$\int_{\mathbb{R}^d}\psi(x)\varphi_N(x)\,\mu_t(dx)\leqslant 2(1+R_N)^{1/2}R_N^{1/2}+2R_N.$$

Finally, letting $N\to\infty$ and using that $a^{ij},b^i\in L^1(\sigma,\mathbb{R}^d\times[0,T])$, we obtain that the integral of ψ with respect to the measure μ_t is nonpositive. Replacing ψ with $-\psi$, we obtain the opposite inequality. Hence for almost all $t\in[0,T]$ this integral vanishes. Since ψ was an arbitrary function in $C_0^\infty(\mathbb{R}^d)$ with $|\psi|\leqslant 1$, we have the equality $\mu_t=0$. Hence $\sigma^1=\sigma^2$. □

9.3.7. Remark. (i) It is easily seen from the proof of the theorem that we can weaken the conditions on $a^{ij},b^i\in L^1(\mu,\mathbb{R}^d\times[0,T])$ that restrict the class of probability solutions in which we proved uniqueness. Let $V\in C^2(\mathbb{R}^d)$, $V>0$, $\lim\limits_{|x|\to\infty}V(x)=+\infty$. Suppose that $a^{ij}\in VMO_{x,\mathrm{loc}}(\mathbb{R}^d\times[0,T])$ and $A=(a^{ij})$ satisfies condition (H1). Then there is at most one measure $\mu\in\mathcal{SP}_\nu$ with

$$\lim_{N\to\infty}\int_0^T\!\!\int_{N\leqslant V\leqslant 2N}\Bigl[\frac{|L_{A,b}V|}{V}+\frac{|\sqrt{A}\nabla V|^2}{V^2}\Bigr]\,d\mu=0.$$

(ii) It is also seen from the proof of the theorem that the condition on A can be weakened in the following way. The theorem remains true if we assume that $A=(a^{ij})$ satisfies condition (H1) and for every ball $U\subset\mathbb{R}^d$ there is a sequence of symmetric positive-definite matrices $A_k = (a_k^{ij})_{1\leqslant i,j\leqslant d}$ and numbers N, p, and q with $p^{-1}+q^{-1}=(d+1)^{-1}$ for which

$$a_k^{ij} \in C^\infty(U\times[0,T]), \quad \sup_k \sup_{U\times[0,T]} |A_k(x,t)| < \infty,$$

$$\lim_{k\to\infty} \|A - A_k\|_{L^q(U\times(0,T))} = 0,$$

$$\|\varphi\|_{W_p^{1,2}(U\times(0,T))} \leqslant N\|\partial_t\varphi + a_k^{ij}\partial_{x_i}\partial_{x_j}\varphi\|_{L^p(U\times(0,T))} \quad \forall\, k\in\mathbb{N}$$

for all $\varphi\in C^\infty(U\times[0,T])$ such that $\varphi(\,\cdot\,,T)=0$ and $\operatorname{supp}\varphi(\,\cdot\,,t)\subset U$ for all $t\in[0,T]$.

9.3.8. Example. Using estimates with Lyapunov functions (see § 7.1), one can obtain the following sufficient conditions for uniqueness. Let (H1) be fulfilled and $a^{ij} \in VMO_{x,\mathrm{loc}}(\mathbb{R}^d\times[0,T])$. Let $\alpha>0$ and $r>2$. Suppose that for all $(x,t)\in\mathbb{R}^d\times[0,T]$ and some positive numbers c_1, c_2, c_3 and $c_4<\alpha$, $c_5<\alpha$ we have

$$\text{(i)}\quad |x|^{r-2}\operatorname{tr} A(x,t) + (r-2)|x|^{r-4}\langle A(x,t)x,x\rangle \\ + \alpha r|x|^{2r-4}\langle A(x,t)x,x\rangle + |x|^{r-2}\langle b(x,t),x\rangle \leqslant c_1,$$

$$\text{(ii)}\quad |a^{ij}(x,t)| \leqslant c_2\exp\bigl(c_4|x|^r\bigr), \quad |b^i(x,t)| \leqslant c_3\exp\bigl(c_5|x|^r\bigr).$$

Let ν be a probability measure on $\mathbb{R}^d$ such that $\exp\bigl(\alpha|x|^r\bigr)\in L^1(\nu)$. Then the set $\mathcal{SP}_\nu$ contains at most one element. Indeed, inequality (i) and our condition on the initial distribution ν enable us to conclude (see § 7.1) that

$$\exp\bigl(\alpha|x|^r\bigr) \in L^1(\mu, \mathbb{R}^d\times[0,T])$$

for every probability solution μ to the Cauchy problem (9.1.1). Condition (ii) ensures the inclusions $a^{ij}, b^i \in L^1(\mu,\mathbb{R}^d\times[0,T])$. Hence we can apply Theorem 9.3.6.

It should be noted that uniqueness of solutions is established under very broad assumptions about the diffusion matrix and drift. However, our global conditions impose restrictions on the whole class of probability solutions for which we prove uniqueness. In addition, verification of our assumptions by means of Lyapunov functions leads to restrictions on the growth of the function $|b(x,t)|$.

9.4. The case of a Lipschitzian diffusion matrix

Here we discuss a method of proving uniqueness of probability solutions that requires some stronger local regularity of the coefficients of the operator L, but enables us to considerably weaken our global assumptions about coefficients and solutions.

We shall assume that along with condition (H1) from § 9.1 also the following condition holds:

(H2) for every ball $U\subset\mathbb{R}^d$ there is $\Lambda=\Lambda(U)>0$ such that for all $x,y\in U$ and $t\in(0,T)$ we have

$$|a^{ij}(x,t) - a^{ij}(y,t)| \leqslant \Lambda|x-y|.$$

It is proved in Chapter 6 that under conditions (H1) and (H2) every solution μ is given by a density ϱ with respect to Lebesgue measure on $\mathbb{R}^d\times(0,T)$. Moreover, if along with conditions (H1) and (H2) we have

$$b\in L^p_{\mathrm{loc}}(\mathbb{R}^d\times(0,T)) \quad\text{and}\quad c\in L^{p/2}_{\mathrm{loc}}(\mathbb{R}^d\times(0,T)) \quad \text{for some } p>d+2,$$

then one can choose a version of ϱ continuous on $\mathbb{R}^d\times(0,T)$, moreover, for almost all $t\in(0,T)$ the function $\varrho(\,\cdot\,,t)$ belongs to $W^{p,1}(U)$ for every ball $U\subset\mathbb{R}^d$. In addition, if for almost all $t \in (0,T)$ the measure $\mu_t(dx) = \varrho(x,t)\,dx$ is a subprobability on $\mathbb{R}^d$ and satisfies the initial condition $\mu|_{t=0} = \nu \neq 0$, then the density ϱ cannot be identically zero in the strip $\mathbb{R}^d\times(0,\tau)$ for any τ. Harnack's inequality yields that for every ball U in $\mathbb{R}^d$ and any compact interval $J \subset (0,T)$ there is a constant $C > 0$ such that $\varrho(x,t) \geqslant C$ for all $(x,t) \in U\times J$. If we work with the continuous version of the density ϱ, then for every $\zeta \in C_0^\infty(\mathbb{R}^d)$ the function

$$t \mapsto \int_{\mathbb{R}^d} \zeta(x)\varrho(x,t)\,dx$$

is continuous on $(0,T)$ and for almost all t coincides with $t \mapsto \displaystyle\int_{\mathbb{R}^d} \zeta\,d\mu_t$. Therefore,

$$\lim_{t\to 0} \int_{\mathbb{R}^d} \zeta(x)\varrho(x,t)\,dx = \int_{\mathbb{R}^d} \zeta\,d\nu.$$

Below we work with the continuous version of the density ϱ.

For every measure μ given by a Sobolev density ϱ with respect to Lebesgue measure, its logarithmic gradient $\beta_{\mu,A}$ with respect to the metric generated by A is defined by the formula

$$\beta^i_{\mu,A} = \sum_{j=1}^d \bigl(\partial_{x_j} a^{ij} + a^{ij}\varrho^{-1}\partial_{x_j}\varrho\bigr).$$

Below in this section we assume that b and $|c|^{1/2}$ are locally integrable to power $p > d+2$ with respect to Lebesgue measure on the strip $\mathbb{R}^d\times(0,T)$, conditions (H1) and (H2) hold, and $c \leqslant 0$.

Suppose that there are two solutions to the Cauchy problem (9.1.1) of class $\mathcal{SP}_\nu$ with a subprobability measure $\nu \neq 0$ defined by densities σ and ϱ with respect to Lebesgue measure. Then these densities are continuous on $\mathbb{R}^d\times(0,T)$. In addition, the functions σ and ϱ are strictly positive. Set $v(x,t) = \sigma(x,t)/\varrho(x,t)$. The function v is continuous and positive on $\mathbb{R}^d\times(0,T)$.

9.4.1. Lemma. *Suppose that for a.e. $t \in (0,T)$ we have*

$$\int_{\mathbb{R}^d} \varrho(x,t)\,dx = \nu(\mathbb{R}^d) + \int_0^t\!\!\int_{\mathbb{R}^d} c(x,s)\varrho(x,s)\,dx\,ds,$$

$$\int_{\mathbb{R}^d} \sigma(x,t)\,dx \leqslant \nu(\mathbb{R}^d) + \int_0^t\!\!\int_{\mathbb{R}^d} c(x,s)\sigma(x,s)\,dx\,ds.$$

Suppose also that for every $\lambda > 0$ the inequality

$$\int_{\mathbb{R}^d} e^{\lambda(1-v(x,t))}\varrho(x,t)\,dx \leqslant 1 \tag{9.4.1}$$

holds for a.e. $t \in (0,T)$. Then $v \equiv 1$, i.e., $\sigma = \varrho$.

PROOF. Let t be such that the densities $\varrho(\,\cdot\,,t)$, $\sigma(\,\cdot\,,t)$ satisfy all relations listed in the formulation of the lemma and (9.4.1) holds for all natural numbers λ. We observe that the set of those t for which this is not true, is a set of Lebesgue measure zero. If there is a ball $U \subset \mathbb{R}^d$ such that $v(x,t) \leqslant 1-\delta$ for every $x \in U$ and some $\delta > 0$, then

$$e^{\lambda\delta}\int_U \varrho\,dx \leqslant \int_U e^{\lambda(1-v(x,t))}\varrho(x,t)\,dx \leqslant 1.$$

Letting $\lambda \to \infty$, we arrive at a contradiction. Hence $v(x,t) \geqslant 1$ and $\sigma \geqslant \varrho$ for all points $(x,t) \in \mathbb{R}^d \times (0,T)$. We observe that

$$\begin{aligned}\nu(\mathbb{R}^d) &= \int_{\mathbb{R}^d} \varrho(x,t)\,dx - \int_0^t\int_{\mathbb{R}^d} c(x,s)\varrho(x,s)\,dx\,ds \\ &\leqslant \int_{\mathbb{R}^d} \sigma(x,t)\,dx - \int_0^t\int_{\mathbb{R}^d} c(x,s)\sigma(x,s)\,dx\,ds \leqslant \nu(\mathbb{R}^d).\end{aligned}$$

Hence the integrals of $\varrho(x,t)$ and $\sigma(x,t)$ in x coincide, whence $v \equiv 1$. □

The next lemma is the key point in our reasoning. In this lemma, one can take for f any of the two functions $z \mapsto e^{\lambda(1-z)}$ and $z \mapsto e^{\lambda(1-z)} - e^{\lambda}$.

9.4.2. Lemma. *Let $\psi \in C_0^\infty(\mathbb{R}^d)$, $\psi \geqslant 0$, $0 < t < T$. Then*

$$\int_{\mathbb{R}^d} f\big(v(x,t)\big)\varrho(x,t)\psi(x)\,dx \leqslant f(1)\int_{\mathbb{R}^d} \psi\,d\nu + \int_0^t\int_{\mathbb{R}^d} \varrho f(v) L\psi\,dx\,ds. \tag{9.4.2}$$

If, in addition, $|b - \beta_{\mu,A}|\varrho \in L^1\big(U \times (0,T)\big)$ for every ball U in $\mathbb{R}^d$, then

$$\begin{aligned}\int_{\mathbb{R}^d} f\big(v(x,t)\big)\varrho(x,t)\psi(x)\,dx &\leqslant f(1)\int_{\mathbb{R}^d} \psi\,d\nu \\ &+ \frac{1}{2}\int_0^t\int_{\mathbb{R}^d} \varrho\langle A\nabla\psi, \nabla\psi\rangle\psi^{-1}|f'(v)|^2 f''(v)^{-1}\,dx\,ds \\ &+ \int_0^t\int_{\mathbb{R}^d} \Big[f(v)\langle b - \beta_{\mu,A}, \nabla\psi\rangle\varrho + \psi f(v)c\varrho\Big]\,dx\,ds.\end{aligned} \tag{9.4.3}$$

A complete proof of the lemma will be given in the next section, but here we give an informal reasoning, which, however, becomes rigorous if we work in the class of smooth functions on $\mathbb{R}^d \times [0,T]$ and assume that the initial condition is given by a strictly positive density.

Proof. Set $h^i = b^i - \sum_{j=1}^d \partial_{x_j} a^{ij}$. We recall some relations following from the Leibniz formula and the chain rule. For any functions ξ, η in $C^\infty\big(\mathbb{R}^d \times (0,T)\big)$ and $\Phi \in C^\infty(\mathbb{R})$ one has the equalities

$$L^*\Phi(\xi) = \Phi'(\xi)L^*\xi + \Phi''(\xi)\langle A\nabla\xi, \nabla\xi\rangle + \big(\xi\Phi'(\xi) - \Phi(\xi)\big)(\mathrm{div} h - c),$$
$$L^*(\xi \cdot \eta) = \eta L^*\xi + \xi L^*\eta + 2\langle A\nabla\xi, \nabla\eta\rangle + \xi\eta(\mathrm{div} h - c).$$

Thus, $\sigma = v\varrho$, $\partial_t\sigma = L^*\sigma$ and also $\partial_t\varrho = L^*\varrho$. Multiplying the equation $\partial_t\varrho = L^*\varrho$ by v and subtracting the result from the equation $\partial_t\sigma = L^*\sigma$, we arrive at the following relation for the function v:

$$\varrho\partial_t v = \varrho L^* v + 2\langle A\nabla\varrho, \nabla v\rangle + \varrho v(\mathrm{div} h - c).$$

Multiplying this relation by the function $f'(v)$ and taking into account the equalities $\partial_t f(v) = f'(v)\partial_t v$ and $\nabla f(v) = f'(v)\nabla v$, we obtain

$$\varrho\partial_t\big(f(v)\big) = \varrho f'(v)L^*v + 2\langle A\nabla\varrho, \nabla f(v)\rangle + \varrho v f'(v)(\mathrm{div} h - c).$$

Since

$$f'(v)L^*v = L^*f(v) - f''(v)\langle A\nabla v, \nabla v\rangle - \big(vf'(v) - f(v)\big)(\mathrm{div} h - c),$$

the term $\varrho\partial_t\big(f(v)\big)$ equals

$$\varrho L^*f(v) + 2\langle A\nabla\varrho, \nabla f(v)\rangle + \varrho f(v)(\mathrm{div} h - c) - \varrho f''(v)\langle A\nabla v, \nabla v\rangle.$$

Summing the last equality and the equality $f(v)\partial_t\varrho = f(v)L^*\varrho$, we find that $\partial_t\bigl(\varrho f(v)\bigr) = L^*\bigl(\varrho f(v)\bigr) - \varrho f''(v)\langle A\nabla v, \nabla v\rangle$. Multiplying this equation by the function ψ and integrating, we have

$$\int_0^t\!\!\int_{\mathbb{R}^d} \partial_t\bigl(f(v)\varrho\bigr)\psi\,dx\,ds + \int_0^t\!\!\int_{\mathbb{R}^d} \varrho\langle A\nabla v, \nabla v\rangle f''(v)\psi\,dx\,ds$$
$$= \int_0^t\!\!\int_{\mathbb{R}^d} \psi L^*\bigl(\varrho f(v)\bigr)\,dx\,ds = \int_0^t\!\!\int_{\mathbb{R}^d} \varrho f(v)L\psi\,dx\,ds.$$

Applying the Newton–Leibniz formula and taking into account the trivial identity $v(x,0)\equiv 1$, we obtain the relation

$$\int_0^t\!\!\int_{\mathbb{R}^d} \partial_t\bigl(f(v)\varrho\bigr)\psi\,dx\,ds = \int_{\mathbb{R}^d} f\bigl(v(x,t)\bigr)\varrho(x,t)\psi(x)\,dx - f(1)\int_{\mathbb{R}^d}\psi\,d\nu.$$

Therefore, we obtain the following equality:

$$\int_{\mathbb{R}^d} f\bigl(v(x,t)\bigr)\varrho(x,t)\psi(x)\,dx + \int_0^t\!\!\int_{\mathbb{R}^d} \varrho\langle A\nabla v, \nabla v\rangle f''(v)\psi\,dx\,ds$$
$$= f(1)\int_{\mathbb{R}^d}\psi\,d\nu + \int_0^t\!\!\int_{\mathbb{R}^d} \varrho f(v)L\psi\,dx\,ds. \tag{9.4.4}$$

Taking into account condition (H2) and the inequalities $f''\geqslant 0$ and $\psi\geqslant 0$, we obtain estimate (9.4.2). In order to derive estimate (9.4.3), some additional transformations of (9.4.4) are needed. Integrating by parts, we obtain

$$\sum_{i,j=1}^d \int_0^t\!\!\int_{\mathbb{R}^d} \varrho f(v)a^{ij}\partial_{x_i}\partial_{x_j}\psi\,dx\,ds$$
$$= -\int_0^t\!\!\int_{\mathbb{R}^d} \varrho f'(v)\langle A\nabla v, \nabla\psi\rangle\,dx\,ds - \int_0^t\!\!\int_{\mathbb{R}^d} \varrho f(v)\langle \beta_{\mu,A}, \nabla\psi\rangle\,dx\,ds.$$

Since

$$|\langle A\nabla v, \nabla\psi\rangle| \leqslant \sqrt{\langle A\nabla v, \nabla v\rangle}\sqrt{\langle A\nabla\psi, \nabla\psi\rangle},$$

the inequality $qr\leqslant 2^{-1}(q^2+r^2)$ with the numbers

$$q = |f''(v)|^{1/2}|\psi|^{1/2}\langle A\nabla v, \nabla v\rangle^{1/2},$$

$$r = |f''(v)|^{-1/2}|\psi|^{-1/2}\langle A\nabla\psi, \nabla\psi\rangle^{1/2}|f'(v)|$$

yields the estimate

$$\int_0^t\!\!\int_{\mathbb{R}^d} f'(v)\langle A\nabla v, \nabla\psi\rangle\varrho\,dx\,ds \leqslant \frac{1}{2}\int_0^t\!\!\int_{\mathbb{R}^d} \varrho\langle A\nabla v, \nabla v\rangle f''(v)\psi\,dx\,ds$$
$$+\frac{1}{2}\int_0^t\!\!\int_{\mathbb{R}^d} \varrho\langle A\nabla\psi, \nabla\psi\rangle\psi^{-1}|f'(v)|^2 f''(v)^{-1}\,dx\,ds.$$

Therefore, we have the following inequality:

$$\int_{\mathbb{R}^d} f\big(v(x,t)\big)\varrho(x,t)\psi(x)\,dx + \frac{1}{2}\int_0^t\!\!\int_{\mathbb{R}^d} \varrho\langle A\nabla v, \nabla v\rangle f''(v)\psi\,dx\,ds$$
$$\leqslant f(1)\int_{\mathbb{R}^d}\psi\,d\nu + \frac{1}{2}\int_0^t\!\!\int_{\mathbb{R}^d}\varrho\langle A\nabla\psi,\nabla\psi\rangle\psi^{-1}|f'(v)|^2 f''(v)^{-1}\,dx\,ds$$
$$+\int_0^t\!\!\int_{\mathbb{R}^d}\big[\varrho f(v)\langle b-\beta_{\mu,A},\nabla\psi\rangle + \psi\varrho f(v)c\big]\,dx\,ds.$$

Since $f''\geqslant 0$ and condition (H2) holds, the last inequality gives estimate (9.4.3). □

Note that it is important in the informal proof above that $v(\,\cdot\,,0)=1$. In the general case we have no this equality in any sense. In addition, under our assumptions the solution density need not be differentiable in the variable t.

9.4.3. Theorem. *Suppose that conditions* (H1) *and* (H2) *hold, that* $c\leqslant 0$ *and that* $b\in L^p_{\text{loc}}\big(\mathbb{R}^d\times(0,T)\big)$, $c\in L^{p/2}_{\text{loc}}\big(\mathbb{R}^d\times(0,T)\big)$ *for some* $p>d+2$. *Assume also that some measure* μ *of class* $\mathcal{SP}_\nu$ *with* ν *satisfies the condition*

$$|a^{ij}|/(1+|x|^2)+|b^i|/(1+|x|)\in L^1\big(\mu,\mathbb{R}^d\times(0,T)\big).$$

Then the set $\mathcal{SP}_\nu$ *consists of exactly one element* μ.

PROOF. We can assume that $\nu\neq 0$. Let ϱ be the density of the measure μ with respect to Lebesgue measure. According to Remark 7.1.9 we have

$$\mu_t(\mathbb{R}^d)=\nu(\mathbb{R}^d)+\int_0^t\!\!\int_{\mathbb{R}^d} c(x,s)\,\mu_s(dx)\,ds \quad \text{for a.e. } t\in(0,T).$$

Suppose that there is another measure in $\mathcal{SP}_\nu$ with density σ. As above, we set $v=\sigma/\varrho$. Let $\psi(x)=\zeta(x/N)$, where a nonnegative function $\zeta\in C_0^\infty(\mathbb{R}^d)$ is such that $\zeta(x)=1$ if $|x|\leqslant 1$ and $\zeta(x)=0$ if $|x|>2$, and there is number $K>0$ such that for all x we have

$$|\zeta(x)|\leqslant K,\quad |\nabla\zeta(x)|\leqslant K,\quad \sum_{i,j=1}^d|\partial_{x_i}\partial_{x_j}\zeta(x)|\leqslant K.$$

Let us set $f(z)=e^{\lambda(1-z)}$. Clearly, $f(z)\leqslant e^\lambda$ if $z\geqslant 0$. Using inequality (9.4.2) in Lemma 9.4.2 and the condition that $c\leqslant 0$, we obtain

$$\int_{\mathbb{R}^d} e^{\lambda(1-v(x,t))}\varrho(x,t)\zeta(x/N)\,dx\leqslant \int_{\mathbb{R}^d}\zeta(x/N)\,\nu(dx)$$
$$+e^\lambda KN^{-2}\sum_{i,j=1}^d\int_0^t\!\!\int_{N<|x|<2N}|a^{ij}(x,t)|\varrho(x,t)\,dx\,ds$$
$$+e^\lambda KN^{-1}\sum_{i=1}^d\int_0^t\!\!\int_{N<|x|<2N}|b^i(x,t)|\varrho(x,t)\,dx\,ds.$$

Since we have $|a^{ij}|(1+|x|)^{-2}\varrho$, $|b^i|(1+|x|)^{-1}\varrho\in L^1\big(\mathbb{R}^d\times(0,T)\big)$, the last two integrals tend to zero as $N\to\infty$, and we arrive at (9.4.1). Lemma 9.4.1 gives the equality $v=1$. □

9.4.4. Remark. Obviously, our condition on the coefficients a^{ij} and b^i is satisfied if

$$a^{ij},\, b^i \in L^1\big(\mu, \mathbb{R}^d\times(0,T)\big).$$

9.4.5. Theorem. *Suppose that conditions* (H1) *and* (H2) *hold and that for some* $p > d+2$ *we have* $b\in L^p_{\mathrm{loc}}\big(\mathbb{R}^d\times(0,T)\big)$, $c\in L^{p/2}_{\mathrm{loc}}\big(\mathbb{R}^d\times(0,T)\big)$, *where* $c\leqslant 0$. *Assume also that for some measure* μ *of class* $\mathcal{SP}_\nu$ *we have*

$$a^{ij}, b^i - \beta^i_{\mu,A} \in L^1\big(\mu, \mathbb{R}^d\times(0,T)\big) \quad \textit{for all } i,j\leqslant d.$$

Then the set $\mathcal{SP}_\nu$ *consists of the single element* μ.

PROOF. Justification is similar to the previous reasoning. Again we can assume that $\nu\neq 0$. The measure μ is given by a density ϱ with respect to Lebesgue measure. According to Remark 7.1.9 we have

$$\mu_t(\mathbb{R}^d) = \nu(\mathbb{R}^d) + \int_0^t\!\!\int_{\mathbb{R}^d} c(x,s)\,\mu_s(dx)\,ds \quad \text{for a.e. } t\in(0,T).$$

Suppose that there is another measure in $\mathcal{SP}_\nu$ with a density σ. As above, we set $v=\sigma/\varrho$, $\psi(x)=\zeta(x/N)$, where a nonnegative function $\zeta\in C_0^\infty(\mathbb{R}^d)$ is such that $\zeta(x)=1$ if $|x|\leqslant 1$ and $\zeta(x)=0$ if $|x|>2$, and there is a number $K>0$ such that for all x we have

$$|\zeta(x)|\leqslant K, \quad |\nabla\zeta(x)|\leqslant K, \quad |\nabla\zeta(x)|^2\zeta^{-1}(x)\leqslant K.$$

Let $f(z) = e^{\lambda(1-z)}$. Then $f(z)\leqslant e^\lambda$ for $z\geqslant 0$. Using inequality (9.4.3) in Lemma 9.4.2 and the condition that $c\leqslant 0$, we obtain

$$\begin{aligned}\int_{\mathbb{R}^d} e^{\lambda(1-v(x,t))}\varrho(x,t)\zeta(x/N)\,dx &\leqslant \int_{\mathbb{R}^d}\zeta(x/N)\,\nu(dx)\\ &+ e^\lambda\frac{K}{N^2}\sum_{i,j=1}^d\int_0^t\!\!\int_{\mathbb{R}^d}|a^{ij}|\varrho\,dx\,ds + e^\lambda\frac{K}{N}\sum_{i=1}^d\int_0^t\!\!\int_{\mathbb{R}^d}|b^i-\beta^i_{\mu,A}|\varrho\,dx\,ds.\end{aligned}$$

By assumption, we have $|a^{ij}|\varrho, |b^i-\beta^i_{\mu,A}|\varrho\in L^1\big(\mathbb{R}^d\times(0,1)\big)$. Letting $N\to\infty$, by Lemma 9.4.1 we obtain that $v=1$. □

Let us now give a condition for uniqueness in terms of Lyapunov functions.

9.4.6. Theorem. *Suppose that conditions* (H1) *and* (H2) *hold,* $c\leqslant 0$ *and* $b\in L^p_{\mathrm{loc}}\big(\mathbb{R}^d\times(0,T)\big)$, $c\in L^{p/2}_{\mathrm{loc}}\big(\mathbb{R}^d\times(0,T)\big)$ *for some* $p>d+2$. *Suppose also that there exists a positive function* $V\in C^2(\mathbb{R}^d)$ *such that* $V(x)\to+\infty$ *as* $|x|\to+\infty$ *and for some number* $C>0$ *and all* $(x,t)\in\mathbb{R}^d\times(0,T)$ *we have*

$$LV(x,t)\leqslant C + CV(x).$$

Then the set $\mathcal{SP}_\nu$ *contains at most one element.*

PROOF. Note that by Theorem 7.1.1 and Proposition 7.1.8 for every measure μ in $\mathcal{SP}_\nu$ we have the equality

$$\mu_t(\mathbb{R}^d) = \nu(\mathbb{R}^d) + \int_0^t\!\!\int_{\mathbb{R}^d} c(x,s)\,\mu_s(dx)\,ds \tag{9.4.5}$$

for almost all $t\in(0,T)$. We can assume that $\nu\neq 0$, since otherwise $\mu=0$. Suppose that the class $\mathcal{SP}_\nu$ contains two measures given by densities σ and ϱ with respect to Lebesgue measure. Set $v=\sigma/\varrho$. Let us take the function $\psi(x)=\zeta\big(N^{-1}V(x)\big)$, where $\zeta \in C_0^\infty(\mathbb{R})$ is a nonnegative function such that $\zeta(0) = 1$, $\zeta(z) = 0$ if $|z| > 1$, $0 \leqslant \zeta \leqslant 1$ and, in addition, $\zeta'(z) \leqslant 0$ and $\zeta''(z) \geqslant 0$ if $z > 0$.

Let $f(z) = e^{\lambda(1-z)} - e^\lambda$. Then $f(z) \leqslant 0$ and $|f(z)| \leqslant 2e^\lambda$ for all $z \geqslant 0$. We observe that

$$f(v)\zeta' LV \leqslant (K + KV) f(v)\zeta',$$

since $f(v)\zeta' \geqslant 0$. Using inequality (9.4.2), we obtain

$$\begin{aligned}\int_D \bigl(e^{\lambda(1-v(x,t))} - e^\lambda\bigr)\varrho(x,t)\zeta\Bigl(\frac{V(x)}{N}\Bigr)\,dx &\leqslant (1-e^\lambda)\int_D \zeta\Bigl(\frac{V}{N}\Bigr)\,d\nu \\ &+ 2e^\lambda M N^{-1}\int_0^t\!\!\int_{V<N}(K+KV)\varrho\,dx\,ds \\ &+ \int_0^t\!\!\int_D \zeta\Bigl(\frac{V(x)}{N}\Bigr) f(v(x,s))c(x,s)\varrho(x,s)\,dx\,ds.\end{aligned}$$

We show that

$$\lim_{N\to+\infty} N^{-1}\int_0^t\!\!\int_{V<N}(K+KV)\varrho\,dx\,ds = 0. \tag{9.4.6}$$

Indeed, let $\gamma \in (0,1)$ and $N > \gamma^{-1}$, then

$$N^{-1}\int_0^t\!\!\int_{V<N}(K+KV)\varrho\,dx\,ds \leqslant \gamma\int_0^t\!\!\int_{V<\gamma N}\varrho\,dx\,ds + K\int_0^t\!\!\int_{\gamma N<V<N}\varrho\,dx\,ds.$$

Therefore, one has the inequality

$$\lim_{N\to+\infty} N^{-1}\int_0^t\!\!\int_{V<N}(K+KV)\varrho\,dx\,ds \leqslant \gamma\int_0^t\!\!\int_{\mathbb{R}^d}\varrho\,dx\,ds.$$

Letting $\gamma \to 0$, we obtain (9.4.6). Letting $N \to +\infty$, we arrive at

$$\begin{aligned}&\int_{\mathbb{R}^d}\bigl(e^{\lambda(1-v(x,t))} - e^\lambda\bigr)\varrho(x,t)\,dx \\ &\qquad\leqslant (1-e^\lambda)\int_{\mathbb{R}^d} d\nu + \int_0^t\!\!\int_{\mathbb{R}^d}\bigl(e^{\lambda(1-v)} - e^\lambda\bigr)c(x,s)\varrho(x,s)\,dx\,ds.\end{aligned}$$

Since $c \leqslant 0$ and for almost all $t \in (0,T)$ equality (9.4.5) holds, we see that for almost all t the integral of $e^{\lambda(1-v(x,t))}\varrho(x,t)$ in x does not exceed 1. Lemma 9.4.1 completes the proof. □

Let us consider an example of application of this theorem.

9.4.7. Example. Let $V(x) = \ln(|x|^2+1)$ if $|x| > 1$. Then the condition $LV \leqslant C + CV$ is equivalent to the inequality

$$\begin{aligned}2\mathrm{tr}A(x,t) - 4\frac{\langle A(x,t)x,x\rangle}{|x|^2+1} &+ c(x,t)(|x|^2+1)\ln(|x|^2+1) \\ &+ 2\langle b(x,t),x\rangle \leqslant C(|x|^2+1) + C(|x|^2+1)\ln(|x|^2+1).\end{aligned}$$

In particular, for unit diffusion matrix and $c = 0$ uniqueness is ensured by the inequality

$$\langle b(x,t),x\rangle \leqslant C_1 + C_2|x|^2\ln(|x|^2+1).$$

Combining Theorems 9.4.6, 6.6.4, 7.1.1 and Proposition 7.1.8, we obtain the following sufficient condition for the existence and uniqueness in the class $\mathcal{SP}_\nu$.

9.4.8. Theorem. *Let conditions* (H1) *and* (H2) *hold, let* $c \leqslant 0$, *and let*

$$c \in L^{p/2}_{\text{loc}}(\mathbb{R}^d \times (0,T)), \quad b \in L^\infty(U \times (0,T))$$

for some $p > d+2$ *and every ball* $U \subset \mathbb{R}^d$. *Suppose that there exists a function* $V \in C^2(\mathbb{R}^d)$ *such that* $\lim\limits_{|x|\to\infty} V(x) = +\infty$ *and for some number* $C > 0$ *and all* $(x,t) \in \mathbb{R}^d \times (0,T)$ *we have*

$$LV(x,t) \leqslant C + CV(x).$$

Then the class $\mathcal{SP}_\nu$, *where* ν *is a probability measure on* $\mathbb{R}^d$, *contains precisely one element* $\mu = (\mu_t)_{t\in(0,T)}$. *Moreover,*

$$\mu_t(\mathbb{R}^d) = 1 + \int_0^t \int_{\mathbb{R}^d} c(x,s)\, \mu_s(dx)\, ds \quad \textit{for a.e. } t \in (0,T).$$

In particular, if $c = 0$, *then for a.e.* t *the measure* μ_t *is a probability measure.*

9.5. Proof of the main lemma

We assume, as above, that b and $|c|^{1/2}$ are locally integrable to power $p > d+2$ with respect to Lebesgue measure on $\mathbb{R}^d \times (0,T)$ and that there are two solutions to the Cauchy problem (9.1.1) in the class $\mathcal{SP}_\nu$, given by continuous and positive densities σ and ϱ with respect to Lebesgue measure on $\mathbb{R}^d \times (0,T)$.

Set $v(x,t) = \sigma(x,t)/\varrho(x,t)$. Let $\omega \in C_0^\infty(\mathbb{R}^d)$ be a probability density such that $|\nabla\omega|^2\omega^{-1} \in C_0^\infty(\mathbb{R}^d)$. For example, we can take $\omega(x) = c_0\zeta(|x|)$, where c_0 is a normalization constant and $\zeta \in C_0^\infty(\mathbb{R})$, $0 \leqslant \zeta \leqslant 1$, $\zeta(y) = 0$ if $|y| \geqslant 1$, $\zeta(y) > 0$ if $|y| < 1$, $\zeta(y) = 1$ if $|y| \leqslant 1/2$, and $\zeta(y) = \exp[(y^2-1)^{-1}]$ near the points -1 and 1. Below we use exactly this function ω.

Let $0 < \varepsilon < 1$ and $\omega_\varepsilon = \varepsilon^{-d}\omega(x\varepsilon^{-1})$. For every locally integrable function g on $\mathbb{R}^d$, let g_ε denote the convolution $g*\omega_\varepsilon$. The following properties of the functions g_ε will be important for us:

1) for almost all x we have $\lim\limits_{\varepsilon\to 0} g_\varepsilon(x) = g(x)$, and if the function g is continuous, then this equality holds for all x;

2) $\lim\limits_{\varepsilon\to 0} \|g_\varepsilon - g\|_{L^1(U)} = 0$ for every ball $U \subset \mathbb{R}^d$;

3) the function g_ε belongs to the class C^∞, and if g is continuous and strictly positive, then g_ε is also strictly positive.

The continuity of σ and ϱ on $\mathbb{R}^d \times (0,T)$ yields that the functions σ_ε and ϱ_ε, obtained by convolution in x, are continuous on $\mathbb{R}^d \times (0,T)$. Moreover,

$$\lim_{\varepsilon\to 0} \sigma_\varepsilon(x,t) = \sigma(x,t), \quad \lim_{\varepsilon\to 0} \varrho_\varepsilon(x,t) = \varrho(x,t).$$

Since $\omega_\varepsilon \in C_0^\infty(\mathbb{R}^d)$, for all $x \in \mathbb{R}^d$, as $t \to 0$, we have

$$\varrho_\varepsilon(x,t) = \int_{\mathbb{R}^d} \varrho(y,t)\omega_\varepsilon(x-y)\, dy \to \int_{\mathbb{R}^d} \omega_\varepsilon(x-y)\, \nu(dy) = \nu * \omega_\varepsilon(x).$$

A similar assertion is true for σ_ε. We define the functions $\varrho_\varepsilon(x,t)$ and $\sigma_\varepsilon(x,t)$ at $t = 0$ by the value $\nu*\omega_\varepsilon(x)$. Therefore, we obtain functions continuous on $\mathbb{R}^d \times (0,T)$ and continuous on $[0,T)$ in t for every fixed x.

Since for almost every $t \in (0,T)$ the functions $\varrho(\,\cdot\,,t)$ and $\sigma(\,\cdot\,,t)$ are subprobability densities, for each fixed ε the functions ϱ_ε and σ_ε are uniformly bounded on the strip $\mathbb{R}^d\times(0,T)$. Hence for every ball U we have

$$\lim_{t\to 0}\|\varrho_\varepsilon(\,\cdot\,,t)-\nu*\omega_\varepsilon\|_{L^1(U)}=0,\quad \lim_{t\to 0}\|\sigma_\varepsilon(\,\cdot\,,t)-\nu*\omega_\varepsilon\|_{L^1(U)}=0.$$

Then $\lim_{t\to 0}\|\varrho_\varepsilon(\,\cdot\,,t)-\sigma_\varepsilon(\,\cdot\,,t)\|_{L^1(U)}=0$. In addition, the functions σ_ε and ϱ_ε are absolutely continuous in t. Indeed, in the sense of generalized functions

$$\partial_t\varrho_\varepsilon=\sum_{i,j=1}^d\partial_{x_i}\partial_{x_j}(a^{ij}\varrho)_\varepsilon-\sum_{i=1}^d\partial_{x_i}(b^i\varrho)_\varepsilon+(c\varrho)_\varepsilon,$$

$$\partial_t\sigma_\varepsilon=\sum_{i,j=1}^d\partial_{x_i}\partial_{x_j}(a^{ij}\sigma)_\varepsilon-\sum_{i=1}^d\partial_{x_i}(b^i\sigma)_\varepsilon+(c\sigma)_\varepsilon,$$

where the right sides of these equalities are integrable on $[0,T]$, since, by the definition of the class $\mathcal{SP}_\nu$ and conditions (H1) and (H2), the functions $b\varrho$, $b\sigma$, $c\varrho$, $c\sigma$, and also $a^{ij}\varrho$ and $a^{ij}\sigma$ are integrable on $U\times[0,T]$ for each ball U.

Set $v^\varepsilon(x,t)=\sigma_\varepsilon(x,t)/\varrho_\varepsilon(x,t)$ if $0<t<1$ and $x\in\mathbb{R}^d$. Then $v^\varepsilon(x,t)$ tends to $v(x,t)=\sigma(x,t)/\varrho(x,t)$ as $\varepsilon\to 0$ for all $x\in\mathbb{R}^d$ and $t\in(0,T)$. Let

$$\mathcal{A}\varphi=a^{ij}\partial_{x_i}\partial_{x_j}\varphi,\ \mathcal{A}^*\varphi=\partial_{x_i}\partial_{x_j}(a^{ij}\varphi),$$

$$R^i_{\varrho,\varepsilon}=\sum_{j=1}^d\bigl(\partial_{x_j}(a^{ij}\varrho)_\varepsilon-\partial_{x_j}(a^{ij}\varrho_\varepsilon)\bigr),\ R^i_{\sigma,\varepsilon}=\sum_{j=1}^d\bigl(\partial_{x_j}(a^{ij}\sigma)_\varepsilon-\partial_{x_j}(a^{ij}\sigma_\varepsilon)\bigr).$$

The following equalities hold:

$$\partial_t\varrho_\varepsilon=\mathcal{A}^*\varrho_\varepsilon-\operatorname{div}\bigl((b\varrho)_\varepsilon-R_{\varrho,\varepsilon}\bigr)+(c\varrho)_\varepsilon,\tag{9.5.1}$$

$$\partial_t\sigma_\varepsilon=\mathcal{A}^*\sigma_\varepsilon-\operatorname{div}\bigl((b\sigma)_\varepsilon-R_{\sigma,\varepsilon}\bigr)+(c\sigma)_\varepsilon.\tag{9.5.2}$$

Let us fix a pair of numbers s,t with $0<s<t<T$. Equation (9.5.2) is equivalent to the integral identity

$$\int_s^t\!\!\int_{\mathbb{R}^d}\bigl[-\partial_t(v^\varepsilon\varrho_\varepsilon)\varphi+\varrho_\varepsilon v^\varepsilon\mathcal{A}\varphi+\langle(b\sigma)_\varepsilon-R_{\sigma,\varepsilon},\nabla\varphi\rangle+(c\sigma)_\varepsilon\varphi\bigr]\,dx\,d\tau=0$$

for every continuous function φ on $\mathbb{R}^d\times(0,T)$, having continuous derivatives $\partial_{x_i}\varphi$, $\partial_{x_i}\partial_{x_j}\varphi$ and vanishing for all x outside some ball $U\subset\mathbb{R}^d$.

Let f be any of the two functions $e^{\lambda(1-z)}-e^\lambda$ and $e^{\lambda(1-z)}$. Let us also take a function $\psi\in C_0^\infty(\mathbb{R}^d)$. Set $\varphi=f'(v^\varepsilon)\psi$. We observe that

$$\partial_t\bigl(v^\varepsilon\varrho_\varepsilon\bigr)f'(v^\varepsilon)=\partial_t\bigl(f(v^\varepsilon)\varrho_\varepsilon\bigr)-\bigl(f(v_\varepsilon)-v_\varepsilon f'(v_\varepsilon)\bigr)\partial_t\varrho_\varepsilon.$$

Substituting φ of the indicated form in the integral identity, we arrive at the following equality:

$$\begin{aligned}\int_s^t\!\!\int_{\mathbb{R}^d}\partial_t(f(v^\varepsilon)\varrho_\varepsilon)\psi\,dx\,d\tau&=\int_s^t\!\!\int_{\mathbb{R}^d}(f(v^\varepsilon)-v^\varepsilon f'(v^\varepsilon))\psi\partial_t(\varrho_\varepsilon)\,dx\,d\tau\\&+\int_s^t\!\!\int_{\mathbb{R}^d}\Bigl[\varrho_\varepsilon v^\varepsilon\mathcal{A}(f'(v^\varepsilon)\psi)+f'(v^\varepsilon)\langle(b\sigma)_\varepsilon-R_{\sigma,\varepsilon},\nabla\psi\rangle\\&\quad+\psi f''(v^\varepsilon)\langle(b\sigma)_\varepsilon-R_{\sigma,\varepsilon},\nabla v^\varepsilon\rangle+(c\sigma)_\varepsilon f'(v^\varepsilon)\psi\Bigr]\,dx\,d\tau.\end{aligned}\tag{9.5.3}$$

Equation (9.5.1) is equivalent to the integral identity

$$\int_s^t\int_{\mathbb{R}^d} \partial_t\varrho_\varepsilon\varphi\,dx\,d\tau = \int_s^t\int_{\mathbb{R}^d} \bigl[\varrho_\varepsilon\mathcal{A}\varphi + \langle(b\varrho)_\varepsilon - R_{\varrho,\varepsilon}, \nabla\varphi\rangle + (c\varrho)_\varepsilon\varphi\bigr]\,dx\,d\tau$$

for every continuous function φ on $\mathbb{R}^d\times(0,T)$, having continuous derivatives $\partial_{x_i}\varphi$, $\partial_{x_i}\partial_{x_j}\varphi$ and vanishing for all x outside some ball $U\subset\mathbb{R}^d$. Set

$$\varphi = \bigl(f(v^\varepsilon) - v^\varepsilon f'(v^\varepsilon)\bigr)\psi.$$

Substituting such a function φ in the integral identity, we have

$$\begin{aligned}\int_s^t\int_{\mathbb{R}^d} \bigl(f(v^\varepsilon) - v^\varepsilon f'(v^\varepsilon)\bigr)\psi\partial_t(\varrho_\varepsilon)\,dx\,d\tau &= \int_s^t\int_{\mathbb{R}^d}\Bigl[\varrho_\varepsilon\mathcal{A}\bigl((f(v^\varepsilon) - v^\varepsilon f'(v^\varepsilon))\psi\bigr)\\ &+ \bigl(f(v^\varepsilon) - v^\varepsilon f'(v^\varepsilon)\bigr)\langle(b\varrho)_\varepsilon - R_{\varrho,\varepsilon}, \nabla\psi\rangle + (c\varrho)_\varepsilon\bigl(f(v^\varepsilon) - v^\varepsilon f'(v^\varepsilon)\bigr)\psi\\ &- \psi v^\varepsilon f''(v^\varepsilon)\langle(b\varrho)_\varepsilon - R_{\varrho,\varepsilon}, \nabla v^\varepsilon\rangle\Bigr]\,dx\,d\tau. \qquad (9.5.4)\end{aligned}$$

For all functions $\xi,\eta\in C^\infty\bigl(\mathbb{R}^d\times(0,T)\bigr)$ and $\Phi\in C^\infty(\mathbb{R})$ we have

$$\mathcal{A}(\xi\eta) = \eta\mathcal{A}\xi + \xi\mathcal{A}\eta + 2\langle A\nabla\xi,\nabla\eta\rangle,\ \ \mathcal{A}\Phi(\xi) = \Phi'(\xi)\mathcal{A}\xi + \Phi''(\xi)\langle A\nabla\xi,\nabla\xi\rangle.$$

Using these relations, we arrive at the equality

$$\mathcal{A}\bigl([f(v^\varepsilon) - v^\varepsilon f'(v^\varepsilon)]\psi\bigr) = f(v^\varepsilon)\mathcal{A}\psi - \langle A\nabla v^\varepsilon, \nabla v^\varepsilon\rangle f''(v^\varepsilon)\psi - v^\varepsilon\mathcal{A}\bigl(f'(v^\varepsilon)\psi\bigr).$$

Summing equalities (9.5.4) and (9.5.3), we obtain

$$\begin{aligned}\int_s^t\int_{\mathbb{R}^d} \partial_t(f(v^\varepsilon)\varrho_\varepsilon)\psi\,dx\,d\tau &+ \int_s^t\int_{\mathbb{R}^d}\langle A\nabla v^\varepsilon, \nabla v^\varepsilon\rangle f''(v^\varepsilon)\varrho_\varepsilon\psi\,dx\,d\tau \qquad (9.5.5)\\ &= \int_s^t\int_{\mathbb{R}^d}\bigl[\varrho_\varepsilon f(v^\varepsilon)\mathcal{A}\psi + f(v^\varepsilon)\langle(b\varrho)_\varepsilon, \nabla\psi\rangle + f(v^\varepsilon)(c\varrho)_\varepsilon\psi\bigr]\,dx\,d\tau\\ &\qquad + K_\varepsilon(s,t) + L_\varepsilon(s,t),\end{aligned}$$

where we use the function

$$K_\varepsilon(s,t) = \int_s^t\int_{\mathbb{R}^d}\bigl[\langle W_\varepsilon, \nabla\psi\rangle f'(v^\varepsilon) + \langle W_\varepsilon, \nabla v^\varepsilon\rangle f''(v^\varepsilon)\psi + Y_\varepsilon f'(v^\varepsilon)\psi\bigr]\,dx\,d\tau,$$

in the definition of which we have set

$$W_\varepsilon = (b\sigma)_\varepsilon - (b\varrho)_\varepsilon v^\varepsilon, \quad Y_\varepsilon = (c\sigma)_\varepsilon - (c\varrho)_\varepsilon v^\varepsilon,$$

and also the function L_ε defined by the formula

$$\begin{aligned}L_\varepsilon(s,t) = -\int_s^t\int_{\mathbb{R}^d}&\bigl[f'(v^\varepsilon)\langle R_{\sigma,\varepsilon}, \nabla\psi\rangle + \bigl(f(v^\varepsilon)\\ &- v^\varepsilon f'(v^\varepsilon)\bigr)(R_{\varrho,\varepsilon}, \nabla\psi) + \psi f''(v^\varepsilon)\langle R_{\sigma,\varepsilon} - v^\varepsilon R_{\varrho,\varepsilon}, \nabla v^\varepsilon\rangle\bigr]\,dx\,d\tau.\end{aligned}$$

We assume further that the parameter ε takes values of the form $1/n$, where n is a natural number.

Let us prove two auxiliary lemmas.

9.5.1. Lemma. *Let $\lambda>0$. Then for every ball $U\subset\mathbb{R}^d$, whenever $0<t<1$, we have*

$$\lim_{\varepsilon\to0}\int_0^t\int_U(|W_\varepsilon| + |Y_\varepsilon|)e^{-\lambda v^\varepsilon}\,dx\,ds = 0,\ \lim_{\varepsilon\to0}\int_0^t\int_U\frac{|W_\varepsilon|^2}{\varrho_\varepsilon}e^{-\lambda v^\varepsilon}\,dx\,ds = 0.$$

PROOF. Let us fix a ball U' such that $U \subset U'$ and for every $y \in U$ and $\varepsilon \in (0,1)$ the support of the function $x \mapsto \omega_\varepsilon(x-y)$ belongs to the ball U'. Let us prove the first equality. We fix s such that the functions $\big[|b(\,\cdot\,,s)| + |b(\,\cdot\,,s)|^2 + c(\,\cdot\,,s)\big]\varrho(\,\cdot\,,s)$ and $\big[|b(\,\cdot\,,s)| + |b(\,\cdot\,,s)|^2 + c(\,\cdot\,,s)\big]\sigma(\,\cdot\,,s)$ belong to $L^1(U')$. By Fubini's theorem, the set of those s where these inclusions fail has measure zero. We observe that for almost all x we have

$$\lim_{\varepsilon\to 0} W_\varepsilon(x,s) = \lim_{\varepsilon\to 0}\big[(b\sigma)_\varepsilon(x,s) - (b\varrho)_\varepsilon(x,s)v^\varepsilon(x,s)\big] = 0.$$

Similarly we have

$$\lim_{\varepsilon\to 0} Y_\varepsilon(x,s) = \lim_{\varepsilon\to 0}\big[(c\sigma)_\varepsilon(x,s) - (c\varrho)_\varepsilon(x,s)v^\varepsilon(x,s)\big] = 0.$$

Therefore, in order to verify convergence it suffices to prove the uniform integrability. Note that $(b\varrho)_\varepsilon$, $(b\sigma)_\varepsilon$, $(c\varrho)_\varepsilon$ and $(c\sigma)_\varepsilon$ are uniformly integrable, since they converge in $L^1(U')$. In addition, the functions $|v^\varepsilon|e^{-\lambda v^\varepsilon}$, $e^{-\lambda v^\varepsilon}$ are bounded uniformly in ε. It is clear that the product of functions that are uniformly bounded in ε by functions that are uniformly integrable is also uniformly integrable. Therefore, for almost all s we have

$$\lim_{\varepsilon\to 0}\int_U (|W_\varepsilon| + |Y_\varepsilon|)e^{-\lambda v^\varepsilon}\,dx = 0.$$

In addition, there exists a constant C independent of ε such that

$$\int_U (|W_\varepsilon| + |Y_\varepsilon|)e^{-\lambda v^\varepsilon}\,dx \leqslant C\Big(\|b(\,\cdot\,,s)\varrho(\,\cdot\,,s)\|_{L^1(U')} + \|b(\,\cdot\,,s)\sigma(\,\cdot\,,s)\|_{L^1(U')}$$
$$+ \|c(\,\cdot\,,s)\varrho(\,\cdot\,,s)\|_{L^1(U')} + \|c(\,\cdot\,,s)\sigma(\,\cdot\,,s)\|_{L^1(U')}\Big).$$

By the Lebesgue dominated convergence theorem we obtain the first assertion. Let us proceed to the second assertion. Let us fix a number s as above. By the Cauchy–Bunyakovsky inequality we have

$$|(b\varrho)_\varepsilon|^2/\varrho_\varepsilon \leqslant (|b|^2\varrho)_\varepsilon, \quad |(b\sigma)_\varepsilon|^2/\varrho_\varepsilon \leqslant (|b|^2\sigma)_\varepsilon v^\varepsilon.$$

Hence

$$|W_\varepsilon|^2 e^{-\lambda v^\varepsilon}/\varrho_\varepsilon \leqslant 2(|b|^2\varrho)_\varepsilon |v^\varepsilon|^2 e^{-\lambda v^\varepsilon} + 2(|b|^2\sigma)_\varepsilon v^\varepsilon e^{-\lambda v^\varepsilon}.$$

Since the families of functions $(|b|^2\sigma)_\varepsilon$ and $(|b|^2\varrho)_\varepsilon$ are uniformly integrable, and the functions $|v^\varepsilon|^2 e^{-\lambda\varepsilon}$ and $v^\varepsilon e^{-\lambda v^\varepsilon}$ are uniformly bounded in ε, the family of functions

$$2(|b|^2\varrho)_\varepsilon |v^\varepsilon|^2 e^{-\lambda v^\varepsilon} + 2(|b|^2\sigma)_\varepsilon v^\varepsilon e^{-\lambda v^\varepsilon}$$

is uniformly integrable. Therefore, the family of functions $|W_\varepsilon|^2 e^{-\lambda v^\varepsilon}/\varrho_\varepsilon$ is uniformly integrable as well. These functions for almost all x tend to zero as $\varepsilon \to 0$. Therefore, for almost all s we have

$$\lim_{\varepsilon\to 0}\int_U \frac{|W_\varepsilon|^2}{\varrho_\varepsilon} e^{-\lambda v^\varepsilon}\,dx = 0.$$

In addition, there exists a constant C independent of ε for which

$$\int_U \frac{|W_\varepsilon|^2}{\varrho_\varepsilon} e^{-\lambda v^\varepsilon}\,dx \leqslant C\Big(\int_{U'} |b|^2\varrho\,dx + \int_{U'} |b|^2\sigma\,dx\Big).$$

By the Lebesgue dominated convergence theorem the integral of the left side over $[0,t]$ tends to zero as $\varepsilon \to 0$. □

9.5.2. Lemma. *Let $\lambda > 0$. Then, for every ball $U \subset \mathbb{R}^d$, whenever $0 < t < 1$, we have*

(i) $\displaystyle\lim_{\varepsilon\to 0}\int_0^t\int_U |R_{\varrho,\varepsilon}|\,dx\,ds = 0, \quad \lim_{\varepsilon\to 0}\int_0^t\int_U |R_{\sigma,\varepsilon}|\,dx\,ds = 0,$

(ii) $\displaystyle\lim_{\varepsilon\to 0}\int_0^t\int_U \frac{|R_{\varrho,\varepsilon}|^2}{\varrho_\varepsilon}\big[|v^\varepsilon|^2+1\big]e^{-\lambda v^\varepsilon}\,dx\,ds = 0.$

PROOF. Let us fix a ball U' such that $U \subset U'$ and for every $y \in U$ and $\varepsilon \in (0,1)$ the support of the function $x \mapsto \omega_\varepsilon(x-y)$ belongs to the ball U'. Let us prove the first assertion. We recall that by assumption the functions a^{ij} are Lipschitzian in the variable x, hence have bounded Sobolev derivatives in x. In addition, for almost all s the function $\varrho(\,\cdot\,,s)$ belongs to the Sobolev class $W^{p,1}(U')$. Let us fix s such that for all i,j the functions $\varrho(\,\cdot\,,s)\partial_{x_j}a^{ij}(\,\cdot\,,s)$ and $a^{ij}(\,\cdot\,,s)\partial_{x_j}\varrho(\,\cdot\,,s)$ belong to $L^1(U)$. The set of those s for which this is not true has measure zero. We observe (see p. 358 for the definition of $R^i_{\varrho,\varepsilon}$) that

$$R^i_{\varrho,\varepsilon} = \sum_{j=1}^d\big[(\varrho\partial_{x_j}a^{ij})_\varepsilon + (a^{ij}\partial_{x_j}\varrho)_\varepsilon - \varrho_\varepsilon\partial_{x_j}a^{ij} - a^{ij}(\partial_{x_j}\varrho)_\varepsilon\big].$$

For almost all x we have $R^i_{\varrho,\varepsilon}(x,s) \to 0$ as $\varepsilon \to 0$. In addition, for any fixed s the family of functions $R^i_{\varrho,\varepsilon}(\,\cdot\,,s)$, $\varepsilon \in (0,1)$ is uniformly integrable. Therefore, for almost all s we have the equality

$$\lim_{\varepsilon\to 0}\|R_{\varrho,\varepsilon}\|_{L^1(U)} = 0.$$

We now write $R_{\varrho,\varepsilon}$ in another form:

$$\begin{aligned}R^i_{\varrho,\varepsilon}(x,s) &= -\varrho_\varepsilon(x,s)\sum_{j=1}^d \partial_{x_j}a^{ij}(x,s)\\ &\quad+\sum_{j=1}^d\int_{\mathbb{R}^d}\big(a^{ij}(y,s)-a^{ij}(x,s)\big)\varrho(y,s)\varepsilon^{-d-1}\partial_{x_j}\omega\Big(\frac{x-y}{\varepsilon}\Big)\,dy.\end{aligned}$$

Using condition (H1), we choose the corresponding number $\Lambda = \Lambda(U') > 0$ and obtain

$$|R^i_{\varrho,\varepsilon}(x,s)| \leqslant \Lambda d\int_{\mathbb{R}^d}\varrho(y,s)\varepsilon^{-d}\frac{|x-y|}{\varepsilon}\Big|\nabla\omega\Big(\frac{x-y}{\varepsilon}\Big)\Big|\,dy + \Lambda d\varrho_\varepsilon(x,s).$$

Integrating in x and taking into account that $\varrho(\,\cdot\,,s)$ is a probability density, we arrive at the inequality

$$\int_U |R^i_{\varrho,\varepsilon}(x,s)|\,dx \leqslant \Lambda d\int_{\mathbb{R}^d}\big[|x|\,|\nabla\omega(x)| + |\omega(x)|\big]\,dx.$$

By the Lebesgue dominated convergence theorem we have

$$\lim_{\varepsilon\to 0}\int_0^t\int_U |R_{\varrho,\varepsilon}|\,dx\,ds = 0.$$

A similar reasoning gives the same for $R_{\sigma,\varepsilon}$. Thus, assertion (i) is proven. Let us prove (ii). Let us fix a number s as above. For almost all x we have

$$\lim_{\varepsilon\to 0}|R_{\varrho,\varepsilon}(x,s)|^2|v^\varepsilon(x,s)|^2e^{-\lambda v^\varepsilon(x,s)}/\varrho_\varepsilon(x,s) = 0.$$

In addition, as already noted above, there holds the inequality

$$|R^i_{\varrho,\varepsilon}(x,s)| \leqslant d\Lambda \int_{\mathbb{R}^d} \varrho(y,s)\varepsilon^{-d}\frac{|x-y|}{\varepsilon}\Big|\nabla\omega\Big(\frac{x-y}{\varepsilon}\Big)\Big|\,dy + d\Lambda\varrho_\varepsilon(x,s).$$

Applying the Cauchy–Bunyakovsky inequality, we obtain

$$\begin{aligned}\int_{\mathbb{R}^d} \varrho(y,s)\varepsilon^{-d}\frac{|x-y|}{\varepsilon}\Big|\nabla\omega\Big(\frac{x-y}{\varepsilon}\Big)\Big|\,dy &= \int_{\mathbb{R}^d} \varrho(x-\varepsilon z,s)|z|\,|\nabla\omega(z)|\,dz\\ &= \int_{\mathbb{R}^d} \big(\varrho(x-\varepsilon z,s)\omega(z)\big)^{1/2}|z|\,|\nabla\omega(z)|\frac{\varrho^{1/2}(x-\varepsilon z,s)}{\omega^{1/2}(z)}\,dz\\ &\leqslant \Big(\int_{\mathbb{R}^d}\varrho(x-\varepsilon z,s)\omega(z)\,dz\int_{\mathbb{R}^d}\varrho(x-\varepsilon z,s)|z|^2\frac{|\nabla\omega(z)|^2}{\omega(z)}\,dz\Big)^{1/2}.\end{aligned}$$

Thus, for some constant C independent of ε we have

$$|R_{\varrho,\varepsilon}(x,s)|^2/\varrho_\varepsilon(x,s) \leqslant C\varrho * \kappa_\varepsilon(x,s),$$

where $\kappa_\varepsilon(x) = \varepsilon^{-d}\kappa(x/\varepsilon)$ and $\kappa(x) = c_d\big(|x|^2|\nabla\omega(x)|^2\omega^{-1}(z) + \omega(x)\big)$, c_d is a normalization constant such that the integral of κ equals 1. Therefore, the function κ is a new averaging kernel. The family of functions $\varrho * \kappa_\varepsilon$, where $\varepsilon \in (0,1)$, is uniformly integrable, since has a limit in L^1 as $\varepsilon \to 0$. Therefore, the functions

$$|R_{\varrho,\varepsilon}(x,s)|^2/\varrho_\varepsilon(x,s), \quad \varepsilon \in (0,1)$$

are uniformly integrable. Since the functions $|v^\varepsilon|^2 e^{-\lambda v^\varepsilon}$ are bounded uniformly in ε, for almost all s we have the equality

$$\lim_{\varepsilon\to 0}\int_U \frac{|R_{\varrho,\varepsilon}(x,s)|^2}{\varrho_\varepsilon(x,s)}|v^\varepsilon(x,s)|^2 e^{-\lambda v^\varepsilon(x,s)}\,dx = 0.$$

In addition, we have the estimate

$$\int_U \frac{|R_{\varrho,\varepsilon}|^2}{\varrho_\varepsilon}|v^\varepsilon|^2 e^{-\lambda v^\varepsilon}\,dx \leqslant C(\lambda)\int_{\mathbb{R}^d}\kappa(x)\,dx.$$

By the Lebesgue dominated convergence theorem

$$\lim_{\varepsilon\to 0}\int_0^t\int_U \frac{|R_{\varrho,\varepsilon}|^2}{\varrho_\varepsilon}|v^\varepsilon|^2 e^{-\lambda v^\varepsilon}\,dx\,ds = 0.$$

Let us prove the analogous equality for σ. We fix a number s such that

$$\sigma(\,\cdot\,,s)\partial_{x_j}a^{ij}(\,\cdot\,,s),\ a^{ij}(\,\cdot\,,s)\partial_{x_j}\sigma(\,\cdot\,,s) \in L^1(U).$$

The set of those points s for which this is not true has measure zero. For almost each x the function

$$|R_{\sigma,\varepsilon}(x,s)|^2 e^{-\lambda v^\varepsilon(x,s)}/\varrho_\varepsilon(x,s)$$

tends to zero as $\varepsilon \to 0$. In addition,

$$|R^i_{\sigma,\varepsilon}(x,s)| \leqslant d\Lambda\int_{\mathbb{R}^d}\sigma(y,s)\varepsilon^{-d}\frac{|x-y|}{\varepsilon}\Big|\nabla\omega\Big(\frac{x-y}{\varepsilon}\Big)\Big|\,dy + d\Lambda\sigma_\varepsilon(x,s).$$

Applying the Cauchy–Bunyakovsky inequality, we obtain that for some constant C independent of ε one has

$$|R_{\sigma,\varepsilon}(x,s)|^2/\varrho_\varepsilon \leqslant Cv^\varepsilon(x,s)(\sigma * \kappa_\varepsilon)(x,s).$$

The family of functions $\sigma * \kappa_\varepsilon$ parametrized by ε is uniformly integrable, and the functions $v^\varepsilon e^{-\lambda v^\varepsilon}$ are bounded uniformly in ε. Therefore, the family of functions

$|R_{\varrho,\varepsilon}(x,s)|^2 e^{-\lambda v^\varepsilon}/\varrho_\varepsilon$ parametrized by ε is uniformly integrable and for almost all s we have

$$\lim_{\varepsilon\to 0}\int_U \frac{|R_{\sigma,\varepsilon}(x,s)|^2}{\varrho_\varepsilon(x,s)} e^{-\lambda v^\varepsilon(x,s)}\,dx = 0.$$

In addition,

$$\int_U \frac{|R_{\sigma,\varepsilon}|^2}{\varrho_\varepsilon} e^{-\lambda v^\varepsilon}\,dx \leqslant C\int_{\mathbb{R}^d}\kappa(x)\,dx.$$

By the Lebesgue dominated convergence theorem

$$\lim_{\varepsilon\to 0}\int_0^t\!\!\int_U \frac{|R_{\sigma,\varepsilon}|^2}{\varrho_\varepsilon} e^{-\lambda v^\varepsilon}\,dx\,ds = 0.$$

Therefore, assertion (ii) is proved. □

Let us continue the proof of Lemma 9.4.2. Let $\delta > 0$ and let U be an arbitrary ball containing the support of the function ψ. Let us estimate the term $K_\varepsilon(s,t)$ in (9.5.5). We recall that $f'' \geqslant 0$. We have

$$\begin{aligned} K_\varepsilon(s,t) \leqslant & \int_0^1\!\!\int_{\mathbb{R}^d} |W_\varepsilon|\,|\nabla\psi|\,|f'(v^\varepsilon)|\,dx\,d\tau \\ & + \int_s^t\!\!\int_{\mathbb{R}^d} |W_\varepsilon|\,|\nabla v^\varepsilon| f''(v^\varepsilon)\psi\,dx\,d\tau + \int_s^t\!\!\int_{\mathbb{R}^d} |Y_\varepsilon|\,|f'(v^\varepsilon)|\psi\,dx\,d\tau. \end{aligned}$$

We can assume that $|\psi| \leqslant 1$, $|\nabla\psi| \leqslant 1$. Applying the inequality $qr \leqslant \delta q^2 + \delta^{-1} r^2$ to the numbers

$$q = |\varrho_\varepsilon|^{1/2}|f''(v^\varepsilon)|^{1/2}|\psi|^{1/2}|\nabla v^\varepsilon| \quad\text{and}\quad r = |\varrho_\varepsilon|^{-1/2}|f''(v^\varepsilon)|^{1/2}|\psi|^{1/2}|W_\varepsilon|,$$

we obtain the following inequality:

$$\begin{aligned} & \int_s^t\!\!\int_{\mathbb{R}^d} |W_\varepsilon|\,|\nabla v^\varepsilon| f''(v^\varepsilon)\psi\,dx\,d\tau \\ & \leqslant \delta\int_s^t\!\!\int_{\mathbb{R}^d} \varrho_\varepsilon |\nabla v^\varepsilon|^2 f''(v^\varepsilon)\psi\,dx\,d\tau + \frac{1}{\delta}\int_s^t\!\!\int_{\mathbb{R}^d} \frac{|W_\varepsilon|^2}{\varrho_\varepsilon} f''(v_\varepsilon)\psi\,dx\,d\tau. \end{aligned}$$

Taking into account the equalities $f'(z) = -\lambda e^{\lambda(1-z)}$ and $f''(z) = \lambda^2 e^{\lambda(1-z)}$, we have

$$\begin{aligned} K_\varepsilon(s,t) \leqslant & \lambda e^\lambda \int_0^t\!\!\int_U |W_\varepsilon| e^{-\lambda v^\varepsilon}\,dx\,d\tau + \lambda e^\lambda \int_0^t\!\!\int_U |Y_\varepsilon| e^{-\lambda v^\varepsilon}\,dx\,d\tau \\ & + \frac{\lambda^2 e^\lambda}{4\delta}\int_0^t\!\!\int_U \frac{|W_\varepsilon|^2}{\varrho_\varepsilon} e^{-\lambda v_\varepsilon}\,dx\,d\tau + \delta\int_s^t\!\!\int_{\mathbb{R}^d} \varrho_\varepsilon |\nabla v^\varepsilon|^2 f''(v^\varepsilon)\psi\,dx\,d\tau. \end{aligned}$$

Let us estimate the term L_ε in (9.5.5). Applying the same inequality with

$$q = |\varrho_\varepsilon|^{1/2}|f''(v^\varepsilon)|^{1/2}|\psi|^{1/2}|\nabla v^\varepsilon|, \quad r = |\varrho_\varepsilon|^{-1/2}|f''(v^\varepsilon)|^{1/2}|\psi|^{1/2}|R_{\sigma,\varepsilon}|$$

in the term with $R_{\sigma,\varepsilon}$ and also with

$$r = |\varrho_\varepsilon|^{-1/2}|f''(v^\varepsilon)|^{1/2}|\psi|^{1/2}|v^\varepsilon|\,|R_{\varrho,\varepsilon}|$$

and the same q in the term $R_{\varrho,\varepsilon}$, we arrive at the estimate

$$\Bigl|\int_s^t\int_{\mathbb{R}^d} \psi f''(v^\varepsilon)\langle R_{\sigma,\varepsilon} - v^\varepsilon R_{\varrho,\varepsilon}, \nabla v^\varepsilon\rangle\,dx\,d\tau\Bigr| \leqslant \frac{\lambda^2}{\delta}e^\lambda \int_0^t\int_U \frac{|R_{\varrho,\varepsilon}|^2}{\varrho_\varepsilon}\bigl[|v^\varepsilon|^2+1\bigr]e^{-\lambda v^\varepsilon}\,dx\,d\tau + 2\delta\int_s^t\int_{\mathbb{R}^d} \varrho_\varepsilon|\nabla v^\varepsilon|^2 f''(v^\varepsilon)\psi\,dx\,d\tau.$$

This gives the estimate

$$L_\varepsilon(s,t) \leqslant 2\lambda e^\lambda \int_0^t\int_U \bigl[|R_{\sigma,\varepsilon}|+|R_{\varrho,\varepsilon}|\bigr]\,dx\,d\tau + \frac{\lambda^2}{\delta}e^\lambda \int_0^t\int_U \frac{|R_{\varrho,\varepsilon}|^2}{\varrho_\varepsilon}\bigl[|v^\varepsilon|^2+1\bigr]e^{-\lambda v^\varepsilon}\,dx\,d\tau + 2\delta\int_s^t\int_{\mathbb{R}^d} \varrho_\varepsilon|\nabla v^\varepsilon|^2 f''(v^\varepsilon)\psi\,dx\,d\tau.$$

Substituting the obtained estimates in (9.5.5), we arrive at the inequality

$$\int_s^t\int_{\mathbb{R}^d} \partial_t(f(v^\varepsilon)\varrho_\varepsilon)\psi\,dx\,d\tau + \int_s^t\int_{\mathbb{R}^d} \langle A\nabla v^\varepsilon, \nabla v^\varepsilon\rangle f''(v^\varepsilon)\varrho_\varepsilon\psi\,dx\,d\tau \leqslant \int_s^t\int_{\mathbb{R}^d} \bigl[\varrho_\varepsilon f(v^\varepsilon)\mathcal{A}\psi + f(v^\varepsilon)\langle(b\varrho)_\varepsilon, \nabla\psi\rangle\bigr]\,dx\,d\tau + 3\delta\int_s^t\int_{\mathbb{R}^d} \varrho_\varepsilon|\nabla v^\varepsilon|^2 f''(v^\varepsilon)\psi\,dx\,d\tau + \Omega(\varepsilon), \tag{9.5.6}$$

where $\Omega(\varepsilon)$ is defined by the following expression:

$$\Omega(\varepsilon) = 2\lambda e^\lambda \int_0^t\int_U \bigl[|R_{\sigma,\varepsilon}|+|R_{\varrho,\varepsilon}|\bigr]\,dx\,d\tau + \frac{\lambda^2}{\delta}e^\lambda \int_0^t\int_U \frac{|R_{\sigma,\varepsilon}|^2}{\varrho_\varepsilon}e^{-\lambda v^\varepsilon}\,dx\,d\tau + \lambda e^\lambda\int_0^t\int_U |W_\varepsilon|e^{-\lambda v^\varepsilon}\,dx\,d\tau + \frac{\lambda^2 e^\lambda}{4\delta}\int_0^t\int_U \frac{|W_\varepsilon|^2}{\varrho_\varepsilon}e^{-\lambda v_\varepsilon}\,dx\,d\tau + \lambda e^\lambda\int_0^t\int_U |Y_\varepsilon|e^{-\lambda v^\varepsilon}\,dx\,d\tau.$$

By Lemmas 9.5.1 and 9.5.2 we have $\lim\limits_{\varepsilon\to 0}\Omega(\varepsilon) = 0$. The Newton–Leibniz formula gives the equality

$$\int_s^t\int_{\mathbb{R}^d} \partial_t(f(v^\varepsilon)\varrho_\varepsilon)\psi\,dx\,d\tau = \int_{\mathbb{R}^d} f\bigl(v^\varepsilon(x,t)\bigr)\varrho_\varepsilon(x,t)\psi(x)\,dx - \int_{\mathbb{R}^d} f\bigl(v^\varepsilon(x,s)\bigr)\varrho_\varepsilon(x,s)\psi(x)\,dx.$$

In addition, by condition (H2), for the ball U there is a number $m=m(U)>0$ such that

$$\langle A\nabla v^\varepsilon, \nabla v^\varepsilon\rangle \geqslant m|\nabla v^\varepsilon|^2 \quad \text{on } U.$$

Since $f''\geqslant 0$, we obtain the following inequality:

$$\int_s^t\int_{\mathbb{R}^d} \langle A\nabla v^\varepsilon, \nabla v^\varepsilon\rangle f''(v^\varepsilon)\varrho_\varepsilon\psi\,dx\,d\tau \geqslant \int_s^t\int_{\mathbb{R}^d} \Bigl[\frac{3}{4}\langle A\nabla v^\varepsilon, \nabla v^\varepsilon\rangle f''(v^\varepsilon)\varrho_\varepsilon\psi + \frac{m}{4}\varrho_\varepsilon|\nabla v^\varepsilon|^2 f''(v^\varepsilon)\psi\Bigr]\,dx\,d\tau.$$

Thus, using (9.5.6) and letting $\delta = m/12$ above, we arrive at the inequality

$$
\begin{aligned}
&\int_{\mathbb{R}^d} f\bigl(v^\varepsilon(x,t)\bigr)\varrho_\varepsilon(x,t)\psi(x)\,dx+\frac{3}{4}\int_s^t\!\int_{\mathbb{R}^d}\varrho_\varepsilon\langle A\nabla v^\varepsilon,\nabla v^\varepsilon\rangle f''(v^\varepsilon)\psi\,dx\,d\tau\\
&\leqslant \int_{\mathbb{R}^d} f\bigl(v^\varepsilon(x,s)\bigr)\varrho_\varepsilon(x,s)\psi(x)\,dx+\Omega(\varepsilon)\\
&\quad+\int_s^t\!\int_{\mathbb{R}^d}\Bigl[\varrho_\varepsilon f(v^\varepsilon)\mathcal{A}\psi+f(v^\varepsilon)\langle(b\varrho)_\varepsilon,\nabla\psi\rangle+f(v^\varepsilon)(c\varrho)_\varepsilon\psi\Bigr]\,dx\,d\tau.
\end{aligned}
\tag{9.5.7}
$$

Dropping the nonnegative term in the left-hand side of the latter inequality, we obtain

$$
\begin{aligned}
&\int_{\mathbb{R}^d} f\bigl(v^\varepsilon(x,t)\bigr)\varrho_\varepsilon(x,t)\psi(x)\,dx\leqslant\int_{\mathbb{R}^d} f\bigl(v^\varepsilon(x,s)\bigr)\varrho_\varepsilon(x,s)\psi(x)\,dx+\Omega(\varepsilon)\\
&\quad+\int_s^t\!\int_{\mathbb{R}^d}\bigl[\varrho_\varepsilon f(v^\varepsilon)\mathcal{A}\psi+f(v^\varepsilon)\langle(b\varrho)_\varepsilon,\nabla\psi\rangle+f(v^\varepsilon)(c\varrho)_\varepsilon\psi\bigr]\,dx\,d\tau.
\end{aligned}
\tag{9.5.8}
$$

We observe that for any fixed ε the function

$$
f(v^\varepsilon)\varrho_\varepsilon\mathcal{A}\psi+f(v^\varepsilon)\langle(b\varrho)_\varepsilon,\nabla\psi\rangle+f(v^\varepsilon)(c\varrho)_\varepsilon\psi
$$

is integrable on $\mathbb{R}^d\times(0,T)$. Hence

$$
\begin{aligned}
&\lim_{s\to 0}\int_s^t\!\int_{\mathbb{R}^d} f(v^\varepsilon)\bigl[\varrho_\varepsilon\mathcal{A}\psi+\langle(b\varrho)_\varepsilon,\nabla\psi\rangle+(c\varrho)_\varepsilon\psi\bigr]\,dx\,d\tau\\
&\qquad=\int_0^t\!\int_{\mathbb{R}^d} f(v^\varepsilon)\bigl[\varrho_\varepsilon\mathcal{A}\psi+\langle(b\varrho)_\varepsilon,\nabla\psi\rangle+(c\varrho)_\varepsilon\psi\bigr]\,dx\,d\tau.
\end{aligned}
\tag{9.5.9}
$$

Let us show that

$$
\lim_{s\to 0}\int_{\mathbb{R}^d} f\bigl(v^\varepsilon(x,s)\bigr)\varrho_\varepsilon(x,s)\psi(x)\,dx=f(1)\int_{\mathbb{R}^d}(\nu*\omega_\varepsilon)\psi\,dx.
\tag{9.5.10}
$$

Since the function f is smooth and its derivative is bounded on $[0,+\infty)$ by the number λe^λ, we have $|f(z)-f(1)|\leqslant\lambda e^\lambda|z-1|$ for all $z\geqslant 0$. We obtain the following inequality:

$$
\int_{\mathbb{R}^d}\bigl|f\bigl(v^\varepsilon(x,s)\bigr)-f(1)\bigr|\varrho_\varepsilon(x,s)\psi(x)\,dx\leqslant\lambda e^\lambda\int_{\mathbb{R}^d}|\sigma_\varepsilon(x,s)-\varrho_\varepsilon(x,s)|\psi(x)\,dx.
$$

As $s\to 0$, the last integral tends to zero, moreover,

$$
\lim_{s\to 0}\int_{\mathbb{R}^d}\varrho_\varepsilon(x,s)\psi(x)\,dx=f(1)\int_{\mathbb{R}^d}(\nu*\omega_\varepsilon)\psi\,dx,
$$

which gives (9.5.10). Passing in inequality (9.5.8) to the limit as $s\to 0$ and using relations (9.5.9), (9.5.10), we obtain that

(9.5.11)

$$
\begin{aligned}
&\int_{\mathbb{R}^d} f\bigl(v^\varepsilon(x,t)\bigr)\varrho_\varepsilon(x,t)\psi(x)\,dx\leqslant f(1)\int_{\mathbb{R}^d}(\nu*\omega_\varepsilon)\psi\,dx\\
&\qquad+\int_0^t\!\int_{\mathbb{R}^d} f(v^\varepsilon)\bigl[\varrho_\varepsilon\mathcal{A}\psi+\langle(b\varrho)_\varepsilon,\nabla\psi\rangle+(c\varrho)_\varepsilon\psi\bigr]\,dx\,d\tau+\Omega(\varepsilon).
\end{aligned}
$$

We observe that, whenever $0<\tau<t$, we have the equality

$$
\lim_{\varepsilon\to 0}\int_{\mathbb{R}^d} f\bigl(v^\varepsilon(x,\tau)\bigr)\varrho_\varepsilon(x,\tau)\mathcal{A}\psi(x)\,dx=\int_{\mathbb{R}^d} f\bigl(v(x,\tau)\bigr)\varrho(x,\tau)\mathcal{A}\psi(x)\,dx
$$

and the estimate

$$\left|\int_{\mathbb{R}^d} f(v^\varepsilon)\varrho_\varepsilon \mathcal{A}\psi\,dx\right| \leqslant 2e^\lambda M \sup_x |D^2\psi(x)|.$$

By the Lebesgue dominated convergence theorem

$$\lim_{\varepsilon\to 0}\int_0^t\int_{\mathbb{R}^d} f(v^\varepsilon)\varrho_\varepsilon \mathcal{A}\psi\,dx\,d\tau = \int_0^t\int_{\mathbb{R}^d} f(v)\varrho\mathcal{A}\psi\,dx\,d\tau.$$

Let U be a ball containing the support of the function ψ. Let us fix a ball U' such that $U\subset U'$ and for every $y\in U$ and $\varepsilon\in(0,1)$ the support of the function $x\mapsto\omega_\varepsilon(x-y)$ belongs to the ball U'. Let us also fix a number $\tau\in(0,t)$ for which

$$b(\,\cdot\,,\tau)\varrho(\,\cdot\,,\tau),\ c(\,\cdot\,,\tau)\varrho(\,\cdot\,,\tau)\in L^1(U).$$

For almost all $x\in U$ the functions $f(v^\varepsilon)(b\varrho)_\varepsilon$ tend to the function $f(v)b\varrho$ and the functions $f(v^\varepsilon)(c\varrho)_\varepsilon$ tend to the function $f(v)c\varrho$ as $\varepsilon\to 0$. In addition, the functions $f(v^\varepsilon)$ are bounded uniformly in ε and the families of functions $(b\varrho)_\varepsilon$ and $(c\varrho)_\varepsilon$ are uniformly integrable. Hence

$$\lim_{\varepsilon\to 0}\int_{\mathbb{R}^d} f(v^\varepsilon)\big[\langle(b\varrho)_\varepsilon,\nabla\psi\rangle+(c\varrho)_\varepsilon\psi\big]\,dx = \int_{\mathbb{R}^d} f(v)\big[\varrho\langle b,\nabla\psi\rangle + c\varrho\big]\,dx.$$

Since $|f(v^\varepsilon)|\leqslant 2e^\lambda$, $|\nabla\psi|\leqslant 1$ and $|\psi|\leqslant 1$, there is a number $C(\lambda)$ such that

$$\left|\int_{\mathbb{R}^d} f(v^\varepsilon)\big[\langle(b\varrho)_\varepsilon,\nabla\psi\rangle+(c\varrho)_\varepsilon\psi\big]\,dx\right| \\ \leqslant C(\lambda)\|b(\,\cdot\,,\tau)\varrho(\,\cdot\,,\tau)\|_{L^1(U')} + C(\lambda)\|c(\,\cdot\,,\tau)\varrho(\,\cdot\,,\tau)\|_{L^1(U')}.$$

The Lebesgue dominated convergence theorem yields

$$\lim_{\varepsilon\to 0}\int_0^t\int_{\mathbb{R}^d} f(v^\varepsilon)\big[\langle(b\varrho)_\varepsilon,\nabla\psi\rangle+(c\varrho)_\varepsilon\psi\big]\,dx\,d\tau \\ = \int_0^t\int_{\mathbb{R}^d} f(v)\varrho\big[\langle b,\nabla\psi\rangle + c\psi\big]\,dx\,d\tau.$$

Letting $\varepsilon\to 0$ in inequality (9.5.11), it is easy to obtain the following estimate:

$$\int_{\mathbb{R}^d} f\big(v(x,t)\big)\varrho(x,t)\psi(x)\,dx \\ \leqslant f(1)\int_{\mathbb{R}^d}\psi\,d\nu + \int_0^t\int_{\mathbb{R}^d}\varrho f(v)\big[\mathcal{A}\psi + \langle b,\nabla\psi\rangle c\psi\big]\,dx\,d\tau.$$

Thus, inequality (9.4.2) is proven. We now obtain (9.4.3). Let us consider the integral of $\varrho_\varepsilon f(v^\varepsilon)\mathcal{A}\psi$ in (9.5.7). We have

$$\int_s^t\int_{\mathbb{R}^d}\varrho_\varepsilon f(v^\varepsilon)\mathcal{A}\psi\,dx\,d\tau = -\int_s^t\int_{\mathbb{R}^d}\big[\varrho_\varepsilon f'(v^\varepsilon)\langle A\nabla\psi,\nabla v^\varepsilon\rangle \\ + f(v^\varepsilon)\langle(\beta_{\mu,A}\varrho)_\varepsilon,\nabla\psi\rangle + f(v^\varepsilon)\langle R_{\varrho,\varepsilon},\nabla\psi\rangle\big]\,dx\,d\tau.$$

We observe that

$$-\varrho_\varepsilon\langle A\nabla v^\varepsilon,\nabla\psi\rangle f'(v^\varepsilon) \leqslant \frac{1}{2}\varrho_\varepsilon\langle A\nabla v^\varepsilon,\nabla v^\varepsilon\rangle f''(v^\varepsilon)\psi \\ + \frac{1}{2}\varrho_\varepsilon\langle A\nabla\psi,\nabla\psi\rangle\psi^{-1}|f'(v^\varepsilon)|^2 f''(v^\varepsilon)^{-1}.$$

Therefore,

$$\begin{aligned}(9.5.12)\quad \int_{\mathbb{R}^d} f\big(v^\varepsilon(x,t)\big)\varrho_\varepsilon(x,t)\psi(x)\,dx \leqslant \int_{\mathbb{R}^d} f\big(v^\varepsilon(x,s)\big)\varrho_\varepsilon(x,s)\psi(x)\,dx \\ +\frac{1}{2}\int_s^t\int_{\mathbb{R}^d} \varrho_\varepsilon\langle A\nabla\psi,\nabla\psi\rangle\psi^{-1}|f'(v^\varepsilon)|^2 f''(v^\varepsilon)^{-1}\,dx\,d\tau \\ -\int_s^t\int_{\mathbb{R}^d}\big[\langle\big((b-\beta_{\mu,A})\varrho\big)_\varepsilon,\nabla\psi\rangle+(c\varrho)_\varepsilon\psi\big]f(v^\varepsilon)\,dx\,d\tau+\widetilde{\Omega}(\varepsilon),\end{aligned}$$

where we set

$$\widetilde{\Omega}(\varepsilon)=\Omega(\varepsilon)+\lambda e^\lambda\sup_x|\nabla\psi(x)|\int_0^t\int_U |R_{\varrho,\varepsilon}|\,dx\,d\tau,\quad \lim_{\varepsilon\to 0}\widetilde{\Omega}(\varepsilon)=0.$$

The passage to the limit as $s\to 0$, and then as $\varepsilon\to 0$, is justified exactly as in the proof of estimate (9.4.2). Hence, letting s and ε go to zero in (9.5.12), we arrive at the following inequality:

$$\begin{aligned}\int_{\mathbb{R}^d} f\big(v(x,t)\big)\varrho(x,t)\psi(x)\,dx \leqslant f(1)\int_{\mathbb{R}^d}\psi(x)\,d\nu \\ +\frac{1}{2}\int_0^t\int_{\mathbb{R}^d}\varrho\langle A\nabla\psi,\nabla\psi\rangle\psi^{-1}|f'(v)|^2 f''(v)^{-1}\,dx\,d\tau \\ +\int_0^t\int_{\mathbb{R}^d}\big[\langle b-\beta_{\mu,A},\nabla\psi\rangle+c\psi\big]f(v)\varrho\,dx\,d\tau.\end{aligned}$$

Thus, we have obtained estimate (9.4.3). The main lemma is proven.

9.5.3. Remark. The conclusion of Lemma 9.4.2 is true not only for the functions $e^{\lambda(1-z)}$ or $e^{\lambda(1-z)}-e^\lambda$ with $\lambda>0$, but also for every function f of class $C^\infty\big([0,+\infty)\big)$ such that $f''>0$ and

$$\sup_{z\in[0,+\infty)}\big(|f(z)|+|z|\,|f'(z)|+|z|^2|f''(z)|\big)<\infty.$$

9.6. Uniqueness of integrable solutions

Here we find sufficient conditions for the uniqueness in the class $\mathcal{I}_\nu$ (its definition is given on p. 338). Suppose that condition (H1) from § 9.1 and condition (H2) from § 9.4 hold and that $c\leqslant 0$. Let $\mu\in\mathcal{I}_\nu$ (where ν can be signed). Then by our assumption

$$b^i\in L^p_{\mathrm{loc}}\big(|\mu|,\mathbb{R}^d\times(0,T)\big),\quad c\in L^{p/2}_{\mathrm{loc}}\big(|\mu|,\mathbb{R}^d\times(0,T)\big).$$

By Corollary 6.5.1 the measure μ has a density $\varrho\in\mathbb{H}^{p,1}_{\mathrm{loc}}\big(\mathbb{R}^d\times(0,1)\big)$ with respect to Lebesgue measure and one can choose a continuous version of ϱ.

As in the case of probability solutions, we need some auxiliary lemmas.

9.6.1. Lemma. *Let a measure $\mu=\varrho\,dx\,dt$ belong to the class $\mathcal{I}_\nu$. Then, for every nonnegative function $\psi\in C_0^\infty(\mathbb{R}^d)$ and any numbers s and t in $(0,T)$, whenever $s<t$, we have*

$$\int_{\mathbb{R}^d}\psi(x)|\varrho(x,t)|\,dx\leqslant\int_{\mathbb{R}^d}\psi(x)|\varrho(x,s)|\,dx+\int_s^t\int_{\mathbb{R}^d}L\psi(x,\tau)|\varrho(x,\tau)|\,dx\,d\tau.$$

Let us give some informal reasoning explaining this assertion; this reasoning becomes rigorous in the case of smooth coefficients. A rigorous proof of this lemma will be given in the next section. Let $f \in C^2([0,+\infty))$, $f \geqslant 0$, $f' \geqslant 0$, $f'' \geqslant 0$, and $\psi \in C_0^\infty(\mathbb{R}^d)$. Set

$$h^i = b^i - \sum_{j=1}^d \partial_{x_j} a^{ij}.$$

We have

$$L^* f(\varrho) = f'(\varrho)L^*\varrho + f''(\varrho)\langle A\nabla\varrho, \nabla\varrho\rangle + \big(\varrho f'(\varrho) - f(\varrho)\big)(\mathrm{div} h - c).$$

Since $f'' \geqslant 0$ and $\partial_t\varrho = L^*\varrho$, we obtain

$$L^* f(\varrho) \geqslant \partial_t\big(f(\varrho)\big) + \big(\varrho f'(\varrho) - f(\varrho)\big)(\mathrm{div} h - c).$$

Multiplying the last inequality by ψ and integrating, we obtain

$$\int_{\mathbb{R}^d} \psi(x) f\big(\varrho(x,t)\big)\,dx \leqslant \int_{\mathbb{R}^d} \psi(x) f\big(\varrho(x,s)\big)\,dx + \int_s^t\!\!\int_{\mathbb{R}^d} L\psi(x,\tau) f\big(\varrho(x,\tau)\big)\,dx\,d\tau$$
$$- \int_s^t\!\!\int_{\mathbb{R}^d} \psi\Big[\varrho(x,\tau) f'\big(\varrho(x,\tau)\big) - f\big(\varrho(x,\tau)\big)\Big](\mathrm{div} h - c)\,dx\,d\tau.$$

This inequality involves no second derivatives of f. Therefore, it remains true for functions that are convex and satisfy the Lipschitz condition. Let us take $f(z) = |z|$. Then we have $zf'(z) - f(z) = 0$, so we obtain the desired inequality.

Our next lemma refines the behavior of solutions as $t \to 0$.

9.6.2. Lemma. *Let $\mu = \varrho(x,t)\,dx\,dt$ belong to $\mathcal{I}_\nu$ with $\nu = 0$. Then*

$$\lim_{t\to 0} \int_U |\varrho(x,t)|\,dx = 0$$

for every ball $U \subset \mathbb{R}^d$.

The proof is given in the next section. We now establish the main result on uniqueness of integrable solutions. Let ν be a bounded Borel measure on $\mathbb{R}^d$, possibly signed.

9.6.3. Theorem. *Suppose that* (H1) *and* (H2) *hold, $c \leqslant 0$ and there exists a positive function $V \in C^2(\mathbb{R}^d)$ such that $V(x) \to +\infty$ as $|x| \to +\infty$ and for some number $C > 0$ and all $(x,t) \in \mathbb{R}^d \times (0,T)$ we have*

$$L_{A,b}V(x,t) \geqslant -CV, \quad |\sqrt{A}\nabla V(x)| \leqslant CV.$$

Then, for every measure $\mu = \varrho(x,t)\,dx\,dt$ in $\mathcal{I}_\nu$, the function

$$t \mapsto \int_{\mathbb{R}^d} |\varrho(x,t)|\,dx$$

decreases on $(0,T)$. Moreover, if $\nu = 0$, then $\mu = 0$. Hence I_ν contains at most one element.

Proof. Let $0 < s < t < T$. Applying Lemma 9.6.1, for every nonnegative function $\psi \in C_0^\infty(\mathbb{R}^d)$ we obtain

$$\int_{\mathbb{R}^d} \psi(x)|\varrho(x,t)|\,dx \leqslant \int_{\mathbb{R}^d} \psi(x)|\varrho(x,s)|\,dx + \int_s^t\!\!\int_{\mathbb{R}^d} L\psi(x,\tau)|\varrho(x,\tau)|\,dx\,d\tau. \tag{9.6.1}$$

Set $\psi(x) = \zeta(N^{-1}V(x))$, where $\zeta \in C^\infty(\mathbb{R})$, $\zeta \geqslant 0$, $\zeta(z) = 1$ if $|z| \leqslant 1$ and $\zeta(z) = 0$ if $|z| > 2$, $\zeta'(z) \leqslant 0$ if $z > 0$ and for some number $M > 0$ and all x we have

$$|\zeta(x)| \leqslant M, \quad |\zeta'(x)| \leqslant M, \quad |\zeta''(x)| \leqslant M.$$

Applying (9.6.1) and the inequality $c \leqslant 0$, we arrive at the inequality

$$\begin{aligned}\int_{\mathbb{R}^d} \psi_N(x)|\varrho(x,t)|\,dx \leqslant \int_{\mathbb{R}^d} \psi(x)|\varrho(x,s)|\,dx \\ + \int_0^t\int_{\mathbb{R}^d} \Bigl[N^{-1}\zeta'(N^{-1}V(x))L_{A,b}V(x,\tau) \\ + N^{-2}\zeta''(N^{-1}V(x))|\sqrt{A(x,t)}\nabla V(x)|^2|\varrho(x,\tau)|\Bigr]\,dx\,d\tau.\end{aligned}$$

As in Proposition 4.3.6, passing to $\ln V$, we can assume that $L_{A,b}V \geqslant -C$ and $|\sqrt{A}\nabla V| \leqslant C$ with a larger C. We observe that $\zeta' L_{A,b}V \leqslant MC$. Therefore,

$$\begin{aligned}\int_{\mathbb{R}^d} \psi_N(x)|\varrho(x,t)|\,dx \leqslant \int_{\mathbb{R}^d} \psi(x)|\varrho(x,s)|\,dx \\ + MC(N^{-1} + CN^{-2}) \int_s^t\int_{\mathbb{R}^d} |\varrho(x,\tau)|\,dx\,d\tau.\end{aligned}$$

Since $\lim\limits_{N\to\infty} \psi_N(x) = 1$, letting $N \to \infty$, we obtain

$$\int_{\mathbb{R}^d} |\varrho(x,t)|\,dx \leqslant \int_{\mathbb{R}^d} |\varrho(x,s)|\,dx. \tag{9.6.2}$$

Thus, the first assertion of the theorem on monotonicity is proved. If $\nu = 0$, then by Lemma 9.6.2 one has the equality

$$\lim_{s\to 0} \int_U |\varrho(x,s)|\,dx = 0,$$

using which along with inequality (9.6.2), we conclude that $\|\varrho(\,\cdot\,,t)\|_1 = 0$ for all points $t \in (0,T)$. □

9.6.4. Example. Let $V(x) = \ln(\ln(1+|x|))$ if $|x| > 1$. Let A and b be locally bounded. Whenever $|x| > 1$, we have

$$|\sqrt{A(x,t)}\nabla V(x)|^2 = \langle A(x,t)\nabla V(x), \nabla V(x)\rangle = \frac{\langle A(x,t)x, x\rangle}{|x|^2(|x|+1)^2\ln^2(|x|+1)}.$$

Let us evaluate $L_{A,b}V(x,t)$ for $|x| > 1$:

$$\begin{aligned}L_{A,b}V(x,t) = -\frac{\langle A(x,t)x, x\rangle}{|x|^2(1+|x|)^2\ln(1+|x|)}\Bigl(1 + \frac{1}{\ln(1+|x|)} + \frac{1+|x|}{|x|}\Bigr) \\ + \frac{\operatorname{tr} A(x,t)}{|x|(1+|x|)\ln(1+|x|)} + \frac{\langle b(x,t), x\rangle}{|x|(|x|+1)\ln(|x|+1)}.\end{aligned}$$

In order to satisfy the hypotheses of the theorem it suffices to have the estimates

$$\begin{aligned}\langle A(x,t)x, x\rangle \leqslant C_1|x|^4 \ln|x| \ln(\ln(1+|x|)), \\ \operatorname{tr} A(x,t) + \langle b(x,t), x\rangle \geqslant -C|x|^2 \ln|x| \ln(\ln(1+|x|))\end{aligned}$$

for all $(x,t) \in \mathbb{R}^d \times (0,T)$ with $|x|$ sufficiently large and some number $C > 0$.

9.7. Proofs of auxiliary lemmas

PROOF OF LEMMA 9.6.1. Let $\omega \in C_0^\infty(\mathbb{R}^d)$, $\omega \geqslant 0$, $\|\omega\|_{L^1} = 1$ and let us set $\omega_\varepsilon(x) = \varepsilon^{-d}\omega(x/\varepsilon)$. For every locally integrable function g let g_ε denote the convolution $g*\omega_\varepsilon$. Let ϱ be the continuous version of the density of the measure μ. The function ϱ_ε is continuous in the variables t and x and infinitely differentiable in the variable x on $\mathbb{R}^d\times(0,T)$ for any fixed $\varepsilon > 0$. Moreover, in the sense of generalized functions

$$\partial_t\varrho_\varepsilon = \partial_{x_i}\partial_{x_j}(a^{ij}\varrho)_\varepsilon - \partial_{x_i}(b^i\varrho)_\varepsilon + (c\varrho)_\varepsilon. \tag{9.7.1}$$

Since by assumption the functions $b\varrho$, $c\varrho$ and ϱ are integrable on $U\times J$ for every ball U and every compact interval $J\subset(0,T)$, the right-hand side of this equality is integrable on J in t (for each fixed x) and the function ϱ_ε is absolutely continuous in the variable t on the interval J. Let

$$\mathcal{A}\varphi = a^{ij}\partial_{x_i}\partial_{x_j}\varphi,\ \ \mathcal{A}^*\varphi = \partial_{x_i}\partial_{x_j}(a^{ij}\varphi),$$

$$R^i_{\varrho,\varepsilon} = \sum_{j=1}^d\bigl(\partial_{x_j}(a^{ij}\varrho)_\varepsilon - \partial_{x_j}(a^{ij}\varrho_\varepsilon)\bigr).$$

Then equation (9.7.1) can be written in the form

$$\partial_t\varrho_\varepsilon = \mathcal{A}^*\varrho_\varepsilon - \operatorname{div}\bigl((b\varrho)_\varepsilon - R_{\varrho,\varepsilon}\bigr) + (c\varrho)_\varepsilon.$$

Let $\delta > 0$. Set $g_\delta(r) = \delta^{-1}r$ if $|r|\leqslant\delta$ and $g_\delta(r) = \operatorname{sign} r$ if $|r|>\delta$. Let

$$f_\delta(t) = \int_0^t g_\delta(s)\,ds.$$

Note that $\lim\limits_{\delta\to 0} f_\delta(t) = |t|$ and $\lim\limits_{\delta\to 0} f'_\delta(t) = \operatorname{sign} t$. Let $\psi\in C_0^\infty(\mathbb{R}^d)$. Integrating by parts, we obtain the following equality:

$$\int_s^t\int_{\mathbb{R}^d} f_\delta(\varrho_\varepsilon)\mathcal{A}\psi\,dx\,d\tau = \int_s^t\int_{\mathbb{R}^d}\Bigl[A(\psi f'_\delta(\varrho_\varepsilon))\varrho_\varepsilon + f''_\delta(\varrho_\varepsilon)\langle A\nabla\varrho_\varepsilon,\nabla\varrho_\varepsilon\rangle\psi\Bigr]\,dx\,d\tau - W_\varepsilon - V_\varepsilon,$$

$$W_\varepsilon = \int_s^t\int_{\mathbb{R}^d}\bigl(\varrho_\varepsilon f'_\delta(\varrho_\varepsilon) - f_\delta(\varrho_\varepsilon)\bigr)\partial_{x_i}a^{ij}\partial_{x_j}\psi\,dx\,dt,$$

$$V_\varepsilon = \int_s^t\int_{\mathbb{R}^d}\psi f''_\delta(\varrho_\varepsilon)\varrho_\varepsilon\partial_{x_i}\varrho_\varepsilon\partial_{x_i}a^{ij}\,dx\,dt.$$

We observe that

$$\begin{aligned}\int_s^t\int_{\mathbb{R}^d}\mathcal{A}(\psi f'_\delta(\varrho_\varepsilon))\varrho_\varepsilon\,dx\,d\tau &= \int_s^t\int_{\mathbb{R}^d}\partial_t(f_\delta(\varrho_\varepsilon))\psi\,dx\,d\tau\\ &- \int_s^t\int_{\mathbb{R}^d}\Bigl[\langle(b\varrho)_\varepsilon - R_{\varrho,\varepsilon},\nabla\psi\rangle f'_\delta(\varrho_\varepsilon) + \langle(b\varrho)_\varepsilon - R_{\varrho,\varepsilon},\nabla\varrho_\varepsilon\rangle f''_\delta(\varrho_\varepsilon)\\ &\qquad - (c\varrho)_\varepsilon f'_\delta(\varrho_\varepsilon)\Bigr]\psi\,dx\,d\tau.\end{aligned}$$

Therefore, we have the following equality:

$$\int_{\mathbb{R}^d} \psi(x) f_\delta\big(\varrho_\varepsilon(x,t)\big)\,dx + \int_s^t\!\!\int_{\mathbb{R}^d} f_\delta''(\varrho_\varepsilon)\langle A\nabla\varrho_\varepsilon, \nabla\varrho_\varepsilon\rangle\psi\,dx\,d\tau$$
$$= \int_{\mathbb{R}^d} \psi(x) f_\delta\big(\varrho_\varepsilon(x,s)\big)\,dx + W_\varepsilon + V_\varepsilon + Z_\varepsilon$$
$$+ \int_s^t\!\!\int_{\mathbb{R}^d} f_\delta(\varrho_\varepsilon)\mathcal{A}\psi + f_\delta'(\varrho_\varepsilon)\langle(b\varrho)_\varepsilon, \nabla\psi\rangle + f_\delta'(\varrho_\varepsilon)(c\varrho)_\varepsilon\,dx\,d\tau,$$

$$Z_\varepsilon = -\int_s^t\!\!\int_{\mathbb{R}^d} \Big[(R_{\varrho,\varepsilon}, \nabla\psi) f_\delta'(\varrho_\varepsilon) + \langle(b\varrho)_\varepsilon - R_{\varrho,\varepsilon}, \nabla\varrho_\varepsilon\rangle f_\delta''(\varrho_\varepsilon)\Big]\psi\,dx\,d\tau.$$

Let U be a ball containing the support of the function ψ. Let $m = m(U) > 0$ and $\Lambda = \Lambda(U) > 0$ be the numbers in conditions (H1) and (H2). We recall that $f_\delta'' \geqslant 0$. Then

$$\int_s^t\!\!\int_{\mathbb{R}^d} f_\delta''(\varrho_\varepsilon)\langle A\nabla\varrho_\varepsilon, \nabla\varrho_\varepsilon\rangle\psi\,dx\,d\tau \geqslant m\int_s^t\!\!\int_{\mathbb{R}^d} f_\delta''(\varrho_\varepsilon)|\nabla\varrho_\varepsilon|^2\psi\,dx\,d\tau.$$

Let us estimate the term W_ε. We observe that $|tf_\delta'(t) - f_\delta(t)| \leqslant \delta$. So we obtain the inequality

$$|W_\varepsilon| \leqslant \Lambda d\delta|U| \max_x |\nabla\psi(x)|.$$

Let us now estimate the term V_ε. Since $f_\delta''(t) = 0$ if $|t| > \delta$ and $|f_\delta''(t)| \leqslant \delta^{-1}$, we have

$$|V_\varepsilon| \leqslant \frac{m}{3}\int_s^t\!\!\int_{\mathbb{R}^d} f_\delta''(\varrho_\varepsilon)|\nabla\varrho_\varepsilon|^2\psi\,dx\,d\tau + 3dm^{-1}\Lambda^2\delta|U|\max_x|\psi|.$$

Finally, let us estimate the term Z_ε. We have

$$|Z_\varepsilon| \leqslant \int_s^t\!\!\int_{\mathbb{R}^d}\Big[|R_{\varrho,\varepsilon}|\,|\nabla\psi| + 3m^{-1}\delta^{-1}\psi|R_{\varrho,\varepsilon}|^2\Big]\,dx\,d\tau$$
$$+ 3m^{-1}\int_s^t\!\!\int_{\mathbb{R}^d} f_\delta''(\varrho_\varepsilon)|(b\varrho)_\varepsilon|^2\psi\,dx\,d\tau + \frac{m}{3}\int_s^t\!\!\int_{\mathbb{R}^d} f_\delta''(\varrho_\varepsilon)|\nabla\varrho_\varepsilon|^2\psi\,dx\,d\tau.$$

Collecting these estimates, we arrive at the inequality

$$\int_{\mathbb{R}^d}\psi(x) f_\delta\big(\varrho_\varepsilon(x,t)\big)\,dx \leqslant \int_{\mathbb{R}^d}\psi(x) f_\delta\big(\varrho_\varepsilon(x,s)\big)\,dx$$
$$+ \int_s^t\!\!\int_{\mathbb{R}^d}\Big[f_\delta(\varrho_\varepsilon)\mathcal{A}\psi + f'(\varrho_\varepsilon)\langle(b\varrho)_\varepsilon, \nabla\psi\rangle + f'(\varrho_\varepsilon)(c\varrho)_\varepsilon\psi\Big]\,dx\,d\tau$$
$$+ \delta C(d,\Lambda,m,\psi) + \int_s^t\!\!\int_{\mathbb{R}^d}\Big[|R_{\varrho,\varepsilon}|\,|\nabla\psi| + 3m^{-1}\delta^{-1}\psi|R_{\varrho,\varepsilon}|^2\Big]\,dx\,d\tau$$
$$+ 3m^{-1}\int_s^t\!\!\int_{\mathbb{R}^d} f_\delta''(\varrho_\varepsilon)|(b\varrho)_\varepsilon|^2\psi\,dx\,d\tau.$$

Letting ε go to zero and taking into account that

$$\lim_{\varepsilon\to 0}\int_s^t\!\!\int_U |R_{\varrho,\varepsilon}|^2\,dx\,d\tau = 0,$$

we arrive at the following inequality:

$$\int_{\mathbb{R}^d} \psi(x) f_\delta\bigl(\varrho(x,t)\bigr)\,dx \leqslant \int_{\mathbb{R}^d} \psi(x) f_\delta\bigl(\varrho(x,s)\bigr)\,dx$$
$$+ \int_s^t\!\!\int_{\mathbb{R}^d} \Bigl[f_\delta(\varrho)\mathcal{A}\psi + f_\delta'(\varrho)\langle b\varrho, \nabla\psi\rangle + f_\delta'(\varrho)\varrho c\psi\Bigr]\,dx\,d\tau + \delta C(d,\Lambda,m,\psi)$$
$$+ 3m^{-1}\int_s^t\!\!\int_{\mathbb{R}^d} f_\delta''(\varrho)|b\varrho|^2\psi\,dx\,d\tau.$$

We observe that

$$\biggl|\int_s^t\!\!\int_{\mathbb{R}^d} f_\delta''(\varrho)|b\varrho|^2\psi\,dx\,d\tau\biggr| \leqslant \int_s^t\!\!\int_{|\varrho|<\delta} |b|^2|\varrho|\psi\,dx\,d\tau.$$

Hence the left-hand side tends to zero as $\delta \to 0$. Therefore, we have

$$\int_{\mathbb{R}^d} \psi(x)|\varrho(x,t)|\,dx$$
$$\leqslant \int_{\mathbb{R}^d} \psi(x)|\varrho(x,s)|\,dx + \int_s^t\!\!\int_{\mathbb{R}^d} \Bigl[\mathcal{A}\psi + \langle b, \nabla\psi\rangle + c\psi\Bigr]|\varrho(x,\tau)|\,dx\,d\tau.$$

Lemma 9.6.1 is proven.

PROOF OF LEMMA 9.6.2. Set $A(x,t) = A(x,-t)$ for $t \in [-T,0]$. Then the functions a^{ij} are continuous on $\mathbb{R}^d \times [-T,T]$ and satisfy conditions (H1) and (H2). Suppose also that $\psi \in C_0^\infty(\mathbb{R}^d)$, $\psi \geqslant 0$ and U is a ball containing the support of the function ψ. One can redefine the functions a^{ij} outside U in such a way that conditions (H1) and (H2) will be fulfilled on $\mathbb{R}^d \times [0,T]$ with the same numbers $\gamma = \gamma(U)$, $M = M(U)$ and $\Lambda = \Lambda(U)$. Set $g_\delta(r) = \delta^{-1}r$ if $|r| < \delta$, $g_\delta(r) = \operatorname{sign} r$ if $|r| \geqslant \delta$. Let $0 < t < T$. According to Porper, Eidelman [**785**, Theorem 1.3], there is a solution Q to the Cauchy problem

$$\partial_s Q + A_0 Q = 0, \quad Q\big|_{s=t} = g_\delta\bigl(\varrho(\,\cdot\,,t)\bigr), \quad A_0 Q = \operatorname{tr}(AD^2Q)$$

on $\mathbb{R}^d \times (-1,t)$. The function $x \mapsto g_\delta\bigl(\varrho(x,t)\bigr)$ is continuous, $|g_\delta| \leqslant 1$. According to [**785**, Theorems 1.1 and 1.3] (see also Friedman [**386**]), we have the inclusion

$$Q \in C_b\bigl(\mathbb{R}^d \times (-T,t]\bigr) \cap C^{2,1}\bigl(\mathbb{R}^d \times (-T,t)\bigr)$$

and the inequality

$$|Q(x,u)| \leqslant C, \ |\nabla Q(x,u)| \leqslant C(t-u)^{-1/2}$$

with some number C depending on γ, Λ and d. Fix $s < t$. Applying Lemma 9.1.1 with $\varphi(x,u) = Q(x,u)\psi(x)$ on $\mathbb{R}^d \times [0,s]$, which is possible, since φ belongs to the corresponding class, after integration by parts (which is also possible due to the indicated properties of Q and the compactness of the support of ψ) we obtain

$$\begin{aligned}\int_{\mathbb{R}^d} Q(x,s)\psi(x)\varrho(x,s)\,dx = \int_0^s\!\!\int_{\mathbb{R}^d} \Bigl[&\langle A(x,\tau)\nabla Q(x,\tau), \nabla\psi(x)\rangle \\ &+ Q(x,\tau)A_0\psi(x) + \langle b(x,\tau), \nabla Q(x,\tau)\rangle\psi(x) \\ &+ \langle b(x,\tau), \nabla\psi(x)\rangle Q(x,\tau) + c(x,\tau)Q(x,\tau)\psi(x)\Bigr]\varrho(x,\tau)\,dx\,d\tau.\end{aligned} \tag{9.7.2}$$

We can estimate the right-hand side of this equality by

$$C \sup_{x\in\mathbb{R}^d} \bigl(|\psi(x)| + |A\nabla\psi(x)| + |A_0\psi(x)|\bigr)$$
$$\times \biggl(\int_0^s\!\!\int_{\mathbb{R}^d} \bigl(1+(t-\tau)^{-1/2}\bigr)\bigl(1+|b(y,\tau)|\bigr)|\varrho(x,\tau)|\,dx\,d\tau$$
$$+ \int_0^s\!\!\int_{\mathbb{R}^d} |c(y,\tau)|\,|\varrho(x,\tau)|\,dx\,d\tau\biggr).$$

We recall that $\varrho \in L^\infty\bigl((0,T), L^1(\mathbb{R}^d)\bigr)$ and $b \in L^p\bigl(|\mu|, \mathbb{R}^d\times(0,T)\bigr)$. By Hölder's inequality for the functions $|b|$ and $(t-\tau)^{-1/2}$ we can estimate the left-hand side of (9.7.2) by

$$C_1 t^{(2-p')/2p'} + C_2 \int_0^t\!\!\int_{\mathbb{R}^d} |c(y,\tau)|\,|\varrho(x,\tau)|\,dx\,d\tau,$$

where the numbers C_1, C_2 do not depend on s and t. Since ϱ and Q are continuous functions on $\mathbb{R}^d\times(0,t]$, letting $s\to t$, we find that

$$\int_{\mathbb{R}^d} \psi(x) g_\delta\bigl(\varrho(x,t)\bigr)\varrho(x,t)\,dx \leqslant C_1 t^{(2-p')/2p'} + C_2 \int_0^t\!\!\int_{\mathbb{R}^d} |c(y,\tau)|\,|\varrho(x,\tau)|\,dx\,d\tau.$$

Letting $\delta\to 0$, we arrive at the inequality

$$\int_{\mathbb{R}^d} \psi(x)|\varrho(x,t)|\,dx \leqslant C_1 t^{(2-p')/2p'} + C_2 \int_0^t\!\!\int_{\mathbb{R}^d} |c(y,\tau)|\,|\varrho(x,\tau)|\,dx\,d\tau.$$

Since $p>2$, we have $p'<2$ and $t^{(2-p')/2p'}\to 0$ as $t\to 0$. Lemma 9.6.2 is proven. This completes the proofs of the results presented in § 9.6.

9.8. Complements, comments, and exercises

9.8(i). Uniqueness for degenerate equations in the class of absolutely continuous measures

Here we turn to the uniqueness problem for solutions to the Cauchy problem for degenerate Fokker–Planck–Kolmogorov equations with singular coefficients. In the case of coefficients of class C^2 the uniqueness of solutions to the Cauchy problem for degenerate Fokker–Planck–Kolmogorov equations is proved, for example, in Stroock, Varadhan [**884**]. In this case, Holmgren's principle, described at the beginning of this chapter, works.

The case of integrable and Sobolev coefficients is studied in Le Bris, Lions [**593**], Figalli [**362**], Wei, Liu [**936**]. We recall that in § 6.7(ii) (see Theorem 6.7.4) we have discussed a construction of a solution given by a density $\varrho \in L^\infty\bigl((0,T), L^p(\mathbb{R}^d)\bigr)$, in the case of a degenerate diffusion matrix.

In the paper DiPerna, Lions [**304**] for the continuity equation (i.e., $A\equiv 0$, see § 9.8(iv)) and in Le Bris, Lions [**593**] for the Fokker–Planck–Kolmogorov equation, a new method of proving uniqueness in the class of absolutely continuous measures was developed, the main idea of which is to obtain the following inequality (in the sense of generalized functions) for the density ϱ of a solution μ to the equation $\partial_t\mu = L^*_{A,b}\mu$ and a convex function $f \in C^2(\mathbb{R})$:

$$\partial_t f(\varrho) - L^*_{A,b} f(\varrho) \leqslant (\partial_{x_i} b^i - \partial_{x_i}\partial_{x_j} a^{ij})\big(f(\varrho) - \varrho f'(\varrho)\big),$$

from which one derives the estimate

$$\frac{d}{dt}\int_{\mathbb{R}^d} \varrho^+(x,t)\,dx \leqslant 0,$$

where $\varrho^+ = \max(0,\varrho)$. This inequality yields that if $\varrho(x,0)=0$, then $\varrho(x,t)=0$, whence the uniqueness follows. If one manages to justify the above inequality for the solution ϱ, then such a solution is called a *renormalized solution* (which is not the same as in § 1.8(iii)). Let us give an example of sufficient conditions for uniqueness obtained in this approach. The next result is [**593**, Proposition 5]. Let us set $A := \sigma\sigma^*$ and $\beta^i := b^i - \partial_{x_j} a^{ij}$.

9.8.1. Theorem. *Suppose that in the natural notation*

$$\sigma \in L^2\big([0,T], W^{2,1}_{\rm loc}(\mathbb{R}^d)\big), \quad \beta \in L^1\big([0,T], W^{1,1}_{\rm loc}(\mathbb{R}^d)\big),$$

$$\operatorname{div}\beta \in L^1\big([0,T], L^\infty(\mathbb{R}^d)\big), \quad \frac{|\beta|}{1+|x|} \in L^1\big([0,T], L^1(\mathbb{R}^d)\big) + L^1\big([0,T], L^\infty(\mathbb{R}^d)\big),$$

$$\frac{\sigma}{1+|x|} \in L^2\big([0,T], L^2(\mathbb{R}^d)\big) + L^2\big([0,T], L^\infty(\mathbb{R}^d)\big).$$

Then, for every initial condition given by a density ϱ_0 from $L^1(\mathbb{R}^d)\cap L^\infty(\mathbb{R}^d)$, there exists a unique solution to the Cauchy problem $\partial_t\varrho = L^*_{A,b}\varrho$, $\varrho|_{t=0} = \varrho_0$ in the class

$$\Big\{\varrho\colon\ \varrho \in L^\infty\big([0,T], L^1(\mathbb{R}^d)\cap L^\infty(\mathbb{R}^d)\big),\ \sigma^*\nabla\varrho \in L^2\big([0,T], L^2(\mathbb{R}^d)\big)\Big\}.$$

Generalizations to the class $L^\infty\big([0,T], L^p(\mathbb{R}^d)\big)$ and various options to weaken conditions on the coefficients can be found in Wei, Liu [**936**] and Luo [**650**].

In the paper Röckner, Zhang [**822**] in the case of bounded A, b and $c=0$, uniqueness is proved by probabilistic methods discussed below. In Blanchard, Röckner, Russo [**118**] in the one-dimensional case and in Belaribi, Russo [**99**] in the multidimensional case, uniqueness has been studied for the equation $\partial_t z = \Delta(\varrho z)$, where no smoothness of ϱ is assumed, but just the measurability. In all these papers the existence of densities of solutions is part of the hypotheses. Note that without this assumption in the case of a degenerate diffusion matrix the uniqueness of solutions can fail: let us consider Girsanov's example [**410**], where $d=1$, $b=0$, $A=1$, $\varrho(x)=|x|$, Dirac's measure at zero is a solution, but there is also another solution (another example with bounded ϱ is discussed in [**118**, Remark 3.11]).

The next results from Bogachev, Röckner, Shaposhnikov [**172**] are generalizations of the results from [**118**] and [**99**].

Let us consider a second order operator of the form

$$Lu = \varrho\operatorname{div}\big(A\nabla u\big) + \sqrt{\varrho}\langle b, \nabla u\rangle,$$

where $\varrho(x,t)\geqslant 0$, $A(x,t)$ is a nonnegative-definite $d\times d$-matrix and $b(x,t)\in\mathbb{R}^d$. We discuss the uniqueness of solutions to the Cauchy problem

$$\partial_t z = L^* z, \quad z|_{t=0} = 0. \tag{9.8.1}$$

As compared to our earlier considered framework, we now slightly change the interpretations of solutions and initial conditions.

We shall say that a function z, belonging to $L^2(\mathbb{R}^d \times [\kappa, T])$ for every number $\kappa \in (0,T)$, satisfies the Fokker–Planck–Kolmogorov equation $\partial_t z = L^* z$ if the function $z\varrho$ is locally square integrable and the following identity holds:

$$\int_0^T \int_{\mathbb{R}^d} \bigl[\partial_t u(x,\tau) + Lu(x,\tau)\bigr] z(x,\tau)\, dx\, d\tau = 0 \quad \forall\, u \in C_0^\infty\bigl(\mathbb{R}^d \times (0,T)\bigr).$$

Therefore, considering this equation with respect to the measure $z\,dx\,dt$, one can say that here for solutions we admit only measures with densities in the indicated class. We shall say that a function z satisfies the initial condition $z|_{t=0} = 0$ if

$$(I - \Delta)^{-1} z(x,t) \in L^1(\mathbb{R}^d) \quad \text{for almost all } t \in (0,T),$$

and

(9.8.2) $$\lim_{n\to\infty} \operatorname{ess\,sup}_{t\in(0,1/n)} \|(I-\Delta)^{-1} z(\,\cdot\,,t)\|_{L^1} = 0,$$

(9.8.3) $$\lim_{n\to\infty} \operatorname{ess\,sup}_{t\in(0,1/n)} \|(I-\Delta)^{-1/2} z(\,\cdot\,,t)\|_{L^2} = 0.$$

The second equality means convergence to zero in the negative Sobolev space $H^{2,-1}(\mathbb{R}^d)$ that is the dual to the space $H^{2,1}(\mathbb{R}^d)$, so that $z(\,\cdot\,,t)$ belongs to this space once it belongs to $L^2(\mathbb{R}^d)$; the first equality corresponds to the norm in the space $H^{1,-2}(\mathbb{R}^d)$, also belonging to the scale of spaces

$$H^{p,r}(\mathbb{R}^d) := (I-\Delta)^{-r/2}\bigl(L^p(\mathbb{R}^d)\bigr), \quad \|(I-\Delta)^{-r/2} f\|_{p,r} = \|f\|_p,$$

where $\|f\|_p$ is the norm in $L^p(\mathbb{R}^d)$. In terms of the new function $g = (I-\Delta)^{-1} z$ condition (9.8.3) means that $\lim\limits_{t\to 0} \|g(\,\cdot\,,t)\|_{2,1} = 0$ provided that we consider t in some set of full measure in $(0,T)$, while condition (9.8.3) is just $\lim\limits_{t\to 0} \|g(\,\cdot\,,t)\|_1 = 0$ with the same interpretation.

Similarly, one can consider the Cauchy problem

(9.8.4) $$\partial_t z = L^* z, \quad z|_{t=0} = \nu$$

in the class of probability densities $z(\,\cdot\,,t)$ for a.e. t (such solutions will be called probability solutions) with the initial condition ν that is a probability measure on $\mathbb{R}^d$. In this case our definition of the initial condition in (9.8.4) is as follows: there exists a set $S_0 \subset (0,T)$ of full measure such that, as $t \to 0$ in S_0, the measures $z(\,\cdot\,,t)dx$ converge weakly to the measure ν, i.e.,

(9.8.5) $$\lim_{t\to 0, t\in S_0} \int_{\mathbb{R}^d} \varphi(x) z(x,t)\, dx = \int_{\mathbb{R}^d} \varphi(x)\, \nu(dx) \quad \forall\, \varphi \in C_b(\mathbb{R}^d).$$

We recall that in the case of probability measures it suffices to have this equality just for all $\varphi \in C_0^\infty(\mathbb{R}^d)$, so that it follows from (9.8.2) or from (9.8.3) (the latter, of course, requires also the membership in L^2).

Let us list our assumptions about the coefficients of the operator L:

(H1) ϱ is a nonnegative Borel function on $\mathbb{R}^d \times (0,T)$ (no boundedness or strict positivity or smoothness is assumed);

(H2) $A(x,t) = \bigl(a^{ij}(x,t)\bigr)_{i,j\leqslant d}$ is a symmetric positive-definite matrix with bounded and Lipschitzian in (x,t) entries a^{ij} and $A(x,t) \geqslant \lambda \cdot \mathrm{I}$ for some $\lambda > 0$;

(H3) $b = (b^i)_{i\leqslant d}$ is a vector mapping, $b^i \in L^\infty\bigl(\mathbb{R}^d \times (0,T)\bigr)$.

We recall that a probability measure satisfying (9.8.4) is given by a density, provided that $\varrho > 0$, otherwise singular solutions may exist.

9.8.2. Theorem. *Let conditions* (H1), (H2) *and* (H3) *be fulfilled. If for every* $\kappa\in(0,T)$ *and every ball* $U\subset\mathbb{R}^d$ *we have* $\varrho|z|\in L^2(U\times[\kappa,T])$ *and*

$$\lim_{N\to\infty}\int_\kappa^T\int_{N\leqslant|x|\leqslant 2N}\biggl[\frac{\sqrt{\varrho(x,t)}|z(x,t)|}{1+|x|} + \frac{\varrho(x,t)|z(x,t)|}{1+|x|}+\frac{\varrho(x,t)^2|z(x,t)|^2}{1+|x|^2}\biggr]\,dx\,dt=0, \tag{9.8.6}$$

then the Cauchy problem (9.8.1) *has the unique solution* $z=0$.

If z_1 *and* z_2 *are two probability solutions to the Cauchy problem* (9.8.4) *such that the function* $z=z_1-z_2$ *satisfies the condition* $z(\,\cdot\,,t)\in L^2(\mathbb{R}^d\times[\kappa,T])$ *for every* $\kappa\in(0,T)$ *and also* (9.8.3) *and* (9.8.6) *hold, then* $z_1=z_2$.

Finally, if (9.8.6) *is fulfilled with* $\kappa=0$, *then for every probability measure* ν *on* $\mathbb{R}^d$ *the Cauchy problem* (9.8.4) *with* (9.8.5) *in place of* (9.8.2), (9.8.3) *has at most one probability solution.*

9.8.3. Remark. (i) Let $V\in C^2(\mathbb{R}^d)$, $V\geqslant 0$, and $\lim\limits_{|x|\to\infty}V(x)=+\infty$. Condition (9.8.6) in this theorem can be replaced with the following one:

$$\lim_{N\to\infty}N^{-1}\int_\kappa^T\int_{\Omega_N}\Bigl((|LV|+\varrho|\sqrt{A}\nabla V|^2)|z|+\varrho^2|z|^2|A\nabla V|^2\Bigr)\,dx\,dt=0$$

for every $\kappa\in(0,T)$, where $\Omega_N=\{N\leqslant V\leqslant 2N\}$.

(ii) Condition (9.8.6) can be replaced with the condition

$$\lim_{N\to\infty}\int_\kappa^T\int_{N\leqslant|x|\leqslant 2N}\varrho^2|z|^2\,dx\,dt=0\quad\forall\,\kappa\in(0,T).$$

9.8.4. Remark. Note also that (9.8.3) is equivalent to the equality

$$\lim_{n\to\infty}\operatorname{ess\,sup}_{t\in(0,1/n)}\|(1+|y|^2)^{-1/2}\widehat{z}(y,t)\|_{L^2}=0,$$

where $\widehat{z}(y,t)$ denotes the Fourier transform in the first argument. In particular, this equality is true in the case of probability solutions on $\mathbb{R}^1$; for $d=2$ it is ensured by the estimate $|\widehat{z}(y,t)|\leqslant C/\ln(|y|+1)$, in any dimension it is ensured by the estimate $|\widehat{z}(y,t)|^2\leqslant(1+|y|)^{3-d}\Phi(y)$ with an integrable function Φ.

If $d=1$, then, as noted above, condition (9.8.3) can be omitted (since it is automatically fulfilled), but the requirement that the admissible solutions belong to L^2 is essential for uniqueness, as the aforementioned example from [**118**] with bounded ϱ shows.

Finally, let us consider an example showing that the multiplication of b and c by suitable powers of ϱ is essential. Let $d=1$ and

$$Lu=\varrho u''+bu',$$

where

$$\varrho(x)=\sqrt{1+x^2},\ b(x)=-(1+x^2)\operatorname{arctg}x=-\varrho^2(x)\operatorname{arctg}x.$$

Let us take the probability measure $\nu=\bigl(\pi(1+x^2)\bigr)^{-1}dx$. In this example it is not enough to scale just the first order term: we would obtain uniqueness by replacing ϱ^2 with $\sqrt{\varrho}$. The Cauchy problem $\partial_t z=L^*z$, $z|_{t=0}=\nu$ has two different solutions, one of which can be written explicitly: $z_1(x,t)=e^t\bigl(\pi(1+x^2)\bigr)^{-1}$, which is verified directly, but there is another solution z_2 that is a probability one. This probability solution exists according to the results in § 6.6, since we have the Lyapunov function $V(x) = |x|^2$. Indeed,

$$L|x|^2 = 2\sqrt{1+x^2} - x(1+x^2)\mathrm{arctg}\, x \leqslant 0$$

for sufficiently large $|x|$. It follows from what we have proved in Chapter 7 that the solution z_2 is a bounded function on $\mathbb{R}^d\times[\kappa, T]$ for all $\kappa > 0$. Therefore, both solutions belong to $L^2(\mathbb{R}^d\times[\kappa, T])$ for every $\kappa > 0$, so all the conditions required in the theorem to guarantee their coincidence are fulfilled (except for a wrong scaling of b, of course). The drift b is nice by itself, but if we write it in the form required in the theorem, we obtain $b(x) = -\sqrt{\varrho(x)}(1+x^2)^{3/4}\mathrm{arctg}\, x$ with the unbounded (although not rapidly growing) mapping $-(1+x^2)^{3/4}\mathrm{arctg}\, x$. We do not know whether $\sqrt{\varrho}$ is the only admissible scaling of b in this theorem.

9.8(ii). Probabilistic methods of proving uniqueness

Let $\mathcal{W}_t$ be the minimal σ-algebra in $C[0, +\infty)$ containing all sets of the form $\{x\colon x(t) \in B\}$, where B is a Borel set in $\mathbb{R}^d$. Given a probability measure ν on $\mathbb{R}^d$, a probability measure P_ν on $\big(C[0, +\infty), \mathcal{W}_T\big)$ is called a solution of the martingale problem with the operator $L = a^{ij}\partial_{x_i}\partial_{x_j} + b^i\partial_{x_i}$ if ν coincides with the distribution of $x(0)$ with respect to P_ν and for every function $f \in C_0^\infty(\mathbb{R}^d)$ the expression

$$f\big(x(t)\big) - f\big(x(0)\big) - \int_0^t Lf\big(x(\tau)\big)\, d\tau$$

is a martingale with respect to $(P_\nu, \mathcal{W}_t)$.

Below in this subsection we assume that the coefficients a^{ij} and b^i are measurable and uniformly bounded. Let $A = \sigma\sigma^*/2$, where an operator-valued mapping σ is measurable, but $\sigma(x, t)$ need not be symmetric.

It is known (see Ikeda, Watanabe [**473**, Chapter 6]) that if there is a solution P_ν of the martingale problem, then there exist a probability space $(\Omega, \mathcal{F}, P, \mathcal{F}_t)$, a continuous random process ξ_t, and a Wiener process w_t on $[0, T]$ adapted to $\mathcal{F}_t$ such that

$$d\xi_t = b(\xi_t, t)\, dt + \sigma(\xi_t, t)\, dw_t$$

and ξ has the distribution P_ν in $C[0, +\infty)$. The collection $(\Omega, \mathcal{F}, P, \mathcal{F}_t, \xi_t, w_t)$ is called a weak solution of the corresponding stochastic equation.

If there exist two weak solutions $(\Omega^i, \mathcal{F}^i, P^i, \mathcal{F}_t^i, \xi_t^i, w_t^i)$, $i = 1, 2$, with the same initial distribution ν, then there exist a probability space $(\Omega, \mathcal{F}, P, \mathcal{F}_t)$, a Wiener process w_t on $[0, T]$, and random processes η_t^1 and η_t^2, adapted to $\mathcal{F}_t$, such that $P(\eta_0^1 = \eta_0^2) = 1$, the collections $(\Omega, \mathcal{F}, P, \mathcal{F}_t, \eta_t^i, w_t)$, $i = 1, 2$, are weak solutions and the distributions of η^1 and η^2 coincide with the distributions of ξ_t^1 and ξ_t^2, respectively. Therefore, if we have two weak solutions, then one can construct two weak solutions with the same distributions on a common probability space with a common filtration and a common Wiener process. This observation enables us to prove uniqueness of weak solutions of stochastic equations.

It is clear that the problem of uniqueness of a solution to the martingale problem, or, which is the same, of a weak solution to the stochastic equation, is close to the problem of uniqueness of a probability solution to the Cauchy problem for the Fokker–Planck–Kolmogorov equation. The proof of uniqueness of solutions to martingale problems consists usually of two steps. At the first step, for a sufficiently large class of functions f, for example, $C_0^\infty(\mathbb{R}^d)$, one proves the equality

$$\mathbb{E}^1_\nu f\bigl(x(t)\bigr) = \mathbb{E}^2_\nu f\bigl(x(t)\bigr) \tag{9.8.7}$$

for every $t \in [0,T]$, where $\mathbb{E}^1_\nu$ and $\mathbb{E}^1_\nu$ are the expectations with respect to two solutions to the martingale problem P^1_ν and P^2_ν. Equality (9.8.7) is equivalent to the coincidence of P^1_ν and P^2_ν on all sets of the form $\{x\colon x(t) \in B\}$, where B is a Borel set and $t \in [0,T]$. The second step is connected with deriving equality $P^1_\nu = P^2_\nu$ from (9.8.7).

We observe that equality (9.8.7) is equivalent to the coincidence of the solutions μ^1 and μ^2 to the Cauchy problem for equation (9.1.1) with initial condition $\mu|_{t=0} = \nu$, corresponding to P^1_ν and P^2_ν. Hence the proof of (9.8.7) often reduces to the proof of uniqueness of a probability solution to the Cauchy problem for the Fokker–Planck–Kolmogorov equation. Moreover, as shown in Figalli [**362**], in the case of bounded coefficients every probability solution (μ_t) to the Cauchy problem (9.1.1) with initial condition ν is generated by some solution P_ν to the martingale problem, i.e.,

$$\int_{\mathbb{R}^d} f(x)\, \mu_t(dx) = \int_{C[0,+\infty)} f\bigl(x(t)\bigr)\, P_\nu(dx) \quad \forall f \in C_0^\infty(\mathbb{R}^d).$$

Therefore, for bounded coefficients, the proof of uniqueness of the solution to the Cauchy problem reduces to the proof of uniqueness of a solution to the corresponding martingale problem, which is especially useful in the case of a degenerate diffusion matrix.

By using Figalli's result, one can extend to Fokker–Planck–Kolmogorov equations the following known result from Stroock, Varadhan [**884**, Theorem 6.3.4].

9.8.5. Theorem. *Let $A = \sigma\sigma^*$. Suppose that σ^{ij}, b^i are measurable and bounded on $\mathbb{R}^d \times (0,T)$ and*

$$|b(x,t) - b(y,t)| + \|\sigma(x,t) - \sigma(y,t)\|^2 \leqslant C|x-y|$$

for all $x, y \in \mathbb{R}^d$, $t \in (0,T)$ and some number $C > 0$. Then a probability solution to the Cauchy problem (10.5.1) *is unique.*

In Röckner, Zhang [**822**] and Luo [**651**], Figalli's result is applied for proving uniqueness in the class of absolutely continuous measures, but in the case of irregular coefficients. This approach requires some new conditions for uniqueness of solutions to the martingale problem and a study of the so-called Lagrangian stochastic flows (see Zhang [**959**]), when one considers solutions P_ν to the martingale problem such that the measures μ_t defined by $\mu_t(B) = P_\nu\bigl(\omega\colon \omega(t) \in B\bigr)$ are absolutely continuous with respect to Lebesgue measure.

The next result is obtained in [**822**]. Let $p \geqslant 1$. Let $\mathcal{M}^p(\mathbb{R}^d)$ denote the class of functions $u \in L^p\bigl((0,T), L^p(\mathbb{R}^d)\bigr)$ such that $x \mapsto u(x,t)$ is a probability density for every $t \in (0,T)$. Let U_R be the ball of radius R centered at the origin.

9.8.6. Theorem. *Suppose that σ and b are measurable and bounded and for some $q \in [0,+\infty]$ for each $R > 0$ there exists a function $f_R \in L^q(U_R \times [0,T])$ such that for almost all $(x,y,t) \in U_R \times U_R \times (0,T)$ we have*

$$2\langle x-y, b(x,t) - b(y,t)\rangle + \|\sigma(x) - \sigma(y)\|^2 \leqslant \bigl(f_R(x) + f_R(y)\bigr)|x-y|^2. \tag{9.8.8}$$

Then, for every initial condition $\nu = \varrho_0\, dx$, where ϱ_0 is a probability density, there exists at most one solution to the Cauchy problem (9.1.1) *in the class $\mathcal{M}^p(\mathbb{R}^d)$, where $p = q/(q-1)$.*

PROOF. Assume we have two solutions $u^1, u^2 \in \mathcal{M}^p(\mathbb{R}^d)$ to the Cauchy problem (9.1.1). Then there exist two solutions to the martingale problem P^1_ν and P^2_ν such that

$$\int_{\mathbb{R}^d} f(x)u^i(x,t)\,dx = \int_{C[0,+\infty)} f\big(x(t)\big)\,P^i_\nu(dx) \quad \forall\, f \in C_0^\infty(\mathbb{R}^d),\ i = 1,2.$$

Hence there are a probability space $(\Omega, \mathcal{F}, P, \mathcal{F}_t)$, a Wiener process w_t on $[0,T]$ and processes η^1_t and η^2_t with the distributions P^1_ν and P^2_ν, adapted to $\{\mathcal{F}_t\}$, such that $P(\eta^1_0 = \eta^2_0) = 1$, and the collections $(\Omega, \mathcal{F}, P, \mathcal{F}_t, \eta^i_t, w_t)$, $i = 1,2$, are weak solutions to the equations $d\eta^i_t = b(\eta^i_t, t)\,dt + \sigma(\eta^i_t, t)\,dw_t$. Set $Z_t = \eta^1_t - \eta^2_t$ and $\tau_R = \inf\{t \in [0,T]\colon |\eta^1_t| > R,\ |\eta^2_R| > R\}$. By Itô's formula for every $\delta > 0$ we have

$$\begin{aligned}
&\ln\Big(\frac{Z_{t\wedge\tau_R}}{\delta^2} + 1\Big) \\
&= \int_0^{t\wedge\tau_R} \Big(2\big\langle Z_s, b(\eta^1_s, s) - b(\eta^2_s, s)\big\rangle + \|\sigma(\eta^1_s, s) - \sigma(\eta^2_s, s)\|^2\Big)\big(|Z_s|^2 + \delta^2\big)^{-1}\,ds \\
&\qquad + 2\int_0^{t\wedge\tau_R} \big(|Z_s|^2 + \delta^2\big)^{-1}\big\langle Z_s, \big(\sigma(\eta^1_s, s) - \sigma(\eta^2_s, s)\big)\,dw_s\big\rangle \\
&\qquad\qquad - 2\int_0^{t\wedge\tau_R} \big(|Z_s|^2 + \delta^2\big)^{-1}\big|\big(\sigma(\eta^1_s, s) - \sigma(\eta^2_s, s)\big)Z_s\big|^2\,ds.
\end{aligned}$$

Approximating the coefficients by smooth ones and applying condition (9.8.8), from the last equality we obtain

$$\begin{aligned}
&\mathbb{E}\ln\Big(\frac{Z_{t\wedge\tau_R}}{\delta^2} + 1\Big) \\
&\qquad \leqslant \|f_R\|_{L^q(B_R\times[0,T])}\big(\|u^1(\,\cdot\,,t)\|_{L^p(B_R\times[0,T])} + \|u^2(\,\cdot\,,t)\|_{L^p(B_R\times[0,T])}\big).
\end{aligned}$$

Letting $\delta \to 0$ we conclude that $Z_{t\wedge\tau_R} = 0$ almost surely. Since σ and b are bounded, we have $\mathbb{E}\sup_{t\in[0,T]}|\eta^i_t| < \infty$, $i = 1,2$. Therefore, $P\big(\lim_{R\to\infty}\tau_R = 1\big) = 1$. Letting $R \to \infty$ in the equality $Z_{t\wedge\tau_R} = 0$, we obtain $Z_t = 0$ almost surely for all $t \in [0,T]$. Thus, $P^1_\nu = P^2_\nu$ and, respectively, $u^1 = u^2$. □

We observe that condition (9.8.8) of the theorem holds if for some $q > d$ we have $b \in L^q\big((0,T), W^{q,1}(U_R)\big)$ and $\sigma \in L^r\big((0,T), W^{r,1}(U_R)\big)$ for every ball U_R, where $r = \min\{q, 2\}$.

If the coefficients are unbounded, the situation becomes more complicated. If we deduce the uniqueness of probability solutions of the Fokker–Planck–Kolmogorov equation from the uniqueness of solutions to the martingale problem, then it is necessary to prove, at least under the same assumptions that ensure uniqueness for the martingale problem, that every solution to the Cauchy problem is generated by a solution to the martingale problem. It is also clear that if we have managed to prove uniqueness for the Cauchy problem under the same assumptions as it is proved for the martingale problem, then the unique probability solution will be automatically generated by the solution to the martingale problem. Finally, let us note that the proof of the well-posedness of the martingale problem with unbounded coefficients often employs the following observation (see Stroock, Varadhan [**884**, Theorem 10.1.1]), which in a sense localizes the uniqueness problem.

Suppose that Ω is a domain in $\mathbb{R}^d$, $z \in \Omega$ and $\tau(x) = \inf\{t\colon x(t) \notin \Omega\}$. Let the martingale problem with bounded coefficients a^{ij}, b^i and initial condition δ_z

be well-posed and let P_z be its solution. Let the functions $\overline{a}^{ij}$ and $\overline{b}^i$ be locally bounded. If $\overline{a}^{ij} = a^{ij}$ and $\overline{b}^i = b^i$ on Ω and the measure $\overline{P}_z$ is some solution to the martingale problem with the coefficients $\overline{a}^{ij}$ and $\overline{b}^i$, then $P_z = \overline{P}_z$ on the σ-algebra $\mathcal{W}_\tau$ consisting of the events $A \in \mathcal{W}_T$ such that $A \cap \{\tau \leqslant t\} \in \mathcal{W}_t$. This property has no analog for probability solutions to the Fokker–Planck–Kolmogorov equation, which complicates finding sufficient conditions for the uniqueness of solutions to the Cauchy problem.

9.8(iii). Uniqueness for degenerate equations in the class of all measures

We have discussed in § 9.8(i) the uniqueness problem for solutions to Fokker–Planck–Kolmogorov equations with degenerating diffusion matrices in the class of absolutely continuous measures. The case where singular solutions are allowed is much more complicated. If the coefficients are globally bounded, then Figalli's result cited above enables us to deduce uniqueness of solutions to Fokker–Planck–Kolmogorov equations from uniqueness of solutions to the corresponding martingale problems. However, in case of unbounded coefficients there might be no probabilistic representations, because the Cauchy problem can have several solutions, while the martingale problem has at most one (see the example of nonuniqueness in § 9.2 and Stroock, Varadhan [**884**, Corollary 10.1.2]). Following Bogachev, Röckner, Shaposhnikov [**175**], we give several recent results, in which we do not assume boundedness of coefficients and absolute continuity of solutions.

As above, we consider the Cauchy problem

$$\partial_t \mu = L_{A,b}^* \mu, \quad \mu|_{t=0} = \nu, \tag{9.8.9}$$

in the class of solutions of the form $\mu(dxdt) = \mu_t(dx)\,dt$, where μ_t is a probability measure on $\mathbb{R}^d) = 1$ for each $t \in (0, T)$. The initial condition ν is also a probability measure. No absolute continuity is assumed.

The first result concerns a partially degenerate diffusion matrix.

Let $(x, y, t) \in \mathbb{R}^n \times \mathbb{R}^m \times (0, T)$, $n + m = d$.

Let also $A(x, y, t) = \bigl(a^{ij}(x, y, t)\bigr)_{1 \leqslant i,j \leqslant d}$ be a symmetric matrix satisfying the following conditions:

(H1) a^{ij} are continuous functions having two continuous derivatives in the variables (x, y) and, for some number $\lambda > 0$, the inequality

$$\langle A(x, y, t)\xi, \xi\rangle \geqslant \lambda(\xi_1^2 + \xi_2^2 + \cdots + \xi_n^2)$$

holds for all $(x, y, t) \in \mathbb{R}^d \times (0, T)$ and $\xi = (\xi_1, \ldots, \xi_n, \ldots, \xi_d) \in \mathbb{R}^d$.

Set $B = (b^1, b^2, \ldots, b^n)$ and $C = (b^{n+1}, b^{n+2}, \ldots, b^d)$. We assume below that also the following condition is fulfilled:

(H2) for every ball $U \subset \mathbb{R}^{n+m}$, the vector field C is bounded and continuous in (x, y) uniformly in t on $U \times (0, T)$ and there exists a number $\Lambda(U) > 0$ such that

$$\langle z - y, C(x, z, t) - C(u, y, t)\rangle \leqslant \Lambda(U)(|z - y|^2 + |x - u|^2),$$

for all $(x, z), (u, y) \in U$ and $t \in (0, T)$.

The main result in this subsection is this.

9.8.7. Theorem. *If conditions* (H1) *and* (H2) *are fulfilled, then the Cauchy problem* (9.8.9) *possesses at most one probability solution* μ *with the properties that*

$b^i \in L^2(U\times(0,T),\mu)$ for every ball $U \subset \mathbb{R}^d$ and all $i \leqslant n$ and one has

$$\lim_{K\to\infty}\int_0^T\int_{K<|x|<2K}\left(\frac{|a^{ij}(x,t)|}{1+|x|^2}+\frac{|b^i(x,t)|}{1+|x|}\right)\mu_t(dx)\,dt=0. \tag{9.8.10}$$

9.8.8. Example. Let $a^{ij}=\delta^{ij}$ if $1\leqslant i,j\leqslant n$ and $a^{ij}=0$ else. Let us consider the Cauchy problem with the operator

$$Lu=\Delta_x u+\langle B,\nabla_x u\rangle+\langle C,\nabla_y u\rangle.$$

Suppose that the coefficient B is bounded and C is Lipschitzian in the variable y on $U\times(0,T)$ for every ball $U\subset\mathbb{R}^d$. Suppose also that

$$\langle B(x,y,t),x\rangle+\langle C(x,y,t),y\rangle\leqslant C_1-C_2(|x|^2+|y|^2)^{k+1},$$

$$|B(x,y,t)|+|C(x,y,t)|\leqslant C_1\exp\bigl[C_3(|x|^2+|y|^2)^k\bigr]$$

for some $C_1,C_3>0$, $C_2>kC_3$, $k\geqslant 1$ and all $(x,y,t)\in\mathbb{R}^n\times\mathbb{R}^m\times(0,T)$. If $\exp\bigl(2kC_3(|x|^2+|y|^2)^k\bigr)\in L^1(\nu)$, then, according to a priori estimates with the Lyapunov function $\exp\bigl(C_3(|x|^2+|y|^2)^k\bigr)$, condition (9.8.10) is fulfilled, so the Cauchy problem (9.8.9) with such initial condition ν has at most one solution in the class of all probability measures.

It is worth noting that stochastic differential equations with partially degenerate diffusion matrices are studied in Levakov, Vas'kovskii [**603**], Veretennikov [**913**], and Guillin, Wang [**429**].

As shown in the previous subsection, in case of bounded coefficients, uniqueness of a probability solution follows from Lipschitzness of σ and b, where $A=\sigma\sigma^*$. Naturally a question arises about the case of unbounded coefficients. The next result gives a partial answer. It is based on the approximate Holmgren method. In this approach the main difficulty is related to the absence of solution densities.

Let now $s_k(\lambda_1,\dots,\lambda_N)$ be the elementary symmetric polynomial of degree $k=1,\dots,N$, i.e., $s_1(\lambda_1,\dots,\lambda_N)=\lambda_1+\cdots+\lambda_N$, $s_2(\lambda_1,\dots,\lambda_N)=\sum_{i,j}\lambda_i\lambda_j$, $\dots$, $s_N(\lambda_1,\dots,\lambda_N)=\lambda_1\cdots\lambda_N$.

Let $\Lambda_k(\lambda_1,\dots,\lambda_N)$ denote the collection of all possible finite linear combinations of the products $\lambda_{i_1}\cdots\lambda_{i_k}$, where $i_1,\dots,i_k$ are distinct numbers between 1 and N.

9.8.9. Theorem. *Let $A=\sigma\sigma^*$. Suppose that for every ball U the coefficient b is bounded and continuous in x uniformly in t on $U\times(0,T)$ and that there exists a number $M(U)>0$ such that*

$$\|\sigma(x,t)-\sigma(y,t)\|^2+\langle x-y,b(x,t)-b(y,t)\rangle\leqslant M(U)|x-y|^2 \tag{9.8.11}$$

for all $x,y\in U$ and all $t\in(0,T)$. Suppose that $A=A_0+A_1$, where A_0 is a constant matrix and the matrix A_1 has entries belonging to $\Lambda_k(g_1,\dots,g_N)$ with some k and some functions $g_1,\dots,g_N$, $N\geqslant k$, on $\mathbb{R}^d\times(0,T)$ that are nonnegative, locally bounded and locally Lipschitzian in x. Then there is at most one solution to the Cauchy problem in the class of measures $\mu=\mu_t\,dt$ given by a family of probability measures $(\mu_t)_{t\in(0,T)}$ on $\mathbb{R}^d$ and having the properties that $s_{k-1}(g_1,\dots,g_N)\mu$ is an absolutely continuous measure with respect to Lebesgue measure and

$$\lim_{K\to\infty}\int_0^T\int_{K<|x|<2K}\left(\frac{|a^{ij}(x,t)|}{1+|x|^2}+\frac{|b^i(x,t)|}{1+|x|}\right)\mu_t(dx)\,dt=0. \tag{9.8.12}$$

In particular, (9.8.12) *holds in the class of solutions with respect to which the functions* $a^{ij}(x,t)/(1+|x|^2)$, $b^i(x)/(1+|x|)$ *are integrable.*

9.8.10. Remark. Since $A = \sigma\sigma^*$ and the matrix σ is locally Lipschitz in x, we have $a^{ij} \in \Lambda_2(\{\sigma^{kl}\}_{1\leqslant k,l\leqslant d})$. Hence uniqueness holds for μ such that the measure $\sigma^{kl}\mu$ is absolutely continuous with respect to Lebesgue measure for all k,l, or, which in the case of a symmetric matrix σ is equivalent, the measure $\sqrt{\mathrm{tr}A}\cdot\mu$ is absolutely continuous.

9.8.11. Corollary. *Suppose that* $A = q^2R$, *where* $R(x) = \bigl(r^{ij}(x)\bigr)_{i,j\leqslant d}$ *is a positive definite symmetric matrix with* r^{ij} *Lipschitzian in* x *uniformly with respect to* t. *Assume that for every ball* U *the coefficient* b *is bounded and continuous in* x *on* $U\times(0,T)$ *uniformly in* t *and that there exists a number* $M(U)$ *such that*

$$|q(x,t)-q(y,t)|^2 + \langle x-y, b(x,t)-b(y,t)\rangle \leqslant M(U)|x-y|^2$$

for all $x,y\in U$ *and all* $t\in(0,T)$. *Then, a solution to the Cauchy problem is unique in the class of measures satisfying condition* (9.8.12).

PROOF. We observe that, by Stroock, Varadhan [**884**, Theorem 5.2.2], the matrix $\sqrt{R}$ is Lipschitzian in x. As shown in Chapter 6, the measure $q^{2d/(d+1)}\mu$ is absolutely continuous and has a density from $L^{(d+1)'}$. Therefore, the measure $q\mu$ is absolutely continuous. Thus, all conditions of Theorem 9.8.9 are fulfilled (see the remark above). Finally, the existence of a subprobability solution follows by the local boundedness and the condition that b is continuous in x uniformly in t. Condition (9.8.12) implies that this is a probability solution. □

9.8.12. Remark. We observe that if $\sqrt{A}$ and b have a linear growth, then condition (9.8.12) is fulfilled automatically. However, also for considerably faster growing coefficients this condition can be verified by using estimates with Lyapunov functions. Let $\alpha>0$ and $r>2$. Suppose that for all $(x,t)\in\mathbb{R}^d\times(0,T)$ and some positive numbers c_1, c_2, c_3 and $c_4<\alpha$, $c_5<\alpha$ we have

$$\text{(i)}\quad |x|^{r-2}\mathrm{tr}\,A(x,t) + (r-2)|x|^{r-4}\langle A(x,t)x,x\rangle + \\ + \alpha r|x|^{2r-4}\langle A(x,t)x,x\rangle + |x|^{r-2}\langle b(x,t),x\rangle \leqslant c_1,$$

$$\text{(ii)}\quad |a^{ij}(x,t)| \leqslant c_2\exp\bigl(c_4|x|^r\bigr),\quad |b^i(x,t)|\leqslant c_3\exp\bigl(c_5|x|^r\bigr).$$

Suppose also that ν is a probability measure on $\mathbb{R}^d$ such that $\exp\bigl(\alpha|x|^r\bigr)\in L^1(\nu)$, which holds, e.g., if ν is of bounded support. Then condition (9.8.12) is fulfilled for every solution μ to the Cauchy problem with the initial condition ν.

The case of identically zero diffusion matrix is considered in the next subsection.

9.8(iv). Uniqueness for the continuity equation

Here we consider the uniqueness problem for the continuity equation

$$\partial_t\mu_t + \mathrm{div}(b\mu_t) = 0$$

with respect to measures on $\mathbb{R}^d$. For the proofs of the presented results, see Bogachev, Da Prato, Röckner, Shaposhnikov [**139**]. We consider solutions given by families of locally bounded Borel measures $(\mu_t)_{t\in(0,T)}$. There is a vast literature devoted to uniqueness and existence problems for the Cauchy problem for such equations. An important problem is to specify a class of measures μ_t in which, under reasonable assumptions about coefficients and initial data, there is a unique solution to the Cauchy problem. Certainly, if the coefficients are sufficiently regular, say, Lipschitzian or satisfy the Osgood type condition, then we can take the whole class of bounded measures μ_t (see, e.g., Ambrosio, Bernard [**40**]).

According to a well-known result of Ambrosio [**39**] (see also Maniglia [**663**] for equations with a potential term) on representations of nonnegative bounded solutions by means of averaging with respect to measures concentrated on solutions to the corresponding ordinary equation $\dot{x} = b(x,t)$. More precisely, if $\mu = \mu_t\,dt$ with probability measures μ_t satisfies the continuity equation $\partial_t\mu_t + \operatorname{div}(b\mu_t) = 0$ and $|b(x,t)|/(1+|x|)$ is μ-integrable, then there exists a nonnegative Borel measure η on the space $\mathbb{R}^d \times C\big((0,T),\mathbb{R}^d\big)$ concentrated on the set of pairs (x,γ) such that γ is an absolutely continuous solution of the integral equation

$$\gamma(t) = x + \int_0^t b\big(\gamma(s),s\big)\,ds$$

and, for each function $\varphi \in C_b(\mathbb{R}^d)$ and each $t \in (0,T)$, one has

$$\int \varphi(x)\,\mu_t(dx) = \int \varphi\big(\gamma(t)\big)\,\eta(dxd\gamma).$$

In other words, the measure μ_t coincides with the image of η under the evaluation mapping $(x,\gamma) \mapsto \gamma(t)$. This representation enables us to transfer conditions for uniqueness of solutions of the ordinary equation to the corresponding continuity equation in the class of nonnegative bounded measures.

However, in the class of signed measures there is no such representation. In the case of nonsmooth coefficients a class convenient in many respects is the class of measures absolutely continuous with respect to Lebesgue measure, which is quite natural, in particular, taking into account existence results. A study of this class initiated by Cruzeiro [**257**]–[**259**] and DiPerna and Lions [**304**] was continued by many researchers. A large number of papers are devoted to the so-called Lagrangian flows and their generalizations (see Ambrosio [**38**], [**39**]). However, this class of absolutely continuous measures is rather narrow, in particular, it does not enable one to deal with singular initial data (and is essentially oriented towards vector fields having at least some minimal regularity such as the existence of divergence or being BV, see Mucha [**727**]). The paper Ambrosio, Trevisan [**47**] develops continuity equations in metric measure spaces, but again only considering solutions absolutely continuous with respect to the underlying fixed measure.

Thus, it is natural to look for other classes of measures, apart from absolutely continuous measures, in which the existence and uniqueness of solutions hold in the case of nonsmooth coefficients. It turns out that even in the one-dimensional case in the present framework new results still can be obtained.

The result presented below can be briefly formulated as follows: uniqueness holds in a certain class of measures with respect to which the given vector field b can be suitably approximated by smooth vector fields (thus, the uniqueness class may depend on b). A precise formulation is given below (see Theorems 9.8.13 and 9.8.15), but we observe that this result is consistent with typical methods of constructing solutions when b is approximated by smooth fields b_k and the solution is obtained as a limit point of the sequence of solutions μ_t^k for b_k. This yields some results for the continuity equation with a merely continuous b. In particular, this improves the result from Crippa [**255**], where the uniqueness is proved in dimension one for absolutely continuous solutions under the assumptions that b is continuous and nonnegative, the trajectories of $\dot{x} = b(x)$ do not blow up in finite time and the set of zeros $Z = \{x\colon b(x) = 0\}$ consists of a finite union of points and closed intervals. Other results with different assumptions can be found in the already cited papers and also in Alberti, Bianchini, Crippa [**17**], Ambrosio, Crippa [**41**], Ambrosio, Crippa, Figalli, Spinolo [**42**], Andreianov, Bénilan, Kruzhkov [**49**], Andreu-Vaillo, Caselles, Mazón [**50**], Bouchut, Crippa [**200**], Bouchut, James [**201**], Cipriano, Cruzeiro [**250**], Crippa, De Lellis [**256**], De Lellis [**287**], [**288**], Desjardins [**297**], Gwiazda, Jamróz, Marciniak-Czochra [**431**], Hauray, Le Bris [**445**], Jabin [**486**], Kruzkov [**544**], and Panov [**762**]. In Panov [**763**] Piccoli, Rossi [**774**], equations with additional nondivergence nonlinear terms $h(\mu)$ are studied.

Let us consider the Cauchy problem

$$\partial_t \mu + \operatorname{div}(b\mu) = 0, \quad \mu|_{t=0} = \nu, \tag{9.8.13}$$

where ν is a locally bounded Borel measure on $\mathbb{R}^d$ (possibly signed). In this subsection we assume that $b\colon \mathbb{R}^d \times (0,T) \to \mathbb{R}^d$ is Borel measurable.

9.8.13. Theorem. *Let $\mu^1 = \mu^1(dx)\,dt$ and $\mu^2 = \mu^2(dx)\,dt$ be two solutions to the Cauchy problem* (9.8.13). *Assume that the measure $\mu = \mu_t(dx)\,dt$, where $\mu_t = \mu_t^1 - \mu_t^2$, satisfies the following conditions: for some number $C_1 \geqslant 0$ and for every ball $U \subset \mathbb{R}^d$, one can find a number $C_2 \geqslant 0$, a sequence of vector fields $b_k \in C^\infty(\mathbb{R}^d \times \mathbb{R}^1, \mathbb{R}^d)$ and a sequence of positive functions $V_k \in C^1(\mathbb{R}^d)$ (all depending on the considered measure μ, the constant C_1 and the ball U) such that*

(i) $|b_k(x,t)| \leqslant C_1 + C_1|x|$,

(ii) $\inf_k \inf_U V_k(x) > 0$ *and*

$$\langle b_k(x,t), \nabla V_k(x)\rangle \leqslant \Bigl(C_2 - 2\max_{|\xi|=1}\langle \mathcal{B}_k(x,t)\xi,\xi\rangle\Bigr)V_k(x)$$

for every k and $(x,t) \in U \times [0,T]$, where $\mathcal{B}_k = (\partial_{x_i} b_k^j)_{i,j\leqslant d}$,

(iii) $\lim\limits_{k\to\infty} \|(b_k - b)\sqrt{V_k}\|_{L^1(|\mu_t|\,dt,\, U\times(0,T))} = 0$.

Then $\mu = 0$, i.e., $\mu^1 = \mu^2$.

Note once again that C_1 is global, but other objects and conditions are local.

We observe that (i) and (iii) yield that b has a $|\mu|$-version bounded by $C_1 + C_1|x|$. Of course, for singular μ this does not say that a certain natural original version of b admits this bound. For smooth b such a bound would be enough, so (ii) and (iii) can be regarded as some compensation for a possible lack of smoothness. These conditions enable us to employ Holmgren's approach combined with certain a priori estimates derived from (ii) and (iii). In case $d = 1$ condition (ii) can be written as $\bigl(b_k\sqrt{V_k}\bigr)' \leqslant 2^{-1}C_2\sqrt{V_k}$. Thus, varying V_k, we obtain diverse conditions on b_k, hence on b. In particular, for $V_k = 1$ we arrive at the classical condition $b_k' \leqslant C$, which is fulfilled if b is one-sided Lipschitzian (the function $b(x) - Cx$ is nondecreasing for some $C \geqslant 0$).

Let us consider the standard example of an ordinary differential equation on the real line without uniqueness and see what kind of uniqueness for the associated continuity equation is offered by our theorem.

9.8.14. Example. Let $d=1$ and $b(x)=\sqrt{|x|}$. Set $b_k(x)=(x^2+k^{-2})^{1/4}$. It is obvious that $|b(x)-b_k(x)|\leqslant k^{-1/2}$. Let us calculate $b_k'(x)$:

$$b_k'(x)=\frac{x}{2(x^2+k^{-2})^{3/4}}.$$

Finally, we take

$$V_k(x)=\frac{1}{b_k(x)^2}=\frac{1}{(x^2+k^{-2})^{1/2}}.$$

Clearly, (ii) is fulfilled. Moreover, we have

$$g_k(x):=|b(x)-b_k(x)|\sqrt{V_k(x)}=|b(x)-b_k(x)|b_k(x)^{-1}.$$

Note that $g_k(x)=1$ if $b(x)=0$, i.e., $x=0$. If $x\neq 0$, then $g_k(x)\leqslant k^{-1/2}|x|^{-1/2}\to 0$ if $k\to\infty$. Thus, $|g_k|\leqslant 1$ and $g_k\to I_{\{0\}}$, where $I_{\{0\}}$ is the indicator of the set $\{b=0\}=\{0\}$. Let $\mu=\mu_t\,dt$ be a locally bounded measure on $\mathbb{R}^1\times[0,T]$. According to the Lebesgue dominated convergence theorem, one has

$$\lim_{k\to\infty}\int_0^T\int_U |b-b_k|\sqrt{V_k}\,d|\mu_t|\,dt=\int_0^T |\mu_t|\bigl(\{x\in U\colon\ b(x)=0\}\bigr)\,dt$$

for every interval U. Hence the uniqueness holds in the class of all locally bounded (possibly signed) solutions $\mu=\mu_t\,dt$ such that $|\mu_t|(\{0\})=0$ for almost all $t\in[0,T]$.

It is interesting that this result is sharp: there exist two solutions $\mu_t^1\equiv\delta_0$ and $\mu_t^2=\delta_{t^2/4}$ to the Cauchy problem with $b(x)=\sqrt{|x|}$ and $\nu=\delta_0$. Note that only μ_t^2 belongs to our uniqueness class and that this solution is not absolutely continuous.

We also give a result without restrictions on the growth of b_k, but with some additional conditions on the solution.

9.8.15. Theorem. *Let $\mu^1=\mu^1(dx)\,dt$ and $\mu^2=\mu^2(dx)\,dt$ be two solutions to the Cauchy problem* (9.8.13). *Assume that the measure $\mu=\mu_t(dx)\,dt$, where $\mu_t=\mu_t^1-\mu_t^2$, satisfies the following conditions:*

(i) $\displaystyle\lim_{N\to\infty}N^{-1}\int_0^T\int_{N<|x|<2N}|b(x,t)|\,|\mu_t|(dx)\,dt=0,$

(ii) *for every ball $U\subset\mathbb{R}^d$, one can find numbers $C>0$ and $\delta>0$, a sequence of vector fields $b_k\in C^\infty(U\times\mathbb{R}^1,\mathbb{R}^d)$ and a sequence of positive functions V_k of class $C^1(U)$ (all depending on the considered measure μ and the ball U) such that*

$$\inf_k\inf_U V_k(x)>0,$$

$$\langle b_k(x,t),\nabla V_k(x)\rangle\leqslant\Bigl(C-\delta|b_k(x,t)|^2-2\max_{|\xi|=1}\langle\mathcal{B}_k(x,t)\xi,\xi\rangle\Bigr)V_k(x)$$

for every k and all $(x,t)\in U\times[0,T]$, where $\mathcal{B}_k=(\partial_{x_i}b_k^j)_{i,j\leqslant d}$,

(iii) $\displaystyle\lim_{k\to\infty}\|(b_k-b)\sqrt{V_k}\|_{L^1(|\mu_t|\,dt,\,U\times(0,T))}=0.$

Then $\mu=0$, i.e., $\mu^1=\mu^2$.

9.8.16. Remark. Note that if the sequence of functions $|b_k|$ on $U\times(0,T)$ is uniformly bounded, by making δ smaller and C larger we can restate the second estimate in (i) as

$$\langle b_k(x,t),\nabla V_k(x)\rangle\leqslant\Bigl(C-2\max_{|\xi|=1}\langle\mathcal{B}_k(x,t)\xi,\xi\rangle\Bigr)V_k(x).$$

It is worth noting that V_k looks like a so-called Lyapunov function, but there is an important difference: its sublevel sets need not be compact.

9.8.17. Corollary. *Let $d = 1$, $b \in C(\mathbb{R}^1)$ and $0 \leqslant b(x) \leqslant C + C|x|$ for every $x \in \mathbb{R}^1$ and some number $C > 0$. Then the corresponding Cauchy problem has at most one solution in the class of all locally bounded (possibly signed) measures μ on $\mathbb{R}^d \times (0,T)$ given by families of locally bounded measures $(\mu_t)_{t\in[0,T]}$ such that*

$$|\mu_t|(\partial Z) = 0 \quad \text{for almost all} \quad t \in (0,T),$$

where ∂Z is the boundary of the set $Z = \{x\colon\ b(x) = 0\}$. In particular, the latter holds for absolutely continuous μ_t provided that $b^{-1}(0)$ has Lebesgue measure zero.

9.8.18. Corollary. *Let $d = 1$. Assume that $b \in C(\mathbb{R}^1)$, $|b(x)| \leqslant C + C|x|$ and $b = g + f$, where $f \geqslant 0$, g is a Lipschitzian function with Lipschitz constant Λ, and*

$$g(x) < 0 \Longrightarrow f(x) = 0.$$

Then the corresponding Cauchy problem (10.5.1) *has at most one solution in the class of all locally bounded measures μ given by families of locally bounded measures $(\mu_t)_{t\in(0,T)}$ such that*

$$|\mu_t|(\partial Z_f) = 0 \quad \text{for almost all} \quad t \in (0,T),$$

where ∂Z_f is the boundary of the set $Z_f = \{x\colon\ f(x) = 0\}$.

Let us now consider the case $d \geqslant 2$. First we mention a known example of uniqueness (see, e.g., [**40**]), in which it is easy to check our hypotheses.

9.8.19. Example. Assume that $b \in C\big(\mathbb{R}^d \times (0,T), \mathbb{R}^d\big)$ and there exist numbers C_1, C_2 and C_3 such that

$$|b(x,t)| \leqslant C_1 + C_2|x|, \quad \langle b(x+\xi,t) - b(x,t), \xi\rangle \leqslant C_3|\xi|^2$$

for all $(x,t) \in \mathbb{R}^d \times (0,T)$ and every $\xi \in \mathbb{R}^d$. Then there exists at most one solution to (9.8.13) in the class of all locally bounded measures μ given by families of locally bounded measures $(\mu_t)_{t\in[0,T]}$.

Let us consider a more specific situation, where

$$b(x) = -\beta(|x|^2)x$$

is a radially symmetric vector field.

9.8.20. Example. Assume that $\beta \geqslant 0$ is bounded and continuous on $[0,+\infty)$ and that for every interval $[0,N]$ there exists a number $\Lambda_N > 0$ such that the function $s \mapsto \beta(s) - \Lambda_N s$ is decreasing on $[0,N]$. Then the Cauchy problem (9.8.13) with $b(x) = -\beta(|x|^2)x$ has at most one solution in the class of all locally bounded measures μ given by families of locally bounded measures $(\mu_t)_{t\in(0,T)}$ such that

$$|\mu_t|(\partial Z \setminus \{0\}) = 0 \quad \text{for almost all} \quad t \in (0,T),$$

where ∂Z is the boundary of the set $Z = \{x\colon\ \beta(|x|^2) = 0\}$.

Finally, let us discuss the existence of solutions possessing our uniqueness properties. We observe that not every approximation of b yields a solution from the considered uniqueness class. Indeed, let $b(0) = 0$ and take the approximations by smooth functions b_k such that $b_k(0) = 0$. Then for $\nu = \delta_0$ we obtain the unique solution $\mu_t^k = \delta_0$ for every k. So, the limit measure of such μ_t^k is again δ_0.

9.8.21. Proposition. *Assume that* $b\colon \mathbb{R}^d\times(0,T)\to\mathbb{R}^d$ *is continuous. Suppose that there exist a sequence of vector fields* $b_k\in C^\infty(\mathbb{R}^d\times\mathbb{R}^1,\mathbb{R}^d)$ *and a sequence of positive functions* $V_k\in C^1(\mathbb{R}^d)$ *such that*

(i) *there exist numbers* C_1, C_2 *and* C_3 *such that* $V_k\leqslant C_1V_m+C_2$ *for every* $k\leqslant m$ *and*

$$\langle b_k(x,t),\nabla V_k(x)\rangle\leqslant C_3V_k(x)\quad\forall\,(x,t)\in\mathbb{R}^d\times[0,T],$$

(ii) *the function* $(1+|x|)^{-1}|b_k(x,t)|$ *is bounded for every* k *and* $b_k\to b$ *uniformly on* $U\times(0,T)$ *for every ball* U.

Then, for any initial nonnegative finite measure ν *with* $\sup_k\|V_k\|_{L^1(\nu)}<\infty$, *there exists a family of nonnegative finite measures* $(\mu_t)_{t\in(0,T)}$ *solving the Cauchy problem* (9.8.13) *with the following property: for every ball* U *one has*

$$\lim_{k\to\infty}\int_0^T\int_U|b-b_k|\sqrt{V_k}\,d\mu_t\,dt=0. \tag{9.8.14}$$

9.8.22. Example. Let us illustrate the previous proposition in the situation of Corollary 9.8.17. Assume that b on $\mathbb{R}^1$ satisfies the following additional condition:

$$(b(x+y)-b(x))y\geqslant -C|y|^2$$

for all x,y and some $C>0$. Then all conditions of the above proposition are fulfilled with the approximations $b_k=b*\varrho_{1/k}+3\omega(1/k)$ and $V_k=b_k^{-2}$. Indeed, one has

$$b_k'\geqslant -C\quad\text{and}\quad b_kV_k'=-2b_k'V_k\leqslant 2CV_k.$$

Let $k>m$. Note that the sequence of numbers $\omega(k^{-1})$ is decreasing and

$$b*\varrho_{1/k}(x)+3\omega(1/k)\geqslant b(x)+2\omega(1/k)\geqslant b(x)+2\omega(1/m)\geqslant b*\varrho_{1/m}(x)+w(1/m).$$

Hence we have $V_k\leqslant 9V_m$. Assume that

$$\int_{\mathbb{R}^1\backslash Z^0}\frac{1}{b^2}\,d\nu<\infty.$$

As above we replace ν by $\nu-\nu_0$, where $\nu_0=\nu|_{Z^0}$ is a stationary solution. Then

$$\sup_k\int_{\mathbb{R}^1}V_k\,d\nu\leqslant\int_{\mathbb{R}^1\backslash Z^0}\frac{1}{(b+2\omega(1/k))^2}\,d\nu\leqslant\int_{\mathbb{R}^1\backslash Z^0}\frac{1}{b^2}\,d\nu<\infty.$$

By the proposition there exists a solution $(\mu_t)_{t\in(0,T)}$ such that equality (9.8.14) holds. Moreover,

$$\int_{\mathbb{R}^1\backslash Z^0}\frac{1}{b^2}\,d\mu_t<\infty$$

and $\mu_t(\partial Z)=0$ for almost all $t\in(0,T)$.

9.8(v). Estimates in the total variation and Kantorovich metrics

In this subsection we present upper bounds for the total variation, entropy and Kantorovich distances between probability solutions to parabolic Fokker–Planck–Kolmogorov equations in the spirit of the estimates discussed in § 4.5(iv), where the necessary definitions and auxiliary results can be found.

We first consider the case of solutions to the same equation having different initial conditions. Let us consider the Cauchy problem

$$\partial_t\mu=\Delta\mu-\operatorname{div}(b\mu),\quad \mu|_{t=0}=\nu, \tag{9.8.15}$$

where b is a Borel vector field that is λ-monotone, i.e.,

$$\langle b(x)-b(y), x-y\rangle \leqslant -\lambda|x-y|^2 \quad \forall\, x,y\in\mathbb{R}^d.$$

Here λ is some real number.

By a solution μ we understand a family of probability measures μ_t on $\mathbb{R}^d$ such that the mapping $t\mapsto\mu_t$ is weakly continuous on $[0,+\infty)$ and

$$\int_0^\infty\int_{\mathbb{R}^d}\bigl[\partial_t u+\Delta u+\langle b,\nabla u\rangle\bigr]\,d\mu_t\,dt=0 \quad \forall\, u\in C_0^\infty\bigl(\mathbb{R}^d\times(0,+\infty)\bigr).$$

We need more general cost functions in the Kantorovich distance than in § 4.5(iv). Let h be a continuous increasing function on the ray $[0,+\infty)$ such that $h(0)=0$. The cost function corresponding to h is defined by

$$C_h(\mu,\nu)=\inf_\pi\int_{\mathbb{R}^d\times\mathbb{R}^d} h(|x-y|)\,\pi(dx\,dy), \tag{9.8.16}$$

where inf is taken over all probability measures π on $\mathbb{R}^d\times\mathbb{R}^d$ with projections μ and ν on the factors. If $h(r)=r^p$, then $C_h^{1/p}$ is the Kantorovich metric W_p.

Let us now use the cost function (9.8.16) with $h_s(r)=h(re^s)$, where $s\in\mathbb{R}$ and $r\geqslant 0$.

The next result is obtained in Natile, Peletier, Savaré [**735**].

9.8.23. Theorem. *Let $\mu^1=(\mu_t^1)_{t\geqslant 0}$ and $\mu^2=(\mu_t^2)_{t\geqslant 0}$ be two solutions to the Cauchy problem (9.8.15) with initial conditions ν^1 and ν^2, respectively. Suppose that for all $0<\tau_0<\tau_1<\infty$ we have*

$$\int_{\tau_0}^{\tau_1}\int_{\mathbb{R}^d}|b(x)+\lambda x|\,(\mu_t^1+\mu_t^2)(dx)\,dt<\infty.$$

Then $C_{h_{\lambda t}}(\mu_t^1,\mu_t^2)\leqslant C_h(\nu^1,\nu^2)$ for $t>0$. In particular, if $\nu^1=\nu^2$, then $\mu^1=\mu^2$ and the uniqueness of solutions holds.

We observe that in the case $h(r)=r^p$, which gives the Kantorovich distance W_p, the following estimate holds:

$$W_p(\mu_t^1,\mu_t^2)\leqslant e^{\lambda t}W_p(\nu_1,\nu_2).$$

Let us explain how to obtain the estimate in the theorem in the simple situation where $\lambda=0$ and the W_1-norm is used (here $h(r)=r$).

Let b_n be a sequence of smooth and bounded vector fields converging to b in $L^1(\mu^1+\mu^2)$ such that

$$\langle b_n(x)-b_n(y), x-y\rangle\leqslant 0.$$

Let $\psi\in C_0^\infty(\mathbb{R}^d)$ and $|\psi|\leqslant 1$, $|\nabla\psi|\leqslant 1$. Consider the Cauchy problem

$$\partial_s f+\Delta f+\langle b_n,\nabla f\rangle=0,\quad f|_{s=t}=\psi.$$

Applying the maximum principle to the functions f and $|\nabla f|^2$, we derive the estimates $|f|\leqslant 1$ and $|\nabla f|\leqslant 1$. We observe that the difference $\mu^1-\mu^2$ satisfies the initial condition $\nu^1-\nu^2$ and the equation

$$\partial_t(\mu^1-\mu^2)=\Delta(\mu^1-\mu^2)-\operatorname{div}\bigl(b(\mu^1-\mu^2)\bigr).$$

Multiplying this equation by f and integrating by parts, we obtain the inequality

$$\int_{\mathbb{R}^d}\psi\,d(\mu_t^1-\mu_t^2)\leqslant\int_{\mathbb{R}^d}f(x,0)\,(\nu^1-\nu^2)(dx)+\int_0^t\int_{\mathbb{R}^d}|b-b_n|\,|\nabla f|\,d(\mu_t^1+\mu_t^2).$$

Therefore, we have

$$\int_{\mathbb{R}^d} \psi \, d(\mu_t^1 - \mu_t^2) \leqslant W_1(\nu^1, \nu^2) + \int_0^t \int_{\mathbb{R}^d} |b - b_n| \, d(\mu_t^1 + \mu_t^2).$$

Letting $n \to \infty$ and taking into account that ψ was an arbitrary function satisfying the conditions $|\psi| \leqslant 1$ and $|\nabla\psi| \leqslant 1$, we obtain $W_1(\mu_t^1, \mu_t^2) \leqslant W_1(\nu^1, \nu^2)$. In the general case the reasoning is much more involved, but the idea is the same.

Finally, note also the following interesting corollary related to stationary solutions, i.e., solutions σ of the equation $\Delta\sigma - \operatorname{div}(b\sigma) = 0$.

9.8.24. Corollary. *Let b be strictly monotone, i.e., $\lambda > 0$. Then there exists at most one stationary probability solution σ such that*

$$\int_{\mathbb{R}^d} |b(x) - \lambda x| \, \sigma(dx) < \infty.$$

Moreover, for the solution $\mu = (\mu_t)$ to the Cauchy problem (9.8.15) *one has*

$$W_p(\mu_t, \sigma) \leqslant e^{-\lambda(t-t_0)} W_p(\mu_{t_0}, \sigma).$$

Note that the uniqueness of probability stationary solutions in Chapters 4 and 5 has been proved under weaker assumptions.

Let us proceed to upper bounds for the total variation, entropy and Kantorovich distances between two probability solutions $\varrho_1(x,t)$ and $\varrho_2(x,t)$ to Fokker–Planck–Kolmogorov equations

$$\partial_t \varrho_k(x,t) = \partial_{x_i}\partial_{x_j}\bigl(a_k^{ij}(x,t)\varrho_k(x,t)\bigr) - \partial_{x_i}\bigl(b_k^i(x,t)\varrho_k(x,t)\bigr), \quad k = 1, 2, \tag{9.8.17}$$

with different diffusion matrices and drifts on $\mathbb{R}^d \times (0,T)$ with fixed $T > 0$. In case of equal initial distributions and identity diffusion matrices, for the entropy of ϱ_2 with respect to ϱ_1 we obtain the estimate

$$\int_{\mathbb{R}^d} \ln \frac{\varrho_2(x,t)}{\varrho_1(x,t)} \, \varrho_2(x,t) \, dx \leqslant \frac{1}{2} \int_{\mathbb{R}^d} |b_1(x,t) - b_2(x,t)|^2 \, \varrho_2(x,t) \, dx,$$

and for the total variation norm we obtain the estimate

$$\|\varrho_1(\,\cdot\,,t) - \varrho_2(\,\cdot\,,t)\|^2 \leqslant \int_0^t \int_{\mathbb{R}^d} |b_1(x,s) - b_2(x,s)|^2 \, \varrho_2(x,s) \, dx \, ds.$$

In the general case we obtain quite comparable estimates under rather broad assumptions about our coefficients: the diffusion matrices are locally uniformly elliptic and locally Lipschitzian in space, the drifts are locally bounded, and either some mild integrability conditions are imposed or a certain Lyapunov function exists (an advantage of the latter condition is that it is expressed entirely in terms of the coefficients). In examples we give a number of effectively verified conditions. The principal difficulty concerns the case of different diffusion matrices (see comments in Remark 9.8.29), but also the simpler case of the same diffusion matrix is interesting. The main result is applied to nonlinear Fokker–Planck–Kolmogorov equations. For the proofs, see Bogachev, Röckner, Shaposhnikov [**176**].

Let us consider a time-dependent second order elliptic operator of the form

$$L_{A,b}u = \sum_{i,j=1}^{d} a^{ij}\partial_{x_i}\partial_{x_j}u + \sum_{i=1}^{d} b^i \partial_{x_i} u,$$

where $A(x,t) = \big(a^{ij}(x,t)\big)_{i,j\leqslant d}$ is a positive symmetric matrix with Borel measurable entries and $b(x,t) = \big(b^i(x,t)\big)_{i=1}^d \colon \mathbb{R}^d\times(0,T) \to \mathbb{R}^d$ is a Borel measurable mapping. Suppose that b is locally bounded and A is locally Lipschitzian in x and locally strictly positive, i.e.,

(H) for every ball $U \subset \mathbb{R}^d$, there exist numbers $\lambda = \lambda(U) \geqslant 0$, $\alpha = \alpha(U) > 0$ and $m = m(U) > 0$ such that

$$|a^{ij}(x,t) - a^{ij}(y,t)| \leqslant \lambda|x-y|, \quad \alpha\cdot \mathrm{I} \leqslant A(x,t) \leqslant m\cdot \mathrm{I}$$

for all $x, y \in U$ and $t \in (0,T)$.

As above, we study solutions to the Cauchy problem

$$\partial_t\mu = L^*_{A,b}\mu, \quad \mu|_{t=0} = \nu, \tag{9.8.18}$$

where ν is a Borel probability measure on $\mathbb{R}^d$. In our model example of transition probabilities of diffusion processes, we obtain a bound on the distance between transition probabilities at a fixed time. As always, we consider solutions of the form $\mu(dxdt) = \mu_t(dx)\,dt$ on $\mathbb{R}^d\times(0,T)$.

Suppose now that $\mu = (\mu_t)_{t\in(0,T)}$ and $\sigma = (\sigma_t)_{t\in(0,T)}$ are two solutions to the Cauchy problem (9.8.18) with coefficients A_μ, b_μ and A_σ, b_σ, respectively, and the same initial condition ν (the case of different initial conditions is addressed in Remark 9.8.27 just to simplify the obtained estimates). The corresponding operators will be denoted by L_μ and L_σ for brevity.

Until the end of this subsection we assume that

A_μ and A_σ satisfy Condition (H) *and b_μ and b_σ are locally bounded Borel measurable.*

Let $\mu = \varrho_\mu(x,t)\,dxdt$ and $\sigma = \varrho_\sigma(x,t)\,dxdt$. Set

$$v(x,t) = \frac{\varrho_\sigma(x,t)}{\varrho_\mu(x,t)}, \quad \text{i.e., } \sigma = v\cdot\mu.$$

Let us introduce vector mappings

$$h_\mu = (h^i_\mu)^d_{i=1}, \quad h_\sigma = (h^i_\sigma)^d_{i=1}, \quad h^i_\mu = b^i_\mu - \sum_{j=1}^d \partial_{x_j} a^{ij}_\mu, \quad h^i_\sigma = b^i_\sigma - \sum_{j=1}^d \partial_{x_j} a^{ij}_\sigma,$$

$$\Phi = \frac{(A_\mu - A_\sigma)\nabla\varrho_\sigma}{\varrho_\sigma} + (h_\mu - h_\sigma).$$

The latter mapping is crucial: the distances between μ_t and σ_t will be estimated through the $L^2(\sigma)$-norm of $A_\mu^{-1/2}\Phi$. Observe that in case of equal diffusion matrices we obtain just the difference of the drifts: $\Phi = b_\mu - b_\sigma$. In case of equal drifts and constant diffusion matrices, only the first term of this mapping appears.

For the definitions of the total variation distance $\|\mu_1 - \mu_2\|$, entropy $H(\mu_1|\mu_2)$, and the Kantorovich distance $W_p(\mu_1,\mu_2)$, see §4.5(iv), p. 169.

9.8.25. Theorem. *Let $|A_\mu^{-1/2}\Phi| \in L^2\big(\mathbb{R}^d\times(0,T),\sigma\big)$. Suppose also that at least one of the following two conditions is fulfilled:*

(a) $(1+|x|)^{-2}|a^{ij}_\mu|$, $(1+|x|)^{-1}|b_\mu| \in L^1\big(\mathbb{R}^d\times(0,T),\mu\big)$, $(1+|x|)^{-1}|\Phi| \in L^1\big(\mathbb{R}^d\times(0,T),\sigma\big)$.

(b) *there exist a nonnegative function $V \in C^2(\mathbb{R}^d)$ and a number $M \geqslant 0$ such that*

$$\lim_{|x|\to\infty} V(x) = +\infty, \quad L_{A_\mu,b_\mu} V \leqslant MV, \quad \frac{\langle \Phi, \nabla V\rangle}{1+V} \in L^1\big(\mathbb{R}^d\times(0,T),\sigma\big).$$

Then

$$H(\sigma_t|\mu_t) = \int_{\mathbb{R}^d} v \ln v \, d\mu_t \leqslant \frac{1}{2}\int_0^t\!\!\int_{\mathbb{R}^d} \big|A_\mu^{-1/2}\Phi\big|^2 \, d\sigma_s \, ds. \tag{9.8.19}$$

9.8.26. Corollary. *Under the assumptions of the theorem, for every nonnegative measurable function φ on $\mathbb{R}^d\times(0,T)$, we have*

$$\|\varphi(\mu_t-\sigma_t)\|^2 \leqslant \big(1+\ln\alpha(t)\big)\int_0^t\!\!\int_{\mathbb{R}^d}\bigg|\frac{A_\mu^{-1/2}(A_\mu-A_\sigma)\nabla\varrho_\sigma}{\varrho_\sigma} + A_\mu^{-1/2}(h_\mu-h_\sigma)\bigg|^2 d\sigma_s\,ds, \tag{9.8.20}$$

where

$$\alpha(t) := \int_{\mathbb{R}^d} e^{\varphi^2(x,t)}\,\mu_t(dx).$$

In particular, if $A_\mu = A_\sigma = A$, then

$$\|\varphi(\mu_t-\sigma_t)\|^2 \leqslant \big(1+\ln\alpha(t)\big)\int_0^t\!\!\int_{\mathbb{R}^d}\big|A^{-1/2}(b_\mu-b_\sigma)\big|^2\,d\sigma_s\,ds,$$

and if $A_\mu = A_\sigma = \mathrm{I}$, then

$$\|\varphi(\mu_t-\sigma_t)\|^2 \leqslant \big(1+\ln\alpha(t)\big)\int_0^t\int_{\mathbb{R}^d}|b_\mu-b_\sigma|^2\,d\sigma_s\,ds,$$

If we have $b_\mu = b_\sigma = b$ and A_μ, A_σ do not depend on x, then

$$\|\varphi(\mu_t-\sigma_t)\|^2 \leqslant \big(1+\ln\alpha(t)\big)\int_0^t\!\!\int_{\mathbb{R}^d}\frac{\big|(A_\mu^{1/2}A_\sigma^{-1/2}-A_\mu^{-1/2}A_\sigma^{1/2})A_\sigma^{1/2}\nabla\varrho_\sigma\big|^2}{\varrho_\sigma}\,dx\,ds.$$

9.8.27. Remark. In the case of different initial conditions ν_μ and ν_σ a bit longer estimates hold. In place of (9.8.19) we have

$$\int_{\mathbb{R}^d} v\ln v\,d\mu_t \leqslant \frac{1}{2}\int_0^t\!\!\int_{\mathbb{R}^d}\big|A_\mu^{-1/2}\Phi\big|^2\,d\sigma_s\,ds + H(\nu_\sigma|\nu_\mu). \tag{9.8.21}$$

The extra term with $H(\nu_\sigma|\nu_\mu)$ will be added also to the integral in the right-hand side of (9.8.20).

In the case where $\varphi = 1$ we obtain the usual total variation distance, hence the classical Pinsker–Csiszár–Kullback inequality (see (4.5.17))

$$\|\mu-\sigma\|^2 \leqslant 2H(\sigma|\mu)$$

can be applied.

The estimate in the theorem can be combined with the estimate

$$W_p(\mu_1,\mu_2) \leqslant C\Big[H(\mu_1|\mu_2)^{1/p} + 2^{-1/(2p)}H(\mu_1|\mu_2)^{1/(2p)}\Big]$$

established in Bolley, Villani [**194**], where C is a number that depends on the integral of $\exp(\kappa|x|^p)$ against μ_2 for any fixed number κ (so that if we fix κ and consider only measures μ_2 such that the integral of $\exp(\kappa|x|^p)$ against μ_2 does not exceed a fixed number M, then we can find C depending only on κ and M).

9.8.28. Remark. The theorem and the corollary involve (through Φ) the logarithmic gradient $\nabla\varrho_\sigma/\varrho_\sigma$ of the measure μ_σ (in the case where A_μ and A_σ are different). If the norms of $A_\mu - A_\sigma$ and A_μ^{-1} are uniformly bounded, then, up to a constant factor, the right-hand side of (9.8.19) is estimated by the $L^2(\sigma)$-norms of $|b_\mu - b_\sigma|$, $|\nabla a_\mu^{ij} - \nabla a_\sigma^{ij}|$, and $|\nabla\varrho_\sigma|/\varrho_\sigma$. Let us recall the estimates of the $L^2(\sigma)$-norm of $\nabla\varrho_\sigma/\varrho_\sigma$ obtained in Chapter 7. Suppose that λ and α in (H1) can be chosen independent of U and that $|b_\sigma| \in L^2(\sigma)$. Assume also that the function $\Lambda(x) := \ln\max(|x|, 1)$ belongs to $L^2(\sigma)$, which is true if, for example, $\langle b_\sigma(x,t), x\rangle \leqslant C_1|x|^2\Lambda(x) + C_2$ with some constants C_1 and C_2 and $\Lambda \in L^2(\nu)$. If the initial distribution ν has finite entropy, i.e., possesses a density ϱ_ν such that $\ln\varrho_\nu \in L^1(\nu)$, then for every $\tau < T$ we have

$$\int_0^\tau\int_{\mathbb{R}^d} \frac{|\nabla\varrho_\sigma(x,t)|^2}{\varrho_\sigma(x,t)}\,dx\,dt \leqslant K < \infty,$$

where K is a number that depends on the $L^2(\sigma)$-norm of $|b_\sigma|$, the entropy of ν, and the bounds on the integrals of Λ against σ_t. More precisely,

(9.8.22)

$$\int_0^\tau\int_{\mathbb{R}^d} \frac{|\nabla\varrho_\sigma|^2}{\varrho_\sigma}\,dx\,dt \leqslant \alpha^{-2}\Bigl(\|b_\sigma\|_{L^2(\sigma)} + \sqrt{3}\lambda d^{5/2}\Bigr)^2$$
$$+ 2\ln 2\alpha^{-1} + 2\alpha^{-1}\int_{\mathbb{R}^d} \varrho_\nu(x)\ln\varrho_\nu(x)\,dx + 2\alpha^{-1}(d+1)\int_{\mathbb{R}^d} \varrho_\sigma(x,\tau)\Lambda(x)\,dx.$$

If the integrals of $\Lambda(x)$ against σ_t over $\mathbb{R}^d$ remain bounded as $t \to T$ (which holds, for example, if $\langle b_\sigma(x,t), x\rangle \leqslant C_1|x|^2 + C_2$ with some constants C_1 and C_2 and $\Lambda \in L^1(\nu)$), then (9.8.22) is true with $\tau = T$.Therefore, there are efficient conditions in terms of the coefficients to verify that the right-hand side of our estimate is finite. Thus, in the previous situation we arrive at the following bound:

(9.8.23)

$$\|\varphi(\mu_t - \sigma_t)\|$$
$$\leqslant C(t) \sup_{x,t,i,j}\Bigl[|b_\mu(x,t) - b_\sigma(x,t)| + |a_\mu^{ij}(x,t) - a_\sigma^{ij}(x,t)| + |\nabla a_\mu^{ij}(x,t) - \nabla a_\sigma^{ij}(x,t)|\Bigr],$$

where $C(t)$ depends also on $d, \lambda, \alpha, \|b_\sigma\|_{L^2(\sigma)}, \|\Lambda\|_{L^1(\sigma)}$, and $\|\ln\varrho_\nu\|_{L^1(\nu)}$.

There is also a modification of (9.8.22): for almost all $\tau \in (0,T)$ one has

$$\int_0^\tau\int_{\mathbb{R}^d} \Bigl|\frac{A_\sigma^{1/2}\nabla\varrho_\sigma}{\varrho_\sigma}\Bigr|^2\,d\sigma$$
$$\leqslant \int_0^\tau\int_{\mathbb{R}^d} |A^{-1/2}h_\sigma|^2\,d\sigma + 2\int_{\mathbb{R}^d} [\varrho_\nu(x)\ln\varrho_\nu(x) - \varrho_\sigma(x,\tau)\ln\varrho_\sigma(x,\tau)]\,dx.$$

9.8.29. Remark. Let us observe that if $d = 1$, $A = 1$, $\varphi = 1$, and there exist diffusion processes ξ_1 and ξ_2 with drifts b_1 and b_2 and initial distribution ν (which is the case, e.g., for bounded drifts), our estimates agree with the estimates obtained in Kabanov, Liptser, Shiryaev [**496**], Liese [**612**], and Liese, Schmidt [**613**], [**614**] for the total variation distance between the distributions of ξ_1 and ξ_2 in the path space $C[0,T]$. Assuming for simplicity that the drifts b_1 and b_2 are bounded and do not depend on t, we obtain by the Girsanov theorem, that for any fixed t the distributions in $C[0,t]$ of the diffusions governed by the stochastic equations

$$d\xi_1(t) = \sqrt{2}dw_t - b_1\bigl(\xi_1(t)\bigr)dt, \quad d\xi_2(t) = \sqrt{2}dw_t - b_2\bigl(\xi_2(t)\bigr)dt$$

with $\xi_1(0)$ and $\xi_2(0)$ having distribution η are equivalent to the Wiener measure P and the corresponding Radon–Nikodym densities are given by

$$\varrho_i(w) = \exp\biggl(\int_0^t b_i(w_s)\,dw_s - \frac{1}{2}\int_0^t |b_i(w_s)|^2\,ds\biggr).$$

This enables one to estimate the $L^1(P)$-norm of the function $\varrho_1(w) - \varrho_2(w)$, which yields an estimate on the total variation norm of the measure $\mu_t^1 - \mu_t^2$, where μ_t^i is the distribution of $\xi_i(t)$. However, in spite of this explicit expression, the derivation of the desired estimate is not trivial, and the Hellinger distance and the associated Hellinger processes are employed in the cited papers. Apparently, this method extends to multidimensional diffusions, but in this way it is impossible to deal with the case where the diffusion processes have different diffusions matrices, because in such a case their distributions in the functional space are typically mutually singular, as happens, e.g., if $A_1 = \mathrm{I}$ and $A_2 = 2\mathrm{I}$ (in the one-dimensional case with different analytic A_1 and A_2, they are always mutually singular, see Bogachev [**126**, Section 4.4]).

9.8.30. Remark. It should be also noted that if we are interested only in W_p-estimates, then a simpler approach is possible. Let $\mu_t\,dt$ and $\sigma_t\,dt$ be solutions on $\mathbb{R}^d\times(0,T)$ to the continuity equations

$$\partial_t\mu_t + \mathrm{div}(b\mu_t) = 0, \quad \partial_t\sigma_t + \mathrm{div}(h\sigma_t) = 0$$

with initial distribution ν. Suppose that b and h are Lipschitzian on $\mathbb{R}^d$ with constant λ. Then $\mu_t = \nu\circ x_t^{-1}$ and $\sigma_t = \nu\circ y_t^{-1}$, where

$$\dot{x}_t(z) = b\bigl(x_t(z)\bigr), \quad x_0(z) = z, \quad \dot{y}_t(z) = h\bigl(y_t(z)\bigr), \quad y_0(z) = z.$$

We observe that

$$\begin{aligned}\frac{d}{dt}\frac{|x_t - y_t|^p}{p} &\leqslant |b(x_t) - b(y_t)|\,|x_t - y_t|^{p-1} + |b(y_t) - h(y_t)|\,|x_t - y_t|^{p-1}\\ &\leqslant \Bigl(\lambda + \frac{p-1}{p}\Bigr)|x_t - y_t|^p + \frac{1}{p}|b(y_t) - h(y_t)|^p,\end{aligned}$$

where we do not indicate dependence of solutions on the initial point z. Therefore,

$$|x_t - y_t|^p \leqslant \int_0^t e^{c_p(t-s)}|b(y_s) - h(y_s)|^p\,ds, \quad c_p = p\lambda + p - 1.$$

From the definition of the metric W_p we find that

$$W_p^p(\mu_t,\sigma_t) \leqslant \int_{\mathbb{R}^d}|x_t(z) - y_t(z)|^p\,\nu(dz) \leqslant \int_0^t e^{c_p(t-s)}\int_{\mathbb{R}^d}\bigl|b\bigl(y_s(z)\bigr) - h\bigl(y_s(z)\bigr)\bigr|^p\,\nu(dz)\,ds.$$

Since $\sigma_t = \nu\circ y_t^{-1}$, we arrive at the inequality

$$W_p(\mu_t,\sigma_t) \leqslant C(\lambda,T)\biggl(\int_0^t\int_{\mathbb{R}^d}|b(z) - h(z)|^p\,\sigma_s(dz)\,ds\biggr)^{1/p},$$

where $C(\lambda,T) = e^{c_pT/p}$.

In a similar manner one can obtain upper bounds for solutions $\mu_t\,dt$ and $\sigma_t\,dt$ to the Fokker–Planck–Kolmogorov equation

$$\partial_t\mu_t = \Delta\mu_t - \mathrm{div}(b(x)\mu_t), \quad \partial_t\sigma_t = \Delta\sigma_t - \mathrm{div}(h(x)\sigma_t)$$

with initial distribution ν. Suppose again that b and h are Lipschitzian with constant λ. Consider the solutions x_t and y_t to the stochastic equations

$$dx_t = \sqrt{2}dw_t + b(x_t)\,dt, \quad dy_t = \sqrt{2}dw_t + h(y_t)\,dt$$

with the same initial distribution ν. We observe that $d(x_t - y_t) = \bigl(b(x_t) - h(y_t)\bigr)\,dt$. Repeating the previous reasoning we obtain

$$W_p^p(\mu_t, \sigma_t) \leqslant \mathbb{E}|x_t - y_t|^p \leqslant C(\lambda, T)\int_0^t\int_{\mathbb{R}^d} |b(z) - h(z)|^p\,\sigma_s(dz)\,ds.$$

Let us give effective conditions to verify the above assumptions (a) or (b).

9.8.31. Corollary. *Let $A_\mu = A_\sigma = A$ be uniformly bounded (and satisfy* (H)*). Suppose that for some numbers $\gamma_1 > 0$ and $\gamma_2 > 0$ we have*

$$\langle b_\mu(x,t), x\rangle \leqslant \gamma_1 + \gamma_2|x|^2.$$

Then

$$\|\mu_t - \sigma_t\|^2 \leqslant 2\int_0^t\int_{\mathbb{R}^d} |A^{-1/2}(b_\mu - b_\sigma)|^2\,d\sigma_s\,ds.$$

Moreover, for any $p \geqslant 1$ and $K > 0$ the following estimate holds:

$$\|(1+|x|^p)(\mu_t - \sigma_t)\|^2$$
$$\leqslant 2K^{-1}\biggl(1 + \ln\biggl(\int_{\mathbb{R}^d} e^{K(1+|x|^p)^2}\,\mu_t(dx)\biggr)\biggr)\int_0^t\int_{\mathbb{R}^d} |A^{-1/2}(b_\mu - b_\sigma)|^2\,d\sigma_s\,ds.$$

9.8.32. Example. In case $A_\mu = A_\sigma$ is uniformly bounded, condition (a) is fulfilled if $|b_\mu(x)| \leqslant C + C|x|$ or if $|b_\mu(x)| \leqslant C + C|x|^m$ and $|x|^{m-1}$ is integrable with respect to μ and σ. The latter can be verified by using Lyapunov functions. Also the assumption that $|b_\mu - b_\sigma|^2$ is σ-integrable can be verified in these terms. Certainly, the case of bounded b_μ and b_σ is covered by both conditions.

9.8.33. Example. Let L_μ be the Ornstein–Uhlenbeck operator

$$\Delta u(x) - \langle x, \nabla u(x)\rangle$$

and let L_σ be its perturbation by a first order term generated by a bounded Borel vector field b_0 on $\mathbb{R}^d$. Then

$$\|\varphi(\mu_t - \sigma_t)\|^2 \leqslant \bigl(1 + \ln\alpha(t)\bigr)\int_0^t\int_{\mathbb{R}^d} |b_0|^2\,d\sigma_s\,ds.$$

In particular, for $\varphi = 1$ we obtain that

$$\|\mu_t - \sigma_t\|^2 \leqslant 2\int_0^t\int_{\mathbb{R}^d} |b_0|^2\,d\sigma_s\,ds.$$

This dimension-free estimate can be extended to the infinite-dimensional case considered in Chapter 10 (see Exercise 10.5.14).

9.8.34. Remark. If A_μ is uniformly bounded and for some $p \geqslant 1$, $K > 0$, $\gamma_1 > 0$ and $\gamma_2 > 2pK$ we have

$$\langle b_\mu(x,t), x\rangle \leqslant \gamma_1 - \gamma_2|x|^{2p}.$$

Hence, by Corollary 7.1.2, for some $C > 0$ and all $t \in (0,T)$ we obtain that

$$\int_{\mathbb{R}^d} e^{K|x|^{2p}}\,\mu_t(dx) \leqslant e^{Ct} - 1 + e^{Ct}\int_{\mathbb{R}^d} e^{K|x|^{2p}}\,\nu(dx).$$

9.8.35. Corollary. *Let A_μ and A_σ satisfy* (H). *Suppose that there are numbers $\lambda_1, \lambda_2 > 0$ such that*

$$\lambda_1\cdot \mathrm{I} \leqslant A_\mu(x,t) \leqslant \lambda_2\cdot \mathrm{I},\quad \lambda_1\cdot \mathrm{I} \leqslant A_\sigma(x,t) \leqslant \lambda_2\cdot \mathrm{I} \quad \textit{for all } (x,t).$$

Assume also that $|x|^m \in L^1(\nu)$, $\nu = \varrho_0\,dx$, $\varrho_0 \ln \varrho_0 \in L^1(\mathbb{R}^d)$ and

$$\langle b_\mu(x,t), x\rangle \leqslant \gamma_1 + \gamma_2|x|^2, \quad |b_\sigma(x,t)| \leqslant \gamma_3 + \gamma_4|x|^m$$

for some numbers $m, \gamma_i \geqslant 0$. Then

$$\|\mu_t - \sigma_t\|^2 \leqslant C(T)\sup_{x,t}\|A_\mu - A_\sigma\|^2 + C(T)\int_0^t\!\!\int_{\mathbb{R}^d} |A_\mu^{-1/2}(h_\mu - h_\sigma)|^2\,d\sigma_s\,ds,$$

where $h_\mu^i - h_\sigma^i = b_\mu^i - b_\sigma^i - \partial_{x_j}(a_\mu^{ij} - a_\sigma^{ij})$ and the number $C(T)$ on the right depends only on $T, m, \lambda_i, \gamma_i$, $\big\|\,|x|^{2m}\big\|_{L^1(\nu)}$, and $\|\varrho_0 \ln \varrho_0\|_{L^1(\mathbb{R}^d)}$.

Suppose now that, for every measure μ on $\mathbb{R}^d\times(0,T)$ given by a family $(\mu_t)_{t\in(0,T)}$ of probability measures on $\mathbb{R}^d$, we are given a locally bounded Borel measurable mapping

$$b(\mu,\cdot,\cdot)\colon\ \mathbb{R}^d\times(0,T) \to \mathbb{R}^d.$$

Then we can consider the Cauchy problem for the nonlinear Fokker–Planck–Kolmogorov equation (see § 6.7(iii), p. 278)

$$\partial_t\mu = \Delta\mu - \operatorname{div}\big(b(\mu,x,t)\mu\big), \quad \mu|_{t=0} = \nu. \tag{9.8.24}$$

By a solution we mean a measure μ given by a family of probability measures $(\mu_t)_{t\in(0,T)}$ such that the corresponding integral identity holds, see § 6.7(iii). Certainly, nonlinear equations with nonconstant second order coefficients $a^{ij}(\mu,x,t)$ can be considered similarly. We deal with $A = \mathrm{I}$ just for simplicity.

Below we use the notation $L_\mu u = \Delta u + \langle b(\mu), \nabla u\rangle$.

Let $C^+[0,T]$ denote the set of nonnegative continuous functions on $[0,T]$. Suppose that $V \in C^2(\mathbb{R}^d)$ and $V \geqslant 1$. For $\alpha \in C^+[0,\tau_0]$ and $\tau \in (0,T)$ we set

$$\mathcal{M}_{\tau,\alpha}(V) = \Big\{\mu(dxdt) = \mu_t(dx)\,dt\colon\ \mu_t \geqslant 0,\ \mu_t(\mathbb{R}^d) = 1,$$
$$\int_{\mathbb{R}^d} V(x)\,\mu_t(dx) \leqslant \alpha(t),\ t \in (0,\tau]\Big\}.$$

If $V(x) = e^{K|x|^{2p}}$, then the corresponding set $\mathcal{M}_{\tau,\alpha}(V)$ will be denoted by the symbol $\mathcal{M}_{\tau,\alpha}^{K,p}$.

Let $\|\mu\|_{p,\tau}$ be the norm defined by

$$\|\mu\|_{p,\tau}^2 := \int_0^\tau \|(1+|x|^p)\mu_s\|^2\,ds$$

on the linear space of signed measures for which it is finite (to see that it is a norm, one can use the triangle inequality for the total variation norm and the Cauchy inequality or apply the fact that here we deal with the L^2-norm of a mapping with values in a normed space). Note that $\mathcal{M}^{K,p}_{\tau,\alpha}$ is a complete metric space with respect to the metric generated by this norm, which follows by the completeness of the space $L^2([0,\tau],X)$ of L^2-mappings with values in a Banach space X.

For shortening notation, we shall write occasionally $b(\mu)$ in place of $b(\mu,x,t)$.

9.8.36. Corollary. *Let $p \geqslant 1$, $K > 0$ and suppose that for every function $\alpha \in C^+[0,T]$ there exist numbers $\gamma_1(\alpha) > 0$ and $\gamma_2(\alpha) > 2pK$ such that for every $\tau \in (0,T)$ and $\mu \in \mathcal{M}^{K,p}_{\tau,\alpha}$ one has*

$$\langle b(\mu,x,t),x\rangle \leqslant \gamma_1(\alpha) - \gamma_2(\alpha)|x|^{2p} \quad \forall\,(x,t) \in \mathbb{R}^d\times(0,\tau].$$

Suppose also that

$$|b(\mu,x,t) - b(\sigma,x,t)| \leqslant Ce^{K|x|^{2p}}\|(1+|x|^p)(\mu_t-\sigma_t)\|.$$

Then, for every probability measure ν on $\mathbb{R}^d$ such that $e^{K|x|^{2p}} \in L^1(\nu)$, there exist $\tau \in (0,T)$ and $\alpha \in C^+[0,T]$ such that a solution to the Cauchy problem (9.8.24) in the class of measures $\mathcal{M}^{K,p}_{\tau,\alpha}$ exists and is unique.

9.8.37. Example. Let

$$b(\mu,x,t) = \beta(x,t) + \int_{\mathbb{R}^d} K(x,y)\,\mu_t(dy),$$

where $\beta\colon \mathbb{R}^d\times[0,T] \to \mathbb{R}^d$ and $K\colon \mathbb{R}^d\times\mathbb{R}^d \to \mathbb{R}^d$ are Borel measurable locally bounded mappings such that there exist numbers $C > 0$, $2p > q > 0$, $\gamma_1 > 0$, $\gamma_2 > 2pK$ for which

$$|K(x,y)| \leqslant C(1+|x|^q)(1+|y|^p), \quad \langle\beta(x,t),x\rangle \leqslant \gamma_1 - \gamma_2|x|^{2p}.$$

Then all conditions of the above corollary are fulfilled.

9.8.38. Corollary. *Suppose that there exists a Lyapunov function $V \in C^2(\mathbb{R}^d)$ such that $\lim\limits_{|x|\to\infty} V(x) = +\infty$ and for every $\tau \in (0,T)$ and $\alpha \in C^+[0,T]$ the following conditions are fulfilled:*

(i) *for any $\mu \in \mathcal{M}_{\tau,\alpha}(V)$ there is a number $M(\alpha)$ such that $L_\mu V \leqslant M(\alpha)V$;*

(ii) *for every μ and σ in $\mathcal{M}_{\tau,\alpha}(V)$ one has $b(\mu) \in L^2(\sigma)$ and, in addition, for every ball $U \subset \mathbb{R}^d$ there exists a number $C(U,\alpha,\tau) > 0$ such that*

$$\sup_{\mu\in\mathcal{M}_{\tau,\alpha}(V)} \|b(\mu)\|_{L^\infty(U\times(0,\tau])} \leqslant C(U,\tau,\alpha);$$

(iii) *if a sequence of measures $\{\mu^n\}$ and a measure μ in $\mathcal{M}_{\tau,\alpha}(V)$ are such that*

$$\lim_{n\to\infty}\int_0^\tau \|\mu^n_t - \mu_t\|^2\,dt = 0,$$

then the mappings $b(\mu_n)$ converge to $b(\mu)$ in $L^2(\mu,\mathbb{R}^d)$.

Then there exists a number $\tau \in (0,\tau_0]$ such that on the interval $[0,\tau]$ the Cauchy problem (9.8.24) has a solution.

Let $\mathcal{M}_T(V)$ denote the set of nonnegative measures $\mu(dxdt) = \mu_t(dx)\,dt$ such that

$$\sup_{t\in(0,T)}\int_{\mathbb{R}^d} V\,d\mu_t < \infty.$$

9.8.39. Corollary. *Let $V > 1$ and $W = \sqrt{\ln V}$. Suppose that for every measure μ in $\mathcal{M}_T(V)$ there exist a positive function $\Psi_\mu \in C^2(\mathbb{R}^d)$ and a number $\beta(\mu)$ such that $\lim\limits_{|x|\to\infty} \Psi_\mu(x) = +\infty$, $|\nabla\Psi_\mu|\Psi_\mu^{-1}V^{-1}$ is bounded and $L_\mu\Psi_\mu \leqslant \beta(\mu)\Psi_\mu$. Suppose also that there exists an increasing continuous function G on $[0,+\infty)$ such that $G(0) = 0$ and*

$$|b(\mu,x,t) - b(\sigma,x,t)| \leqslant \sqrt{V(x)}G(\|W(\mu_t - \sigma_t)\|) \quad \forall\, \mu,\sigma,x,t.$$

If we have

$$\int_{0+} \frac{du}{G^2(\sqrt{u})} = +\infty,$$

then the Cauchy problem (9.8.24) has at most one solution in the class $\mathcal{M}_T(V)$.

The estimate from Theorem 9.8.25 can also be used for proving the differentiability of solutions to the Cauchy problem for linear Fokker–Planck–Kolmogorov equations with respect to a parameter. For a different approach to this, see Pardoux, Veretennikov [**764**].

9.8.40. Corollary. *Suppose that for every $\alpha \in [0,1]$ there exists a mapping*

$$b(\alpha,\,\cdot\,,\,\cdot\,)\colon \mathbb{R}^d\times(0,T) \to \mathbb{R}^d$$

such that b is continuously differentiable in α and for every ball U there exists a number $C(U)$ such that

$$\|b(\alpha,\,\cdot\,,\,\cdot\,)\|_{L^\infty(U\times(0,T))} + \|\partial_\alpha b(\alpha,\,\cdot\,,\,\cdot\,)\|_{L^\infty(U\times(0,T))} \leqslant C(U).$$

Suppose that for every $\alpha \in [0,1]$ there exist numbers $\gamma_1(\alpha)$ and $\gamma_2(\alpha)$ such that

$$|b(\alpha,x,t)| \leqslant \gamma_1(\alpha) + \gamma_2(\alpha)|x|\ln(1+|x|).$$

Let μ^α be a probability solution to the Cauchy problem (9.8.18) with $b(\alpha,x,t)$ and $A = \mathrm{I}$. Suppose that for every $\alpha_0 \in [0,1]$

$$\lim_{r\to 0}\int_0^T\int_{\mathbb{R}^d}\Bigl|\frac{b(\alpha_0+r,x,t) - b(\alpha_0,x,t)}{r} - \partial_\alpha b(\alpha_0,x,t)\Bigr|^2 \mu_t^{\alpha_0}(dx)\,dt = 0.$$

Then the density $\varrho(\alpha,x,t)$ of the measure μ^α is differentiable in α.

Another approach to proving uniqueness of probability solutions to nonlinear Fokker–Planck–Kolmogorov equations, based on a modification of Holmgren's principle, is developed in Manita, Romanov, Shaposhnikov [**664**], [**665**].

9.8(vi). Mean field games

Mean field games is a recent area at the border line of statistical physics, stochastic and nonlinear PDEs, initiated in Huang, Malhame, Caines [**467**], [**468**] and Lasry, Lions [**586**], [**587**], [**588**] and concerned with a study of the limiting behavior of systems involving very large numbers of rational agents which play differential games under partial information. For recent surveys, see Bensoussan, Frehse, Yam [**106**], Cardaliaguet [**218**], Gomes, Saude [**417**], Guéant, Lasry, Lions [**427**], and Lasry, Lions [**587**]. A typical model for mean field games is the following system:

$$\begin{cases} -\partial_t u - \Delta u + H(x,\mu_t,\nabla u) = F(x,\mu_t), & \mathbb{R}^d\times[0,T),\\ \partial_t\mu_t - \Delta\mu_t - \operatorname{div}\bigl(\frac{\partial}{\partial p}H(x,\mu_t,\nabla u)\mu_t\bigr) = 0, & \mathbb{R}^d\times(0,T], \end{cases}$$

with the initial-terminal conditions $u(x,T) = G(x,\mu_T)$, $\mu_0 = \nu$ of suitable classes.

The heuristic interpretation is this: an average agent controls the stochastic differential equation $dx_t = \sqrt{2}d\,w_t + b(x_t,t)\,dt$, where w_t is a standard Winer processes. Letting $\mu_t(B) = P(x_t \in B)$, he aims at minimizing the quantity

$$\mathbb{E}\Bigl(\frac{1}{2}\int_0^T L(x_t,\mu_t,b(x_t,t))\,dt + G(x_T,\mu_T)\Bigr),$$

where L is the Fenchel conjugate of H with respect to the last variable. The function u is the value function of our average player. His optimal control is heuristically given in feedback form by $b(x,t) = -\frac{\partial}{\partial p}H\bigl(x,\mu_t,\nabla u(x,t)\bigr)$.

As an example, consider $H(x,m,p) = |p|^2/2$. Then the systems of equations takes the form

$$\left\{\begin{array}{ll} -\partial_t u - \Delta u + \frac{1}{2}|\nabla u|^2 = F(x,\mu_t), & \mathbb{R}^d\times[0,T),\\ \partial_t\mu_t - \Delta\mu_t - \operatorname{div}(\mu_t\nabla u) = 0, & \mathbb{R}^d\times(0,T],\end{array}\right. \tag{9.8.25}$$

with initial-terminal conditions $u(x,T) = G(x,\mu_T)$, $\mu_0 = \nu$. Note that if u is known, then the second equation is exactly the Fokker–Planck–Kolmogorov equation.

Let $\mathcal{P}_1(\mathbb{R}^d)$ be the space of probability measures on $\mathbb{R}^d$ with finite first moment equipped with the Kantorovich metric $W_1(\mu,\sigma)$.

Let us give two results from Cardaliaguet [**218**] and Gomes, Saude [**417**].

9.8.41. Theorem. *Suppose that ν is a probability measure on $\mathbb{R}^d$ with finite second moment, $F,G\colon \mathbb{R}^d\times\mathcal{P}_1(\mathbb{R}^d)\to\mathbb{R}$ are bounded and Lipschitz. Then, there exist $(\mu_t)_{t\in[0,T]}\in C\bigl([0,T],\mathcal{P}_1(\mathbb{R}^d)\bigr)$ and $u\in C^{2,1}\bigl(\mathbb{R}^d\times[0,T)\bigr)\bigcap C\bigl(\mathbb{R}^d\times[0,T]\bigr)$ satisfying the system of equations* (9.8.25) *with $u(x,T) = G(x,\mu_T)$, $\mu_0=\nu$.*

Proof. We observe that $\Delta e^{-u/2} = 2^{-1}e^{-u/2}\Bigl[-\Delta u + 2^{-1}|\nabla u|^2\Bigr]$. The first equation in the system can be written as $\partial_t v + \Delta v - 2Fv = 0$, where $v = e^{-u/2}$. Since F and G are bounded and Lipschitz, the Cauchy problem for this linear equation with terminal condition $v(x,T) = e^{-G(x,\mu_T)/2}$ has a unique positive bounded solution of class $C^{2,1}\bigl(\mathbb{R}^d\times[0,T)\bigr)\bigcap C\bigl(\mathbb{R}^d\times[0,T]\bigr)$. By using the maximum principle and a priori estimates, one can readily show that the solution v, hence also u, continuously depends on F and G. In addition, $|\nabla u|$ is bounded on $\mathbb{R}^d\times[0,T]$ by a constant depending on F and G.

Let $(\sigma_t)_{t\in[0,T]}\in C\bigl([0,T],\mathcal{P}_1(\mathbb{R}^d)\bigr)$ and $\sigma_0=\nu$. Substituting σ_t in F and G, we find a solution u to the first equation of the system. We now solve the second equation, which is the Fokker–Planck–Kolmogorov equation with initial condition $\mu_0=\nu$. We observe that the solution (μ_t) satisfies the estimates

$$\sup_{t\in[0,T]}\int_{\mathbb{R}^d}|x|^2\,\mu_t(dx)\leqslant C_1,\quad W_1(\mu_t,\mu_s)\leqslant C_2\sqrt{|t-s|}$$

where C_1, C_2 depend only on F, G and ν, but not on (σ_t). The first estimate is a standard estimate obtained by means of the Lyapunov function $V(x) = |x|^2$. The second one is readily deduced from a probabilistic representation of the solution $\mu_t(B) = P(x_t\in B)$, where $dx_t = \sqrt{2}dw_t - \nabla u(x_t,t)\,dt$, w_t is a Wiener process. It suffices to note that

$$W_1(\mu_t,\mu_s)\leqslant\mathbb{E}|x_t-x_s|\leqslant\frac{1}{2}\sqrt{|t-s|}+\int_s^t\!\!\int_{\mathbb{R}^d}|\nabla u(x,\tau)|\,\mu_\tau(dx)\,d\tau.$$

Let us introduce the set

$$\mathcal{K} = \Big\{\mu_t \in C\big([0,T], \mathcal{P}_1(\mathbb{R}^d)\big):$$
$$\mu_0 = \nu, \sup_{t\in[0,T]} \int_{\mathbb{R}^d} |x|^2\, \mu_t(dx) \leqslant C_1, \quad W_1(\mu_t, \mu_s) \leqslant C_2\sqrt{|t-s|}\Big\}.$$

It is clear that $\mathcal{K}$ is a convex compact set in $C\big([0,T], \mathcal{P}_1(\mathbb{R}^d)\big)$. We define a mapping of this compact set into itself as follows: $\mu_t = \Psi(\sigma_t)$, where (μ_t) is a probability solution to the second equation constructed for the coefficient ∇u, where u is a solution to the first equation of the system with $F(x,\sigma_t)$ and $G(x,\sigma_t)$. The mapping Ψ is continuous and by the Schauder theorem has a fixed point $\mu_t = \Psi(\mu_t)$. The pair (μ_t, u) is the desired solution to the system. □

9.8.42. Theorem. *Suppose that in the previous theorem we have*

$$\int_{\mathbb{R}^d} \big[F(x,\mu) - F(x,\sigma)\big]\,(\mu-\sigma)(dx) > 0, \ \int_{\mathbb{R}^d} \big[G(x,\mu) - G(x,\sigma)\big]\,(\mu-\sigma)(dx) \geqslant 0$$

for all $\mu, \sigma \in \mathcal{P}_1(\mathbb{R}^d)$ with $\mu \neq \sigma$. Then (9.8.25) has a unique solution.

PROOF. Let (u, μ_t) and (v, σ_t) be two solutions. We observe that

$$-\partial_t(u-v) - \Delta(u-v) + \frac{1}{2}|\nabla u|^2 - \frac{1}{2}|\nabla v|^2 = F(x,\mu) - F(x,\sigma),$$
$$\partial_t(\mu_t - \sigma_t) - \Delta(\mu_t - \sigma_t) - \operatorname{div}(\mu\nabla u - \sigma\nabla v) = 0.$$

Multiplying the second equation by $u - v$ and integrating by parts, we obtain

$$\int_0^T\!\!\int_{\mathbb{R}^d} \big[F(x,\mu_t) - F(x,\sigma_t)\big]\,(\mu_t - \sigma_t)(dx)\,dt$$
$$= -\int_{\mathbb{R}^d} \big[G(x,\mu_T) - G(x,\sigma_T)\big]\,(\mu_T - \sigma_T)(dx) - \frac{1}{2}\int_0^T\!\!\int_{\mathbb{R}^d} |\nabla u - \nabla v|^2\, d(\mu_t - \sigma_t)\,dt.$$

If we interchange μ_t and σ_t, then the second term on the right changes sign, while the first term and the left side do not change sign. Hence the second term vanishes, which gives the equality

$$\int_0^T\!\!\int_{\mathbb{R}^d} \big[F(x,\mu_t) - F(x,\sigma_t)\big]\,(\mu_t - \sigma_t)(dx)\,dt$$
$$= -\int_{\mathbb{R}^d} \big[G(x,\mu_T) - G(x,\sigma_T)\big]\,(\mu_T - \sigma_T)(dx),$$

which is only possible if $\mu_t = \sigma_t$. □

Note that analogous systems can be studied on bounded sets such as $[0,1]^d$ with periodic boundary conditions, see Lasry, Lions [**586**], [**587**]. Concerning properties of solutions, see Gomes, Saude [**417**] and Porretta [**787**].

Comments

In relation to uniqueness problems, let us quote Kolmogorov from [**527**] (see also [**528**]), changing his notation for the diffusion matrix and drift to our notation and noting that (133) is the forward Kolmogorov equation, (86) is the Chapman–Kolmogorov equation (5) from p. xi, (143) means that the initial distribution is

Dirac's measure at the point x, (85) and (142) express the condition that we consider probability measures:

The principle question concerning uniqueness of solutions is the following: under what conditions can one assert that, for any fixed s *and* x, *a unique nonnegative function* $f(s,x,t,y)$ *of variables* t,y *defined for all values of* y *and* $t>s$ *satisfying equation* (133) *and conditions* (142)–(143) *can exist? In some important partial cases this question can be given a positive answer; for example, such are all cases considered in the next two sections.*

Suppose now that the functions $b(t,y)$ *and* $A(t,y)$ *are given in advance; one can pose the question whether there is a nonnegative function* $f(s,x,t,y)$ *that, on the one hand, would satisfy equations* (85) *and* (86) (*as it was explained in* § 11, *these requirements are necessary in order that* $f(s,x,t,y)$ *could define a stochastic system*), *and, on the other hand, after passing to the limit by formulae* (122) *and* (124), *would lead to the given functions* $b(t,y)$ *and* $A(t,y)$.

For solving such a problem, one could, for example, define some nonnegative solution to our second differential equation (133) *satisfying conditions* (142)–(143), *and then investigate whether it is indeed a solution to our problem. The following two general questions arise here:*

1. *Under what conditions does there exist such a solution to equation* (133)*?*

2. *Under what conditions can one assert that this solution satisfies indeed equations* (85) *and* (86)*?*

There are all grounds to believe that these conditions have a sufficiently general character.

Existence and uniqueness of integrable solutions to the Cauchy problem for Fokker–Planck–Kolmogorov equations are investigated in the papers Figalli [**362**], Le Bris, Lions [**593**], Lemle [**599**], Wei, Liu [**936**], Wu, Zhang [**943**]. Surveys of results on uniqueness of nonnegative solutions to the Cauchy problem for parabolic equations can be found in Ishige, Murata [**480**], Pinchover [**777**]. Uniqueness of probability solutions to the Cauchy problem has been studied in Bogachev, Da Prato, Röckner, Stannat [**142**] and Röckner, Zhang [**822**]. The main results of this chapter are obtained in Shaposhnikov [**849**], [**850**], [**851**], and also in Bogachev, Röckner, Shaposhnikov [**170**]. In the proofs of uniqueness results the smoothness of solutions of degenerate equations proved by Oleĭnik [**754**], [**755**] is employed.

In addition to the methods of obtaining sufficient conditions for uniqueness listed in § 9.1 one should mention the papers Wu, Zhang [**943**] and Lemle [**599**], where the L^∞- and L^1-uniqueness for operators of the form $L=\Delta+\langle b,\nabla\rangle$ and L^* is studied. Their main idea is to introduce a topology on the space $L^\infty(\mathbb{R}^d)$ such that $L^1(\mathbb{R}^d)$ becomes the dual space (say, $*$-weak topology is such a topology). In the cited papers, sufficient conditions for the uniqueness of solutions to the Cauchy problem in the class $C\bigl([0,1],L^1(\mathbb{R}^d)\bigr)$ are obtained in terms of the behavior of $\langle b(x),x\rangle$ under the assumption that the coefficient b is locally bounded and does not depend on t. The uniqueness of solutions to the Cauchy problem for the Fokker–Planck–Kolmogorov equation is studied also in Gray [**421**], Luo [**650**]. On some symmetries of solutions, see Kozlov [**538**].

A number of essential specific features arise in the study of degenerate equations for measures. Such equations usually cannot be reduced to equations for functions due to the absence of densities, although in the case of smooth coefficients a moderate degeneracy of the diffusion matrix can be compensated by the drift and can even lead to existence of smooth densities. The latter phenomenon — hypoellipticity — is investigated by methods of the theory of partial differential equations (see Hörmander [**461**]) and methods of the Malliavin calculus (see, e.g., Bell [**101**], Ikeda, Watanabe [**473**], Nualart [**750**], and Shigekawa [**856**]).

There is a vast literature on degenerate Fokker–Planck–Kolmogorov equations and degenerate diffusions; we note the classical papers Feller [**354**], [**356**], [**357**], and also more modern ones Bensoussan, Turi [**107**], Bramanti, Brandolini, Lanconelli, Uguzzoni [**202**], Cinti, Pascucci, Polidoro [**243**], Cinti, Polidoro [**244**], Da Prato, Lunardi [**271**], [**273**], Da Prato, Frankowska [**268**], [**269**], Di Francesco, Polidoro [**300**], Elst, Robinson, Sikor [**330**], Epstein, Mazzeo [**337**], Figalli [**362**], Guillin, Wang [**429**], Helffer, Nier [**448**], Hérau, Nier [**449**], Ichihara, Kunita [**472**], Ivasyshen, Medynsky [**485**], Le Bris, Lions [**593**], Konakov, Menozzi, Molchanov [**532**], Kushner [**568**], Lanconelli, Polidoro [**579**], Lanconelli, Uguzzoni [**580**], Lorenzi [**640**], Montanari [**722**], Nier [**744**], Polidoro [**782**], [**783**], Polidoro, Ragusa [**784**], Quastel, Varadhan [**796**], Sakbaev [**833**], Stasi [**875**], [**876**], Stroock, Varadhan [**883**], Tamura, Watanabe [**893**], Wang, Zhang [**933**], Wei, Liu [**936**]. On uniqueness problems connected with degenerate elliptic operators, see Robinson [**809**], Robinson, Sikora [**811**], [**812**], [**813**]. Degenerate diffusions in manifolds are studied in Kliemann [**519**].

Various estimates for transition probabilities of diffusions involving the total variation distance or Kantorovich-type distances have become popular in the last decade. There are many works on this topic, see, e.g., Ambrosio, Savaré, Zambotti [**46**], Arnold, Markowich, Toscani, Unterreiter [**69**], Bakry, Cattiaux, Guillin [**84**], Bolley, Gentil [**188**], Bolley, Gentil, Guillin [**191**], Bolley, Gentil, Guillin [**189**], Bolley, Guillin, Villani [**193**], Carrillo, McCann, Villani [**223**], Carrillo, Toscani [**224**], Dolbeault, Nazaret, Savaré [**307**], Guillin, Léonard, Wu, Yao [**428**], Marton [**670**], Natile, Peletier, Savaré [**735**], Otto, Westdickenberg [**760**], von Renesse, Sturm [**804**], and references therein. The principal feature of the presented estimates is that they compare diffusions with different drifts or even different diffusion matrices, not with different initial distributions.

Exercises

9.8.43. Give an example of a probability measure on $\mathbb{R}\times(0,1)$ with a smooth unbounded density that satisfies the parabolic equation $\partial_t\mu = L^*_{\mathrm{I},b}\mu$ with a smooth drift b and an initial condition given by a bounded density.

9.8.44. Let b be a locally bounded Borel vector field on the space $\mathbb{R}^d$ such that $b(-x) = -b(x)$. Suppose that the equation $\partial_t\mu = L^*_{\mathrm{I},b}\mu$ has a unique probability solution $\mu = \mu_t\,dt$ with a centrally symmetric initial condition. Prove that μ possesses a density centrally symmetric in x.

HINT: Show that the image of the measure μ under the mapping $x \mapsto -x$ satisfies the same equation.

9.8.45. Let b be a locally bounded Borel vector field on $\mathbb{R}^d$ and $Ub(x) = b(Ux)$ for every orthogonal operator U. Suppose that the equation $\partial_t\mu = L^*_{\mathrm{I},b}\mu$ has a unique probability solution $\mu = \mu_t\,dt$ on $\mathbb{R}^d\times(0,1)$ with a spherically symmetric initial condition. Prove that μ possesses a density spherically symmetric in x.

Apply this assertion to $b(x) = f(|x|)x$, where f is a locally bounded Borel function such that $f(|x|)|x|^2 \to -\infty$ as $|x| \to \infty$.

HINT: Show that for every orthogonal operator U on $\mathbb{R}^d$ the transformed measure $\mu\circ U^{-1} = \mu_t\circ U^{-1}\,dt$ satisfies the same equation.

9.8.46. If in place of the strip $\mathbb{R}^d\times(0,1)$ we consider $\mathbb{R}^d\times\mathbb{R}$ and do not assume any conditions as $t\to-\infty$, then there are simple examples of equations $\partial_t\mu = L^*\mu$ having several probability solutions. Let $d=1$, $A(x)=1$ and $b(x,t)=-x$. Verify that, for every number α, the measure $\mu^\alpha = \mu_t^\alpha(dx)\,dt$, where

$$\mu_t^\alpha(dx) = (2\pi)^{-1/2}\exp\bigl(-(x+e^{-t}\alpha)^2/2\bigr)\,dx,$$

is a probability solution of our equation.

9.8.47. Let $d=1$ and $A(x)=1$,

$$b(x) = -2x(1+x^2)^{-1} - (1+x^2)\mathrm{arctg}\,x, \quad u(x) = \bigl(\pi(1+x^2)\bigr)^{-1}.$$

Prove that the Cauchy problem for the Fokker–Planck–Kolmogorov equation with the operator $L_{1,b}$ with the initial condition $\nu = u\,dx$ has a unique subprobability solution that is not a unique integrable and unique nonnegative solution.

HINT: The Cauchy problem for the measure $\mu = v(x,t)\,dx\,dt$ with the initial condition $\nu = u\,dx$ can be written as the Cauchy problem for the density v as follows: $\partial_t v = \partial_x(\partial_x v - bv)$, $v(x,0) = u(x)$. We are looking for a nonnegative solution v of the form $e^t\Phi'(x)$. Substituting in the equation we obtain the equality $\Phi' = (\Phi'' - b\Phi')'$, integrating which we arrive at the relation $\Phi = \Phi'' - b\Phi' + \text{const}$. Set $\Phi(x) = \mathrm{arctg}\,x$. Then we have $b(x) = [\Phi''(x) - \Phi(x)]/\Phi'(x)$. Since $b(x)x \leqslant 0$, by Theorem 9.4.8 there exists a unique subprobability solution to the Cauchy problem. However, there is another integrable and nonnegative solution $e^t u(x)\,dx\,dt$, which is not a subprobability solution.

9.8.48. Construct an example of a smooth vector field b on $\mathbb{R}^3$ such that for $A = \mathrm{I}$ the Cauchy problem for the Fokker–Planck–Kolmogorov equation has two different probability solutions.

HINT: According to Exercises 4.5.17 and 5.6.49, there is a smooth function B on $\mathbb{R}$ such that the probability solution ν of the equation $L_{1,B}^*\nu = 0$ is not an invariant measure for the corresponding semigroup $\{T_t\}_{t\geqslant 0}$, but only subinvariant. Let now $C(y) = \bigl(C^1(y), C^2(y)\bigr)$ be a smooth vector field on $\mathbb{R}^2$ for which there are two different probability solutions σ^1 and σ^2 to the equation $L_{\mathrm{I},C}^*\sigma = 0$. Set

$$\mu_t^1 \equiv \nu\otimes\sigma_1, \quad \mu_t^2 = (\nu - T_t^*\nu)\otimes(\sigma_2 - \sigma_1) + \nu\otimes\sigma_1.$$

It is readily seen that μ_t^1 and μ_t^2 are probability solutions to the Cauchy problem

$$\partial_t\mu = L_{\mathrm{I},b}^*\mu, \quad \mu|_{t=0} = \nu\otimes\sigma_1,$$

where $b(x,y) = \bigl(B(x), C^1(y), C^2(y)\bigr)$, $x\in\mathbb{R}$, $y\in\mathbb{R}^2$.

9.8.49. Give an example of a smooth function b on the real line and a smooth probability density u_0 for which the Cauchy problem for the parabolic equation $\partial_t u = u'' - (bu)'$ with the initial condition u_0 has exactly one integrable solution which is a subprobability, but not probability solution.

HINT: See the hint to Exercise 9.8.47.

CHAPTER 10

The Infinite-Dimensional Case

The problems considered in the previous chapters arise also in the infinite-dimensional case; as we noted in the introduction, the infinite-dimensional case even served as one of motivations for the study of such problems. Since a thorough discussion of this area would substantially increase the size of this book, here we only briefly discuss the setting of the main problems and some model examples (mostly in the elliptic case). Moreover, we confine ourselves to the questions directly connected with the finite-dimensional case, so that the material of this chapter should be considered as applications of the results obtained in the finite-dimensional case.

In § 10.1 we explain the infinite-dimensional framework and § 10.2 is devoted to certain simple properties of solutions to infinite-dimensional equations, which mostly follow from the fact that finite-dimensional projections of such equations satisfy equations whose coefficients are the conditional expectations of the original coefficients with respect to the solution and the σ-algebra generated by the projection. In § 10.3 and § 10.4 some simple cases of solvability of elliptic and parabolic equations in infinite dimensions are discussed.

10.1. Equations in infinite-dimensional spaces

First of all, it should be noted that many of the principal objects considered above, such as stochastic equations, transition probabilities, stationary distributions, Fokker–Planck and Kolmogorov equations, are defined on very general spaces and bear no finite-dimensional features. However, in the infinite-dimensional case new phenomena arise. Partly this is due to the absence of exact analogs of Lebesgue measure, but there are also other reasons. One of the starting points is again a Markov process with a state space X whose transition semigroup $\{T_t\}_{t\geqslant 0}$ is defined on a suitable function space $\mathcal{F}$ (such as $C_b(X)$ or $L^p(\mu)$). If this process has an invariant probability measure μ on X, i.e.,

$$\int_X f(x)\,\mu(dx) = \int_X T_t f(x)\,\mu(dx) \quad \forall\, f\in\mathcal{F},$$

then, under broad assumptions, $\{T_t\}_{t\geqslant 0}$ extends to a strongly continuous semigroup on $L^1(\mu)$ which has a generator L on some domain $\mathcal{D}$, and μ satisfies the equation

$$L^*\mu = 0 \tag{10.1.1}$$

in the sense that the integral of Lf vanishes for all $f\in\mathcal{D}$. Typically, there is a smaller class $\mathcal{D}_0\subset\mathcal{D}$ of functions for which Lf is defined explicitly without any reference to the semigroup and the domain of its generator. Then we can consider equation (10.1.1) as the identity

$$\int_X Lf(x)\,\mu(dx) = 0 \quad \forall\, f \in \mathcal{D}_0. \tag{10.1.2}$$

Therefore, questions arise about the relation between (10.1.2) and the invariance of μ with respect to the semigroup, as well as about the properties of solutions to (10.1.2), i.e., the same questions that we have discussed in the finite-dimensional case. For example, let μ be an invariant measure of a diffusion process in l^2 or in the space $\mathbb{R}^\infty$ of all real sequences defined by the stochastic differential equation

$$d\xi_t = \sqrt{2}dw_t + b(\xi_t)dt, \tag{10.1.3}$$

where b is a Borel vector field and W_t is a Wiener process (in a suitable sense, see [**124**, Chapter 7]; e.g., in case of $\mathbb{R}^\infty$ this can be just $w_t = (w_{t,n})$, where $w_{n,t}$ are independent real Wiener processes, and in case of l^2 this can be $w_t = (\alpha_n w_{t,n})$, where $\alpha_n > 0$ and $(\alpha_n)_{n=1}^\infty \in l^2$). Then one can take for $\mathcal{D}_0$ the class $\mathcal{F}C_b^\infty$ of all functions of the form $f(x_1, \ldots, x_n)$, where $f \in C_b^\infty(\mathbb{R}^n)$. Then L has the form

$$Lf = \sum_i [\partial_{x_i}^2 f + b^i \partial_{x_i} f].$$

However, as we shall see below in concrete examples, in the infinite-dimensional case it is often desirable to have a broader setting where b cannot be interpreted as a vector field with values in the original space (e.g., when dealing with l^2 it may happen that b takes values in $\mathbb{R}^\infty$). For this reason, the following more general framework is used.

Suppose that X is a locally convex space with the topological dual X^* and there is a sequence $\{l_i\} \subset X^*$ separating the points in X; such a sequence exists if X is a separable metrizable space or a Souslin space.

Let $\mathcal{F}C_b^\infty(\{l_i\})$ denote the class of all functions of the form

$$f(x) = f_0\big(l_1(x), \ldots, l_n(x)\big), \quad \text{where } f_0 \in C_b^\infty(\mathbb{R}^n), \tag{10.1.4}$$

and let $\mathcal{F}C_0^\infty(\{l_i\})$ denote the class of all functions of the form (10.1.4) with functions $f \in C_0^\infty(\mathbb{R}^n)$; finally, let $\mathcal{F}C_b^\infty(X)$ be the class of functions of the form (10.1.4) with functionals $l_i \in X^*$, $n \in \mathbb{N}$. Note that $\mathcal{F}C_0^\infty(\{l_i\})$ is not a linear space.

Suppose that we are given Borel functions a^{ij}, b^i on X. The infinite matrix with entries $a^{ij}(x)$ will be denoted by $A(x)$ and the collection of scalar functions b^i will be denoted by b, although we do not assume that $A(x)$ and $b(x)$ correspond to some operator or vector. Then we can consider the elliptic operator

$$L_{A,b}f(x) = \sum_{i,j} a^{ij}(x)\partial_{e_i}\partial_{e_j} f(x) + \sum_i b^i(x)\partial_{e_i} f(x), \tag{10.1.5}$$

where e_i are some vectors in X, ∂_{e_i} denotes the partial derivative along e_i, i.e., $\partial_{e_i} f(x) = \lim\limits_{t\to 0}[f(x+te_i) - f(x)]/t$.

We shall say that a Radon measure μ on X satisfies equation (10.1.1) with respect to the class $\mathcal{F}C_b^\infty(\{l_i\})$ if $a^{ij}, b^i \in L^1(\mu)$ for all i,j and (10.1.2) is fulfilled with $\mathcal{D}_0 = \mathcal{F}C_b^\infty(\{l_i\})$. Sometimes it is more convenient to use the smaller class $\mathcal{D}_0 = \mathcal{F}C_0^\infty(\{l_i\})$, which in this case gives an equivalent definition due to the integrability of coefficients, or some other classes of test functions.

The parabolic operators and Fokker–Planck–Kolmogorov equations are defined similarly in complete analogy with the finite-dimensional case. In the parabolic case the functions a^{ij} and b^i may depend also on $t \in (0,T)$ and the equation

$$\partial_t \mu = L^* \mu \tag{10.1.6}$$

with respect to a measure μ of the form $\mu = \mu_t\, dt$ on $X_T = X \times (0,T)$ with Radon measures μ_t on X (regarding this form of μ, see p. 242; moreover, here we require the integrability of $t \mapsto \|\mu_t\|$) is understood as the identity

$$\int_{X_T} [\partial_t \varphi + L\varphi]\, d\mu = 0 \quad \forall \varphi \in \mathcal{FC}_b^\infty(X_T), \tag{10.1.7}$$

where $\mathcal{FC}_b^\infty(X_T)$ is the class of all functions on X_T of the form

$$\psi(l_1(x), \ldots, l_n(x), t), \quad \text{where } \psi \in C_b^\infty(\mathbb{R}^n \times (0,T)).$$

When we consider the Cauchy problem, the initial condition $\mu_0 = \nu$ is understood as the relation

$$\lim_{t\to 0} \int_X \varphi\, d\mu_t = \int_X \varphi\, d\nu \quad \forall \varphi \in \mathcal{FC}_b^\infty(\{l_i\}).$$

Measures satisfying equation (10.1.1) are called infinitesimally invariant for the operator L, since under broad assumptions this equation holds for the true invariant measures of the diffusion generated by L (certainly, provided that such a diffusion exists and has a stationary distribution). An advantage of dealing with infinitesimally invariant measures (as in the finite-dimensional case) is that the equation can be meaningful and possess solutions under much weaker assumptions than needed for the existence of the associated diffusion. Moreover, the existence of solutions to the elliptic equation often enables us to construct the diffusion (this subject is studied in Eberle [**320**], Stannat [**872**], [**873**], [**874**]).

10.1.1. Remark. If we apply (10.1.2) or in the parabolic case (10.1.7) to the function $\varphi(x) = \exp(il(x))$, where $l \in X^*$, then on account of the equality $L_{A,b}\varphi(x) = g(x)\varphi(x)$, where the function g is given by some explicit expression, we obtain an equation with respect to the Fourier transform of the measure μ or the measure μ_t in the parabolic case, defined by the formula

$$\widetilde{\mu}(l) = \int_X \exp(il)\, d\mu.$$

For example, if $X = \mathbb{R}^\infty$, $a^{ii} = \alpha_i > 0$, $a^{ij} = 0$ whenever $i \neq j$, and the functions b^i do not depend on t, then $g = \sum_{j=1}^\infty [-\alpha_j l_j^2 + i l_j b^j]$. This integro-differential equation is called Hopf's equation [**460**], who introduced it in his study of statistical solutions of the Navier – Stokes equations (see Vishik, Fursikov [**922**]).

Infinite-dimensional equations for measures have been considerably less studied so far as compared to the finite-dimensional case. There are sufficient conditions for the existence and sometimes for the uniqueness of solutions, but not much is known about their properties and connections between infinitesimal invariance and proper invariance with respect to the associated semigroups (the very existence of such semigroups has also been less studied). The most difficult problems are connected with uniqueness. Standard methods of proving the existence of stationary distributions are based on Prohorov's theorem and Lyapunov functions combined with a priori estimates or on convergence of transition probabilities (which in turn employs various assumptions of dissipativity and Lyapunov functions). The same techniques are used for the proof of existence of solutions to elliptic or parabolic equations for measures. Generally speaking, existence results for elliptic equations can be obtained under broader assumptions on the coefficients, since they do not assume the existence of the corresponding diffusions (the latter is usually more stringent). However, there are cases where there is no direct proof of the solvability of the elliptic equation for measures, and one has to construct a process and analyze its transition semigroup.

For solutions to elliptic and parabolic equations for measures on infinite-dimensional spaces we can consider various properties of their finite-dimensional projections, their mutual absolute continuity or singularity, and directional properties such as continuity or differentiability. We recall that a Radon measure μ on a locally convex space X is continuous along a vector $h \in X$ if $\lim\limits_{t\to 0} \|\mu_{th} - \mu\| = 0$, where $\mu_{th}(B) := \mu(B + th)$ for every Borel set B. If, for every Borel set $B \subset X$, the function $t \mapsto \mu(B + th)$ is differentiable, then the measure μ is called Fomin differentiable along h. This is equivalent to the existence of a function $\beta_h^\mu \in L^1(\mu)$, called the logarithmic derivative of μ along h, such that the following integration by parts formula holds for all functions $f \in \mathcal{FC}_b^\infty(X)$:

$$\int_X \partial_h f(x)\, \mu(dx) = -\int_X f(x)\beta_h^\mu(x)\, \mu(dx).$$

If $X = \mathbb{R}^d$, then a measure μ is Fomin differentiable along all vectors precisely when it has a density $\varrho \in W^{1,1}(\mathbb{R}^d)$; then $\beta_h^\mu = \langle \nabla\varrho, h\rangle$. In the infinite-dimensional case, only the zero measure is differentiable along all vectors.

For example, if the measure μ equals the countable product of probability measures on the real line with differentiable densities p_n with integrable derivatives p_n', then μ is differentiable along every vector of the form $h = (h_1, \ldots, h_n, 0, 0, \ldots)$ and we have $\beta_h^\mu(x) = \sum_{i=1}^n h_i p_i'(x_i)/p_i(x_i)$.

There is also a reasonable infinite-dimensional analog of the logarithmic gradient $\nabla\varrho/\varrho$, although neither ϱ nor $\nabla\varrho$ exist separately. Namely, suppose that we a have a continuously and densely embedded Hilbert space $H \subset X$ with the inner product $(u, v)_H$, generating the embedding $j = j_H : X^* \to X$ in the following way: for every $l \in X^*$ there is a unique element $j(l) \in H$ for which

$$l(h) = \big(j(l), h\big)_H \quad \forall\, h \in H.$$

The standard example is $X = \mathbb{R}^\infty$, $H = l^2$. Then X^* is the space $\mathbb{R}_0^\infty$ of finite sequence, naturally embedded into $\mathbb{R}^\infty$.

If there exists a μ-measurable mapping $\beta_H \colon X \to X$ such that for each $l \in X^*$ the measure μ is differentiable along $j(l)$ and

$$\beta_{j(h)}^\mu = \langle l, \beta_H \rangle,$$

then β_H is called a logarithmic gradient of μ associated with H.

It can happen that the measure μ is differentiable along all vectors $j(l)$, where $l \in X^*$, but there is no logarithmic gradient. However, for the aforementioned product-measure $\mu = \bigotimes_{i=1}^\infty p_i\, dx_i$ on $\mathbb{R}^\infty$ and for $H = l^2$ the logarithmic gradient obviously exists and is given by the formula $\beta_H^\mu(x) = \big(p_i'(x_i)/p(x_i)\big)$.

A detailed account of the theory of differentiable measures is given in Bogachev [**126**].

Let $\{l_i\} \subset X^*$ be such that for the vectors $e_i := j(l_i)$ we have $l_j(e_i) = \delta_{ij}$. We shall assume additionally that the functionals l_i separate the points in X. Then, under mild assumptions about X (see Bogachev [**124**, Chapter 7]) there is a process W_t in X, called a Wiener process in X associated with H, such that the scalar processes $\langle l_i, W_t\rangle$ are independent Wiener processes. It is known (see Albeverio, Röckner [**31**], [**33**], [**126**, Chapter 12]) that, under broad assumptions, for a given probability measure μ possessing a logarithmic gradient β_H, there is a diffusion process in X governed by the stochastic differential equation (10.1.3) with $b = \beta_H$ such that μ is its invariant measure and the corresponding transition semigroup $\{T_t\}_{t\geqslant 0}$ is symmetric on $L^2(\mu)$; its generator is given by

$$Lf = \sum_i [\partial_{e_i}^2 f + l_i(\beta_H)\partial_{e_i} f]$$

on functions $f \in \mathcal{F}C_b^\infty(\{l_i\})$. An efficient method of constructing more general processes is based on perturbations of the drift β_H. This method is based on the symmetry of the operator with drift β_H in $L^2(\mu)$. Similarly to the finite-dimensional case, the symmetry of the operator $L_{\mathrm{I},b}$ is equivalent to the property that b is a logarithmic gradient.

10.1.2. Proposition. *Let $b^i \in L^2(\mu)$. The operator*

$$Lf = \sum_{i=1}^\infty [\partial_{e_i}^2 f + b^i\partial_{e_i} f]$$

is symmetric on $\mathcal{F}C_b^\infty(\{l_i\}) \subset L^2(\mu)$ precisely when the measure μ is differentiable along each vector $e_n = j(l_n)$ and $\beta_{e_n}^\mu = b^n$ μ-a.e.

PROOF. If $\beta_{e_n}^\mu$ exists, then

$$\int_X g(x)\partial_{e_n}^2 f(x)\,\mu(dx) = -\int_X [\partial_{e_n} g(x)\partial_{e_n} f(x) + g(x)\partial_{e_n} f(x)\beta_{e_n}^\mu(x)]\,\mu(dx),$$

whence we immediately obtain the equality of the integrals of gLf and fLg for all functions $f, g \in \mathcal{F}C_b^\infty(\{l_i\})$. Conversely, suppose that the operator L is symmetric on the domain $\mathcal{F}C_b^\infty(\{l_i\})$. Let $f = \exp(il)$, $g = \exp(itl_n)$, where l is a finite linear combination of the functionals l_i. Let $\langle l, b\rangle := \sum_{k=1}^\infty l(e_k)b^k$, where the sum is finite, since only finitely many numbers $l(e_k)$ are nonzero. Then

$$Lf = -\sum_{k=1}^\infty l(e_k)^2 e^{il} + i\langle l, b\rangle e^{il}, \quad Lg = -t^2 e^{itl_n} + itb^n e^{itl_n},$$

whence by the symmetry of L we obtain the equality

$$\int_X e^{il+itl_n}\Bigl[-\sum_{k=1}^\infty l(e_k)^2 + i\langle l, b\rangle\Bigr]\,\mu(dx) = \int_X e^{il+itl_n}\Bigl[-t^2 + itb^n\Bigr]\,\mu(dx).$$

Letting $\xi := l + tl_n$ and replacing l by $\xi - tl_n$, we can write this as follows:

$$\int_X e^{i\xi}\Bigl[-\sum_{k=1}^\infty \xi(e_k)^2 + i\langle \xi, b\rangle\Bigr]\,\mu(dx) = 2\int_X e^{i\xi}\Bigl[-t\xi(e_n) + itb^n\Bigr]\,\mu(dx).$$

Since the left-hand side is independent of t, the right-hand side must vanish. Therefore, we obtain

$$-i\xi(e_n)\int_X e^{i\xi}\,\mu(dx) = \int_X e^{i\xi}b^n\,\mu(dx)$$

for any functional ξ that is a finite linear combination of the functionals l_i. Since μ is a Radon measure and $\{l_i\}$ separates the points in X, it follows that the same is true for any $\xi \in X^*$. This identity yields that μ is differentiable along e_n and $\beta_{e_n}^\mu = b^n$ (see Bogachev [**126**, Theorem 3.6.7]). Note that without the assumption that $\{l_i\}$ separates the points in X we could obtain the same assertion for μ on the σ-algebra generated by $\{l_i\}$. □

The case of nonconstant A is studied similarly in Long [**636**].

The simplest (but also one of the most important) infinite-dimensional examples of solutions to the stationary Fokker–Planck–Kolmogorov equation is the standard Gaussian measure γ on $\mathbb{R}^\infty$ that is the countable power of the standard Gaussian measure on the real line and satisfies the equation $L^*_{\mathrm{I},b}\gamma = 0$ with the drift $b(x) = -x$. This operator on the class of smooth functions in finitely many variables has the form

$$Lf(x) = \sum_{i=1}^{n} [\partial^2_{x_i} f(x) - x_i \partial_{x_i} f(x)]$$

and is called the infinite-dimensional Ornstein–Uhlenbeck operator. The closure of the operator L in $L^2(\gamma)$ coincides with the generator of the Ornstein–Uhlenbeck semigroup $\{T_t\}_{t\geqslant 0}$ on $L^2(\gamma)$ given by the same formula (5.1.1) as in the finite-dimensional case. It is readily seen (Exercise 10.5.13) that γ is the unique probability solution of the equation $L^*\gamma = 0$ with the Ornstein–Uhlenbeck operator L. It is easy to show that the mapping $-x$ is the logarithmic gradient of the measure γ for the space $H = l^2$, called the Cameron–Martin space of the measure γ. For every $h = (h_i) \in l^2$ one has $\beta^\gamma_h(x) = -\sum_{i=1}^\infty h_i x_i$, where the series converges in $L^2(\gamma)$ and almost everywhere; outside l^2 there are no vectors of differentiability for γ.

On an arbitrary locally convex space X a Gaussian measure is a Radon probability measure μ such that for every functional $l \in X^*$ the induced measure $\mu\circ l^{-1}$ on the real line is Gaussian, i.e., l is a Gaussian random variable. If all functionals l have zero integrals, then μ is called centered. The celebrated Tsirelson theorem (see Bogachev [**124**, § 3.4], [**126**], [**127**]) says that every centered Radon Gaussian measure on X not concentrated on a finite-dimensional subspace is the image of the standard Gaussian measure γ on $\mathbb{R}^\infty$ under some measurable linear mapping (moreover, this linear mapping is a Borel isomorphism of some Borel linear subspaces $X_1 \subset \mathbb{R}^\infty$ and $X_2 \subset X$ of full measure). The exact analog of the space l^2 for the measure μ is its Cameron–Martin space $H(\mu)$ consisting of all vectors h such that the measures μ_h and μ are equivalent. This space is separable Hilbert with norm $|\cdot|_{H(\mu)}$ defined by the formula

$$|h|_{H(\mu)} = \sup\{l(h)\colon\ l \in X^*,\ \|l\|_{L^2(\mu)} \leqslant 1\}.$$

Moreover, $H(\mu)$ consists of all vectors h with $|h|_{H(\mu)} < \infty$. It is known that the closure of $H(\mu)$ in X has measure 1, so one can assume that $H(\mu)$ is dense in X, but $\mu(H(\mu))=0$ if $\dim H(\mu)=\infty$. It is readily verified that in this case the logarithmic gradient of μ along $H = H(\mu)$ is the mapping $-x$. Also the Ornstein–Uhlenbeck operator L naturally arises. Taking an orthonormal basis $\{e_i\}$ in H of the form $e_n = j(l_n)$ with $l_n \in X^*$, we set

$$Lf = \sum_i [\partial^2_{e_i} f - l_i \partial_{e_i} f]$$

for $f \in \mathcal{F}C_b^\infty(\{l_i\})$. As above, the closure of L in $L^2(\mu)$ coincides with the generator of the Ornstein–Uhlenbeck semigroup $\{T_t\}_{t\geqslant 0}$ on $L^2(\gamma)$.

The case of a gradient type drift has deep and interesting connections with the study of Gibbs measures, i.e., measures on infinite products of spaces with given conditional distributions. For example, for a broad class of models in statistical physics, constructing Gibbs measures is equivalent to constructing measures with given logarithmic gradients; the latter problem is studied in Albeverio, Kondratiev, Pazurek, Röckner [**23**], Albeverio, Kondratiev, Röckner [**26**], [**27**], Albeverio, Kondratiev, Röckner, Tsikalenko [**28**], Bogachev [**123**], Bogachev [**126**], Kirillov [**511**], [**513**], [**514**], [**516**]; in particular, a method of reconstructing measures with a given logarithmic gradient by means of Lyapunov functions was suggested by A.I. Kirillov. For general elliptic equations for measures this method was developed in the papers Bogachev, Röckner [**158**], [**161**], Bogachev, Röckner, Wang [**179**]. New sufficient conditions that the corresponding solutions, i.e., the infinitesimally invariant measures, are symmetrizing, hence Gibbs, are identified in Bogachev, Röckner, Wang [**181**].

Let us give examples in which operators of the form (10.1.5) and the corresponding infinite-dimensional Fokker–Planck–Kolmogorov equations arise in the study of stochastic partial differential equations such as stochastic porous media equations, reaction–diffusion equations, Burgers equations, and Navier–Stokes equations. Details can be found in Bogachev, Da Prato, Röckner [**130**], Bogachev, Da Prato, Röckner, Shaposhnikov [**138**], Bogachev, Röckner [**161**].

Let us illustrate a general plan of studying such equations in the case of stochastic porous media equations. The primary object is then a nonlinear partial differential equation of the form

$$\partial_t u = \Delta\big[\Psi(u)\big] + \Phi(u, \nabla u),$$

where Ψ and Φ are some functions on $\mathbb{R}$ and $\mathbb{R}^2$, e.g., polynomials. The analysis of this equation usually turns out to be very complicated. However, it turns out that adding a stochastic noise in the right-hand side leads to simplifications (it is even possible that the stochastic equation has at least the same physical significance). This stochastic partial differential equation is heuristically written as

$$d\xi_t = \sqrt{2}dw_t + \big(\Delta[\Psi(\xi_t)] + \Phi(\xi_t, \nabla\xi_t)\big)\, dt,$$

where w_t is a Wiener process with a suitable covariance operator Q. But a rigorous interpretation in the case of nonlinear functions Ψ and Φ is not obvious. One possible approach to this problem is to consider the associated infinite-dimensional elliptic operator L on a suitable domain, find an infinitesimally invariant measure μ for L, construct a Markov semigroup on $L^2(\mu)$ with μ as an invariant measure in such a way that the generator of this semigroup extends L, and finally construct a Markov process solving the martingale problem corresponding to this operator. One can also try to solve the parabolic equation for the transition probabilities of the expected process. The case $\Psi(s) = s|s|^{m-1}$ and $\Phi = 0$ corresponds to the porous media equation and the case $\Psi(s) = s$ and $\Phi(s, s') = -s|s|^{m-1}$ to the reaction-diffusion equation.

Let us see how we can associate to these objects well-defined infinite-dimensional Fokker–Planck–Kolmogorov equations.

10.1.3. Example. Let $\Omega \subset \mathbb{R}^d$ be a bounded open domain with smooth boundary, let $\{e_n\}$ be the orthonormal basis in $L^2(\Omega)$ formed by the eigenfunctions of the Laplacian Δ with Dirichlet boundary condition, and let λ_1 be the minimal eigenvalue of $-\Delta$. Let Ψ be a C^1-function with $\Psi(0) = 0$ such that for some positive numbers κ_0, C_0, κ_1, and $r \geqslant 1$ we have

$$\kappa_0|s|^{r-1} \leqslant \Psi'(s) \leqslant C_0 + \kappa_1|s|^{r-1} \quad \text{for all } s \in \mathbb{R}^1,$$

and let Φ be a continuous function satisfying the condition

$$|\Phi(s)| \leqslant C + \delta|s|^r,$$

where $0 < \delta < 4\kappa_0\lambda_1(r+1)^{-2}$ and C is a constant. For example, it suffices that $|\Phi(s)| \leqslant \kappa_2 + \kappa_3|s|^q$, where $q \in (0,r)$, $\kappa_2, \kappa_3 \in (0,+\infty)$. We are interested in the existence of infinitesimally invariant measures for the infinite-dimensional elliptic operator L which is informally given by

$$Lf := \Delta_Q f + \langle b, \nabla f\rangle, \quad b(x) = \Delta[\Psi(x)] + \Phi(x)$$

on smooth cylindrical functions defined on $X := L^2(\Omega)$ or on the negative Sobolev space $W^{2,-1}(\Omega)$. A rigorous interpretation is as follows. Elements of X will be denoted by u in place of x. Let

$$b^i(u) := \int_\Omega \Bigl[\Psi\bigl(u(\omega)\bigr)\Delta e_i(\omega) + \Phi\bigl(u(\omega)\bigr)e_i(\omega)\Bigr]\, d\omega, \quad u \in L^r(\Omega).$$

Let $q_i > 0$ be such that $S := \sum_{i=1}^\infty q_i < \infty$. The operator

$$Lf := \sum_{i=1}^\infty \bigl[q_i\partial_{e_i}^2 f + b^i\partial_{e_i} f\bigr]$$

where ∂_{e_i} is the partial derivative along e_i, is well-defined on $\mathcal{FC}_b^\infty(\{l_i\})$, where $l_i(x) = (x, e_i)_2$ and $(x,y)_2$ is the inner product in $L^2(\Omega)$. The second order part in L can be regarded as $\operatorname{tr}(QD^2 f)$, where Q is the operator on X defined by $Qe_i = q_i e_i$, which is just the covariance operator of the Wiener process w_t in the indicated stochastic equation. A minor nuance is that the functions b^i above, hence also Lf, are defined not on all of X, but only on $L^r(\Omega)$.

As we shall see below (§ 10.3 and Exercise 10.5.18), under these assumptions there is a Borel probability measure μ on $L^r(\Omega)$ that is infinitesimally invariant for L.

For example, if $\Psi(u) = u^r$, where r is an odd number, then one can take for Φ any polynomial of degree r with a sufficiently small leading coefficient.

10.1.4. Example. The stochastic Burgers type equation

$$du(x,t) = \sqrt{2}dw(x,t) + \bigl[\Delta u(x,t) - \psi\bigl(u(x,t)\bigr)\partial_x u(x,t) + f(x)\bigr]\, dt$$

is defined in the space $X = L^2[0,1]$, where Δ is the Laplace operator with zero boundary condition and eigenbasis $\{\eta_n\}$ with eigenvalues λ_n, w is a Wiener process in X of the form $w(t) = \sum_{n=1}^\infty \alpha_n w_n(t)\eta_n$, $\{w_n\}$ is a sequence of independent real Wiener processes, $\alpha_n > 0$, $\sum_{n=1}^\infty \alpha_n^2 < \infty$, ψ is a locally bounded Borel function, and $f \in L^\infty[0,1]$. Let

$$\Psi(y) := \int_0^y \psi(s)\, ds, \quad \Psi(y) \leqslant C + |y|^\kappa, \ \kappa < 6,$$

$$b^n(u) = \lambda_n u_n + \bigl(\Psi(u), \eta_n'\bigr)_2 + (f, \eta_n)_2, \quad u_n := (u, \eta_n)_2.$$

To these coefficients we associate the operator

$$L\varphi = \sum_{n=1}^\infty [\alpha_n^2\partial_{\eta_n}^2\varphi + b^n\partial_{\eta_n}\varphi], \quad \varphi \in \mathcal{FC}_b^\infty(\{u_n\}).$$

As shown in § 10.3, there exists a Borel probability measure μ on X satisfying the equation $L^*\mu = 0$. If $\kappa \leqslant 2$, then the function $\exp(\varepsilon\|u\|_2^2)$ is integrable with respect to μ for some $\varepsilon > 0$.

10.1.5. Example. Another similar example is related to the one-dimensional stochastic "reaction–diffusion" equation

$$du(x,t) = \sqrt{2}dw(x,t) + \bigl[\partial_x^2 u(x,t) + \Phi\bigl(u(x,t)\bigr)\bigr]\,dt,$$

where the function $\Phi \in C(\mathbb{R})$ on the real line is such that $|\Phi(s)| \leqslant C + Cs^\kappa$, $s\Phi(s) \leqslant C + \varepsilon s^2$ and $0 \leqslant \varepsilon < 1$, $0 \leqslant \kappa < 6$. For example, one can use the functions $\Phi(s) = -s^2\operatorname{sign} s$ and $\Phi(s) = -s^4\operatorname{sign} s$. The functions b^n have the form

$$b^n(u) = \lambda_n u_n + \bigl(\Phi(u), \eta_n\bigr)_2, \quad u_n = (u, \eta_n)_2.$$

The technical conditions imposed on Φ are needed to apply Theorem 10.3.7.

10.1.6. Example. The stochastic Navier–Stokes type equation is considered in the space X of all $\mathbb{R}^d$-valued mappings $\xi = (\xi^1, \ldots, \xi^d)$ on a bounded domain $\Omega \subset \mathbb{R}^d$ with smooth boundary such that $\xi^j \in W_0^{2,1}(\Omega)$ and $\operatorname{div}\xi = 0$. The space X is equipped with a Hilbert norm $\|\xi\|_0$ defined by

$$\|\xi\|_0^2 := \|\nabla\xi^1\|_{L^2}^2 + \cdots + \|\nabla\xi^d\|_{L^2}^2.$$

The equation is formally written in the form

$$d\xi(x,t) = \sqrt{2}dw(x,t) + \bigl[\Delta_x\xi(x,t) - \xi^j(x,t)\partial_{x_j}\xi(x,t) + F\bigl(x, \xi(x,t), t\bigr)\bigr]dt,$$

where w is a suitable Wiener process in X (the requirements on it will be specified below) and $F\colon \Omega\times\mathbb{R}^d\times(0,T) \to \mathbb{R}^d$ is a bounded continuous mapping. Since the Laplacian Δ is not defined on all of X, this equation needs some interpretation. Our approach suggests the following procedure. Let $\{\eta_n\}$ be an orthonormal basis in the closure of X in $L^2(\Omega, \mathbb{R}^d)$ and let $\eta_n \in X$ be eigenfunctions of the Laplacian: $\Delta\eta_n = -\lambda_n\eta_n$, $\lambda_n > 0$. Let us introduce the functions

$$b^n(\xi, t) := (\xi, \Delta\eta_n)_2 - \sum_{j=1}^d (\partial_{x_j}\xi, \xi^j\eta_n)_2 + \bigl(F(\,\cdot\,, \xi(\,\cdot\,), t), \eta_n\bigr)_2.$$

These functions are defined already on all of X. It is readily seen that they are continuous on all balls in X with respect to the topology from $L^2(\Omega, \mathbb{R}^d)$. Taking the Wiener process of the form $w_t = \sum_{n=1}^\infty \alpha_n w_{t,n}\eta_n$, where $\{w_{t,n}\}$ is a sequence of independent real Wiener processes, $\sum_{n=1}^\infty \alpha_n^2 < \infty$, we arrive at a parabolic operator $\partial_t + L$, where

$$Lf(\xi, t) = \sum_{n=1}^\infty \alpha_n^2\partial_{\eta_n}^2 f(\xi, t) + \sum_{n=1}^\infty b^n(\xi, t)\partial_{\eta_n} f(\xi, t).$$

Here we also have a probability measure μ on $X\times[0,1)$ satisfying the parabolic equation $\partial_t\mu = L^*\mu$ with any initial distribution ν such that $\|\xi\|_0^2\|\xi\|_2^k \in L^1(\nu)$ for all k (see § 10.4). If F does not depend on t, then similarly we consider the elliptic equation and establish the existence of infinitesimally invariant measures for the stochastic Navier–Stokes equation.

In lattice models in statistical physics, the study of stochastic dynamics of interacting spins systems leads to infinite systems of stochastic equations. The spins take values in $\mathbb{R}^d$, and their sites are indexed by elements i of some countable set S. Their interactions are described by potentials of self-interaction V_i and potentials $W_{i,j}$ of interaction with other particles with indices j. This leads to systems of the form

$$dx_t^i = \sqrt{2}dw_t^i + \nabla V_i(x^i)dt + \sum_j \nabla_{x^i} W_{i,j}(x_t^i, x_t^j)dt.$$

In the case of finite radius of interaction only one and the same fixed number of potentials $W_{i,j}$ may differ from zero for each i. In the general case certain decay conditions are required so that these sums are defined. It is convenient to consider this system in the space $\mathbb{R}^S$. Investigation of the stationary Fokker–Planck–Kolmogorov equation leads to the elliptic operator with the drift $b = (b^i)$, where $b^i(x) = \nabla V_i(x^i) + \sum_j \nabla_{x^i} W_{i,j}(x^i, x^j)$. In order to apply the results presented below on existence of solutions to this equation, we have to impose certain technical restrictions on the functions b^i. For example, for $d = 1$ a typical case is this: $b^i(x) = -(x^i)^3 + f^i(x)$, where f^i satisfies appropriate estimates. Let us give precise conditions.

Suppose that S is a countable set, $J = (J_{i,j})_{i,\,j\in S}$ is a possibly infinite symmetric matrix, $J_{i,j} \geqslant 0$, $q = (q_i)$, $q_i > 0$, $\sum_{i\in S} q_i < \infty$. We write $Jq \leqslant Cq$ with $C \in [0,\infty)$ if $\sum_{i\in S} q_i J_{i,j} \leqslant Cq_t$ for all $j \in S$. Let $l^1(q)$ denote the weighted l^1-space of all collections $x = (x_i)_{i\in S}$ with $\|x\|_{l^1(q)} = \sum_{i\in S} q_i|x_i| < \infty$. The condition $Jq \leqslant Cq$ holds if $J_{i,j} = b_{i,j}c_{i,j}$, where $b_{j,l}q_l \leqslant C_1 q_j$ and $\sum_{l\in S} c_{j,l} \leqslant C_2$ for all $j, l \in S$. For example, this is true if $S = \mathbb{Z}^d$ and $J_{n,j} = a(n-j)$, where $a \geqslant 0$ is an even function with $a(n) \leqslant \text{const}\, q_n^2$ and $q_j q_{n-j} \leqslant \text{const}\, q_n$ for all $n, j \in \mathbb{Z}^d$. The latter holds if $q_n \sim |n|^{-r}$, $r > d$, and $a(n) \leqslant \text{const}|n|^{-2r}$.

Suppose that J generates a bounded operator on $l^1(q)$, i.e.,

$$\sum_{i,j\in S} q_j J_{j,i}|x_i| \leqslant \lambda \sum_{j\in S} q_j|x_j|$$

for all $x \in l^1(q)$ and some $\lambda \geqslant 0$. The minimal possible λ is the operator norm $\|J\|$ of the operator J on $l^1(q)$. Clearly, whenever $Jq \leqslant Cq$, we have $\|J\| \leqslant C$.

As an example of suitable J and q take the integer lattice $S = \mathbb{Z}^d$ in $\mathbb{R}^d$ and for $n \in \mathbb{Z}^d$ set $q_n = (|n|+1)^{-r}$, where $r > d$, so that $\sum_{n\in S} q_n < \infty$. Let

$$\|J\|_p = \sup_{n\in\mathbb{Z}^d} \sum_{j\in\mathbb{Z}^d} (1+|j|)^{2p} J_{n,j+n}^2 < \infty \quad \forall\, p \in \mathbb{N}.$$

Then $\|J\| < \infty$. In addition, since $J_{n,j} \leqslant \|J\|_r(1+|n-j|)^{-r}$, one has

$$\sum_{j\in S} J_{n,j}q_j \leqslant C_1\bigl(|n/2|+1\bigr)^{-r} + \sum_{|j|<|n|/2} J_{n,j}q_j$$
$$\leqslant C_1\bigl(|n/2|+1\bigr)^{-r} + C_2\bigl(|n/2|+1\bigr)^{-r} \leqslant C_3 q_n.$$

10.1.7. Example. Let $X = \mathbb{R}^S$, $\alpha \geqslant 2$, let $q = (q_i)_{i\in S} \in l_1$, and let X_0 be the weighted Banach space of sequences with finite norm

$$|x|_0 = \Bigl(\sum_{i\in S} q_i|x_i|^\alpha\Bigr)^{1/\alpha} < \infty.$$

Suppose that $J = (J_{i,j})_{i,j\in S}$ is the same matrix as above,

$$\|J\| \leqslant \lambda, \quad \sum_{j\in S} J_{i,j} \leqslant \lambda.$$

Let $b = (b^i)_{i\in S}$ be a collection of Borel functions on X_0 such that

$$x_i b^i(x) \leqslant c - (\lambda+\varepsilon)|x_i|^\alpha + \sum_{j\in S} J_{i,j}|x_j|^\alpha \quad \forall x\in X_0,\ i\in S,$$

$$|b^i(x)| \leqslant C_i + K_i\|x\|^{d_i} \quad \forall x\in X_0,\ i\in S,$$

where $c>0$, $\varepsilon>0$, $C_i, K_i, d_i \geqslant 0$. Then, as we shall see in § 10.3, there exists a probability measure μ on $\mathbb{R}^S$ satisfying the equation $L^*_{\mathrm{I},b}\mu = 0$ with respect to the class $\mathcal{FC}_b^\infty(\mathbb{R}^S)$, moreover, $\mu(X_0)=1$ and

$$\int_{X_0} |x_i|^\alpha\,\mu(dx) \leqslant \frac{c+1}{\varepsilon} \quad \forall i\in S, \qquad \int_{X_0} |x|_0^\alpha\,\mu(dx) \leqslant \frac{c+1}{\varepsilon}\sum_{i\in S} q_i.$$

This result applies to many models considered in the literature (see, e.g., Albeverio, Kondratiev, Pasurek, Röckner [**23**]).

10.2. Properties of solutions

Here we discuss those properties of finite-dimensional projections of a measure μ satisfying an infinite-dimensional elliptic or parabolic equation that are direct corollaries of the results presented in Chapters 1–6.

We shall assume that we are given a dense continuous embedding $H\subset X$ generating an embedding $j\colon X^*\to H$, and the functionals l_i are such that the vectors $e_i := j(l_i)$ have the property $l_i(e_j)=\delta_{ij}$. Let μ be a probability measure satisfying the equation $L^*_{A,b}\mu = 0$ with $A=(a^{ij})_{i,j\geqslant 1}$ and $b=(b^i)_{i\geqslant 1}$ in the sense explained above. We consider projections $P_n x = \bigl(l_1(x),\dots,l_n(x)\bigr)$. Let $\mathbb{E}_n$ denote the conditional expectation with respect to the σ-field generated by P_n and let

$$a_n^{ij} := \mathbb{E}_n a^{ij}, \quad b_n^i := \mathbb{E}_n b^i.$$

On $\mathbb{R}^n$ we consider the elliptic operator

$$L_n f := \sum_{i,j=1}^n a_n^{ij}(x)\partial_{x_i}\partial_{x_j} f(x) + \sum_{i=1}^n b_n^i(x)\partial_{x_i} f(x).$$

For the measures $\mu_n := \mu\circ P_n^{-1}$ on $\mathbb{R}^n$, $\mu\circ P_n^{-1}(B) := \mu\bigl(P_n^{-1}(B)\bigr)$, we obtain

$$L_n^*\mu_n = 0.$$

Indeed, if $f(x) = f\bigl(l_1(x),\dots,l_n(x)\bigr)$, then

$$\begin{aligned}\int_{\mathbb{R}^n} L_n\varphi(y)\,\mu_n(dy) &= \int_X \Bigl[\sum_{i,j\leqslant n} a_n^{ij}(x)\partial_{e_i}\partial_{e_j}\varphi\bigl(l_1(x),\dots,l_n(x)\bigr) \\ &\qquad + \sum_{i\leqslant n} b_n^i(x)\partial_{e_i}\varphi\bigl(l_1(x),\dots,l_n(x)\bigr)\Bigr]\,\mu(dx) \\ &= \int_X \Bigl[\sum_{i,j\leqslant n} a^{ij}(x)\partial_{e_i}\partial_{e_j}\varphi\bigl(l_1(x),\dots,l_n(x)\bigr) \\ &\qquad + \sum_{i\leqslant n} b^i(x)\partial_{e_i}\varphi\bigl(l_1(x),\dots,l_n(x)\bigr)\Bigr]\,\mu(dx) = 0.\end{aligned}$$

We draw the reader's attention to the following circumstance: the coefficients of the equation satisfied by the measure μ_n depend on the original measure μ (since the conditional expectation is taken with respect to it). Hence in the general case the projections of two different solutions of the infinite-dimensional equation will satisfy different finite-dimensional equations! For this reason the infinite-dimensional uniqueness problem does not reduce to the finite-dimensional one. Say, if the functions b^i are bounded and $a^{ij} = \delta_{ij}$, then the projections will satisfy the equation with unit diffusion matrix and a bounded drift, which in the finite-dimensional case implies uniqueness of a probability solution, but this does not help to obtain the coincidence of the projections of different solutions, since for them the finite-dimensional drifts may be different. Though, there is a simple case where one can assert the coincidence of the finite-dimensional coefficients: this is the case where a^{ij} and b^i with $i, j \leqslant n$ depend only on $l_1, \ldots, l_n$ for arbitrarily large numbers n. Here we obtain uniqueness of a probability solution in the infinite-dimensional case under the following assumptions: the functions a^{ij} are constant and the matrices $(a^{ij})_{i,j\leqslant n}$ are positive-definite, the functions b^i are bounded and b^i depend only on $l_1, \ldots, l_n$ for $i \leqslant n$. In this case, the conditional expectation of $b^i(x) = b^i\bigl(l_1(x), \ldots, l_k(x)\bigr)$ with respect to the σ-algebra generated by $l_1, \ldots, l_k$ is b^i itself for any measure.

10.2.1. Theorem. *Suppose that the functions a^{ij} are constant, the finite-dimensional matrices $(a^{ij})_{i,j\leqslant n}$ are positive-definite and $b^i \in L^p(\mu)$ for all $p < \infty$. Then the measure μ_n has a bounded continuous density ϱ_n of class $W^{p,1}(\mathbb{R}^n)$ for all n. If $\exp\bigl(\kappa_i|b^i|\bigr) \in L^1(\mu)$ for some numbers $\kappa_i > 0$, then the continuous density ϱ_n is positive. If only the inclusion $b^i \in L^2(\mu)$ is given, then μ_n possesses a density ϱ_n of class $W^{1,1}(\mathbb{R}^n)$ and $|\nabla\varrho_n/\varrho_n|^2 \in L^2(\mu_n)$.*

PROOF. Let $\kappa = \min(\kappa_1, \ldots, \kappa_n)$. According to Jensen's inequality for conditional expectations (see § 1.3), we have $\exp(\kappa|b_n|) \in L^1(\mu_n)$, since we have $\exp(\kappa|b^i|) \in L^1(\mu)$. Hence we can apply the finite-dimensional theorems. □

Similarly, for the solution $\mu = \mu_t\,dt$ of the parabolic equation on $X\times(0,T)$ given by a family of probability measures μ_t on X, the projection of the measure $\mu_n = \mu_{t,n}\,dt$ on $\mathbb{R}^n\times(0,T)$, where $\mu_{t,n}$ is the projection of μ_t, satisfies the parabolic equation $\partial_t\mu_n = L_n^*\mu_n$, where $L_n = L_{A_n,b_n}$, $A_n = (a_n^{ij})_{i,j\leqslant n}$, $b_n = (b_n^i)_{i=1}^n$, a_n^{ij} and b_n^i are the conditional expectations of a^{ij} and b^i. If μ solves the Cauchy problem with initial condition ν, its projection will solve the corresponding Cauchy problem with the initial condition equal to the projection of ν on $\mathbb{R}^n$.

10.2.2. Theorem. *If a Borel measure $\mu = \mu_t\,dt$ on $X\times(0,T)$ given by a family of probability measures μ_t on X satisfies equation* (10.1.6) *and there exist numbers $\kappa_i > 0$ such that $\exp\bigl(\kappa_i|b^i|^2\bigr) \in L^1(\mu)$ for every $i \geqslant 1$, then, for every projection $P_n\colon (x,t) \mapsto (x_1, \ldots, x_n, t)$, $X\times(0,T) \to \mathbb{R}^n\times(0,T)$, the measure μ_n on $\mathbb{R}^n\times(0,T)$ has a continuous strictly positive density.*

The proof is similar to the justification of the previous theorem.

For some special equations there are other results, for example, estimates of measures of balls uniform in dimension, see Agrachev, Kuksin, Sarychev, Shirikyan [**7**], Barbu, Bogachev, Da Prato, Röckner [**88**], Shirikyan [**858**]. The following question is open. Let μ be a probability measure on l^2 such that $L_{\mathrm{I},b}^*\mu = 0$ with respect to $\mathcal{F}C_b^\infty(\{l_i\})$, where $\{l_i\}$ are coordinate functions, the drift b is continuous (or even bounded continuous); is μ positive on all balls of nonzero radius? For example, this is true if μ is equivalent to a nondegenerate Gaussian measure.

In the paper Bogachev, Röckner [**161**], the method of Lyapunov functions is applied to obtain sufficient conditions of exponential integrability of the norm with respect to the measure μ satisfying equation (10.1.1), which ensures also the condition required in Theorem 10.2.1. These conditions use certain coercivity of the drift. Let us give a typical example.

10.2.3. Example. Let $q_n > 0$, $\sum_{n=1}^\infty q_n < \infty$. Let X be the weighted Hilbert space of sequences $x = (x_n)$ with the square of the norm $\|x\|_0^2 := \sum_{n=1}^\infty q_n x_n^2 < \infty$. Suppose that we are given functions b^n on X, $n \in \mathbb{N}$, that are continuous in the weak topology on all balls and for some numbers $C_n > 0$ satisfy the estimates $|b^n(x)| \leqslant C_n + C_n\|x\|_0^2$. Suppose that there exists a number $C > 0$ such that

$$\sum_{n=1}^\infty q_n x_n b^n(x) \leqslant C - C\|x\|_0^2, \quad \text{if } x = (x_1, \ldots, x_n, 0, 0, \ldots).$$

Assume that $a^{nk} = 0$ if $n \neq k$, $0 < a^{nn} \leqslant M$, and a^{nn} does not depend on x. As we shall see in § 10.3, on X there is a Borel probability solution μ of the equation $L_{A,b}^*\mu = 0$ such that the function $\exp(\varepsilon\|x\|_0^2)$ is integrable with respect to μ for any positive $\varepsilon < (2M \sup_n q_n)^{-1}C$.

Then $\exp(\varepsilon_n|b^n|) \in L^1(\mu)$ if $\varepsilon_n < \varepsilon C_n^{-1}$, which yields the existence of continuous strictly positive densities of the finite-dimensional projections of the measure μ. If for some $r > 1$ we have the estimate $|b^n(x)| \leqslant C_n + C_n\|x\|_0^r$, then it suffices to have the estimate $\sum_{n=1}^\infty q_n x_n b^n(x) \leqslant C - C\|x\|_0^r$ on finite sequences.

Let us consider an example with a parabolic equation for the transition probabilities of a diffusion process.

10.2.4. Example. Let ξ_t be a diffusion process in a Banach space X defined by the stochastic equation

$$d\xi_t = dw_t + b(\xi_t, t)dt,$$

where w_t is some Wiener process in X (see Bogachev [**124**, Chapter 7] for the definition) and $b\colon X\times[0,T] \to X$ is a Borel mapping such that $\|b(x,t)\| \leqslant C + C\|x\|$. This equation is understood as the integral equation

$$\xi_t(\omega) = \xi_0(\omega) + w_t(\omega) + \int_0^t b\bigl(\xi_s(\omega), s\bigr)\, ds.$$

Suppose that for some $\varepsilon_0 > 0$ the function $\exp(\varepsilon_0\|x\|^2)$ is integrable with respect to the distribution of ξ_0 (which holds, for example, if ξ_0 is a nonrandom point). Using Gronwall's lemma, we obtain

$$\|\xi_t(\omega)\| \leqslant \bigl[\|\xi_0(\omega)\| + \|w_t(\omega)\| + C\bigr]e^{Ct}, \quad t \in [0, T].$$

By Fernique's theorem (see Bogachev [**124**, § 2.6]) for some $\varepsilon_1 > 0$ the random variables $\exp\bigl(\varepsilon_1\|w_t\|^2\bigr)$ with $t \in [0,T]$ have uniformly bounded expectations. Hence there exists $\varepsilon_2 > 0$ such that the random variables $\exp\bigl(\varepsilon_2\|\xi_t\|^2\bigr)$ for $t \in [0,T]$ also have uniformly bounded expectations. This means that for the law $P(t, \cdot)$ of ξ_t, the function $\exp\bigl(\varepsilon_2\|x\|^2\bigr)$ is integrable with respect to the measure $P(t, \cdot)\, dt$ on $X\times[0,T]$, which gives also the integrability of the function $\exp\bigl(C^{-2}\varepsilon_2\|b\|^2\bigr)$ with respect to this measure and implies the positivity of densities of finite-dimensional projections (for the latter property, of course, it is important that w_t is a Wiener process).

It is more difficult to obtain infinite-dimensional properties of solutions which could be regarded as infinite-dimensional analogs of absolute continuity and differentiability. There are results (a typical example is the theorem below) asserting that, under rather special assumptions, a stationary distribution of an infinite-dimensional diffusion or a solution to an elliptic equation is absolutely continuous with respect to a given measure (typically Gaussian) and its density belongs to some Sobolev class; see Bogachev, Da Prato, Röckner [**128**], Bogachev, Krylov, Röckner [**149**], Bogachev, Röckner [**157**], [**158**], Da Prato, Debussche [**263**], Da Prato, Debussche, Goldys [**264**], Da Prato, Lunardi [**274**], Da Prato, Zabczyk [**280**], [**281**], Es-Sarhir [**338**], Gaveau, Moulinier [**402**], Goldys, Maslowski [**415**], Hino [**455**] and references in Bogachev [**126**, Chapter 12]. It should be noted that Tolmachev [**896**] constructed an example of an infinite-dimensional diffusion with smooth coefficients and bounded and uniformly nondegenerate diffusion coefficient such that its transition probabilities and stationary distribution have no directions of continuity (in the sense defined above). A long-standing open problem is whether this can happen if $A = \mathrm{I}$. Some positive results for special drifts b can be found in the works cited above. Here we just mention one typical example, which is a result in [**158**] solving a problem raised by Shigekawa [**855**]. For notational simplicity we formulate this result for the space $\mathbb{R}^\infty$, although it holds for general locally convex spaces.

10.2.5. Theorem. *Let $v = (v^i)$ be a Borel vector field on $X = \mathbb{R}^\infty$ with values in the space $H = l^2$ and let μ be a Borel probability measure on $\mathbb{R}^\infty$ satisfying equation* (10.1.1) *with $A = \mathrm{I}$ and $b(x) = -x + v(x)$, where $|v|_H \in L^2(\mu)$ and $l_i \in L^2(\mu)$, $l_i(x) = x_i$. Then μ is absolutely continuous with respect to the Gaussian measure γ that is the countable power of the standard Gaussian measure on the real line.*

The role of the Gaussian measure γ is that it is a solution for $v=0$, and a moderate perturbation of the drift yields a solution that is absolutely continuous with respect to γ. However, it is important here that the perturbation takes values in H; say, if we take $v(x) = -\varepsilon x$, then the solution will be another Gaussian measure (also a countable power of a Gaussian measure on the real line), but this time mutually singular with γ (see Exercise 10.5.11).

10.3. Existence in the elliptic case

In this section, $\{l_n\}$ is a sequence of continuous linear functionals separating the points in X and $\{e_n\} \subset X$ is such that $l_n(e_k) = \delta_{nk}$. As above, one can assume for simplicity that we deal with the countable power $\mathbb{R}^\infty$ of the real line with the coordinate functions l_n, e_n is the vector with 1 as its nth and only nonzero component.

We shall study the existence of solutions to the equation

$$L^*_{A,b}\mu = 0 \tag{10.3.1}$$

in the special case where $a^{ij} = 0$ if $i \neq j$ and $a^{nn} = a^n$, i.e., we consider the operator $L_{A,b}$ heuristically given by $L_{A,b} = \sum_{n=1}^\infty (a^n \partial^2_{e_n} + b^n \partial_{e_n})$. We note, however, that in the theorems below we deal with classes of cylindrical functions, on which $L_{A,b}$ is meaningful as a finite sum.

We recall that a function $G\colon X \to [0, +\infty]$ on a topological space X is called compact if the sets $\{G \leqslant c\}$, where $c \in \mathbb{R}^1$, are compact.

10.3.1. Theorem. *Suppose that* $\Theta\colon X \to [0,+\infty]$ *is compact and is finite on the finite dimensional spaces* E_n *spanned by* $e_1,\ldots,e_n$. *Let* $a_n \geqslant 0$ *and* a_n *be functions on* X *which are continuous on the sets* $\{\Theta \leqslant c\}$, $c \in \mathbb{R}^1$, *as well as on the subspaces* E_j. *Assume that there exist a number* $C > 0$ *and a nonnegative function* V *on* X *such that, for every* n, *the restriction of* V *to* E_n *is compact and twice continuously differentiable and*

$$\sum_{j=1}^{n}\bigl[a^j(x)\partial_{e_j}^2 V(x) + \partial_{e_j}V(x)b^j(x)\bigr] \leqslant C - \Theta(x), \quad x \in E_n. \tag{10.3.2}$$

Finally, let us assume that

$$a_n(x) + |b^n(x)| \leqslant C_n + \delta_n\bigl(\Theta(x)\bigr)\Theta(x), \quad x \in \{\Theta < +\infty\}. \tag{10.3.3}$$

where $C_n \in (0,+\infty)$ *and* δ_n *is a nonnegative bounded Borel function on* $[0,+\infty)$ *with* $\lim\limits_{r\to+\infty}\delta_n(r) = 0$. *Then, there is a probability measure* μ *on* X *such that* $L_{A,b}^*\mu = 0$ *with respect to the class* $\mathcal{F}C_b^\infty(X,\{l_n\})$. *In addition,*

$$\int_X \Theta(x)\,\mu(dx) \leqslant C. \tag{10.3.4}$$

PROOF. Let E_n be equipped with the inner product making $e_1,\ldots,e_n$ an orthonormal basis. Then the sets $\{x \in E_n\colon \Theta(x) \leqslant c\}$ are compact in E_n. Hence $\Theta(x) \to +\infty$ and $V(x) \to +\infty$ as $|x|_{E_n} \to +\infty$. According to the finite-dimensional case, there exists a probability measure μ_n on E_n such that $L_n^*\mu_n = 0$ with respect to $C_0^\infty(E_n)$, where

$$L_n\psi(x) = \sum_{j=1}^{n}\bigl[a^j(x)\partial_{e_j}^2\psi(x) + \partial_{e_j}\psi(x)b^j(x)\bigr], \quad \psi \in C_0^\infty(E_n).$$

Clearly, we have also $L_n^*\mu_n = 0$ with respect to the class $C_0^2(E_n)$. The functions $a^j\partial_{e_j}^2 V$ and $b^j\partial_{e_j}V$ are bounded on the compact sets $\{\Theta \leqslant c\} \cap E_n$. According to (10.3.2) the function Θ is bounded on all balls in E_n. Since $\varphi\circ V \in C_0^2(E_n)$ for every $\varphi \in C_0^\infty(\mathbb{R}^1)$, it follows from (10.3.2) and Theorem 2.3.2 that

$$\int_{E_n} \Theta(x)\,\mu_n(dx) \leqslant C. \tag{10.3.5}$$

Let $K_n = \sup_r \delta_n(r)$. By (10.3.3) we obtain

$$\int_{E_n} \bigl[a^j(x) + |b^j(x)|\bigr]\,\mu_n(dx) \leqslant C_j + CK_j \quad \forall n, j \in \mathbb{N}. \tag{10.3.6}$$

In particular, we conclude that the equality $L_n^*\mu_n = 0$ holds with respect to the class $C_b^\infty(E_n) = \mathcal{F}C_b^\infty(E_n,\{l_j\}_{j=1}^n)$. We shall consider μ_n as a measure on X (i.e., we extend μ_n to X by setting $\mu_n(X\backslash E_n) = 0$). Since the sets $\{\Theta \leqslant c\}$ are compact, the sequence $\{\mu_n\}$ is uniformly tight. Since compact sets in X are metrizable, because the sequence $\{l_n\}$ is separating (see Bogachev [**125**, Exercise 6.10.24]), this yields that there is a subsequence $\{\mu_{n_i}\}$ which converges weakly to some Radon probability measure μ on X (see Bogachev [**125**, Theorem 8.6.7]). We may assume that the whole sequence $\{\mu_n\}$ converges weakly to μ. It is readily seen that the measure μ is concentrated on the union of the compact sets $\{\Theta \leqslant m\}$, $m \in \mathbb{N}$. We observe that we have not used so far that δ_n in (10.3.3) tends to zero at infinity. Note that, for every function ψ of the form $\psi(x) = \psi_0(l_1(x), \ldots, l_m(x))$, $\psi_0 \in C_b^\infty(\mathbb{R}^m)$, and every $n \geqslant m$, we have

$$\int_X L_{A,b}\psi \, d\mu_n = \int_X \sum_{j=1}^{m} \bigl[a^j \partial_{e_j}^2 \psi + \partial_{e_j}\psi b^j\bigr] \, d\mu_n$$
$$= \int_X \sum_{j=1}^{n} \bigl[a^j \partial_{e_j}^2 \psi + \partial_{e_j}\psi b^j\bigr] \, d\mu_n = 0.$$

Let us show that (10.3.4) holds and $L_{A,b}^*\mu = 0$ with respect to $\mathcal{FC}_b^\infty(X, \{l_n\})$. Clearly, it suffices to show that, for every $\psi \in \mathcal{FC}_b^\infty(X, \{l_n\})$ and every fixed j, one has a^j, $b^j \in L^1(\mu)$ and

$$\lim_{n\to\infty} \int_{E_n} a^j \psi \, d\mu_n = \int_X a^j \psi \, d\mu, \quad \lim_{n\to\infty} \int_{E_n} b^j \psi \, d\mu_n = \int_X b^j \psi \, d\mu.$$

We verify the second equality, with the first one we can proceed completely analogously. Let $R > 0$ and $\Omega_R = \{\Theta \leqslant R\}$. Let

$$\varepsilon_j(R) := R^{-1} \sup_{t\in[0,R]} t\delta_j(t).$$

Then ε_j is a nonnegative function on $[0, +\infty)$ such that

$$\lim_{R\to+\infty} \varepsilon_j(R) = 0 \quad \text{and} \quad \sup_{t\in[0,R]} \delta_j(t)t \leqslant \varepsilon_j(R)R.$$

Hence $\sup_{\Omega_R} |b^j| \leqslant C_j + \varepsilon_j(R)R$. By (10.3.5) we have

$$\mu_n(X\backslash\Omega_R) \leqslant CR^{-1}.$$

By weak convergence and the compactness of Ω_R we have

$$\mu(X\backslash\Omega_R) \leqslant \liminf_{n\to\infty} \mu_n(X\backslash\Omega_R) \quad \forall R \geqslant 0,$$

and since Θ is lower semicontinuous we obtain that

$$\int_X \Theta(x)\, \mu(dx) \leqslant \liminf_{n\to\infty} \int_X \Theta(x)\, \mu_n(dx) \leqslant C, \quad \mu(X\backslash\Omega_R) \leqslant CR^{-1}.$$

By (10.3.3) we obtain that a^j, $b^j \in L^1(\mu)$. Since b^j is continuous on the compact set Ω_R, there is a continuous function G_R on X such that $G_R = b^j$ on Ω_R and $|G_R| \leqslant C_j + \varepsilon_j(R)R$. By weak convergence we have

$$\lim_{n\to\infty} \int_{E_n} G_R(x)\psi(x)\, \mu_n(dx) = \int_X G_R(x)\psi(x)\, \mu(dx).$$

By the above estimates and the equality $G_R = B_j$ on Ω_R we obtain

$$\int_{E_n} |G_R(x)\psi(x) - b^j(x)\psi(x)|\, \mu_n(dx) \leqslant \bigl[C_j + \varepsilon_j(R)R\bigr] \sup_x |\psi(x)|\, \mu_n(X\backslash\Omega_R)$$
$$+ \sup_x |\psi(x)| \int_{X\backslash\Omega_R} \bigl[C_j + \delta_j(\Theta(x))\Theta(x)\bigr]\, \mu_n(dx)$$
$$\leqslant C\Bigl[C_j R^{-1} + \varepsilon_j(R) + C_j R^{-1} + \sup_{t\geqslant R} \delta_j(t)\Bigr] \sup_x |\psi(x)|.$$

The right-hand side of this estimate goes to zero as $R \to +\infty$. The same is true for μ in place of μ_n, which completes the proof. □

We shall also employ the following modification of Theorem 10.3.1 proved by a similar method.

10.3.2. Theorem. *Suppose that in the situation of Theorem* 10.3.1 *condition* (10.3.3) *is replaced by the following conditions:*

$$a^n(x) + |b^n(x)| \leqslant C_n + K_n V(x)^{d_n}\Bigl(1 + \delta_n\bigl(\Theta(x)\bigr)\Theta(x)\Bigr), \quad x \in \{\Theta < +\infty\},$$

$$\sum_{j=1}^{n} a_j(x)|\partial_{e_j} V(x)|^2 \leqslant C + \delta\bigl(\Theta(x)\bigr)\Theta(x)V(x), \quad x \in E_n,$$

where $C_n, K_n, d_n \geqslant 0$ are numbers, δ_n and δ are nonnegative bounded Borel functions with $\lim\limits_{r\to+\infty} \delta_n(r) = \lim\limits_{r\to+\infty} \delta(r) = 0$. Assume, in addition, that V is bounded on the sets $\{\Theta \leqslant c\}$, $c \in [0, +\infty)$. Then there exists a probability measure μ on X such that a^n, $b^n \in L^1(\mu)$, $L^*_{A,b}\mu = 0$ with respect to the class $\mathcal{F}C_b^\infty(X, \{l_n\})$, and (10.3.4) *holds. Moreover, if V is continuous on the set $\{\Theta < \infty\}$ (or, more generally, the functions $V^m\Theta$ are lower semicontinuous), then*

$$\int_X V^m[1 + \Theta]\, d\mu < \infty \quad \forall m \in \mathbb{N}.$$

Let us prove yet another closely related result.

10.3.3. Theorem. *Let X, $\{q_n\}$, and X_0 be the same as in Example* 10.2.3. *Assume, in addition, that the embedding $X_0 \subset X$ is compact. Let $b^n\colon X_0 \to \mathbb{R}^1$ be continuous on all balls in X_0 with respect to the topology of X and satisfy the estimates*

$$|b^n(x)| \leqslant C_n + K_n |x|_0^{d_n} \quad \forall x \in X_0. \tag{10.3.7}$$

Assume that

$$\sum_{n=1}^{\infty} q_n l_n(x) b^n(x) \leqslant C - \varrho(|x|_0) \tag{10.3.8}$$

on the linear span of the vectors e_n, where ϱ is a nonnegative bounded Borel function on $[0, +\infty)$ such that $\lim\limits_{t\to\infty} \varrho(t) = +\infty$. Then there exists a probability measure μ on X_0 satisfying the equation $L^*_{A,b}\mu = 0$ with respect to $\mathcal{F}C_b^\infty(X, \{l_n\})$ such that

$$\int_X \varrho(|x|_0)\, \mu(dx) \leqslant C + \sum_{n=1}^{\infty} q_n, \quad \int_X |x|_0^m [1 + \varrho(|x|_0)]\, \mu(dx) < \infty \quad \forall m \in \mathbb{N}.$$

PROOF. Let $V(x) = \sum_{n=1}^\infty q_n l_n(x)^2 = |x|_0^2$. We have

$$\sum_{j=1}^{n} \bigl[a^j \partial_{e_j}^2 V(x) + b^j(x)\partial_{e_j} V(x)\bigr] \leqslant 2\sum_{j=1}^{n} a^j q_j + 2C - 2\varrho(|x|_0), \quad x \in E_n.$$

Let $\Theta(x) = 2\varrho(|x|_0)$. The functions b^n and V are bounded and continuous on compact sets $\{\Theta \leqslant c\}$, which are closed balls in X_0. In addition, we have $a^n = 1$ for all n and also

$$\sum_{j=1}^{\infty} |\partial_{e_j} V|^2 \leqslant 4(\sup_j q_j)V.$$

Now we can use Theorem 10.3.2 with $\delta(t) = (|t| + 1)^{-1}$ and $\delta_n = 0$. □

Using the function $V(x) = \exp\bigl(\lambda |x|_0^2\bigr)$, we obtain the following fact.

10.3.4. Corollary. *Assume that in Theorem* 10.3.3 *we have* $\varrho(t) = kt^2$ *and that* $k > 2\lambda \sup_n q_n$. *Then*

$$\int_X \exp\bigl(\lambda |x|_0^2\bigr)\, \mu(dx) < \infty.$$

Moreover, for the existence of μ *it suffices to replace the power estimates on* b^n *by the estimates*

$$|b^n(x)| \leqslant C_n + K_n \exp\bigl(d_n |x|_0\bigr).$$

If $\varrho(t) = kt^{2+\delta}$, *where* k, $\delta > 0$, *then it is enough to have the estimates*

$$|b^n(x)| \leqslant C_n + K_n \exp\bigl(d_n |x|_0^2\bigr).$$

We observe that Theorem 10.3.3 can be equivalently reformulated in terms of a Hilbert space.

10.3.5. Theorem. *Let* Z *be a separable Hilbert space with an orthonormal basis* $\{\eta_n\}$ *and let* $b^n\colon Z \to \mathbb{R}^1$ *be continuous on all balls with respect to the weak topology. Let* P_n *be the orthogonal projection to the linear span* E_n *of* $\eta_1, \ldots, \eta_n$. *Suppose that there exist constants* C, C_n, K_n, d_n *and a locally bounded nonnegative Borel function* ϱ *on* $[0, +\infty)$ *such that* $\lim_{R\to+\infty} \varrho(R) = +\infty$ *and for all* $n \in \mathbb{N}$

$$\sup_{x\colon |P_n x|_Z \geqslant R} \bigl(b^1(P_n x)\eta_1 + \ldots + b^n(P_n x)\eta_n, P_n x\bigr)_Z \leqslant C - \varrho(R),$$

$$|b^n(x)| \leqslant C_n + K_n |x|_Z^{d_n}.$$

If $t_n > 0$ *and* $\sum_{n=1}^\infty t_n^2 < \infty$, *then there exists a probability measure* μ *on* Z *such that*

$$\sum_{n=1}^\infty \int_Z \bigl[q_n^2 \partial_{\eta_n}^2 f + b^n \partial_{\eta_n} f\bigr]\, d\mu = 0 \quad \forall f \in \mathcal{F}C_b^\infty(\{\eta_n\}). \tag{10.3.9}$$

If $\sum_{n=1}^\infty \|b^n\|_{L^2(\mu)}^2 < \infty$, *then* (10.3.9) *holds for all* $f \in \mathcal{F}C_b^\infty(Z)$.

Let us apply our existence result to the situation in Example 10.1.7.

10.3.6. Theorem. *Let* $X = \mathbb{R}^S$, *let* S *be a countable set, and let* X_0 *be the weighted space from Example* 10.1.7. *Suppose that* $b = (b^i)_{i\in S}$ *is a collection of continuous functions on* $(X_0, |\cdot|_0)$ *satisfying the same conditions as in Example* 10.1.7. *Then there exists a probability measure* μ *on* X_0 *such that* (10.3.1) *is satisfied with respect to* $\mathcal{F}C_b^\infty(X)$.

Proof. We may assume that $S = \mathbb{N}$. As in the proof of Theorem 10.3.3, we find probability measures μ_n on the n-dimensional linear subspaces $\mathbb{R}^n$ in $\mathbb{R}^\infty$ satisfying the elliptic equations $L_n^*\mu_n = 0$. It follows from the results in § 2.3 that

$$\sup_{n,j} \int |x_j|^p\, \mu_n(dx) < \infty \quad \forall p \in \mathbb{N}.$$

We can find numbers $\widetilde{q}_n > 0$ such that $\sum_{n=1}^\infty \widetilde{q}_n < \infty$ and $\lim_{n\to\infty} q_n/\widetilde{q}_n = 0$. Hence, considering the function $\Psi := \sum_{n=1}^\infty \widetilde{q}_n |x_n|^\alpha$ and using that $\sup_n \|\Psi\|_{L^1(\mu_n)} < \infty$, we see that the sequence $\{\mu_n\}$ is uniformly tight on X_0. Let us take for $\{l_n\}$ the natural coordinate functions on $\mathbb{R}^S$. Now the same reasoning as in Theorem 10.3.3 completes the proof. □

Let us return to the situation considered in Example 10.1.4.

10.3.7. Theorem. *Let* $|\Psi(y)|\leqslant c+c|y|^\kappa$, $\kappa<6$. *Then on* $X = L^2[0,1]$ *there exists a probability solution* μ *of the equation* $L^*_{A,b}\mu=0$ *with respect to* $\mathcal{F}C_b^\infty(X,\{l_n\})$, *where* $l_n(u) = u_n$. *For the Sobolev class* $X_0 = W_0^{2,1}[0,1]$ *we have* $\mu(X_0) = 1$ *and also* $\exp(\varepsilon\|u'\|_2^2) \in L^1(\mu)$ *for any* $\varepsilon < (2\sup_j \alpha_j^2)^{-1}$.

PROOF. We apply Theorem 10.3.3 and Corollary 10.3.4 with numbers $a^n=\alpha_n^2$, $e_n=\eta_n$, and $\Theta(u) = \|u'\|_2^2$, taking on X_0 the equivalent norm $\|u'\|_2$. The functions b^n are continuous on any ball S in X_0 in the norm topology from $X = L^2[0,1]$, since if $u^k \to u$ in X and $u^k \in S$, then the functions u^k uniformly converge to u. If u belongs to the linear span of $\{\eta_n\}$, then

$$\sum_{n=1}^\infty b^n(u)u_n = \sum_{n=1}^\infty \lambda_n u_n^2 + \int_0^1 fu\,dx$$
$$= -\int_0^1 (u')^2\,dx + \int_0^1 fu\,dx \leqslant \|f\|_2^2 - \frac{1}{2}\|u'\|_2^2.$$

Indeed, let $G(x)$ be the integral of $y\psi(y)$ over $[0,x]$. Then

$$\sum_{n=1}^\infty \bigl(\psi(u)u',\eta_n\bigr)_2(u,\eta_n)_2 = \int_0^1 \psi(u)u'u\,dx = \int_0^1 [G(u)]'\,dx = 0, \ u\in X_0,$$

since $u(0) = u(1)$. In addition, $\sum_{n=1}^\infty \lambda_n u_n^2 = -\|u'\|_2^2$. Since

$$\int_0^1 u(x)^6\,dx = \int_0^1 u^2(x)\biggl(\int_0^x 2uu'\,dt\biggr)^2 dx \leqslant 4\|u\|_2^4\|u'\|_2^2, \quad u \in X_0,$$

the following inequality holds:

$$\int_0^1 |\Psi(u)|\,dx \leqslant c+c\int_0^1 |u|^\kappa\,dx \leqslant c+c\|u\|_6^\kappa \leqslant c+c4^{\kappa/6}\|u\|_2^{4\kappa/6}\|u'\|_2^{2\kappa/6}.$$

Hence the hypotheses of Theorem 10.3.2 are fulfilled with numbers $d_n = \max(\kappa/3,1)$ and $s = (1-\kappa/6) > 0$ and functions $\delta_n(r) = r^{-s}$, $\delta(r) = r^{-1}$. □

If $\kappa \leqslant 2$, then the densities of finite-dimensional projections are positive, since $\exp(\varepsilon_n|b^n|) \in L^1(\mu)$ for sufficiently small ε_n (see § 10.2).

The same reasoning applies in Example 10.1.5, where the functions b^n have the form $b^n(u) = -n^2u_n + (\Phi\circ u,\eta_n)_2$, $u \in X_0$. These functions are still continuous on all balls in X_0 with the topology from X, and one can use the same Lyapunov function $V(u) = \|u\|_2^2$. Thus, we obtain a probability solution μ of the equation $L^*_{A,b}\mu = 0$ with the same $a^n = \alpha_n^2$ as above. For $\kappa \leqslant 2$ we also obtain the positivity of the densities of finite-dimensional projections.

In the situation of Example 10.1.6 (the stochastic Navier–Stokes equation), analyzed in detail in the parabolic case in Example 10.4.3, there exists also a probability solution μ of the elliptic equation $L^*_{A,b}\mu = 0$ such that $\exp(\varepsilon\|u\|_0^2) \in L^1(\mu)$ for small ε. As one can see from the calculations in the parabolic case, here on the linear span of $\{u_n\}$ we have the estimate $\sum_{n=1}^\infty b^n(u)u_n \leqslant C_1 - C_2\|u\|_0^2$. The functions b^n are continuous on all balls in X with the topology from $L^2(\Omega,\mathbb{R}^d)$, since if $u_i \to u$ in $L^2(\Omega,\mathbb{R}^d)$ and the norms $\|u_i\|_0$ are uniformly bounded, then by the compactness of the embedding $W^{2,1}(\Omega) \to L^2(\Omega)$ one has convergence $\partial_{x_j}u_i \to \partial_{x_j}u$ in

$L^2(\Omega, \mathbb{R}^d)$. Hence the presented results are applicable (and the densities of finite-dimensional projections are positive). Exercise 10.5.19 gives some reinforcement. There is a similar justification for Example 10.1.3. But there is a nuance discussed in Exercise 10.5.18. In all these examples, conditions for uniqueness remain unclear.

10.4. Solvability of the Cauchy problem

Here we prove two theorems on existence of solutions to the Cauchy problem for infinite-dimensional Fokker–Planck–Kolmogorov equations. For simplicity we consider $X = \mathbb{R}^\infty$. Let $\{e_n\}$ be the standard orthonormal basis in l^2, $e_{nk} = \delta_{nk}$. The linear span of $e_1, \ldots, e_n$ will be denoted by E_n. Let us fix $T > 0$. Let $a^{ij}\colon \mathbb{R}^\infty \times [0,T] \to \mathbb{R}^1$ and $b^i\colon \mathbb{R}^\infty \times [0,T] \to \mathbb{R}^1$ be Borel functions. Suppose that the matrices $(a^{ij})_{i,j\leqslant n}$ are symmetric and nonnegative-definite for all n. Set

$$L\varphi(x,t) := \sum_{i,j=1}^{n} a^{ij}(x,t)\partial_{e_i}\partial_{e_j}\varphi(x,t) + \sum_{i=1}^{n} b^i(x,t)\partial_{e_i}\varphi(x,t), \ (x,t) \in \mathbb{R}^\infty \times [0,T]$$

for any function φ that is a smooth function in finitely many variables $x_1, \ldots, x_n, t$.

Let $B_n := (b^1, \ldots, b^n)$ and $P_n x = (x_1, \ldots, x_n)$.

10.4.1. Theorem. *Let $\Theta\colon \mathbb{R}^\infty \to [0,+\infty]$ be a compact function, finite on every space E_n, such that the functions a^{ij} and b^i are continuous in x on all sets $\{\Theta \leqslant R\} \times [0,T]$ with $R < \infty$. Suppose that there exist numbers $M_0, C_0 \geqslant 0$ and a Borel function $V\colon \mathbb{R}^\infty \to [1,+\infty]$ with compact sublevel sets $\{V \leqslant R\}$ for $R < \infty$ and its restrictions on E_n of class C^2 such that for all $x \in E_n$, $n \geqslant 1$ we have*

$$\begin{aligned} &\sum_{i,j=1}^{n} a^{ij}(x,t)\partial_{e_i}V(x)\partial_{e_j}V(x) \leqslant M_0 V(x)^2, \\ &LV(x,t) \leqslant C_0 V(x) - \Theta(x). \end{aligned} \tag{10.4.1}$$

Suppose also that there are constants $C_i \geqslant 0$ and $k_i \geqslant 0$ such that for all i and $j \leqslant i$ we have

$$|a^{ij}(x,t)| + |b^i(x,t)| \leqslant C_i V(x)^{k_i}\bigl[1 + \delta_i(\Theta(x))\Theta(x)\bigr] \tag{10.4.2}$$

if $(x,t) \in \mathbb{R}^\infty \times [0,T]$, where δ_i is a bounded nonnegative Borel function on $[0,+\infty)$ with $\lim_{s\to\infty} \delta_i(s) = 0$. Then, for every Borel probability measure ν on $\mathbb{R}^\infty$ satisfying the condition $W_k := \sup_n \|V^k \circ P_n\|_{L^1(\nu)} < \infty$ for all $k \in \mathbb{N}$, the Cauchy problem $\partial_t \mu = L^*\mu$ with initial condition ν has a solution of the form $\mu = \mu_t\, dt$ with Borel probability measures μ_t on $\mathbb{R}^\infty$ such that for all $t \in (0,T)$ and $k \in \mathbb{N}$ we have

$$\int_{\mathbb{R}^\infty} V^k\, d\mu_t + k\int_0^t\!\!\int_{\mathbb{R}^\infty} V^{k-1}\Theta\, d\mu_s\, ds \leqslant N_k W_k, \tag{10.4.3}$$

where $N_k := M_k e^{M_k} + 1$ and $M_k = k\bigl(C_0 + (k-1)M_0\bigr)$. In particular, we have $\mu_t(V < \infty) = 1$ for all t and $\mu_t(\Theta < \infty) = 1$ for almost all t.

Proof. For every n let a_n^{ij} denote the restriction of a^{ij} to $E_n \times (0,T)$ and let $A_n := (a_n^{ij})_{i,j\leqslant n}$. Let ν_n be the projection of ν on E_n. We show that there exist Borel probability measures $\mu_{t,n}$ on E_n such that the measure $\mu_n := \mu_{t,n}\, dt$ satisfies the Cauchy problem with the coefficients A_n and B_n on $E_n \times (0,T)$ and initial distribution ν_n. To this end, we consider the Lyapunov function $V_m(x) = V(x)^m$ on the subspace E_n, where $m \geqslant 1$. Letting $M_m := m\bigl(C_0 + (m-1)M_0\bigr)$, we obtain

$$
\begin{aligned}
LV_m &= mV^{m-1}\Bigl(LV + (m-1)V^{-1}\sum_{i,j=1}^{n} a^{ij}\partial_{e_i}V\partial_{e_j}V\Bigr)\\
&\leqslant mV^{m-1}\bigl(C_0V - \Theta + (m-1)M_0V\bigr) \leqslant M_mV^m - mV^{m-1}\Theta.
\end{aligned}
$$

Since the function V_m is ν_n-integrable, one can apply the existence theorem from § 6.6 and obtain the desired probability measures $\mu_{t,n}$ on E_n, for which the function

$$
t \mapsto \int_{E_n} \zeta(x)\,\mu_{t,n}(dx)
$$

is continuous on $[0,T)$ for every $\zeta \in C_0^\infty(E_n)$, where $\mu_{0,n} = \nu_n$. Moreover, according to a priori estimates in § 7.1, for every $m \geqslant 1$ and the numbers

$$
N_m := M_m e^{M_m} + 1, \quad M_m = m\bigl(C_0 + (m-1)M_0\bigr),
$$

for almost all $t \in (0,T)$ the following estimate holds:

$$
\begin{aligned}
\text{(10.4.4)}\quad \int_{H_n} V_m(x)\,\mu_{t,n}(dx) + m\int_0^t\!\!\int_{H_n} V_{m-1}(x)\Theta(x)\,\mu_{s,n}(dx)\,ds&\\
\leqslant N_m\int_{E_n} V_m(x)\,\nu_n(dx) \leqslant N_m + N_mW_m.&
\end{aligned}
$$

Therefore, the Fatou theorem and the aforementioned continuity of $t \mapsto \mu_{t,n}$ yield that (10.4.4) is true for all $t \in [0,T)$. Indeed, replacing the functions V_m and ΘV_{m-1} in the left-hand side with $\min(k, V_m)$ and $\min(k, \Theta V_{m-1})$, we obtain the required estimate for all $t \in [0,T)$ for each fixed k, and then we let $k \to \infty$.

Suppose now that $\zeta \in C_0^\infty(\mathbb{R}^d)$. Let us identify E_n with the space $\mathbb{R}^n$. If $n \geqslant d$, then ζ as a function on $\mathbb{R}^n$ belongs to the class $C_b^\infty(\mathbb{R}^n)$. Let $m = \max(k_1, \ldots, k_d)$ and $\delta = \delta_1 + \cdots + \delta_d$. Then we have

$$
\text{(10.4.5)}\qquad |L\zeta(x,t)| \leqslant K + KV_m(x) + KV_m(x)\delta\bigl(\Theta(x)\bigr)\Theta(x)
$$

for $(x,t) \in \mathbb{R}^n\times[0,T]$, where K is a number that depends on ζ (but does not depend on n, since ζ is a function of $x_1, \ldots, x_d$). Hence using approximations, inequality (10.4.4) and the Lebesgue dominated convergence theorem, we obtain

$$
\text{(10.4.6)}\qquad \int_{E_n} \zeta(x)\,\mu_{t,n}(dx) = \int_0^t\!\!\int_{E_n} L\zeta(x,s)\,\mu_{s,n}(dx)\,ds + \int_{E_n} \zeta(x)\,\nu_n(dx),
$$

since by the finite-dimensional case this identity is true for all $\zeta \in C_0^\infty(\mathbb{R}^n)$, so in our situation it remains true also for all $\zeta \in C_b^\infty(\mathbb{R}^n)$. Letting now

$$
\varphi_n(t) := \int_{H_n} \zeta(x)\,\mu_{t,n}(dx), \; t \in [0,T),
$$

we conclude by (10.4.4) and (10.4.6) that the function φ_n is Lipschitzian (one can show that it is everywhere differentiable in $(0,T)$), moreover, inequality (10.4.5) gives the estimate

$$
|\varphi_n'(t)| \leqslant \int_{E_n} |L\zeta(x,t)|\,\mu_{t,n}(dx) \leqslant K_\zeta\int_{E_n} [1 + V_{m-1}(x)\Theta(x)]\,\mu_{t,n}(dx)
$$

with some number K_ζ independent of n (but depending only on ζ). Therefore, by (10.4.4) the functions φ_n possess uniformly bounded variations, hence there is a subsequence of $\{\varphi_n\}$ converging pointwise on $[0,T)$. One can assume that this is true for the whole sequence. Moreover, one can do this in such a way that the pointwise convergence will take place for every function ζ in a fixed countable collection $\mathcal{F}$ with the following property: weak convergence of a uniformly tight sequence of probability measures on $\mathbb{R}^\infty$ follows from convergence of their integrals of every function in $\mathcal{F}$.

It follows by (10.4.4) and the compactness of the sets $\{V_m \leqslant R\}$ and $\{\Theta \leqslant R\}$ that, for every fixed $t \in (0,T)$, the sequence of measures $\mu_{t,n}$ is uniformly tight on $\mathbb{R}^\infty$ (see Bogachev [**125**, Example 8.6.5]). Hence one can find a subsequence, denoted for simplicity by the same indices n, such that $\{\mu_{t,n}\}$ converges weakly on $\mathbb{R}^\infty$ for every rational $t \in [0,T)$. However, since we have convergence of $\varphi_n(t)$ at every point $t \in [0,T)$ for every $\zeta \in \mathcal{F}$, we see that $\{\mu_{t,n}\}$ converges weakly for each $t \in [0,T)$.

Estimate (10.4.3) follows from (10.4.4) taking into account that the functions $V \geqslant 1$ and $\Theta \geqslant 0$ are lower semicontinuous, hence V^k and $V^{k-1}\Theta$ are also lower semicontinuous.

The family of measures μ_t obtained in this way is the desired solution. Indeed, let us fix $\zeta \in C_0^\infty(\mathbb{R}^d)$. We have to show that the integrals of $L\zeta(x,t)$ over the set $\mathbb{R}^\infty \times [0,T']$, where $T' < T$, with respect to μ_n converge to the integral with respect to $\mu = \mu_t\,dt$. This reduces to the proof of such convergence for all functions $f = \partial_{x_i}\zeta b^i$ and $f = a^{ij}\partial_{x_j}\partial_{x_i}\zeta$. Suppose that we have managed to show this for the functions $f_N = \max\bigl(\min(f,N), -N\bigr)$. Then (10.4.2) and (10.4.4) enable us to extend the same to the original functions f, since for every $\varepsilon > 0$ these estimates give a number N such that the integral of $|f|I_{|f|>N}$ with respect to $\mu_{t,n}\,dt$ is less than ε. Indeed, it suffices to show that the integral of $G := V^k\bigl(1+\delta_i(\Theta)\Theta\bigr)$ over the set $\{G \geqslant N\}$ with respect to $\mu_{t,n}\,dt$ does not exceed ε for sufficiently large N. Let us take n_1 such that $1/n_1 + \delta_i(s) < c\varepsilon$ for all $s \geqslant n_1$, where the number $c > 0$ is so small that $cN_{k+1}W_{k+1} < 1/2$. One can assume that $\delta_i \leqslant 1$. We have

$$\begin{aligned}\int_0^T\int_{\{\Theta\geqslant n_1\}} G\,d\mu_{t,n}\,dt &= \int_0^T\int_{\{\Theta\geqslant n_1\}} [\Theta^{-1} + \delta_i(\Theta)]V^k\Theta\,d\mu_{t,n}\,dt \\ &\leqslant c\varepsilon\int_0^T\int_{H_n} V^k\Theta\,d\mu_{t,n}\,dt \leqslant \varepsilon/2.\end{aligned}$$

For any $N \geqslant n_1$ and $t < T$ the integral of G against the measure $\mu_{t,n}$ over the set $\{G \geqslant N, \Theta \leqslant n_1\}$ does not exceed

$$(1+n_1)\int_{V^k\geqslant N/(1+n_1)} V^k\,d\mu_{t,n} \leqslant N^{-1}(1+n_1)^2 N_k W_k,$$

which can be made smaller than $\varepsilon/2$ uniformly in $t < T$ for all sufficiently large N.

Thus, it remains to justify the desired convergence in the case of a single function f_N, which from now on will be denoted by f. We recall that the restrictions of such a function f to the sets $\{\Theta \leqslant R\}\times[0,T]$ are continuous in the first variable. Dividing by N one can assume that $|f| \leqslant 1$. If the function f were continuous in x on the whole space, this would follow at once by weak convergence of $\mu_{t,n}$ for every fixed t. Our situation reduces to this in a standard way: given $\varepsilon > 0$ we find R so large that the set $\{\Theta > R\}\times[0,T]$ has measure less than ε with respect to all measures $\mu_{t,n}\,dt$ and $\mu_t\,dt$. By our assumption the set $\Omega = \{\Theta \leqslant R\}$ is compact in $\mathbb{R}^\infty$. The mapping $t \mapsto f(\,\cdot\,,t)$ from $[0,T]$ to $C(\Omega)$ is Borel measurable. By the

Dugundji theorem (see Borsuk [**197**, Chapter III, Section 7]), there exists a linear extension operator $E\colon C(\Omega)\to C_b(\mathbb{R}^\infty)$ such that

$$E\varphi(x)=\varphi(x)\quad \forall\,\varphi\in C(\Omega),\, x\in\Omega \quad\text{and}\quad \|E\varphi\|_\infty=\|\varphi\|_\infty.$$

Letting $g(x,t)=Ef(\,\cdot\,,t)(x)$, we obtain a Borel function (since it is Borel measurable in t and continuous in x, see Bogachev [**125**, Lemma 6.4.6]) such that $|g|\leqslant 1$ and $g(x,t)=f(x,t)$ for all $x\in\Omega$. The integrals of g with respect to the measures $\mu_{t,n}\,dt$ converge to the integral of g with respect to the measure $\mu_t\,dt$, and the integrals of $|f-g|$ with respect to these measures do not exceed ε. Hence the measure $\mu=\mu_t\,dt$ satisfies our equation with initial distribution ν. □

The condition that $V\geqslant 1$ is taken just for simplifying our estimates: it can be easily replaced with the condition $V\geqslant 0$ if we add constants in the right-hand sides of (10.4.1) and (10.4.2).

In typical examples V and Θ are quadratic functions (with added constants). For example, one can use $V(x)=\sum_{i=1}^\infty \beta_i x_i^2+1$ and $\Theta(x)=\sum_{i=1}^\infty \alpha_i x_i^2$. There is also a version of this theorem applicable to exponents of quadratic functions (replacing the first inequality in (10.4.1), which is not suitable for such functions).

10.4.2. Theorem. *Suppose that in Theorem* 10.4.1 *condition* (10.4.1) *is replaced with the condition*

$$LV(x,t)\leqslant V(x)-V(x)\Theta(x) \tag{10.4.7}$$

and that (10.4.2) *is replaced with the condition*

$$|a^{ij}(x,t)|+|b^i(x,t)|\leqslant C_i\big[1+\delta_i\big(V(x)\Theta(x)\big)V(x)\Theta(x)\big] \tag{10.4.8}$$

for all $(x,t)\in\mathbb{R}^\infty\times[0,T]$, $j\leqslant i$. *Then, for every Borel probability measure* ν *on* $\mathbb{R}^\infty$ *with* $W_1:=\sup_n\|V\circ P_n\|_{L^1(\nu)}<\infty$, *the Cauchy problem* $\partial_t\mu=L^*\mu$ *with initial distribution* ν *has a solution of the form* $\mu=\mu_t\,dt$ *with Borel probability measures* μ_t *on* $\mathbb{R}^\infty$ *such that for all* $t\in[0,T)$ *we have*

$$\int_{\mathbb{R}^\infty} V\,d\mu_t+\int_0^t\!\int_{\mathbb{R}^\infty} V\Theta\,d\mu_s\,ds\leqslant 4W_1. \tag{10.4.9}$$

PROOF. We can apply the same reasoning as in the proof of the previous theorem, but we have to use only one Lyapunov function V and apply (10.4.7) in place of (10.4.4) for obtaining the estimates

$$\int_{E_n} V(x)\,\mu_{t,n}(dx)+\int_0^t\!\int_{E_n} V(x)\Theta(x)\,\mu_{s,n}(dx)\,ds$$
$$\leqslant (e+1)\int_{E_n} V(x)\,\nu_n(dx)\leqslant 4W_1.$$

One more place where some difference appears is connected with estimating the integral of $fI_{|f|>N}$, where $|f|$ is estimated by $C[1+\delta_i(V\Theta)V\Theta]$. This is easily done by using the previous inequality and the condition that $\delta_i(s)\to 0$ as $s\to\infty$. □

10.4.3. Example. Let us return to Example 10.1.6, where we discussed the stochastic Navier–Stokes equation. We apply the previous results to show the existence of a solution to the Cauchy problem for the corresponding Fokker–Planck–Kolmogorov equation. We recall that the functions b^n introduced there are defined on the whole space

$$X = \{u \in W_0^{2,1}(\Omega, \mathbb{R}^d)\colon \operatorname{div} u = 0\},$$

equipped with the norm $\|\cdot\|_0$ given by $\|u\|_0^2 = \|\nabla u^1\|_2^2 + \cdots + \|\nabla u^d\|_2^2$, are continuous on all balls in X with respect to the topology of $L^2(\Omega, \mathbb{R}^d)$, which follows by the compactness of the embedding $W_0^{2,1}(\Omega) \to L^2(\Omega)$. Since for every u in the linear span of $\{\eta_n\}$ we have

$$\sum_{n=1}^{\infty}\sum_{j=1}^{d}(u,\eta_n)_2(\partial_{x_j}u, u^j\eta_n)_2 = \sum_{j=1}^{d}(u, u^j\partial_{x_j}u)_2 = -\frac{1}{2}\int_\Omega |u(\omega)|^2 \operatorname{div} u(\omega)\, d\omega = 0$$

and $(\Delta u, u)_2 = -\|u\|_0^2$, we arrive at the estimate

$$\sum_{n=1}^{N}(u,\eta_n)_2 b^n(u,t) \leqslant C_1 - C_1\|u\|_0^2$$

for all u in the linear span of $\eta_1, \ldots, \eta_N$, where C_1 is a number that does not depend on N. It is clear that $|b^n(u,t)| \leqslant C_2(n) + C_2(n)\|u\|_2^2$. Therefore, by Theorem 10.4.1 applied with the functions $\Theta(u) = C_1\|u\|_0^2$, $V(u) = \|u\|_2^2 + 1$ (the estimate above along with convergence of the series of α_n means that we have (10.4.1)), for every initial distribution ν with $\|u\|_2^k \in L^1(\nu)$ for all k, there exists a probability measure $\mu = \mu_t\, dt$ on $L^2(\Omega, \mathbb{R}^d) \times [0,T)$ satisfying (10.5.1) such that $\mu_t(X) = 1$ for almost all t. Exercise 10.5.19 contains an important refinement.

10.5. Complements, comments, and exercises

10.5(i). Uniqueness in the infinite-dimensional case

The problem of uniqueness of solutions to elliptic and parabolic Fokker–Planck–Kolmogorov equations in infinite-dimensional spaces is much more complicated than in finite-dimensional ones. Moreover, one can hardly hope for getting such broad sufficient conditions as we have seen in $\mathbb{R}^d$. For the proof of the following assertion, see Bogachev [**126**, Theorem 7.6.7], Bogachev, Röckner [**158**].

10.5.1. Theorem. *There is a bounded linear operator b on a separable Hilbert space X such that the equation $L_{\mathrm{I},b}^*\mu = 0$ has for solutions two different centered Gaussian measures μ_1 and μ_2 for which the operator $L_{\mathrm{I},b}$ in the respective L^2-spaces is symmetric and serves as the logarithmic gradient for both measures with respect to some densely embedded Hilbert space H. Hence this equation has also a non-Gaussian probability solution $(\mu_1 + \mu_2)/2$.*

In the finite-dimensional case this is impossible, since there is only one Gaussian measure whose logarithmic gradient equals a given linear operator. In the framework of the theory of Gibbs measures such examples arise in case of phase transitions, i.e., different Gibbs distributions with equal local conditional distributions. A trivial (but of little use) condition for uniqueness for the equation (elliptic or parabolic) in the space $\mathbb{R}^\infty$ with constant coefficients a^{ij} for which the matrices $(a^{ij})_{i,j\leqslant n}$ are invertible is this: there exist arbitrarily large numbers n such that b^i with $i \leqslant n$ is a locally bounded function of $x_1, \ldots, x_n$ of at most linear growth.

Under such conditions, the corresponding finite-dimensional projections of all solutions satisfy the same equations, for which the uniqueness is established. More interesting uniqueness theorems are obtained for equations whose drift coefficients are in a suitable sense small perturbations of drifts of the indicated trivial form. Results of this type can be found in Albeverio, Bogachev, Röckner [**18**], Bogachev, Röckner, Zhang [**182**], Da Prato, Zabczyk [**281**]. We mention just one result from Hino [**455**] giving a condition for uniqueness in the situation of Theorem 10.2.5; in the paper the case of a nonconstant diffusion coefficient is considered.

10.5.2. Theorem. *If in Theorem* 10.2.5 *we have* $\exp(\theta|v|_H^2) \in L^1(\gamma)$ *for some number* $\theta > 2$, *then a probability solution is unique.*

Let us mention a uniqueness theorem in the parabolic case from Bogachev, Da Prato, Röckner, Shaposhnikov [**138**]. For simplicity we assume that a^{ij} are constant, $a^{ii} = \alpha_i > 0$, $a^{ij} = 0$ for $i \neq j$, $T = 1$. Suppose we are given Borel functions b^k on $\mathbb{R}^\infty \times [0,1]$ and a Borel probability measure ν on $\mathbb{R}^\infty$. Let $\mathbb{P}_\nu$ be some *convex* set of solutions $\mu = \mu_t\,dt$ to the Cauchy problem for the equation $\partial_t \mu = L^*_{A,b}\mu$ with the initial condition ν such that μ_t are probability measures, $|b^k| \in L^2(\mu)$ for all k and the following condition is fulfilled:

for every $\varepsilon > 0$ and every $d \in \mathbb{N}$ there exist a natural number $N \geqslant d$ and a $C_b^{2,1}$-map $(b_\varepsilon^k)_{k=1}^N \colon \mathbb{R}^N \times [0,1] \to \mathbb{R}^N$ such that

$$\int_0^{T_0} \int_{\mathbb{R}^\infty} \sum_{k=1}^N \alpha_k^{-1} |b^k(x,t) - b_\varepsilon^k(x_1,\ldots,x_N,t)|^2\, \mu_t(dx)\, dt < \varepsilon.$$

10.5.3. Theorem. *Under the indicated conditions,* $\mathbb{P}_\nu$ *contains at most one element.*

Let us give some examples.

10.5.4. Example. (i) If b^k depends only on $t, x_1, x_2, \ldots, x_k$, then it suffices to have just the inclusion $|b^k| \in L^2(\mu)$ for all $k \geqslant 1$.

(ii) Let

$$l_A^2 = \Bigl\{(z_k)\colon \sum_{k=1}^\infty \alpha_k^{-1} z_k^2 < \infty\Bigr\}, \quad \|x\|_A = \Bigl(\sum_{k=1}^\infty \alpha_k^{-1} z_k^2\Bigr)^{1/2},$$

where $\bigl(b^k(x,t)\bigr) \in l_A^2$ for μ-a.e. (x,t) and $\|b\|_A \in L^2(\mu)$. For every $\varepsilon > 0$ and every natural number d we find a number $M > d$ such that

$$\sum_{k=M+1}^\infty \int_0^1 \int_{\mathbb{R}^\infty} \alpha_k^{-1} |b^k|^2\, d\mu_t\, dt < \varepsilon/2.$$

Next we take a smooth function b_ε^k of the first n_k variables such that

$$\int_0^1 \int_{\mathbb{R}^\infty} \alpha_k^{-1} |b^k - b_\varepsilon^k|^2\, d\mu_t\, dt < \varepsilon(2M)^{-1}, \quad k = 1, \ldots, M.$$

Set $N = \max\{M, n_1, n_2, \dots, n_M\}$ and $b_\varepsilon^k \equiv 0$ if $k > N$. Then

$$\sum_{k=1}^{N} \int_0^1 \int_{\mathbb{R}^\infty} \alpha_k^{-1} |b^k - b_\varepsilon^k|^2 \, d\mu_t \, dt$$
$$= \sum_{k=1}^{M} \int_0^1 \int_{\mathbb{R}^\infty} \alpha_k^{-1} |b^k - b_\varepsilon^k|^2 \, d\mu_t \, dt + \sum_{k=M+1}^{N} \int_0^1 \int_{\mathbb{R}^\infty} \alpha_k^{-1} |b^k|^2 \, d\mu_t \, dt < \varepsilon.$$

(iii) Finally, one can combine both previous cases. Let $b = G + F$, where $G^k, F^k \in L^2(\mu)$, $G^k(x,t) = G^k(x_1, \dots, x_k, t)$, $F(x,t) \in l_A^2$, $\|F\|_A \in L^2(\mu)$. It is clear that for given b^k of such a type the set of all solutions $\mu = \mu_t \, dt$ of our equation with probability measures μ_t satisfying the previous conditions of integrability is convex.

Let us mention another uniqueness theorem from Bogachev, Da Prato, Röckner, Shaposhnikov **[138]**. Let $a^{ij} = 0$ if $i \neq j$, let $a^{ii} = \alpha_i \geqslant 0$ be constant, and let b^i have smooth restrictions to $\mathbb{R}^n$. Let ν be a Borel probability measure on $\mathbb{R}^\infty$ and let $\widetilde{\mathbb{P}}_\nu$ be some *convex* set of solutions $\mu = \mu_t \, dt$ to our equation with initial condition ν for which μ_t are probability measures, $|b^k| \in L^2(\mu)$ for all $k \in \mathbb{N}$ and the following condition holds:

there are a μ-integrable function $V \geqslant 1$ on $\mathbb{R}^\infty$ with smooth restrictions to $\mathbb{R}^n$ and a number $C_0 \geqslant 0$ such that for every $\varepsilon > 0$ and $d \in \mathbb{N}$ there exist a number $N \geqslant d$ and number $\delta_\varepsilon > 0$ for which

$$\text{(i)} \quad \int_0^1 \int_{\mathbb{R}^\infty} |B_N(x,t) - B_N(P_N x, t)| \sqrt{V(P_N x)} \, \mu_t(dx) \, dt < \varepsilon,$$

where we have set $B_N = (b^1, \dots, b^N)$ and $|\cdot|$ is the norm on $\mathbb{R}^N$;

(ii) the matrix $\mathcal{B} = (\partial_{x_j} b^i)_{i,j \leqslant N}$ and the operator

$$L_N \varphi(x,t) = \sum_{i \leqslant N} [a^{ii} \partial_{x_i} \partial_{x_j} \varphi(x,t) + b^i(x,t) \partial_{x_i} \varphi(x,t)]$$

satisfy the estimate $L_N V(x,t) \leqslant \bigl(C_0 - \Lambda(x,t)\bigr) V(x)$, $(x,t) \in \mathbb{R}^N \times [0,1]$, where $\Lambda(x,t) := \delta_\varepsilon (1 + |x|^2)^{-1} |B_N(x,t)|^2 + 2 \sup_{|h| \leqslant 1} \langle \mathcal{B}(x,t) h, h \rangle$.

10.5.5. Theorem. *Under the stated assumptions, $\widetilde{\mathbb{P}}_\nu$ contains at most one element.*

10.5.6. Example. Let us return to Example 10.1.6, where we discussed the stochastic Navier–Stokes equation, consideration of which was continued in Example 10.4.3. Let now $d = 2$ and $F = 0$. Then

$$b^n(u) = -\lambda_n u_n + f^n(u), \quad f^n(u) = -(\partial_{x_1} u, u^1 \eta_n)_2 - (\partial_{x_2} u, u^2 \eta_n)_2, \quad u \in X.$$

We recall that

$$\sum_{n=1}^{\infty} \sum_{j=1}^{2} (u, \Delta \eta_n)_2 (\partial_{x_j} u, u^j \eta_n)_2 = 0$$

for all u in the linear span of $\{\eta_n\}$. Set $V(u) = \exp\bigl(\kappa \|u\|_0^2\bigr)$. For u in the linear span of $\{\eta_n\}$ the following equality holds:

$$LV(u) = 2\kappa \Bigl(\sum_n \alpha_n \lambda_n + 2\kappa \sum_n \alpha_n \lambda_n^2 u_n^2 - \sum_n \lambda_n^2 u_n^2 \Bigr) V(u).$$

Suppose also that $\sum_{n=1}^{\infty}\alpha_n\lambda_n \leqslant 1$, $\alpha_n \leqslant 1$. Then for $\kappa = 1/4$ we have

$$LV(u) \leqslant \bigl(1-\Theta(u)\bigr)V(u), \quad \Theta(u) = \frac{1}{8}\|\Delta u\|_2^2,$$

where $\Theta(u) = +\infty$ for $u^i \notin W^{2,2}(\Omega)$. By Theorem 10.4.2 and Exercise 10.5.19, for every initial condition ν with $\exp\bigl(\kappa\|u\|_0^2\bigr) \in L^1(\nu)$, there is a solution $\mu = (\mu_t)$ to the Cauchy problem $\partial_t\mu = L^*\mu$, $\mu|_{t=0} = \nu$ given by probability measures μ_t concentrated on the space $H = W^{2,2}(\Omega, \mathbb{R}^d)$ such that

$$\int_0^T\int_H (1+\|\Delta u\|_2^2)e^{\kappa\|u\|_0^2}\,\mu_t(du)\,dt < \infty.$$

Let us verify the hypotheses of the previous theorem for the class of all solutions $\mu = \mu_t\,dt$ with probability measures μ_t such that $\bigl(1+\|\Delta u\|_2^2\bigr)e^{\kappa\|u\|_0^2} \in L^1(\mu)$. Since $f^n(u) = \sum_{j=1,2}(u, u^j\partial_{x_j}\eta_n)_2$ by the condition $\operatorname{div} u = 0$, one has the estimate $|f^n(u)| \leqslant C_1(n) + C_2(n)\|u\|_2^2$. It is known (see (1.2.1) that there exists a number $C_1 > 0$ such that

$$\|g\|_{2,2} \leqslant C_1\|\Delta g\|_2 \quad \forall g \in W_0^{2,1}(\Omega)\cap W^{2,2}(\Omega).$$

Moreover, for all $g \in W^{2,2}(\Omega)$ and some $C_2 > 0$ we have $\|g\|_4 \leqslant C_2\|g\|_{2,1}$ by the Sobolev embedding theorem (here any $d \leqslant 4$ is suitable). Hence

$$\begin{aligned}\sum_{n=1}^{\infty}|f^n(u)|^2 &\leqslant \int_\Omega |\nabla u(\omega)|^2|u(\omega)|^2\,d\omega \\ &\leqslant \Bigl(\int_\Omega|\nabla u(\omega)|^4\,d\omega\Bigr)^{1/2}\Bigl(\int_\Omega|u(\omega)|^4\,d\omega\Bigr)^{1/2} \leqslant C_1^2C_2^4(1+\|\Delta u\|_2^2)\|u\|_0^2.\end{aligned}$$

Let $P_N u = u_1\eta_1 + \cdots + u_N\eta_N$. Then we obtain

$$\lim_{N\to\infty}\int_0^T\int_X\sum_{n=1}^{\infty}|f^n(u) - f^n(P_N u)|\,e^{\kappa\|u\|_0^2/2}\,\mu_t(du)\,dt = 0.$$

Hence for any $\varepsilon > 0$ there exists N such that the integral in condition (i) in front of the theorem will be less than ε. It is known that, in the considered case where $d = 2$, the equality $\bigl(f(P_N u), \Delta P_N u\bigr)_2 = 0$ holds, i.e., $\sum_{n=1}^{\infty}\lambda_n f^n(u)u_n = 0$ (see Da Prato [**261**, Proposition 6.3]). In addition, for every $\gamma \in (0,1)$ and $h = (h_1, \ldots, h_N)$, we have

$$\sum_{i,j\leqslant N}\partial_{\eta_i}f^j(P_N u)h_ih_j \leqslant (C_\gamma + \gamma\|\Delta P_N u\|_2^2)|h|^2 + \sum_{i\leqslant N}\lambda_i h_i^2, \ |h|^2 = h_1^2+\cdots+h_N^2.$$

Indeed, let $u = P_N u$. Hölder's inequality with $q = 3/2$ gives the estimate

$$\sum_{i,j}\partial_{\eta_i}f^j(u)h_ih_j \leqslant \int_\Omega|\nabla u|\,|h|^2\,dx \leqslant \Bigl(\int_\Omega|\nabla u|^3\,dx\Bigr)^{1/3}\Bigl(\int_\Omega|h|^3\,dx\Bigr)^{2/3},$$

where h denotes also $h_1\eta_1 + \cdots + h_N\eta_N$. The first factor is estimated by $C\|\Delta u\|_2$. In order to estimate the second factor we recall that by the Sobolev embedding theorem $\|h\|_3 \leqslant C_3\|h\|_{2,1/3}$, which by the interpolation inequality does not exceed $C_3\|h\|_2^{2/3}\|h\|_{2,1}^{1/3}$. Finally, by Young's inequality we estimate $\sum_{i,j}\partial_{\eta_i}f^j(u)h_ih_j$ by

$$\|h\|_{2,1}^2 + C_1\|\Delta u\|_2^q|h|^2 \leqslant \sum_{i\geqslant 1}\lambda_i h_i^2 + (C_\gamma + \gamma\|\Delta u\|_2)|h|^2.$$

Let us take $\gamma < 1/32$ and $C_0 = C_\gamma + 1$, which gives the estimate required in condition (ii) in front of the last theorem. Thus, μ is the unique solution satisfying the condition $(1 + \|\Delta u\|_2^2)e^{\kappa\|u\|_0^2} \in L^1(\mu)$.

10.5(ii). Infinite-dimensional nonlinear equations for measures

Here we present some results of the papers Bogachev, Röckner, Shaposhnikov [**168**], Bogachev, Da Prato, Röckner, Shaposhnikov [**137**] on existence of solutions to the following nonlinear evolution equation ("the continuity equation") for probability measures on a separable Hilbert space X with an orthonormal basis $\{e_i\}$:

$$\partial_t \mu_t + \sum_{i=1}^{\infty} \partial_{e_i}\big(b^i(\mu, \cdot\,, \cdot\,)\mu_t\big) = 0, \quad \mu|_{t=0} = \nu, \tag{10.5.1}$$

where for every $i \in \mathbb{N}$ the function $b^i\colon \mathcal{P}(X\times[0,1])\times X \times [0,1] \to \mathbb{R}^1$ is Borel measurable and the space $\mathcal{P}(X\times[0,1])$ of Borel probability measures on $X\times[0,1]$ is equipped with the weak topology (in the first theorem we consider the weak topology on the space of measures corresponding to the weak topology w on X, i.e., we deal with (X, w), and in the two other theorems the usual weak topology on the space of measures is used). In fact, in place of $\mathcal{P}(X\times[0,1])$ we need only its subspace formed by measures of the form $\mu = \mu_t\,dt$ with $\mu_t \in \mathcal{P}(X)$.

We shall say that a family $\mu := (\mu_t)_{t\in[0,1)}$ of Borel probability measures on X (regarded also as a measure on $X\times[0,1]$ by means of the identification $\mu := \mu_t\,dt$) satisfies equation (10.5.1) if

$$b^i(\mu, \cdot\,, \cdot\,) \in L^1(\mu),$$

and for all $t \in [0, 1)$ we have

$$\int_X \varphi\,d\mu_t - \int_X \varphi\,d\nu = \int_0^t\int_X \sum_{i=1}^{m} b^i(\mu, x, s)\partial_{e_i}\varphi(x)\,\mu_s(dx)\,ds \tag{10.5.2}$$

for every function φ of the form

$$\varphi(x) = \varphi_0(x_1, \ldots, x_m), \ x_i = (x, e_i), \ \varphi_0 \in C_b^\infty(\mathbb{R}^m), \ m \in \mathbb{N}.$$

The class of all such functions will be denoted by $\mathcal{FC}^\infty(\{e_i\})$.

Set

$$P_n x := \sum_{i=1}^{n} x_i e_i, \quad X_n := P_n(X).$$

Let us identify X_n with $\mathbb{R}^n$. We introduce the following two conditions on b:

(B1) for every fixed measure $\mu \in \mathcal{P}(X\times[0,1])$ and every fixed i, the function $x \mapsto b^i(\mu, x, t)$ is weakly continuous on all balls for a.e. t and one has uniform convergence $b^i(\mu_j, x, t) \to b^i(\mu, x, t)$ on bounded sets in $X\times[0,1]$ if $\mu_j \to \mu$ weakly, provided that the Hilbert space X is equipped with the weak topology;

(B2) there exist numbers $\alpha > 0$, $c_i > 0$, and $\kappa \geqslant 2$ such that for all points $(x,t) \in X_n\times[0,1]$, $n \in \mathbb{N}$, and all $\mu \in \mathcal{P}(X\times[0,1])$ we have

$$\sum_{i=1}^{n} b^i(\mu, x, t)x_i \leqslant \alpha(1 + |x|^2), \quad |b^i(\mu, x, t)| \leqslant c_i(1 + |x|^\kappa).$$

The first inequality in (B2) actually means that one can use the Lyapunov function $V(x) = (x,x)$ (or its power) for obtaining the estimate $LV \leqslant c + cV$, where L is the first order operator generated by b. Such estimates ensure some a priori bounds for solutions, on which our results on existence of solutions are based.

We recall that the requirement of the weak continuity on balls means simply that $b^i(\mu, x^j, t) \to b^i(\mu, x, t)$ if $x^j \to x$ weakly.

Note also that weak convergence $\nu_j \to \nu$ in $\mathcal{P}(X)$ with respect to the weak topology on X is weaker than the usual weak convergence (associated with the norm topology on X) and is equivalent to the following two conditions: for every bounded continuous function f in finitely many variables, the integrals of f with respect to ν_j converge to the integral with respect to ν and for every $\varepsilon > 0$ there is a ball $U_\varepsilon \subset X$ with

$$\nu_j(X \backslash U_\varepsilon) < \varepsilon \quad \forall j.$$

The situation is similar with weak convergence $\mu_j \to \mu$ in $\mathcal{P}(X \times [0,1])$ with respect to the weak topology on X, but the latter condition is replaced with

$$\mu_j\bigl((X \backslash U_\varepsilon) \times [0,1]\bigr) < \varepsilon \quad \forall j.$$

It is worth recalling that we *do not assume* that $b = (b^i)$ corresponds to a vector field on X: this is just a collection of scalar functions b^i.

10.5.7. Theorem. *Let ν be a Borel probability measure on X such that for some $p > \kappa$ we have*

$$\int_X |x|^p \, \nu(dx) < \infty.$$

Let a collection $b = (b^i)$ satisfy conditions (B1) *and* (B2) *above. Then there exists a family $\mu = (\mu_t)_{t \in [0,1)}$ of probability measures satisfying* (10.5.1). *Moreover,*

$$\sup_{t \in [0,1)} \int_X |x|^p \, \mu_t(dx) < \infty.$$

A certain drawback of our assumptions is the weak continuity of the functions b^i on balls. For example, this excludes functions depending on the norm, since the latter is not weakly sequentially continuous. So far, it has not been clarified whether this assumption is really necessary, but it is used in the proof in the cited papers when we check that the weak limit of finite-dimensional solutions satisfies our equation. In addition, this assumption is naturally connected with another assumption that the functions b^i are weakly continuous in μ with respect to the weak topology, which is again stronger than the continuity associated with the norm topology. The second main result in this subsection weakens these two assumptions to, apparently, more natural continuities connected with the norm topology, but at the expense of some stronger dissipativity of the drift.

Let us consider the Borel function

$$V(x) = \sum_{n=1}^{\infty} \lambda_i x_i^2, \quad \text{where } \lambda_i > 0, \ \lambda_i \to +\infty,$$

defined on the compactly embedded weighted Hilbert space X_V of all sequences $x = (x_i)$ with finite norm $V^{1/2}$.

Let us modify the previous assumptions (B1) and (B2) as follows:

(B3) for every fixed measure $\mu \in \mathcal{P}(X\times[0,1])$ and every fixed index i the function $x \mapsto b^i(\mu, x, t)$ is defined and continuous on all compact sets $\{V \leqslant R\}$, $R > 0$, with respect to the norm on X for a.e. t; moreover, one has uniform convergence $b^i(\mu_j, x, t) \to b(\mu, x, t)$ on all sets $\{V \leqslant R\}\times[0,1]$, $R>0$, if $\mu_j \to \mu$ weakly with respect to the norm topology on X;

(B4) there exist numbers $\alpha > 0$, $c_i > 0$, and $\kappa \geqslant 1$ such that for all points $(x,t)\in X_n\times[0,1]$, $n\in\mathbb{N}$, and all $\mu\in\mathcal{P}(X\times[0,1])$ we have

$$\sum_{i=1}^n \lambda_i b^i(\mu, x, t)x_i \leqslant \alpha\bigl(1+V(x)\bigr), \quad |b^i(\mu,x,t)| \leqslant c_i\bigl(1+V(x)^\kappa\bigr).$$

As compared to (B2), condition (B4) admits a stronger growth of b^i.

10.5.8. Theorem. *Let ν be a Borel probability measure on X with $V\in L^p(\nu)$ for some $p>\kappa$. Then, under assumptions* (B3) *and* (B4), *there exists a family $\mu=(\mu_t)_{t\in[0,1)}$ of probability measures on X satisfying* (10.5.1). *Moreover, we have $\mu_t(X_V)=1$ for all t and also $\sup_{t\in[0,1)}\|V\|_{L^p(\mu_t)}<\infty$.*

In some situations (as in the example below) assumption (B4) is less convenient than (B2) (in the first part), since the Lyapunov function behind this assumption is not the inner product in X, but the inner product in X_V. However, one can modify this part of (B4) as follows:

(B5) there exist numbers $\alpha > 0$, $c_i > 0$, and $\kappa \geqslant 1$ such that for all points $(x,t)\in X_n\times[0,1]$, $n\in\mathbb{N}$, and all $\mu\in\mathcal{P}(X\times[0,1])$ we have

$$\sum_{i=1}^n b^i(\mu, x, t)x_i \leqslant \alpha - \alpha V(x), \quad |b^i(\mu,x,t)| \leqslant c_i\bigl(1+V(x)\|x\|^\kappa\bigr).$$

10.5.9. Theorem. *Let ν be a Borel probability measure on X such that for some $p>\kappa$ we have*

$$\int_X V(x)\|x\|^{2p}\,\nu(dx)<\infty.$$

Then, under assumptions (B3) *and* (B5), *there exists a family $\mu=(\mu_t)_{t\in[0,1)}$ of probability measures satisfying* (10.5.1). *Moreover, $\mu_t(X_V)=1$ for all t and*

$$\sup_{t\in[0,1)}\int_X V(x)\|x\|^{2p}\,\mu_t(dx)<\infty.$$

10.5.10. Example. Let Ω be a bounded domain in $\mathbb{R}^2$ with smooth boundary, and let Δ be the Laplace operator with zero boundary condition and eigenbasis $\{e_i\}$ with the corresponding eigenvalues $\{\lambda_i\}$. Let $X=L^2(\Omega)$, where the elements of X will be denoted by u. Set

$$V(u)=\int_\Omega |\nabla u(\omega)|^2\,d\omega.$$

Then X_V is the Sobolev class $W_0^{2,1}(\Omega)$. Finally, let b be given by the heuristic expression

$$b(\mu,u,t)=\Delta u+\alpha_3(\mu,u,t)u^3+\alpha_2(\mu,u,t)u^2+\alpha_1(\mu,u,t)u+\alpha_0(\mu,u,t),$$

where the functions $\alpha_3,\dots,\alpha_0$ are Borel measurable, uniformly bounded, continuous in u on all balls in $W_0^{2,1}(\Omega)$ with respect to the L^2-norm and α_3 satisfies the inequality

$$\alpha_3(\mu,u,t)\leqslant -M, \quad \text{where } M>0 \text{ is a constant.}$$

Suppose also that if $\mu_j \to \mu$ weakly, then

$$\alpha_k(\mu_j, u, t) \to \alpha_k(\mu, u, t), \quad 0 \leqslant k \leqslant 3$$

uniformly in $t \in [0, 1]$ and u from every ball in $W_0^{2,1}(\Omega)$.

The corresponding functions b^i are defined by the equality

$$b^i(\mu, u, t) := \lambda_i \int_\Omega u(\omega) e_i(\omega)\, d\omega + \int_\Omega \bigl[\alpha_3(\mu, u, t) u^3(\omega) + \cdots + \alpha_0(\mu, u, t)\bigr] e_i(\omega)\, d\omega.$$

Then there exists a family $\mu = (\mu_t)_{t\in[0,1]}$ of probability measures satisfying (10.5.1). Let us show how this can be derived from the last theorem.

Let us verify conditions (B3) and (B5). The multiplicative Sobolev inequality (see Adams, Fournier [**3**, p. 139]) ensures the estimate

$$\int_\Omega |u(\omega)|^4\, d\omega \leqslant C \int_\Omega |\nabla u(\omega)|^2\, d\omega \int_\Omega |u(\omega)|^2\, d\omega,$$

which shows that the second inequality in (B5) is true with $\kappa = 2$. The first estimate required in (B5) follows by the equality

$$\begin{aligned} &\int_\Omega \bigl[u(\omega)\Delta u(\omega) + \alpha_3(\mu, u, t) u(\omega)^4 + \cdots + \alpha_0(\mu, u, t) u(\omega)\bigr]\, d\omega \\ &= -\int_\Omega |\nabla u(\omega)|^2\, du + \int_\Omega \bigl[\alpha_3(\mu, u, t) u(\omega)^4 + \cdots + \alpha_0(\mu, u, t) u(\omega)\bigr]\, d\omega \end{aligned}$$

on all of X_n, where the right-hand side is majorized by the quantity

$$C_1 - \int_\Omega |\nabla u(\omega)|^2\, d\omega$$

with some constant C_1, since $\alpha_3(\mu, u, t) \leqslant -M$ and $\alpha_0, \alpha_1, \alpha_2$ are bounded. The continuity of b^i on all balls in $W_0^{2,1}(\Omega)$ with respect to the L^2-norm follows by our assumption and the Sobolev embedding theorem combined with the compactness of the embedding of $W_0^{2,1}(\Omega)$ into $L^2(\Omega)$.

It is important in this example that the term with the Sobolev norm appears when we apply the operator L to the inner product in L^2. Theorem 10.5.8 does not work here, since the growth of $|b^i|$ is not controlled by powers of the L^2-norm. But as soon as we use bigger functions, stronger dissipativity is needed to control their moments.

Comments

There is an extensive literature on stationary distributions of infinite-dimensional diffusions, see, for example, the books Da Prato, Zabczyk [**281**], Liu, Röckner [**634**], and Prévôt, Röckner [**790**]. There are particularly many works related to stochastic partial differential equations, see Agrachev, Kuksin, Sarychev, Shirikyan [**7**], Assing [**76**], Assing, Manthey [**77**], Barbu, Da Prato, Röckner [**89**], Blömker, Flandoli, Romito [**120**], Brzezniak, Li [**210**], Caraballo, Kloeden, Real [**217**], Chojnowska-Michalik [**245**], Da Prato [**261**], [**262**], Da Prato, Flandoli, Röckner [**266**], Da Prato, Röckner, Rozovskii, Wang [**276**], Debussche [**289**], Debussche, Romito [**290**], E, Mattingly, Sinai [**319**], Eckmann, Hairer [**324**], Es-Sarhir, Stannat [**340**], Ferrario [**358**]–[**360**], Flandoli [**363**]–[**366**], Flandoli, Gatarek [**367**], Flandoli, Maslowski [**368**], Gatarek, Goldys [**401**], Goldys, Maslowski [**413**], [**414**], Hairer, Mattingly [**434**], [**436**], Jacquot [**487**], Jacquot,

Royer [**488**], Kim [**508**]–[**510**], Kuksin, Shirikyan [**559**], [**560**], Maslowski [**673**], Manthey, Maslowski [**669**], Maslowski, Seidler [**676**], Maslowski, Simão [**678**], Mueller [**729**], Otto, Weber, Westdickenberg [**759**], Röckner, Wang [**818**], Röckner, Zhang [**821**], Röckner, Zhu, Zhu [**823**], Romito [**825**], Scherbakov [**835**], Seidler [**839**], Shirikyan [**858**], Sowers [**867**], Wiesinger [**940**].

Another important component is infinite gradient systems in statistical physics, Gibbs measures, and stochastic quantization, see Albeverio, Ferrario [**19**], Albeverio, Haba, Russo [**20**], Albeverio, Kondratiev, Pasurek, Röckner [**23**], Albeverio, Kondratiev, Röckner, Tsikalenko [**28**], Albeverio, Ma, Röckner [**29**], Albeverio, Röckner [**33**], Cattiaux, Roelly, Zessin [**233**], Doss, Royer [**312**], Eckmann, Hairer [**323**], Föllmer [**378**], Fritz [**387**], [**388**], Fritz, Funaki, Lebowitz [**389**], Holley, Stroock [**458**], Jona-Lasinio, Mitter [**494**], Künsch [**562**], [**563**], Lang [**582**], Namiki [**733**], Roelly, Zessin [**824**], Shiga [**854**], Stroock [**880**], Stroock, Zegarlinski [**885**]. Results of general character about invariant measures in infinite-dimensional spaces can be found in Es-Sarhir [**339**], Hino [**455**], Maslowski [**672**], Maslowski, Seidler [**675**], Miyahara [**718**], von Vintschger [**921**], Zegarlinski [**952**].

Prohorov's theorem and Lyapunov functions combined with a priori estimates are used in many works. This is a standard techniques, see, for example, Chow, Khasminskii [**247**], Leha, Maslowski, Ritter [**595**], Leha, Ritter [**596**], [**597**], Leha, Ritter, Wakolbinger [**598**], Liu, Mandrekar [**631**], [**632**], and also the papers cited above.

The strong Feller property of diffusion semigroups in the infinite-dimensional case requires much more restrictive conditions than in the finite-dimensional case, see Da Prato, Elworthy, Zabczyk [**265**], Da Prato, Zabczyk [**281**], Maslowski, Seidler [**677**], Peszat, Zabczyk [**772**].

Other applications of the results of Chapters 6–9 to finite-dimensional projections of solutions of infinite-dimensional equations for measures are given in Bogachev, Röckner, Shaposhnikov [**166**].

There are many works on equations connected with Ornstein–Uhlenbeck type operators and semigroups in infinite-dimensional spaces, see Bogachev, Röckner [**159**], Bogachev, Röckner, Schmuland [**163**], Chojnowska-Michalik, Goldys [**246**], Goldys, van Neerven [**416**], Jegaraj [**490**], Maslowski, Pospíŝil [**674**], van Neerven, Weis [**739**].

We have discussed above only Kolmogorov equations, but not the existence of infinite-dimensional diffusions whose transition or stationary probabilities could satisfy these equations. In the infinite-dimensional case this question becomes even more difficult, see Da Prato, Zabczyk [**279**]. The construction of diffusions by the method of Dirichlet forms is discussed in Albeverio, Høegh-Krohn [**21**], Albeverio, Høegh-Krohn, Streit [**22**], Albeverio, Röckner [**31**]–[**33**], and Ma, Röckner [**653**].

On existence and uniqueness of solutions to infinite-dimensional parabolic equations for measures, see also Ambrosio, Savaré, Zambotti [**46**], Bogachev, Da Prato, Röckner [**134**]–[**136**], Bogachev, Da Prato, Röckner, Shaposhnikov [**138**], Lemle, Wang, Wu [**600**], [**601**], and Manca [**659**]–[**662**]. The results in § 10.4 generalize some results of the paper Bogachev, Da Prato, Röckner [**132**] and are taken partly from [**138**]. Infinite-dimensional analogs of results on connections between flow solutions to ordinary equations and continuity equations for measures are obtained in Cruzeiro [**258**], [**259**], where this direction was initiated, and in Ambrosio, Figalli [**43**], Bogachev, Mayer-Wolf [**156**], Peters [**773**], Üstünel, and Zakai [**908**]. For other results on infinite-dimensional continuity equations, see Bogachev, Da Prato, Röckner, Shaposhnikov [**138**], Da Prato, Flandoli, Röckner [**267**], and Kolesnikov, Röckner [**526**].

Differentiable measures introduced by S.V. Fomin have been studied in many works, the first surveys are given in Averbuh, Smolyanov, Fomin [**79**] and Dalecky, Fomin [**282**]; the modern theory is presented in Bogachev [**126**], where also the Malliavin calculus is discussed, including the theory of Sobolev classes on infinite-dimensional spaces with measures.

On infinite-dimensional Harnack inequalities, see Bass, Gordina [**93**], Da Prato, Röckner, Wang [**277**], Liu [**633**], Röckner, Wang [**815**], and Wang [**929**], [**932**].

About diffusion semigroups in the infinite-dimensional case, see Aida, Kawabi [**10**], Dragoni, Kontis, Zegarliński [**313**], Lugiewicz, Zegarliński [**645**].

Kolmogorov operators, problems of uniqueness (or essential self-adjointness in the symmetric case) and the corresponding semigroups on infinite-dimensional spaces are studied in Albeverio, Kondratiev, Röckner [**24**], [**25**], Bakry, Ledoux [**87**], Da Prato, Röckner [**275**], Kondratiev, Tsycalenko [**536**], Liskevich, Röckner [**621**], Liskevich, Röckner, Sobol, Us [**622**], Liskevich, Sobol, Röckner [**625**], Long, Simão [**637**], van Neerven [**738**], Röckner, Sobol [**814**], Takeda [**890**], see also Bogachev [**126**, Chapter 12] and the references in these works.

Time-reversal for infinite-dimensional diffusions is studied in Föllmer [**378**], Föllmer, Wakolbinger [**379**], Handa [**439**], see also Bogachev [**122**] about connections with logarithmic gradients.

Various problems related to the subject of this chapter in the case of infinite-dimensional manifolds are studied in Airault [**11**], Airault, Ouerdiane [**12**], Belopol'skaya, Dalecky [**102**], Bogachev, Röckner, Wang [**179**], [**181**], [**185**], Gliklikh [**411**], Holley, Stroock [**459**], Ramírez [**799**], [**800**], Ramirez, Varadhan [**801**].

The equations for measures we considered are Fokker–Planck–Kolmogorov equations by the aforementioned terminology. We do not discuss closely connected Kolmogorov equations for functions (see, for example, Bogachev, Da Prato, Röckner [**133**], [**135**], Cerrai [**234**], Da Prato [**261**], Dalecky, Fomin [**282**]).

Equations on loop spaces are considered in Fang, Shao [**350**].

Nonlinear parabolic equations for measures on infinite-dimensional spaces have become an object of study relatively recently, see Bogachev, Da Prato, Röckner, Shaposhnikov [**137**].

Exercises

10.5.11. Let γ be the measure on $\mathbb{R}^\infty$ that is the countable power of the standard Gaussian measure on the real line. Prove that it is mutually singular with all its images under mappings $x \mapsto kx$, where $|k| \neq 1$, and also with the countable power of any probability measure on the real line different from the standard Gaussian measure. It follows that the transition probabilities of the infinite-dimensional Wiener process are mutually singular at all different times.

10.5.12. Let γ_1 and γ_2 be two distinct Gaussian measures on a locally convex space. Show that the measure $(\gamma_1 + \gamma_2)/2$ is not Gaussian.

10.5.13. Show that the standard Gaussian measure on $\mathbb{R}^\infty$ is the unique probability solution of the equation $L^*\gamma = 0$ with the Ornstein–Uhlenbeck L operator.

HINT: Consider the equations satisfied by the finite-dimensional projections of the solution and apply the finite-dimensional uniqueness result.

10.5.14. Extend the estimate from Example 9.8.33 to the infinite-dimensional case where μ is the countable power of the standard Gaussian measure on $\mathbb{R}$ and the perturbation b_0 is a bounded Borel vector field with values in l^2, i.e., in the Cameron–Martin space of μ.

10.5.15. Suppose that μ is a Borel probability measure on the space $\mathbb{R}^\infty$ satisfying the elliptic equation $L^*_{\mathrm{I},b}\mu = 0$ with respect to the class $\mathcal{FC}_b^\infty(\mathbb{R}^\infty)$ with $b(x) = -x + b_0(x)$, where $b_0\colon \mathbb{R}^\infty \to l^2$ is a bounded Borel mapping. Prove that μ is positive on all nonempty open sets, moreover, the same is true for the restriction of μ to the weighted Hilbert space of sequences with $\sum_{n=1}^\infty \alpha_n^2 x_n^2 < \infty$ for any sequence of numbers $\alpha_n > 0$ such that $(\alpha_n) \in l^2$.

10.5.16. Suppose that μ is a Borel probability measure on the space $\mathbb{R}^\infty$ satisfying the elliptic equation $L^*_{\mathrm{I},b}\mu = 0$ with respect to the class $\mathcal{FC}_b^\infty(\mathbb{R}^\infty)$ with $b(x) = -x + b_0(x)$, where $b_0\colon \mathbb{R}^\infty \to l^2$ is a bounded Borel mapping such that $b_0(-x) = -b_0(x)$. Prove that μ is centrally symmetric.

10.5.17. Suppose that μ is a Borel probability measure on the space $\mathbb{R}^\infty$ satisfying the elliptic equation $L^*_{\mathrm{I},b}\mu = 0$ with respect to the class $\mathcal{FC}_b^\infty(\mathbb{R}^\infty)$ with $b(x) = -x + b_0(x)$, where $b_0\colon \mathbb{R}^\infty \to \mathbb{R}^n \subset l^2$ is a bounded Borel mapping such that $b_0(x) = b_0(x_1, \ldots, x_n)$. Prove that μ is the product of a probability measure on $\mathbb{R}^n$ and the countable power of the standard Gaussian measure on the real line.

10.5.18. Prove the existence of an infinitesimally invariant probability measure on $L^r(\Omega)$ in the situation of Example 10.1.3, and also the existence of a solution to the corresponding Cauchy problem.

HINT: Use $V(u) = \|u\|_2^2$, $\Theta(u) = \|\nabla(\zeta\circ u)\|_2^2$, $\zeta(u) = |u|^{(r+1)/2}\mathrm{sign}\, u$; since

$$\sum_{i=1}^n u_i b^i(u) = \bigl(\Delta u, \Psi(u)\bigr)_2 + \bigl(u, \Phi(u)\bigr)_2$$

on E_n and

$$\bigl(\Delta u, \Psi(u)\bigr)_2 = -\int_\Omega \Psi'(u)|\nabla u|^2\, d\omega \leqslant -\kappa_0 \int_\Omega \Psi'(u)|\nabla u|^2\, d\omega \leqslant -C_r\Theta(u),$$

one obtains an estimate necessary for applying the results in § 10.3; in addition, the sets $\{u\colon \Theta(u) \leqslant R\}$ are compact in $L^2(\Omega)$; see also Bogachev, Da Prato, Röckner [**130**], [**134**].

10.5.19. Using the function $V(u) = \exp(\kappa\|u\|_0^2)$ from Example 10.5.6, show that in Example 10.1.3 the constructed solution is concentrated on $W^{2,2}(\Omega, \mathbb{R}^d)$ and $\|\Delta u\|_2^2 \exp(\kappa\|u\|_0^2) \in L^1(\mu)$ for small κ if the series of $\alpha_n\lambda_n$ converges. In the situation of Example 10.4.3 with initial condition ν such that $\exp(\kappa\|u\|_0^2) \in L^1(\nu)$, prove a parabolic analog.

10.5.20. Suppose that μ is a Borel probability measure on l^2 satisfying the elliptic equation $L^*_{A,b}\mu = 0$ with respect to the class $\mathcal{FC}_b^\infty(l^2)$, where $b\colon l^2 \to l^2$ is a bounded Lipschitz mapping and A is a diagonal operator with eigenvalues $\alpha_n > 0$ such that $\sum_{n=1}^\infty \alpha_n < \infty$. Find sufficient conditions on A and b under which μ is positive on all open balls, but need not be absolutely continuous with a respect to a Gaussian measure.

Bibliography

[1] Abourashchi, N., Veretennikov, A. Yu., *On stochastic averaging and mixing.* Theory Stoch. Process. 2010. V. 16, №1. P. 111–129. [*235*]

[2] Adams, R. A., *Sobolev spaces.* Academic Press, New York, 1975; 268 p. [*51*]

[3] Adams, R. A., Fournier, J. J. F., *Sobolev spaces.* 2nd ed. Academic Press, New York, 2003; xiii+305 p. [*6, 51, 433*]

[4] Agafontsev, B. V., Bogachev, V. I., Shaposhnikov S. V., *A condition for the positivity of the density of an invariant measure.* Dokl. Ross. Akad. Nauk. 2011. V. 438, №3. P. 295–299 (in Russian); English transl.: Dokl. Math. 2011. V. 83, №3. P. 332–336. [*127*]

[5] Agmon S., *Lectures on elliptic boundary value problems.* AMS Chelsea Publ., Providence, Rhode Island, 2010; x+216 p. [*51*]

[6] Agmon, S., Douglis A., Nirenberg, L., *Estimates near the boundary for solutions of elliptic partial differential equations satisfying general boundary conditions. I, II.* Comm. Pure Appl. Math. 1959. V. 12. P. 623–727; 1964. V. 17. P. 35–92. [*51*]

[7] Agrachev, A., Kuksin, S., Sarychev, A., Shirikyan, A., *On finite-dimensional projections of distributions for solutions of randomly forced 2D Navier–Stokes equations.* Ann. Inst. H. Poincaré, Probab. Stat. 2007. V. 43, №4. P. 399–415. [*414, 433*]

[8] Agueh M., *Existence of solutions to degenerate parabolic equations via the Monge–Kantorovich theory.* Adv. Differ. Equ. 2005. V. 10, №3. P. 309–360. [*285*]

[9] Ahmed, N. U., Ding, X., *On invariant measures of nonlinear Markov processes.* J. Appl. Math. Stoch. Anal. 1993. V. 6, №4. P. 385–406. [*235*]

[10] Aida S., Kawabi, H., *Short time asymptotics of certain infinite dimensional diffusion process.* In: Stochastic Analysis and Related Topics, VII (Kusadasi, 1998); Progr. Probab. V. 48, pp. 77–124. Birkhäuser Boston, Boston, 2001. [*435*]

[11] Airault, H., *Projection of the infinitesimal generator of a diffusion.* J. Funct. Anal. 1989. V. 85. P. 353–391. [*435*]

[12] Airault H., Ouerdiane, H., *Ornstein–Uhlenbeck operators and unitarizing measures in the Poincaré disk.* Bull. Sci. Math. 2009. V. 133, №7. P. 671–692. [*435*]

[13] Akhiezer, A. I., Peletminskii, S. V., *Methods of statistical physics.* Nauka, Moscow, 1977; 368 p. (in Russian); English transl.: Pergamon Press, Oxford – New York, 1981; xv+450 p. [*283*]

[14] Albanese A., Lorenzi L., Mangino, E., *L^p-uniqueness for elliptic operators with unbounded coefficients in $\mathbb{R}^N$.* J. Funct. Anal. 2009. V. 256, №4. P. 1238–1257. [*234*]

[15] Albanese, A. A., Mangino, E., *Cores of second order differential linear operators with unbounded coefficients on $\mathbb{R}^N$.* Semigroup Forum. 2005. V. 70, №2. P. 278–295. [*234*]

[16] Albanese, A. A., Mangino, E. M., *Cores for Feller semigroups with an invariant measure.* J. Differ. Equ. 2006. V. 225, №1. P. 361–377; correction: ibid. 2008. V. 244, №11. P. 2980–2982. [*234*]

[17] Alberti, G., Bianchini, S., Crippa, G., *A uniqueness result for the continuity equation in two dimensions.* J. Eur. Math. Soc. 2014. V. 16, №2. P. 201–234. [*384*]

[18] Albeverio, S., Bogachev, V., Röckner, M., *On uniqueness of invariant measures for finite- and infinite-dimensional diffusions.* Comm. Pure Appl. Math. 1999. V. 52. P. 325–362. [*209, 234, 427*]

[19] Albeverio, S., Ferrario, B., *Uniqueness results for the generators of the two-dimensional Euler and Navier–Stokes flows. The case of Gaussian invariant measures.* J. Funct. Anal. 2002. V. 193, №1. P. 77–93. [*434*]

[20] Albeverio, S., Haba, Z., Russo, F., *Stationary solutions of stochastic parabolic and hyperbolic Sine–Gordon equations.* J. Phys. A. 1993. V. 26. P. 711–718. [*434*]

[21] Albeverio, S., Høegh-Krohn, R., *Dirichlet forms and diffusion processes on rigged Hilbert spaces.* Z. Wahr. theor. verw. Geb. 1977. B. 40. S. 1–57. [x, *434*]

[22] Albeverio, S., Høegh-Krohn, R., Streit, L., *Energy forms, Hamiltonians, and distorted Brownian paths.* J. Math. Phys. 1977. V. 18, № 5. P. 907–917. [*434*]

[23] Albeverio, S., Kondratiev, Yu. G., Pazurek, T., Röckner, M., *Existence of and a priori estimates for Euclidean Gibbs states.* Trudy Mosk. Matem. Ob. 2006. V. 67. P. 3–103 (in Russian); English transl.: Trans. Moscow Math. Soc. 2006. P. 1–85. [*409, 413, 434*]

[24] Albeverio, S., Kondratiev, Yu. G., Röckner, M., *An approximate criterium of essential self-adjointness of Dirichlet operators.* Potential Anal. 1992. V. 1, №3. P. 307–317; Addendum: ibid. 1993. V. 2, №2. P. 195–198. [*435*]

[25] Albeverio, S., Kondratiev, Yu. G., Röckner, M., *Dirichlet operators via stochastic analysis.* J. Funct. Anal. 1995. V. 128, №1. P. 102–138. [*435*]

[26] Albeverio, S., Kondratiev, Yu. G., Röckner, M., *Ergodicity of L^2-semigroups and extremality of Gibbs states.* J. Funct. Anal. 1997. V. 144, №2. P. 394–423. [*409*]

[27] Albeverio, S., Kondratiev, Yu. G., Röckner, M., *Ergodicity for the stochastic dynamics of quasi-invariant measures with applications to Gibbs states.* J. Funct. Anal. 1997. V. 149, №2. P. 415–469. [*409*]

[28] Albeverio, S., Kondratiev, Yu. G., Röckner, M., Tsikalenko, T. V., *A priori estimates for symmetrizing measures and their applications to Gibbs states.* J. Funct. Anal. 2000. V. 171, №2. P. 366–400. [*409, 434*]

[29] Albeverio, S., Ma, Z., Röckner, M., *Quasi-regular Dirichlet forms and the stochastic quantization problem.* Festschrift Masatoshi Fukushima on the occasion of his Sanju (Chen, Z.Q., Jacob, N., Takeda, M., Vemura, T., eds.), pp. 27–58. World Sci., Singapore, 2015. [*434*]

[30] Albeverio, S., Marinelli, C., *Reconstructing the drift of a diffusion from partially observed transition probabilities.* Stoch. Process. Appl. 2005. V. 115, №9. P. 1487–1502. [*236*]

[31] Albeverio, S., Röckner, M., *Classical Dirichlet forms on topological vector spaces – construction of an associated diffusion process.* Probab. Theory Related Fields. 1989. V. 83. P. 405–434. [*407, 434*]

[32] Albeverio, S., Röckner, M., *Classical Dirichlet forms on topological vector spaces – closability and a Cameron-Martin formula.* J. Funct. Anal. 1990. V. 88. P. 395–436. [*434*]

[33] Albeverio, S., Röckner, M., *Stochastic differential equations in infinite dimensions: solutions via Dirichlet forms.* Probab. Theory Related Fields. 1991. V. 89. P. 347–386. [*407, 434*]

[34] Aleksandrov, A. D., *Uniqueness conditions and estimates for the solution of the Dirichlet problem.* Vestnik Leningrad. Univ. Ser. Mat. Meh. Astronom. 1963. №13, issue 3. P. 5–29 (in Russian); English transl.: In: Ten papers on differential equations and functional analysis. Amer. Math. Soc. Trans. Ser. 2. V. 68. P. 89–119. Amer. Math. Soc., Providence, Rhode Island, 1968. [*52*]

[35] Aleksandrov, A. D., *Majorants of solutions of linear equations of order two.* Vestnik Leningrad. Univ. 1966. №1, issue 1. P. 5–25 (in Russian); English transl. in [**37**]. [*52*]

[36] Aleksandrov, A. D., *The impossibility of general estimates of solutions and of uniqueness conditions for linear equations with norms weaker than in L_n.* Vestnik Leningrad. Univ. 1966. №13, issue 3. P. 5–10 (in Russian); English transl.: In: Ten papers on differential equations and functional analysis. Amer. Math. Soc. Trans. Ser. 2. V. 68. P. 162–168. Amer. Math. Soc., Providence, Rhode Island, 1968. [*52*]

[37] Alexandrov, A. D., *Selected works. Part 1*: Selected scientific papers. Ed. by Yu. G. Reshetnyak and S. S. Kutateladze, Gordon and Breach, Amsterdam, 1996. [*52*]

[38] Ambrosio, L., *Transport equation and Cauchy problem for BV vector fields.* Invent. Math. 2004. V. 158, №2. P. 227–260. [*284, 383*]

[39] Ambrosio, L., *Transport equation and Cauchy problem for non-smooth vector fields.* Lecture Notes in Math. 2008. V. 1927. P. 2–41. [*284, 383*]

[40] Ambrosio, L., Bernard, P., *Uniqueness of signed measures solving the continuity equation for Osgood vector fields.* Atti Accad. Naz. Lincei Cl. Sci. Fis. Mat. Natur. Rend. Lincei (9) Mat. Appl. 2008. V. 19, №3. P. 237–245. [*383*]

[41] Ambrosio, L., Crippa, G., *Existence, uniqueness, stability and differentiability properties of the flow associated to weakly differentiable vector fields.* In: Transport Equations and Multi-D Hyperbolic Conservation Laws, Lecture Notes Unione Mat. Ital. 5, pp. 3–57. Springer, Berlin, 2008. *[384]*

[42] Ambrosio, L., Crippa, G., Figalli, A., Spinolo, L. V., *Some new well-posedness results for continuity and transport equations, and applications to the chromatography system.* SIAM J. Math. Anal. 2009. V. 41, №5. P. 1890–1920. *[384]*

[43] Ambrosio, L., Figalli, A., *On flows associated to Sobolev vector fields in Wiener spaces: an approach à la DiPerna–Lions.* J. Funct. Anal. 2009. V. 256, №1. P. 179–214. *[434]*

[44] Ambrosio, L., Gigli, N., *A user's guide to optimal transport.* Lecture Notes in Math. 2013. V. 2062. P. 1–155. *[285]*

[45] Ambrosio, L., Gigli N., Savaré G., *Gradient flows in metric spaces and in the Wasserstein spaces of probability measures.* 2nd ed. Birkhäuser, Basel, 2008; x+334 p. *[273, 275]*

[46] Ambrosio, L., Savaré, G., Zambotti, L., *Existence and stability for Fokker–Planck equations with log-concave reference measure.* Probab. Theory Related Fields. 2009. V. 145, №3-4. P. 517–564. *[401, 434]*

[47] Ambrosio, L., Trevisan, D., *Well-posedness of Lagrangian flows and continuity equations in metric measure spaces.* Analysis & PDE. 2014. V. 7, №5. P. 1179–1234. *[383]*

[48] Ancona, A., *Elliptic operators, conormal derivatives and positive parts of functions (with an appendix by Haïm Brezis).* J. Funct. Anal. 2009. V. 257. P. 2124–2158. *[35]*

[49] Andreianov, B. P., Bénilan, Ph., Kruzhkov, S. N., *L^1-theory of scalar conservation law with continuous flux function.* J. Funct. Anal. 2000. V. 171, №1. P. 15–33. *[384]*

[50] Andreu-Vaillo, F., Caselles, V., Mazón, J. M., *Parabolic quasilinear equations minimizing linear growth functionals.* Birkhäuser, Basel, 2004; xiv+340 p. *[384]*

[51] Angiuli, L., Lorenzi, L., *Compactness and invariance properties of evolution operators associated with Kolmogorov operators with unbounded coefficients.* J. Math. Anal. Appl. 2011. V. 379, №1. P. 125–149. *[314]*

[52] Angiuli, L., Lorenzi, L., *On the Dirichlet and Neumann evolution operators in $\mathbb{R}^d_+$.* Potential Anal. 2014. V. 41, №4. P. 1079–1110. *[314]*

[53] Angiuli, L., Lorenzi, L., Lunardi, A., *Hypercontractivity and asymptotic behavior in nonautonomous Kolmogorov equations.* Comm. Partial Differ. Equ. 2013. V. 38, №12. P. 2049–2080. *[314]*

[54] Angst, J., *Trends to equilibrium for a class of relativistic diffusions.* J. Math. Phys. 2011. V. 52, №11. 9 pp. *[235]*

[55] Arapostathis, A., Borkar, V. S., *Uniform recurrence properties of controlled diffusions and applications to optimal control.* SIAM J. Control Optim. 2010. V. 48, №7. P. 4181–4223. *[52, 235]*

[56] Arapostathis, A., Borkar, V. S., *A relative value iteration algorithm for nondegenerate controlled diffusions.* SIAM J. Control Optim. 2012. V. 50, №4. P. 1886–1902. *[235]*

[57] Arapostathis, A., Borkar, V. S., Ghosh, M. K., *Ergodic control of diffusion processes.* Cambridge University Press, Cambridge, 2012; xvi+323 p. *[234, 235]*

[58] Arapostathis, A., Borkar, V. S., Kumar, K. S., *Convergence of the relative value iteration for the ergodic control problem of nondegenerate diffusions under near-monotone costs.* SIAM J. Control Optim. 2014. V. 52, №1. P. 1–31. *[235]*

[59] Araujo, M. T., Drigo Filho, E., *A general solution of the Fokker–Planck equation.* J. Stat. Phys. 2012. V. 146, №3. P. 610–619. *[284]*

[60] Arendt, W., *The abstract Cauchy problem, special semigroups and perturbation.* Lecture Notes in Math. 1986. V. 1184. P. 25–47. *[189]*

[61] Arendt, W., *Positive semigroups of kernel operators.* Positivity. 2008. V. 12. P. 25–44. *[214, 216]*

[62] Arendt, W., Biegert, M., ter Elst, A. F. M., *Diffusion determines the manifold.* J. Reine Angew. Math. 2012. B. 667. S. 1–25. *[236]*

[63] Arendt, W., Metafune, G., Pallara, D., *Gaussian estimates for elliptic operators with unbounded drift.* J. Math. Anal. Appl. 2008. V. 338, №1. P. 505–517. *[229]*

[64] Arnold, A., Carlen, E., Ju, Q., *Large-time behavior of non-symmetric Fokker–Planck type equations.* Commun. Stoch. Anal. 2008. V. 2, №1. P. 153–175. *[314]*

[65] Arnold, A., Markowich, P., Toscani, G., Unterreiter, A., *On convex Sobolev inequalities and the rate of convergence to equilibrium for Fokker–Planck type equations.* Comm. Partial Differ. Equ. 2001. V. 26, №1-2. P. 43–100. [*314*]

[66] Arnold, L., Eizenberg, A., Wihstutz, V., *Large noise asymptotics of invariant measures with applications to Lyapunov exponents.* Stochastics Stoch. Rep. 1996. V. 59, №1-2. P. 71–142. [*235*]

[67] Arnold, L., Kliemann, W., *Qualitative theory of stochastic systems.* In: Probabilistic analysis and related topics, Vol. 3 (A.T. Bharucha-Reid ed.), pp. 1–79, Academic Press, New York, 1983. [*235*]

[68] Arnold, L., Kliemann, W., *On unique ergodicity for degenerate diffusions.* Stochastics. 1987. V. 21. P. 41–61. [*235*]

[69] Arnold, A., Markowich, P., Toscani, G., Unterreiter, A., *On convex Sobolev inequalities and the rate of convergence to equilibrium for Fokker–Planck type equations.* Comm. Partial Differ. Equ. 2001. V. 26, №1-2. P. 43–100. [*401*]

[70] Aronson, D. G., *Uniqueness of positive weak solutions of second order parabolic equations.* Ann. Polon. Math. 1965. V. 16. P. 285–303. [*339*]

[71] Aronson, D. G., *Bounds for the fundamental solutions of a parabolic equation.* Bull. Am. Math. Soc. 1967. V. 73. P. 890–896. [*311*]

[72] Aronson, D. G., *Non-negative solutions of linear parabolic equations.* Ann. Sc. Norm. Sup. Pisa Cl. Sci. 1968. V. 22. P. 607–694. [*339*]

[73] Aronson, D. G., Besala, P., *Uniqueness of solutions of the Cauchy problem for parabolic equations.* J. Math. Anal. Appl. 1966. V. 13. P. 516–526. [*339*]

[74] Aronson, D. G., Serrin, J., *Local behavior of solutions of quasilinear parabolic equations.* Arch. Rational Mech. Anal. 1967. V. 25. P. 81–122. [*250, 316, 317, 334*]

[75] Arsen'ev, A. A., *Lectures on kinetic equations.* Nauka, Moscow, 1992; 216 p. (in Russian). [*283*]

[76] Assing, S., *A pregenerator for Burgers equation forced by conservative noise.* Comm. Math. Phys. 2002. V. 225, №3. P. 611–632. [*433*]

[77] Assing, S., Manthey, R., *Invariant measures for stochastic heat equations with unbounded coefficients.* Stoch. Process. Appl. 2003. V. 103, №2. P. 237–256. [*433*]

[78] Auscher, P., Qafsaoui, M., *Observations on $W^{1,p}$ estimates for divergence elliptic equations with VMO coefficients.* Boll. Unione Mat. Ital. Sez. B Artic. Ric. Mat. (8). 2002. V. 5, №2. P. 487–509. [*8*]

[79] Averbuh, V. I., Smolyanov, O. G., Fomin, S. V., *Generalized functions and differential equations in linear spaces.* Trans. Moscow Math. Soc. 1971. V. 24. P. 140–184. [*435*]

[80] Azéma, J., Kaplan-Duflo, M., Revuz, D., *Mesure invariante sur les classes récurrentes des processus de Markov.* Z. Wahr. theor. verw. Geb. 1967. B. 8. S. 157–181. [*235*]

[81] Azencott, R., *Behavior of diffusion semi-groups at infinity.* Bull. Soc. Math. France. 1974. V. 102. P. 193–240. [*234*]

[82] Bakry, D., *L'hypercontractivité et son utilisation en théorie des semigroupes.* Lecture Notes in Math. 1994. V. 1581. P. 1–114. [*228*]

[83] Bakry, D., Bolley, F., Gentil, I., *Dimension dependent hypercontractivity for Gaussian kernels.* Probab. Theory Related Fields. 2012. V. 154, №3-4. P. 845–874. [*228*]

[84] Bakry, D., Cattiaux, P., Guillin, A., *Rate of convergence for ergodic continuous Markov processes: Lyapunov versus Poincaré.* J. Funct. Anal. 2008. V. 254, №3. P. 727–759. [*228, 401*]

[85] Bakry, D., Emery, M., *Diffusions hypercontractives.* Lecture Notes in Math. 1985. V. 1123. P. 77–206. [*228*]

[86] Bakry, D., Gentil, I., Ledoux, M., *Analysis and geometry of Markov diffusion operators.* Springer, Berlin, 2013; 490 p. [*228*]

[87] Bakry, D., Ledoux, M. *Lévy–Gromov's isoperimetric inequality for an infinite-dimensional diffusion generator.* Invent. Math. 1996. V. 123, №2. P. 259–281. [*435*]

[88] Barbu, V., Bogachev, V. I., Da Prato G., Röckner, M., *Weak solutions to the stochastic porous media equation via Kolmogorov equations: the degenerate case.* J. Funct. Anal. 2006. V. 237, №1. P. 54–75. [*414*]

[89] Barbu, V., Da Prato, G., Röckner, M., *Some results on stochastic porous media equations.* Boll. Unione Mat. Ital. (9). 2008. V. 1, №1. P. 1–15. [*433*]

[90] Barlow, M. T., Grigor'yan A., Kumagai, T., *On the equivalence of parabolic Harnack inequalities and heat kernel estimates.* J. Math. Soc. Japan. 2012. V. 64, №4. P. 1091–1146. [*335*]

[91] Bass, R. F., *Diffusions and elliptic operators.* Springer-Verlag, New York, 1998; xiv+232 p. [*52*]

[92] Bass, R. F., Burdzy, K., Chen, Zh.-Q., Hairer, M., *Stationary distributions for diffusions with inert drift.* Probab. Theory Related Fields. 2010. V. 146, №1-2. P. 1–47. [*236*]

[93] Bass, R. F., Gordina M., *Harnack inequalities in infinite dimensions.* J. Funct. Anal. 2012. V. 263, №11. P. 3707–3740. [*435*]

[94] Bauman, P., *Positive solutions of elliptic equations in nondivergent form and their adjoints.* Ark. Mat. 1984. V. 22. P. 153–173. [*27, 49, 52*]

[95] Bauman P., *Equivalence of the Green's functions for diffusion operators in* R^n*: a counterexample.* Proc. Amer. Math. Soc. 1984. V. 91. P. 64–68. [*30, 52*]

[96] Baur, B., Grothaus, M., *Construction and strong Feller property of distorted elliptic diffusion with reflecting boundary.* Potential Anal. 2014. V. 40, №4. P. 391–425. [*284*]

[97] Baur, B., Grothaus, M., Stilgenbauer, P., *Construction of* L^p*-strong Feller processes via Dirichlet forms and applications to elliptic diffusions.* Potential Anal. 2013. V. 38, №4. P. 1233–1258. [*284*]

[98] Baxendale, P., *T. E. Harris's contributions to reccurent Markov processes and stochastic flows.* Ann. Probab. 2011. V. 39, №2. P. 417–428. [*235*]

[99] Belaribi, N., Russo, F., *Uniqueness for Fokker–Planck equations with measurable coefficients and applications to the fast diffusion equation.* Electr. J. Probab. 2013. V. 17, №84. P. 1–28. [*374*]

[100] Belgacem, F. B., Jabin, P.-E., *Compactness for nonlinear continuity equations.* J. Funct. Anal. 2013. V. 264, №1. P. 139–168. [*284*]

[101] Bell, D. R., *Degenerate stochastic differential equations and hypoellipticity.* Longman, Harlow, 1995; xii+114 p. [*401*]

[102] Belopol'skaya, Ya. I., Dalecky, Yu. L., *Stochastic equations and differential geometry.* Kluwer, Dordrecht, 1990; xvi+260 p. (Russian ed.: Kiev, 1990). [*435*]

[103] Beneš, V. E., *Existence of finite invariant measures for Markov processes.* Proc. Amer. Math. Soc. 1967. V. 18. P. 1058–1061. [*235*]

[104] Beneš, V. E., *Finite regular invariant measures for Feller processes.* J. Appl. Probab. 1968. V. 5. P. 203–209. [*235*]

[105] Bensoussan, A., *Perturbation methods in optimal control.* Wiley/Gauthier-Villars, Paris, 1988; 573 p. [*52, 79*]

[106] Bensoussan, A., Frehse, J., Yam, Ph., *Mean field games and mean field type control theory.* Springer, New York, 2013; x+128 p. [*397*]

[107] Bensoussan, A., Turi, J., *Degenerate Dirichlet problems related to the invariant measure of elasto-plastic oscillators.* Appl. Math. Optim. 2008. V. 58, №1. P. 1–27. [*401*]

[108] Bergh, J., Löfström, J., *Interpolation spaces. An introduction.* Springer-Verlag, Berlin – New York, 1976; x+207 p. [*152, 224, 240*]

[109] Bers, L., John F., Schechter, M., *Partial differential equations.* Interscience Publ., John Wiley & Sons, New York – London – Sydney, 1964; xiii+343 p. [*51*]

[110] Bertoldi, M., Lorenzi, L., *Estimates of the derivatives for parabolic operators with unbounded coefficients.* Trans. Amer. Math. Soc. 2005. V. 357, №7. P. 2627–2664. [*314*]

[111] Bertozzi, A. L., Carrillo, J. A., Laurent, T., *Blow-up in multidimensional aggregation equations with mildly singular interaction kernels.* Nonlinearity. 2009. V. 22. P. 683–710. [*282*]

[112] Besov, O. V., Il'in, V. P., Nikolskiĭ, S. M., *Integral representations of functions and imbedding theorems.* Vols. I, II. Winston & Sons, Washington; Halsted Press, New York – Toronto – London, 1978, 1979; viii+345 p., viii+311 p. (Russian ed.: Moscow, 1975). [*51, 246, 248*]

[113] Bhatt, A. G., Karandikar, R. L., *Invariant measures and evolution equations for Markov processes characterized via martingale problems.* Ann. Probab. 1993. V. 21, №4. P. 2246–2268. [*235*]

[114] Bhatt, A. G., Karandikar, R. L., *Evolution equations for Markov processes: application to the white-noise theory of filtering.* Appl. Math. Optim. 1995. V. 31, №3. P. 327–348. [*235*]

[115] Bhattacharya, R. N., *Criteria for recurrence and existence of invariant measures for multidimensional diffusions.* Ann. Probab. 1978. V. 6. P. 541–553; Correction: ibid. 1980. V. 8. P. 1194–1195. [*235*]

[116] Bhattacharya, R. N., Ramasubramanian, S., *Recurrence and ergodicity of diffusions.* J. Multivar. Anal. 1982. V. 12, №1. P. 95–122. [*235*]

[117] Birrell, J., Herzog, D. P., Wehr, J., *The transition from ergodic to explosive behavior in a family of stochastic differential equations.* Stoch. Process. Appl. 2012. V. 122, №4. P. 1519–1539. [*79*]

[118] Blanchard, P., Röckner, M., Russo, F., *Probabilistic representation for solutions of an irregular porous media type equation.* Ann. Probab. 2010. V. 38, №5. P. 1870–1900. [*374*]

[119] Blanchet, A., Carlen, E. A., Carrillo, J. A., *Functional inequalities, thick tails and asymptotics for the critical mass Patlak–Keller–Segel model.* J. Funct. Anal. 2012. V. 262, №5. P. 2142–2230. [*284*]

[120] Blömker, D., Flandoli, F., Romito, M., *Markovianity and ergodicity for a surface growth PDE.* Ann. Probab. 2009. V. 37, №1. P. 275–313. [*433*]

[121] Bouchut, F., *Existence and uniqueness of a global smooth solution for the Vlasov–Poisson–Fokker–Planck system in three dimensions.* J. Funct. Anal. 1993. V. 111, №1. P. 239–258. [*284*]

[122] Bogachev, V. I., *Remarks on invariant measures and reversibility of infinite dimensional diffusions.* In: Probab. Theory and Math. Statist. (Proc. Conf. on Stoch. Anal., Euler Math. Inst., St.-Peterburg, 1993), I.A. Ibragimov et al., eds., pp. 119–132. Gordon and Breach, Amsterdam, 1996. [*435*]

[123] Bogachev, V. I., *Differentiable measures and the Malliavin calculus.* J. Math. Sci. 1997. V. 87, №5. P. 3577–3731. [*409*]

[124] Bogachev, V. I., *Gaussian measures.* Amer. Math. Soc., Providence, Rhode Island, 1998; 433 p. [*404, 406, 408, 415*]

[125] Bogachev, V. I., *Measure theory.* Vols. 1, 2. Springer, Berlin, 2007; xvii+500 p., xiii+575 p. [*3, 54, 76, 102, 129, 170, 210, 219, 222, 239, 276, 279, 306, 417, 424, 425*]

[126] Bogachev, V. I., *Differentiable measures and the Malliavin calculus.* Amer. Math. Soc., Providence, Rhode Island, 2010; xvi+488 p. (Russian ed.: Moscow, 2008). [*4, 51, 70, 80, 84, 110, 156, 393, 406, 408, 409, 416, 426, 435*]

[127] Bogachev, V. I., *Gaussian measures on infinite-dimensional spaces.* In: Real and Stochastic Analysis. Current Trends (M.M. Rao ed.), pp. 1–83. World Sci., Singapore, 2014. [*408*]

[128] Bogachev, V. I., Da Prato, G., Röckner, M., *Regularity of invariant measures for a class of perturbed Ornstein–Uhlenbeck operators.* Nonlinear Diff. Equat. Appl. 1996. V. 3, №2. P. 261–268. [*416*]

[129] Bogachev, V. I., Da Prato, G., Röckner, M., *On weak parabolic equations for probability measures.* Dokl. Ross. Akad. Nauk. 2002. V. 386, №3. P. 295–299 (in Russian); English transl.: Dokl. Math. 2002. V. 66, №2. P. 192–196. [*284*]

[130] Bogachev, V. I., Da Prato, G., Röckner, M., *Invariant measures of generalized stochastic porous medium equations.* Dokl. Ross. Akad. Nauk. 2004. V. 396, №1. P. 7–11 (in Russian); English transl.: Dokl. Math. 2004. V. 69, №3. P. 321–325. [*409, 436*]

[131] Bogachev, V. I., Da Prato, G., Röckner, M., *Existence of solutions to weak parabolic equations for measures.* Proc. London Math. Soc. 2004. V. 88, №3. P. 753–774. [*284, 434*]

[132] Bogachev, V. I., Da Prato, G., Röckner, M., *On parabolic equations for measures.* Comm. Partial Differ. Equ. 2008. V. 33, №1-3. P. 397–418. [*284*]

[133] Bogachev, V. I., Da Prato, G., Röckner, M., *Infinite-dimensional Kolmogorov operators with time dependent drift coefficients.* Dokl. Ross. Akad. Nauk. 2008. V. 419, №5. P. 587–591; English transl.: Dokl. Math. 2008. V. 77, №2. P. 276–280. [*435*]

[134] Bogachev, V. I., Da Prato, G., Röckner, M., *Parabolic equations for measures on infinite-dimensional spaces.* Dokl. Ross. Akad. Nauk. 2008. V. 421, №4. P. 439–444 (in Russian); English transl.: Dokl. Math. 2008. V. 78, №1. P. 544–549. [*434, 436*]

[135] Bogachev, V. I., Da Prato G., Röckner, M., *Fokker–Planck equations and maximal dissipativity for Kolmogorov operators with time dependent singular drifts in Hilbert spaces.* J. Funct. Anal. 2009. V. 256, №3. P. 1269–1298. [*434*]

[136] Bogachev, V., Da Prato, G., Röckner, M., *Existence results for Fokker–Planck equations in Hilbert spaces.* In: Proceedings of the conference “Stochastic Analysis, Random Fields and Applications VI” (Ascona, May 19–23, 2008; R. Dalang, M. Dozzi, F. Russo eds.), pp. 23–35. Progress in Probability, V. 63. Birkhäuser, 2011. [*434*]

[137] Bogachev, V. I., Da Prato, G., Röckner, M., Shaposhnikov, S. V., *Nonlinear evolution equations for measures on infinite dimensional spaces.* In: Stochastic Partial Differential

Equations and Applications. P. 51–64. Quaderni di Matematica, V. 25, Series edited by Dipartimento di Matematica Seconda Università di Napoli, 2010. [*430, 435*]

[138] Bogachev V. I., Da Prato, G., Röckner, M., Shaposhnikov, S. V., *An analytic approach to infinite-dimensional continuity and Fokker–Planck–Kolmogorov equations.* Ann. Sc. Norm. Sup. Pisa. 2015. V. 15, №3. P. 583–1023. [*409, 427, 428, 434*]

[139] Bogachev, V. I., Da Prato, G., Röckner, M., Shaposhnikov, S. V., *On the uniqueness of solutions to continuity equations.* J. Differ. Equ. 2015. V. 259, №8. P. 3854–3873. [*382*]

[140] Bogachev, V. I., Da Prato, G., Röckner, M., Sobol, Z., *Global gradient bounds for dissipative diffusion operators.* C. R. Math. Acad. Sci. Paris. 2004. T. 339, №4. P. 277–282. [*229*]

[141] Bogachev, V. I., Da Prato, G., Röckner, M., Sobol, Z., *Gradient bounds for solutions of elliptic and parabolic equations.* Stochastic partial differential equations and applications-VII, pp. 27–34, Lect. Notes Pure Appl. Math., V. 245, Chapman & Hall/CRC, Boca Raton, Florida, 2006. [*229*]

[142] Bogachev, V. I., Da Prato, D., Röckner M., Stannat, W., *Uniqueness of solutions to weak parabolic equations for measures.* Bull. Lond. Math. Soc. 2007. V. 39, №4. P. 631–640. [*284, 340, 400*]

[143] Bogachev, V. I., Kirillov, A. I., Shaposhnikov, S. V., *Invariant measures of diffusions with gradient drift.* Dokl. Ross. Akad. Nauk. 2010. V. 434, №6. P. 730–734 (in Russian); English transl.: Dokl. Math. 2010. V. 82, №2. P. 790–793. [*174*]

[144] Bogachev, V. I., Kirillov, A. I., Shaposhnikov, S. V., *On probability and integrable solutions to the stationary Kolmogorov equation.* Dokl. Ross. Akad. Nauk. 2011. V. 438, №2. P. 154–159 (in Russian); English transl.: Dokl. Math. 2011. V. 83, №3. P. 309–313. [*79, 174, 175*]

[145] Bogachev, V. I., Kirillov, A. I., Shaposhnikov, S. V., *Integrable solutions of the stationary Kolmogorov equation.* Dokl. Ross. Akad. Nauk. 2012. V. 444, №1. P. 11–16 (in Russian); English transl.: Dokl. Math. 2012. V. 85, №3. P. 309–314. [*174*]

[146] Bogachev, V. I., Kirillov, A. I., Shaposhnikov, S. V., *The stationary Fokker–Planck–Kolmogorov equation with a potential.* Dokl. Ross. Akad. Nauk. 2014. V. 454, №2. P. 131–136 (in Russian); English transl.: Dokl. Math. 2014. V. 89, №1. P. 24–29. [*174*]

[147] Bogachev, V. I., Kirillov, A. I., Shaposhnikov, S. V., *The Kantorovich and variation distances between invariant measures of diffusions and nonlinear stationary Fokker–Planck–Kolmogorov equations.* Math. Notes. 2014. V. 96, №6. P. 17–25. [*169*]

[148] Bogachev, V. I., Kolesnikov, A. V., *The Monge–Kantorovich problem: achievements, connections, and perspectives.* Uspehi Matem. Nauk. 2012. V. 67, №5. P. 3–110 (in Russian); English transl.: Russian Math. Surveys. 2012. V. 67, №5. P. 785–890. [*169, 170, 285*]

[149] Bogachev, V. I., Krylov, N. V., Röckner, M., *Regularity of invariant measures: the case of non-constant diffusion part.* J. Funct. Anal. 1996. V. 138. P. 223–242. [*52, 127, 416*]

[150] Bogachev, V. I., Krylov, N. V., Röckner, M., *Elliptic regularity and essential self-adjointness of Dirichlet operators on* $\mathbb{R}^n$. Ann. Sc. Norm. Sup. Pisa Cl. Sci. (4). 1997. V. 24, №3. P. 451–461. [*208*]

[151] Bogachev, V. I., Krylov, N. V., Röckner, M., *Differentiability of invariant measures and transition probabilities of singular diffusions.* Dokl. Ross. Akad. Nauk. 2001. V. 376, №2. P. 151–154 (in Russian); English transl.: Dokl. Math. 2001. V. 63, №1. P. 18–21. [*284*]

[152] Bogachev, V. I., Krylov, N. V., Röckner, M., *On regularity of transition probabilities and invariant measures of singular diffusions under minimal conditions.* Comm. Partial Differ. Equ. 2001. V. 26, №11–12. P. 2037–2080. [*52, 284*]

[153] Bogachev, V. I., Krylov, N. V., Röckner, M., *Regularity and global bounds of densities of invariant measures of diffusion processes.* Dokl. Ross. Akad. Nauk. 2005. V. 405, №5. P. 583–587 (in Russian); English transl.: Dokl. Math. 2005. V. 72, №3. P. 934–938. [*127*]

[154] Bogachev, V. I., Krylov, N. V., Röckner, M., *Elliptic equations for measures: regularity and global bounds of densities.* J. Math. Pures Appl. 2006. V. 85, №6. P. 743–757. [*127*]

[155] Bogachev, V. I., Krylov, N. V., Röckner, M., *Elliptic and parabolic equations for measures.* Uspehi Matem. Nauk. 2009. V. 64, №6. P. 5–116 (in Russian); English transl.: Russian Math. Surveys. 2009. V. 64, №6. P. 973–1078. [*70, 79*]

[156] Bogachev, V. I., Mayer-Wolf, E., *Absolutely continuous flows generated by Sobolev class vector fields in finite and infinite dimensions.* J. Funct. Anal. 1999. V. 167, №1. P. 1–68. [*108, 130, 434*]

[157] Bogachev, V. I., Röckner, M., *Hypoellipticity and invariant measures for infinite dimensional diffusions.* C. R. Acad. Sci. Paris. 1994. T. 318. P. 553–558. [*52, 416*]

[158] Bogachev, V. I., Röckner, M., *Regularity of invariant measures on finite and infinite dimensional spaces and applications.* J. Funct. Anal. 1995. V. 133. P. 168–223. [*52, 409, 416, 426*]

[159] Bogachev, V. I., Röckner, M., *Mehler formula and capacities for infinite dimensional Ornstein–Uhlenbeck processes with general linear drift.* Osaka J. Math. 1995. V. 32, №2. P. 237–274. [*434*]

[160] Bogachev, V. I., Röckner, M., *A generalization of Khasminskii's theorem on the existence of invariant measures for locally integrable drifts.* Teor. Verojatn. i Primen. 2000. V. 45, №3. P. 417–436; correction: ibid. 2001. V. 46, №3. P. 600 (in Russian); English transl.: Theory Probab. Appl. 2000. V. 45, №3. P. 363–378. [*53, 79, 80*]

[161] Bogachev, V. I., Röckner, M., *Elliptic equations for measures on infinite dimensional spaces and applications.* Probab. Theory Related Fields. 2001. V. 120. P. 445–496. [*409, 415*]

[162] Bogachev, V. I., Röckner, M., *On L^p-uniqueness of symmetric diffusion operators on Riemannian manifolds.* Matem. Sbornik. 2003. V. 194, №7. P. 15–24 (in Russian); English transl.: Sb. Math. 2003. V. 194, №7-8. P. 969–978. [*209*]

[163] Bogachev, V. I., Röckner, M., Schmuland, B., *Generalized Mehler semigroups and applications.* Probab. Theory Related Fields. 1996. V. 105, №2. P. 193–225. [*434*]

[164] Bogachev, V. I., Röckner, M., Shaposhnikov, S. V., *Global regularity and bounds for solutions of parabolic equations for probability measures.* Teor. Verojatn. Primen. 2005. V. 50, №4. P. 652–674 (in Russian); English transl.: Theory Probab. Appl. 2006. V. 50, №4. P. 561–581. [*313*]

[165] Bogachev, V. I., Röckner, M., Shaposhnikov, S. V., *Estimates of densities of stationary distributions and transition probabilities of diffusion processes.* Teor. Verojatn. i Primen. 2007. V. 52, №2. P. 240–270 (in Russian); English transl.: Theory Probab. Appl. 2008. V. 52, №2. P. 209–236. [*52, 127, 316*]

[166] Bogachev, V. I., Röckner M., Shaposhnikov, S. V., *Positive densities of transition probabilities of diffusion processes.* Teor. Verojatn. i Primen. 2008. V. 53, №2. P. 213–239 (in Russian); English transl.: Theory Probab. Appl. 2009. V. 53, №2. P. 194–215. [*130, 334, 335*]

[167] Bogachev, V. I., Röckner M., Shaposhnikov, S. V., *Lower estimates of densities of solutions of elliptic equations for measures.* Dokl. Ross. Akad. Nauk. 2009. V. 426, №2. P. 156–161 (in Russian); English transl.: Dokl. Math. 2009. V. 79, №3. P. 329–334. [*127*]

[168] Bogachev, V. I., Röckner, M., Shaposhnikov, S. V., *Nonlinear evolution and transport equations for measures.* Dokl. Ross. Akad. Nauk. 2009. V. 429, №1. P. 7–11 (in Russian); English transl.: Dokl. Math. 2009. V. 80, №3. P. 785–789. [*284, 430*]

[169] Bogachev, V. I., Röckner, M., Shaposhnikov, S. V., *On uniqueness problems related to elliptic equations for measures.* J. Math. Sci. (New York). 2011. V. 176, №6. P. 759–773. [*174*]

[170] Bogachev, V. I., Röckner M., Shaposhnikov, S. V., *On uniqueness problems related to the Fokker–Planck–Kolmogorov equations for measures.* J. Math. Sci. (New York). 2011. V. 179, №1. P. 759–773. [*400*]

[171] Bogachev, V. I., Röckner, M., Shaposhnikov, S. V., *On positive and probability solutions of the stationary Fokker–Planck–Kolmogorov equation.* Dokl. Ross. Akad. Nauk. 2012. V. 444, №3. P. 245–249 (in Russian); English transl.: Dokl. Math. 2012. V. 85, №3. P. 350–354. [*79, 174*]

[172] Bogachev, V. I., Röckner, M., Shaposhnikov, S. V., *On uniqueness of solutions to the Cauchy problem for degenerate Fokker–Planck–Kolmogorov equations.* J. Evol. Equ. 2013. V. 13, №3. P. 577–593. [*374*]

[173] Bogachev, V. I., Röckner, M., Shaposhnikov, S. V., *On parabolic inequalities for generators of diffusions with jumps.* Probab. Theory Related Fields. 2014. V. 158, №1-2. P. 465–476. [*285*]

[174] Bogachev, V. I., Röckner, M., Shaposhnikov, S. V., *On existence of Lyapunov functions for a stationary Kolmogorov equation with a probability solution.* Dokl. Ross. Akad. Nauk. 2014. V. 457, №2. P. 136–140 (in Russian); English transl.: Dokl. Math. 2014. V. 90, №1. P. 424–428. [*234*]

[175] Bogachev, V. I., Röckner, M., Shaposhnikov, S. V., *Uniqueness problems for degenerate Fokker–Planck–Kolmogorov equations.* J. Math. Sci. (New York). 2015. V. 207, №2. P. 147–165. [*380*]

[176] Bogachev, V. I., Röckner, M., Shaposhnikov, S. V., *Distances between transition probabilities of diffusions and applications to nonlinear Fokker–Planck–Kolmogorov equations.* Preprint CRC 701 N 15021. 18 pp. Bielefeld Univ. 2015. [*389*]

[177] Bogachev, V. I., Röckner, M., Stannat, W., *Uniqueness of invariant measures and maximal dissipativity of diffusion operators on L^1.* In: "Infinite Dimensional Stochastic Analysis" (Proceedings of the Colloquium, Amsterdam, 11–12 February, 1999), Ph. Clément, F. den Hollander, J. van Neerven and B. de Pagter eds., Royal Netherlands Academy of Arts and Sciences, pp. 39–54, Amsterdam, 2000. [*140, 174*]

[178] Bogachev, V. I., Röckner, M., Stannat, W., *Uniqueness of solutions of elliptic equations and uniqueness of invariant measures of diffusions.* Matem. Sbornik. 2002. V. 197, №7. P. 3–36 (in Russian); English transl.: Sb. Math. 2002. V. 193, №7. P. 945–976. [*140, 174*]

[179] Bogachev, V. I., Röckner, M., Wang, F.-Y., *Elliptic equations for invariant measures on finite and infinite dimensional manifolds.* J. Math. Pures Appl. 2001. V. 80. P. 177–221. [*72, 79, 87, 119, 160, 236, 409, 435*]

[180] Bogachev, V. I., Röckner, M., Wang, F.-Y., *Elliptic equations for invariant measures on Riemannian manifolds: existence and regularity of solutions.* C. R. Acad. Sci. Paris, Sér. I Math. 2001. T. 332, №4. P. 333–338. [*160*]

[181] Bogachev, V. I., Röckner, M., Wang, F.-Y., *Invariance implies Gibbsian: some new results.* Comm. Math. Phys. 2004. V. 248. P. 335–355. [*119, 160, 409, 435*]

[182] Bogachev, V. I., Röckner, M., Zhang, T. S., *Existence of invariant measures for diffusions with singular drifts.* Appl. Math. Optim. 2000. V. 41. P. 87–109. [*73, 79, 427*]

[183] Bogachev, V. I., Shaposhnikov, S. V., Veretennikov, A. Yu., *Differentiability of solutions of stationary Fokker–Planck–Kolmogorov equations with respect to a parameter.* Discr. Contin. Syst. A (to appear). [*122, 127*]

[184] Bogachev, V. I., Veretennikov, A. Yu., Shaposhnikov, S. V., *Differentiability of invariant measures of diffusions with respect to a parameter.* Dokl. Ross. Akad. Nauk. 2015. V. 460, №5. P. 507–511 (in Russian); English transl.: Dokl. Math. 2015. V. 91, №1. P. 76–79. [*122*]

[185] Bogachev, V., Wang, F.-Y., Röckner, M., *Elliptic equations connected with invariant measures of diffusions on finite and infinite dimensional manifolds.* Dokl. Ross. Akad. Nauk. 2001. V. 378, №4. P. 439–442 (in Russian); English transl.: Dokl. Math. 2001. V. 63. [*236, 435*]

[186] Bogachev, V. I., Wang, F.-Yu., Röckner M., *Invariant measures of stochastic gradient systems on Riemannian manifolds, and Gibbs measures.* Dokl. Ross. Akad. Nauk. 2002. V. 386, №2. P. 151–155 (in Russian); English transl.: Dokl. Math. 2002. V. 66, №2. P. 179–183. [*160, 236*]

[187] Bogolyubov, N. N., *Collection of scientific works. Statistical mechanics. Vol. 5. Nonequilibrium statistical mechanics.* 1939–1980. Nauka, Moscow, 2006; 805 p. (in Russian). [*284*]

[188] Bolley, F., Gentil, I., *Phi-entropy inequalities for diffusion semigroups.* J. Math. Pures Appl. (9). 2010. V. 93, №5. P. 449–473. [*228, 401*]

[189] Bolley, F., Gentil, I., Guillin, A., *Convergence to equilibrium in Wasserstein distance for Fokker–Planck equations.* J. Funct. Anal. 2012. V. 263, №8. P. 2430–2457. [*314, 401*]

[190] Bolley, F., Gentil, I., Guillin, A., *Uniform convergence to equilibrium for granular media.* Arch. Ration. Mech. Anal. 2013. V. 208, №2. P. 429–445. [*285*]

[191] Bolley, F., Gentil, I., Guillin, A., *Dimensional contraction via Markov transportation distance.* J. London Math. Soc. (2). 2014. V. 90, №1. P. 309–332. [*401*]

[192] Bolley, F., Guillin, A., Malrieu, F., *Trend to equilibrium and particle approximation for a weakly selfconsistent Vlasov–Fokker–Planck equation.* Math. Model. Numer. Anal. 2010. V. 44, №5. P. 867–884. [*285*]

[193] Bolley, F., Guillin, A., Villani, C., *Quantitative concentration inequalities for empirical measures on non-compact spaces.* Probab. Theory Related Fields. 2007. V. 137, №3-4. P. 541–593. [*285, 401*]

[194] Bolley, F., Villani, C., *Weighted Csiszar–Kullback–Pinsker inequalities and applications to transportation inequalities.* Ann. Fac. Sci. Toulouse Math. (6). 2005. V. 14, №3. P. 331–352. [*170, 391*]

[195] Bony, J. M., *Principe du maximum, inégalité de Harnack et unicité du problème de Cauchy pour les opérateurs elliptiques dégénérés.* Ann. Inst. Fourier. 1969. V. 19. P. 277–304. [*79*]

[196] Borovkov, A. A., *Ergodicity and stability of stochastic processes.* John Wiley & Sons, Chichester, 1998; xxiv+585 p. [*235*]

[197] Borsuk, K., *Theory of retracts.* Polish Scientific Publ., Warszawa, 1967; 251 p. [*425*]

[198] Borsuk, M., *Transmission problems for elliptic second-order equations in non-smooth domains.* Birkhäuser/Springer Basel, Basel, 2010; xii+218 p. [*51*]

[199] Borsuk, M., Kondratiev, V., *Elliptic boundary value problems of second order in piecewise smooth domains.* Elsevier, Amsterdam, 2006; vi+531 p. [*51*]

[200] Bouchut, F., Crippa, G., *Lagrangian flows for vector fields with gradient given by a singular integral.* J. Hyperb. Differ. Equ. 2013. V. 10, №2. P. 235–282. [*384*]

[201] Bouchut, F., James, F., *One-dimensional transport equations with discontinuous coefficients.* Nonlinear Anal. 1998. V. 32, №7. P. 891–933. [*384*]

[202] Bramanti, M., Brandolini, L., Lanconelli, E., Uguzzoni, F., *Non-divergence equations structured on Hörmander vector fields: heat kernels and Harnack inequalities.* Mem. Amer. Math. Soc. 2010. V. 204, №961; vi+123 p. [*401*]

[203] Bramanti, M., Cupini, G., Lanconelli, E., Priola, E., *Global L^p estimates for degenerate Ornstein–Uhlenbeck operators.* Math. Z. 2010. B. 266, №4. S. 789–816. [*234*]

[204] Bramanti, M., Cupini, G., Lanconelli, E., Priola, E., *Global L^p estimates for degenerate Ornstein–Uhlenbeck operators with variable coefficients.* Math. Nachr. 2013. B. 286, №11-12. P. 1087–1101. [*234*]

[205] Bratteli, O., Robinson, D. W., *Subelliptic operators on Lie groups: variable coefficients.* Acta Appl. Math. 1996. V. 42, №1. 104 pp. [*128*]

[206] Brezis, H., *Opérateurs maximaux monotones et semi-groupes de contractions dans les espaces de Hilbert.* North-Holland, Amsterdam – London, 1973; vi+183 p. [*232*]

[207] Brezis, H., *On a conjecture of J. Serrin.* Rend. Lincei Mat. Appl. 2008. V. 19. P. 335–338. [*35*]

[208] Brezis, H., *Functional analysis, Sobolev spaces and partial differential equations.* Springer, New York, 2011; xiii+599 p. [*51*]

[209] Brunel, A., Horowitz, S., Lin, M., *On subinvariant measures for positive operators in L_1.* Ann. Inst. H. Poincaré Probab. Statist. 1993. V. 29, №1. P. 105–117. [*235*]

[210] Brzezniak, Z., Li, Y., *Asymptotic compactness and absorbing sets for 2D stochastic Navier–Stokes equations on some unbounded domains.* Trans. Amer. Math. Soc. 2006. V. 358, №12. P. 5587–5629 . [*433*]

[211] Bukhvalov, A. V., Gutman, A. E., Korotkov, V. B., Kusraev, A. G., Kutateladze, S. S., Makarov, B. M., *Vector lattices and integral operators.* Kluwer, Dordrecht, 1996; x+462 p. [*210, 239*]

[212] Burenkov, V. I., *Sobolev spaces on domains.* Teubner, Stuttgart, 1998; 312 p. [*51*]

[213] Butkovsky, O., *Subgeometric rates of convergence of Markov processes in the Wasserstein metric.* Ann. Appl. Probab. 2014. V. 24, №2. P. 526–552. [*235*]

[214] Byun, S.-S., *Elliptic equations with BMO coefficients in Lipschitz domains.* Trans. Amer. Math. Soc. 2005. V. 357, № 3. P. 1025–1046. [*8*]

[215] Cabré, X., *Nondivergent elliptic equations on manifolds with nonnegative curvature.* Comm. Pure Appl. Math. 1997. V. 50, №7. P. 623–665. [*52*]

[216] Calogero, S., *Exponential convergence to equilibrium for kinetic Fokker–Planck equations.* Comm. Partial Differ. Equ. 2012. V. 37, №8. P. 1357–1390. [*314*]

[217] Caraballo, T., Kloeden, P. E., Real, J., *Invariant measures and statistical solutions of the globally modified Navier–Stokes equations.* Discrete Contin. Dyn. Syst. Ser. B. 2008. V. 10, №4. P. 761–781. [*433*]

[218] Cardaliaguet, P., *Notes on mean field games.* Preprint. 2013. 59 pp. https://www.ceremade.dauphine.fr/ cardalia/MFG20130420.pdf. [*398*]

[219] Carlen, E. A., Kusuoka, S., Stroock, D. W., *Upper bounds for symmetric Markov transition functions.* Ann. Inst. H. Poincaré Probab. Statist. 1987. V. 23, №2, suppl. P. 245–287. [*228*]

[220] Carmona, R., Klein, A., *Exponential moments for hitting times of uniformly ergodic Markov processes.* Ann. Probab. 1983. V. 11, №3. P. 648–655. [*235*]

[221] Carrillo, J. A., Difrancesco, M., Figalli, A., Laurent, T., Slepčev, D., *Global-in-time weak measure solutions and finite-time aggregation for non-local interaction equations.* Duke Math. J. 2011. V. 156, №2. P. 229–271. [*282, 284, 285*]

[222] Carrillo, J. A., Jüngel, A., Markowich, P. A., Toscani, G., Unterreiter, A., *Entropy dissipation methods for degenerate parabolic problems and generalized Sobolev inequalities.* Monatsh. Math. 2001. V. 133, №1. P. 1–82. [*285*]

[223] Carrillo, J. A., McCann, R. J., Villani, C., *Kinetic equilibration rates for granular media and related equations: entropy, dissipation and mass transportation estimates.* Rev. Math. Iberoamer. 2003. V. 19. P. 971–1018. [*284, 285, 401*]

[224] Carrillo, J. A., Toscani, G., *Exponential convergence toward equilibrium for homogeneous Fokker–Planck-type equations.* Math. Methods Appl. Sci. 1998. V. 21, №13. P. 1269–1286. [*314, 401*]

[225] Carrillo, J. A., Toscani, G., *Contractive probability metrics and asymptotic behavior of dissipative kinetic equations.* Riv. Mat. Univ. Parma (7). 2007. V. 6. P. 75–198. [*284*]

[226] Cattiaux, P., Fradon, M., *Entropy, reversible diffusion processes, and Markov uniqueness.* J. Funct. Anal. 1996. V. 138, №1. P. 243–272. [*234*]

[227] Cattiaux, P., Guillin, A., *Trends to equilibrium in total variation distance.* Ann. Inst. H. Poincaré Probab. Stat. 2009. V. 45, №1. P. 117–145. [*235*]

[228] Cattiaux, P., Guillin, A., Roberto, C., *Poincaré inequality and the L^p convergence of semigroups.* Electron. Commun. Probab. 2010. V. 15. P. 270–280. [*228*]

[229] Cattiaux, P., Guillin, A., Wang, F.-Y., Wu, L., *Lyapunov conditions for super Poincaré inequalities.* J. Funct. Anal. 2009. V. 256, №6. P. 1821–1841. [*228*]

[230] Cattiaux, P., Léonard, C., *Minimization of the Kullback information of diffusion processes.* Ann. Inst. H. Poincaré. 1994. V. 30, №1. P. 83–132; correction: ibid. 1995. V. 31, №4. P. 705–707. [*235, 254*]

[231] Cattiaux, P., Léonard, C., *Large deviations and Nelson processes.* Forum Math. 1995. V. 7, №1. P. 95–115. [*235*]

[232] Cattiaux, P., Léonard C., *Minimization of the Kullback information for some Markov processes.* Lecture Notes in Math. 1996. V. 1626. P. 288–311. [*235*]

[233] Cattiaux, P., Roelly S., Zessin H., *Une approche gibbsienne des diffusions browniennes infini-dimensionnelles.* Probab. Theory Related Fields. 1996. V. 104, №2. P. 147–179. [*434*]

[234] Cerrai, S., *Second order PDE's in finite and infinite dimension.* A probabilistic approach. Lecture Notes in Math. V. 1762. Springer-Verlag, Berlin, 2001; x+330 p. [*435*]

[235] Chapman, S., *On the Brownian displacements and thermal diffusion of grains suspended in a non-uniform fluid.* Proceedings Royal Soc. London (A). 1928. V. 119. P. 34–54. [ix, *52*]

[236] Chen, M. F., *From Markov chains to non-equilibrium particle systems.* 2nd ed. World Sci., River Edge, New Jersey, 2004; xii+597 p. [*235*]

[237] Chen, Y.-Zh., Wu, L.-Ch., *Second order elliptic equations and elliptic systems.* Amer. Math. Soc., Providence, Rhode Island, 1998; xiv+246 p. [*51*]

[238] Chen, Zh.-Q., Qian, Zh., Hu Y., Zheng, W., *Stability and approximations of symmetric diffusion semigroups and kernels.* J. Funct. Anal. 1998. V. 152, №1. P. 255–280. [*234*]

[239] Chicco, M., *Principio di massimo generalizzato e valutazione del primo autovalore per problemi ellittici del secondo ordine di tipo variazionale.* Ann. Mat. Pura Appl. (4). 1970. V. 87. P. 1–9. [*79*]

[240] Chicco, M., *Solvability of the Dirichlet problem in $H^{2,p}(\Omega)$ for a class of linear second order elliptic partial differential equations.* Boll. Unione Mat. Ital., IV. Ser. 1971. V. 4. P. 374–387. [*191, 195*]

[241] Chill, R., Fašangovà, E., Metafune, G., Pallara, D., *The sector of analyticity of the Ornstein–Uhlenbeck semigroup on L^p spaces with respect to invariant measure.* J. London Math. Soc. (2). 2005. V. 71, №3. P. 703–722. [*234*]

[242] Chill, R., Fašangovà, E., Metafune, G., Pallara, D., *The sector of analyticity of nonsymmetric submarkovian semigroups generated by elliptic operators.* C. R. Math. Acad. Sci. Paris. 2006. V. 342, №12. P. 909–914. [*234*]

[243] Cinti, Ch., Pascucci, A., Polidoro S., *Pointwise estimates for a class of non-homogeneous Kolmogorov equations.* Math. Ann. 2008. B. 340, №2. S. 237–264. [*401*]

[244] Cinti, Ch., Polidoro, S., *Bounds on short cylinders and uniqueness in Cauchy problem for degenerate Kolmogorov equations.* J. Math. Anal. Appl. 2009. V. 359, №1. P. 135–145. [*401*]

[245] Chojnowska-Michalik, A., *Transition semigroups for stochastic semilinear equations on Hilbert spaces.* Disser. Math. (Rozprawy Mat.). 2001. V. 396. 59 p. [*433*]

[246] Chojnowska-Michalik, A., Goldys, B., *On regularity properties of nonsymmetric Ornstein–Uhlenbeck semigroup in L^p spaces.* Stochastics Stoch. Rep. 1996. V. 59, №3-4. P. 183–209. [*434*]

[247] Chow, P.L., Khasminskii, R. Z., *Stationary solutions of nonlinear stochastic evolution equations.* Stoch. Anal. Appl. 1997. V. 15, №5. P. 671–699. [*434*]

[248] Chung, L. O., *Existence of harmonic L^1 functions in complete Riemannian manifolds.* Proc. Amer. Math. Soc. 1983. V. 88. P. 531–532. [*72, 119*]

[249] Chupin, L., *Fokker–Planck equation in bounded domain.* Ann. Inst. Fourier (Grenoble). 2010. V. 60, №1. P. 217–255. [*284*]

[250] Cipriano, F., Cruzeiro, A. B., *Flows associated with irregular $\mathbb{R}^d$-vector fields.* J. Differ. Equ. 2005. V. 219, №1. P. 183–201. [*384*]

[251] Clément, Ph., Heijmans, H. J. A. M., Angenent, S., van Duijn, C. J., de Pagter, B., *One-parameter semigroups.* North-Holland, Amsterdam, 1987; x+312 p. [*234*]

[252] Coffey, W. T., Kalmykov, Yu. P., *The Langevin equation with applications to stochastic problems in physics, chemistry and electrical engineering.* 3d ed. World Sci., Singapore, 2012; xxii+827 p. [*284*]

[253] Cohen de Lara, M., *Geometric and symmetry properties of a nondegenerate diffusion process.* Ann. Probab. 1995. V. 23, №4. P. 1557–1604. [*235*]

[254] Coifman, R. R., Fefferman, C., *Weighted norm inequalities for maximal functions and singular integrals.* Studia Math. 1974. V. 51. P. 241–250. [*29*]

[255] Crippa, G., *Lagrangian flows and the one-dimensional Peano phenomenon for ODEs.* J. Differ. Equ. 2011. V. 250, №7. P. 3135–3149. [*383*]

[256] Crippa, G., De Lellis, C., *Estimates and regularity results for the DiPerna–Lions flow.* J. Reine Angewandte Math. 2008. B. 616. S. 15–46. [*384*]

[257] Cruzeiro, A. B., *Équations différentielles ordinaires: non explosion et mesures quasi-invariantes.* J. Funct. Anal. 1983. V. 54, №2. P. 193–205. [*383*]

[258] Cruzeiro, A. B., *Équations différentielles sur l'espace de Wiener et formules de Cameron–Martin non-linéaires.* J. Funct. Anal. 1983. V. 54, №2. P. 206–227. [*383, 434*]

[259] Cruzeiro, A. B., *Unicité de solutions d'équations différentielles sur l'espace de Wiener.* J. Funct. Anal. 1984. V. 58, №3. P. 335–347. [*383, 434*]

[260] Cruzeiro, A.-B., Malliavin, P., *Nonperturbative construction of invariant measure through confinement by curvature.* J. Math. Pures Appl. (9). 1998. V. 77, №6. P. 527–537. [*383*]

[261] Da Prato, G., *Kolmogorov equations for stochastic PDEs.* Birkhäuser, Basel, 2004; x+182 p. [*429, 433, 435*]

[262] Da Prato, G., Debussche, A., *Ergodicity for the 3D stochastic Navier–Stokes equations.* J. Math. Pures Appl. (9). 2003. V. 82, №8. P. 877–947. [*433*]

[263] Da Prato, G., Debussche, A., *Absolute continuity of the invariant measures for some stochastic PDEs.* J. Stat. Phys. 2004. V. 115, №1-2. P. 451–468. [*416*]

[264] Da Prato, G., Debussche, A., Goldys, B., *Some properties of invariant measures of non symmetric dissipative stochastic systems.* Probab. Theory Related Fields. 2002. V. 123, №3. P. 355–380. [*416*]

[265] Da Prato, G., Elworthy, K. D., Zabczyk, J., *Strong Feller property for stochastic semilinear equations.* Stoch. Anal. Appl. 1995. V. 13, №1. P. 35–45. [*434*]

[266] Da Prato, G., Flandoli, F., Röckner, M., *Fokker–Planck equations for SPDE with non-trace-class noise.* Commun. Math. Stat. 2013. V. 1, №3. P. 281–304. [*433*]

[267] Da Prato, G., Flandoli, F., Röckner, M., *Uniqueness for continuity equations in Hilbert spaces with weakly differentiable drift.* Stoch. Partial Differ. Equ. Anal. Comput. 2014. V. 2, №2. P. 121–145. [*435*]

[268] Da Prato, G., Frankowska, H., *Existence, uniqueness, and regularity of the invariant measure for a class of elliptic degenerate operators.* Differ. Integ. Equ. 2004. V. 17, №7-8. P. 737–750. [*401*]

[269] Da Prato, G., Frankowska, H., *Invariant measure for a class of parabolic degenerate equations.* Nonlin. Differ. Eq. Appl. 2005. V. 12, №4. P. 481–501. [*401*]

[270] Da Prato, G., Goldys, B., *Elliptic operators on $\mathbb{R}^d$ with unbounded coefficients.* J. Differ. Equ. 2001. V. 172, №2. P. 333–358. [*234*]

[271] Da Prato, G., Lunardi, A., *Maximal dissipativity of a class of elliptic degenerate operators in weighted L^2 spaces.* Discrete Contin. Dyn. Syst. Ser. B. 2006. V. 6, №4. P. 751–760. [*401*]

[272] Da Prato, G., Lunardi, A., *On a class of self-adjoint elliptic operators in L^2 spaces with respect to invariant measures.* J. Differ. Equ. 2007. V. 234, №1. P. 54–79. [*234*]

[273] Da Prato, G., Lunardi, A., *On a class of degenerate elliptic operators in L^1 spaces with respect to invariant measures.* Math. Z. 2007. V. 256, №3. P. 509–520. [*401*]

[274] Da Prato, G., Lunardi, A., *Sobolev regularity for a class of second order elliptic PDE's in infinite dimension.* Ann. Probab. 2014. V. 42, №5. P. 2113–2160. [*416*]

[275] Da Prato, G., Röckner, M., *Well posedness of Fokker–Planck equations for generators of time-inhomogeneous Markovian transition probabilities.* Atti Accad. Naz. Lincei Cl. Sci. Fis. Mat. Natur. Rend. Lincei (9) Mat. Appl. 2012. V. 23, №4. P. 361–376. [*435*]

[276] Da Prato, G., Röckner, M., Rozovskii, B. L., Wang, F.-Y., *Strong solutions of stochastic generalized porous media equations: existence, uniqueness, and ergodicity.* Comm. Partial Differ. Equ. 2006. V. 31. P. 277–291. [*433*]

[277] Da Prato, G., Röckner, M., Wang, F.-Y., *Singular stochastic equations on Hilbert spaces: Harnack inequalities for their transition semigroups.* J. Funct. Anal. 2009. V. 257, №4. P. 992–1017. [*435*]

[278] Da Prato, G., Vespri, V., *Maximal L^p regularity for elliptic equations with unbounded coefficients.* Nonlinear Anal. Ser. A. 2002. V. 49, №6. P. 747–755. [*234*]

[279] Da Prato, G., Zabczyk, J., *Stochastic equations in infinite dimensions.* Cambridge University Press, Cambridge, 1992; xviii+454 p. [*434*]

[280] Da Prato, G., Zabczyk, J., *Regular densities of invariant measures in Hilbert spaces.* J. Funct. Anal. 1995. V. 130, №2. P. 427–449. [*416*]

[281] Da Prato, G., Zabczyk, J., *Ergodicity for infinite-dimensional systems.* Cambridge University Press, Cambridge, 1996; xii+339 p. [*218, 416, 427, 433*]

[282] Dalecky (Daletskiĭ), Yu. L., Fomin, S. V., *Measures and differential equations in infinite-dimensional space.* Kluwer, Dordrecht, 1991; xvi+337 p. (Russian ed.: Moscow, 1983). [*435*]

[283] Davies, E. B., *L^1 properties of second order elliptic operators.* Bull. London Math. Soc. 1992. V. 17, №5. P. 417–436. [*234*]

[284] Davies, E. B., *Heat kernels and spectral theory.* Cambridge University Press, Cambridge, 1989; x+197 p. [*313, 335*]

[285] Davies, P. L., *Rates of convergence to the stationary distribution for k-dimensional diffusion processes.* J. Appl. Probab. 1986. V. 23. P. 370–384. [*235*]

[286] De Giorgi, E., *Sulla differenziabilità e l'analiticità delle estremali degli integrali multipli regolari.* Mem. Accad. Sci. Torino. CI. Sci. Fis. Mat. Nat. (3). 1957. V. 3. P. 25–43. [*35*]

[287] De Lellis, C., *Notes on hyperbolic systems of conservation laws and transport equations.* In: Handbook of differential equations: evolutionary equations. Vol. III, pp. 277–382, Handb. Differ. Equ., Elsevier/North-Holland, Amsterdam, 2007. [*384*]

[288] De Lellis, C., *ODEs with Sobolev coefficients: the Eulerian and the Lagrangian approach.* Discrete Contin. Dyn. Syst. Ser. S. 2008. V. 1, №3. P. 405–426. [*384*]

[289] Debussche, A., *Ergodicity results for the stochastic Navier–Stokes equations: an introduction.* Lecture Notes in Math. 2013. V. 2073. P. 23–108. [*433*]

[290] Debussche, A., Romito, M., *Existence of densities for the 3D Navier–Stokes equations driven by Gaussian noise.* Probab. Theory Related Fields. 2014. V. 158, №3-4. P. 575–596. [*433*]

[291] Dechant, A., Lutz, E., Barkai, E., Kessler, D. A., *Solution of the Fokker–Planck equation with a logarithmic potential.* J. Stat. Phys. 2011. V. 145, №6. P. 1524–1545. [*284*]

[292] Delbrouck, L. E. N., *On stochastic boundedness and stationary measures for Markov processes.* J. Math. Anal. Appl. 1971. V. 33. P. 149–162. [*235*]

[293] Demengel, F., Demengel, G., *Functional spaces for the theory of elliptic partial differential equations.* Springer, London; EDP Sciences, Les Ulis, 2012; xviii+465 p. [*51*]

[294] Denisov, V. N., *On the behaviour of solutions of parabolic equations for large values of time.* Uspehi Matem. Nauk. 2005. V. 60, №4. P. 145–212 (in Russian); English transl.: Russian Math. Surveys. 2005. V. 60, №4. P. 721–790. [*313*]

[295] Denisov, V. N., *Stabilization of solution of Cauchy problem for a non-divergent parabolic equation.* J. Math. Sci. (New York). 2013. V. 189, №2. P. 188–222. [*313*]

[296] Dermoune, A., Filali, S., *Estimates of the transition density of a gas system.* J. Math. Pures Appl. (9). 2004. V. 83, №11. P. 1353–1371. [*313*]

[297] Desjardins, B., *A few remarks on ordinary differential equations.* Comm. Partial Differ. Equ. 1996. V. 21, №11-12. P. 1667–1703. [*384*]

[298] Desvillettes, L., Villani, C., *On the trend to global equilibrium in spatially inhomogeneous entropy-dissipating systems: the linear Fokker–Planck equation.* Comm. Pure Appl. Math. 2001. V. 54, №1. P. 1–42. [*285*]

[299] Deuschel, J.-D., Stroock, D. W., *Hypercontractivity and spectral gap of symmetric diffusions with applications to the stochastic Ising models.* J. Funct. Anal. 1990. V. 92, №1. P. 30–48. [*228*]

[300] Di Francesco, M., Polidoro, S., *Schauder estimates, Harnack inequality and Gaussian lower bound for Kolmogorov-type operators in non-divergence form.* Adv. Differ. Equ. 2006. V. 11, №11. P. 1261–1320. [*401*]

[301] DiBenedetto, E., Gianazza, U., Vespri, V., *Harnack's inequality for degenerate and singular parabolic equations.* Springer, New York, 2012; xiv+278 p. [*127*]

[302] DiPerna, R. J., Lions, P. L., *On the Fokker–Planck–Boltzmann equation.* Comm. Math. Phys. 1988. V. 120. P. 1–23. [*340*]

[303] DiPerna, R. J., Lions, P. L., *On the Cauchy problem for Boltzmann equations: global existence and weak stability.* Ann. Math. 1989. V. 130, №2. P. 321–366. [*340*]

[304] DiPerna, R. J., Lions, P. L., *Ordinary differential equations, transport theory and Sobolev spaces.* Invent. Math. 1989. V. 98. P. 511–547. [*285, 340, 373, 383*]

[305] Dobrushin, R. L., *Vlasov equations.* Funk. Anal. i Pril. 1979. V. 13, №2. P. 48–58 (in Russian); English transl.: Funct. Anal. Appl. 1979. V. 13. P. 115–123. [*284, 285*]

[306] Doeblin, W., *Sur les propriétés asymptotiques de mouvement régis par certains types de chaines simples.* Bull. Math. Soc. Roumanie Sci. 1937. V. 39, №1. P. 57–115; №2. P. 3–61. [*235*]

[307] Dolbeault, J., Nazaret, B., Savaré, G., *From Poincaré to logarithmic Sobolev inequalities: a gradient flow approach.* SIAM J. Math. Anal. 2012. V. 44, №5. P. 3186–3216. [*401*]

[308] Dong, H., *Solvability of second-order equations with hierarchically partially BMO coefficients.* Trans. Amer. Math. Soc. 2012. V. 364, №1. P. 493–517. [*8*]

[309] Dong, H., *Parabolic equations with variably partially VMO coefficients.* St. Petersburg Math. J. 2012. V. 23, №3. P. 521–539. [*8*]

[310] Doob, J. L., *Asymptotic properties of Markoff transitions probabilities.* Trans. Amer. Math. Soc. 1948. V. 63. P. 394–421. [*218, 235*]

[311] Doob, J. L., *Stochastic processes.* Wiley, New York, 1953; viii+654 p. [*235*]

[312] Doss, H., Royer, G., *Processus de diffusion associés aux mesures de Gibbs sur* R^{Z^d}. Z. Wahr. theor. verw. Geb. 1979. B. 46. S. 125–158. [*434*]

[313] Dragoni, F., Kontis, V., Zegarliński, B., *Ergodicity of Markov semigroups with Hörmander type generators in infinite dimensions.* Potential Anal. 2012. V. 37, №3. P. 199–227. [*435*]

[314] Dreyer, W., Huth, R., Mielke, A., Rehberg, J., Winkler, M., *Global existence for a nonlocal and nonlinear Fokker–Planck equation.* Z. Angew. Math. Phys. 2015. V. 66, №2. P. 293–315. [*284*]

[315] Dubins, L. E., Freedman, D. A., *Invariant probabilities for certain Markov processes.* Ann. Math. Statist. 1966. V. 37. P. 837–848. [*235*]

[316] Dunford, N., Schwartz, J. T., *Linear operators, I. General Theory.* Interscience, New York, 1958; xiv+858 p. [*216, 219*]

[317] Durrett, R., *Reversible diffusion processes.* Probability theory and harmonic analysis, Papers of Mini-Conference (Chao J.-A. and Woyczynski W.A., eds., Cleveland/Ohio 1983). P. 67–89. Pure Appl. Math., Marcel Dekker, 1986. [*236*]

[318] Dym, H., *Stationary measures for the flow of a linear differential equation driven by white noise.* Trans. Amer. Math. Soc. 1966. V. 123. P. 130–164. [*235*]

[319] E, W., Mattingly, J. C., Sinai, Ya., *Gibbsian dynamics and ergodicity for the stochastically forced Navier–Stokes equation.* Commun. Math. Phys. 2001. V. 224, №1. P. 83–106. [*433*]

[320] Eberle, A., *Uniqueness and non-uniqueness of singular diffusion operators.* Lecture Notes in Math. V. 1718, Springer, Berlin, 1999; 262 p. [*208, 405*]

[321] Eberle, A., L^p *uniqueness of non-symmetric diffusion operators with singular drift coefficients. I. The finite-dimensional case.* J. Funct. Anal. 2000. V. 173, №2. P. 328–342. [*208*]

[322] Echeverria, P., *A criterion for invariant measures of Markov processes.* Z. Wahr. theor. verw. Geb. 1982. B. 61, №1. P. 1–16. [*235*]

[323] Eckmann, J.-P., Hairer, M., *Non-equilibrium statistical mechanics of strongly anharmonic chains of oscillators.* Commun. Math. Phys. 2000. V. 212, №1. P. 105–164. [*434*]

[324] Eckmann, J.-P., Hairer, M., *Invariant measures for stochastic partial differential equations in unbounded domains.* Nonlinearity. 2001. V. 14, №1. P. 133–151. [*433*]

[325] Egorov, Y., Kondratiev, V., *On spectral theory of elliptic operators.* Birkhäuser, Basel, 1996; x+328 p. [*51*]

[326] Èĭdel'man, S. D., *Parabolic systems.* London North-Holland, Amsterdam – London; Wolters-Noordhoff, Groningen, 1969; v+469 p. (Russian ed.: Moscow, 1964). [*283*]

[327] Eidelman, S. D., Ivasyshen, S. D., Kochubei, A. N., *Analytic methods in the theory of differential and pseudo-differential equations of parabolic type.* Birkhäuser, Basel, 2004; ix+387 p. *[283]*

[328] Eidelman, S. D., Zhitarashu, N. V., *Parabolic boundary value problems.* Birkhäuser, Basel, 1998; x+298 p. *[283, 313, 335]*

[329] Einstein, A., *Zur Theorie der Brownschen Bewegung.* Ann. Phys. 1906. B. 19. S. 371–381. *[51]*

[330] ter Elst, A. F. M., Robinson, D. W., Sikora, A., *Flows and invariance for degenerate elliptic operators.* J. Aust. Math. Soc. 2011. V. 90, №3. P. 317–339. *[401]*

[331] Elworthy, K. D., *Stochastic differential equations on manifolds.* Cambridge University Press, Cambridge – New York, 1982; xiii+326 p. *[236]*

[332] Elworthy, K. D., Le Jan, Y., Li, X.-M., *On the geometry of diffusion operators and stochastic flows.* Lecture Notes in Math. V. 1720. Springer-Verlag, Berlin, 1999; ii+118 p. *[236]*

[333] Elworthy, K. D., Le Jan, Y., Li, X.-M., *The geometry of filtering.* Birkhäuser, Basel, 2010; xii+169 p. *[236]*

[334] Elworthy, K. D., Li, X.-M., *Formulae for the derivatives of heat semigroups.* J. Funct. Anal. 1994. V. 125, №1. P. 252–286. *[236]*

[335] Engelking, P., *General topology.* Polish Sci. Publ., Warszawa, 1977; 626 p. *[211]*

[336] Englefield, M. J., *Exact solutions of a Fokker–Planck equation.* J. Statist. Phys. 1988. V. 52, №1-2. P. 369–381. *[284]*

[337] Epstein, Ch. L., Mazzeo, R., *Degenerate diffusion operators arising in population biology.* Princeton University Press, Princeton, 2013; 320 p. *[401]*

[338] Es-Sarhir, A., *Sobolev regularity of invariant measures for generalized Ornstein–Uhlenbeck operators.* Infin. Dimens. Anal. Quantum Probab. Related Top. 2006. V. 9, №4. P. 595–606. *[416]*

[339] Es-Sarhir, A., *Existence and uniqueness of invariant measures for a class of transition semigroups on Hilbert spaces.* J. Math. Anal. Appl. 2009. V. 353, №2. P. 497–507. *[434]*

[340] Es-Sarhir, A., Stannat, W., *Invariant measures for semilinear SPDE's with local Lipschitz drift coefficients and applications.* J. Evol. Equ. 2008. V. 8, №1. P. 129–154. *[433]*

[341] Escauriaza, L., *Weak type-$(1,1)$ inequalities and regularity properties of adjoint and normalized adjoint solutions to linear nondivergence form operators with VMO coefficients.* Duke Math. J. 1994. V. 74. P. 177–201. *[49, 52]*

[342] Escauriaza, L., *Bounds for the fundamental solution of elliptic and parabolic equations in nondivergence form.* Comm. Partial Differ. Equ. 2000. V. 25, №5-6. P. 821–845. *[49, 52]*

[343] Escauriaza, E., Kenig, C. E., *Area integral estimates for solutions and normalized adjoint solutions to nondivergence form elliptic equations.* Ark. Mat. 1993. V. 31. P. 275–296. *[52]*

[344] Evans, C., Gariepy, R. F., *Measure theory and fine properties of functions.* CRC Press, Boca Raton – London, 1992; viii+268 p. *[51, 156, 167]*

[345] Fabes, E. B., Garofalo, N., *Parabolic B.M.O. and Harnack's inequality.* Proc. Amer. Math. Soc. 1985. V. 95, №1. P. 63–69. *[334]*

[346] Fabes, E. B., Kenig, C. E., *Examples of singular parabolic measures and singular transition probability densities.* Duke Math. J. 1981. V. 48. P. 845–856. *[254]*

[347] Fabes, E. B., Kenig, C. E., Serapioni, R. P., *The local regularity of solutions of degenerate elliptic equations.* 1982. V. 7, №1. P. 77–116. *[29]*

[348] Fabes, E. B., Stroock, D. W., *The L^p-integrability of Green's functions and fundamental solutions for elliptic and parabolic equations.* Duke Math. J. 1984. V. 51. P. 977–1016. *[52]*

[349] Fabes, E. B., Stroock, D. W., *A new proof of Moser's parabolic Harnack inequality using the old ideas of Nash.* Arch. Rational Mech. Anal. 1986. V. 96, №4. P. 327–338. *[311, 335]*

[350] Fang, S., Shao, J., *Fokker–Planck equation with respect to heat measures on loop groups.* Bull. Sci. Math. 2011. V. 135, №6-7. P. 775–794. *[435]*

[351] Farkas, B., Lunardi, A., *Maximal regularity for Kolmogorov operators in L^2 spaces with respect to invariant measures.* J. Math. Pures Appl. (9). 2006. V. 86, №4. P. 310–321. *[234]*

[352] Fattler, T., Grothaus, M., *Strong Feller properties for distorted Brownian motion with reflecting boundary condition and an application to continuous N-particle systems with singular interactions.* J. Funct. Anal. 2007. V. 246, №2. P. 217–241. *[235]*

[353] Fattler, T., Grothaus, M., *Construction of elliptic diffusions with reflecting boundary condition and an application to continuous N-particle systems with singular interactions.* Proc. Edinb. Math. Soc. (2). 2008. V. 51, №2. P. 337–362. *[235]*

[354] Feller, W., *Two singular diffusion problems.* Ann. Math. (2). 1951. V. 54. P. 173–182. [*401*]

[355] Feller, W., *The parabolic differential equations and the associated semigroups of transformations.* Ann. Math. (2). 1952. V. 55. P. 468–519. [*284*]

[356] Feller, W., *Diffusion processes in one dimension.* Trans. Amer. Math. Soc. 1954. V. 77. P. 1–31. [*401*]

[357] Feller, W., *Generalized second order differential operators and their lateral conditions.* Illinois J. Math. 1957. V. 1. P. 459–504. [*401*]

[358] Ferrario, B., *Ergodic results for stochastic Navier–Stokes equation.* Stochastics Stoch. Rep. 1997. V. 60, №3–4. P. 271–288. [*433*]

[359] Ferrario, B., *The Bénard problem with random perturbations: dissipativity and invariant measures.* Nonlin. Differ. Equ. Appl. 1997. V. 4, №1. P. 101–121. [*433*]

[360] Ferrario, B., *Invariant measures for a stochastic Kuramoto–Sivashinsky equation.* Stoch. Anal. Appl. 2008. V. 26, №2. P. 379–407. [*433*]

[361] Ferretti, E., Safonov, M. V., *Growth theorems and Harnack inequality for second order parabolic equations.* Harmonic analysis and boundary value problems. Contemp. Math., V. 277. P. 87–112. Amer. Math. Soc., Providence, Rhode Island, 2001. [*335*]

[362] Figalli, A., *Existence and uniqueness of martingale solutions for SDEs with rough or degenerate coefficients.* J. Funct. Anal. 2008. V. 254, №1. P. 109–153. [*277, 340, 373, 378, 400*]

[363] Flandoli, F., *Dissipativity and invariant measures for stochastic Navier–Stokes equations.* Nonlin. Differ. Equ. Appl. 1994. V. 1, №4. P. 403–423. [*433*]

[364] Flandoli, F., *Regularity theory and stochastic flows for parabolic SPDEs.* Gordon and Breach Sci. Publ., Yverdon, 1995; x+79 p. [*433*]

[365] Flandoli, F., *Irreducibility of the 3-D stochastic Navier–Stokes equation.* J. Funct. Anal. 1997. V. 149, №1. P. 160–177. [*433*]

[366] Flandoli, F., *Random perturbation of PDEs and fluid dynamic models.* Lecture Notes in Math. V. 2015. Springer, Heidelberg, 2011; x+176 p. [*433*]

[367] Flandoli, F., Gatarek, D., *Martingale and stationary solutions for stochastic Navier–Stokes equations.* Probab. Theory Related Fields. 1995. V. 102, №3. P. 367–391. [*433*]

[368] Flandoli, F., Maslowski, B., *Ergodicity of the 2-D Navier–Stokes equation under random perturbations.* Comm. Math. Phys. 1995. V. 172, №1. P. 119–141. [*433*]

[369] Foguel, S. R., *Existence of invariant measures for Markov processes.* Proc. Amer. Math. Soc. 1962. V. 13. P. 833–838. [*235*]

[370] Foguel, S. R., *Markov processes with stationary measure.* Pacific J. Math. 1962. V. 12. P. 505–510. [*235*]

[371] Foguel, S. R., *An L_p theory for a Markov process with a sub-invariant measure.* Proc. Amer. Math. Soc. 1965. V. 16. P. 398–406. [*235*]

[372] Foguel, S. R., *Existence of invariant measures for Markov processes. II.* Proc. Amer. Math. Soc. 1966. V. 17. P. 387–389. [*235*]

[373] Foguel, S. R., *Positive operators on $C(X)$.* Proc. Amer. Math. Soc. 1969. V. 22. P. 295–297. [*235*]

[374] Foguel, S. R., *The ergodic theory of Markov processes.* Van Nostrand, 1969; v+102 p. [*235*]

[375] Foguel, S. R., *The ergodic theory of positive operators on continuous functions.* Ann. Sc. Norm. Sup. Pisa (3). 1973. V. 27. P. 19–51. [*235*]

[376] Foguel, S. R., *A new approach to the study of Harris type Markov operators.* Rocky Mountain J. Math. 1989. V. 19, №2. P. 491–512. [*235*]

[377] Fokker, A. D., *Die mittlere Energie rotierender elektrischer Dipole im Strahlungsfeld.* Ann. Phys. 1914. B. 348 (4. Folge 43). S. 810–820. [ix, *51, 52*]

[378] Föllmer, H., *Random fields and diffusion processes.* Lect. Notes in Math. 1988. V. 1362. P. 101–203. [*236, 434, 435*]

[379] Föllmer, H., Wakolbinger, A., *Time reversal of infinite-dimensional diffusions.* Stoch. Process. Appl. 1986. V. 22. P. 59–77. [*435*]

[380] Fornaro, S., Fusco, N., Metafune, G., Pallara, D., *Sharp upper bounds for the density of some invariant measures.* Proc. Roy. Soc. Edinburgh Sect. A. 2009. V. 139, №6. P. 1145–1161. [*127*]

[381] Fornaro, S., Vespri, V., *Harnack estimates for non-negative weak solutions of a class of singular parabolic equations.* Manuscripta Math. 2013. V. 141, №1-2. P. 85–103. [*335*]

[382] Foss, M., Hrusa, W. J., Mizel, V. J., *The Lavrentiev gap phenomenon in nonlinear elasticity.* Arch. Rational Mech. Anal. 2003. V. 167, №4. P. 337–365. [*164*]

[383] Fraenkel, L. E., *An introduction to maximum principles and symmetry in elliptic problems.* Cambridge University Press, Cambridge, 2000; x+340 p. [*78*]

[384] Frank, T. D., *Nonlinear Fokker–Planck equations. Fundamentals and applications.* Springer-Verlag, Berlin, 2005; xii+404 p. [*284*]

[385] Friedman, A., *On the uniqueness of the Cauchy problem for parabolic equations.* Amer. J. Math. 1959. V. 81. P. 503–511. [*339*]

[386] Friedman, A., *Partial differential equations of parabolic type.* Prentice-Hall, Englewood Cliffs, New Jersey, 1964; xiv+347 p. [*266, 283, 339, 340, 372*]

[387] Fritz, J., *Stationary measures of stochastic gradient systems, infinite lattice models.* Z. Wahr. theor. verw. Geb. 1982. B. 59. S. 479–490. [*434*]

[388] Fritz, J., *On the stationary measures of anharmonic systems in the presence of a small thermal noise.* J. Statist. Phys. 1986. V. 44, №1-2. P. 25–47. [*434*]

[389] Fritz, J., Funaki, T., Lebowitz, J. L., *Stationary states of random Hamiltonian systems.* Probab. Theory Related Fields. 1994. V. 99, №2. P. 211–236. [*434*]

[390] Fujita, Y., *On a critical role of Ornstein–Uhlenbeck operators in the Poincaré inequality.* Differ. Integ. Equ. 2006. V. 19, №12. P. 1321–1332. [*228*]

[391] Fukushima, M., *Dirichlet forms and Markov processes.* Amsterdam – New York, North Holland, 1980; x+196 p. [*234*]

[392] Fukushima, M., *Energy forms and diffusion processes.* In: Mathematics and Physics (Streit L., ed.). V. 1. P. 65–97. World Sci., Singapore, 1984. [*235*]

[393] Fukushima, M., Oshima, Y., Takeda, M., *Dirichlet forms and symmetric Markov processes.* 2nd ed. Walter de Gruyter, Berlin – New York, 2011; x+489 p. [*235*]

[394] Fuller, A. T., *Analysis of nonlinear stochastic systems by means of the Fokker–Planck equation.* Internat. J. Control. 1969. V. 9, №6. P. 603–655. [*52*]

[395] Funaki, T., *A certain class of duffusion processes associated with nonlinear parabolic equations.* Z. Wahr. theor. verw. Geb. 1984. B. 67. S. 331–348. [*284*]

[396] Galkin, V. A., *The Smoluchowski equation.* Fizmatlit, Moscow, 2001; 336 p. (in Russian). [*284*]

[397] Gallot, S., Hulin, D., Lafontaine, J., *Riemannian geometry.* 3d ed. Springer–Verlag, Berlin, 2004; xvi+322 p. [*72*]

[398] Ganidis, H., Roynette, B., Simonot, F., *Convergence rate of some semi-groups to their invariant probability.* Stoch. Processes Appl. 1999. V. 79, №2. P. 243–263. [*235*]

[399] Gardiner, C., *Stochastic methods. A handbook for the natural and social sciences.* 4th ed. Springer-Verlag, Berlin, 2009; xviii+447 p. [*284*]

[400] Garroni, M. G., Menaldi, J. L., *Second order elliptic integro-differential problems.* Chapman & Hall/CRC, Boca Raton, Florida, 2002; xvi+221 p. [*51*]

[401] Gatarek, D., Goldys, B., *On invariant measures for diffusions on Banach spaces.* Potential Anal. 1997. V. 7, №2. P. 539–553. [*433*]

[402] Gaveau, B., Moulinier, J.-M., *Régularité des mesures et perturbations stochastiques de champs de vecteurs sur des espaces de dimension infinie.* Publ. Res. Inst. Math. Sci. 1985. V. 21, №3. P. 593–616. [*416*]

[403] Gehring, F. W., *The L^p-integrability of the partial derivatives of a quasiconformal mapping.* Acta Math. 1973. V. 130. P. 265–277. [*28*]

[404] Geissert, M., Lunardi, A., *Invariant measures and maximal L^2 regularity for nonautonomous Ornstein–Uhlenbeck equations.* J. London Math. Soc. (2). 2008. V. 77, №3. P. 719–740. [*234*]

[405] Gerlach, M., Nittka, R., *A new proof of Doob's theorem.* J. Math. Anal. Appl. 2012. V. 388, №2. P. 763–774. [*218*]

[406] Getoor, R. K., *Transience and recurrence of Markov processes.* Lecture Notes in Math. 1980. V. 784. P. 397–409. [*235*]

[407] Gihman, I. I., Skorohod, A. V., *Stochastic differential equations and their applications.* Naukova Dumka, Kiev, 1982; 611 p. (in Russian). [*51, 284*]

[408] Gikhman, I. I., Skorokhod, A. V., *The theory of stochastic processes. Vol. II.* Springer-Verlag, Berlin, 2004, vii+441 p. (Russian ed.: Moscow, 1973). [*13*]

[409] Gilbarg, D., Trudinger, N. S., *Elliptic partial differential equations of second order.* Springer-Verlag, Berlin – New York, 1977; x+401 p. [*6, 7, 11, 23, 49, 51, 57, 100, 143*]

[410] Girsanov, I. V., *On transforming a certain class of stochastic processes by absolutely continuous substitution of measures.* Teor. Veroyatn. i Primenen. 1961. V. 5, №3. P. 314–330 (in Russian); English transl.: Theory Probab. Appl. 1960. V. 5. P. 285–301. [*374*]

[411] Gliklikh, Y. E., *Global and stochastic analysis with applications to mathematical physics.* Springer-Verlag London, London, 2011; xxiv+436 p. [*435*]

[412] Gol'dshteĭn, V. M., Reshetnyak, Yu. G., *Quasiconformal mappings and Sobolev spaces.* Kluwer, Dordrecht, 1990; xx+371 p. (Russian ed.: Moscow, 1983). [*51*]

[413] Goldys, B., Maslowski, B., *Uniform exponential ergodicity of stochastic dissipative systems.* Czechoslovak Math. J. 2001. V. 51 (126), №4. P. 745–762. [*433*]

[414] Goldys, B., Maslowski, B., *Exponential ergodicity for stochastic Burgers and 2D Navier–Stokes equations.* J. Funct. Anal. 2005. V. 226, №1. P. 230–255. [*433*]

[415] Goldys, B., Maslowski, B., *Lower estimates of transition densities and bounds on exponential ergodicity for stochastic PDE's.* Ann. Probab. 2006. V. 34, №4. P. 1451–1496. [*416*]

[416] Goldys, B., van Neerven, J. M. A. M., *Transition semigroups of Banach space-valued Ornstein–Uhlenbeck processes.* Acta Appl. Math. 2003. V. 76, №3. P. 283–330. [*434*]

[417] Gomes, D. A., Saude, J., *Mean field games models — a brief survey.* Dyn. Games Appl. 2014. V. 4. P. 110–154. [*398, 399*]

[418] Gong, G., *The finite invariant measures and the reversible measures of the diffusions with the Ventzel's boundary conditions in the circular disc.* Acta Math. Sci. 1984. V. 4. P. 451–461. [*236*]

[419] Gozlan, N., *Integral criteria for transportation-cost inequalities.* Electron. Commun. Probab. 2006. V. 11. P. 64–77. [*170*]

[420] Grasman, J., van Herwaarden, O. A., *Asymptotic methods for the Fokker-Planck equation and the exit problem in applications.* Springer-Verlag, Berlin, 1999; x+220 p. [*284*]

[421] Gray, A. H., Jr., *Uniqueness of steady-state solutions to the Fokker–Planck equation.* J. Math. Phys. 1965. V. 6. P. 644–647. [*400*]

[422] Greiner, G., *Spektrum und Asymptotik stark stetiger Halbgruppen positiver Operatoren.* Sitzungsber. Heidelb. Akad. Wiss. Math.-Natur. Kl. 1982, №3. S 55–80. [*214*]

[423] Grigoryan, A. A., *Stochastically complete manifolds and summable harmonic functions.* Izv. Akad. Nauk SSSR. Ser. Mat. 1988. V. 52, №5. P. 1102–1108 (in Russian); English transl.: Math. USSR Izvestiya. 1989. V. 33. P. 425–432. [*72, 119, 160*]

[424] Grigor'yan, A., *Analytic and geometric background of recurrence and non-explosion of the Brownian motion on Riemannian manifolds.* Bull. Amer. Math. Soc. (N.S.). 1999. V. 36. P. 135–249. [*161, 236*]

[425] Grigor'yan, A., *Heat kernel and analysis on manifolds.* Amer. Math. Soc., Providence, Rhode Island, 2009; xvii+482 p. [*313, 335*]

[426] Gross, L., *Logarithmic Sobolev inequalities and contractivity properties of semigroups.* Lecture Notes in Math. 1993. V. 1563. P. 54–88. [*228*]

[427] Guéant, O., Lasry, J.-M., Lions, P.-L., *Mean field games and applications.* Lecture Notes in Math. 2011. V. 2003. P. 205–266. [*397*]

[428] Guillin, A., Léonard, C., Wu, L.-M., Yao, N., *Transportation-information inequalities for Markov processes.* Probab. Theory Related Fields. 2009. V. 144, №3-4. P. 669–695. [*401*]

[429] Guillin, A., Wang, F.-Y., *Degenerate Fokker–Planck equations: Bismut formula, gradient estimate and Harnack inequality.* J. Diff. Eq. 2012. V. 253, №1. P. 20–40. [*381, 401*]

[430] Gushchin, A. K., *On the interior smoothness of solutions of second-order elliptic equations.* Sib. Mat. Zh. 2005. V. 46, №5. P. 1036–1052 (in Russian); English transl.: Siberian Math. J. 2005. V. 46, №5. P. 826–840. [*52*]

[431] Gwiazda, P., Jamróz, G., Marciniak-Czochra, A., *Models of discrete and continuous cell differentiation in the framework of transport equation.* SIAM J. Math. Anal. 2012. V. 44, №2. P. 1103–1133. [*384*]

[432] Gyöngy, I., Krylov, N., *Existence of strong solutions for Itô's stochastic equations via approximations.* Probab. Theory Related Fields. 1996. V. 105, №2. P. 143–158. [*16*]

[433] Gyrya, P., Saloff-Coste, L., *Neumann and Dirichlet heat kernels in inner uniform domains.* Asterisque №336. 2011; viii+144 p. [*335*]

[434] Hairer, M., Mattingly, J. C., *Spectral gaps in Wasserstein distances and the 2D stochastic Navier–Stokes equations.* Ann. Probab. 2008. V. 36, №6. P. 2050–2091. [*433*]

[435] Hairer, M., Mattingly, J. C., *Yet another look at Harris' ergodic theorem for Markov chains.* In: Seminar on Stochastic Analysis, Random Fields and Applications VI, pp. 109–117, Progr. Probab., V. 63. Birkhäuser – Springer Basel, Basel, 2011. [*235*]

[436] Hairer, M., Mattingly, J. C., *A theory of hypoellipticity and unique ergodicity for semilinear stochastic PDEs.* Electron. J. Probab. 2011. V. 16, №23. P. 658–738. [*433*]

[437] Hairer, M., Mattingly, J. C., Scheutzow, M., *Asymptotic coupling and a general form of Harris' theorem with applications to stochastic delay equations.* Probab. Theory Related Fields. 2011. V. 149, №1-2. P. 223–259. [*235*]

[438] Han, Q., Lin, F., *Elliptic partial differential equations.* 2nd ed. Courant Inst. Math. Sci., New York; Amer. Math. Soc., Providence, Rhode Island, 2011; x+147 p. [*51*]

[439] Handa, K., *Quasi-invariance and reversibility in the Fleming–Viot process.* Probab. Theory Related Fields. 2002. V. 122, №4. P. 545–566. [*435*]

[440] Haroske, D., Triebel, H., *Distributions, Sobolev spaces, elliptic equations.* European Math. Soc., Zürich, 2008; ix+294 p. [*51*]

[441] Harris, T. E., *The existence of stationary measures for certain Markov processes.* In: Proc. 3rd Berkeley Symp. Math. Stat. Probab., 1954–1955, vol. II, pp. 113–124, University of California Press, Berkeley and Los Angeles, 1956. [*235*]

[442] Harris, T. E., *Transient Markov chains with stationary measures.* Proc. Amer. Math. Soc. 1957. V. 8. P. 937–942. [*235*]

[443] Harrison, J. M., Williams, R. J., *Multidimensional reflected Brownian motions having exponential stationary distributions.* Ann. Probab. 1987. V. 15, №1. P. 115–137. [*236*]

[444] Hasminskii, R. Z., *Ergodic properties of reccurent diffusion processes and stabilization of the solution of the Cauchy problem for parabolic equations.* Teor. Verojatn. Primen. 1960. V. 5, №2. P. 196–214 (in Russian); English tranls.: Theory Probab. Appl. 1960. V. 5. P. 179–196. [*79*]

Hasminskii=Khasminskii

[445] Hauray, M., Le Bris, C., *A new proof of the uniqueness of the flow for ordinary differential equations with BV vector fields.* Ann. Mat. Pura Appl. (4). 2011. V. 190, №1. P. 91–103. [*384*]

[446] Haussmann, U. G., Pardoux, E., *Time reversal of diffusions.* Ann. Probab. 1986. V. 14, №4. P. 1188–1205. [*236*]

[447] Helffer, B., *Spectral theory and its applications.* Cambridge University Press, Cambridge, 2013; vi+255 p. [*51*]

[448] Helffer, B., Nier, F., *Hypoelliptic estimates and spectral theory for Fokker–Planck operators and Witten Laplacians.* Lecture Notes in Math. V. 1862. Springer-Verlag, Berlin, 2005; x+209 p. [*401*]

[449] Hérau, F., Nier, F., *Isotropic hypoellipticity and trend to equilibrium for the Fokker–Planck equation with a high-degree potential.* Arch. Ration. Mech. Anal. 2004. V. 171, №2. P. 151–218. [*401*]

[450] Hernandez-Lerma, O., Lasserre, J. B., *Invariant probabilities for Feller–Markov chains.* J. Appl. Math. Stoch. Anal. 1995. V. 8, №4. P. 341–345. [*235*]

[451] Hernandez-Lerma, O., Lasserre, J. B., *Existence of bounded invariant probability densities for Markov chains.* Statist. Probab. Lett. 1996. V. 28, №4. P. 359–366. [*235*]

[452] Hervé, R. M., *Recherches axiomatiques sur la théorie des fonctions surharmoniques et du potentiel.* Ann. Inst. Fourier (Grenoble). 1962. V. 12. P. 415–571. [*52*]

[453] Hieber, M., Lorenzi, L., Prüss, J., Rhandi, A., Schnaubelt, R., *Global properties of generalized Ornstein–Uhlenbeck operators on $L^p(\mathbb{R}^N, \mathbb{R}^N)$ with more than linearly growing coefficients.* J. Math. Anal. Appl. 2009. V. 350, №1. P. 100–121. [*234*]

[454] Hille, E., Phillips, R. S., *Functional analysis and semi-groups.* Amer. Math. Soc., Providence, Rhode Island, 1974; xii+808 p. [*179*]

[455] Hino, M., *Existence of invariant measures for diffusion processes on a Wiener space.* Osaka J. Math. 1998. V. 35, №3. P. 717–734. [*416, 427, 434*]

[456] Hino, M., *Exponential decay of positivity preserving semigroups on L^p.* Osaka J. Math. 2000. V. 37, №3. P. 603–624; Correction: ibid. 2002. V. 39, №3. P. 771. [*228*]

[457] Hino, M., Ramírez, J. A., *Small-time Gaussian behavior of symmetric diffusion semigroups.* Ann. Probab. 2003. V. 31, №3. P. 1254–1295. [*228*]

[458] Holley, R. A., Stroock, D. W., *In one and two dimensions, every stationary measure for a stochastic Ising model is a Gibbs state.* Comm. Math. Phys. 1977. V. 55, №1. P. 37–45. [*434*]

[459] Holley, R., Stroock, D., *Diffusions on an infinite dimensional torus.* J. Funct. Anal. 1981. V. 42. P. 29–63. [*435*]

[460] Hopf, E., *Statistical hydromechanics and functional calculus.* J. Rational Mech. Anal. 1952. V. 1. P. 87–123. [*405*]

[461] Hörmander, L., *The analysis of linear partial differential operators. I–IV.* Springer-Verlag, Berlin – New York, 1983, 1985; ix+391 p., viii+391 p., viii+525 p., vii+352 p. [*51, 283, 401*]

[462] Horowitz, S., *Some limit theorems for Markov processes.* Israel J. Math. 1968. V. 6. P. 107–118. [*235*]

[463] Horowitz, S., *Transition probabilities and contractions of L_∞.* Z. Wahr. theor. verw. Geb. 1972. B. 24. S. 263–274. [*235*]

[464] Horowitz, S., *Semi-groups of Markov operators.* Ann. Inst. H. Poincaré Sect. B (N.S.). 1974. V. 10. P. 155–166. [*235*]

[465] Horowitz, S., *Pointwise convergence of the iterates of a Harris-recurrent Markov operator.* Israel J. Math. 1979. V. 33, №3-4. P. 177–180. [*235*]

[466] Hostinský, B., *Application du calcul des probabilités à la théorie du mouvement brownien.* Annales Inst. H. Poincaré. 1932. V. 3. P. 1–74. [*52*]

[467] Huang, M., Malhame, R. P., Caines, P. E., *Large population stochastic dynamic games: closed-loop McKean–Vlasov systems and the Nash certainty equivalence principle.* Commun. Inf. Syst. 2006. V. 6, №3. P. 221–251. [*397*]

[468] Huang, M., Malhame, R. P., Caines, P. E., *Large-population cost-coupled LQG problems with nonuniform agents: individual-mass behavior and decentralized ε-Nash equilibria.* IEEE Trans. Autom. Control. 2007. V. 52, №9. P. 1560–1571. [*397*]

[469] Hwang, C.-R., Hwang-Ma, S.-Y., Sheu, S.-J., *Accelerating diffusions.* Ann. Appl. Probab. 2005. V. 15, №2. P. 1433–1444. [*236*]

[470] Ichihara, K., *Some global properties of symmetric diffusion processes.* Publ. Res. Inst. Math. Sci. 1978. V. 14, №2. P. 441–486. [*235*]

[471] Ichihara, K., *Explosion problems for symmetric diffusion processes.* Trans. Amer. Math. Soc. 1986. V. 298, №2. P. 515–536. [*235*]

[472] Ichihara, K., Kunita, H., *A classification of the second order degenerate elliptic operators and its probabilistic characterization.* Z. Wahr. theor. Verw. Geb. 1974. B. 30. S. 235–254; suppl. and corr.: ibid. 1977. B. 39, №1. S. 81–84. [*401*]

[473] Ikeda, N., Watanabe, S., *Stochastic differential equations and diffusion processes.* 2nd ed. North-Holland, Amsterdam, 1989; xvi+555 p. [*16, 377, 401*]

[474] Il'in, A. M., Has'minskiĭ, R. Z., *On an ergodic property of non-homogeneous diffusion processes.* Dokl. Akad. Nauk SSSR. 1962. V. 145. P. 986–988 (in Russian). [*314*]

[475] Il'in, A. M., Has'minskiĭ, R. Z., *Asymptotic behavior of solutions of parabolic equations and an ergodic property of non-homogeneous diffusion processes.* Matem. Sbornik. 1963. V. 60. P. 366–392 (in Russian); English transl. in: Ten papers on functional analysis and measure theory. Amer. Math. Soc. Transl. Ser. 2, V. 49. Amer. Math. Soc., Providence, Rhode Island, 1966. [*314*]

[476] Il'in, A. M., Khasminskii, R. Z., Yin, G., *Singularly perturbed switching diffusions: rapid switchings and fast diffusions.* J. Optim. Theory. Appl. 1999. V. 102. P. 555–591. [*284*]

[477] Isaac, R., *Markov processes and unique stationary probability measures.* Pacific J. Math. 1962. V. 12. P. 273–286. [*235*]

[478] Isaac, R., *A uniqueness theorem for stationary measures of ergodic Markov processes.* Ann. Math. Statist. 1964. V. 35. P. 1781–1786. [*235*]

[479] Isaac, R., *Non-singular recurrent Markov processes have stationary measures.* Ann. Math. Statist. 1964. V. 35. P. 869–871. [*235*]

[480] Ishige, K., Murata, M., *Uniqueness of nonnegative solutions of the Cauchy problem for parabolic equations on manifolds or domains.* Ann. Sc. Norm. Sup. Pisa Cl. Sci. 2001. V. 30, №1. P. 171–223. [*400*]

[481] Isihara, A., *Statistical physics.* Academic Press, New York – London, 1971; xv+438 p. [*284*]

[482] Itô, K., Nisio, M., *On stationary solutions of a stochastic differential equation.* J. Math. Kyoto Univ. 1964. V. 4. P. 1–75. [*235*]

[483] Ito, Y., *Invariant measures for Markov processes.* Trans. Amer. Math. Soc. 1964. V. 110. P. 152–184. [*235*]

[484] Ivanov, A. V., *The Harnack inequality for generalized solutions of second order quasilinear parabolic equations.* Trudy Mat. Inst. Steklov. 1967. V. 102. P. 51–84. (in Russian); English transl. in: Proc. Steklov Inst. Math. 1967. V. 102: Boundary value problems of mathematical physics. V. P. 55–94. Amer. Math. Soc., Providence, Rhode Island, 1970. [*334*]

[485] Ivasyshen, S. D., Medynsky, I. P., *The Fokker–Planck–Kolmogorov equations for some degenerate diffusion processes.* Theory Stoch. Process. 2010. V. 16, №1. P. 57–66. [*401*]

[486] Jabin, P.-E., *Differential equations with singular fields.* J. Math. Pures Appl. (9). 2010. V. 94, №6. P. 597–621. [*384*]

[487] Jacquot, S., *Strong ergodicity results on Wiener space.* Stochastics Stoch. Rep. 1994. V. 51, №1-2. P. 133–154. [*433*]

[488] Jacquot, S., Royer, G., *Ergodicity of stochastic plates.* Probab. Theory Related Fields. 1995. V. 102, №1. P. 19–44. [*433*]

[489] Jain, N., Krylov, N., *Large deviations for occupation times of Markov processes with L_2 semigroups.* Ann. Probab. 2008. V. 36, №5. P. 1611–1641. [*235*]

[490] Jegaraj, T., *Small time asymptotics for Ornstein–Uhlenbeck densities in Hilbert spaces.* Electron. Commun. Probab. 2009. V. 14. P. 552–559. [*434*]

[491] Jikov, V. V., Kozlov, S. M., Oleinik, O. A., *Homogenization of differential operators and integral functionals.* Springer-Verlag, Berlin, 1994; xii+570 p. (Russian ed.: Moscow, 1993). [*285*]

[492] Jin, T., Maz'ya, V., Van Schaftingen, J., *Pathological solutions to elliptic problems in divergence form with continuous coefficients.* C. R. Acad. Sci. Paris. 2009. T. 347, №13-14. P. 773–778. [*35*]

[493] John, F., Nirenberg, L., *On functions of bounded mean oscillation.* Comm. Pure Appl. Math. 1961. V. 14. P. 415–426. [*120*]

[494] Jona-Lasinio, G., Mitter, P. K., *On the stochastic quantization of field theory.* Commun. Math. Phys. 1985. V. 101. P. 406–436. [*434*]

[495] Jordan, R., Kinderlehrer, D., Otto, F., *The variational formula of the Fokker–Planck equation.* SIAM J. Math Anal. 1998. V. 29, №1. P. 1–17. [*275, 285*]

[496] Kabanov, Yu. M., Liptser, R. Sh., Shiryaev, A. N., *On the variation distance for probability measures defined on a filtered space.* Probab. Theory Relat. Fields. 1986. V. 71, №1. P. 19–35. [*392*]

[497] Kakutani, S., Yosida, K., *Operator-theoretical treatment of Markoff's process and mean ergodic theorem.* Ann. Math. 1941. V. 42. P. 188–228. [*235*]

[498] Kantorovich, L. V., Akilov, G. P., *Functional analysis.* Pergamon Press, Oxford – Elmsford, New York, 1982; xiv+589 p. [*210*]

[499] Kassmann, M., *Harnack inequalities: an introduction.* Bound. Value Probl. 2007. Art. ID 81415, 21 pp. [*335*]

[500] Keizer, J., *Statistical thermodynamics of nonequilibrium processes.* Springer-Verlag, Berlin, 1987; 526 p. [*284*]

[501] Kenig, C. E., *Harmonic analysis techniques for second order elliptic boundary value problems.* Amer. Math. Soc., Providence, Rhode Island, 1994; xii+146 p. [*51*]

Khasminskii=Hasminskii

[502] Khasminskii, R. Z., *Stochastic stability of differential equations.* 2nd ed. Springer, Heidelberg, 2012; xviii+339 p. (Russian ed.: Moscow, 1969). [*79, 235*]

[503] Khasminskii, R. Z., Yin, G., *Asymptotic series for singularly perturbed Kolmogorov–Fokker–Planck equations.* SIAM J. Appl. Math. 1996. V. 56, №6. P. 1766–1793. [*284*]

[504] Khasminskii, R. Z., Yin, G., *On transition densities of singularly perturbed diffusions with fast and slow components.* SIAM J. Appl. Math. 1996. V. 56, №6. P. 1794–1819. [*284*]

[505] Khasminskii, R. Z., Yin, G., *Asymptotic behavior of parabolic equations arising from one-dimensional null-recurrent diffusions.* J. Differ. Equ. 2000. V. 161, №1. P. 154–173. [*284*]

[506] Khasminskii, R. Z., Yin, G., *On averaging principles: an asymptotic expansion approach.* SIAM J. Math. Anal. 2004. V. 35, №6. P. 1534–1560. [*284*]

[507] Khasminskii, R. Z., Yin, G., *Limit behavior of two-time-scale diffusions revisited.* J. Differ. Equ. 2005. V. 212, №1. P. 85–113. [*284*]

[508] Kim, J. U., *Invariant measures for the stochastic von Karman plate equation.* SIAM J. Math. Anal. 2005. V. 36, №5. P. 1689–1703 [*434*]

[509] Kim, J. U., *On the stochastic porous medium equation.* J. Differ. Equ. 2006. V. 220, №1. P. 163–194. [*434*]

[510] Kim, J. U., *On the stochastic Benjamin–Ono equation.* J. Differ. Equ. 2006. V. 228, №2. P. 737–768. [*434*]

[511] Kirillov, A. I., *Two mathematical problems of canonical quantization. I, II, III, IV.* Teoret. Mat. Fiz. 1991. V. 87, №1. P. 22–33; 1991. V. 87, №2. P. 163–172; 1992. V. 91, №3. P. 377–395; 1992. V. 93, №2. P. 249–263 (in Russian); English transl.: Theoret. Math. Phys. 1991. V. 87, №1. P. 345–353; 1991. V. 87, №2. P. 447–454; 1992. V. 91, №3. P. 591–603; 1992. V. 93, №2. P. 1251–1261. [x, *409*]

[512] Kirillov, A. I., *Brownian motion with drift in a Hilbert space and its application in integration theory.* Teor. Verojat. i Primen. 1993. V. 38, №3. P. 629–634 (in Russian); Theory Probab. Appl. 1993. V. 38, №3. P. 529–533. [x]

[513] Kirillov, A. I., *Infinite-dimensional analysis and quantum theory as semimartingale calculi.* Uspehi Mat. Nauk. 1994. V. 49, №3. P. 43–92 (in Russian); English transl.: Russian Math. Surveys. 1994. V. 49, №3. P. 43–95. [x, *409*]

[514] Kirillov, A. I., *A field of sine-Gordon type in space-time of arbitrary dimension: the existence of the Nelson measure.* Teoret. Mat. Fiz. 1994. V. 98. №1. P. 12–28 (in Russian); English transl.: Theoret. Math. Phys. 1994. V. 98. №1. P. 8–19. [x, *409*]

[515] Kirillov, A. I., *A field of sine-Gordon type in space-time of arbitrary dimension. II. Stochastic quantization.* Teoret. Mat. Fiz. 1995. V. 105, №2. P. 179–197 (in Russian); English transl.: Theoret. Math. Phys. 1995. V. 105, №2. P. 1329–1345. [x]

[516] Kirillov, A. I., *On the reconstruction of measures from their logarithmic derivatives.* Izv. Ross. Akad. Nauk Ser. Mat. 1995. V. 59, №1. P. 121–138 (in Russian); English transl.: Izv. Math. 1995. V. 59, №1. P. 121–139. [x, *409*]

[517] Kleius, E. M., Frankfurt. U. I., *Max Planck (1858–1947).* Nauka, Moscow, 1980; 392 p. (in Russian). [*50*]

[518] Kliemann, W., *Qualitative theory of stochastic dynamical systems – applications to life sciences.* Bull. Math. Biol. 1983. V. 45. P. 483–506. [*235*]

[519] Kliemann, W., *Recurrence and invariant measures for degenerate diffusions.* Ann. Probab. 1987. V. 15. P. 690–707. [*401*]

[520] Klimontovich, Yu. L., *Statistical physics.* Harwood Acad. Publ., Chur, 1986; xxvi+734 p. [*284*]

[521] Klyatskin, V. I., *Stochastic equations through the eye of the physicist. Basic concepts, exact results and asymptotic approximations.* Elsevier, Amsterdam, 2005; xviii+538 p. [*284*]

[522] Knerr, B., *Parabolic interior Schauder estimates by the maximum principle.* Arch. Rational Mech. Anal. 1980. V. 75. P. 51–58. [*247*]

[523] Knowles, J., *Measures on topological spaces.* Proc. London Math. Soc. 1967. V. 17. P. 139–156. [*219*]

[524] Kobayashi, Sh., Nomizu, K., *Foundations of differential geometry.* V. I, II. John Wiley & Sons, New York, 1996; xii+329 p., xvi+468 p. [*72*]

[525] Kolb, M., *On the strong uniqueness of some finite dimensional Dirichlet operators.* Infin. Dimens. Anal. Quantum Probab. Related Top. 2008. V. 11, №2. P. 279–293. [*208*]

[526] Kolesnikov, A. V., Röckner, M., *On continuity equations in infinite dimensions with non-Gaussian reference measure.* J. Funct. Anal. 2014. V. 266, №7. P. 4490–4537. [*435*]

[527] Kolmogoroff, A. N., *Über die analytischen Methoden in der Wahrscheinlichkeitsrechnung.* Math. Ann. 1931. B. 104. S. 415–458; see also [**530**]. [ix, xi, *52, 399*]

[528] Kolmogoroff, A. N., *Zur Theorie der stetigen zufälligen Prozesse.* Math. Ann. 1933. B. 104. S. 149–160; see also [**530**]. [ix, *52, 399*]

[529] Kolmogoroff, A. N., *Zur Umkehrbarkeit der statistischen Naturgesetze.* Math. Ann. 1937. B. 113. S. 766–772; see also [**530**]. [*52, 174, 236*]

[530] Kolmogorov, A. N., *Selected works.* V. II. Probability theory and mathematical statistics. Kluwer, Dordrecht, 1992; xvi+597 p. [*50*]

[531] Kolokoltsov, V. N., *Markov processes, semigroups and generators.* Walter de Gruyter, Berlin, 2011; xviii+430 p. [*235*]

[532] Konakov, V., Menozzi, S., Molchanov, S., *Explicit parametrix and local limit theorems for some degenerate diffusion processes.* Ann. Inst. H. Poincaré Probab. Stat. 2010. V. 46, №4. P. 908–923. [*401*]

[533] Kondrat'ev, V. A., Landis, E. M., *Qualitative theory of second-order linear partial differential equations. Partial differential equations*, V. 3. P. 99–215. Itogi Nauki i Tekhniki, Akad. Nauk SSSR, Vsesoyuz. Inst. Nauchn. i Tekhn. Inform., Moscow, 1988 (in Russian); English transl.: *Partial Differential Equations III.* Encycl. Math. Sci. 1991. V. 32. P. 87–192. [*51, 283*]

[534] Kondratiev, V., Liskevich, V., Sobol, Z., Us, O., *Estimates of heat kernels for a class of second-order elliptic operators with applications to semi-linear inequalities in exterior domains.* J. London Math. Soc. (2). 2004. V. 69, №1. P. 107–127. [*314*]

[535] Kondratiev, Yu. G., Konstantinov, A. Yu., Röckner, M., *Uniqueness of diffusion generators for two types of particle systems with singular interactions.* J. Funct. Anal. 2004. V. 212, №2. P. 357–372. [*235*]

[536] Kondratiev, Ju. G., Tsycalenko, T. V., *Infinite-dimensional Dirichlet operators. I. Essential selfadjointness and associated elliptic equations.* Potential Anal. 1993. V. 2, №1. P. 1–21. [*435*]

[537] Koshelev, A. I., *Regularity of solutions of elliptic equations and systems.* Nauka, Moscow, 1986; 240 p. (in Russian). [*51*]

[538] Kozlov, R., *On symmetries of the Fokker–Planck equation.* J. Engrg. Math. 2013. V. 82. P. 39–57. [*400*]

[539] Kozlov, V. V., *The generalized Vlasov kinetic equation.* Uspehi Matem. Nauk. 2008. V. 63, №4. P. 93–130 (in Russian); English transl.: Russian Math. Surveys. 2008. V. 63, №4. P. 691–726. [*75, 78, 284, 285*]

[540] Kozlov, V. V., *The Vlasov kinetic equation, dynamics of continuum and turbulence.* Regul. Chaotic Dyn. 2011. V. 16, №6. P. 602–622. [*75, 284, 285*]

[541] Kozlov, V. A., Maz'ya, V. G., Rossmann, J., *Elliptic boundary value problems in domains with point singularities.* Amer. Math. Soc., Providence, Rhode Island, 1997; x+414 p. [*51*]

[542] Kresin, G., Maz'ya, V., *Maximum principles and sharp constants for solutions of elliptic and parabolic systems.* Amer. Math. Soc., Providence, Rhode Island, 2012; viii+317 p. [*51, 78, 283*]

[543] Kružkov, S. N., *A priori bounds and some properties of solutions of elliptic and parabolic equations.* Matem. Sbornik. 1964. V. 65. №4. P. 522–570 (in Russian); English transl.: In: Ten papers on differential equations and functional analysis. Amer. Math. Soc. Trans. Ser. 2. V. 68. P. 169–220. Amer. Math. Soc., Providence, Rhode Island, 1968. [*120*]

[544] Kruzkov, S. N., *First order quasilinear equations with several independent variables.* Matem. Sbornik. 1970. V. 81. P. 228–255 (in Russian); English transl.: Math. USSR Sb. 1970. V. 10. P. 217–243. [*384*]

[545] Kryloff, N., Bogoliouboff, N., *Sur les propriétés ergodiques de l'équation de Smoluchovsky.* Bull. Soc. Math. France. 1936. V. 64. P. 49–56. [*235*]

[546] Kryloff, N., Bogolioùboff, N., *Sur les équations de Focker–Planck déduites dans la théorie des perturbations à l'aide d'une méthode basée sur les propriétés spectrales de l'hamiltonien perturbateur.* (*Application à la mécanique classique et à la mécanique quantique*). Ann. Chaire Phys. Math. Kiev. 1939. V. 4. P. 5–157. [*284*]

[547] Krylov, N. V., *A certain estimate from the theory of stochastic integrals.* Teor. Verojatnost. i Primenen. 1971. V. 16. P. 446–457 (in Russian); English transl.: Theor. Probability Appl. 1971. V. 16. P. 438–448. [*52*]

[548] Krylov, N. V., *Sequences of convex functions, and estimates of the maximum of the solution of a parabolic equation.* Sibirsk. Mat. Z. 1976. V. 17, №2. P. 290–303 (in Russian); English transl.: Siberian Math. J. 1976. V. 17, №2. P. 226–236. [*250, 284*]

[549] Krylov, N. V., *Controlled diffusion processes.* Springer-Verlag, New York, 1980; 308 p. (Russian ed.: Moscow, 1977). [*51, 254, 284*]

[550] Krylov, N. V., *Nonlinear elliptic and parabolic equations of the second order.* Reidel, Dordrecht, 1987; xiv+462 p. (Russian ed.: Moscow, 1985). [*250*]

[551] Krylov, N. V., *Introduction to the theory of diffusion processes.* Amer. Math. Soc., Rhode Island, Providence, 1995; xii+271 p. [*16*]

[552] Krylov, N. V., *Lectures on elliptic and parabolic equations in Hölder spaces.* Amer. Math. Soc., Rhode Island, Providence, 1996; 164 p. [*7, 51, 193, 247, 250, 283*]

[553] Krylov, N. V., *An analytic approach to SPDEs.* In: Stochastic partial differential equations: six perspectives; R. Carmona, B. Rozovskii, eds., pp. 185–241. Amer. Math. Soc., Rhode Island, Providence, 1999. [*246, 247*]

[554] Krylov, N. V., *Some properties of traces for stochastic and deterministic parabolic weighted Sobolev spaces.* J. Funct. Anal. 2001. V. 183, no. 1. P. 1–41. [*246, 248*]

[555] Krylov, N. V., *Parabolic and elliptic equations with VMO coefficients.* Comm. Partial Differ. Equ. 2007. V. 32. P. 453–475. [*8, 344, 345*]

[556] Krylov, N. V., *Lectures on elliptic and parabolic equations in Sobolev spaces*, Amer. Math. Soc., Rhode Island, 2008; xviii+357 p. [*7, 51, 283*]

[557] Krylov, N. V., *Second-order elliptic equations with variably partially VMO coefficients.* J. Funct. Anal. 2009. V. 257, №6. P. 1695–1712. [*8*]

[558] Kufner, A., Sändig, A.-M., *Some applications of weighted Sobolev spaces.* Teubner, Leipzig, 1987; 268 p. [*51*]

[559] Kuksin, S., Shirikyan, A., *Stochastic dissipative PDEs and Gibbs measures.* Comm. Math. Phys. 2000. V. 213, №2. P. 291–330. [*434*]

[560] Kuksin, S., Shirikyan, A., *Mathematics of two-dimensional turbulence.* Cambridge University Press, Cambridge, 2012; xvi+320 p. [*434*]

[561] Kulik, A. M., *Asymptotic and spectral properties of exponentially φ-ergodic Markov processes.* Stoch. Process. Appl. 2011. V. 121, №5. P. 1044–1075. [*235*]

[562] Künsch, H., *Nonreversible stationary measures for infinite interacting particle systems.* Z. Wahr. theor. verw. Geb. 1984. B. 66, №3. S. 407–424. [*434*]

[563] Künsch, H., *Time reversal and stationary Gibbs measures.* Stoch. Process. Appl. 1984. V. 17, №1. P. 159–166. [*434*]

[564] Kunze, M., Lorenzi, L., Lunardi, A., *Nonautonomous Kolmogorov parabolic equations with unbounded coefficients.* Trans. Amer. Math. Soc. 2010. V. 362, №1. P. 169–198. [*284*]

[565] Kuo, H.-J., Trudinger, N. S., *New maximum principles for linear elliptic equations.* Indiana Univ. Math. J. 2007. V. 56, №5. P. 2439–2452. [*50, 52*]

[566] Kushner, H. J., *Stochastic stability and control.* Academic Press, New York, 1967; xiv+161 p. [*235*]

[567] Kushner, H., *Converse theorems for stochastic Liapunov functions.* SIAM J. Control. 1967. V. 5. P. 228–233. [*234*]

[568] Kushner, H. J., *The Cauchy problem for a class of degenerate parabolic equations and asymptotic properties of the related diffusion processes.* J. Differ. Equ. 1969. V. 6. P. 209–231. [*401*]

[569] Kushner, H. J., *Stability and existence of diffusions with discontinuous or rapidly growing drift terms.* J. Differ. Equ. 1972. V. 11. P. 156–168. [*235*]

[570] Kushner, H. J., *Probability methods for approximations in stochastic control and for elliptic equations.* Academic Press, New York – London, 1977; xvii+243 p. [*51*]

[571] Kushner, H. J., *Asymptotic distributions of solutions of ordinary differential equations with wide band noise inputs: approximate invariant measures.* Stochastics. 1981/82. V. 6, №3-4. P. 259–277. [*235*]

[572] Kushner, H. J., Yu, C. F., *The approximate calculation of invariant measures of diffusions via finite difference approximations to degenerate elliptic partial differential equations.* J. Math. Anal. Appl. 1975. V. 51, №2. P. 359–367. [*235*]

[573] Kusuoka, S. (Jr.), *Hölder continuity and bounds for fundamental solutions to nondivergence form parabolic equations.* Analysis & PDE. 2015. V. 8, №1. P. 1–32. [*313, 335*]

[574] Kusuoka, S., Stroock, D., *Some boundedness properties of certain stationary diffusion semigroups.* J. Funct. Anal. 1985. V. 60, №2. P. 243–264. [*234*]

[575] Kuz'menko, Ju. T., Molĉanov, S. A., *Counterexamples to theorems of Liouville type.* Vestnik Moskov. Univ. Ser. I Mat. Mekh. 1979, №6. P. 39–43 (in Russian); English transl.: Vestnik Moscow Univ. Math. 1979. V. 34, №6. P. 35–39. [*72*]

[576] Ladyz'enskaya, O. A., Solonnikov, V. A., Ural'tseva, N. N., *Linear and quasilinear equations of parabolic type.* Amer. Math. Soc., Rhode Island, Providence, 1968; 648 p. (Russian ed.: Moscow, 1967). [*247, 250, 257, 271, 283, 346*]

[577] Ladyzhenskaya, O. A., Ural'tseva, N. N., *Linear and quasilinear elliptic equations.* Academic Press, New York, 1968; xviii+495 p. (Russian 2nd ed.: Moscow, 1973). [*35, 51, 191*]

[578] Lanconelli, E., *Global L^p estimates for degenerate Ornstein–Uhlenbeck operators: a general approach.* Rev. Un. Mat. Argentina. 2011. V. 52, №1. P. 57–72. [*234*]

[579] Lanconelli, E., Polidoro, S., *On a class of hypoelliptic evolution operators.* Rend. Sem. Mat. Univ. Pol. Torino, Partial Diff. Eqs. 1994. V. 52. P. 29–63. [*401*]

[580] Lanconelli, E., Uguzzoni, F., *Potential analysis for a class of diffusion equations: a Gaussian bounds approach.* J. Differ. Equ. 2010. V. 248, №9. P. 2329–2367. [*401*]

[581] Landis, E. M., *Second order equations of elliptic and parabolic type.* Amer. Math. Soc., Providence, Rhode Island, 1998; xii+203 p. (Russian ed.: Moscow, 1971). [*51, 283*]

[582] Lang, R., *On the asymptotic behaviour of infinite gradient systems.* Comm. Math. Phys. 1979. V. 65, №2. P. 129–149. [*434*]

[583] Lasota, A., Myjak, J., Szarek, T., *Markov operators with a unique invariant measure.* J. Math. Anal. Appl. 2002. V. 276, №1. P. 343–356. [*238*]

[584] Lasota, A., Szarek, T., *Lower bound technique in the theory of a stochastic differential equation.* J. Differ. Equ. 2006. V. 231, №2. P. 513–533. [*238*]

[585] Lasota, A., Yorke, J. A., *Lower bound technique for Markov operators and iterated function systems.* Random Comput. Dynam. 1994. V. 2, №1. P. 41–77. [*235*]

[586] Lasry, J.-M., Lions, P.-L., *Jeux a champ moyen. I. Le cas stationnaire.* C. R. Acad. Sci. Paris. Math. 2006. T. 343, №9. P. 619–625. [*397, 399*]

[587] Lasry, J.-M., Lions, P.-L., *Jeux a champ moyen. II. Horizon fini et controle optimal.* C. R. Acad. Sci. Paris. Math. 2006. T. 343, №10. P. 679–684. [*397, 399*]

[588] Lasry, J.-M., Lions, P.-L., *Mean field games.* Japan. J. Math. 2007. V. 2, №1. P. 229–260. [*397*]

[589] Lasserre, J. B., *Existence and uniqueness of an invariant probability for a class of Feller Markov chains.* J. Theoret. Probab. 1996. V. 9, №3. P. 595–612. [*235*]

[590] Lasserre, J. B., *Invariant probabilities for Markov chains on a metric space.* Statist. Probab. Lett. 1997. V. 34, №3. P. 259–265. [*235*]

[591] Lasserre, J. B., *A Lyapunov criterion for invariant probabilities with geometric tail.* Probab. Engrg. Inform. Sci. 1998. V. 12, №3. P. 387–391. [*235*]

[592] Lavrentiev, M., *Sur quelques problèmes du calcul des variations.* Ann. Mat. Pura Appl. 1926. V. 4. P. 107–124. [*164*]

[593] Le Bris, C., Lions, P. L., *Existence and uniqueness of solutions to Fokker–Planck type equations with irregular coefficients.* Comm. Partial Differ. Equ. 2008. V. 33. P. 1272–1317. [*277, 340, 373, 374, 400*]

[594] Ledoux, M., *On an integral criterion for hypercontractivity of diffusion semigroups and extremal functions.* J. Funct. Anal. 1992. V. 105, №2. P. 444–465. [*228*]

[595] Leha, G., Maslowski, B., Ritter, G., *Stability of solutions to semilinear stochastic evolution equations.* Stoch. Anal. Appl. 1999. V. 17, №6. P. 1009–1051. [*434*]

[596] Leha, G., Ritter, G., *Lyapunov-type conditions for stationary distributions of diffusion processes in Hilbert spaces.* Stochastics. 1994. V. 48. P. 195–225. [*434*]

[597] Leha, G., Ritter, G., *Lyapunov functions and stationary distributions of stochastic evolution equations.* Stoch. Anal. Appl. 2003. V. 21, №4. P. 763–799. [*434*]

[598] Leha, G., Ritter, G., Wakolbinger, A., *An improved Lyapunov-function approach to the behavior of diffusion processes in Hilbert spaces.* Stoch. Anal. Appl. 1997. V. 15, №1. P. 59–89. [*434*]

[599] Lemle, L. D., *$L^1(\mathbb{R}^d, dx)$-uniqueness of weak solutions for the Fokker–Planck equation associated with a class of Dirichlet operators.* Elect. Research Announc. Math. Sci. 2008. V. 15. P. 65–70. [*400*]

[600] Lemle, L. D., Wang, R., Wu, L., *Uniqueness of Fokker–Planck equations for spin lattice systems* (I): *compact case.* Semigroup Forum. 2013. V. 86, №3. P. 583–591. [*434*]

[601] Lemle, L. D., Wang, R., Wu, L. M., *Uniqueness of Fokker–Planck equations for spin lattice systems* (II): *non-compact case.* Sci. China Math. 2014. V. 57, №1. P. 161–172. [*434*]

[602] Leoni, G., *A first course in Sobolev spaces.* Amer. Math. Soc., Providence, Rhode Island, 2009; xvi+607 p. [*51*]

[603] Levakov, A. A., Vas'kovskii, M. M., *Existence of weak solutions of stochastic differential equations with discontinuous coefficients and a partially degenerate diffusion operator.* Differ. Uravn. 2007. V. 43, №8. P. 1029–1042 (in Russian); English transl.: Differ. Equ. 2007. V. 43, №8. P. 1051–1066. [*381*]

[604] Li, H., Toscani, G., *Long-time asymptotics of kinetic models of granular flows.* Arch. Rational Mech. Anal. 2004. V. 172. P. 407–428. [*282, 285*]

[605] Li, P., Schoen, R., *L^p and mean value properties of subharmonic functions on Riemannian manifolds.* Acta Math. 1984. V. 153, N 3-4. P. 279–301. [*72, 119, 160*]

[606] Li, X.-D., *Liouville theorems for symmetric diffusion operators on complete Riemannian manifolds.* J. Math. Pures Appl. (9). 2005. V. 84, №10. P. 1295–1361. [*128*]

[607] Li, X.-D., *Perelman's W-entropy for the Fokker–Planck equation over complete Riemannian manifolds.* Bull. Sci. Math. 2011. V. 135, №6-7. P. 871–882. [*284*]

[608] Liboff, R. L., *Introduction to the theory of kinetic equations.* Wiley, New York – Toronto, 1969; viii+397 p. [*284*]

[609] Liberzon, D., Brockett, R. W., *Nonlinear feedback systems perturbed by noise: steady-state probability distributions and optimal control.* IEEE Trans. Automat. Control. 2000. V. 45, №6. P. 1116–1130. [*235*]

[610] Lieberman, G. M., *Intermediate Schauder theory for second order parabolic equations IV: time irregularity and regularity.* Diff. and Integr. Equations. 1992. V. 5, №6. P. 1219–1236. [*247*]

[611] Lieberman, G. M., *Second order parabolic differential equations.* World Sci., Singapore, 1996; 439 p. [*122, 247, 283, 334*]

[612] Liese, F., *Hellinger integrals of diffusion processes.* Statistics. 1986. V. 17, №1. P. 63–78. [*392*]

[613] Liese, F., Schmidt, W., *A note on the convergence of integral functionals of diffusion processes. An application to strong convergence.* Math. Nachr. 1993. B. 161. S. 283–289. [*392*]

[614] Liese, F., Schmidt, W., *On the strong convergence, contiguity and entire separation of diffusion processes.* Stochastics Stoch. Rep. 1994. V. 50, №3-4. P. 185–203. [*392*]

[615] Lifshitz, E. M., Pitaevskii, L. P., *Course of theoretical physics ("Landau–Lifshits").* V. 10. Pergamon Press, Oxford – Elmsford, New York, 1981: xi+452 p. (Russian ed.: Moscow, 1979). [*284*]

[616] Lin, T. F., Huang, M. J., *Poincaré–Chernoff type inequalities for reversible probability measures of diffusion processes.* Soochow J. Math. 1990. V. 16, №1. P. 109–122. [*228*]

[617] Lin, W.-T., Ho, C.-L., *Similarity solutions of the Fokker–Planck equation with time-dependent coefficients.* Ann. Physics. 2012. V. 327, №2. P. 386–397. [*284*]

[618] Lions, J.-L., Magenes, E., *Non-homogeneous boundary value problems and applications.* V. I–III. Springer-Verlag, Heidelberg, 1972, 1973; xvi+357 p., xi+242 p., xii+308 p. [*51, 283*]

[619] Lions, P.-L., Perthame, B., Tadmor E., *A kinetic formulation of multidimensional scalar conservation laws and related equations.* J. Amer. Math. Soc. 1994. V. 7, №1. P. 169–191. [*284*]

[620] Liskevich, V., *On the uniqueness problem for Dirichlet operators.* J. Funct. Anal. 1999. V. 162, №1. P. 1–13. [*208*]

[621] Liskevich, V., Röckner, M., *Strong uniqueness for certain infinite-dimensional Dirichlet operators and applications to stochastic quantization.* Ann. Sc. Norm. Sup. Pisa Cl. Sci. (4). 1998. V. 27, №1. P. 69–91 [*435*]

[622] Liskevich, V., Röckner, M., Sobol, Z., Us, O., *L^p-uniqueness for infinite-dimensional symmetric Kolmogorov operators: the case of variable diffusion coefficients.* Ann. Sc. Norm. Sup. Pisa Cl. Sci. (4). 2001. V. 30, №2. P. 285–309. [*435*]

[623] Liskevich, V. A., Semenov, Yu. A., *Dirichlet operators: a priori estimates and the uniqueness problem.* J. Funct. Anal. 1992. V. 109. P. 199–213. [*208*]

[624] Liskevich, V., Sobol, Z., *Estimates of integral kernels for semigroups associated with second-order elliptic operators with singular coefficients.* Potential Anal. 2003. V. 18, №4. P. 359–390. [*234*]

[625] Liskevich, V., Sobol, Z., Röckner, M., *Dirichlet operators with variable coefficients in L^p-spaces of functions of infinitely many variables coefficients.* Infin. Dimens. Anal. Quantum Probab. Related Top. 1999. V. 2, №4. P. 487–502. [*435*]

[626] Liskevich, V., Sobol, Z., Vogt, H., *On the L_p-theory of C_0-semigroups associated with second-order elliptic operators. II.* J. Funct. Anal. 2002. V. 193, №1. P. 55–76. [*234*]

[627] Liskevich, V., Us, O., *L^p-uniqueness for Dirichlet operators with singular potentials.* J. Evol. Equ. 2002. V. 2, №3. P. 275–298. [*208*]

[628] Littman, W., *A strong maximum principle for weakly L-subharmonic functions.* J. Math. Mech. 1959. V. 8. P. 761–770. [*52*]

[629] Littman, W., *Generalized subharmonic functions: monotonic approximations and an improved maximum principle.* Ann. Sc. Norm. Super. Pisa Cl. Sci. 1963. V. 17. P. 207–222. [*52*]

[630] Liu, L., Shen, Y., *Sufficient and necessary conditions on the existence of stationary distribution and extinction for stochastic generalized logistic system.* Adv. Difference Equ. 2015. N 10. 13 pp. [*79*]

[631] Liu, R., Mandrekar, V., *Ultimate boundedness and invariant measures of stochastic evolution equations.* Stochastics Stoch. Rep. 1996. V. 56, №1-2. P. 75–101. [*434*]

[632] Liu, R., Mandrekar, V., *Stochastic semilinear evolution equations: Lyapunov function, stability, and ultimate boundedness.* J. Math. Anal. Appl. 1997. V. 212, №2. P. 537–553. [*434*]

[633] Liu, W., *Harnack inequality and applications for stochastic evolution equations with monotone drifts.* J. Evol. Equ. 2009. V. 9, №4. P. 747–770. [*435*]

[634] Liu, W., Röckner, M., *Stochastic partial differential equations: an introduction.* Springer, Berlin, 2015; 238 p. [*433*]

[635] Lo, C. F., *Exactly solvable Fokker–Planck equation with time-dependent nonlinear drift and diffusion coefficients — the Lie-algebraic approach.* Eur. Phys. J. B. 2011. V. 84, №1. P. 131–136. [*284*]

[636] Long, H., *Necessary and sufficient conditions for the symmetrizability of differential operators over infinite dimensional state spaces.* Forum Math. 2000. V. 12, №2. P. 167–196. [*408*]

[637] Long, H., Simão, I., *A note on the essential self-adjointness of Ornstein–Uhlenbeck operators perturbed by a dissipative drift and a potential.* Infin. Dimens. Anal. Quantum Probab. Related Top. 2004. V. 7, №2. P. 249–259. [*435*]

[638] Lorenz, T., *Mutational analysis. A joint framework for Cauchy problems in and beyond vector spaces.* Lecture Notes in Math. V. 1996. Springer-Verlag, Berlin, 2010; xiv+509 p. [*284*]

[639] Lorenz, T., *Radon measures solving the Cauchy problem of the nonlinear transport equation.* IWR Preprint http://www.ub.uni-heidelberg.de/archiv/7252, 2007. [*284, 285*]

[640] Lorenzi, L., *Estimates of the derivatives for a class of parabolic degenerate operators with unbounded coefficients in* $\mathbb{R}^N$. Ann. Sc. Norm. Sup. Pisa Cl. Sci. (5). 2005. V. 4, №2. P. 255–293. [*401*]

[641] Lorenzi, L., *Nonautonomous Kolmogorov equations in the whole space: a survey on recent results.* Discrete Contin. Dyn. Syst. Ser. S. 2013. V. 6, №3. P. 731–760. [*314*]

[642] Lorenzi, L., Bertoldi, M., *Analytical methods for Markov semigroups.* Boca Raton, Florida, Chapman & Hall/CRC, 2007; xxxi+526 p. [*208, 234*]

[643] Lorenzi, L., Lunardi, A., *Elliptic operators with unbounded diffusion coefficients in* L^2 *spaces with respect to invariant measures.* J. Evol. Equ. 2006. V. 6, №4. P. 691–709. [*234*]

[644] Lorenzi, L., Zamboni, A., *Cores for parabolic operators with unbounded coefficients.* J. Differ. Equ. 2009. V. 246, №7. P. 2724–2761. [*234*]

[645] Ługiewicz, P., Zegarliński, B., *Coercive inequalities for Hörmander type generators in infinite dimensions.* J. Funct. Anal. 2007. V. 247, №2. P. 438–476. [*435*]

[646] Lunardi, A., *Analytic semigroups and optimal regularity in parabolic problems.* Birkhäuser Verlag, Basel, 1995; xviii+424 p. [*283*]

[647] Lunardi, A., *On the Ornstein–Uhlenbeck operator in* L^2 *spaces with respect to invariant measures.* Trans. Amer. Math. Soc. 1997. V. 349, №1. P. 155–169. [*234*]

[648] Lunardi, A., Metafune, G., *On the domains of elliptic operators in* L^1. Differ. Integral Equ. 2004. V. 17, №1-2. P. 73–97. [*234*]

[649] Lunardi, A., Metafune, G., Pallara, D., *Dirichlet boundary conditions for elliptic operators with unbounded drift.* Proc. Amer. Math. Soc. 2005. V. 133, №9. P. 2625–2635; Erratum: ibid. 2006. V. 134, №8. P. 2479–2480. [*234*]

[650] Luo, D., *Fokker–Planck type equations with Sobolev diffusion coefficients and BV drift coefficients.* Acta Math. Sin. (Engl. Ser.). 2013. V. 29, №2. P. 303–314. [*374, 400*]

[651] Luo, D., *Uniqueness of degenerate Fokker–Planck equations with weakly differentiable drift whose gradient is given by a singular integral.* Electron. Commun. Probab. 2014. V. 19, №43. 14 pp. [*378*]

[652] Lyons, T., *A simple criterion for transience of a reversible Markov chain.* Ann. Probab. 1983. V. 11, №2. P. 393–402. [*235*]

[653] Ma, Z.-M., Röckner, M., *Introduction to the theory of (non-symmetric) Dirichlet forms.* Springer, 1992; 209 p. [*235, 434*]

[654] Makov, Ju. N., Hohlov, R. V., *A class of solutions of the Fokker–Planck equation.* Dokl. Akad. Nauk SSSR. 1976. V. 227, №2. P. 315–317 (in Russian); English transl.: Soviet Physics Dokl. 1976. V. 21, №3. P. 138–139. [*284*]

[655] Malanin, V. V., Poloskov, I. E., *Random processes in nonlinear dymanic systems. Analytic and numeric methods of study. Regular and Chaotic Dynamics*, Moscow – Izhevsk, 2001; 160 p. (in Russian). [*284*]

[656] Malliavin, P., Nualart, E., *Density minoration of a strongly non-degenerated random variable.* J. Funct. Anal. 2009. V. 256, №12. P. 4197–4214. [*110*]

[657] Malyshkin, M. N., *Subexponential estimates for the rate of convergence to the invariant measure for stochastic differential equations.* Teor. Veroyat. i Primenen. 2000. V. 45, №3. P. 489–504 (in Russian); English transl.: Theory Probab. Appl. 2002. V. 45, №3. P. 466–479. [*235*]

[658] Mamontov, E., *Nonstationary invariant distributions and the hydrodynamics-style generalization of the Kolmogorov-forward/Fokker–Planck equation.* Appl. Math. Lett. 2005. V. 18, №9. P. 976–982. [*284*]

[659] Manca, L., *Kolmogorov equations for measures.* J. Evol. Equ. 2008. V. 8, №2. P. 231–262. [*434*]

[660] Manca, L., *Kolmogorov operators in spaces of continuous functions and equations for measures.* Tesi. Scuola Norm. Sup. Pisa. 10. Ed. della Normale, Pisa, 2008; xiv+127 p. [*434*]

[661] Manca, L., *Fokker–Planck equation for Kolmogorov operators with unbounded coefficients.* Stoch. Anal. Appl. 2009. V. 27, №4. P. 747–769. [*434*]

[662] Manca, L., *The Kolmogorov operator associated to a Burgers SPDE in spaces of continuous functions.* Potential Anal. 2010. V. 32, №1. P. 67–99. [*434*]

[663] Maniglia, S., *Probabilistic representation and uniqueness results for measure-valued solutions of transport equation.* J. Math. Pures Appl. 2007. V. 87. P. 601–626. [*284, 383*]

[664] Manita, O. A., Romanov, M. S., Shaposhnikov, S. V., *Uniqueness of solutions to nonlinear Fokker–Planck–Kolmogorov equations.* Dokl. Ross. Akad. Nauk. 2015. V. 461, №1. P. 18–22 (in Russian); English transl.: Dokl. Math. 2015. V. 91, №2. P. 142–146. [*397*]

[665] Manita, O. A., Romanov, M. S., Shaposhnikov, S. V., *On uniqueness of solutions to nonlinear Fokker–Planck–Kolmogorov equations.* ArXiv:1407.8047. 26 pp. 2014. [*397*]

[666] Manita, O. A., Shaposhnikov, S. V., *Nonlinear parabolic equations for measures.* Dokl. Ross. Akad. Nauk. 2012. V. 447, №6. P. 610–614 (in Russian); English transl.: Dokl. Math. 2012. V. 86, №3. P. 857–860. [*278*]

[667] Manita, O. A., Shaposhnikov, S. V., *Nonlinear parabolic equations for measures.* Algebra i Analiz. 2013. V. 25, №1. P. 64–93 (in Russian); English transl.: St.-Petersburg Math. J. 2014. V. 25, №1. P. 43–62. [*278, 280, 282, 286*]

[668] Manita, O. A., Shaposhnikov, S. V., *On the Cauchy problem for Fokker–Planck–Kolmogorov equations with potential terms on arbitrary domains.* J. Dyn. Diff. Equ. 2015. DOI 10.1007/s10884-015-9453-y [*284*]

[669] Manthey, R., Maslowski, B., *Qualitative behaviour of solutions of stochastic reaction-diffusion equations.* Stoch. Process. Appl. 1992. V. 43, №2. P. 265–289. [*434*]

[670] Marton, K., *An inequality for relative entropy and logarithmic Sobolev inequalities in Euclidean spaces.* J. Funct. Anal. 2013. V. 264, №1. P. 34–61. [*401*]

[671] Maruyama, G., Tanaka, H., *Ergodic property of N-dimensional recurrent Markov processes.* Mem. Fac. Sci. Kyushu Univ. Ser. A. 1959. V. 13. P. 157–172. [*235*]

[672] Maslowski, B., *Uniqueness and stability of invariant measures for stochastic differential equations in Hilbert spaces.* Stochastics Stoch. Rep. 1989. V. 28, №2. P. 85–114. [*434*]

[673] Maslowski, B., *Stability of semilinear equations with boundary and pointwise noise.* Ann. Sc. Norm. Sup. Pisa Cl. Sci. (4). 1995. V. 22, №1. P. 55–93. [*434*]

[674] Maslowski, B., Pospíšil, J., *Ergodicity and parameter estimates for infinite-dimensional fractional Ornstein–Uhlenbeck process.* Appl. Math. Optim. 2008. V. 57, №3. P. 401–429. [*434*]

[675] Maslowski, B., Seidler, J., *Ergodic properties of recurrent solutions of stochastic evolution equations.* Osaka J. Math. 1994. V. 31, №4. P. 965–1003. [*434*]

[676] Maslowski, B., Seidler, J., *Invariant measures for nonlinear SPDE's: uniqueness and stability.* Arch. Math. (Brno). 1998. V. 34, №1. P. 153–172. [*434*]

[677] Maslowski, B., Seidler, J., *Probabilistic approach to the strong Feller property.* Probab. Theory Related Fields. 2000. V. 118, №2. P. 187–210. [*434*]

[678] Maslowski, B., Simão, I., *Asymptotic properties of stochastic semilinear equations by the method of lower measures.* Colloq. Math. 1997. V. 72, №1. P. 147–171. [*434*]

[679] Matthes, D., Jüngel, A., Toscani, G., *Convex Sobolev inequalities derived from entropy dissipation.* Arch. Ration. Mech. Anal. 2011. V. 199, №2. P. 563–596. [*228*]

[680] Mauceri, G., Noselli, L., *Riesz transforms for a non-symmetric Ornstein–Uhlenbeck semigroup.* Semigroup Forum. 2008. V. 77, №3. P. 380–398. [*234*]

[681] Maugeri, A., Palagachev, D. K., Softova, L. G., *Elliptic and parabolic equations with discontinuous coefficients.* Wiley-VCH Verlag, Berlin, 2000; 256 p. [*51, 283*]

[682] Maz'ja, V. G., *Sobolev spaces.* Springer-Verlag, Berlin, 1985; xix+486 p. [*51*]

[683] Maz'ya, V., *Sobolev spaces with applications to elliptic partial differential equations.* Springer, Heidelberg, 2011; xxviii+866 p. [*51*]

[684] Maz'ya, V., McOwen, R., *Asymptotics for solutions of elliptic equations in double divergence form.* Comm. Partial Differ. Equ. 2007. V. 32, №1-3. P. 191–207. [*52*]

[685] Maz'ya, V., Rossmann, J., *Elliptic equations in polyhedral domains.* Amer. Math. Soc., Providence, Rhode Island, 2010; viii+608 p. [*51*]

[686] McCauley, J. L., *Stochastic calculus and differential equations for physics and finance.* Cambridge University Press, Cambridge, 2013; xii+206 p. [*284*]

[687] McKean, H. P., Jr., *A class of Markov processes associated with nonlinear parabolic equations.* Proc. Nat. Acad. Sci. U.S.A. 1966. V. 56. P. 1907–1911. [*284*]

[688] McKean, H. P., Jr., *Propagation of chaos for a class of non-linear parabolic equations.* In: Stochastic Differential Equations (Lecture Series in Differential Equations, Session 7, Catholic Univ., 1967), pp. 41–57, Air Force Office Sci. Res., Arlington, Virginia. [*284*]

[689] Metafune, G., *L^p-spectrum of Ornstein–Uhlenbeck operators.* Ann. Sc. Norm. Sup. Pisa Cl. Sci. (4). 2001. V. 30, №1. P. 97–124. [*234*]

[690] Metafune, G., Ouhabaz, E. M., Pallara, D., *Long time behavior of heat kernels of operators with unbounded drift terms.* J. Math. Anal. Appl. 2011. V. 377, №1. P. 170–179. [*229*]

[691] Metafune, G., Pallara, D., *Discreteness of the spectrum for some differential operators with unbounded coefficients in* R^n. Atti Accad. Naz. Lincei Cl. Sci. Fis. Mat. Natur. Rend. Lincei (9) Mat. Appl. 2000. V. 11, №1. P. 9–19. [*234*]

[692] Metafune, G., Pallara, D., Priola, E., *Spectrum of Ornstein–Uhlenbeck operators in L^p spaces with respect to invariant measures.* J. Funct. Anal. 2002. V. 196, №1. P. 40–60. [*234*]

[693] Metafune, G., Pallara, D., Prüss, J., Schnaubelt, R., *L^p-theory for elliptic operators on $\mathbb{R}^d$ with singular coefficients.* Z. Anal. Anwendungen. 2005. V. 24, №3. S. 497–521. [*234*]

[694] Metafune, G., Pallara, D., Rhandi, A., *Global regularity of invariant measures.* J. Funct. Anal. 2005. V. 223. P. 396–424. [*127*]

[695] Metafune, G., Pallara, D., Rhandi, A., *Global properties of transition probabilities of singular diffusions.* Theory Probab. Appl. 2009. V. 54, №1. P. 116–148. [*228, 246, 292*]

[696] Metafune, G., Pallara, D., Vespri, V., *L^p-estimates for a class of elliptic operators with unbounded coefficients in* $\mathbf{R}^N$. Houston J. Math. 2005. V. 31, №2. P. 605–620 [*234*]

[697] Metafune, G., Pallara, D., Wacker, M., *Compactness properties of Feller semigroups.* Studia Math. 2002. V. 153, №2. P. 179–206. [*234*]

[698] Metafune, G., Pallara, D., Wacker, M., *Feller semigroups on* $\mathbf{R}^N$. Semigroup Forum. 2002. V. 65, №2. P. 159–205. [*234*]

[699] Metafune, G., Priola, E., *Some classes of non-analytic Markov semigroups.* J. Math. Anal. Appl. 2004. V. 294, №2. P. 596–613. [*234*]

[700] Metafune, G., Prüss, J., Rhandi, A., Schnaubelt, R., *The domain of the Ornstein–Uhlenbeck operator on an L^p-space with invariant measure.* Ann. Sc. Norm. Sup. Pisa Cl. Sci. (5). 2002. V. 1, №2. P. 471–485. [*234*]

[701] Metafune, G., Prüss, J., Schnaubelt, R., Rhandi, A., *L^p-regularity for elliptic operators with unbounded coefficients.* Adv. Differ. Equ. 2005. V. 10, №10. P. 1131–1164. [*234*]

[702] Metafune, G., Spina, Ch., *Elliptic operators with unbounded diffusion coefficients in L^p spaces.* Ann. Sc. Norm. Sup. Pisa Cl. Sci. (5). 2012. V. 11, №2. P. 303–340. [*234*]

[703] Metafune, G., Spina, Ch., *Heat kernel estimates for some elliptic operators with unbounded diffusion coefficients.* Discrete Contin. Dyn. Syst. 2012. V. 32, №6. P. 2285–2299. [*229*]

[704] Metafune, G., Spina, C., Tacelli, C., *Elliptic operators with unbounded diffusion and drift coefficients in L^p spaces.* Adv. Differential Equat. 2014. V. 19, №5-6. P. 473–526. [*234*]

[705] Metafune, G., Spina, C., Tacelli, C., *On a class of elliptic operators with unbounded diffusion coefficients.* Evol. Equ. Control Theory. 2014. V. 3, №4. P. 671–680. [*234*]

[706] Meyer, J., Schröter, J., *Proper and normal solutions of the Fokker–Planck equation.* Arch. Rational Mech. Anal. 1981. V. 76, №3. P. 193–246. [*284*]

[707] Meyer, P.-A., Zheng, W. A., *Construction de processus de Nelson réversibles.* Lecture Notes in Math. 1985. V. 1123. P. 12–26. [*235*]

[708] Meyn, S., Tweedie, R. L., *Markov chains and stochastic stability.* Prologue by Peter W. Glynn. 2nd ed. Cambridge University Press, Cambridge, 2009; xviii+594 p. [*235*]

[709] Mikami, T., *Dynamical systems in the variational formulation of the Fokker–Planck equation by the Wasserstein metric.* Appl. Math. Optim. 2000. V. 42. P. 203–227. [*314*]

[710] Mikami, T., *Semimartingales from the Fokker–Planck equation.* Appl. Math. Optim. 2006. V. 53, №2. P. 209–219. [*314*]

[711] Millet, A., Nualart, D., Sanz, M., *Integration by parts and time reversal for diffusion processes.* Ann. Probab. 1989. V. 17, №1. P. 208–238. [*236*]

[712] Millet, A., Nualart, D., Sanz, M., *Time reversal for infinite-dimensional diffusions.* Probab. Theory Related Fields. 1989. V. 82, №3. P. 315–347. [*236*]

[713] Miranda, C., *Partial differential equations of elliptic type.* 2nd ed. Springer-Verlag, New York – Berlin, 1970; xii+370 p. [*51*]

[714] Mitidieri, È., Pokhozhaev, S. I., *A priori estimates and the absence of solutions of nonlinear partial differential equations and inequalities.* Trudy Matem. Inst. Steklova, 2001. V. 234. P. 1–384 (in Russian); English transl.: Proc. Steklov Inst. Math. 2001. №3 (234). P. 1–362. [*282*]

[715] Mitidieri, È, Pokhozhaev, S. I., *Liouville theorems for some classes of nonlinear nonlocal problems.* Trudy Mat. Inst. Steklova. 2005. V. 248. P. 164–184 (in Russian); English transl.: Proc. Steklov Inst. Math. 2005. V. 248, №1. P. 158–178. [*282*]

[716] Miyahara, Y., *Ultimate boundedness of the systems governed by stochastic differential equations.* Nagoya Math. J. 1972. V. 47. P. 111–144. [*79*]

[717] Miyahara, Y., *Invariant measures of ultimately bounded stochastic processes.* Nagoya Math. J. 1973. V. 49. P. 149–153. [*235*]

[718] Miyahara, Y., *Infinite-dimensional Langevin equation and Fokker–Planck equation.* Nagoya Math. J. 1981. V. 81. P. 177–223. [*434*]

[719] Miyazawa, T., *Theory of the one-variable Fokker–Planck equation.* Phys. Rev. A (3). 1989. V. 39, №3. P. 1447–1468. [*284*]

[720] Modica, L., Mortola, S., *Construction of a singular elliptic-harmonic measure.* Manuscr. Math. 1980. V. 33. P. 81–98. [*30*]

[721] Molčanov, S. A., *Diffusion processes and Riemannian geometry.* Uspehi Matem. Nauk. 1975. V. 30, №1. P. 3–59 (in Russian); English transl.: Russian Math. Surveys. 1975. V. 30, №1. P. 1–63. [*236*]

[722] Montanari, A., *Harnack inequality for totally degenerate Kolmogorov–Fokker–Planck operators.* Boll. Un. Mat. Ital. B. (7). 1996. V. 10, №4. P. 903–926. [*401*]

[723] Morrey, C. B., *Multiple integrals in the calculus of variations.* Springer-Verlag, Berlin – Heidelberg – New York, 1966; x+506 p. [*47, 52*]

[724] Moser, J., *On Harnack's theorem for elliptic differential equations.* Comm. Pure Appl. Math. 1961. V. 14. P. 577–591; Correction: ibid. 1967. V. 20. P. 231–236. [*127*]

[725] Moser, J., *A Harnack inequality for parabolic differential equations.* Comm. Pure Appl. Math. 1964. V. 17. P. 101–134. [*334*]

[726] Moser, J., *On a pointwise estimate for parabolic differential equations.* Comm. Pure Appl. Math. 1971. V. 24. P. 727–740. [*334*]

[727] Mucha, P. B., *Transport equation: extension of classical results for* $\operatorname{div} b \in BMO$. J. Differ. Equ. 2010. V. 249, №8. P. 1871–1883. [*383*]

[728] Muckenhoupt, B., *The equivalence of two conditions for weight functions.* Studia Math. 1973/74. V. 49. P. 101–106. [*29*]

[729] Mueller, C., *Coupling and invariant measures for the heat equation with noise.* Ann. Probab. 1993. V. 21, №4. P. 2189–2199. [*434*]

[730] Nadirashvili, N. S., *Nonuniqueness in the martingale problem and Dirichlet problem for uniformly elliptic operators.* Ann. Sc. Norm. Sup. Pisa Cl. Sci (4). 1997. V. 24. P. 537–550. [*174*]

[731] Nagasawa, M., *Time reversions of Markov processes.* Nagoya Math. J. 1964. V. 24. P. 177–204. [*236*]

[732] Nagasawa, M., *Schrödinger equations and diffusion theory.* Birkhäuser, Basel, 1993; xii+319 p. [*236*]

[733] Namiki, N., *Stochastic quantization.* Lecture Notes in Physics. Springer-Verlag, Berlin, 1992; x+217 p. [*434*]

[734] Nash, J., *Continuity of solutions of parabolic and elliptic equations.* Amer. J. Math. 1958. V. 80. P. 931–954. [*311, 335*]

[735] Natile, L., Peletier, M. A., Savaré, G., *Contraction of general transportation costs along solutions to Fokker–Planck equations with monotone drifts.* J. Math. Pures Appl. (9). 2011. V. 95, №1. P. 18–35. [*314, 388, 401*]

[736] Nazarov, A. I., Ural'tseva, N. N., *The Harnack inequality and related properties of solutions of elliptic and parabolic equations with divergence-free lower-order coefficients.* Algebra i Analiz. 2011. V. 23, №1. P. 136–168 (in Russian); English transl.: St. Petersburg Math. J. 2012. V. 23, №1. P. 93–115. [*160*]

[737] Nazarov, S. A., Plamenevsky, B. A., *Elliptic problems in domains with piecewise smooth boundaries.* Walter de Gruyter, Berlin, 1994; viii+525 p. [*51*]

[738] van Neerven, J. M. A. M., *Second quantization and the L^p-spectrum of nonsymmetric Ornstein–Uhlenbeck operators.* Infin. Dimens. Anal. Quantum Probab. Related Top. 2005. V. 8, №3. P. 473–495. [*435*]

[739] van Neerven, J. M. A. M., Weis, L., *Invariant measures for the linear stochastic Cauchy problem and R-boundedness of the resolvent.* J. Evol. Equ. 2006. V. 6, №2. P. 205–228. [*434*]

[740] Nelson, E., *The adjoint Markoff process.* Duke Math. J. 1958. V. 25. P. 671–690. [*235*]

[741] Nevel'son, M. B., *The behaviour of an invariant measure of a diffusion process with small diffusion on a circle.* Teor. Verojatn. i Primen. 1964. V. 9, №1. P. 139–146 (in Russian); English transl.: Theory Probab. Appl. 1964. V. 9, №1. P. 125–131. [*236*]

[742] Neveu, J., *Mathematical foundations of the calculus of probability.* Holden-Day, San Francisco, 1965; 231 p. [*54*]

[743] Neveu, J., *Existence of bounded invariant measures in ergodic theory.* In: Proc. 5th Berkeley Symp. Math. Stat. Prob., v. 2, pp. 461–472, Berkeley, 1967. [*235*]

[744] Nier, F., *Hypoellipticity for Fokker–Planck operators and Witten Laplacians.* Lectures on the analysis of nonlinear partial differential equations. Part 1, pp. 31–84, Morningside Lect. Math., 1, Internat. Press, Somerville, Massachusetts, 2012. [*401*]

[745] Noarov, A. I., *On a sufficient condition for the existence of a stationary solution of the Fokker–Planck equation.* Zh. Vychisl. Mat. Mat. Fiz. 1997. V. 37, №5. P. 587–598 (in Russian); English transl.: Comput. Math. Math. Phys. 1997. V. 37, №5. P. 572–583. [*52*]

[746] Noarov, A. I., *Numerical investigation of the Fokker–Planck equation.* Zh. Vychisl. Mat. Mat. Fiz. 1999. V. 39, №8. P. 1337–1347 (in Russian); English transl.: Comput. Math. Math. Phys. 1999. V. 39, №8. P. 1283–1292. [*52*]

[747] Noarov, A. I., *On some diffusion processes with stationary distributions.* Teor. Veroyatn. Primen. 2009. V. 54, №3. P. 589–598 (in Russian); English transl.: Theory Probab. Appl. 2010. V. 54, №3. P. 525–533. [*52, 65*]

[748] Noarov, A. I., *On the justification of a projection method for the stationary Fokker–Planck equation.* Zh. Vychisl. Mat. Mat. Fiz. 2011. V. 51, №4. P. 647–653 (in Russian); English transl.: Comput. Math. Math. Phys. 2011. V. 51, №4. P. 602–608. [*52*]

[749] Noarov, A. I., *Stationary diffusion processes with discontinuous drift coefficients.* Algebra i Analiz. 2012. V. 24, №5. P. 141–164 (in Russian); English transl.: St. Petersburg Math. J. 2013. V. 24, №5. P. 795–809. [*52, 66*]

[750] Nualart, D., *The Malliavin calculus and related topics.* 2nd ed. Springer-Verlag, Berlin, 2006; xiv+382 p. [*401*]

[751] Nualart, E., *Exponential divergence estimates and heat kernel tail.* C. R. Math. Acad. Sci. Paris. 2004. T. 338, №1. P. 77–80. [*110*]

[752] Neummelin, E., *General irreducible Markov chains and non-negative operators.* Cambridge University Press, Cambridge, 1984; xi+156 p. [*235*]

[753] O'Connell N., Ortmann, J., *Product-form invariant measures for Brownian motion with drift satisfying a skew-symmetry type condition.* ALEA Lat. Am. J. Probab. Math. Stat. 2014. V. 11, №1. P. 307–329. [*236*]

[754] Oleĭnik, O. A., *On the smoothness of solutions of degenerating elliptic and parabolic equations.* Dokl. Akad. Nauk SSSR. 1965. V. 163. P. 577–580 (in Russian); English transl.: Soviet Math. Dokl. 1965. V. 6. P. 972–976. [*400*]

[755] Oleĭnik, O. A., *Alcuni risultati sulle equazioni lineari e quasi lineari ellittico-paraboliche a derivate parziali del secondo ordine.* Atti Accad. Naz. Lincei Rend. Cl. Sci. Fis. Mat. Natur. (8). 1966. V. 40, P. 775–784. [*400*]

[756] Oleĭnik, O. A., Kružkov, S. N., *Quasi-linear second-order parabolic equations with many independent variables.* Uspehi Mat. Nauk. 1961. V. 16, №5. P. 115–155 (in Russian); English transl.: Russian Math. Surveys. 1961. V. 16, №5. P. 105–146. [*120*]

[757] Oleĭnik, O.A., Radkevič, E. V., *Second order equations with nonnegative characteristic form.* Plenum Press, New York – London, 1973; vii+259 p. [*51, 283*]

[758] Otto, F., Villani, C., *Generalization of an inequality by Talagrand and links with the logarithmic Sobolev inequality.* J. Funct. Anal. 2000. V. 173, №2. P. 361–400. [*170*]

[759] Otto, F., Weber H., Westdickenberg, M. G., *Invariant measure of the stochastic Allen–Cahn equation: the regime of small noise and large system size.* Electron. J. Probab. 2014. V. 19, №23. P. 1–76. [*434*]

[760] Otto, F., Westdickenberg, M., *Eulerian calculus for the contraction in the Wasserstein distance.* SIAM J. Math. Anal. 2005. V. 37. P. 1227–1255. [*401*]

[761] Ouédraogo, A., De Dieu Zabsonré, J., *Continuous dependence of renormalized solution for nonlinear degenerate parabolic problems in the whole space.* Mediterr. J. Math. 2014. V. 11, №3. P. 873–880. [*284*]

[762] Panov, E. Yu., *On existence and uniqueness of entropy solutions to the Cauchy problem for a conservation law with discontinuous flux.* J. Hyperbolic Differ. Equ. 2009. V. 6, №3. P. 525–548. [*384*]

[763] Panov, E. Yu., *Renormalized entropy solutions of the Cauchy problem for a first-order nonhomogeneous quasilinear equation.* Matem. Sbornik. 2013. V. 204, №10. P. 91–126 (in Russian); English transl.: Sb. Math. V. 204, №9-10. P. 1480–1515. [*384*]

[764] Pardoux, E., Veretennikov, A. Yu., *On the Poisson equation and diffusion approximation. I.* Ann. Probab. 2001. V. 29, №3. P. 1061–1085; II. Ibid. 2003. V. 31, №3. P. 1166–1192; III. Ibid. 2005. V. 33, №3. P. 1111–1133. [*125, 127, 235, 397*]

[765] Pardoux, E., Williams, R. J., *Symmetric reflected diffusions.* Ann. Inst. H. Poincaré Probab. Statist. 1994. V. 30, №1. P. 13–62. [*236*]

[766] Pavliotis, G. A., *Stochastic processes and applications. Diffusion processes, the Fokker–Planck and Langevin equations.* Springer, New York, 2014; xiv+339 p. [*284*]

[767] Pazy, A., *Semigroups of linear operators and applications to partial differential equations.* Springer-Verlag, New York, 1983; viii+279 p. [*183, 234*]

[768] Peeters, A. G., Strintzi, D., *The Fokker–Planck equation, and its application in plasma physics.* Ann. Phys. (8). 2008. V. 17, №2-3. P. 142–157. [*284*]

[769] Peetre, J., Rus, I. A., *Sur la positivité de la fonction de Green.* Math. Scand. 1967. V. 21. P. 80–89. [*79*]

[770] Peletier, M. A., Renger, D. R. M., Veneroni, M., *Variational formulation of the Fokker–Planck equation with decay: a particle approach.* Commun. Contemp. Math. 2013. V. 15, №5. 43 pp. [*275, 285*]

[771] Perthame, B., *Kinetic formulation of conservation laws.* Oxford University Press, Oxford, 2002; xii+198 p. [*284*]

[772] Peszat, S., Zabczyk, J., *Strong Feller property and irreducibility for diffusions on Hilbert spaces.* Ann. Probab. 1995. V. 23, №1. P. 157–172. [*434*]

[773] Peters, G., *Anticipating flows on the Wiener space generated by vector fields of low regularity.* J. Funct. Anal. 1996. V. 142, №1. P. 129–192. [*434*]

[774] Piccoli, B., Rossi, F., *Generalized Wasserstein distance and its application to transport equations with source.* Arch. Ration. Mech. Anal. 2014. V. 211, №1. P. 335–358. [*384*]

[775] Pichór K., Rudnicki, R., *Stability of Markov semigroups and applications to parabolic systems.* J. Math. Anal. Appl. 1997. V. 215. P. 56–74. [*235*]

[776] Pichór K., Rudnicki, R., *Asymptotic behaviour of Markov semigroups and applications to transport equations.* Bull. Polish Acad. Sci. Math. 1997. V. 45, №4. P. 379–397. [*235*]

[777] Pinchover, Y., *On uniqueness and nonuniqueness of positive Cauchy problem for parabolic equations with unbounded coefficients.* Math. Z. 1996. B. 233. S. 569–586. [*400*]

[778] Pinsky, M., Pinsky, R. G., *Transience/recurrence and central limit theorem behavior for diffusions in random temporal environments.* Ann. Probab. 1993. V. 21, №1. P. 433–452. [*235*]

[779] Pinsky, R., *A classification of diffusion processes with boundaries by their invariant measures.* Ann. Probab. 1985. V. 13. P. 693–697. [*236*]

[780] Pinsky, R. G., *Positive harmonic functions and diffusion.* Cambridge University Press, Cambridge, 1995; xvi+474 p. [*52, 235, 236*]

[781] Planck, M., *Über einen Satz der statistischen Dynamik und seine Erweiterung in der Quantentheorie.* Sitzungber.Preussischen Akad. Wissenschaften. 1917. S. 324–341. [ix, *52*]

[782] Polidoro, S., *Uniqueness and representation theorems for solutions of Kolmogorov–Fokker–Planck equations.* Rend. Mat. Appl. (7). 1995. V. 15, №4. P. 535–560. [*401*]

[783] Polidoro, S., *A global lower bound for the fundamental solution of Kolmogorov–Fokker–Planck equations.* Arch. Rational Mech. Anal. 1997. V. 137, №4. P. 321–340. [*401*]

[784] Polidoro, S., Ragusa, M. A., *Harnack inequality for hypoelliptic ultraparabolic equations with a singular lower order term.* Rev. Mat. Iberoam. 2008. V. 24, №3. P. 1011–1046. [*401*]

[785] Porper, F. O., Eidelman, S. D., *Two-sided estimates of fundamental solutions of second-order parabolic equations, and some applications.* Uspehi Matem. Nauk. 1984. V. 39, №3. P. 107–156 (in Russian); English transl.: Russian Math. Surveys. 1984. V. 39, №3. P. 119–178. [*313, 335, 372*]

[786] Porper, F. O., Eidelman, S. D., *Properties of solutions of second order parabolic equations with lower order terms.* Trudy Mosk. Matem. Ob. 1992. V. 54. P. 118–159 (in Russian); English transl.: Trans. Mosc. Math. Soc. 1993. P. 101–137. [*313, 335*]

[787] Porretta, A., *Weak solutions to Fokker-Planck equations and mean field games.* Arch. Ration. Mech. Anal. 2015. V. 216, №1, P. 1–62. [*399*]

[788] Portenko, N. I., *Generalized diffusion processes.* Amer. Math. Soc., Providence, Rhode Island, 1990; x+180 p. (Russian ed.: Kiev, 1982). [*254*]

[789] Portenko, N. I., Skorokhod, A. V., Shurenkov, V. M., *Markov processes.* Itogi Nauki i Tekhniki, Akad. Nauk SSSR. V. 46. P. 5–245. VINITI, Moscow, 1989 (in Russian). [*235*]

[790] Prévôt, C., Röckner, M., *A concise course on stochastic partial differential equations.* Lecture Notes in Math. V. 1905. Springer, Berlin, 2007; 144 p. [*433*]

[791] Priola, E., Wang, F.-Y., *Gradient estimates for diffusion semigroups with singular coefficients.* J. Funct. Anal. 2006. V. 236, №1. P. 244–264. [*234*]

[792] Protter, M. H., Weinberger, H. F., *Maximum principles in differential equations.* Prentice-Hall, Englewood Cliffs, New Jersey, 1967; x+261 p. [*78*]

[793] Prüss, J., Rhandi, A., Schnaubelt, R., *The domain of elliptic operators on $L^p(\mathbb{R}^d)$ with unbounded drift coefficients.* Houston J. Math. 2006. V. 32, №2. P. 563–576. [*234*]

[794] Pucci, P., Serrin, J., *The maximum principle.* Birkhäuser, Basel, 2007; x+235 p. [*78*]

[795] Qian, Zh., *On conservation of probability and the Feller property.* Ann. Probab. 1996. V. 24, №1. P. 280–292. [*236*]

[796] Quastel, J., Varadhan, S. R. S., *Diffusion semigroups and diffusion processes corresponding to degenerate divergence form operators.* Comm. Pure Appl. Math. 1997. V. 50, №7. P. 667–706. [*401*]

[797] Quittner, P., Souplet, Ph., *Superlinear parabolic problems. Blow-up, global existence and steady states.* Birkhäuser, Basel, 2007; xii+584 p. [*284*]

[798] Radkevich, E. V., *Equations with nonnegative characteristics form. I, II.* J. Math. Sci. (New York). 2009. V. 158, №3. P. 297–452; №4. P. 453–604. [*51*]

[799] Ramírez, A. F., *Relative entropy and mixing properties of infinite dimensional diffusions.* Probab. Theory Related Fields. 1998. V. 110, №3. P. 369–395. [*435*]

[800] Ramírez, A. F., *Uniqueness of invariant product measures for elliptic infinite dimensional diffusions and particle spin systems.* ESAIM, Probab. Stat. 2002. V. 6. P. 147–155. [*435*]

[801] Ramirez, A. F., Varadhan, S. R. S., *Relative entropy and mixing properties of interacting particle systems.* J. Math. Kyoto Univ. 1996. V. 36, №4. P. 869–875. [*435*]

[802] Reed, M., Simon, B., *Methods of modern mathematical physics. I. Functional analysis.* 2nd ed. Academic Press, New York, 1980; xv+400 p. [*217*]

[803] Rempel, S., Schulze, B.-W., *Index theory of elliptic boundary problems.* Akademie-Verlag, Berlin, 1982; 393 p. [*51*]

[804] von Renesse, M.-K., Sturm, K.-T., *Transport inequalities, gradient estimates, entropy and Ricci curvature.* Comm. Pure Appl. Math. 2005. V. 68. P. 923–940. [*401*]

[805] Revuz, D., *Lois du tout ou rien et comportement asymptotique pour les probabilités de transition des processus de Markov.* Ann. Inst. H. Poincaré Sect. B. 1983. V. 19, №1. P. 9–24. [*235*]

[806] Revuz, D., *Markov chains* . 2nd ed. North-Holland, Amsterdam, 1984; xi+374 p. [*235*]

[807] Risken, H., *The Fokker–Planck equation: methods of solutions and applications.* 2nd ed., Springer, Berlin, 1989; xiv+472 p. [*284*]

[808] Robinson, D. W., *Elliptic operators and Lie groups.* The Clarendon Press, Oxford University Press, New York, 1991; xii+558 p. [*128*]

[809] Robinson, D. W., *Uniqueness of diffusion operators and capacity estimates.* J. Evol. Equ. 2013. V. 13, №1. P. 229–250. [*401*]

[810] Robinson, D. W., *Gaussian bounds, strong ellipticity and uniqueness criteria.* Bull. Lond. Math. Soc. 2014. V. 46, №5. P. 1077–1090. [*335*]

[811] Robinson, D. W., Sikora, A., *Degenerate elliptic operators in one dimension.* J. Evol. Equ. 2010. V. 10, №4. P. 731–759. [*401*]

[812] Robinson, D. W., Sikora, A., *Markov uniqueness of degenerate elliptic operators.* Ann. Sc. Norm. Sup. Pisa Cl. Sci. (5). 2011. V. 10, №3. P. 683–710. [*401*]

[813] Robinson, D. W., Sikora, A., *L^1-uniqueness of degenerate elliptic operators.* Studia Math. 2011. V. 203, №1. P. 79–103. [*401*]

[814] Röckner, M., Sobol, Z., *Kolmogorov equations in infinite dimensions: well-posedness and regularity of solutions, with applications to stochastic generalized Burgers equations.* Ann. Probab. 2006. V. 34, №2. P. 663–727. [*435*]

[815] Röckner, M., Wang, F.-Y., *Harnack and functional inequalities for generalized Mehler semigroups.* J. Funct. Anal. 2003. V. 203, №1. P. 237–261. [*435*]

[816] Röckner, M., Wang, F.-Y., *Supercontractivity and ultracontractivity for (non-symmetric) diffusion semigroups on manifolds.* Forum Math. 2003. V. 15, №6. P. 893–921. [*171, 227*]

[817] Röckner, M., Wang, F.-Y., *On the spectrum of a class of non-sectorial diffusion operators.* Bull. London Math. Soc. 2004. V. 36, №1. P. 95–104. [*234*]

[818] Röckner, M., Wang, F.-Y., *Concentration of invariant measures for stochastic generalized porous media equations.* Infin. Dimens. Anal. Quantum Probab. Related Top. 2007. V. 10, №3. P. 397–409. [*434*]

[819] Röckner, M., Zhang, T. S., *Uniqueness of generalized Schrödinger operators and applications.* J. Funct. Anal. 1992. V. 105, №1. P. 187–231. [*208*]

[820] Röckner, M., Zhang, T. S., *Uniqueness of generalized Schrödinger operators. Part II.* J. Funct. Anal. 1994. V. 119. P. 455–467. [*208*]

[821] Röckner, M., Zhang, X., *Stochastic tamed 3D Navier-Stokes equations: existence, uniqueness and ergodicity.* Probab. Theory Related Fields. 2009. V. 145, №1-2. P. 211–267. [*434*]

[822] Röckner, M., Zhang, X., *Weak uniqueness of Fokker–Planck equations with degenerate and bounded coefficients.* C. R. Math. Acad. Sci. Paris. 2010. T. 348, №7-8. P. 435–438. [*374, 378, 400*]

[823] Röckner, M., Zhu, R., Zhu, X., *A note on stochastic semilinear equations and their associated Fokker–Planck equations.* J. Math. Anal. Appl. 2014. V. 415, №1. P. 83–109. [*434*]

[824] Roelly, S., Zessin, H., *Une caractérisation des mesures de Gibbs sur $C(0,1)^{\mathbb{Z}^d}$ par le calcul des variationes stochastiques.* Ann. Inst. H. Poincaré. 1993. V. 29. P. 327–338. [*434*]

[825] Romito, M., *Analysis of equilibrium states of Markov solutions to the 3D Navier–Stokes equations driven by additive noise.* J. Stat. Phys. 2008. V. 131, №3. P. 415–444. [*434*]

[826] Rosenblatt, M., *Markov processes. Structure and asymptotic behavior.* Springer-Verlag, Berlin – New York, 1971; xiii+268 p. [*235*]

[827] Royer, G., *Initiation to logarithmic Sobolev inequalities.* Amer. Math. Soc., Providence, Rhode Island, 2007; 119 p. [*228*]

[828] Runst, Th., Sickel, W., *Sobolev spaces of fractional order, Nemytskij operators, and nonlinear partial differential equations.* Walter de Gruyter, Berlin, 1996; x+547 p. [*51*]

[829] Rytov, S. M., Kravtsov, Yu. A., Tatarskii, V. I., *Principles of statistical radiophysics. Vol. 3. Elements of random fields.* Springer-Verlag, Berlin, 1989; x+239 p. [*284*]

[830] Safonov, M. V., *Harnack's inequality for elliptic equations and Ho"lder property of their solutions.* Zap. Nauchn. Sem. Leningrad. Otdel. Mat. Inst. Steklov. (LOMI). 1980. V. 96. P. 272–287 (in Russian); English transl.: J. Math. Sci. (New York). 1983. V. 21, №5. P. 851–863. [*127*]

[831] Safonov, M. V., *An example of a diffusion process with the singular distribution at a fixed moment.* In: Abstracts Communications of the Third International Vilnius Conference on Probability Theory and Mathematical Statistics, Vol. II, pp. 133–134. Vilnius, 1981. [*254*]

[832] Safonov, M., *Nonuniqueness for second order elliptic equations with measurable coefficients.* SIAM J. Math. Anal. 1999. V. 30. P. 879–895. [*174*]

[833] Sakbaev, V. Zh., *On the Cauchy problem for the Fokker–Planck equation that is degenerate on the half-line.* Differ. Uravn. 2007. V. 43, №8. P. 1127–1143 (in Russian); English transl.: Differ. Equ. 2007. V. 43, №8. P. 1153–1171. [*401*]

[834] Sauvigny, F., *Partial differential equations 2. Functional analytic methods.* Springer-Verlag, Berlin – Heidelberg, 2006; xiv+388 p. [*51*]

[835] Scherbakov, V. V., *Full-space exponential convergence for Itô stochastic partial differential equations.* Selecta Math. Soviet. 1993. V. 12, №3. P. 223–249. [*434*]

[836] Scheutzow, M., Weizsäcker, H. von., *Which moments of a logarithmic derivative imply quasiinvariance?* Doc. Math. 1998. V. 3. P. 261–272. [*108, 110*]

[837] Schrödinger, E., *Über die Umkehrung der Naturgesetze.* Sitzungsber. Preuss. Akad. Wiss. Berlin, Phys.-Math. Kl., 1931, 12. Marz, S. 144–153. [*174, 236*]

[838] Schuss, Z., *Brownian dynamics at boundaries and interfaces. In physics, chemistry, and biology.* Springer, New York, 2013; xx+322 p. [*284*]

[839] Seidler, J., *Ergodic behaviour of stochastic parabolic equations.* Czechoslovak Math. J. 1997. V. 47, №2. P. 277–316. [*434*]

[840] Semenov, Yu. A., *On perturbation theory for linear elliptic and parabolic operators; the method of Nash.* In: Applied analysis (Baton Rouge, LA, 1996), pp. 217–284, Contemp. Math., V. 221, Amer. Math. Soc., Providence, Rhode Island, 1999. [*314*]

[841] Semenov, Yu. A., *Dirichlet operators: a priori estimates and uniqueness problems* II. J. Funct. Anal. 2014. V. 267, №12. P. 4567–4634. [*208*]

[842] Serrin, J., *Pathological solutions of elliptic differential equations.* Ann. Sc. Norm. Sup. Pisa. 1965. V. 19. P. 593–608. [*35*]

[843] Shaposhnikov, S. V., *On Morrey's estimate for the Sobolev norms of solutions of elliptic equations.* Mat. Zametki. 2006. V. 79, №3. P. 450–469 (in Russian); English transl.: Math. Notes. 2006. V. 79, №3-4. P. 413–430. [*52*]

[844] Shaposhnikov, S. V., *Positiveness of invariant measures of diffusion processes.* Dokl. Ross. Akad. Nauk. 2007. V. 415, №2. P. 174–179 (in Russian); English transl.: Dokl. Math. 2007. V. 76, №1. P. 533–538. [*127*]

[845] Shaposhnikov, S. V., *On nonuniqueness of solutions to elliptic equations for probability measures.* J. Funct. Anal. 2008. V. 254, №10. P. 2690–2705. [*174*]

[846] Shaposhnikov, S. V., *On interior estimates for the Sobolev norms of solutions of elliptic equations.* Mat. Zametki. 2008. V. 83, №2. P. 316–320 (in Russian); English transl.: Math. Notes. 2008. V. 83, №1-2. P. 285–289. [*52*]

[847] Shaposhnikov, S. V., *On the nonuniqueness of solutions of elliptic equations for probability measures.* Dokl. Ross. Akad. Nauk. 2008. V. 420, №3. P. 320–323 (in Russian); English transl.: Dokl. Math. 2008. V. 77, №3. P. 401–403. [*174*]

[848] Shaposhnikov, S. V., *Lower bounds for the densities of solutions of parabolic equations for measures.* Dokl. Ross. Akad. Nauk. 2009. V. 429, №5. P. 600–604 (in Russian); English transl.: Dokl. Math. 2009. V. 80, №3. P. 877–881. [*334*]

[849] Shaposhnikov, S. V., *On the uniqueness of the probabilistic solution of the Cauchy problem for the Fokker–Planck–Kolmogorov equation.* Teor. Veroyatn. Primen. 2011. V. 56, №1. P. 77–99 (in Russian); English transl.: Theory Probab. Appl. 2012. V. 56, №1. P. 96–115. [*400*]

[850] Shaposhnikov, S. V., *Regularity and qualitative properties of solutions of parabolic equations for measures.* Teor. Veroyatn. Primen. 2011. V. 56, №2. P. 318–350 (in Russian); English transl.: Theory Probab. Appl. 2012. V. 56, №2. P. 252–279. [*313, 400*]

[851] Shaposhnikov, S. V., *On the uniqueness of integrable and probabilistic solutions of the Cauchy problem for the Fokker–Planck–Kolmogorov equation.* Dokl. Ross. Akad. Nauk. 2011. V. 439, №3. P. 323–328 (in Russian); English transl.: Dokl. Math. 2011. V. 84, №1. P. 565–570. [*400*]

[852] Shaposhnikov, S. V., *Fokker–Planck–Kolmogorov equations with a potential and a non-uniformly elliptic diffusion matrix.* Trudy Mosk. Matem. Ob. 2013. V. 74, №1. P. 1–18 (in Russian); English transl.: Trans. Moscow Math. Soc. 2013. P. 15–29. [*284*]

[853] Sheu, S. J., *Some estimates of the transition density of a nondegenerate diffusion Markov process.* Ann. Probab. 1991. V. 19, №2. P. 538–561. [*314*]

[854] Shiga, T., *Ergodic theorems and exponential decay of sample paths for certain interacting diffusion systems.* Osaka J. Math. 1992. V. 29, №4. P. 789–807. [*434*]

[855] Shigekawa, I., *Existence of invariant measures of diffusions on an abstract Wiener space.* Osaka J. Math. 1987. V. 24, №1. P. 37–59. [*416*]

[856] Shigekawa, I., *Stochastic analysis.* Amer. Math. Soc., Providence, Rhode Island, 2004; xii+182 p. [*401*]

[857] Shimakura, N., *Partial differential operators of elliptic type.* Amer. Math. Soc., Providence, Rhode Island, 1992; xiv+288 p. [*51*]

[858] Shirikyan, A., *Qualitative properties of stationary measures for three-dimensional Navier–Stokes equations.* J. Funct. Anal. 2007. V. 249, №2. P. 284–306. [*414, 434*]

[859] Shishmarev, I. A., *Introduction to the theory of elliptic equations.* Moskov. Gos. Univ., Moscow, 1979; 184 p. (in Russian). [*51*]

[860] Shurenkov, V. M., *Ergodic Markov processes.* Nauka, Moscow, 1989; 336 p. (in Russian). [*235*]

[861] Sjögren, P., *On the adjoint of an elliptic linear differential operator and its potential theory.* Ark. Mat. 1973. V. 11. P. 153–165. [*29, 52*]

[862] Skorohod, A. V., *Asymptotic methods in the theory of stochastic differential equations.* Amer. Math. Soc., Rhode Island, 1989; xvi+339 p. [*235*]

[863] von Smoluchowski, M., *Über Brownsche Molekularbewegung unter Einwirkung äußerer Kräfte und den Zusammenhang mit der verallgemeinerten Diffusionsgleichung.* Ann. Phys. 1915. B. 353 (4. Folge 48). S. 1103–1112. [ix, *52*]

[864] Snyders, J., *Stationary probability distributions for linear time-invariant systems.* SIAM J. Control Optimization. 1977. V. 15, №3. P. 428–437. [*235*]

[865] Soize, C., *The Fokker–Planck equation for stochastic dynamical systems and its explicit steady state solutions.* World Sci., River Edge, New Jersey, 1994; xvi+321 p. [*51, 284*]

[866] Soloveitchik, M. R., *Focker–Planck equation on a manifold. Effective diffusion and spectrum.* Potential Anal. 1995. V. 4, №6. P. 571–593. [*284*]

[867] Sowers, R., *Large deviations for the invariant measure of a reaction-diffusion equation with non-Gaussian perturbations.* Probab. Theory Related Fields. 1992. V. 92, №3. P. 393–421. [*434*]

[868] Sperb, R. P., *Maximum principles and their applications.* Academic Press, New York – London, 1981; ix+224 p. [*78*]

[869] Spina, Ch., *Kernel estimates for a class of Kolmogorov semigroups.* Arch. Math. 2008. V. 91, №3. P. 265–279. [*292*]

[870] Stampacchia, G., *Équations elliptiques du second ordre à coefficients discontinus.* Les Presses de l'Université de Montréal, 1966; 326 p. [*51*]

[871] Stannat, W., *First order perturbations of Dirichlet operators: existence and uniqueness.* J. Funct. Anal. 1996. V. 141, №1. P. 216–248. [*235*]

[872] Stannat, W., (*Nonsymmetric*) *Dirichlet operators on L^1: existence, uniqueness and associated Markov processes.* Ann. Sc. Norm. Sup. Pisa Cl. Sci. (4). 1999. V. 28, №1. P. 99–140. [*188, 208, 234, 238, 405*]

[873] Stannat, W., *The theory of generalized Dirichlet forms and its applications in analysis and stochastics.* Mem. Amer. Math. Soc. 1999. V. 142, №678; viii+101 p. [*405*]

[874] Stannat, W., *Time-dependent diffusion operators on L^1.* J. Evol. Equ. 2004. V. 4, №4. P. 463–495. [*405*]

[875] Stasi, R., *Essential self-adjointness and domain characterization for some degenerate gradient systems.* Dynam. Systems Appl. 2003. V. 12, №3-4. P. 259–273. [*401*]

[876] Stasi, R., *Maximal dissipativity for degenerate Kolmogorov operators.* Nonlin. Differ. Equ. Appl. 2005. V. 12, №4. P. 419–436. [*401*]

[877] Stein, E., *Singular integrals and differentiability properties of functions.* Princeton University Press, Princeton, 1970; xiv+290 p. [*51, 166*]

[878] Stettner, L., *Large deviations of invariant measures for degenerate diffusions.* Probab. Math. Statist. 1989. V. 10, №1. P. 93–105. [*236*]

[879] Stettner, L., *Remarks on ergodic conditions for Markov processes on Polish spaces.* Bull. Polish Acad. Sci. Math. 1994. V. 42. P. 103–114. [*235*]

[880] Stroock, D. W., *Lectures on infinite interacting systems.* Lectures in Mathematics, Kyoto Univ., №11. Kinokuniya Book-Store Co., Tokyo, 1978; i+72 p. [*434*]

[881] Stroock, D. W., *On the spectrum of Markov semigroups and the existence of invariant measures.* Lecture Notes in Math. 1982. V. 923. P. 286–307. [*235*]

[882] Stroock, D. W., *Partial differential equations for probabilists.* Cambridge University Press, Cambridge, 2008; xvi+215 p. [*51*]

[883] Stroock, D., Varadhan, S. R. S., *On degenerate elliptic-parabolic operators of second order and their associated diffusions.* Comm. Pure Appl. Math. 1972. V. 25. P. 651–713. [*401*]

[884] Stroock, D. W., Varadhan, S. R. S., *Multidimensional diffusion processes.* Springer-Verlag, Berlin – New York, 1979; xii+338 p. [*51, 254, 284, 340, 344, 373, 378, 380, 382*]

[885] Stroock, D., Zegarlinski, B., *On the ergodic properties of Glauber dynamics.* J. Statist. Phys. 1995. V. 81, №5-6. P. 1007–1019. [*434*]

[886] Sunyach, C., *Une classe de chaînes de Markov récurrentes sur un espace métrique complet.* Ann. Inst. H. Poincaré Sect. B. (N.S). 1975. V. 11, №4. P. 325–343. [*235*]

[887] Szarek, T., *Invariant measures for nonexpensive Markov operators on Polish spaces.* Dissertationes Math. (Rozprawy Mat.). 2003. V. 415, 62 p. [*235*]

[888] Szarek, T., *Feller processes on nonlocally compact spaces.* Ann. Probab. 2006. V. 34, №5. P. 1849–1863. [*235*]

[889] Szarek, T., *The uniqueness of invariant measures for Markov operators.* Studia Math. 2008. V. 189, №3. P. 225–233. [*235*]

[890] Takeda, M., *On the uniqueness of Markovian selfadjoint extension of diffusion operators on infinite-dimensional spaces.* Osaka J. Math. 1985. V. 22, №4. P. 733–742. [*435*]

[891] Takeda, M., *The maximum Markovian selfadjoint extensions of generalized Schrödinger operators.* J. Math. Soc. Japan. 1992. V. 44, №1. P. 113–130. [*208*]

[892] Takeda, M., *Two classes of extensions for generalized Schrödinger operators.* Potential Anal. 1996. V. 5, №1. P. 1–13. [*435*]

[893] Tamura, T., Watanabe, Y., *Hypoellipticity and ergodicity of the Wonham filter as a diffusion process.* Appl. Math. Optim. 2011. V. 64, №1. P. 13–36. [*401*]

[894] Taylor, M., *Pseudodifferential operators.* Princeton University Press, Princeton, New Jersey, 1981; xi+452 p. [*20, 43*]

[895] Tian, G.-J., Wang, X.-J., *Moser–Trudinger type inequalities for the Hessian equation.* J. Funct. Anal. 2010. V. 259. P. 1974–2002. [*50*]

[896] Tolmachev, N. A., *On the smoothness and singularity of invariant measures and transition probabilities of infinite-dimensional diffusions.* Teor. Veroyatn. Primen. 1998. V. 43, №4. P. 798–808 (in Russian); English transl.: Theory Probab. Appl. 1998. V. 43, №4. P. 655–664. [*416*]

[897] Tonoyan, L. G., *Nonlinear elliptic equations for measures.* Dokl. Ross. Akad. Nauk. 2011. V. 439, №2. P. 174–177 (in Russian); English transl.: Dokl. Math. 2011. V. 84, №1. P. 558–561. [*79*]

[898] Trèves, F., *Introduction to pseudodifferential and Fourier integral operators.* Vols. 1, 2. Plenum Press, New York – London, 1980; xxvii+299+xi p., xiv+301 p. [*21*]

[899] Triebel, H., *Interpolation theory. Function spaces. Differential operators.* Deutscher Verlag des Wissenschaften, Berlin, 1978; 528 p. [*51*]

[900] Triebel, H., *Theory of function spaces.* Birkhäuser/Springer Basel, Basel, 2010; 285 p. [*51*]

[901] Troianiello, G. M., *Elliptic differential equations and obstacle problems.* Plenum Press, New York, 1987; xiv+353 p. [*51*]

[902] Trudinger, N. S., *Pointwise estimates and quasilinear parabolic equations.* Comm. Pure Appl. Math. 1968. V. 21. P. 205–226. [*250, 334*]

[903] Trudinger, N. S., *On the regularity of generalized solutions of linear, non-uniformly elliptic equations.* Arch. Rational Mech. Anal. 1971. V. 42. P. 50–62. [*122*]

[904] Trudinger, N. S., *Linear elliptic operators with measurable coefficients.* Ann. Sc. Norm. Sup. Pisa (3). 1973. V. 27. P. 265–308. [*35, 57*]

[905] Trudinger, N. S., *Maximum principles for linear, non-uniformly elliptic operators with measurable coefficients.* Math. Z. 1977. B. 156. S. 291–301. [*58*]

[906] Trutnau, G., *On a class of non-symmetric diffusions containing fully nonsymmetric distorted Brownian motions.* Forum Math. 2003. V. 15, №3. P. 409–437. [*236*]

[907] Tychonoff, A. N., *A uniqueness theorem for the heat equation.* Matem. Sbornik. 1935. V. 42. P. 199–216. [*339*]

[908] Üstünel, A.-S., Zakai, M., *Transformation of measure on Wiener space.* Springer-Verlag, Berlin, 2000; xiv+296 p. [*434*]

[909] Varadhan, S. R. S., *Lectures on diffusion problems and partial differential equations.* Tata Institute of Fundamental Research, Bombay, 1980; iii+315 p. [*234*]

[910] Varopoulos, N. Th., *Hardy–Littlewood theory for semigroups.* J. Funct. Anal. 1985. V. 63, №2. P. 240–260. [*228*]

[911] Varopoulos, N. Th., Saloff-Coste, L., Coulhon, T., *Analysis and geometry on groups.* Cambridge University Press, Cambridge, 1992; xii+156 p. [*228*]

[912] Vedenyapin, V. V., *The Bolzman and Vlasov kinetic equations.* Fizmatlit, Moscow, 2001; 111 p. (in Russian). [*284*]

[913] Veretennikov, A. Yu., *On stochastic equations with degenerate diffusion with respect to some of the variables.* Izv. Akad. Nauk SSSR Ser. Mat. 1983. V. 47, №1. P. 188–196 (in Russian); English transl.: Math. USSR Izvestiya. 1984. V. 22, №1. P. 173–180. [*381*]

[914] Veretennikov, A. Yu., *On polynomial mixing bounds for stochastic differential equations.* Stoch. Process. Appl. 1997. V. 70. P. 115–127. [*235*]

[915] Veretennikov, A. Yu., *On polynomial mixing and convergence rate for stochastic difference and differential equations.* Teor. Veroyatn. Primen. 1999. V. 44, №2. P. 312–327 (in Russian); English transl.: Theory Probab. Appl. 2000. V. 44, №2. P. 361–374. [*235*]

[916] Veretennikov, A. Yu., *On Sobolev solutions of Poisson equations in* $\mathbb{R}^d$ *with a parameter.* J. Math. Sci. (New York). 2011. V. 179, №1. P. 48–79. [*125, 235*]

[917] Veretennikov, A. Yu., Klokov, S. A., *On subexponential mixing rate for Markov processes.* Teor. Veroyatn. Primen. 2004. V. 49, №1. P. 21–35 (in Russian); English transl.: Theory Probab. Appl. 2005. V. 49, №1. P. 110–122. [*235*]

[918] Villani, C., *Topics in optimal transportation.* Amer. Math. Soc., Rhode Island, 2003; 370 p. [*170, 285*]

[919] Villani, C., *Optimal transport. Old and new.* Springer-Verlag, Berlin, 2009; xxii+973 p. [*170*]

[920] Villani, C., *Hypocoercivity.* Mem. Amer. Math. Soc. 2009. V. 202, №950, iv+141 pp. [*285*]

[921] von Vintschger, R., *The existence of invariant measures for* $C[0,1]$*-valued diffusions.* Probab. Theory Related Fields. 1989. V. 82. P. 307–313. [*434*]

[922] Vishik, M. I., Fursikov, A. V., *Mathematical problems of statistical hydromechanics.* Kluver Acad. Publ., Dordrecht, Boston, London, 1988; 473 p. (Russian ed.: Moscow, 1980). [*405*]

[923] Vlasov, A. A., *Statistical distribution functions.* Nauka, Moscow, 1966; 356 p. (in Russian). [*284*]

[924] Volpert, V., *Elliptic partial differential equations. Volume 1: Fredholm theory of elliptic problems in unbounded domains.* Birkhäuser/Springer Basel, Basel, 2011; xviii+639 p. [*51*]

[925] Wallstrom, T. C., *Ergodicity of finite-energy diffusions.* Trans. Amer. Math. Soc. 1990. V. 318, №2. P. 735–747. [*235*]

[926] Wang, F.-Y., *Gradient estimates of Dirichlet heat semigroups and application to isoperimetric inequalities.* Ann. Probab. 2004. V. 32, №1A. P. 424–440. [*234*]

[927] Wang, F.-Y., L^1*-convergence and hypercontractivity of diffusion semigroups on manifolds.* Studia Math. 2004. V. 162, №3. P. 219–227. [*236*]

[928] Wang, F.-Y., *Functional inequalities, Markov semigroups and spectral theory.* Elsevier, Beijing, 2006; vii+305 p. [*228*]

[929] Wang, F.-Y., *Harnack inequality and applications for stochastic generalized porous media equations.* Ann. Probab. 2007. V. 35, №4. P. 1333–1350. [*435*]

[930] Wang, F.-Y., *Entropy-cost inequalities for diffusion semigroups with curvature unbounded below.* Proc. Amer. Math. Soc. 2008. V. 136, №9. P. 3331–3338. [*234*]

[931] Wang, F.-Y., *From super Poincaré to weighted log-Sobolev and entropy-cost inequalities.* J. Math. Pures Appl. 2008. V. 90, №3. P. 270–285. [*170*]

[932] Wang, F.-Y., *Harnack inequalities for stochastic partial differential equations.* Springer, New York, 2013; x+125 p. [*435*]

[933] Wang, F.-Y., Zhang, X., *Derivative formula and applications for degenerate diffusion semigroups.* J. Math. Pures Appl. (9). 2013. V. 99, №6. P. 726–740. [*401*]

[934] Wang, X.-J., *The k-Hessian equation.* Lecture Notes in Math. 2009. V. 1977. P. 177–252. [*49*]

[935] Watson, N. A., *Parabolic equations on an infinite strip.* Marcel Dekker, New York, 1989; xiv+289 p. [*283*]

[936] Wei, J., Liu, B., L^p*-solutions of Fokker–Planck equations.* Nonlinear Anal. 2013. V. 85. P. 110–124. [*373, 400, 401*]

[937] Wentzell, A. D., *A course in the theory of stochastic processes.* McGraw-Hill International Book, New York, 1981; x+304 p. (Russian ed.: Moscow, 1975). [*12, 15, 16*]

[938] Widder, D. V., *Positive temperatures on the infinite rod.* Trans. Amer. Math. Soc. 1944. V. 55, №1. P. 85–95. [*339*]

[939] Wielens, N., *The essential selfadjointness of generalized Schrödinger operators.* J. Funct. Anal. 1985. V. 61, №1. P. 98–115. [*208*]

[940] Wiesinger, S., *Uniqueness for solutions of Fokker–Planck equations related to singular SPDE driven by Lévy and cylindrical Wiener noise.* J. Evol. Equ. 2013. V. 13, №2. P. 369–394. [*434*]

[941] Williams, R. J., *Recurrence classification and invariant measure for reflected Brownian motion in a wedge.* Ann. Probab. 1985. V. 13, №3. P. 758–778. [*236*]

[942] Wonham, W. M., *Liapunov criteria for weak stochastic stability.* J. Differ. Equ. 1966. V. 2. P. 195–207. [*235*]

[943] Wu, L. M., Zhang, Y., *A new topological approach to the L^∞-uniqueness of operators and the L^1-uniqueness of Fokker–Planck equations.* J. Funct. Anal. 2006. V. 241, №2. P. 557–610. [*157, 234, 400*]

[944] Wu, Z., Yin, J., Wang, Ch., *Elliptic & parabolic equations.* World Sci., Hackensack, New Jersey, 2006; xvi+408 p. [*51, 283*]

[945] Yosida, K., *On the integration of diffusion equations in Riemannian spaces.* Proc. Amer. Math. Soc. 1952. V. 3. P. 864–873. [*237, 284*]

[946] Yosida, K., *Functional analysis.* Springer-Verlag, Berlin, 1995; xii+501 p. [*216*]

[947] Zaharopol, R., *Invariant probabilities of Markov–Feller operators and their supports.* Birkhäuser, Basel, 2005; xiv+108 p. [*235*]

[948] Zaharopol, R., *An ergodic decomposition defined by transition probabilities.* Acta Appl. Math. 2008. V. 104, №1. P. 47–81. [*235*]

[949] Zakai, M., *On the ultimate boundedness of moments associated with solutions of stochastic differential equations.* SIAM J. Control. 1967. V. 5. P. 588–593. [*79*]

[950] Zakai, M., *A Lyapunov criterion for the existence of stationary probability distributions for systems perturbed by noise.* SIAM J. Control. 1969. V. 7. P. 390–397. [*79, 235*]

[951] Zakai, M., Snyders, J., *Stationary probability measures for linear differential equations driven by white noise.* J. Differ. Equ. 1970. V. 8. P. 27–33. [*235*]

[952] Zegarlinski, B., *Ergodicity of Markov semigroups.* In: Stochastic partial differential equations (Edinburgh, 1994), pp. 312–337, London Math. Soc. Lecture Note Ser., 216, Cambridge University Press, Cambridge, 1995. [*434*]

[953] Zeitouni, O., *On the nonexistence of stationary diffusions which satisfy the Beneš condition.* Systems Control Lett. 1983. V. 3, №6. P. 329–330. [*54*]

[954] Zelenyak, T. I., Lavrentiev, M. M. (Jr.), Vishnevskii, M. P., *Qualitative theory of parabolic equations. Part 1.* VSP, Utrecht, 1997; ii+417 p. [*283*]

[955] Zhang, W., Bao, J., *Regularity of very weak solutions for elliptic equation of divergence form.* J. Funct. Anal. 2012. V. 262, №4. P. 1867–1878. [*35*]

[956] Zhang, X., *Exponential ergodicity of non-Lipschitz stochastic differential equations.* Proc. Amer. Math. Soc. 2009. V. 137, №1. P. 329–337. [*235*]

[957] Zhang, X. S., *Existence and uniqueness of invariant probability measure for uniformly elliptic diffusion.* In: Dirichlet forms and stochastic processes (Beijing, 1993), pp. 417–423. Walter de Gruyter, Berlin, 1995. [*235*]

[958] Zhang, X., *Variational approximation for Fokker–Planck equation on Riemannian manifold.* Probab. Theory Related Fields. 2007. V. 137, №3-4. P. 519–539. [*284*]

[959] Zhang, X., *Well-posedness and large deviation for degenerate SDEs with Sobolev coefficients.* Rev. Mat. Iberoam. 2013. V. 29, №1. P. 25–52. [*378*]

[960] Zhikov, V. V., *On Lavrentiev's phenomenon.* Russian J. Math. Physics. 1995. V. 3. P. 249–269. [*164*]

[961] Zhikov, V. V., *Weighted Sobolev spaces.* Matem. Sbornik. 1998. V. 189, №8. P. 27–58 (in Russian); English transl.: Sb. Math. 1998. V. 189, №8. P. 1139–1170. [*51*]

[962] Zhikov, V. V., *Remarks on the uniqueness of the solution of the Dirichlet problem for a second-order elliptic equation with lower order terms.* Funk. Anal. i Pril. 2004. V. 38, №3. P. 15–28 (in Russian); English transl.: Funct. Anal. Appl. 2004. V. 38, №3. P. 173–183. [*162, 174*]

[963] Zhikov, V. V., *On the density of smooth functions in a weighted Sobolev space.* Dokl. Ross. Akad. Nauk. 2013. V. 453, №3. P. 247–251 (in Russian); English transl.: Dokl. Math. 2013. V. 88, №3. P. 669–673. [*165*]

[964] Zhikov, V. V., Estimates of Nash–Aronson type for degenerate parabolic equations. J. Math. Sci. (New York). 2013. V. 190, №1. P. 66–79. [*312*]

[965] Ziemer, W., *Weakly differentiable functions.* Springer-Verlag, New York – Berlin, 1989; xvi+308 p. [*51*]

Subject Index

Notation

Selected Published Titles in This Series

207 **Vladimir I. Bogachev, Nicolai V. Krylov, Michael Röckner, and Stanislav V. Shaposhnikov,** Fokker–Planck–Kolmogorov Equations, 2015

206 **Bennett Chow, Sun-Chin Chu, David Glickenstein, Christine Guenther, James Isenberg, Tom Ivey, Dan Knopf, Peng Lu, Feng Luo, and Lei Ni,** The Ricci Flow: Techniques and Applications: Part IV: Long-Time Solutions and Related Topics, 2015

205 **Pavel Etingof, Shlomo Gelaki, Dmitri Nikshych, and Victor Ostrik,** Tensor Categories, 2015

204 **Victor M. Buchstaber and Taras E. Panov,** Toric Topology, 2015

203 **Donald Yau and Mark W. Johnson,** A Foundation for PROPs, Algebras, and Modules, 2015

202 **Shiri Artstein-Avidan, Apostolos Giannopoulos, and Vitali D. Milman,** Asymptotic Geometric Analysis, Part I, 2015

201 **Christopher L. Douglas, John Francis, André G. Henriques, and Michael A. Hill, Editors,** Topological Modular Forms, 2014

200 **Nikolai Nadirashvili, Vladimir Tkachev, and Serge Vlăduţ,** Nonlinear Elliptic Equations and Nonassociative Algebras, 2014

199 **Dmitry S. Kaliuzhnyi-Verbovetskyi and Victor Vinnikov,** Foundations of Free Noncommutative Function Theory, 2014

198 **Jörg Jahnel,** Brauer Groups, Tamagawa Measures, and Rational Points on Algebraic Varieties, 2014

197 **Richard Evan Schwartz,** The Octagonal PETs, 2014

196 **Silouanos Brazitikos, Apostolos Giannopoulos, Petros Valettas, and Beatrice-Helen Vritsiou,** Geometry of Isotropic Convex Bodies, 2014

195 **Ching-Li Chai, Brian Conrad, and Frans Oort,** Complex Multiplication and Lifting Problems, 2014

194 **Samuel Herrmann, Peter Imkeller, Ilya Pavlyukevich, and Dierk Peithmann,** Stochastic Resonance, 2014

193 **Robert Rumely,** Capacity Theory with Local Rationality, 2013

192 **Messoud Efendiev,** Attractors for Degenerate Parabolic Type Equations, 2013

191 **Grégory Berhuy and Frédérique Oggier,** An Introduction to Central Simple Algebras and Their Applications to Wireless Communication, 2013

190 **Aleksandr Pukhlikov,** Birationally Rigid Varieties, 2013

189 **Alberto Elduque and Mikhail Kochetov,** Gradings on Simple Lie Algebras, 2013

188 **David Lannes,** The Water Waves Problem, 2013

187 **Nassif Ghoussoub and Amir Moradifam,** Functional Inequalities: New Perspectives and New Applications, 2013

186 **Gregory Berkolaiko and Peter Kuchment,** Introduction to Quantum Graphs, 2013

185 **Patrick Iglesias-Zemmour,** Diffeology, 2013

184 **Frederick W. Gehring and Kari Hag,** The Ubiquitous Quasidisk, 2012

183 **Gershon Kresin and Vladimir Maz'ya,** Maximum Principles and Sharp Constants for Solutions of Elliptic and Parabolic Systems, 2012

182 **Neil A. Watson,** Introduction to Heat Potential Theory, 2012

181 **Graham J. Leuschke and Roger Wiegand,** Cohen-Macaulay Representations, 2012

180 **Martin W. Liebeck and Gary M. Seitz,** Unipotent and Nilpotent Classes in Simple Algebraic Groups and Lie Algebras, 2012

179 **Stephen D. Smith,** Subgroup Complexes, 2011

178 **Helmut Brass and Knut Petras,** Quadrature Theory, 2011

For a complete list of titles in this series, visit the
AMS Bookstore at **www.ams.org/bookstore/survseries/**.